IET TELECOMMUNICATIONS SERIES 107

Physical Layer Security for 6G Networks

Other volumes in this series:

Volume 9 **Phase Noise in Signal Sources** W.P. Robins
Volume 12 **Spread Spectrum in Communications** R. Skaug and J.F. Hjelmstad
Volume 13 **Advanced Signal Processing** D.J. Creasey (Editor)
Volume 19 **Telecommunications Traffic, Tariffs and Costs** R.E. Farr
Volume 20 **An introduction to Satellite Communications** D.I. Dalgleish
Volume 26 **Common-Channel Signalling** R.J. Manterfield
Volume 28 **Very Small Aperture Terminals (VSATs)** J.L. Everett (Editor)
Volume 29 **ATM: The broadband telecommunications solution** L.G. Cuthbert and J.C. Sapanel
Volume 31 **Data Communications and Networks, 3rd Edition** R.L. Brewster (Editor)
Volume 32 **Analogue Optical Fibre Communications** B. Wilson, Z. Ghassemlooy and I.Z. Darwazeh (Editors)
Volume 33 **Modern Personal Radio Systems** R.C.V. Macario (Editor)
Volume 34 **Digital Broadcasting** P. Dambacher
Volume 35 **Principles of Performance Engineering for Telecommunication and Information Systems** M. Ghanbari, C.J. Hughes, M.C. Sinclair and J.P. Eade
Volume 36 **Telecommunication Networks, 2nd Edition** J.E. Flood (Editor)
Volume 37 **Optical Communication Receiver Design** S.B. Alexander
Volume 38 **Satellite Communication Systems, 3rd Edition** B.G. Evans (Editor)
Volume 40 **Spread Spectrum in Mobile Communication** O. Berg, T. Berg, J.F. Hjelmstad, S. Haavik and R. Skaug
Volume 41 **World Telecommunications Economics** J.J. Wheatley
Volume 43 **Telecommunications Signalling** R.J. Manterfield
Volume 44 **Digital Signal Filtering, Analysis and Restoration** J. Jan
Volume 45 **Radio Spectrum Management, 2nd Edition** D.J. Withers
Volume 46 **Intelligent Networks: Principles and applications** J.R. Anderson
Volume 47 **Local Access Network Technologies** P. France
Volume 48 **Telecommunications Quality of Service Management** A.P. Oodan (Editor)
Volume 49 **Standard Codecs: Image compression to advanced video coding** M. Ghanbari
Volume 50 **Telecommunications Regulation** J. Buckley
Volume 51 **Security for Mobility** C. Mitchell (Editor)
Volume 52 **Understanding Telecommunications Networks** A. Valdar
Volume 53 **Video Compression Systems: From first principles to concatenated codecs** A. Bock
Volume 54 **Standard Codecs: Image compression to advanced video coding, 3rd Edition** M. Ghanbari
Volume 59 **Dynamic Ad Hoc Networks** H. Rashvand and H. Chao (Editors)
Volume 60 **Understanding Telecommunications Business** A Valdar and I. Morfett
Volume 65 **Advances in Body-Centric Wireless Communication: Applications and state-of-the-art** Q.H. Abbasi, M.U. Rehman, K. Qaraqe and A. Alomainy (Editors)
Volume 67 **Managing the Internet of Things: Architectures, theories and applications** J. Huang and K. Hua (Editors)
Volume 68 **Advanced Relay Technologies in Next Generation Wireless Communications** I. Krikidis and G. Zheng
Volume 69 **5G Wireless Technologies** A. Alexiou (Editor)
Volume 70 **Cloud and Fog Computing in 5G Mobile Networks** E. Markakis, G. Mastorakis, C.X. Mavromoustakis and E. Pallis (Editors)
Volume 71 **Understanding Telecommunications Networks, 2nd Edition** A. Valdar
Volume 72 **Introduction to Digital Wireless Communications** Hong-Chuan Yang
Volume 73 **Network as a Service for Next Generation Internet** Q. Duan and S. Wang (Editors)
Volume 74 **Access, Fronthaul and Backhaul networks for 5G & Beyond** M.A. Imran, S.A.R. Zaidi and M.Z. Shakir (Editors)
Volume 76 **Trusted Communications with Physical Layer Security for 5G and Beyond** T.Q. Duong, X. Zhou and H.V. Poor (Editors)
Volume 77 **Network Design, Modelling and Performance Evaluation** Q. Vien
Volume 78 **Principles and Applications of Free Space Optical Communications** A.K. Majumdar, Z. Ghassemlooy and A.A.B. Raj (Editors)
Volume 79 **Satellite Communications in the 5G Era** S.K. Sharma, S. Chatzinotas and D. Arapoglou
Volume 80 **Transceiver and System Design for Digital Communications, 5th Edition** Scott R. Bullock
Volume 81 **Applications of Machine Learning in Wireless Communications** R. He and Z. Ding (Editors)
Volume 83 **Microstrip and Printed Antenna Design, 3rd Edition** R. Bancroft
Volume 84 **Low Electromagnetic Emission Wireless Network Technologies: 5G and Beyond** M.A. Imran, F. Héliot and Y.A. Sambo (Editors)
Volume 86 **Advances in Communications Satellite Systems Proceedings of the 36th International Communications Satellite Systems Conference (ICSSC-2018)** I. Otung, T. Butash and P. Garland (Editors)
Volume 87 **Real Time Convex Optimisation for 5G Networks and Beyond** T.Q. Duong, L.D. Nguyen and H.D. Tuan
Volume 89 **Information and Communication Technologies for Humanitarian Services** M.N. Islam (Editor)
Volume 90 **Communication Technologies for Networked Smart Cities** S.K. Sharma, N. Jayakody, S. Chatzinotas and A. Anpalagan (Editors)

Volume 91	**Green Communications for Energy-Efficient Wireless Systems and Networks** Himal Asanga Suraweera, Jing Yang, Alessio Zappone and John S. Thompson (Editors)
Volume 92	**Flexible and Cognitive Radio Access Technologies for 5G and Beyond** H. Arslan and E. Başar (Editors)
Volume 93	**Antennas and Propagation for 5G and Beyond** Q. Abbasi, S.F. Jilani, A. Alomainy and M.A. Imran (Editors)
Volume 94	**Intelligent Wireless Communications** G. Mastorakis, C.X. Mavromoustakis, J.M. Batalla and E. Pallis (Editors)
Volume 95	**ISDN Applications in Education and Training** R. Mason and P.D. Bacsich
Volume 96	**Edge Caching for Mobile Networks** H. Vincent Poor and Wei Chen (Editors)
Volume 98	**Artificial Intelligence Applied to Satellite-based Remote Sensing Data for Earth Observation** M.P. Del Rosso, A. Sebastianelli and S.L. Ullo (Editors)
Volume 99	**Metrology for 5G and Emerging Wireless Technologies** T.H. Loh (Editor)
Volume 101	**Wireless Mesh Networks for IoT and Smart Cities: Technologies and applications** L. Davoli and Gi. Ferrari (Editors)
Volume 104	**Enabling Technologies for Social Distancing: Fundamentals, concepts and solutions** D.N. Nguyen, D.T. Hoang, T.X. Vu, E. Dutkiewicz, S. Chatzinotas and B. Ottersten
Volume 105	**Non-Geostationary Satellite Communications Systems** E. Lagunas, S. Chatzinotas, K. An and B.F. Beidas
Volume 108	**The Role of 6G and Beyond on the Road to Net-Zero Carbon** M. Ali Imran, A. Taha, S. Ansari, M. Usman and Q.H. Abbasi (Editors)

Physical Layer Security for 6G Networks

Edited by
Trung Q. Duong, Junqing Zhang, Nan Yang,
Xiangyun Zhou and Vishal Sharma

The Institution of Engineering and Technology

Published by The Institution of Engineering and Technology, London, United Kingdom

The Institution of Engineering and Technology is registered as a Charity in England & Wales (no. 211014) and Scotland (no. SC038698).

First published 2024

The Institution of Engineering and Technology
Futures Place
Kings Way, Stevenage
Hertfordshire, SG1 2UA, United Kingdom

www.theiet.org

British Library Cataloguing in Publication Data
A catalogue record for this product is available from the British Library

ISBN 978-1-83953-691-5 (hardback)

ISBN 978-1-83953-692-2 (PDF)

Typeset in India by MPS Limited

Cover Image: Yuichiro Chino / Moment via Getty Images

Contents

About the editors

Trung Q. Duong is a Canada Excellence Research Chair at Memorial University of Newfoundland, Canada, and an adjunct professor at Queen's University Belfast, UK. He also holds the prestigious Research Chair of the Royal Academy of Engineering. His current research interests include physical layer security, 5G–6G networks (small-cell networks, ultra-dense networks, UAV, URLLC, caching, energy-harvesting), real-time optimization, data analytics, and machine learning. He has authored or co-authored several books. He is currently serving as an editor for *IEEE Transactions on Vehicular Technology* and *IEEE Wireless Communications Letters*. He is the recipient of the prestigious Royal Academy of Engineering Research Fellowship (2016–20), RAEng Research Chair (2021–25) and won the Newton Prize in 2017. He holds a PhD degree in telecommunications systems from Blekinge Institute of Technology, Sweden.

Junqing Zhang is a senior lecturer in the Department of Electrical Engineering and Electronics at the University of Liverpool, UK. His research interests include the Internet of Things, wireless security, physical-layer security, key generation, and radio frequency fingerprint identification. He has published 35 journal articles in the wireless security area. He serves as a TPC member for major IEEE conferences. He received his PhD degree in electronics and electrical engineering from Queen's University Belfast, UK.

Nan Yang is an associate professor in the School of Engineering at the Australian National University, Australia. His research interests include cyber-physical security, terahertz communications, ultra-reliable low-latency communications, and molecular communications. He serves on the editorial board of *IEEE Transactions on Molecular, Biological, and Multi-Scale Communications* and *IEEE Communications Letters*. He received the IEEE ComSoc Asia-Pacific Outstanding Young Researcher Award in 2014. He received his PhD degree in electronic engineering from Beijing Institute of Technology, China.

Xiangyun Zhou is an associate professor at the School of Engineering, the Australian National University, Australia. His research interests cover communication theory and wireless networks, with over 180 publications to date. He has served as an editor of *IEEE Transactions on Wireless Communications*, *IEEE Transactions on*

Communications, and *IEEE Wireless Communications Letters* and as an area editor of *IEEE Communications Letters*. He is a fellow of the IEEE.

Vishal Sharma is a senior lecturer in the School of Electronics, Electrical Engineering and Computer Science (EEECS) at Queen's University Belfast (QUB), Northern Ireland, UK. His areas of interest include defense and security, UAVs, distributed ledger technology, and digital twins. He has authored over 100 journal and conference articles and book chapters, co-edited two books, and won six best paper awards. He served/serves on the editorial board of *IEEE Communications Magazine, CAAI Transactions on Intelligence Technology, IET Wireless Communications and Mobile Computing, IET Networks, ICT Express* and was the section editor-in-chief of *Drones* journal. Before coming to QUB, he was a research fellow in the Information Systems Technology and Design (ISTD) Pillar at the Singapore University of Technology and Design (SUTD), Singapore, with Prof. Paweł Szałachowski and Prof. Jianying Zhou, under a project entitled, future-proof blockchain, funded by MoE-Tier 2 grant, Singapore. Before joining SUTD, he worked as a research assistant professor under Industry-Academia Cooperation Foundation in the MobiSec Lab at Soonchunhyang University with research sponsored by ETRI-Korea, NRF-Korea, IITP-Korea, MSIT-Korea, and AFOSR-USA. He was a postdoctoral researcher with Prof. Ilsun You in the Information Security Engineering Department at Soonchunhyang University in association with Soongsil University, South Korea. Since 2022, he has been affiliated with the Global Innovation Institute as a fellow and works with the Centre for Secure Information Technologies (CSIT) and Centre for Data Science and Scalable Computing (DSSC). He is the director of the Innovation-by-design lab at QUB. He leads the British Computer Society (BCS) Accreditation for QUB and is the Chair of the Computer Science Programme Review Working Group. He is the Northern Ireland Advanced Research and Engineering Centre (ARC) co-investigator. He is a senior member of IEEE and a professional member of ACM.

Preface

The evolution of wireless technologies has been a key driver in the digital transformation and seamless connectivity of human society. While the world's telecommunications service providers are busy deploying 5G networks and services to major metropolitan areas, extensive discussions and research on the next-generation wireless technologies, namely 6G, are well underway. A wide range of 6G-enabled use cases has been envisioned, from holographic teleportation, and digital twins, to autonomous and intelligence-embedded manufacturing. These use cases require a level of data rate and end-to-end latency to be one-to-two orders of magnitude better than what 5G can support. As a result, a fast-growing effort has been devoted to improved and new wireless access and networking technologies from both academia and key industry players.

One major consequence of realizing the 6G-enabled applications is the exponentially growing amount of data, a big portion of which is private and sensitive, being communicated through wireless links. For example, holographic teleportation enables high-resolution and ultra-reliable remote diagnosis and surgery in healthcare, where personal information is sent through the air. The modeling and controlling of the physical systems and environments through their digital twins require a huge volume of sensing and control data to be generated, often in a distributed manner, and wirelessly transmitted. People and organizations would be highly skeptical about adopting such technologies if there is no strong security protection in denying unauthorized access to the data or preventing unexpected interruption of the service. With machine intelligence being envisioned to be an innate feature in many communication and computing infrastructures, which form the core of tomorrow's digital society, the security requirement needs to be extended from data to the machine-learning models and parameters, particularly when the updates of such models and parameters rely on wireless links.

It is, therefore, of paramount importance to carefully consider the security implications of potential 6G technologies and provide solutions to respond to security threats. This requires efforts through different layers of the networking protocol stack. Relatively speaking, the physical layer is the least investigated layer for security protection. Nevertheless, the research in the past two decades has demonstrated a great potential for leveraging the characteristics of wireless transceivers, smart signal processing techniques, as well as the randomness and interference in the wireless propagation medium, for security enhancements. This motivated a large amount of recent research attempting to tackle the security issues of the envisioned 6G technologies at the physical layer.

The book chapters are arranged into four parts according to their topics. In **Part I**, there are four chapters covering the fundamental theories of physical layer security. *Chapter 1* considers the confidential data transmission over the wiretap channels. First, new wiretap code designs by machine learning are proposed and compared to several recent designs. Second, the resulting secrecy outage capacity is studied for statistically dependent channels to the legitimate receiver and eavesdropper. The results enable practical wiretap designs for future secure communication systems. *Chapter 2* describes the fundamental system model, challenges, and technical solutions to terahertz (THz) physical layer security. Several THz secure or covert communication schemes are introduced and elaborated. Open challenges are presented to point out some possible research directions for physical layer security. *Chapter 3* also studies physical layer security with THz communications. It discusses challenges in 6G security, especially due to environmental factors like scattering and eavesdropping. Drawing from extensive research, the chapter outlines the evolution and potential pitfalls in securing next-generation networks. *Chapter 4* explores the potential of intelligent reflecting surface (IRS), an appealing cutting-edge technique in 6G, in enhancing physical layer security for wireless communication networks. Four scenarios of IRS-aided physical layer security are investigated, including the multi-input multi-output system, the pattern index modulation system, the integrated sensing and communication system, and the artificial intelligence-driven system, with a design framework established for each case. Several interesting and challenging open issues are also presented for future research works.

Part II consists of three papers on physical layer authentication using channel features and hardware impairments. *Chapter 5* overviews the physical layer authentication technique and its challenges. Two adaptive multi-dimensional physical layer authentication schemes are proposed, including a non-parametric model based on the back-propagation neural network and a parametric model based on fuzzy learning. The proposed physical layer authentication schemes relying on time-varying features are modeled as an intelligent process, which can achieve more reliable and efficient authentication performance. *Chapter 6* designs a radio frequency fingerprint identification system, which authenticates wireless devices by analyzing their inherent hardware imperfections. In order to enhance identification performance, deep learning techniques are employed for feature extraction. The chapter employs LoRa as a practical example and includes experiments involving off-the-shelf LoRa devices and a USRP SDR receiver. *Chapter 7* also studies radio frequency fingerprint identification. The concept, origin of radio frequency fingerprint (RFF), typical RFF extraction, and identification methods are elaborated. An example is presented for Bluetooth low energy (BLE) devices to demonstrate the application for wireless device authentication.

Key generation is an important branch of physical layer security, which exploits channel randomness as cryptographic keys. **Part III** provides good coverage of the recent research effort on key generation. *Chapter 8* presents an overview of secret key generation from wireless channels. The five stages of processing (channel sounding, feature extraction, quantization, information reconciliation, and privacy amplification) to produce secret keys are described and explained with some examples.

Chapter 9 studies the effects of reconfigurable intelligent surface (RIS) on physical layer key generation (PKG) from both constructive and destructive perspectives. Static and wave-blockage environments are identified as two RIS-empowered-PKG applications in future wireless systems, while RIS could be utilized by an attacker to perform new jamming and leakage attacks. *Chapter 10* explores key generation within LoRa-enabled networks, presenting different schemes and their practical implications. As LoRa forms a critical part of LPWAN technology, the chapter emphasizes the growing importance and anticipated research interest in enhancing physical layer security in LoRa networks. *Chapter 11* discusses general directions for the incorporation of physical layer security and more precisely secret key generation in future networks, starting with 6G. Aspects arising from the use of finite and short-blocklength encoders in particular are discussed. Furthermore, the majority of known attacks on SKG and countermeasures are presented. In particular, eavesdropping (on-the-shoulder) attacks, as well as man-in-the-middle, denial of service, and spoofing attacks are addressed and SKG is shown to be a mature technology for being incorporated in 6G systems. *Chapter 12* presents multi-user secret key generation in massive MIMO wireless communications. Several parameters of secret key generation are designed and the security is discussed, which can help wireless system engineers to apply the physical layer of secret key generation in wireless communication systems and research to further study the secret key generation in the future.

Finally, the applications of physical layer security to various networks and areas are presented in **Part IV**. *Chapter 13* designs a secure transmission scheme for unmanned aerial vehicle (UAV) wireless networks in which the source transmits artificial noise, along with information signals, to interfere with the eavesdropper. The chapter also presents an analytical framework to evaluate the performance achieved by our proposed scheme, thus providing valuable insights into the design of novel secure transmission schemes for UAV wireless networks. *Chapter 14* designs a physical layer-based secure transmission protocol for a multiuser non-orthogonal multiple access (NOMA) radio frequency energy harvesting (RF EH) relaying mobile edge computing system. Furthermore, to evaluate the secrecy performance, closed-form expressions of the existence probability of secrecy capacity and secrecy outage probability for selected users and relay are derived. This chapter also presents an optimization algorithm based on particle swarm optimization to obtain an optimal parameter set. It is proved that the secrecy performance under the proposed protocol and optimal algorithm is superior to that of the traditional orthogonal multiple access solution. *Chapter 15* investigates two key features for 6G location security, i.e., the security associated with localization services, aiming at ensuring that the reportedly derived location is correct, and location privacy techniques, aiming at preventing the disclosure of the position information to other entities. It also discusses new solutions based on quantum security to improve location privacy and verification. *Chapter 16* discusses a physical-layer keyless secure wireless transmission technology, namely directional modulation, from its origin to its recent development. Its mathematical principle has been revealed, which boosts the development of both analog and digital variants. The energy aspect is also investigated, which leads to a few new energy-efficient implementations. *Chapter 17* presents a brief introduction about how NOMA interplays with many enabling techniques of 6G in the area of PHY security. The corresponding open problems and future research directions are discussed. Taking RIS-NOMA as an example, how the security advantage brought by RIS to NOMA can be exploited is presented respectively for the typical external- or internal-eavesdropping scenario. *Chapter 18* discusses the decentralized security of Industrial Cyber-Physical Systems (ICPSs), focusing on unpredictable adversaries that can be detected using blockchain. The chapter explains how digital warfare and vandalism are critical motivators for adversaries that can fool the detection using intelligent techniques and prevent themselves from being logged in the attack corpus. It further explains using decentralized technologies to help tame these adversaries and several open issues relevant to ICPSs. *Chapter 19* employs a support vector machine (SVM) for detecting active spoof attacks. The training process considers both active attackers with known attack channels and passive eavesdroppers without channel information. The authors carry out a comprehensive simulation evaluation, where four kernel functions of SVM are studied.

The research and development of physical layer security technologies have been extensively conducted over the past more than two decades. The priceless efforts from our global telecommunications community have greatly deepened our understanding of harnessing the unique properties of wireless propagation environment to safeguard various emerging communication systems, ranging from large-scale multi-antenna systems, to millimeter wave and THz systems, to UAV-aided systems, to RIS-enabled systems, and to ultra-reliable low latency communication (URLLC) systems. Such advancement stands as a grand leap forward, compared to the fundamental work on physical layer security at the early stage, which has focused on basic multiuser transmission and multi-antenna systems. The technical contributions presented in the chapters of this book collectively narrate the tale of how physical layer security plays a pivotal role in securing information and data transmitted across next-generation wireless networks, achieved through an exploration of physical layer characteristics. Despite so, it is important to acknowledge that physical layer security technologies are still in a relatively nascent stage of theoretical development, while their incorporation into wireless standards has not been realized. Also, a comprehensive investigation into the precise alignment of physical layer security technologies with the key quality-of-service requirements anticipated by 6G wireless networks remains lacking. To propel physical layer security forward and ultimately realize it, our telecommunications community must continue to commit unwavering and persistent efforts to its research, development, and implementation.

Anticipating the upcoming decade, it is imperative for the community to meticulously assess the potential held by physical layer security technologies in addressing the novel security challenges and opportunities that 6G will bring. Notably, the ITU-R has recently finalized the recommendation framework for IMT-2030, which outlines the global vision for 6G. Within this vision, 6G wireless networks are projected to support six distinct usage scenarios, where three scenarios are extensions of those in 5G, while the remaining three are entirely new. Particularly, 6G will enable:

i Immersive communication, which is the enhancement of enhanced Mobile Broadband (eMBB), aiming to deliver enhanced accessibility to a wide range of multimedia content, services, and data, and ensure a seamless user experience.

ii Massive communication, which is the enhancement of massive machine-type communications (mMTC), aiming to support a massive number of low-power and low-complexity Internet of Things (IoT) devices.
iii Hyper-reliable and low-latency communication, which is the enhancement of URLLC, aiming to offer real-time communication services with extremely low latency and high reliability.
iv Ubiquitous connectivity, which is a new usage scenario, aiming to ensure that communication networks cover remote and under-served areas with reliable connectivity.
v Integrated artificial intelligence (AI) and communication, which is a new usage scenario, aiming to create an efficient, adaptive, and intelligent network to deliver a highly customized and context-aware user experience.
vi Integrated sensing and communication (ISAC), which is a new usage scenario, aiming to underpin real-time data acquisition from diverse sensors and create an interconnected and data-driven network.

In order to seamlessly integrate physical layer security technologies into 6G wireless networks, the community must pioneer innovative disruptions that enable flexible, scalable, and adaptive security mechanisms tailored to the security requirements within each distinct usage scenario. To achieve this goal, the development of physical layer security technologies should be rooted in a deep and accurate comprehension of the communication medium in 6G, complemented by demonstrated adaptiveness and agility by monitoring the communication environment in specific wireless systems. Moreover, the multifaceted attributes of 6G usage scenarios, such as the characteristics of signal propagation at the THz frequency band, a huge number of IoT devices, ultra-low-latency transmission, edge- and device-level artificial intelligence, and dual-functional integrated sensing and communication waveforms, are required to be fully incorporated into the design of 6G-oriented physical layer security technologies. Furthermore, the cross-layer testing and implementation of physical layer security technologies in tandem with cryptographic solutions need to be accelerated. This will allow standardization bodies and industry collaborators to seamlessly incorporate physical layer security within the operational guidelines and protocols that will pave the way for 6G realization.

The creation of this book has been made possible through the assistance and support of many people. The Editors extend their heartfelt gratitude to the chapter authors for their valuable contributions. Also, the Editors express their appreciation for the diligent efforts of many reviewers whose input greatly enhanced the quality of each chapter. Finally, the Editors would like to convey their deep appreciation for the steadfast support offered by the IET staff throughout the preparation process of this book.

Trung Q. Duong, Junqing Zhang, Nan Yang,
Xiangyun (Sean) Zhou, and Vishal Sharma
Editors

Part I

Fundamental theories of physical layer security

Chapter 1

Learned codes for dependent wiretap channels*

Karl-Ludwig Besser[1,2], *Pin-Hsun Lin*[1,2] *and Eduard A. Jorswieck*[1,2]

1.1 Introduction

The sixth generation of mobile communications, which is currently investigated in the research departments at industrial and academic institutes, will continue in the following directions: higher frequency bands (sub-THz), a larger number of antennas [extreme multiple-input multiple-output (MIMO)], and nodes (massive multiple access), and enable integrated sensing and communications with learning capabilities distributed among all nodes in the network, including edge and mobile nodes. These technologies will allow consumers as well as industrial nodes to exchange their data with higher data rates, lower latency, and higher reliability [3].

On the one hand, these vast amounts of data, both from consumers and machines, will contain sensitive information which needs to be secured. On the other hand, these technologies will enable new intelligent, adaptive, and aware physical layer security (PHYSEC) schemes. Therefore, PHYSEC is considered an important technology to secure the 6G communication links [4–6].

In the recent survey papers [7–10], the role of PHYSEC for 6G and vice versa is highlighted. Open security issues in 5G and research challenges ahead of 6G are reviewed in [7]. Furthermore, a roadmap to address these challenges is presented and viable solutions to address specific security vulnerabilities in 5G and 6G are outlined. The authors of [8] provide a vision of 6G security and security key performance indicator (KPI) with the tentative threat landscape based on the foreseen 6G network architecture. Security considerations with 6G-enabling technologies such as distributed ledger technology (DLT), PHYSEC, distributed AI/ML, visible light communication (VLC), THz, and quantum computing are discussed. A systematic overview of security and privacy issues is provided in [9] based on future technologies for 6G in the physical, connection, and service layers, as well as through lessons learned from the failures of existing security architectures and state-of-the-art defenses. A review of fundamental contributions to information security can be

*Parts of this chapter have been reported partly in [1] and [2].

[1]Department of Electrical and Computer Engineering, Princeton University, USA
[2]Technische Universität Braunschweig, Institute for Communications Technology, Germany

found in [10]. The tutorial makes the case that information-theoretic (IT) secrecy and privacy metrics have well-grounded cryptography properties, reviews how models for IT security have evolved since the beginning of the field, and emphasizes the meaningful connections between the foundational results and the tools that would accurately reflect the foundational insights in system design. Finally, the paper [6] reviews a subset of promising PHYSEC techniques which can enable and support a range of crucial key value indicators (KVIs).

Besides cryptography, PHYSEC [11] is a promising approach to enable secure data transmission. In contrast to cryptography, no shared key is required, but the physical properties of the wireless channel are exploited.

Additionally, PHYSEC provides information-theoretic security, i.e., the attacker does not obtain sufficient information to attack the system. This establishes post-quantum security by information-theoretic methods and offers better protection than many common cryptography schemes, which rely on the complexity of certain computations, that can be solved fast on quantum computers [12].

One critical challenge in the application of PHYSEC technologies is the uncertainty with respect to the physical parameters of the attacker. If confidentiality is considered, the uncertainty comprises the location and the channel state and received signal-to-noise ratio (SNR) at the wiretapper. Therefore, the scenario and its parameters impact the decision to apply PHYSEC. If there exists an advantage of the legitimate link compared to the eavesdropper, PHYSEC can be applied to generate a covert or stealth communication link [13,14], a confidential link [15], or a secret key [16]. For scenarios, in which the transmitter does not know if there will be an advantage or not, a combination of both cryptography and PHYSEC to achieve hybrid security is proposed in [7]. In situations where the physical layer parameters cannot contribute to improved security, it is then always possible to fall back to the classical cryptography solution.

In scenarios where PHYSEC can be applied, the main benefits compared to and in addition to cryptography algorithms are the following:

1. Often the PHYSEC approaches do not require infrastructure, e.g., both wiretap coding as well as secret key generation work without another node that distributes secret keys or performs public key management.
2. The overhead and latency associated with PHYSEC procedures are in general smaller than their crypto counterpart. The estimation of parameters on the physical layer is part of the communications protocol anyway. Therefore, no additional communications or signaling is required. If additional signaling is required, then only between the two nodes which establish the secure link.
3. The PHYSEC solutions scale well with the number of devices that participate in the communication model, e.g., in multi-user scenarios PHYSEC-based security can be efficiently established for individual pairs of users whenever they are able to communicate. This includes scenarios with many devices like Internet of Things (IoT), where key management among the devices could be very challenging. Furthermore, the multiple access interference can be additionally applied as physical layer parameter to benefit the secure links.

In this work, we focus on confidential data transmission between one transmitter and one receiver with a passive eavesdropper, but no active attacks,† DoS attacks, and other threats. We consider the construction of wiretap codes by a machine-learning approach and its performance for dependent wiretap channels. The material presented in the chapter is based on the publications in [1,2,18–21].

1.1.1 State of the art

In the following, we give an overview of using neural networks (NNs) for designing channel codes without and with secrecy constraint.

In [22], the authors use deep NNs for one-shot decoding of random and structured codes, such as polar codes. They observe that structured codes are easier to learn and the NN can generalize to codewords that it has never seen during training for structured, but not for random codes. These results show that NNs can learn a form of decoding algorithm, rather than only a simple classifier. In [23], the authors interpret a communication system as an autoencoder and develop a new way to think about communication system design as an end-to-end reconstruction task that seeks to jointly optimize transmitter and receiver components in a single process. They also show how this idea can be extended to networks of multiple transmitters and receivers. In [24], the authors show that over-the-air transmissions are possible: a complete communication system can be built, trained, and run, solely composed of NNs using non-synchronized off-the-shelf software-defined radios and open-source deep-learning software libraries. They also extend the existing ideas toward continuous data transmission, which not only eases their current restriction to short block lengths but also entails the issue of receiver synchronization. In [25], the authors present a learning algorithm which alleviates the problem that an end-to-end learning of communication systems through NN-based autoencoders requires a differentiable channel model. Their proposed algorithm iterates between supervised training of the receiver and reinforcement learning (RL)-based training of the transmitter. In [26], the authors use deep NNs to develop an end-to-end wireless communication system, where NNs are applied to all signal-related functionalities, including encoding, decoding, modulation, and equalization. In [27], the authors present an approach that utilizes a recently proposed neural estimator of mutual information. In particular, they use this estimator to optimize the encoder for maximized mutual information, only relying on channel samples.

1.1.1.1 Wiretap code design by NN

The works discussed above only consider a peaceful communication system. Therefore, we give a brief overview of the state-of-the-art works on using NNs to design wiretap codes in a PHYSEC scenario.

In [28], the authors develop the code separately in a secrecy layer and a reliability layer. The reliability layer only deals with the reliability constraint by an autoencoder

†Active attacks have been considered in the physical layer security under the theme of arbitrarily varying wiretap channels [17].

via feed-forward neural networks (FF-NNs). The secrecy layer only deals with the secrecy constraint by a hash function. This approach allows a simplified code design, e.g., if only one of the two layers needs to be modified. Note that the information leakage is measured by a neural network estimator invented by [29]. In particular, mutual information can be equivalently expressed by Kullback–Leibler (KL) divergence, while Donsker–Varadhan representation is a functional lower bound of the Kullback-Leibler (KL) divergence. The optimal solution can reach the KL divergence with equality. Mutual information neural estimator (MINE) was invented in [29], where the authors propose to use an NN to solve the optimal function based on a measured pair of sequences. However, using MINE to estimate the information leakage may underestimate the leakage since MINE leads to a lower bound. Nevertheless, it should be mentioned that MINE is proved to be a consistent estimator [29]. In contrast to the lower bound of KL divergence derived by [29], the authors in [30] investigate the way of finding the upper bound of KL divergence also by NN.

In [31], the authors use an autoencoder to model the processing between Alice and Bob. Instead of considering any processing at Eve in the theory of physical layer security, another NN is used at Eve, trying to decode the transmitted messages. The objective function of training the whole system is a weighted difference between two cross-entropies: one is between the channel input distribution and the distribution of the decoded messages at Bob, and another is between the channel input distribution and the distribution of the decoded messages at Eve.

In [32], the authors study an RL approach for secure encoding. In particular, to derive the NN parameters, a per sample loss function, which is a weighted sum of per sample losses derived from reliability as well as secrecy, is used to update the gradient of a policy function. The secure encoding approach uses the altered one-hot input distribution for model-aware training. The authors show that the resulting method can learn to cluster the codewords into co-sets as the binning structure, enabling secrecy with appropriate encoding without model knowledge.

In contrast to the above works, in [33], the authors consider a model of transmitting images through a binary wiretap channel. In particular, Alice and Bob use a variational autoencoder (VAE) and Eve uses a classifier based on a neural joint source-channel coding from [34]. More specifically, Alice and Bob share data like images, while they keep the latent variable inside the images unknown to Eve. The authors' goal is to minimize the weighted sum of the reconstruction distortion (of the image) and the information leakage, meanwhile, to maximize the mutual information between the transmitted image and the received signal at Bob. Then they invoke the above neural network setting to solve the encoder and the decoder.

The overall comparison of wiretap code design by NN is summarized in Table 1.1. Note that the papers [31,32] do not provide the measured leakage. We use the goodput at Eve as the information leakage (IL) here, i.e., the product of the number of message bits and the probability of correct decoding at Eve.

1.1.1.2 Correlated wiretap channels and secrecy outage capacity

When we transmit a wiretap codeword through a slow fading channel, while the transmitter only knows the fading distribution, a well-accepted performance metric is

Table 1.1 Comparison of related works

Aspects	References [2,18]	[28]	[31]	[32]	[33]
Type of wiretap channel	G-WTC	G-WTC	G-WTC	G-WTC	BS-WTC
Message type	Uniform	Uniform	Uniform	Uniform	Image
ENC	AE	AE/Hash	AE	RL	VAE
DEC	AE	AE/Hash	AE	NN	VAE
Codeword field	$\mathbb{R}$	$\mathbb{R}$	$\mathbb{R}$	$\mathbb{R}$	{0, 1}
Eve's operation	NA	NA	NN	NN	Classifier [34]
Reliability constraint derivation	MSE	Cross entropy	Cross entropy	Cross entropy	Reconstruction distortion – MI between Alice and Bob
Secrecy constraint derivation	Boundings of IL	MINE [29]	Weighted sum of cross entropies at Bob and Eve	Policy gradient method	Boundings of IL by [35]
Training SNR (Bob/Eve)	0/−5 dB	9/−5 dB	10/7 dB	diff = 12 dB	$\epsilon_B = 0.1, \epsilon_E = 0.3$
Max block length	64	16	32	32	200
Secrecy code rate	$\frac{4}{64}$	$\frac{2}{16}/\frac{2}{13}$	$\frac{2}{32}$	$\frac{2}{32}$	NA
Reliability measure	$P_e = 0.08 \sim 0.23$	$P_e = 10^{-4} \sim 10^{-5}$	$P_e = 10^{-1} \sim 10^{-4}$	$P_e = 10^{-1} \sim 10^{-5}$	NA
Secrecy measure (IL)	$0.7 \sim 1.4$ bit	$10^{-1} \sim 10^{-3}$ bit	$0.012 \sim 0.5$ bit	NA	$0.6 \sim 1.1$ bit
Binning structure	✓	–	✓	✓	–

the secrecy outage probability given a specific secrecy outage capacity (or the secrecy outage capacity given a specific secrecy outage probability) [11].

While there exist multiple definitions for an outage event, the most common definition is that a secrecy outage occurs, if the instantaneous secrecy capacity C_S of the channel is less than the secrecy rate R_S that is used for the transmission. Based on this, the secrecy outage probability is given as [11, Def. 5.1]

$$\varepsilon = \mathbb{P}\left(C_S < R_S\right). \tag{1.1}$$

If the transmitter has perfect channel state information (CSI) about the channel realizations, the rate of the used code can be adjusted such that no outage occurs. However, in practice, this knowledge might not be available but only statistical CSI. If there is no diversity between Alice and Bob, an outage cannot be avoided in this case, i.e., the outage probability is strictly positive. In order to reduce the outage probability, a diversity can be established, e.g., by using multiple antennas at the receiver. In such a setup, the secrecy outage probability depends on the joint distribution of all fading links. The marginal distributions of the fading coefficients for different positions (or users) can easily be measured. However, the dependency between them is usually unclear because the attacker might choose position and orientation according to the legitimate transmitter and receiver. On the other hand, relaying technologies as well as new emerging technologies like reconfigurable intelligent surfaces [36] might enable a proactive design of the radio environment and, therefore, a control of the dependency structure.

In the literature, usually, independent legitimate and eavesdropper channels are considered [37–41]. However, real measurements demonstrate that the commonly used assumption of independent channels is not always justified [42,43]. In particular, if the eavesdropper chooses its position and orientation carefully. Therefore, some work has already studied the influence of the correlation between channels on the secrecy performance. The outage probability and outage secrecy capacity for correlated wiretap channels at high SNR are investigated in [44]. In [45], bounds on the secrecy capacity for correlated fading channels are provided. However, the results also hold only for the high SNR regime and only for the specific case of Rayleigh fading. Additionally, the not well-justified assumption of full channel-state information at the transmitter (CSI-T) about both channels is assumed. The results of [45] are extended in [46] for general SNR. A different fading distribution, namely log-normal fading, is considered in [47]. Additional differences are that the author of [47] focuses on the outage probability instead of the secrecy capacity and no CSI-T about the eavesdropper's channel is assumed. The secrecy outage probability of correlated channels with small-scale fading and shadowing has been investigated in [48]. In all previous works, only (positive) linear correlation is assumed. A work that considers more general dependencies described by a specific copula is [49].

However, when including all possible joint distributions for given marginals, it can be shown that there can exist a significant performance gap compared to only linear dependencies [50]. In particular, it is shown in [21] that it is possible to achieve zero secrecy outages for dependent fading channels when at least two connections to Bob exist. This implies that a positive zero-outage secrecy capacity (ZOSC) can be achieved for dependent channels with only statistical CSI-T, i.e., one can transmit at a positive rate without any outages without requiring perfect CSI-T. The development of the positive ZOSC is motivated by the notion of a positive ZOC for dependent channels without secrecy constraint [51,52]. The main idea is that, given marginal fading gain distributions, it is possible to find joint fading distributions, such that the corresponding support leads to a positive value of the ZOC/ZOSC.

1.1.2 Outline of chapter

First, we introduce the system model of the considered Gaussian wiretap channel in the next section. We then describe an extension of this model to a two-state channel with a possible correlation of Bob's and Eve's channel states. There already exist results on various performance metrics for this general scenario of dependent fading channels. Some of these results for the outage probability are restated at the end of Section 1.2.

After introducing the necessary preliminaries, we describe a framework for designing finite block length wiretap codes using NN autoencoders in Section 1.3.

This approach is numerically evaluated in Section 1.4. After demonstrating the effectiveness of the basic Gaussian wiretap channel, we determine the outage probability of the learned code for the previously described two-state channel with dependency.

Finally, the chapter is concluded in Section 1.5, and a summary of open research directions is provided.

1.2 Preliminaries

In this section, we introduce the basic system model along with some preliminaries from the area of physical layer security. The considered model for fixed channel parameters matches the one in [2].

1.2.1 System model

In the following, we consider a Gaussian wiretap channel model, where the transmitter Alice wants to confidentially transmit a message $\boldsymbol{M}$ from the set of possible messages $\mathcal{M}$ to the legitimate user Bob. The communication is overheard through a second channel by a passive eavesdropper, Eve. Both the main channel between Alice and Bob and the eavesdropper's channel between Alice and Eve are assumed to have independent additive white Gaussian noises (AWGN) $\boldsymbol{N}_{\mathrm{B}}$ and $\boldsymbol{N}_{\mathrm{E}}$ with zero mean and variances σ_B^2 and σ_E^2, respectively. The received signals at Bob and Eve are then given as

$$\boldsymbol{Y} = \boldsymbol{X} + \boldsymbol{N}_{\mathrm{B}}, \tag{1.2}$$

$$\boldsymbol{Z} = \boldsymbol{X} + \boldsymbol{N}_{\mathrm{E}}, \tag{1.3}$$

respectively. The output $\widehat{\boldsymbol{M}}_{\mathrm{B}}$ of Bob's decoder represents his estimation of the sent message $\boldsymbol{M}$. The eavesdropper's estimation of the sent message is denoted as $\widehat{\boldsymbol{M}}_{\mathrm{E}}$.

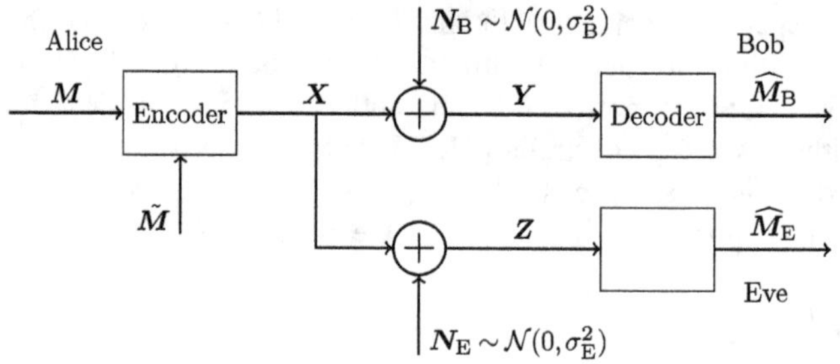

Figure 1.1 System model of a Gaussian wiretap channel

However, we do not assume a specific decoding operation at Eve in order to not underestimate the attacker. The system model is depicted in Figure 1.1.

To protect the information, Alice uses a wiretap code with a binning structure [11], where secret messages $\boldsymbol{M} \in \mathcal{M}$ of length k together with confusion messages $\tilde{\boldsymbol{M}} \in \tilde{\mathcal{M}}$ of r random bits are mapped to codewords $\boldsymbol{X} \in \mathcal{C}$. Thus, for each secret message m, 2^r different codewords $\mathcal{C}_m$ exist. The stochastic encoder function γ and the encoder function γ_m for message m are given as

$$\gamma : \mathcal{M} \times \tilde{\mathcal{M}} \to \mathcal{C}, \tag{1.4}$$

$$\gamma_m : \tilde{\mathcal{M}} \to \mathcal{C}_m, \tag{1.5}$$

where the messages and random confusion bits are from the sets $\mathcal{M} = \{0, 1\}^k$ and $\tilde{\mathcal{M}} = \{0, 1\}^r$, respectively. The codebook is denoted as $\mathcal{C} \subset \mathbb{R}^n$. The subset $\mathcal{C}_m \subset \mathcal{C}$ contains only the codewords embedding the secret message m. The secrecy rate of such a wiretap code is given as $R_S = k/n$.

Throughout this work, we assume that the secure and confusion messages are uniformly and independently distributed. As usual, there is a power constraint for wireless transmissions, i.e., the variance of the transmitted codewords is not allowed to exceed a certain value P_T

$$\frac{1}{n}\sum_{i=1}^{n} x_i(j)^2 \leq P_T, \ \forall j \in \mathcal{M} \times \tilde{\mathcal{M}}. \tag{1.6}$$

Remark 1. *Note that the codewords are over the real numbers, which implicitly includes the modulation. Therefore, the codewords are directly set as the channel input without further processing.*

1.2.2 Reliability and confidentiality

The point-to-point communication model includes one transmitter (Alice) who wants to reliably transmit a message to one receiver (Bob). In addition, Alice wants to ensure the secrecy of the message against the eavesdropper (Eve). Thus, there exist two performance metrics in parallel.

First, the reliability of the transmission to the legitimate receiver is measured in terms of block error rate (BLER). This is defined as the average probability that the decoded message at Bob $\widehat{M}_{\mathrm{B}}$ is not identical to the message m transmitted by Alice,

$$\mathrm{BLER} = \frac{1}{|\mathcal{M}|} \sum_{m \in \mathcal{M}} \mathbb{P}\left(\widehat{M}_{\mathrm{B}} \neq m | M = m\right). \tag{1.7}$$

Next, we need to quantify the leaked information to Eve. In order to do this, we use the mutual information $I(M;Z)$ between the message M and the received signal at the eavesdropper Z. In order to ensure a secure transmission, the leakage should go to zero for an increasing block length n, i.e.,

$$\lim_{n \to \infty} I(M;Z) = 0, \tag{1.8}$$

which corresponds to the notion of strong secrecy [11].

1.2.3 Transmissions over multiple blocks in slow two-state fading

Next, we consider the transmission of multiple codewords over B blocks. Within each block, $b \in \{1, \dots, B\}$, the properties of the two channels, the legitimate and the wiretap channel, stay constant. This means that the noise variances σ_B^2 and σ_E^2 are constant for the n channel uses within one codeword and then change to a new state.

We model the noise variances as random variables and the codebook has to be learned based only on statistical knowledge. It is reasonable to assume that the legitimate link does not know the channel properties of the wiretapper because it is usually not part of the active communication system. However, it might be possible to make assumptions about the statistical properties of the channel of the wiretapper, i.e., to build a subjective probability distribution about the unknown parameters.

We will assume that the realizations change from one block to the next block in an i.i.d. fashion, however, they are distributed according to an underlying joint probability distribution $p_{\sigma_B,\sigma_E}(\sigma_B, \sigma_E)$. It is difficult to know the joint distribution both theoretically and practically. Theoretically, it would be necessary to develop an underlying model of the propagation environment and derive from their statistical model. In practice, both legitimate receiver and wiretapper would need to transmit orthogonal pilots which are then received at the transmit side to estimate the joint channels.

Hence, we will assume that the transmitter only knows the marginal distribution of the channel to the wiretapper $p_{\sigma_E}(\sigma_E)$. For the legitimate channel, we have two scenarios, one where we know the channel realization σ_B^2 perfectly and one where we have only knowledge about the marginal distribution $p_{\sigma_B}(\sigma_B)$.

This model corresponds to a slow-fading environment and the correct performance metric is the outage probability [53], i.e., the probability that within the next block, the confidential transmission fails.

1.2.3.1 Single links to Bob and Eve

Within this chapter, we will consider a simple bi-variate Bernoulli model with channels with two states, a good and a bad state corresponding to small noise variance σ_{low}^2 (high

Table 1.2 Joint probability distribution for bi-variate Bernoulli noise variances

$\sigma_{\mathrm{B}}^2, \sigma_E^2$	$\sigma^2_{\mathrm{E_{high}}}$	$\sigma^2_{\mathrm{E,low}}$	
$\sigma^2_{\mathrm{B_{high}}}$	p_{00}	p_{01}	$1-p_{\mathrm{B}}$
$\sigma^2_{\mathrm{B_{low}}}$	p_{10}	p_{11}	p_{B}
	$1-p_{\mathrm{E}}$	p_{E}	

SNR) and high noise variance σ^2_{high} (low SNR), respectively. Such binary models with a high and low SNR state could arise in modern millimeter wave (mmWave) systems. Such high-frequency communication systems typically rely on a strong line-of-sight (LoS) connection which is highly susceptible to blockage [54,55]. This effect can be modeled by a binary random variable where we identify the two states with 0 (bad) and 1 (good). Then the joint probability distribution [56, Section 7.1] of the two noise variances can then be described by the following matrix in Table 1.2.

The marginal distributions of the noise variances at Bob and Eve are given by

$$p_{\mathrm{B}} = \mathbb{P}\left(\sigma_{\mathrm{B}}^2 = \sigma^2_{\mathrm{B_{low}}}\right) = p_{10} + p_{11}, \tag{1.9}$$

$$p_{\mathrm{E}} = \mathbb{P}\left(\sigma_{\mathrm{E}}^2 = \sigma^2_{\mathrm{E_{low}}}\right) = p_{01} + p_{11}, \tag{1.10}$$

respectively. While we assume that the marginal probabilities p_B and p_E are known and fixed, the correlation between σ_{B}^2 and σ_{E}^2 may vary. By definition, the correlation coefficient ρ is given for the considered two-state channel as

$$\rho = \frac{p_{11} - p_{\mathrm{B}}p_{\mathrm{E}}}{\sqrt{p_{\mathrm{B}}p_{\mathrm{E}}(1-p_{\mathrm{B}})(1-p_{\mathrm{E}})}}. \tag{1.11}$$

In general, the correlation coefficient is within the interval $[\rho_{\min}, \rho_{\max}]$,

$$-1 \le \rho_{\min} \le \rho \le \rho_{\max} \le 1,$$

where the boundary values depend on the marginal distributions [57, Thm. 3]. For the case of two correlated binary variables, the bounds on the correlation coefficient are given as [56, Chap. 7]

$$\rho_{\min} = \frac{\min\{p_B, p_E\} - p_Bp_E}{\sqrt{p_Bp_E(1-p_B)(1-p_E)}} \le \rho \le$$

$$\frac{\max\{p_B + p_E - 1, 0\} - p_Bp_E}{\sqrt{p_Bp_E(1-p_B)(1-p_E)}} = \rho_{\max}. \tag{1.12}$$

Remark 2. *Depending on the operating point, the communication system may only not be in an outage, if Bob's channel is in the high SNR state and Eve's channel is in* *the low SNR state simultaneously. In this case, the outage probability depends on the joint distribution of the channel states, and thus, it also depends on the correlation coefficient. A numerical example of this relation will be investigated in detail in Section 1.4.1.*

1.2.3.2 Multiple links to bob

When we fix the number of links to Eve as one but increase the number of links to Bob to more than one, the situation will be different from what has been discussed above. Let us take two links to Bob as an example in the following. Denote the equivalent noise variances of the first and second link to Bob as $\sigma_{\mathrm{B}_1^2}$ and $\sigma_{\mathrm{B}_2^2}$, respectively. Note that $\sigma_{\mathrm{B}_1^2}$, $\sigma_{\mathrm{B}_2^2}$ and σ_{E}^2 are assumed to be random two-state variables in the following discussion. Denote the correlation between $\sigma_{\mathrm{B}_1^2}$ and $\sigma_{\mathrm{B}_2^2}$ by ρ_{12}; denote the correlation between $\sigma_{\mathrm{B}_1^2}$ and σ_{E}^2 by ρ_{13}; denote the correlation between $\sigma_{\mathrm{B}_2^2}$ and σ_{E}^2 by ρ_{23}. Similar to the notation in Table 1.2, we use 1 in the subscript to denote the channel is good (low noise power, or equivalently, high SNR) and use 0 in the subscript to denote the channel is bad (high noise power, or equivalently, low SNR). In the considered scenario with given marginal probabilities (when each state is 1) $p_{\mathrm{B}_1}, p_{\mathrm{B}_2}, p_{\mathrm{E}}$, we want to discuss the relation between the joint distribution and the three cross-correlations. In particular, we have a joint distribution described by

$$p = \left(p_{000}\ p_{001}\ p_{010}\ p_{011}\ p_{100}\ p_{101}\ p_{110}\ p_{111}\right)^T. \tag{1.13}$$

By the given marginal probabilities, we know that

$$p_{100} + p_{101} + p_{110} + p_{111} = p_{\mathrm{B}1}, \tag{1.14}$$

$$p_{010} + p_{011} + p_{011} + p_{111} = p_{\mathrm{B}2}, \tag{1.15}$$

$$p_{001} + p_{011} + p_{101} + p_{111} = p_{\mathrm{E}}. \tag{1.16}$$

Feasible joint distribution with fixed marginals

We first consider the case where only marginal distributions are given. We can formulate a new linear system as

$$\mathbf{A}p = c, \tag{1.17}$$

where

$$\mathbf{A} := \begin{pmatrix} 0 & 0 & 0 & 0 & 1 & 1 & 1 & 1 \\ 0 & 0 & 1 & 1 & 0 & 0 & 1 & 1 \\ 0 & 1 & 0 & 1 & 0 & 1 & 0 & 1 \\ 1 & 1 & 1 & 1 & 1 & 1 & 1 & 1 \end{pmatrix} \tag{1.18}$$

and

$$c := (p_{B1}, p_{B2}, p_E, 1)^T. \tag{1.19}$$

Based on (1.17), we can formulate an optimization problem as

$$\begin{aligned} \text{P1}: \quad \min \quad & g(\boldsymbol{p}, p_{B1}, p_{B2}, p_E) && (1.20) \\ \text{s.t.} \quad & (1.17) \\ & p_{ijk} \geq 0,\ i, j, k \in \{0, 1\}, && (1.21) \end{aligned}$$

where $g(\cdot)$ is the objective function depending on the scenario considered. Later in this chapter, we will consider a specific case of g and solve for the optimal solution.

Feasible joint distribution with fixed marginals and cross correlations

When ρ_{12}, ρ_{13}, and ρ_{23} are additionally known/fixed, based on (1.11), we have the following additional constraints

$$c_1 := p_{110} + p_{111} = \rho_{12}\sqrt{p_{B1}p_{B2}(1-p_{B1})(1-p_{B2})} + p_{B1}p_{B2}, \tag{1.22}$$

$$c_2 := p_{101} + p_{111} = \rho_{13}\sqrt{p_{B1}p_E(1-p_{B1})(1-p_E)} + p_{B1}p_E, \tag{1.23}$$

$$c_3 := p_{011} + p_{111} = \rho_{23}\sqrt{p_{B2}p_E(1-p_{B2})(1-p_E)} + p_{B2}p_E. \tag{1.24}$$

Based on (1.14), (1.15), (1.16), (1.21), (1.22), (1.23), and (1.24), we can rearrange these seven constraints as follows:

$$\tilde{\mathbf{A}}\boldsymbol{p} = \tilde{\boldsymbol{c}}, \tag{1.25}$$

where

$$\tilde{\mathbf{A}} := \begin{pmatrix} 0&0&0&0&1&1&1&1 \\ 0&0&1&1&0&0&1&1 \\ 0&1&0&1&0&1&0&1 \\ 0&0&0&0&0&0&1&1 \\ 0&0&0&0&0&1&0&1 \\ 0&0&0&1&0&0&0&1 \\ 1&1&1&1&1&1&1&1 \end{pmatrix} \tag{1.26}$$

and

$$\tilde{\boldsymbol{c}} := (p_{B1}, p_{B2}, p_E, c_1, c_2, c_3, 1)^T. \tag{1.27}$$

Similar to P1, we can formulate an optimization problem. Besides, from (1.25) we know that there are eight variables with seven equations. Therefore, it is possible that given the three marginal distributions and the three cross-correlations, we can have different joint distributions. Note also that, depending on the given marginal distributions and the cross-correlations, it may be possible that there exists no feasible joint distribution. For correlation matrices with specific structures like moving average with order 1 or auto-regression with order 1, it is possible to specify the relation between the joint distribution and the correlation matrix as shown in [58].

1.2.4 General distributions of the channels

As mentioned above, the correlation between the states of Bob's and Eve's channels can have a (significant) impact on the resulting performance, e.g., in terms of the outage probability. Thus, when facing uncertainty about the joint distribution of the channels, it is relevant to determine bounds on the performance with respect to all possible joint distributions, e.g., for worst-case design. This is especially important for modern applications which require an ultra-high reliability [59]. We will, therefore, take a closer look at the influence of the dependency between the channel states in the following.

We denote the receiver SNRs at Bob and Eve as $\xi_{\mathrm{B}} = P_T/\sigma_{\mathrm{B}}^2$ and $\xi_{\mathrm{E}} = P_T/\sigma_{\mathrm{E}}^2$, respectively. For one realization of the slow fading coefficients, the instantaneous secrecy capacity C_S is given by [60, Theorem 1 specialized to SIMOME wiretap channels]

$$C_S = \left[\log_2\left(1 + \sum_{i=1}^{n_B} \xi_{\mathrm{B}} \boldsymbol{X}_i\right) - \log_2\left(1 + \sum_{i=1}^{n_E} \xi_{\mathrm{E}} \boldsymbol{Y}_i\right)\right], \tag{1.28}$$

where $\boldsymbol{X}_i$ and $\boldsymbol{Y}_i$ are the squared magnitudes of the channel gains, and n_B and n_E are the number of antennas at Bob and Eve, respectively.

For the transmission, we consider a constant secrecy rate R_S since the transmitter does not have perfect information about the SNRs. With this, it follows that a secrecy outage event happens, if the secrecy rate is greater than the instantaneous secrecy capacity, i.e., $C_S < R_S$. This leads to the definition of the secrecy outage probability ε as [11, Definition 5.1]

$$\varepsilon = \mathbb{P}\left(C_S < R_S\right). \tag{1.29}$$

Based on the secrecy outage probability, we can define the ε-outage secrecy capacity R_S^ε as the highest rate, such that the secrecy outage probability is at most ε, i.e.,

$$R_S^\varepsilon = \sup_{R \geq 0} \left\{R \mid \mathbb{P}\left(C_S < R\right) \leq \varepsilon\right\}. \tag{1.30}$$

Remark 3. *It should be noted that we have $\boldsymbol{X}_i = \boldsymbol{Y}_i = 1$ for the simple two-state channel model described above, where no additional channel fading is considered.*

1.2.4.1 Single links to Bob and Eve

We start with the case that both Bob and Eve only have a single antenna, i.e., $n_B = n_E = 1$.

For the described two-state channel model with one link to Bob and one link to Eve, the correlation coefficient ρ fully determines the joint distribution for given marginals p_{B} and p_{E}. However, in the general case with additional given (continuous) marginal fading distributions F_X and F_Y, the linear correlation coefficient does not capture all possible dependencies between the two fading channels. A mathematical framework to describe general dependency structures is copula theory [61]. Based on this, bounds on the secrecy outage probability are derived in [20]. They are valid for general fading channels with continuous marginal probability density functions (PDFs) f_X and f_Y and an arbitrary dependency structure.

Theorem 1 (Lower bound on the secrecy outage probability [20, Theorem 2]). *Let X and Y be random variables over the non-negative real numbers representing the* *squared magnitude of the channel gains to Bob and Eve, respectively. The transmitter has perfect CSI only about the main channel to Bob and no CSI about the channel to Eve. The secrecy outage probability is then lower bounded by*

$$\underline{\varepsilon} = \max_{\tilde{y}^\star \in \mathcal{Y}} \begin{cases} F_{\tilde{X}}(s) & \text{if } f'_{\tilde{\imath}}(\tilde{y}^\star) + f'_{\tilde{X}}(s - \tilde{y}^\star) \geq 0 \\ g\,(\tilde{y}^\star) & \text{if } f'_{\tilde{Y}}(\tilde{y}^\star) + f'_{\tilde{X}}(s - \tilde{y}^\star) < 0 \end{cases}, \tag{1.31}$$

with

$$g(\tilde{y}) = F_{\tilde{X}}(s - \tilde{y}) + F_{\tilde{Y}}(\tilde{y}) - 1, \tag{1.32}$$

and where the maximum is over all $\tilde{y}^\star \leq 0$. *The set of feasible* $\tilde{y}^\star$ *is the following:*

$$\mathcal{Y} = \left\{\tilde{y}^\star \mid \tilde{y}^\star < 0 \wedge f_{\tilde{Y}}(\tilde{y}^\star) = f_{\tilde{X}}(s - \tilde{y}^\star)\right\} \cup \{0\}. \tag{1.33}$$

The used shorthands are $\tilde{\boldsymbol{X}} = \xi_{\mathrm{B}}\boldsymbol{X}$, $\tilde{\boldsymbol{Y}} = -2^{R_S}\xi_{\mathrm{E}}\boldsymbol{Y}$ *and* $s = 2^{R_S} - 1$.

The lower bound on the secrecy outage probability provided by Theorem 1 represents the best case. While this is not required for worst-case design, it can serve as a benchmark and design guideline, if active dependency control is available.

The worst-case outage probability over all possible joint distributions for fixed marginal distributions is given by the upper bound in the following theorem.

Theorem 2 (Upper bound on the secrecy outage probability [20, Theorem 3]). *Let* $\boldsymbol{X}$ *and* $\boldsymbol{Y}$ *be random variables over the non-negative real numbers representing the squared magnitude of the channel gains to Bob and Eve, respectively. The transmitter has perfect CSI about the main channel to Bob. The secrecy outage probability is then upper bounded by*

$$\overline{\varepsilon} = \min_{\tilde{y}^\star \in \mathcal{Z}} \begin{cases} h(\tilde{y}^\star) & \text{if } f'_{\tilde{Y}}(\tilde{y}^\star) + f'_{\tilde{X}}(s - \tilde{y}^\star) \geq 0 \\ 1 & \text{if } f'_{\tilde{Y}}(\tilde{y}^\star) + f'_{\tilde{X}}(s - \tilde{y}^\star) < 0 \end{cases}, \tag{1.34}$$

with

$$h(\tilde{y}) = F_{\tilde{X}}(s - \tilde{y}) + F_{\tilde{Y}}(\tilde{y}) \tag{1.35}$$

and where the minimum is over all $\tilde{y}^\star$ *which are from the following feasible set:*

$$\mathcal{Z} = \left\{\tilde{y}^\star \mid \tilde{y}^\star < 0 \wedge f_{\tilde{Y}}(\tilde{y}^\star) = f_{\tilde{X}}(s - \tilde{y}^\star)\right\} \cup \{-\infty, 0\}. \tag{1.36}$$

Besides the bounds from Theorems 1 and 2, one of the key takeaways from [20] is the mathematical confirmation of the intuitive assumption that the lowest outage probability is achieved for a negative correlation of Bob's and Eve's channel, i.e., if Bob's channel is in a high-SNR state, Eve's channel is in a low-SNR state, and vice versa. In contrast, the worst case, i.e., the highest outage probability, occurs for a strong positive correlation of the channel states.

1.2.4.2 Two links to Bob

Next, we want to extend the results for the multi-antenna scenario where two links are established to Bob simultaneously, i.e., $n_B = 2$. In [52], it is shown that this diversity allows for a positive ZOC when the channels are dependent, i.e., there exists a joint distribution such that it can be transmitted at a positive rate without any outages. From this, the natural question arises, if this is also possible for a PHYSEC system with an eavesdropper, i.e., if a positive ZOSC can be achieved.

Such results about achievable secrecy rates and zero-outage secrecy rates are presented in [21]. For the scenario with two links to Bob, i.e., $n_B = 2$, and one link to Eve, i.e., $n_E = 1$, the following result characterizes an achievable zero-outage secrecy rate for Rayleigh fading.

Theorem 3 ([21, Theorem 3]). *The wiretap channel with two channels to Bob, both with marginal standard exponential distribution, and one channel to Eve exponentially distributed with mean $\mu \geq 0$ can support a zero-outage secrecy rate*

$$\mathrm{ZOSR} = \begin{cases} 0 & \mu \geq 1 \\ \log_2(1/\mu) & 1 \geq \mu \geq \frac{1}{1+\ln(2)} \\ \min\limits_{0\leq U\leq 1} \log_2\left(\frac{\ln(1/4-1/U^2)-1}{\ln(1-U)\mu-1}\right) & \frac{1}{1+\ln(2)} > \mu. \end{cases} \tag{1.37}$$

The result implies that the specific dependency structure where the two SNRs of the two links to Bob are negatively dependent and their sum is positively dependent with the SNR to Eve can support positive zero-outage secrecy rates.

In [21], the authors were not able to show that this dependency structure is optimal. Therefore, one interesting question is about the optimal dependency structure given the discrete Bernoulli distributions of the SNR with their marginals as defined above.

1.3 Autoencoder-based wiretap code design

In this section, we first introduce the problem formulation of learning wiretap codes for a Gaussian wiretap channel. Next, we propose loss functions for a neural network that are suitable for such a learning task. Finally, we present an implementation to solve the problem by an autoencoder as shown in [2].

1.3.1 Problem formulation

The goal of the code design is to minimize both the BLER and the leakage. This is a multi-objective programming problem (MOP) which can be solved by scalarization approaches [62]. Here we adopt the weighted sum and the scalarized objective is expressed as

$$\min_{C} \; w_B \cdot \mathrm{BLER} + w_E \cdot I(\boldsymbol{M};\boldsymbol{Z}), \tag{1.38}$$

with a specific pair of positive objective weights $w_B, w_E \in \mathbb{R}^+$. The solution to this problem yields one point in the BLER-IL tradeoff plane. By varying the weights, the operating point is moved along the Pareto boundary of the performance region.

1.3.2 Feed-forward neural networks

In the following, we want to apply FF-NNs to solve the problem in (1.38). Therefore, we first give a short overview of the fundamental aspects of FF-NN. Further details on NNs can be found, e.g., in [63].

The basic building block of FF-NNs is a neuron. The output y of a single neuron is a weighted sum of its k inputs x (with an optional bias b) which is transformed by a nonlinear activation function f

$$y = f\left(w^{\mathrm{T}}x + b\right), \quad x, w \in \mathbb{R}^k,\ y, b \in \mathbb{R}. \tag{1.39}$$

Multiple neurons can be grouped into a layer. By doing so, the individual vectors containing the weights w_j are stacked to a weighting matrix $W^{(i)} \in \mathbb{R}^{L_i \times k_i}$ of layer i, $i = 1, 2, \ldots, D$, where D represents the total number of layers in the NN. The output $y^{(i)}$ of layer i with L_i neurons is given as,

$$y^{(i)} = f^{(i)}\left(W^{(i)}x^{(i)} + b^{(i)}\right),\ x^{(i)} \in \mathbb{R}^{k_i},\ y^{(i)}, b^{(i)} \in \mathbb{R}^{L_i}. \tag{1.40}$$

To get a so-called *deep neural network*, multiple layers can be concatenated. In this case, the input of layer i is the output of layer $i-1$. The final output $y^{(D)}$ of the deep NN with D layers with input x is given as a composition of (1.40) as

$$\begin{aligned} y^{(D)} &= f^{(D)}\left(W^{(D)}f^{(D-1)}\left(W^{(D-1)}f^{(D-2)}\left(\ldots\right.\right.\right. \\ &\left.\left.\left. f^{(1)}\left(W^{(1)}x + b^{(1)}\right)\ldots\right) + b^{(D-1)}\right) + b^{(D)}\right). \end{aligned} \tag{1.41}$$

Often used activation functions are the rectified linear unit (ReLU) $g(x) = \max(0, x)$, sigmoid $\sigma(x) = \frac{1}{1+\exp(-x)}$, and the softmax function $\mathrm{softmax}(x)_i = \frac{\exp(x_i)}{\sum_{j=1}^{k}\exp(x_j)}$ [64]. Note that among these activation functions, only the softmax function is applied to the vector output of a layer. Sigmoid and ReLU are used component-wise.

This work uses NNs as a supervised learning algorithm [64]. Therefore, a labeled dataset is needed for training, i.e., the training set $\mathcal{T}$ consists of tuples (x_j, y_j) which contain the inputs (observations) $x_j \in \mathbb{R}^{k_1}, j = 1, \ldots, |\mathcal{T}|$ with the corresponding desired output $y_j \in \mathbb{R}^{L_D}$.

The process of applying NNs consists of two phases. In the first phase, namely the *training phase*, a loss function which calculates a loss between the input and the desired output of the training samples, is minimized by modifying the weights and biases in the network. This usually happens by using a gradient-based optimization algorithm. In the second phase, the *prediction phase*, the trained system is used to evaluate new samples which are not part of the training set.

One special way of using NNs is the autoencoder. In this case, the inputs are used as the output labels during training. The network tries to reconstruct the input by first finding an internal representation of the data and decoding this afterwards [63, Chapter 14]. Commonly, this internal representation of the hidden layers can be seen as an encoded version of the data and is often used to extract the relevant information of the data, i.e., for compression. In contrast, for code design in our work, the representation of the data is a redundant version of the message, where the redundancy is used to protect the messages, i.e., it generates codewords.

1.3.3 Implementation by autoencoders

After introducing the basic concept of FF-NNs, we now want to apply them to solve the optimization problem from (1.38). In order to do this, we implement the wiretap system model from Figure 1.1 as an autoencoder. The basic structure of the implementation is shown in Figure 1.2. The highly similar structures of autoencoders and the encoder/decoder in communications make them a natural choice for modeling the communication system. The individual components can be directly translated into components of the NN as follows.‡ The autoencoder has two vector inputs and two vector outputs. The first input $\boldsymbol{M}$ contains k message bits $m_i \in \{0,1\}^k$. The second input $\tilde{\boldsymbol{M}}$ contains r uniformly random bits $\tilde{m}_i \in \{0,1\}^r$ which are used to introduce the confusion messages to the wiretap code. The two outputs of the network are the estimated bits $\hat{m}_i$ at Bob and the codewords x_i. The latter is used to calculate the loss function at Eve, e.g., a leakage estimation. The NN of the autoencoder consists of an encoder part at Alice and a decoder part at Bob. Both parts are implemented as FF-NNs [63]. On top of the last layer of the NN at Alice, a normalization layer is added which centers and scales the data, to be zero mean and unit variance. This is in order to fulfill the power constraint on the codewords.

The two parts are separated by a noise layer which adds Gaussian noise with a variance σ_B^2 to model the AWGN channel between Alice and Bob. The second output of the system, which is used for training, are the channel inputs, i.e., the codewords. At this output, the found codewords from the encoder at Alice and the noise variance of the wiretap channel σ_E^2 are used to calculate L_E.

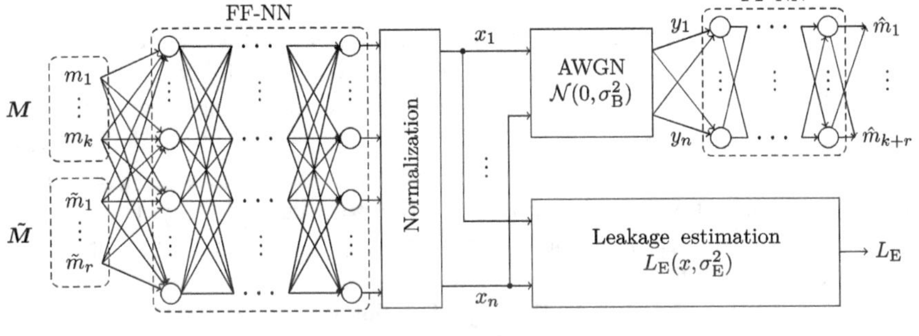

Figure 1.2 Implementation structure of the autoencoder for designing finite block length wiretap codes (adapted from [1]). Both the encoder and decoder at Alice and Bob, respectively, are implemented by FF-NNs, between which is a noise layer. © IEEE 2020. Reprinted with permission from [2].

‡Autoencoder is by no means the only possible choice and using other methods/algorithms is also possible and interesting to investigate. However, it is out of scope for this work and can be considered in future work.

During training, all $k+r$ bits are used for minimizing the loss function L_B at Bob's output. This is to force the network to design a codebook with a binning structure. Otherwise, multiple codewords for one secret message might have a vanishing Euclidean distance in the n-dimensional codeword space, i.e., the NN would design a regular point-to-point code. The training process results in an encoder and a decoder system implicitly including the modulation and demodulation steps.

However, the loss functions in (1.38) cannot be directly solved by NN in practice due to two reasons: first, the estimation of the BLER relies on a sufficiently large number of realizations in a Monte Carlo simulation [63, Ch. 17]. Because in each training epoch the BLER should be estimated again, the Monte Carlo scheme will cause the complexity of the training to be unacceptably high due to a large number of training epochs in our setting [22]. Second, due to the fact that the received signal $\boldsymbol{Z}$ at Eve is a Gaussian mixture (GM), an analytical closed form of $I(\boldsymbol{M};\boldsymbol{Z})$ is still unknown in general [65].

Therefore, we choose the following loss functions at Bob and Eve when training the autoencoder.

1.3.3.1 Choice of loss functions

At Bob, the mean squared error (MSE) is used instead of the BLER, i.e., we set

$$L_{\mathrm{B}} = \mathbb{E}_{\mathcal{T}}\left[\frac{1}{k+r}\left\|\hat{m} - m'\right\|^2\right], \tag{1.42}$$

where $\hat{m}$ is the $k+r$-dimensional output of the NN, m' is the desired output, i.e., the message input $(m, \tilde{m})$, and the expectation is taken over the whole training set $\mathcal{T}$. It is shown in [22] that this loss function and the error rate are related such that reducing this loss function reduces the error rate. The idea behind this is the following. During the testing phase, the soft-output of the NN is rounded to bits. To get a correct prediction, the output of the value should be close to the desired value. Thus, the loss function should penalize more the farther it is away from the desired output. This is fulfilled by the MSE loss function. Note that other loss functions like the binary cross-entropy showed similar results [22]. Note that this MSE loss function is only used during training. Once the training is finished, we run Monte Carlo simulations to estimate the BLER at Bob. During training, Gaussian noise is added in Bob's channel to consider the same model during the training and prediction phase. This also helps during the training of the network to reduce the decoding error probability, cf. [18,22].

At Eve's output, we need a way to efficiently estimate the IL $I(\boldsymbol{M};\boldsymbol{Z})$. For this, we propose using an approximation based on the Taylor series as a loss function L_{E}.

The IL is defined as the mutual information between $\boldsymbol{M}$ and $\boldsymbol{Z}$ which can equivalently be expressed as [66]

$$I(\boldsymbol{M};\boldsymbol{Z}) = h(\boldsymbol{Z}) - h(\boldsymbol{Z}|\boldsymbol{M}), \tag{1.43}$$

where h denotes the differential entropy. In the described AWGN setting, the distribution of Eve's received signal $\boldsymbol{Z}$ is a GM which can be expressed as

$$\boldsymbol{Z} \sim \sum_{i=1}^{|\mathcal{C}|} \omega_i \mathcal{N}(\mu_i, \sigma_{\mathrm{E}}^2 I_n) \tag{1.44}$$

with the codewords of length n, $\mu_i \in \mathcal{C}$ is the ith codeword as the means of each Gaussian component and the weights $\omega_i \triangleq \mathbb{P}(\mu_i \text{ is selected}), i \in \{1, \ldots, 2^{k+r}\}$. In the case of uniformly distributed messages, $\omega_i = |\mathcal{C}|^{-1}$ for all codewords $\mu_i \in \mathcal{C}$.

For a given transmitted message $\boldsymbol{M} = m$, the conditional distribution of $\boldsymbol{Z}$ given $\boldsymbol{M} = m$ is also a GM. In particular,

$$\boldsymbol{Z}|_{\boldsymbol{M}=m} \sim \sum_{j=1}^{|\mathcal{C}_m|} \omega_j \mathcal{N}(\mu_j, \sigma_{\mathrm{E}}^2 I_n), \tag{1.45}$$

where $j \in \{1, \ldots, 2^r\}$. Recall that the set $\mathcal{C}_m \subset \mathcal{C}$ contains only the codewords embedding the secret message m.

Due to the GMs, there is no closed-form expression for the information leakage. Therefore, we resort to an approximation based on the Taylor expansion [65]. Such a Taylor series approach to approximate the logarithm inside the differential entropy function has been considered in [65,67]. The precision of this scheme can be arbitrarily high when a sufficiently high order is used. However, considering the number of training epochs needed to converge and also the block length of the codewords, we use the second-order expansion for the feasibility of implementation. The zeroth and the second term of the Taylor expansion on the logarithm of the differential entropies of GMs and the approximated differential entropy can be found in [65] as

$$H_0(\boldsymbol{Z}) \triangleq 2^{-(k+r)} \sum_{j=1}^{2^{k+r}} \log g(\mu_j), \tag{1.46}$$

$$H_2(\boldsymbol{Z}) \triangleq -2^{-(k+r+1)} \sigma_{\mathrm{E}}^2 \sum_{j=1}^{2^{k+r}} \operatorname{tr}\big(\mathbf{F}(\mu_j)\big), \tag{1.47}$$

$$h(\boldsymbol{Z}) \approx h_T(\boldsymbol{Z}) = H_0(\boldsymbol{Z}) + H_2(\boldsymbol{Z}), \tag{1.48}$$

where the function g denotes the PDF of the GM. The covariance matrices are scaled identity matrices for all components of the GM, the matrix function $\mathbf{F}$ [65] is defined as

$$\mathbf{F}(x) = \frac{1}{g(x)\sigma_{\mathrm{E}}^2} 2^{-(k+r)} \sum_{j=1}^{2^{k+r}} \mathcal{N}(x; \mu_j, \sigma_{\mathrm{E}}^2 I) \cdot \left(\frac{1}{g(x)} (x - \mu_j)(\nabla g(x))^\top + \sigma_{\mathrm{E}}^{-2} (x - \mu_j)(x - \mu_j)^\top - I \right).$$

The approximation of the differential entropy from (1.48) is then applied to both terms in (1.43). Then the loss function at Eve is approximated by

$$L_{\mathrm{E},T} = h_T(\boldsymbol{Z}) - h_T(\boldsymbol{Z}|\boldsymbol{M}). \tag{1.49}$$

Finally, the loss functions are combined into a weighted sum of the individual losses

$$L \triangleq w_B L_{\mathrm{B}} + w_E L_{\mathrm{E}}. \tag{1.50}$$

Remark 4. *There exist other methods of combining different loss functions than the one used in* (1.38). *Another method proposed in [68] is*

$$\min_{C} \; -\left(\log\left(w_B - L_B\right) + \log\left(w_E - L_E\right)\right). \tag{1.51}$$

1.3.3.2 Hyperparameter selection

The performance of machine learning (ML) algorithms highly depends on the choice of *hyperparameters*. These are parameters of the ML model chosen before training the system and are not changed during the training phase, e.g., the number of layers and nodes in a NN. In this section, we provide guidelines and justifications for certain hyperparameter choices. However, the general selection of well-performing NN hyperparameters is still unknown and relies on trial-and-error [63].

Both the encoder and the decoder NN use the ReLU activation function on all layers but the last one. The final output layer at Bob uses the *sigmoid* activation function to ensure a soft output between 0 and 1. It should be noted that the minimal required size of the NNs is a single layer with n nodes at the encoder and a single layer of size $k + r$ at the decoder. However, both parts may also have additional hidden layers for improved performance.

1.4 Numerical assessments

For the numerical simulations, we use the parameters shown in Table 1.3. The code parameters for the simulations are fixed to block length $n = 16$, number of secure bits $k = 4$, and number of random bits $r = 3$. The short block length of 16 has been chosen due to the high complexity of the NNs. However, such short block lengths are relevant in applications where an ultra-low latency is required [69]. The SNRs at Bob and Eve are set to 0 dB and -5 dB, respectively.

The above code parameters are applied to the autoencoder implementation described in Section 1.3.3 to generate wiretap codes for the Gaussian wiretap channel. The varied hyperparameters are the structures of the encoder and decoder NNs of the autoencoder, and the weights of the loss functions for Bob and Eve. The detailed settings of the different autoencoder structures can be found in Table 1.4. The NN layers are described by a tuple, where each entry represents the number of nodes in the respective layer, e.g., (128, 64, 16) means that there are three layers, each has 128, 64, and 16 nodes, respectively. The used training algorithm is Adam [70] with a learning rate of 0.001. After training, the system is tested with 10^5 messages from which the BLER at Bob is determined using only k secure message bits. The leakage of the found codebook is estimated by Monte Carlo simulations using $2 \cdot 10^5$ samples.

Table 1.3 System parameters for implementing the autoencoder (Adapted from [2, Table II]. © IEEE 2020. Reprinted with permission from [2].)

Parameters	**Values/functions**
SNR at Bob	0 dB (E_b/N_0)
SNR at Eve	-5 dB (E_b/N_0)
Block length n	16
Number of secure message bits k	4
Number of confusion message bits r	3
NN type	Feed-forward
Optimizer	Adam [70] with a learning rate of 0.001
Number of layers and nodes in NN	See Table 1.4
Loss function at Bob	L_{B} in (1.42)
Loss function at Eve	L_{E} in (1.49)
Number of training epochs	$7.5 \cdot 10^5$
Weight w_B	$w_B = 0.9, \ldots, 1$
Weight w_E	$w_E = 1 - w_B$

Table 1.4 The evaluated autoencoders for a $(16, 4, 3)$ *wiretap code. © IEEE 2020. Reprinted with permission from [2].*

Name	**Encoder layers**	**Decoder layers**
AE1	(16)	(7)
AE2	(128, 64, 16)	(128, 64, 7)
AE3	(256, 16)	(256, 7)

All of the numerical results shown in the following were obtained by a TensorFlow implementation of the described architectures running on an off-the-shelf NVIDIA GTX 1080 Ti GPU [2].

Since there exists a tradeoff for finite block length wiretap codes between block length, rate, reliability, and secrecy [71], the variation of the weights allows a flexible operating point of the system for fixed code parameters. By increasing w_E relative to w_B, secrecy constraints are considered more important, which reduces the leakage with the cost of the reduced reliability. This tradeoff between the leaked information and the error rate can be shown in the BLER-IL plane, which will be used in the following to present the simulation results. Note that the performance of a wiretap code is a Pareto improvement, if it is closer to the origin (zero errors at Bob while leaking no information). Our goal is to find the Pareto boundary in the BLER-IL plane as close to the origin as possible.

The tradeoff between the IL and BLER for the specified parameters can be found in Figure 1.3.

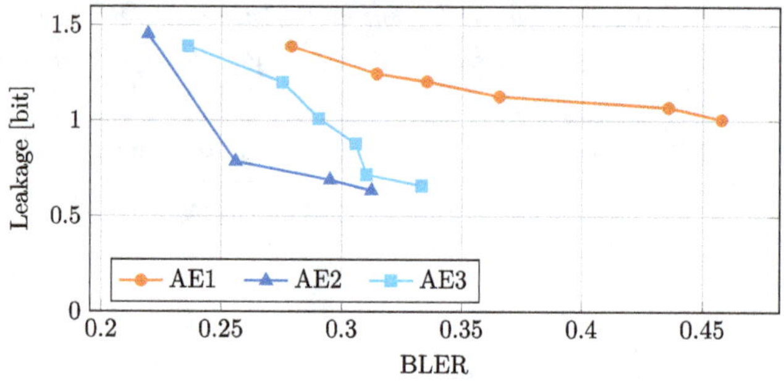

Figure 1.3 *IL versus BLER for different network structures when the second-order Taylor expansion-based loss L_E is used at Eve for designing a* (16, 4, 3) *wiretap code. © IEEE 2020. Reprinted with permission from [2].*

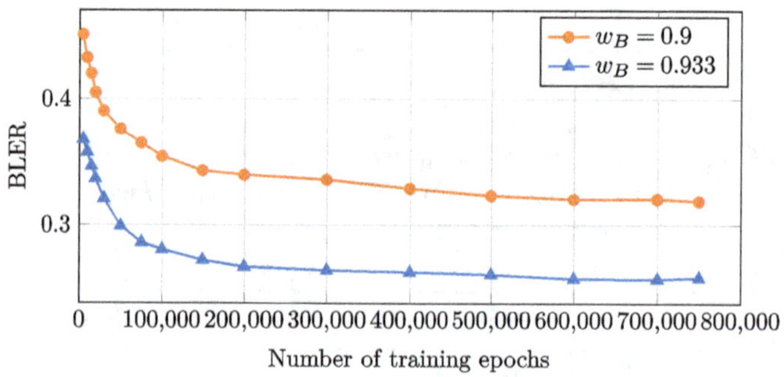

Figure 1.4 *The learning curve of the achieved BLER for AE2 from Table 1.4 with different weight combinations over the number of training epochs. © IEEE 2020. Reprinted with permission from [2].*

It can be seen that the performance of the system improves with an increasing size of the NN. The smallest network (AE1) has the worst performance, since it is too small. The larger networks AE2 and AE3 achieve better performance. In particular, AE2 achieves a leakage of around 0.78 bit at a BLER of 25.6%.

In Figure 1.4, we present the learning curve for AE2 for two different values of w_B. This demonstrates the influence of the number of training epochs on the achieved BLER. It can be seen that the curves saturate with an increasing number of training epochs, i.e., there is a limit on how much the BLER can be reduced for a certain network structure. In many NN applications, the number of training epochs is chosen by monitoring the loss function and the training is stopped, when a certain stabilization is reached [63]. In our simulations, the loss function was not decreasing significantly after around 5000 training epochs. However, as can be seen in Figure 1.4, the network was still learning and further reducing the BLER. For this reason, we continued the training to $7.5 \cdot 10^5$ epochs until also the BLER reached a stable value.

1.4.1 Dependent fading channels with a single connectivity

After investigating the performance of the learned wiretap codes for a wiretap channel with a single channel state, we now want to focus on the two-state channel described in Section 1.2.3. In addition to the SNR values from the previous example, both Bob's and Eve's channels may now also be in a state of a higher SNR. In particular, we have $\xi_{\mathrm{B_{low}}} = 0$ dB, $\xi_{\mathrm{B_{high}}} = 5$ dB, $\xi_{\mathrm{E_{low}}} = -5$ dB, and $\xi_{\mathrm{E_{high}}} = 0$ dB.

Since the configuration of AE2 performs best, we limit the simulations in the following to the wiretap codes previously found by it. The resulting performance of these codes for the described two-state channel is shown in Figure 1.5. Since the BLER at Bob only depends on the SNR ξ_{B} and the IL only depends on the SNR of Eve's channel ξ_{E}, a grid pattern is clearly visible.

For a fixed operating point ($\mathrm{BLER}_O, \mathrm{IL}_O$), which specifies a tolerated BLER (BLER_O) and IL (IL_O), an outage occurs, if either of the values is larger than the one of the operating points. Therefore, the outage probability is given as

$$\varepsilon = \mathbb{P}\left(\mathrm{BLER} > \mathrm{BLER}_O \vee \mathrm{IL} > \mathrm{IL}_O\right) \tag{1.52}$$

or vice versa

$$1 - \varepsilon = \mathbb{P}\left(\mathrm{BLER} < \mathrm{BLER}_O \wedge \mathrm{IL} < \mathrm{IL}_O\right).$$

The second expression corresponds to the joint distribution of BLER and IL. This can be explicitly evaluated for the considered two-state model.

In the following, we fix the operating point to (0.05, 1), i.e., the application can tolerate a BLER of up to 5% at the legitimate receiver and a leakage of at most 1 bit to an eavesdropper. This operating point is also indicated in Figure 1.5. From the figure, it can clearly be seen that only two points do not cause an outage. This corresponds to two weights w_B at the SNRs $\sigma_B = 5$ dB and $\sigma_E = -5$ dB. Hence, the success

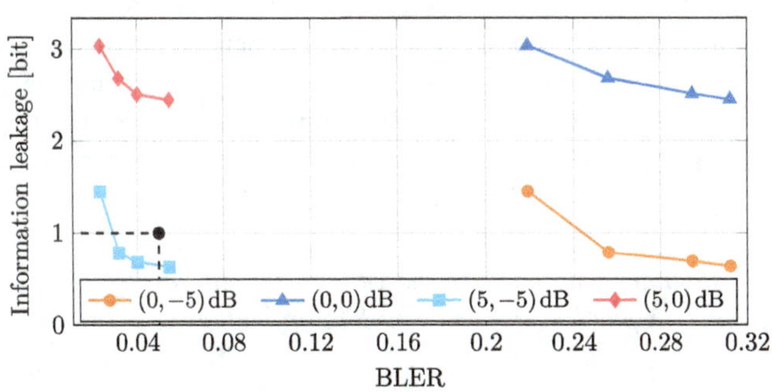

Figure 1.5 BLER-IL trade-off of a (16, 4, 3) *wiretap code for different SNRs* ($\xi_{\mathrm{B}}, \xi_{\mathrm{E}}$) *dB according to the two-state channel model. The wiretap code is the one designed by the autoencoder structure AE2. Additionally, we show the considered operating point* ($BLER_O, IL_O$) = (0.05, 1)*. The dashed lines indicate the outage region.*

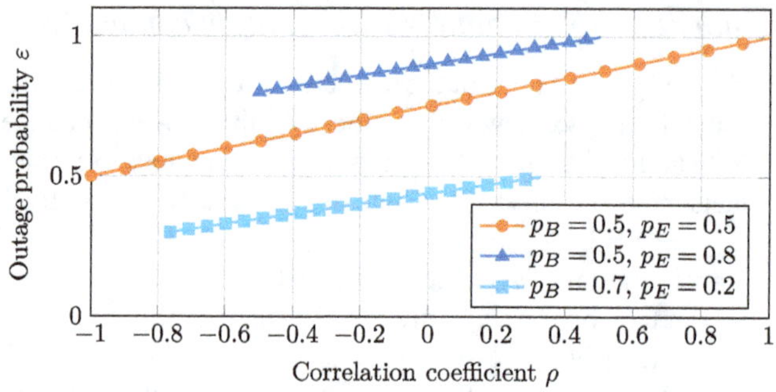

Figure 1.6 Outage probability for the (16, 4, 3) wiretap code designed by autoencoder AE2 in the correlated two-state channel. The probability of the high-SNR state of Bob's (resp. Eve's) channel is p_B (resp. p_E). The operating point is set to $(BLER_O, IL_O) = (0.05, 1)$.

probability, i.e., $1 - \varepsilon$, is given by the probability p_{11} as described in Section 1.2.3. For a given correlation coefficient ρ, this can be calculated according to (1.11) as

$$1 - \varepsilon = p_{11} = \rho\sqrt{p_B p_E (1 - p_B)(1 - p_E)} + p_B p_E.$$

The resulting outage probability ε over the correlation coefficient ρ is shown for different values of p_B and p_E in Figure 1.6. The range of ρ is determined by (1.12). First, it can be seen from Figure 1.6 that the outage probability overall increases for increasing p_E and decreasing p_B, i.e., when the probability of a high SNR channel to Bob is small or the probability of a high SNR channel to Eve is large. Next, it can be seen that the outage probability increases linearly with the correlation coefficient ρ. Hence, for fixed marginal probabilities p_B and p_E, it is smallest for the minimum ρ. The smallest value of ρ corresponds to negatively dependent channels, i.e., it is more likely to have a high SNR channel state to Bob when Eve's channel SNR is small and vice versa. This observation confirms the more general results from previous works [20,72] that negative dependencies of the individual channels improve the performance, e.g., in terms of a lower outage probability.

Remark 5. *It should be noted that the outage definition from* (1.52) *is not the only possible one. Depending on the application, different definitions could also be appropriate.*

Remark 6 (Longer block lengths). *While the results above are for a short block length* $n = 16$*, the idea presented in this chapter can also be applied to longer block lengths. However, this increases the complexity of the training significantly [2]. In order to demonstrate that it is still possible to apply our proposed framework to longer block lengths, we show the results for block length* $n = 64$ *in Figure 1.7. Similar to*

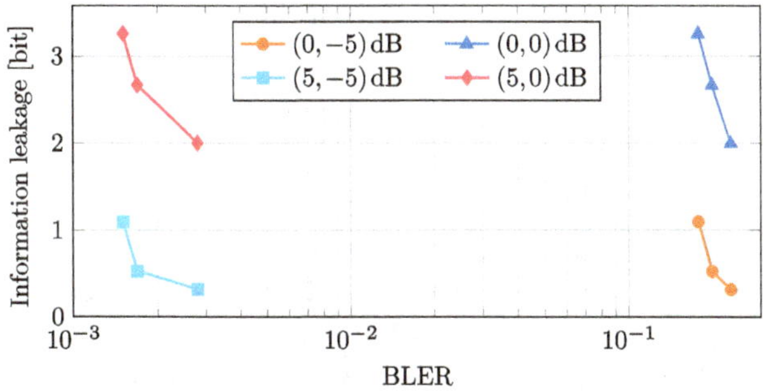

Figure 1.7 BLER-IL trade-off of a (64, 4, 3) *wiretap code for different SNRs* $(\xi_\mathrm{B}, \xi_\mathrm{E})$ *dB according to the two-state channel model*

Figure 1.5 shows the BLER-IL trade-off of a learned wiretap code in the two-state channel model.

1.4.2 Dependent fading channels with multiple connections

We now consider an additional link to Bob, i.e., $n_B = 2$. All the individual links follow the two-state channel model as described in Section 1.2.3.2. The probabilities that Bob's links are in the high-SNR state are p_{B_1} and p_{B_2}. The probability that Eve's link is in the high-SNR state is p_E. As the diversity combining scheme, Bob performs selection combining (SC), i.e., he always uses the better link.

Assuming the same operating point and parameters as before, the outage probability ε is then given via

$$1 - \varepsilon = \mathbb{P}\left(\min\left(\sigma_{B_1}{}^2, \sigma_{B_2}{}^2\right) = \sigma_{B^2_{low}}, \ \sigma_E^2 = \sigma_{E^2_{high}}\right) \tag{1.53}$$

$$= p_{110} + p_{100} + p_{010} \tag{1.54}$$

$$= 1 - p_E - p_{000}, \tag{1.55}$$

where $\min(\sigma_{B_1^2}, \sigma_{B_2^2})$ corresponds to the better of Bob's link due to the SC. Equivalently, we have

$$\varepsilon = p_E + p_{000}.$$

For the best-case scenario, comparable to the ZOSC, we are interested in the following optimization problem:

$$\min \varepsilon = \min p_E + p_{000} = p_E + \min p_{000}, \tag{1.56}$$

where the second equality is due to the fact that p_E is fixed and known.

Based on (1.56) and (1.17), we formulate the following new linear programming problem:

$$\min_{p} \quad p_{000} \tag{1.57}$$

$$\text{s.t.} \quad (1.17),\ (1.21). \tag{1.58}$$

Note that the above linear programming problem can be solved efficiently numerically. However, since it is a linear problem over a convex set with only affine constraints, any point that fulfills the Karush–Kuhn–Tucker (KKT) conditions, is also a global minimum [73]. Based on this, we can determine the solution of (1.57) as follows.

Proposition 1. *The minimum outage probability with respect to the joint distribution of the channel states is given as*

$$\min \varepsilon = p_E + \min p_{000} = p_E + [1 - p_{B_1} - p_{B_2} - p_E]^+. \tag{1.59}$$

Proof. As discussed above, we need to show that there exists at least one Karush-Kuhn-Tucker (KKT) point with the solution from (1.59). In order to show this for all possible combinations, we consider the following three cases: (1) $p_{B_1} + p_{B_2} + p_E \leq 1$, (2) $1 < p_{B_1} + p_{B_2} + p_E < 2$, and (3) $2 \leq p_{B_1} + p_{B_2} + p_E$.

Let $\boldsymbol{\mu} = (\mu_{000}, \mu_{001}, \dots, \mu_{111})$ be the Lagrange multiplier for the inequality constraint $\boldsymbol{p} \geq 0$, and let $\boldsymbol{\lambda} = (\lambda_{B_1}, \lambda_{B_2}, \lambda_E, \lambda_S)$ be the Lagrange multiplier for the equality constraints (1.17).

For the first case with $p_{B_1} + p_{B_2} + p_E \leq 1$, the following solution $(\boldsymbol{p}^\star, \boldsymbol{\mu}^\star, \boldsymbol{\lambda}^\star)$ fulfills the KKT conditions:

$$\boldsymbol{p}^\star = \begin{pmatrix} 1 - p_E - p_{B_1} - p_{B_2} & p_E & p_{B_2} & 0 & p_{B_1} & 0 & 0 & 0 \end{pmatrix}, \tag{1.60}$$

$$\boldsymbol{\mu}^\star = \begin{pmatrix} 0 & 0 & 0 & 1 & 0 & 1 & 1 & 2 \end{pmatrix}, \tag{1.61}$$

$$\boldsymbol{\lambda}^\star = \begin{pmatrix} 1 & 1 & 1 & -1 \end{pmatrix}. \tag{1.62}$$

Hence, we obtain the minimum value of p_{000} as

$$\min p_{000} = 1 - p_E - p_{B_1} - p_{B_2} > 0.$$

The third case $p_{B_1} + p_{B_2} + p_E \geq 2$ corresponds to the scenario where we can simply exchange p_i for $1 - p_i$. With this, we obtain the KKT point

$$\boldsymbol{p}^\star = \begin{pmatrix} 0 & 0 & 0 & 1 - p_{B_1} & 0 & 1 - p_{B_2} & 1 - p_E & p_{B_1} + p_{B_2} + p_E - 2 \end{pmatrix},$$

$$\boldsymbol{\mu}^\star = \begin{pmatrix} 1 & 0 & 0 & 0 & 0 & 0 & 0 & 0 \end{pmatrix},$$

$$\boldsymbol{\lambda}^\star = \begin{pmatrix} 0 & 0 & 0 & 0 \end{pmatrix}.$$

Similarly, a KKT point with $p_{000} = 0$ can be obtained for the last case where $1 < p_{B_1} + p_{B_2} + p_E < 2$. However, the expressions are lengthy since they include additional case definitions based on the relative sizes of p_{B_1}, p_{B_2}, and p_E. For the sake of brevity, we omit these expressions.

Combining the solutions of the KKT points for the three different cases yields the final solution

$$\min p_{000} = [1 - p_E - p_{B_1} - p_{B_2}]^+,$$

which then yields (1.59). □

In order to illustrate the solution from Proposition 1, we take a closer look at some numerical examples in the following. Based on the derived KKT points, we are also able to explicitly state a joint probability distribution that achieves the minimum outage probability.

Example 1. For the first numerical example, assume $p_{B_1} = p_{B_2} = p_E = 0.1$. Based on (1.59) in Proposition 1, we calculate the outage probability to be $\varepsilon = 0.1 + 0.7 = 0.8$. From the KKT point corresponding to the case $p_{B_1} + p_{B_2} + p_E \leq 1$, we get an optimal joint probability distribution which achieves the minimum as

$$\boldsymbol{p}^\star = \begin{pmatrix} 0.7 & 0.1 & 0.1 & 0 & 0.1 & 0 & 0 & 0 \end{pmatrix}.$$

Example 2. For a second example, we assume $p_{B1} = p_{B2} = p_E = 0.4$. This corresponds to the case $1 < p_{B_1} + p_{B_2} + p_E < 2$. Based on the KKT conditions, we determine an optimal joint distribution as

$$\boldsymbol{p}^\star = \begin{pmatrix} 0 & \frac{4}{15} & \frac{4}{15} & \frac{1}{15} & \frac{4}{15} & \frac{1}{15} & \frac{1}{15} & 0 \end{pmatrix}.$$

Since $p_{000} = 0$, we find that $\varepsilon = p_E = 0.4$, which confirms the general solution from (1.59).

Example 3. For a final example, we assume $p_{B_1} = 0.6$, $p_{B_2} = 0.7$, and $p_E = 0.8$. From Proposition 1, we expect to find a joint distribution such that $p_{000} = 0$. Using the KKT point for the case $p_{B_1} + p_{B_2} + p_E \geq 2$, we determine an optimal joint distribution as

$$\boldsymbol{p}^\star = \begin{pmatrix} 0 & 0 & 0 & 0.4 & 0 & 0.3 & 0.2 & 0.1 \end{pmatrix}.$$

The overall outage probability is again only determined by p_E as $\varepsilon = p_E = 0.8$.

From the last two examples, it can be seen that the minimum with $p_{000} = 0$ can be achieved when $p_{B_1} + p_{B_2} + p_E \geq 1$. This corresponds to the best case possible since p_{000} needs to have a non-negative value. The overall outage probability ε from (1.59) is shown for different values of the marginal probabilities p_{B_1}, p_{B_2}, and p_E in Figure 1.8. First, it can be seen that the outage probability increases with increasing p_E, since this corresponds to an increasing probability that Eve's channel is in the high SNR state. Similarly, the outage probability increases when p_B decreases, since this corresponds to a decreasing probability that Bob's channel is in the high SNR state. Additionally, a clear change of behavior at the line $p_{B_1} + p_{B_2} + p_E = 1$ is visible in Figure 1.8. From Proposition 1, we know that above this line, i.e., when $p_{B_1} + p_{B_2} + p_E \geq 1$, we get a minimum of $p_{000} = 0$ and thus, the outage probability ε is only determined by p_E in this region.

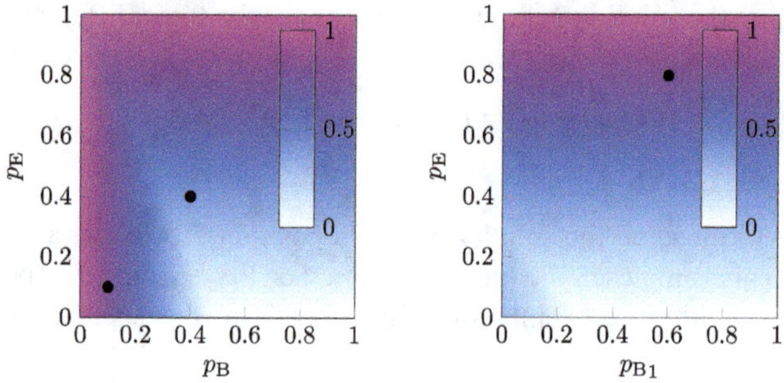

(a) $p_{B1} = p_{B2} = p_B$ (b) $p_{B2} = 0.7$

Figure 1.8 Minimum outage probability from (1.59) *for different marginal probabilities p_{B_1}, p_{B_2}, and p_E. The black markers indicate the parameters from Examples 1–3.*

1.5 Conclusions and future work

In this chapter, we demonstrated how to leverage ML techniques to design finite-block length wiretap codes for the Gaussian wiretap channel. In particular, we formulate the tradeoff between BLER at the legitimate receiver and IL to the eavesdropper as a MOP. This problem is solved by NN autoencoders, which additionally provide a flexible code design along the tradeoff curve.

The results for the basic Gaussian wiretap channel are extended to a two-state channel model with a possible correlation between the channels. The trained wiretap code works for different dependencies between the legitimate and illegitimate channels. The Pareto boundary of the BLER-IL curve depends significantly on the joint distribution of the channels. For this scenario, the outage probability is calculated and it shows that a negative dependency between the channel to Bob and the channel to Eve is beneficial.

In the finite block length regime, we are not able to achieve vanishing leakage to Eve and vanishing error probability at Bob, simultaneously. Even to solely achieve vanishing leakage or error probability, when the block length is not sufficiently long, it will lead to a huge cost, i.e., unacceptably high error probability or leakage, respectively. As a result, to get a better reliability-secrecy tradeoff, which is suitable for practical use, we need to be able to design a longer wiretap code. The main issue of designing the code by NN is the increasing training complexity when the number of messages increases. One of the possible directions for future works is that we may consider to concatenate capacity-achieving codes with the NN code. Especially, considering the robustness of the capacity-achieving code on noisy channels, we may apply the capacity-achieving codes as the inner code, while the NN code works as the outer code. Efficient schemes to adapt the hyper-parameters are critical due to the aforementioned training complexity. The main tasks of measuring the reliability and leakage may possibly be attained by another two NNs, for estimating the mutual information between Alice and Bob and also Alice and Eve, respectively.

To design NN wiretap codes for more practical fading channels as mentioned in this chapter and fully utilize the variation of the channels, we need to consider the fading channels with more states rather than two, or even more, with continuous states. In the latter case, under the asymptotic assumption, we may design the NN wiretap codes for achieving a zero-outage secrecy rate. However, as mentioned, the non-vanishing leakage and error probability should be additionally considered, due to the impact of the finite block length. In particular, to design the code much more precisely in the finite block length regime under fading channels, we need to consider the second-order secrecy rate followed by the ZOSC analysis, in which we can apply the tools from copula theory to the second-order secrecy rate. This way may provide us with a better prediction of the system performance in the finite block length regime.

Another direction for future work is to incorporate channel feedback into the NN to improve the performance of the network [74–76].

References

[1] Besser KL, Janda CR, Lin PH, *et al.* Flexible design of finite blocklength wiretap codes by autoencoders. In: *ICASSP 2019 – 2019 IEEE International Conference on Acoustics, Speech and Signal Processing (ICASSP)*. IEEE; 2019. p. 2512–2516. Available from: https://ieeexplore.ieee.org/document/8683409/.

[2] Besser KL, Lin PH, Janda CR, *et al.* Wiretap code design by neural network autoencoders. *IEEE Transactions on Information Forensics and Security*. 2020;15:3374–3386. Available from: https://ieeexplore.ieee.org/document/8859269/.

[3] Saad W, Bennis M, and Chen M. A vision of 6G wireless systems: applications, trends, technologies, and open research problems. *IEEE Network*. 2019; 1–9:134–142. Available from: https://doi.org/10.1109/mnet.001.1900287.

[4] Dang S, Amin O, Shihada B, *et al.* What should 6G be? *Nature Electronics*. 2020;3(1):20–29.

[5] Gui G, Liu M, Tang F, *et al.* 6G: opening new horizons for integration of comfort, security, and intelligence. *IEEE Wireless Communications*. 2020;27(5):126–132.

[6] Jorswieck E, Lin PH, and Janda C. Physical layer security based enabling technologies for 6G communications values. In: *IEEE Future Networks World Forum*. IEEE; 2022.

[7] Chorti A, Barreto AN, Köpsell S, *et al.* Context-aware security for 6G wireless: the role of physical layer security. *IEEE Communications Standards Magazine*. 2022;6(1):102–108.

[8] Porambage P, Gür G, Osorio DPM, *et al.* The roadmap to 6G security and privacy. *IEEE Open Journal of the Communications Society*. 2021;2: 1094–1122.

[9] Nguyen VL, Lin PC, Cheng BC, *et al.* Security and privacy for 6G: a survey on prospective technologies and challenges. *IEEE Communications Surveys & Tutorials*. 2021;23(4):2384–2428.

[10] Bloch M, Günlü O, Yener A, *et al.* An overview of information-theoretic security and privacy: metrics, limits and applications. *IEEE Journal on Selected Areas in Information Theory*. 2021;2(1):5–22.

[11] Bloch M and Barros J. *Physical-Layer Security*. Cambridge University Press; 2011.

[12] Gidney C and Ekerå M. How to factor 2048 bit RSA integers in 8 hours using 20 million noisy qubits. *Quantum*. 2021;5:433. Available from: http://dx.doi.org/10.22331/q-2021-04-15-433.

[13] Hou J and Kramer G. Effective secrecy: reliability, confusion and stealth. In: *2014 IEEE International Symposium on Information Theory*. IEEE; 2014. p. 601–605. Available from: https://doi.org/10.1109/isit.2014.6874903.

[14] Bloch MR. Covert communication over noisy channels: a resolvability perspective. *IEEE Transactions on Information Theory*. 2016;62(5):2334–2353.

[15] Shakiba-Herfeh M, Chorti A, and Poor HV. Physical layer security: authentication, integrity, and confidentiality. In: Le KN, editor. *Physical Layer Security*. Springer International Publishing; 2021. p. 129–150.

[16] Li G, Sun C, Zhang J, *et al.* Physical layer key generation in 5G and beyond wireless communications: challenges and opportunities. *Entropy*. 2019;21(5):497. Available from: https://www.mdpi.com/1099-4300/21/5/497.

[17] Nötzel J, Wiese M, and Boche H. The arbitrarily varying wiretap channel—secret randomness, stability, and super-activation. *IEEE Transactions on Information Theory*. 2016;62(6):3504–3531.

[18] Besser KL, Lin PH, Janda CR, *et al.* Machine learning assisted wiretapping. In: *2018 52nd Asilomar Conference on Signals, Systems, and Computers*. IEEE; 2018. p. 489–493. Available from: https://ieeexplore.ieee.org/document/8645088/.

[19] Besser KL, Lonnstrom A, and Jorswieck EA. Neural network wiretap code design for multi-mode fiber optical channels. In: *ICASSP 2020 – 2020 IEEE International Conference on Acoustics, Speech and Signal Processing (ICASSP)*. IEEE; 2020. p. 8738–8742. Available from: https://ieeexplore.ieee.org/document/9053933/.

[20] Besser KL and Jorswieck EA. Bounds on the secrecy outage probability for dependent fading channels. *IEEE Transactions on Communications*. 2021;69(1):443–456.

[21] Jorswieck EA, Lin PH, and Besser KL. On the zero-outage secrecy-capacity of dependent fading wiretap channels. *Entropy*. 2022;24(1):99.

[22] Gruber T, Cammerer S, Hoydis J, *et al.* On deep learning-based channel decoding. In: *2017 51st Annual Conference on Information Sciences and Systems (CISS)*. IEEE; 2017.

[23] O'Shea T and Hoydis J. An introduction to deep learning for the physical layer. *IEEE Transactions on Cognitive Communications and Networking*. 2017;3(4):563–575.

[24] Dörner S, Cammerer S, Hoydis J, *et al.* Deep learning based communication over the air. *IEEE Journal of Selected Topics in Signal Processing.* 2018;12(1):132–143.

[25] Aoudia FA and Hoydis J. End-to-end learning of communications systems without a channel model. In: *2018 52nd Asilomar Conference on Signals, Systems, and Computers*. IEEE; 2018. p. 298–303.

[26] Ye H, Li GY, Juang BHF, *et al.* Channel agnostic end-to-end learning based communication systems with conditional GAN. In: *2018 IEEE Globecom Workshops (GC Wkshps)*. IEEE; 2018.

[27] Fritschek R, Schaefer RF, and Wunder G. Deep learning for channel coding via neural mutual information estimation. In: *2019 IEEE 20th International Workshop on Signal Processing Advances in Wireless Communications (SPAWC)*. IEEE; 2019.

[28] Rana V and Chou RA. Short Blocklength Wiretap Channel Codes via Deep Learning: Design and Performance Evaluation. arXiv. 2022 6. Available from: https://arxiv.org/abs/2206.03477.

[29] Belghazi MI, Baratin A, Rajeshwar S, *et al.* Mutual information neural estimation. In: Dy J, Krause A, editors. *Proceedings of the 35th International Conference on Machine Learning*. vol. 80 of *Proceedings of Machine Learning Research*. PMLR; 2018. p. 531–540. Available from: https://proceedings.mlr.press/v80/belghazi18a.html.

[30] Cheng P, Hao W, Dai S, *et al.* CLUB: a contrastive log-ratio upper bound of mutual information. In: *ICML*; 2020.

[31] Fritschek R, Schaefer RF, and Wunder G. Deep Learning for the Gaussian wiretap channel. In: *ICC 2019 – 2019 IEEE International Conference on Communications (ICC)*; 2019.

[32] Fritschek R, Schaefer RF, and Wunder G. Reinforce security: a model-free approach towards secure wiretap coding. In: *ICC 2021 – IEEE International Conference on Communications*; 2021.

[33] Erdemir E, Dragotti PL, and Gunduz D. Privacy-aware communication over a wiretap channel with generative networks. In: *ICASSP 2022 – 2022 IEEE International Conference on Acoustics, Speech and Signal Processing (ICASSP)*. IEEE; 2022.

[34] Choi K, Tatwawadi K, Grover A, *et al.* Neural joint source-channel coding. In: Chaudhuri K, Salakhutdinov R, editors. *Proceedings of the 36th International Conference on Machine Learning*. vol. 97 of *Proceedings of Machine Learning Research*. PMLR; 2019. p. 1182–1192.

[35] Barber D and Agakov FV. The IM algorithm: a variational approach to information maximization. In: *Advances in Neural Information Processing Systems*. The MIT Press, USA. vol. 16; 2003. p. 201–208.

[36] Di Renzo M, Debbah M, Phan-Huy DT, *et al.* Smart radio environments empowered by reconfigurable AI meta-surfaces: an idea whose time has come. *EURASIP Journal on Wireless Communications and Networking*. 2019; 2019(1). Available from: https://jwcn-eurasipjournals.springeropen.com/articles/10.1186/s13638-019-1438-9.

[37] Bloch M, Barros J, Rodrigues MRD, *et al.* Wireless information-theoretic security. *IEEE Transactions on Information Theory*. 2008;54(6):2515–2534. Available from: http://ieeexplore.ieee.org/document/4529264/.

[38] Gopala PK, Lai L, and El Gamal H. On the secrecy capacity of fading channels. *IEEE Transactions on Information Theory*. 2008;54(10):4687–4698.

[39] Bhargav N, Cotton SL, and Simmons DE. Secrecy capacity analysis over κ–μ fading channels: theory and applications. *IEEE Transactions on Communications*. 2016;64(7):3011–3024.

[40] Wang JY, Liu C, Wang JB, *et al.* Secrecy outage probability analysis over Malaga–Malaga fading channels. In: *2018 IEEE International Conference on Communications (ICC)*. IEEE; 2018.

[41] Zhao H, Liu Y, Sultan-Salem A, *et al.* A simple evaluation for the secrecy outage probability over generalized-*K* fading channels. *IEEE Communications Letters*. 2019;23(9):1479–1483.

[42] Lee WCY. Effects on correlation between two mobile radio base-station antennas. *IEEE Transactions on Vehicular Technology*. 1973;22(4):130–140. Available from: http://ieeexplore.ieee.org/document/1622173/.

[43] Sang-Bin R and Zysman G. Results of suburban base station spatial diversity measurements in the UHF band. *IEEE Transactions on Communications*. 1974;22(10):1630–1636. Available from: http://ieeexplore.ieee.org/document/1092096/.

[44] Zhu J, Takahashi O, Jiang X, *et al.* Outage secrecy capacity over correlated fading channels at high SNR. In: *Proceedings of the 2012 International Conference on Mobile Computing and Ubiquitous Networking*; 2012. p. 92–97.

[45] Hyoungsuk J, Namshik K, Jinho C, *et al.* Bounds on secrecy capacity over correlated ergodic fading channels at high SNR. *IEEE Transactions on Information Theory*. 2011;57(4):1975–1983. Available from: http://ieeexplore.ieee.org/document/5730598/.

[46] Sun X, Wang J, Xu W, *et al.* Performance of secure communications over correlated fading channels. *IEEE Signal Processing Letters*. 2012;19(8): 479–482. Available from: http://ieeexplore.ieee.org/document/6213074/.

[47] Liu X. Outage probability of secrecy capacity over correlated log-normal fading channels. *IEEE Communications Letters*. 2013;17(2):289–292. Available from: http://ieeexplore.ieee.org/document/6397537/.

[48] Alexandropoulos GC and Peppas KP. Secrecy outage analysis over correlated composite Nakagami-*m*/Gamma fading channels. *IEEE Communications Letters*. 2018;22(1):77–80. Available from: http://ieeexplore.ieee.org/document/8060499/.

[49] Ghadi FR and Hodtani GA. Copula-based analysis of physical layer security performances over correlated Rayleigh fading channels. *IEEE Transactions on Information Forensics and Security*. 2021;16:431–440. Available from: https://ieeexplore.ieee.org/document/9159617/.

[50] Besser KL and Jorswieck EA. Copula-based bounds for multi-user communications – Part II: outage performance. *IEEE Communications Letters*. 2021;25(1):8–12.

[51] Jorswieck E and Lin PH. Ultra-reliable multi-connectivity with negatively dependent fading channels. In: *2019 16th International Symposium on Wireless Communication Systems (ISWCS)*; 2019. p. 373–378.

[52] Besser KL, Lin PH, and Jorswieck EA. On fading channel dependency structures with a positive zero-outage capacity. *IEEE Transactions on Communications*. 2021;69(10):6561–6574.

[53] Tse D and Viswanath P. *Fundamentals of Wireless Communications*. Cambridge University Press; 2005.

[54] Bai T, Vaze R, and Heath RW. Analysis of blockage effects on urban cellular networks. *IEEE Transactions on Wireless Communications*. 2014;13(9): 5070–5083.

[55] Akdeniz MR, Liu Y, Samimi MK, *et al.* Millimeter wave channel modeling and cellular capacity evaluation. *IEEE Journal on Selected Areas in Communications*. 2014;32(6):1164–1179.

[56] Joe H. *Multivariate Models and Multivariate Dependence Concepts*. CRC Press; 1997.

[57] Embrechts P. Copulas: a personal view. *Journal of Risk and Insurance*. 2009;76(3):639–650.

[58] Chaganty NR and Joe H. Range of correlation matrices for dependent Bernoulli random variables. *Biometrika*. 2006;93(1):197–206. Available from: http://www.jstor.org/stable/20441270.

[59] Park J, Samarakoon S, Shiri H, *et al.* Extreme ultra-reliable and low-latency communication. *Nature Electronics*. 2022;5(3):133–141. Available from: http://arxiv.org/abs/2001.09683 https://www.nature.com/articles/s41928-022-00728-8.

[60] Khisti A and Wornell GW. Secure transmission with multiple antennas—Part II: the MIMOME wiretap channel. *IEEE Transactions on Information Theory*. 2010;56(11):5515–5532.

[61] Nelsen RB. *An Introduction to Copulas*. 2nd ed. Springer Series in Statistics. Springer, New York, NY; 2006.

[62] Björnson E, Jorswieck EA, Debbah M, *et al.* Multiobjective signal processing optimization: the way to balance conflicting metrics in 5G systems. *IEEE Signal Processing Magazine*. 2014;31(6):14–23.

[63] Goodfellow I, Bengio Y, and Courville A. *Deep Learning*. MIT Press; 2016. Available from: http://www.deeplearningbook.org/.

[64] LeCun Y, Bengio Y, and Hinton G. Deep learning. *Nature*. 2015;521(7553):436–444. Available from: http://www.nature.com/articles/nature14539.

[65] Huber MF, Bailey T, Durrant-Whyte H, *et al.* On entropy approximation for Gaussian mixture random vectors. In: *2008 IEEE International Conference on Multisensor Fusion and Integration for Intelligent Systems*. IEEE; 2008. p. 181–188. Available from: http://ieeexplore.ieee.org/document/4648062/.

[66] Cover TM and Thomas JA. *Elements of Information Theory*. 2nd ed. John Wiley & Sons, Inc.; 2006.

[67] Moshksar K and Khandani AK. Arbitrarily tight bounds on differential entropy of Gaussian mixtures. *IEEE Transactions on Information Theory*. 2016;62(6):3340–3354. Available from: https://doi.org/10.1109/tit.2016.2553147.

[68] Miranda CS and Von Zuben FJ. Multi-Objective Optimization for Self-Adjusting Weighted Gradient in Machine Learning Tasks. arXiv. 2015 6. Available from: http://arxiv.org/abs/1506.01113.

[69] Ji H, Park S, Yeo J, *et al*. Ultra-reliable and low-latency communications in 5G downlink: physical layer aspects. *IEEE Wireless Communications*. 2018;25(3):124–130.

[70] Kingma DP and Ba J. Adam: A Method for Stochastic Optimization. arXiv. 2014. Available from: http://arxiv.org/abs/1412.6980.

[71] Yang W, Schaefer RF, and Poor HV. Finite-blocklength bounds for wiretap channels. In: *2016 IEEE International Symposium on Information Theory (ISIT)*. IEEE; 2016. p. 3087–3091.

[72] Besser KL and Jorswieck EA. Reliability bounds for dependent fading wireless channels. *IEEE Transactions on Wireless Communications*. 2020;19(9): 5833–5845.

[73] Boyd S and Vandenberghe L. *Convex Optimization*. Cambridge University Press; 2004.

[74] Kim H, Jiang Y, Kannan S, *et al*. Deepcode: feedback codes via deep learning. *IEEE Journal on Selected Areas in Information Theory*. 2020;1(1):194–206.

[75] Li S, Tuninetti D, and Devroye N. Deep learning-aided coding for the fading broadcast channel with feedback. In: *ICC 2022 – IEEE International Conference on Communications*. IEEE; 2022. p. 3874–3879.

[76] Ahlswede R and Cai N. Transmission, identification and common randomness capacities for wire-tape channels with secure feedback from the decoder. In: *General Theory of Information Transfer and Combinatorics*. No. 4123 in *Lecture Notes in Computer Science*. Springer, Berlin, Heidelberg; 2006. p. 258–275.

Chapter 2

Physical layer security in the terahertz band

Weijun Gao[1] and Chong Han[1]

In next-generation wireless communication networks, the stringent link requirement on ultra-high throughput catalyzes the exploration of unused and unlicensed spectrum. Following the trend from previous generations, the carrier frequency evolves from megahertz, gigahertz to terahertz (THz).* THz band communications, with a frequency from 0.1 to 10 THz and a wavelength from 0.03 to 3 mm, have been envisioned as promising to realize multi-Tbps wireless data links [4]. Besides the increased bandwidth, high propagation attenuation and strong directivity of THz communications put forward intrinsic advantages for physical layer security [5–7]. Specifically, high transmit beamforming gain increases the received power at a legitimate receiver. In contrast, the received power at a passive eavesdropper is not enhanced. With the equipment of ultra-massive multiple-input–multiple-output (MIMO) array, the THz beamwidth is on the order of 1°. There is a high probability that the eavesdropper is outside the beam sector region. Therefore, THz communications have naturally good security.

However, there are several challenges to be addressed. First, when the eavesdropper happens to be located inside the beam sector region, security is jeopardized since the legitimate user and the eavesdropper can equally benefit from the high directivity [8]. Second, when the eavesdropper is nearer than the legitimate user to the transmitter, i.e., in close proximity to Alice, its channel capacity can exceed that of the legitimate user. In this case, the coded symbols received by the legitimate user can be decoded by the eavesdropper. Third, the covertness of THz wireless communications, i.e., the ability of communication systems to conceal the existence of wireless transmissions, is required, as a stricter extent of security. Covert communication is an important topic for future military and defense applications. For example, the geographic information of an unmanned aerial vehicle (UAV) is vulnerable to being exposed to enemies, even though they cannot decode the message and intercept the specific command.

In this chapter, we will start with a description of the fundamental system model for THz physical layer security, including the THz channel model, performance metric, and challenges for wireless information security. Covertness of the THz wireless

*This chapter has been a part of [1–3].

[1]Terahertz Wireless Communications (TWC) Laboratory, Shanghai Jiao Tong University, China

communication is also modeled and discussed which prevents the existence of wireless links rather than information. In light of these, three novel THz secure and covert communication schemes are described. First, the distance-adaptive absorption peak modulation (DA-APM) utilizes the increased path loss of molecular absorption effect of the eavesdropper's link to enhance the security of THz wireless communications. It is further stated that this scheme is effective in improving the covertness of THz wireless communications. Second, the self-interference cancellation (SIC)-free receiver artificial noise (RAN) scheme takes the advantage of the temporal broadening effect caused by the molecular absorption effect to improve information security. This scheme can combat eavesdroppers located nearer than the legitimate receiver to the transmitter. Third, we describe the widely spaced antenna array (WSA) and beamforming scheme as a novel and practical transmit antenna array design. The WSA structure has an impractically large array size in the microwave band. However, it is envisioned to be useful and of reasonable size in the THz band [9]. This scheme can also combat near eavesdropping to guarantee range and angular security, without relying on the aid of the receiver side like the SIC-free RAN scheme.

2.1 Fundamental system model for THz secure communications

The fundamental THz secure communication system consists of a transmitter Alice (A), a legitimate receiver Bob (B), and one or several eavesdroppers Eve (E). Alice sends the confidential message to Bob in the presence of randomly distributed Eve(s). Since Alice and Bob cannot predict how much information about the legitimate transmission the Eve can obtain, such as whether they can know the carrier frequency of the legitimate link, we usually overestimate Eve's detection ability. One basic assumption is that Eve knows exactly how Alice and Bob cooperatively transmit symbols. With the equipment of UM-MIMO of Alice, the directional beam created by the antenna array is directed towards Bob to combat the high path loss.

2.1.1 THz channel model

Since the carrier frequency of the Alice–Bob and Alice–Eve links are in the THz band, the THz channel characteristics and the corresponding beamforming strategy significantly affect its security performance. Compared with microwave and millimeter-wave channels, the THz channel has two unique channel properties, i.e., the molecular absorption effect [1] and the high channel sparsity.

2.1.1.1 Molecular absorption effect

A point source transmits an electromagnetic (EM) and the wave spreads spherically and experiences free-space path loss. During this process, certain molecules inside the medium could resonate with and be excited by the EM wave. In quantum mechanics, different types of molecules have different and intrinsic rotational energy levels. Only energy transitions between different energy levels are allowed. A molecule at a high energy level can spontaneously transit into the low energy levels and emit a photon for the conservation of energy. The energy carried by the photon is equal to the energy level difference. This process also occurs inversely. According to the wave–particle duality, the energy of a photon is equal to hf, where h denotes the Planck constant, and f represents the frequency of the EM wave. When external photons with proper energy coincide with molecules, the molecules can transit from low energy to high energy levels by absorbing the photons. Therefore, only EM waves with resonate frequencies can interact with certain types of molecules. For communications, the absorbance of a photon implies the attenuation of the EM wave, and the photons emitted in this process cause an additional noise to THz channels. This phenomenon is referred to as the molecular absorption effect, which is illustrated in Figure 2.1.

In addition to the ITU recommendation P.676-12's reference model [10], which addresses the molecular absorption loss within the 0.1–1 THz range, a more comprehensive model for both molecular absorption loss and molecular absorption noise spanning the entire 0.1–10 THz range is presented in [11]. In general, the model proposed by Jornet *et al.* in [11] calculates the molecular absorption loss by summing up the power absorbed by all the molecules along the propagation path:

$$PL_{\text{mol}} = \exp\left[\sum_{i,g} \kappa^{i,g}(f)d\right], \tag{2.1}$$

where $\kappa^{i,g}(f)$ denotes the frequency-dependent molecular absorption coefficient of the isotopologue i of gas g. d represents the propagation distance. Within the THz range, water vapor has the most substantial impact on molecule absorption compared to other typical air components. In quantitative terms, water vapor overwhelmingly dominates this absorption phenomenon, being six orders of magnitude greater than the second most influential contributor, which is oxygen [11].

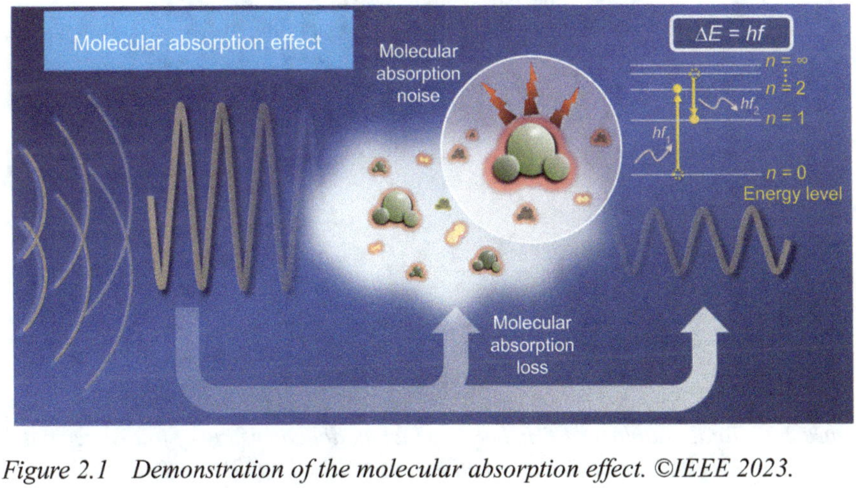

Figure 2.1 Demonstration of the molecular absorption effect. ©IEEE 2023. Reprinted with permission from [1].

The total molecular absorption coefficient can be expressed as

$$\kappa(f) = \sum_{i,g} \kappa^{i,g}(f) \approx \kappa^{H_2O}(f). \tag{2.2}$$

It is shown that the coefficient $\kappa(f)$ has *triple selectivity* [1]. The strength of this effects relies heavily on several factors, including frequency, distance, and various environmental conditions.

- **Frequency selectivity:** The molecular absorption effect depends on the intensity of the resonance between molecules and EM waves with a certain frequency. This makes the molecular absorption highly frequency selective. At resonant frequencies, the molecular absorption effect is maximized, and lots of molecular absorption peaks are created in the THz band. As shown in Figure 2.2, the total path loss of the THz LoS channel versus frequency in the THz band is plotted.
- **Distance dependence:** The molecular absorption loss increases exponentially with the distance. When the propagation distance increases from zero, the molecular absorption loss first appears negligibly weak. Then the loss increases explosively when the distance exceeds a certain threshold. This phenomenon strictly limits the coverage of THz communications unless the transmission scheme adapts to the distance.
- **Environment impact:** Since the water vapor concentration along the propagation path dominates the molecular absorption effect, its impact on THz communications significantly varies under different communication environments. As altitude increases, less water vapor leads to a weaker molecular absorption

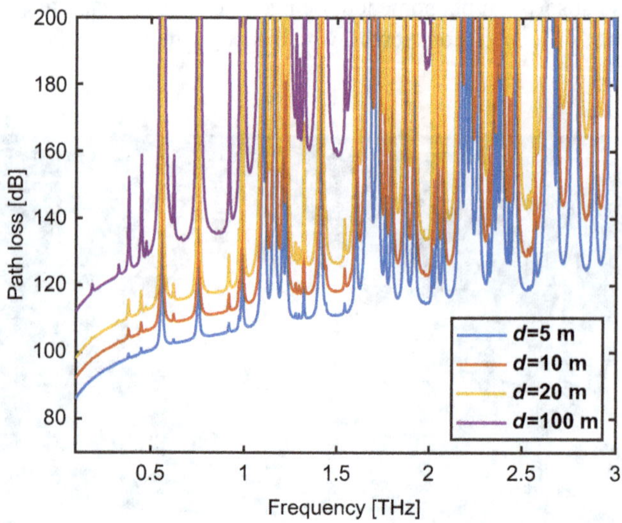

Figure 2.2 Total path loss of THz LoS channel model consisting of free-space path loss and molecular absorption loss. ©IEEE 2023. Reprinted with permission from [3].

effect. In terrestrial indoor and outdoor environments, the water vapor density is maximized.

2.1.1.2 Channel sparsity

Besides the molecular absorption effect, the sparsity of the THz channel is one of the key channel characteristics, where the number of THz multi-paths is limited. The reflection, diffraction, and scattering effect of the THz EM wave is negligibly weak. Traditional Rayleigh or Rician models are not feasible to model the THz channel due to the lack of rich scattering in the environment. Instead, one dominant path is widely adopted to model the THz channel [11,12]. The one dominant path can be the LoS path, if it exists, or the strongest single-reflecting (SR) path. Moreover, to combat the high propagation loss of THz communications, ultra-massive MIMO is usually equipped at the THz transmitter, which creates high-gain and high-directional beams aligning to the strongest path.

2.1.1.3 Line-of-sight and single-reflecting channel model

The THz secure system model for the LoS and SR scenarios is shown in Figures 2.3 and 2.4. By considering the free-space path loss and molecular absorption loss, the THz channel for the LoS and SR case can be expressed by

$$H_{LoS}(f,d) = \frac{c}{4\pi fd}\exp\left[-\frac{1}{2}\kappa(f)d - \frac{j2\pi fd}{c}\right], \tag{2.3}$$

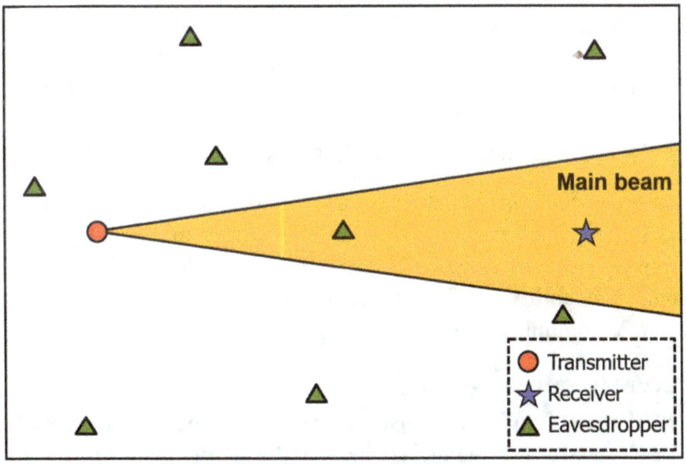

Figure 2.3 THz secure communication with multiple eavesdroppers. Scenario I: line-of-sight (LoS). ©IEEE 2023. Reprinted with permission from [2].

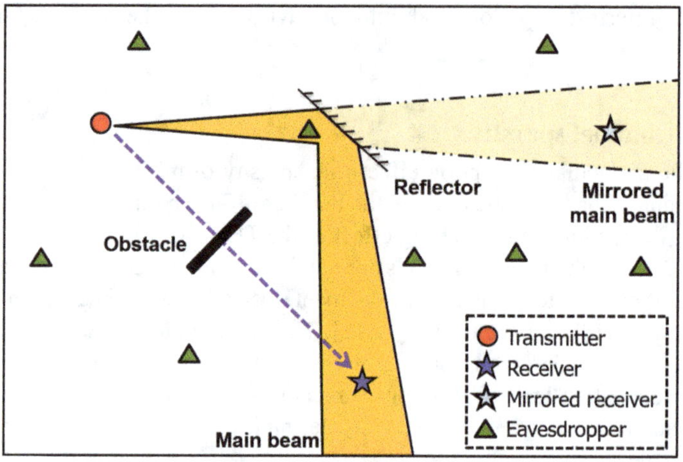

Figure 2.4 THz secure communication with multiple eavesdroppers. Scenario II: single reflection (SR). ©IEEE 2023. Reprinted with permission from [2].

$$H_{SR}(f,d) = \frac{c}{4\pi fd}\exp\left[-\frac{1}{2}\kappa(f)d - \frac{j2\pi fd}{c} - \left(\frac{-2\cos(\theta_{\mathrm{t}})}{\sqrt{n_{\mathrm{t}}^2-1}} - \frac{8\pi^2 f^2\sigma_{\mathrm{ref}}^2\cos^2(\theta_{\mathrm{t}})}{c^2} - j\pi\right)\right], \tag{2.4}$$

where c is the speed of light. The free-space path loss is formulated according to Friis' law. For the SR scenario, the signal experiences an additional reflection loss, which depends on the surface roughness and the incident angle with the reflector. n_{t} is the refractive index of the reflector, θ_{t} denotes the incident angle, and σ_{ref} represents the reflector surface standard deviation indicating the roughness of the surface. The reflection parameters are summarized by [13] and the references therein.

2.1.2 Performance metrics for THz secure communications

Different performance metrics are applied for THz secure communication to characterize the secrecy performance of the investigated communication system. This subsection introduces and discusses secrecy rate, eavesdroppable region (ER), and secrecy outage probability (SOP) as three widely used secrecy performance metrics for THz secure communications.

2.1.2.1 Secrecy rate

The most fundamental and widely used secrecy performance metric is the secrecy rate, which is defined as the channel capacity of the legitimate link minus the channel capacity of the eavesdropper link, given by [14]

$$R_s \triangleq [R_{\mathrm{B}} - R_{\mathrm{E}}]^+, \tag{2.5}$$

where R_{B} and R_{E} denote the capacity of the legitimate channel and the eavesdroppable channel. Remarkably, the intuitive idea to enhance the secrecy rate is to increase the legitimate channel capacity while mitigating the eavesdropper's channel capacity. Moreover, suppose the eavesdropper's channel capacity is larger than the legitimate one. In that case, the secrecy rate is zero, and secure communication is compromised. The drawback of the secrecy rate is its necessity to know the channel capacity of the eavesdroppers. Unlike a legitimate channel where Bob can send feedback information to Alice, the eavesdropper's location, fading, and even existence is unknown to Alice. Therefore, it is impractical for real THz systems to compute the secrecy rate without knowing the eavesdropper's channel.

2.1.2.2 Eavesdroppable region

To handle the challenge of an unknown eavesdropper's location, we use the eavesdroppable region (ER) as a secrecy performance metric (also known as the insecure region). The idea is that we can assume the eavesdropper to be located at a specific position and compute the secrecy rate. By doing this, we define the ER ($\mathscr{A}_{\text{ER}}$) as the area inside which an eavesdropper can compromise secure communication, i.e., the secrecy rate is lower than a pre-decided secrecy rate threshold R_{th}, as

$$\mathscr{A}_{\text{ER}} = \{\mathbb{R}^2 | R_s < R_{\text{th}}\}. \tag{2.6}$$

Taking the eavesdroppable region as the security performance metric avoids knowing the eavesdropper's specific position. However, it cannot characterize the secrecy performance with the presence of multiple eavesdroppers.

2.1.2.3 SOP

Secrecy outage probability (SOP) is a widely used secrecy performance metric that can be applied for secure communications with the presence of one or multiple eavesdroppers. Since we do not know the specific position of each eavesdropper, we assume that the eavesdroppers are randomly distributed in the communication scenario following the point Poisson process (PPP) with density λ_{E}. The secrecy outage event for a single eavesdropper is defined as the case that the secrecy rate with the presence of that eavesdropper is lower than a secrecy rate requirement R_{th}. The SOP is defined as the probability that such a secrecy outage event happens. For non-colluding eavesdroppers, we assume the secrecy outage event for multiple eavesdroppers only occurs when the minimum secrecy rate among all eavesdroppers is smaller than a secrecy rate requirement R_{th}. Therefore, the SOP p_{out} is given by

$$p_{\text{out}} \triangleq \mathbb{P}\left\{\min_i R_s^i < R_{\text{th}}\right\} = 1 - \prod_i \mathbb{P}\left\{R_s^i > R_{th}\right\}, \tag{2.7}$$

where $\mathbb{P}(\cdot)$ denotes the probability measure, i represents the eavesdropper index, and R_s^i is the secrecy rate with only the ith eavesdropper. By considering the definition of the ER, we can expand the expression for SOP in (2.7) as

$$\begin{aligned} p_{\text{out}} &= 1 - \prod_i \sum_{\Omega \in \{\mathscr{A}_{\text{ER}}, \mathscr{A}_{\text{ER}}^{\complement}\}} \mathbb{P}\left\{R_s^i \geq R_{th} | \mathrm{E}_i \in \Omega\right\} \cdot \mathbb{P}\left\{\mathrm{E}_i \in \Omega\right\} \\ &= 1 - \prod_i \mathbb{P}\left\{\mathrm{E}_i \in \mathscr{A}_{\text{ER}}^{\complement}\right\} \\ &= 1 - e^{-\lambda_{\mathrm{E}} |\mathscr{A}_{\text{ER}}|} \approx \lambda_{\mathrm{E}} |\mathscr{A}_{\text{ER}}|, \end{aligned} \tag{2.8}$$

where E_i denotes a random variable characterizing the position of the ith eavesdropper. Equation (2.8) is obtained from (2.7) according to the law of sum probability. The approximation (2.8) applies when $\lambda_{\mathrm{E}} |\mathscr{A}_{\text{ER}}| \to 0$. As implied by (2.8), for PPP-distributed non-colluding eavesdroppers, the SOP depends only on the ER area and the density of the eavesdroppers.

2.1.3 Challenges of THz physical layer security

Unlike the conventional communication requirement of achieving high data rates, physical layer security targets ensure the received signal-to-noise ratio (SNR) at Bob and reduce the SNR of the signal detected by Eve. Maintaining a high-quality legitimate channel and suppressing a low eavesdropper capacity lead to the key security design methodology: widening the capacity gap between Bob and the eavesdropper. Aligning with this key security design methodology, the THz spectrum is envisioned to possess the enticing potential to support security and covertness. Since THz transmissions experience a high spreading loss according to Friis' law, a high-gain and narrow beam is usually created by ultra-massive multiple-input multiple-output (MIMO) to realize directional transmission towards Bob. Thus, an eavesdropper residing outside the narrowbeam is hard to overhear the transmission, while Bob benefited from the antenna gain [8]. However, this directivity-aided mechanism is not always reliable. The remaining challenges of THz physical layer security are three-fold:

- **Eve is inside the beam sector but is located farther than Bob from Alice:** In contrast to the good security performance with eavesdroppers outside the beam sector, eavesdroppers inside the beam become even more devastating. Both Bob and Eve benefit from the high directivity, and traditional secure beamforming methods transmitting such artificial noise generation [15–17] become valid due to the THz channel sparsity.
- **Eve is inside the beam sector and is close proximity to Alice:** When the eavesdropper is nearer than the legitimate user to the transmitter, its channel capacity can exceed that of the legitimate user, and the secrecy rate is zero. Moreover, some traditional physical layer security method such as transmit artificial noise scheme [18] cannot apply when Eve is not only inside the beam sector but also nearer than Bob. THz physical layer security schemes applied under this condition need investigation.

- **Covertness of THz wireless communications is needed as a more stringent requirement:** Covertness is a very challenging security requirement since the legitimate receiver needs to decode the message but the eavesdropper only needs to determine its existence. This implies that traditional jamming-based security schemes are ineffective in providing covertness. For example, the artificial noise generation sends jamming signals to the eavesdroppers to decrease the received signal-to-interference-plus-noise ratio (SINR). This does not provide covertness because eavesdroppers do not need to distinguish information and noise, and the artificial noise also exposes the existence of its source. Since THz communications have potentially good security performance than microwave and millimeter-wave communications, new covertness schemes in the THz band are waiting to be explored.

2.2 Fundamental system model for THz covert communications

Covert communications avert the existence of wireless transmission rather than the information from eavesdropping [19,20]. One of the take-away lessons in traditional research on covert communications is that only $\mathcal{O}(\sqrt{n})$ covert bits can be transmitted reliable and perfectly for n channel uses, where $\mathcal{O}(\cdot)$ represents the complexity measure [21,22]. This principle is known as square root law. Therefore, the average data rate approaches zero as n goes to infinity due to $\lim_{n\to\infty}\mathcal{O}(\sqrt{n})/n \to 0$, which implies that the covert data rate is bounded by zero. The uncertainty of the noise power or noise distribution is used to achieve a positive-rate covert channel capacity to break through this limitation [23]. Specifically, by considering the noise uncertainty in the channel, we can derive a maximum SNR at the eavesdropper such that the eavesdropper is not able to detect the existence of the transmission. This threshold is defined as the SNR wall, which is derived by the hypothesis testing analysis for covert communications.

2.2.1 *Hypothesis testing for covert communications*

When Eve performs eavesdropping on the communication, he/she monitors a frequency band $\mathscr{B}$ and decides on whether the *received band-limited symbol* $y_{\mathrm{E}}^{\mathscr{B}}[n]$ is pure noise or noise plus signal from Alice. The statistical hypotheses $\mathscr{H}_0$ and $\mathscr{H}_1$ on received signal $y_{\mathrm{E}}^{\mathscr{B}}[n]$ can be written as:

$$\mathscr{H}_0 \;: y_{\mathrm{E}}^{\mathscr{B}}[n] = n_{\mathrm{E}}^{\mathscr{B}}[n]; \tag{2.9}$$

$$\mathscr{H}_1 \;: y_{\mathrm{E}}^{\mathscr{B}}[n] = \sum_i h_{\mathrm{E}}^{\mathscr{B}}[i]x_{\mathrm{E}}^{\mathscr{B}}[n-i] + n_{\mathrm{E}}^{\mathscr{B}}[n], \tag{2.10}$$

where $\mathscr{H}_0$ represents no transmit signals existing, and $\mathscr{H}_1$ implies that Alice transmits signal with a positive rate. Note that we use discrete signals to denote the band-limited received symbols. $h_{\mathrm{E}}^{\mathscr{B}}[i]$, $x_{\mathrm{E}}^{\mathscr{B}}[n]$, and $n_{\mathrm{E}}^{\mathscr{B}}[n]$ refer to the band-limited frequency-selective channel gain, band-limited transmitting symbol, and the band-limited noise symbol with respect to frequency band $\mathscr{B}$, respectively. We define

$P_{\mathrm{E}} \triangleq E\left[\sum \left|h_{\mathrm{E}}^{\mathscr{B}}[i]x_{\mathrm{E}}^{\mathscr{B}}[n-i]\right|^2\right]$ as the transmitted signal power at Eve. With the target to detect Alice's signals, Eve calculates a *test statistic* $T(\mathbf{y}_{\mathrm{E}}^{\mathscr{B}})$, based on the receive signal vector of size N, i.e., $\mathbf{y}_{\mathrm{E}}^{\mathscr{B}} = [y_{\mathrm{E}}^{\mathscr{B}}[i]]$, $i \in \{1, \cdots, N\}$. The choice of test statistics depends on the receiving mechanism of the receiver, which relies on the features of noise. For a Gaussian-distributed noise, the radiometer is widely adopted at the receiver [20], which measures the receiving power of the signal, with test statistic given by $T(\mathbf{y}_{\mathrm{E}}^{\mathscr{B}}) = \frac{1}{N}\sum_{i=1}^{N} |y_{\mathrm{E}}^{\mathscr{B}}[i]|^2$. For a non-Gaussian noise, the fractional low order detector (FLOD) is used for a near-optimal detection with test statistic given by $T(\mathbf{y}_{\mathrm{E}}^{\mathscr{B}}) = \frac{1}{N}\sum_{i=1}^{N} |y_{\mathrm{E}}^{\mathscr{B}}[i]|^p, p \in (0, 2)$. Decision on whether Alice is transmitting or not is made by comparing the test statistic with a pre-determined *detection threshold* η, i.e., $T(\mathbf{y}_{\mathrm{E}}^{\mathscr{B}}) \underset{\mathscr{H}_0}{\overset{\mathscr{H}_1}{\gtrless}} \eta$. Since we assume the infinite detection capability of Eve (infinite-long monitoring period), $N \to \infty$ is assumed. By applying the law of large numbers, the test statistics can converge to their *test means*, denoted by μ_0 and μ_1, respectively, i.e.,

$$\begin{aligned} &\lim_{N\to\infty} \Pr\left\{|T(\mathbf{y}_{\mathrm{E}}^{\mathscr{B}}) - \mu_0|<\varepsilon \middle| \mathscr{H}_0\right\} = 1, \\ &\lim_{N\to\infty} \Pr\left\{|T(\mathbf{y}_{\mathrm{E}}^{\mathscr{B}}) - \mu_1|<\varepsilon \middle| \mathscr{H}_1\right\} = 1, \end{aligned} \tag{2.11}$$

for any $\varepsilon > 0$. Let P_{FA} and P_{MD} represent the false alarm probability and missed detection probability, i.e.,

$$\begin{aligned} P_{\mathrm{FA}} &\triangleq \Pr\left\{T(\mathbf{y}_{\mathrm{E}}^{\mathscr{B}}) > \eta|\mathscr{H}_0\right\}, \\ P_{\mathrm{MD}} &\triangleq \Pr\left\{T(\mathbf{y}_{\mathrm{E}}^{\mathscr{B}}) < \eta|\mathscr{H}_1\right\}, \end{aligned} \tag{2.12}$$

where η is the detection threshold. Then, to ensure the covertness of the confidential communications, Eve's detection error probability ξ should satisfy that

$$\xi(\eta) \triangleq P_{\mathrm{FA}} + P_{\mathrm{MD}} \geq 1 - \varepsilon \tag{2.13}$$

for a sufficiently small $\varepsilon \in (0, 1)$. By substituting (2.12) into (2.13) with the assumption $N \to \infty$, the total error probability $\xi(\eta)$ can be simplified as

$$\begin{aligned} \lim_{N\to\infty} \xi(\eta) &\overset{(A)}{=} \lim_{N\to\infty} \Pr\left\{T(\mathbf{y}_{\mathrm{E}}^{\mathscr{B}}) > \eta|\mathscr{H}_0\right\} + \lim_{N\to\infty} \Pr\left\{T(\mathbf{y}_{\mathrm{E}}^{\mathscr{B}}) < \eta|\mathscr{H}_1\right\} \\ &\overset{(B)}{=} \mathbf{1}(\mu_0 > \eta) + \mathbf{1}(\mu_1 < \eta). \end{aligned} \tag{2.14}$$

2.2.2 Covert outage probability

The *covert outage probability* is adopted as the covertness measure, which is defined as the probability that the covertness of the communication is compromised [20], i.e., the probability that (2.13) is violated, defined as

$$p_{\mathrm{out}} \triangleq \max_{\eta} \Pr\{\xi(\eta) < 1 - \varepsilon\}, \quad \varepsilon \in (0, 1), \tag{2.15}$$

where the worst-case performance is considered since η is determined by Eve. Based on the covert outage probability, we set a small *probability threshold* $\delta \in (0, 1]$ and define that the communication is covert if $p_{\text{out}} < \delta$. According to (2.14), if Eve sets the detection threshold value η between μ_0 and μ_1, covertness is compromised ($\xi = 0$). Otherwise, covertness is ensured since $\xi = 1$. If Eve has perfect knowledge of noise distribution and statistics, he knows the test means μ_0 and μ_1. Then, Eve can choose an arbitrary $\eta \in [\mu_0, \mu_1]$ to compromise covertness and therefore the covertness cannot be achieved, i.e., $p_{\text{out}} = 1$.

Due to the failure of covertness when Eve has perfect information on noise, the noise uncertainty is considered based on the fact that the noise variance estimation process unavoidably introduces uncertainty due to calibration error or environmental change. When Eve has uncertainty on noise variance, the test means depend on the estimated noise variance, i.e., $\mu_0(\hat{\psi})$ and $\mu_1(\hat{\psi})$, where $\hat{\psi}$ denotes the estimated noise variance; $\mu_0(\hat{\psi})$ and $\mu_1(\hat{\psi})$ denote the test means under the two hypotheses given that the noise variance is $\hat{\psi}$, respectively. As a result, the probability that Eve sets $\eta \in [\mu_0(\hat{\psi}), \mu_1(\hat{\psi})]$ is not 1, i.e., $p_{\text{out}} < 1$, and therefore covert communication is possible and an SNR wall occurs.

2.2.3 *Signal-to-noise-ratio wall*

In this subsection, a general model is proposed to compute the SNR wall denoted by $\gamma_{\text{E,th}}$ with various noise distributions. In this model, we denote $f_{\hat{\psi}}(\hat{\psi})$ as the probability density function (PDF) of noise estimation. The SNR wall can be computed as follows. Suppose the estimated noise strength lies in the range $\hat{\psi} \in [\psi_{\min}, \psi_{\max}]$, then if there exists a signal power P_{E} exceeds a certain value such that

$$\mu_1(\psi_{\min}) > \mu_0(\psi_{\max}), \tag{2.16}$$

then $p_{\text{out}} = 1$, i.e., covertness is compromised. Otherwise, the resulting outage probability is

$$p_{\text{out}} = \max_{\eta} \int_{\mu_1^{-1}(\eta)}^{\mu_0^{-1}(\eta)} f_{\hat{\psi}}(\hat{\psi}) d\hat{\psi}, \tag{2.17}$$

and Eve's optimal detection threshold is

$$\eta^{\text{opt}} = \arg\max_{\eta} \int_{\mu_1^{-1}(\eta)}^{\mu_0^{-1}(\eta)} f_{\hat{\psi}}(\hat{\psi}) d\hat{\psi}. \tag{2.18}$$

By combining (2.17) and $p_{\text{out}} < \delta$, we obtain the minimal SNR that defines the SNR wall. We further investigate the SNR walls under different levels of Eve's prior knowledge of noise strength and distribution as listed follows.

Gaussian noise with variance uncertainty: With noise variance uncertainty, Eve does not know the exact noise variance. However, we assume the noise variance distribution is prior knowledge. In particular, we assume that the estimated noise variance follows a log-uniform distribution $\hat{\psi} \in [\frac{\psi_0}{\rho}, \rho\psi_0]$, where ψ_0 is the nominal noise variance and uncertainty parameter $\rho \in (1, \infty)$ characterizes the size of the uncertainty. The probability density function (PDF) of the estimated noise variance is given by

$$f_{\hat{\psi}}(\hat{\psi}) = \begin{cases} \frac{1}{2\ln(\rho)\hat{\psi}}, & \frac{\psi_0}{\rho} \le \hat{\psi} \le \rho\psi_0, \\ 0, & \text{otherwise.} \end{cases} \tag{2.19}$$

Proposition 1: *If the noise is Gaussian-distributed and the estimated noise variance is log-uniformly distributed in the uncertainty range* $\hat{\psi} \in [\frac{\psi_0}{\rho}, \rho\psi_0]$*, then the SNR with covert probability* δ *should satisfy*

$$\frac{P_E}{\psi_0} \le \rho^{2\delta-1} - \frac{1}{\rho} = \gamma_{E,th}, \tag{2.20}$$

where the SNR wall is equal to the right-hand side of the inequality.

Proof: When $\mu_1(\psi_{\min}) > \mu_0(\psi_{\max})$, $P_{\mathrm{E}} > \left(\rho - \frac{1}{\rho}\right)\psi_0$, $p_{\mathrm{out}} = 1$, which implies that outage always occurs. To reach a certain outage probability level δ, it is required that $P_{\mathrm{E}} \le \left(\rho - \frac{1}{\rho}\right)\psi_0$. For radiometers, $\mu_0(\psi) = \psi$, $\mu_1(\psi) = P_{\mathrm{E}} + \psi$. The inverse function of $\mu_0(\psi)$ and $\mu_1(\psi)$ is given by $\mu_0^{-1}(\eta) = \eta$ and $\mu_1^{-1}(\eta) = \eta - P_{\mathrm{E}}$, respectively.

The optimal detection threshold for Eve can be calculated as

$$\eta^{\mathrm{opt}} = \arg\max_{\eta} \int_{\eta - P_{\mathrm{E}}}^{\eta} f_{\hat{\psi}}(\hat{\psi}) d\hat{\psi} = P_{\mathrm{E}} + \frac{\psi_0}{\rho}. \tag{2.21}$$

Substituting it into (2.17) gives the covert outage probability as

$$p_{\mathrm{out}} = \int_{\frac{\psi_0}{\rho}}^{P_{\mathrm{E}} + \frac{\psi_0}{\rho}} \frac{1}{2\ln(\rho)\psi} d\psi < \delta. \tag{2.22}$$

Solving it gives $P_{\mathrm{E}} < \left(\rho^{2\delta-1} - \frac{1}{\rho}\right)\psi_0$, which completes the proof.

Equation (2.20) suggests that when the SNR at Eve, i.e., $\frac{P_{\mathrm{E}}}{\psi_0}$, is lower than the SNR wall, covert outage probability, i.e., the probability of communication being compromised, is less than δ. Hence, the communication from Alice to Bob is covert.

Non-Gaussian noise with variance uncertainty: Considering external natural or man-made signal sources like atmospheric noise, the noise cannot be modeled as Gaussian distributed. Instead, the zero-mean symmetric α-stable (SαS) distribution is widely used to model these non-Gaussian noises, where some abnormally large impulses occur. The closed-form probability density function (PDF) of the SαS distributions does not exist. Its characteristic function $\phi(t)$ is expressed as

$$\phi(t) = \exp\left\{-\psi_\phi |t|^\alpha\right\}, \tag{2.23}$$

where $\alpha \in (0, 2]$ denotes the characteristic exponent that determines how frequently the impulses occur. Smaller α implies more and stronger impulses in the noise signal. ψ_ϕ is the dispersion parameter that measures the scale of the non-Gaussian noise.

Note that when $\alpha < 2$, SαS distribution does not have a second-order finite moment, i.e., if X_α follows an SαS distribution, then $\mathbb{E}|X_\alpha|^2 \to \infty$. Therefore, a radiometer based on second-order moment statistic is not suitable for signal detection under a non-Gaussian noise environment since both tests means converge to infinity. Instead, based on the fact that its moment of order $p < \alpha/2$ is finite, *fractional lower order detector* (FLOD) is used for detection, which has the test statistic

$$T_{\text{FLOD}}(\mathbf{y}_{\text{E}}^{\mathscr{B}}) = \frac{1}{N}\sum_{i=1}^{N} |y_{\text{E}}^{\mathscr{B}}[i]|^p \underset{\mathscr{H}_0}{\overset{\mathscr{H}_1}{\gtrless}} \eta. \tag{2.24}$$

Similar to the noise variance uncertainty for Gaussian noise, we assume that the estimated dispersion parameter follows a uniform distribution in the uncertainty range $\hat{\psi}_\phi \in [\frac{\hat{\psi_{\phi 0}}}{\rho}, \rho\hat{\psi_{\phi 0}}]$, where $\rho \in (1, \infty)$ characterizes the size of the uncertainty. Moreover, we define the generalized SNR $\gamma_{\text{E}}^{\text{g}}$ as

$$\gamma_{\text{E}}^{\text{g}} = \frac{P_{\text{E}}}{\psi_\phi^{2/\alpha}}. \tag{2.25}$$

Proposition 2: *If the noise is SαS-distributed and the estimated noise dispersion parameter* $\hat{\psi}_\phi$ *follows a log-uniform distribution in the uncertainty range* $\hat{\psi}_\phi \in [\frac{\psi_{\phi 0}}{\rho}, \rho\psi_{\phi 0}]$*, then the maximal generalized SNR with covertness level* δ *is given by*

$$\frac{P_E}{\psi_\phi^{2/\alpha}} < \frac{C(p,\alpha)}{D(p,\alpha)}(\rho^{\frac{2(\delta p-1)}{\alpha}} - \frac{1}{\rho^{2/\alpha}}) = \gamma_{E,th}^{g}, \tag{2.26}$$

where the right-hand side of the inequality is the generalized SNR wall with

$$C(p,\alpha) = \frac{2^{p+1}\Gamma(\frac{p+1}{2})\Gamma(-p/\alpha)}{\alpha\sqrt{\pi}\Gamma(-p/2)}, D(p,\alpha) = \frac{p(p-1)C(p-2,\alpha)}{2}, \tag{2.27}$$

where $\Gamma(t) \triangleq \int_0^\infty x^{t-1}e^{-x}dx$ *denotes the Gamma function.*

Proof: Under hypothesis $\mathscr{H}_0$ and $\mathscr{H}_1$, the test means of test statistic with FLOM detector and infinite sample size are given by

$$\begin{aligned} \mu_0(\psi_\phi) &= C(p,\alpha)\psi_\phi^{p/\alpha}, \\ \mu_1(\psi_\phi) &= C(p,\alpha)\psi_\phi^{p/\alpha} + D(p,\alpha)P_{\text{E}}\psi_\phi^{(p-2)/\alpha}. \end{aligned} \tag{2.28}$$

Then, we calculate the maximal received power P_{E} such that the worst-case covertness is satisfied, i.e., $\delta = 1$. By substituting the two statistics into (2.16), the condition for $p_{\text{out}} = 1$ becomes

$$C(p,\alpha)(\rho\psi_{\phi 0})^{p/\alpha} > C(p,\alpha)\left(\frac{\psi_{\phi 0}}{\rho}\right)^{\frac{p}{\alpha}} + D(p,\alpha)P_{\text{E}}\left(\frac{\psi_{\phi 0}}{\rho}\right)^{\frac{(p-2)}{\alpha}}, \tag{2.29}$$

which can be simplified into

$$P_{\text{E}} < \frac{C(p,\alpha)}{D(p,\alpha)}\left(\rho^{\frac{2(p-1)}{\alpha}} - \frac{1}{\rho^{2/\alpha}}\right)\psi_{\phi 0}^{2/\alpha}. \tag{2.30}$$

Otherwise, by following Theorem 1, the optimal detection threshold is

$$\eta^{\text{opt}} = \arg\max_{\eta} \int_{\mu_1^{-1}(\eta)}^{\mu_0^{-1}(\eta)} f_{\hat{\psi}_\phi}(\hat{\psi}_\phi) d\hat{\psi}_\phi = \mu_1(\frac{\psi_{\phi 0}}{\rho}). \tag{2.31}$$

Substituting it into (2.17) gives

$$p_{\text{out}} = \int_{\frac{\psi_{\phi 0}}{\rho}}^{\mu_0^{-1}(\mu_1(\frac{\psi_{\phi 0}}{\rho}))} f_{\hat{\psi}_\phi}(\hat{\psi}_\phi) d\hat{\psi}_\phi. \tag{2.32}$$

Solving $p_{\text{out}} < \delta$, we obtain the desired result (2.26), which completes the proof.

Equation (2.26) suggests that when the generalized SNR is lower than the generalized SNR wall, i.e., $P_{\text{E}}/\psi_{\phi 0}^{2/\alpha}$, covert outage probability is less than δ and the communication is covert.

2.3 Distance-adaptive molecular absorption peak modulation scheme

Distance-adaptive molecular absorption peak modulation (DA-APM) is a novel secure modulation scheme to address the security problem in the presence of eavesdroppers inside the beam. The key idea is to utilize the molecular absorption effect in the THz band. The molecular absorption effect is generally considered an unfavored feature in THz communications since it increases the path loss and decreases the channel capacity. However, from the perspective of communication security, it can be beneficial to THz communications. Molecular absorption loss is exponentially dependent on distance, which indicates that it can slightly degrade legitimate while greatly mitigating the eavesdropper's link capacity if Bob is nearer to Alice than Eve. The main principle of DA-APM is to dynamically modulate signals at molecular absorption peaks to enhance security. This scheme can be applied to secure or covert communications.

The design guideline is summarized as follows. First, the molecular absorption loss at Bob should be less than a certain threshold to support reliable transmission. Second, for security and covertness purposes, the molecular absorption loss should be as large as possible, such that the received SNR at Eve is significantly degraded. Thus, there is a trade-off between introducing a large molecular absorption loss for security and a small molecular absorption loss to ensure the reliability of the legitimate link [24]. Note that the DA-APM scheme can be applied on top of other modulation schemes since it only introduces the distance-dependency of carrier frequency. When selecting the frequency bands around molecular absorption peaks, the carrier frequency of the system, which determines the molecular absorption loss, is optimized for the scenario where THz communication is applied by jointly considering the legitimate link capacity and the security requirement. The drawback of the DA-APM scheme is due to the exponential function of the molecular absorption loss. It can only work effectively when the eavesdropper is farther than the receiver. Moreover, the data rate of Bob is partially sacrificed by introducing an additional molecular absorption loss. Therefore, this scheme cannot be applied when the legitimate link is weak.

2.3.1 DA-APM design guideline

The key design objective is to control the molecular absorption loss so that the reliability and covertness requirements can be maintained simultaneously. Since molecular absorption loss is frequency selective, this idea can be realized by carrier frequency tuning. THz band can be divided into many sub-bands [25,26]. When determining the sub-band for signal transmission in the THz band, we focus on the following requirements:

2.3.1.1 Reliability requirement

The molecular absorption effect can increase the path loss of the legitimate channel, degrading the Alice–Bob link's reliability. To make sure reliability, we require the received SNR at Bob to exceed a certain SNR threshold, i.e., $\gamma_B \geq \gamma_{B,th}$. The received SNR at Bob can be expressed according to the link budget analysis of the Alice–Bob link, as

$$\gamma_B[\mathrm{dB}] = P_{Tx} + G_A + G_B - N_B - 20\log\left(\frac{4\pi f d_{AB}}{c}\right) - 2.171\kappa(f)d_{AB}, \quad (2.33)$$

should be greater than or equal to $\gamma_{B,th}$, where P_{Tx} denotes the transmit power; G_A and G_B represent the transmit and receive antenna gain, respectively. N_B represents the AWGN noise power at receiver Bob.

2.3.1.2 Covertness or security requirement

The molecular absorption effect can increase the path loss of the eavesdropper's channel, enhancing the communication system's covertness. We require that the received SNR at Eve should be less than the SNR wall, i.e., $\gamma_E \leq \gamma_{E,th}$. The received SNR at Eve can be expressed according to the link budget analysis of the Alice–Eve link, as

$$\gamma_E[\mathrm{dB}] = P_{Tx} + G_A + G_E - N_E - 20\log\left(\frac{4\pi f d_{AE}}{c}\right) - 2.171\kappa(f)d_{AE}, \quad (2.34)$$

should be smaller than or equal to $\gamma_{E,th}$, where G_E represents the receive antenna gain at Eve, respectively. N_E represents the AWGN noise power at Eve. If we assume that $G_B = G_E$ and $N_B = N_E$, by combining (2.33) and (2.34), the following relationship should be satisfied:

$$\begin{aligned} \kappa(f) &\geq \frac{\gamma_{B,th} - \gamma_{E,th} + 20\log\frac{d_{AB}}{d_{AE}}}{2.171(d_{AB} - d_{AE})}, \quad \text{if } d_{AE} > d_{AB}, \\ \kappa(f) &\leq \frac{\gamma_{E,th} - \gamma_{B,th} + 20\log\frac{d_{AE}}{d_{AB}}}{2.171(d_{AB} - d_{AE})}, \quad \text{if } d_{AE} < d_{AB}. \end{aligned} \quad (2.35)$$

The requirement for covertness is more stringent for reliability, i.e., $\gamma_{B,th} \gg \gamma_{E,th}$, so we require that $d_{AB} < d_{AE}$. This implies that the DA-APM scheme can only work when Eve is located farther than Bob.

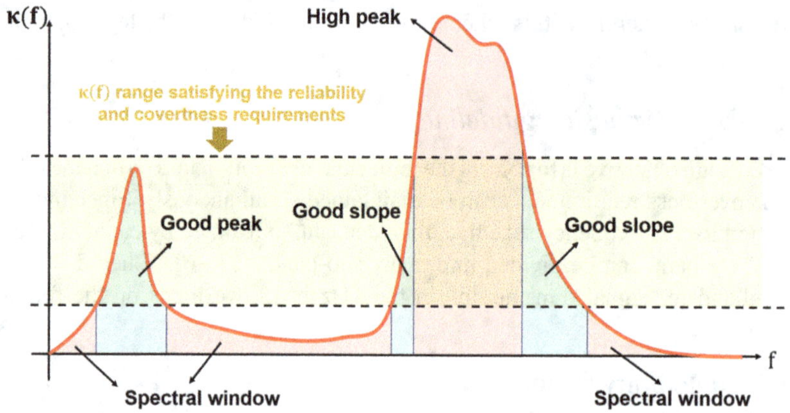

Figure 2.5 Illustration of the feasible region of the molecular absorption coefficient

The range of molecular absorption coefficient is shown in Figure 2.5, where (2.33) and (2.34) poses two constraints on $\kappa(f)$. As a result, the molecular absorption coefficient curve is divided into several areas. The slope and peak areas with feasible $\kappa(f)$ can be selected for secure and covert transmission. Transmissions in spectral window regions have good reliability but bad covertness. In contrast, transmissions in high peaks have good covertness but bad reliability. Moreover, the molecular absorption coefficient is highly dependent on the propagation distances d_{AB} and d_{AE}. After determining a proper molecular absorption coefficient, the carrier frequency can be selected from the $\kappa(f)$ curve. The communications with the presence of proper molecular absorption have better covertness and security than spectral-window communications, and better reliability than communications in high peaks.

2.4 Molecular absorption-assisted SIC-free receiver artificial noise scheme

Eavesdroppers in close proximity are challenging due to the higher eavesdropper's channel capacity than the legitimate one. A widely used counter-method is to transmit artificial noise (AN) signals, which do not bear information, to confuse the eavesdroppers. Most existing studies have suggested the transmitter to generate AN signals, which nevertheless are not applicable for THz line-of-sight transmission since the LoS path is shared between Bob and Eve. An alternative method is allowing the full-duplex receiver to transmit AN signals while receiving information, known as receiver AN (RAN) [27,28]. In the traditional implementation of RAN, the self-interference (SI) caused by the AN signal at Bob is not negligible. To address this problem, the receiver is required to cancel the SI by deploying some SI cancellation mechanisms, which brings high hardware difficulty and is impractical for state-of-the-art THz technologies [29,30].

Against this background, a novel molecular absorption-aided RAN scheme was proposed in the THz band [31], where SIC-free transmission is realized through an interesting waveform design based on the temporal broadening effect (TBE) caused by molecular absorption. The TBE refers to the widening of transmitted pulses that experiences frequency-selective attenuation [31]. An important phenomenon is observed that is when pulses are transmitted in frequency bands with large molecular absorption, the TBE increases considerably with distance. In light of this observation, the SI can be easily removed without incurring any complicated compensation-based SIC methods. Specifically, at Bob, since the SI signal caused by transmitting AN signals is very close to its receive antennas, the SI experiences negligible TBE. Motivated by this, we can separate information signals and AN signals in the time domain while keeping these signals overlapping at eavesdroppers after experiencing TBE since the duration of received SI is extremely short at Bob. As a result, Bob can detect information signals without any influence from AN signals. Meanwhile, the information signals at Eve are heavily polluted by the AN signals. Therefore, accounting for proximal eavesdroppers, the proposed RAN scheme can guarantee THz secure communications without the need for SIC. We present the design guideline for the SIC-free transmission scheme for THz secure communications. The modeling of TBE caused by the molecular absorption effect is first described. Then, the novel SIC-free signal detection strategy is presented by utilizing this unique effect in the THz band. Finally, the waveform parameter design realizing the SIC-free receiver AN scheme is elaborated. By adopting the molecular absorption-assisted SIC-free RAN scheme, the secrecy rate with eavesdroppers in close proximity can reach about 40% of that with SIC-based schemes but with significantly reduced complexity. If neither the SIC-based RAN scheme nor our proposed SIC-free scheme is used, the secrecy rate is zero with eavesdroppers in close proximity.

2.4.1 Temporal broadening effect

The temporal broadening effect stands for the effect of the widening of signal duration after it propagates through the frequency-selective channels. Let $g(t)$ and $h(t)$ represent the normalized waveform with bandwidth B_g and the channel impulse response. The normalized received waveform $g_r(t)$ can be expressed by convolving the transmit waveform with the channel impulse response, which can be expressed by

$$g_r(t) = \mathscr{F}^{-1}\left\{\frac{\mathscr{F}\{g(t)\}\cdot H(f-f_c)}{|\mathscr{F}\{g(t)\}\cdot H(f-f_c)|}\right\} = \frac{\int_{f_u} G(f)e^{-\frac{1}{2}\kappa(f-f_c)d}e^{j2\pi ft}df}{\sqrt{\int_{f_u}|G(f)|^2e^{-\kappa(f-f_c)d}df}}, \tag{2.36}$$

where $H(f)$ represents the frequency response and $G(f)$ represents the Fourier transform of the time-domain signal $g(t)$. f_u represents the frequency range of the transmit waveform, and f_c is the center frequency of f_u. According to (2.36), if the received waveform experiences frequency-selective attenuation, its waveform shape in the frequency domain changes, the time-domain pulse shape distorts, and the pulse duration increases. This phenomenon is known as the temporal broadening effect (TBE) [13]. To characterize the TBE, we define the pulse duration of waveform $g(t)$ as

$$\mathbf{D}(g) \triangleq \sqrt{\frac{\int_{-\infty}^{\infty} |tg(t)|^2 dt}{\int_{-\infty}^{\infty} |g(t)|^2 dt}}. \tag{2.37}$$

Furthermore, we define the temporal broadening factor β_{br} as the ratio of the pulse width of the received pulse shape after passing through the channel $h(t)$ to the transmitted pulse shape $g_r(t)$, as

$$\beta_{\mathrm{br}} = \mathbf{D}(g_t * h)/\mathbf{D}(g_t). \tag{2.38}$$

By considering g_t as Gaussian pulse and $h(t)$ as the THz channel expressed in (2.4), the relationship between β_{br} and the signal propagation distance is elaborated as follows.

Theorem 1 (Temporal broadening factor). *β_{br} versus f_c and bandwidth B_g is approximately linear with transmission distance, as $\beta_{br} \approx 1 + \eta_{br} d$, where η_{br} is given by*

$$\eta_{br} = \frac{f_g \sqrt{\int_{-\infty}^{\infty} \left(e^{-2\left(\frac{f-f_c}{f_g}\right)^2 - \kappa(f)d}\right)^2 \left(\frac{\mathrm{d}}{\mathrm{d}f}\kappa(f)\right)^2 df}}{2\sqrt{\int_{-\infty}^{\infty} \left(e^{-2\left(\frac{f-f_c}{f_g}\right)^2 - \kappa(f)d}\right) df}} \tag{2.39}$$

$$\approx \frac{f_g}{2^{3/4}} \sqrt{8\kappa'(f_c)^2 + \kappa''(f_c)^2 f_g^2},$$

where $f_g = 1/\xi_g$. Moreover, $\kappa'(\cdot)$ and $\kappa''(\cdot)$ denote the 1st and 2nd derivatives of $\kappa(\cdot)$, respectively.

Proof: According to the definition of temporal broadening factor, β_{br} with respect to frequency f_c and distance d can be expressed by

$$\beta_{\mathrm{br}}(f,d) \overset{(A)}{=} \sqrt{\frac{\frac{1}{4\pi^2}\int_{-\infty}^{\infty} \left|\frac{\mathrm{d}}{\mathrm{d}f} G_r(f)\right|^2 df}{\int_{-\infty}^{\infty} |G_r(f)|^2 df}} \Bigg/ \sqrt{\frac{\frac{1}{4\pi^2}\int_{-\infty}^{\infty} |\frac{\mathrm{d}}{\mathrm{d}f} G_t(f)|^2 df}{\int_{-\infty}^{\infty} |G_t(f)|^2 df}} \tag{2.40}$$

$$\beta_{\mathrm{br}}(f,d) \overset{(B)}{=} f_g \sqrt{\frac{\int_{-\infty}^{\infty} (e^{-2\left(\frac{f-f_c}{f_g}\right) - \kappa(f)d})(\frac{2(f-f_c)}{f_g} + \frac{1}{2}\frac{\mathrm{d}}{\mathrm{d}f}\kappa(f)d)^2 df}{\int_{-\infty}^{\infty} e^{-2\frac{f-f_c}{f_g} - \kappa(f)d} df}} \tag{2.41}$$

$$\beta_{\mathrm{br}}(f,d) \overset{(C)}{\approx} 1 + \frac{f_g \sqrt{\int_{-\infty}^{\infty} (e^{-2(\frac{f-f_c}{f_g})^2 - \kappa(f)d})^2 (\frac{\mathrm{d}}{\mathrm{d}f}\kappa(f))^2 df}}{2\sqrt{\int_{-\infty}^{\infty} e^{-2(\frac{f-f_c}{f_g})^2 - \kappa(f)d} df}} d, \tag{2.42}$$

where step (A) is obtained by invoking Parseval's theorem. Then in step (B), by expanding the square term $(\frac{2(f-f_c)}{f_g} + \frac{1}{2}\frac{\mathrm{d}}{\mathrm{d}f}\kappa(f)d)^2$, we have $\beta_{\mathrm{br}} = 1 + \eta_{\mathrm{br}} \cdot d$, and η_{br} is given by the square root of the quadratic coefficient of the expression as shown in step (C).

In (2.39), the temporal broadening parameter η_{br} indicates the increasing trend of the pulse duration with distance, defined as *temporal broadening coefficient*. When $\kappa(f) \approx 0$, $\eta_{\mathrm{br}} = 0$ and $\beta_{\mathrm{br}} = 1$. This case indicates that in the THz band, the temporal broadening effect is led by the frequency selectivity due to molecular absorption effect.

2.4.2 SIC-free receiver AN scheme

The SIC-free receiver AN scheme based on the different TBEs at Bob and Eve is shown in Figures 2.6 and 2.7, where the different receiving methods are demonstrated. In Figure 2.6, the main signal at Bob consists of a lot of narrow pulses. The frame time which is defined as the time duration between consecutive pulses is denoted as T_f. Although Bob's interference power is significant compared with the main signal, the TBE of the AN signal is weak and the AN pulse duration is limited. Thus, Bob can receive the information signal without any interference in part of the time duration. The *detection ratio* at both sides, denoted by α_i where $i \in \{\mathrm{B}, \mathrm{E}\}$ is defined as the ratio of the detection period of node i to the frame time T_f, i.e., $\alpha_{\mathrm{B}} = \frac{T_{\mathrm{B}}^d}{T_f}$. By contrast, the received signal at Eve is shown in Figure 2.7. It can be observed that the TBE of AN is severer due to the longer transmission distance of the AN signals from Bob to Eve. Similarly, we define the detection ratio at Eve $\alpha_{\mathrm{E}} = \frac{T_{\mathrm{E}}^d}{T_f}$. As a result, despite the AN signal having an attenuated power at Eve, the information pulses with larger durations are inevitably "polluted" by the widened AN pulses, which significantly

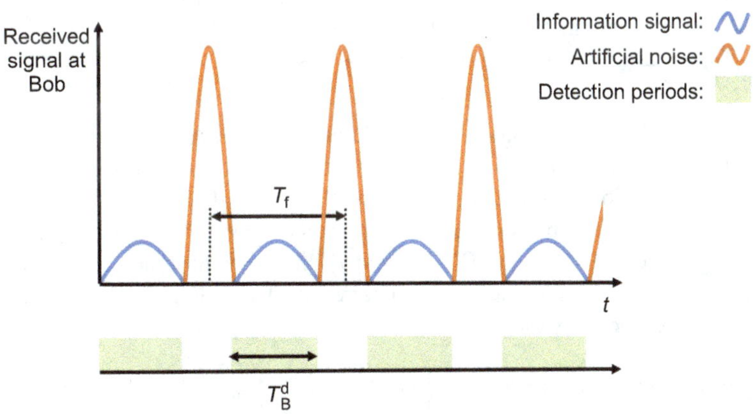

Figure 2.6 Illustration of the SIC-free receiver AN scheme. The received signal at Bob is untouched. ©IEEE 2023. Reprinted with permission from [2].

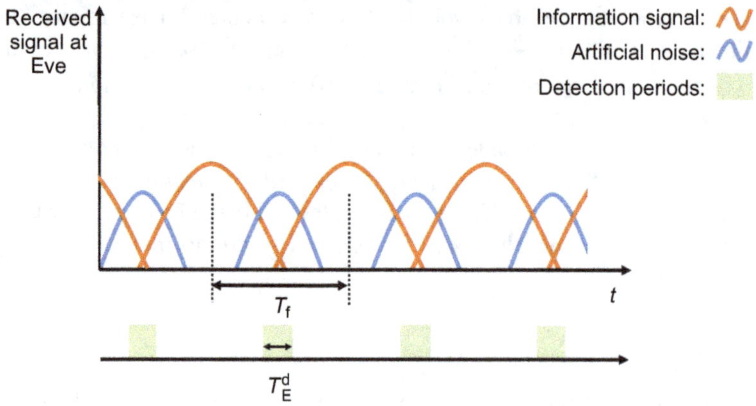

Figure 2.7 Illustration of the SIC-free receiver AN scheme. Received signal at Eve is polluted by AN. ©IEEE 2023. Reprinted with permission from [2].

degrades Eve's wiretap rate. At the transmitter side, the main signal $x_A(t)$ and the AN signal $x_B(t)$ are given by

$$x_A(t) = \sqrt{P_A} \sum_i b_i \cdot g(t - iT_f), \tag{2.43}$$

$$x_B(t) = \sqrt{P_B} \sum_i r_i \cdot g(t - iT_f - \tau_B), \tag{2.44}$$

where i represents the symbol index, P_A and P_B denote the transmit power at Alice and Bob, respectively. T_f represents the frame time as the time duration between two symbols. τ_B is the relative time delay of the AN signal with respect to $x_A(t)$, expressed by $\tau_B = (\frac{d_{AB} - d_{BB}}{c} - \frac{1}{2}T_f) \bmod T_f$. We assume the transmit pulse shape $g(t)$ is a Gaussian-shaped pulse, as

$$g(t) = \sqrt{\pi}/\xi_g \cdot e^{-\pi^2 t^2/\xi_g^2}, \tag{2.45}$$

where $\xi_g = \sqrt{-\ln\sqrt{0.5}}/B_g$, and the pulse duration is $T_p = \xi_g/\pi$. For the main signal in (2.43), the parameter b_i represents the modulated information symbol carrying the confidential information, which satisfies that $E[|b_i|^2] = 1$. For the AN signals expressed in (2.44), r_i represents a unit-power random variable randomly produced by a local generator. Parameters P_A, P_B, and T_f should be designed which represent the power and rate allocation strategies. The received signals from Bob and Eve can be expressed, respectively, as

$$y_B(t) = \sqrt{G_{AB}P_{AB}} \sum_i b_i p_i g_{AB}\left(t - iT_f - \frac{d_{AB}}{c}\right) + \sqrt{G_{BB}P_{BB}} \sum_m r_m \\ g_{BB}\left(t - mT_f - \frac{d_{BB}}{c} - \tau_b\right) + n_B(t), \tag{2.46}$$

$$y_E(t) = \sqrt{G_{\mathrm{AE}}P_{\mathrm{AE}}}\sum_i b_i p_i g_{\mathrm{AE}}\left(t - iT_f - \frac{d_{\mathrm{AE}}}{c}\right) + \sqrt{G_{\mathrm{BE}}P_{\mathrm{BE}}}\sum_m r_m$$
$$g_{\mathrm{BE}}\left(t - mT_f - \frac{d_{\mathrm{BE}}}{c} - \tau_b\right) + n_E(t), \tag{2.47}$$

where $g_{ij}(t)$ denotes the received waveform from i to j. P_{ij} represents the received signal power from i to j, which is given by

$$P_{ij} = P_i \int_{-\infty}^{\infty} |G(f)H_{ij}(f - f_c)|^2 df, \tag{2.48}$$

where H_{ij} stands for the channel frequency response.

The signal detection of our SIC-free RAN scheme is realized by a correlation-based receiver, where a synchronized template signal is correlated with the received signal $y(t)$. The correlation value for each frame is used for symbol detection. Here the template signal should be the normalized received pulse shape of the information-bearing pulse during the detection period $[0, \alpha_i T_f]$, which is expressed as,

$$g_r^i(t) = g_{ai}^*(t) \cdot \Pi_{\alpha_i T_f}(t), \tag{2.49}$$

where $(\cdot)^*$ denotes the conjugate operation, and the rectangular function $\Pi_T(t)$ is defined as

$$\Pi_T(t) \triangleq \begin{cases} 1, & |t| < T/2, \\ 0, & \text{otherwise,} \end{cases} \tag{2.50}$$

where T is the shaping parameter of the rectangular pulse. For AN-free signal detection, $\alpha_i = 1$ is the optimal choice, and all power is fully utilized for symbol detection. However, when the received signal is composed of the main and interference signal, detecting the entire frame time is not the optimal choice since the AN signal degrades the SINR. Instead, partially detecting the signals in the time domain is a better way to detect the main signals.

In practice, the proposed receiver can be implemented by truncating a fraction of the received signal, i.e., only receiving α_{B} of the entire symbol duration. This process does not require additional computational processes and therefore has low complexity.

2.4.3 Eavesdroppable region analysis

We adopt the ER as the security metric for randomly distributed eavesdroppers with density λ_{E}. To theoretically demonstrate the security improvements, we partition the entire eavesdropping region into four sub-regions, according to whether the area is inside the main beam and AN beam or not, denoted by $\{\mathscr{A}_1, \mathscr{A}_2, \mathscr{A}_3, \mathscr{A}_4\}$. Thus, we have $|\mathscr{A}_{\mathrm{ER}}| = \sum_{i=1}^4 |\mathscr{A}_i|$, and the four non-overlapping sub-regions are described in Figure 2.8 and elaborated as follows:

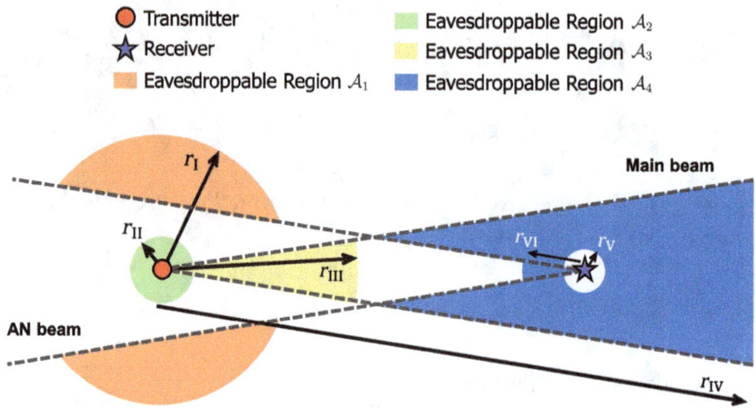

Figure 2.8 The eavesdroppable region model with four sub-regions. ©IEEE 2023. Reprinted with permission from [2].

- $\mathscr{A}_1$: Outside the main beam and outside the AN beam. This is a circular area except for the intersection with the AN beam region, with $r_{\rm I}$ being the radius. The eavesdropping sub-region area can be calculated as

$$|\mathscr{A}_1| = \left[2\arccos\left(\frac{d_{\rm AB}\sin(\theta_{\rm n}/2)}{r_{\rm I}}\right)r_{\rm I}^2 - d_{\rm AB}\sin(\theta_{\rm n}/2) \times\sqrt{r_{\rm I}^2-(d_{\rm AB}\sin(\theta_{\rm n}/2))^2}\right]^+. \tag{2.51}$$

- $\mathscr{A}_2$: Outside the main beam but inside the AN beam. Let $r_{\rm II}$ denote the radius of this area. The eavesdropping sub-region area can be calculated as $|\mathscr{A}_2| = \left(\pi - \frac{\theta_{\rm s}}{2}\right)r_{\rm II}^2$.
- $\mathscr{A}_3$: Inside the main beam and inside the AN beam. This area is the left area of the diamond-shaped beam intersection. Let $r_{\rm III}$ denote the distance from Alice to the right-most point inside this sub-ER. The eavesdropping sub-region area can be calculated as

$$|\mathscr{A}_3| = \begin{cases} \frac{1}{2}\theta_s r_{\rm III}^2, & r_{\rm III}\le r_{\rm th}, \\ \frac{1}{2}\theta_s r_{\rm th}^2 + \frac{1}{2}\theta_n(d_{\rm AB}-r_{\rm th})^2 - \frac{1}{2}\theta_n(d_{\rm AB}-r_{\rm III})^2, & r_{\rm III} > r_{\rm th}, \end{cases} \tag{2.52}$$

where $r_{\rm th} = \dfrac{d_{\rm AB}\theta_n}{\theta_n+\theta_s}$.
- $\mathscr{A}_4$: Inside the main beam but outside the AN beam. $r_{\rm IV}$ denotes the maximal distance from Alice that Eve can eavesdrop the communications. $r_{\rm V}$ is the radius

of the inner circle, and r_{VI} represents the eavesdroppable distance from Bob. The eavesdropping sub-region area can be calculated as

$$|\mathscr{A}_4| = \frac{1}{2}\theta_{\mathrm{s}} r_{\mathrm{IV}}^2 - \frac{d_{\mathrm{AB}}^2}{\cot\theta_{\mathrm{s}} + \cot\theta_{\mathrm{n}}} - \pi r_{\mathrm{V}}^2 + \frac{1}{2}\theta_{\mathrm{n}} r_{\mathrm{VI}}^2. \tag{2.53}$$

Distances $r_{\mathrm{I}}, \ldots, r_{\mathrm{VI}}$ represent the threshold distances of d_{AE} corresponding to each sub-region such that the secrecy rate equals to R_{th}.

2.4.4 Waveform parameter design guideline

Design of waveform parameters including frame time T_f, detection ratios α_{B} and α_{E}, are described and summarized as follows. Based on the proposed SIC-free detection scheme, the instantaneous secrecy rate can be computed as

$$R_s = \frac{1}{B_g T_f}\big[\log(1+\gamma_{\mathrm{B}}) - \max_{\alpha_{\mathrm{E}}\in(0,1]} \log(1+\gamma_{\mathrm{E}})\big], \tag{2.54}$$

where the SINRs at Bob and Eve are, respectively, expressed as

$$\gamma_{\mathrm{B}} = \frac{G_{\mathrm{AB}}P_{\mathrm{AB}}\left(\int_{-\frac{\alpha_{\mathrm{B}}T_f}{2}}^{\frac{\alpha_{\mathrm{B}}T_f}{2}} |g_{\mathrm{AB}}(t) g_{r,\mathrm{norm}}^{\mathrm{B}}(t)| dt\right)^2}{4G_{\mathrm{BB}}P_{\mathrm{BB}}\left(\int_{-\frac{\alpha_{\mathrm{B}}T_f}{2}}^{\frac{\alpha_{\mathrm{B}}T_f}{2}} |g_{\mathrm{BB}}\left(t-\frac{T_f}{2}\right) g_{r,\mathrm{norm}}^{\mathrm{B}}(t)| dt\right)^2 + N_{\mathrm{B}}}, \tag{2.55}$$

$$\gamma_{\mathrm{E}} = \frac{G_{\mathrm{AE}}P_{\mathrm{AE}}\left(\int_{-\frac{\alpha_{\mathrm{E}}T_f}{2}}^{\frac{\alpha_{\mathrm{E}}T_f}{2}} |g_{\mathrm{AE}}(t) g_{r,\mathrm{norm}}^{\mathrm{E}}(t)| dt\right)^2}{4G_{\mathrm{BE}}P_{\mathrm{BE}}\left(\int_{-\frac{\alpha_{\mathrm{E}}T_f}{2}}^{\frac{\alpha_{\mathrm{E}}T_f}{2}} |g_{\mathrm{BE}}(t - T_f/2) g_{r,\mathrm{norm}}^{\mathrm{E}}(t)| dt\right)^2 + N_{\mathrm{E}}}, \tag{2.56}$$

where $N_i = \alpha_i B_g N_0$, $g_{r,\mathrm{norm}}^i(t) \triangleq g_r^i(t)/||g_r^i(t)||, i \in \{\mathrm{B}, \mathrm{E}\}$ is the normalized template signal at node i. Since Bob cannot know Eve's detection ratio α_{E}, the worst situation for α_{E} is considered, i.e., the minimum secrecy rate with varying α_{E} is analyzed in (2.54). The feasible solution of the frame time is found based on the following theorem.

Theorem 2 (Feasible region of the frame time). *To achieve a non-zero secrecy rate, the frame time should be chosen within the range*

$$T_f \in \Big(2x_{apr}\frac{T_{p2,b}}{T_{p1,b}}\sqrt{T_{p1,b}^2 + T_{p2,b}^2}/B_g, 2x_{apr}\frac{T_{p1,w}}{T_{p2,w}}\sqrt{T_{p1,w}^2 + T_{p2,w}^2}/B_g\Big), \tag{2.57}$$

where $T_{p1,i} = (1+\eta_{br}d_{ai})T_p$, $T_{p2,i} = (1+\eta_{br}d_{bi})T_p$, x_{apr} *denotes the truncation threshold of the approximated error function, i.e.,*

$$\mathrm{erf}(x) \approx \mathrm{erf}_{\mathrm{apr}}(x) = \begin{cases} \mathrm{erf}(x), & |x| < x_{apr}, \\ 0, & |x| > x_{apr}, \end{cases} \tag{2.58}$$

and x_{apr} is given by

$$x_{apr} \approx \text{erf}^{-1}\left[1 - \left(\frac{d_{BB}}{d_{AB}}\right)^2\right]. \tag{2.59}$$

Theorem 3 (Optimal detection ratio at Bob). *To achieve a non-zero secrecy rate, the optimal detection ratio at Bob is given by*

$$\alpha_B^{opt} = \frac{T_f T_{p1,b}^2 - x_{apr} * 2T_{p1,b}T_{p2,b}\sqrt{T_{p1,b}^2 + T_{p2,b}^2}}{T_f(T_{p1,b}^2 + T_{p2,b}^2)}, \tag{2.60}$$

and the resulting rate at Bob is given by

$$R_b^{opt} = \frac{1}{B_g T_f}\log_2\left(1 + \frac{P_{AB}\cdot \text{erf}\left[\frac{\alpha_B^{opt} T_f}{\sqrt{2}T_{p1,b}}\right]}{\alpha_B^{opt} B_g N_0}\right). \tag{2.61}$$

2.5 Widely spaced antenna communications for THz physical layer security

The widely spaced antenna array is an array structure where the spacing between antennas is much larger than half wavelength. The ground-truth model of the EM transmission is the spherical-wave propagation. In traditional MIMO antenna arrays, the spacing between different antennas is half the wavelength, which is much smaller than the propagation distance. Therefore, an approximation is applied where the spherical wave can be viewed as a planar wave. The threshold distance that determines whether this approximation applies is known as the Fraunhofer distance [32,33]. The Fraunhofer distance for a 2×2 critically spaced antenna array ($d = \lambda/2$) can be calculated as $\lambda/4$, where λ represents the wavelength. The area with a distance farther than the Fraunhofer distance from the transmitter is the far-field region, and the remaining part is referred to as the near-field region. For a microwave communication system where the frequency is 300 MHz, the Fraunhofer distance is 1 m. Therefore, the planar-wave assumption is valid for microwave communications since the propagation distance is much larger than larger than 1 m. However, the traditional far-field propagation experiences limitation in the secrecy performance when Eve is inside the beam sector. In this section, we investigate how the WSA structure can safeguard the THz range security. The WSA means the antenna spacing is widened such that Bob and Eve are located in the near-field regions of Alice, as illustrated in Figure 2.9. Since the spherical-wave propagation model is the ground-truth model and the planar-wave propagation is an approximated method, our analysis can still be applied when either of Bob or Eve or both of them are in far field. We first develop the WSA model with hybrid beamforming and formulate a secrecy rate maximization

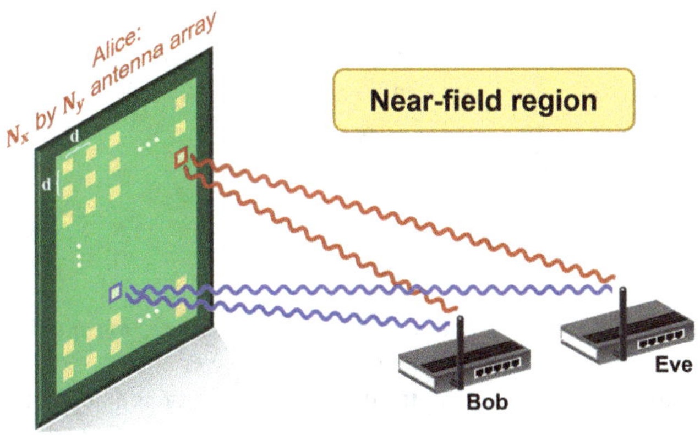

Figure 2.9 THz widely spaced antenna array

problem for the WSA communications. Then, a beamforming design strategy for our WSA transmission scheme is proposed based on a new algorithm. Finally, we discuss the trade-off between the antenna array size and the secrecy rate. This demonstrates that the proposed scheme is realistic but unique for the THz band.

Thanks to the short wavelength, THz communications can maintain a small array size even when the antenna spacing is wide enough to enhance the secrecy rate. A sample comparison between millimeter-wave and THz WSA design is provided as follows. For a 4×4 communication system where the wavelength is on the order of sub-meters, e.g., 0.3 m for the 1 GHz system, the spherical-wave propagation should be used when $3d \approx 4.0$ m. This has an impractically large antenna array. In contrast, for the THz band with a wavelength on the order of millimeters, e.g., 1 mm for a 300 GHz system, the spherical-wave propagation should be used when $3d \approx 0.08$ m. Therefore, a 0.08 m $\times$ 0.08 m THz antenna array can well meet the near-field communication requirement.

2.5.1 Widely spaced array and beamforming

A universal model for the both spherical-wave and planar-wave propagation is described as follows. The beamformer can be represented by a matrix $\mathbf{W} = [w_1, \cdots, w_{N_t}]$ with complex gains w_i satisfying $\sum_i |w_i|^2 = 1, i \in 1, 2, \cdots, N_t$, where N_t represents the number of antennas. For example, for a beamformer based on phase shifter, the complex gain for each antenna can be represented by $w_i = \frac{1}{\sqrt{N_t}} e^{j\phi_i}$, where ϕ_i denotes the phase. The array factor for the antenna array is given by

$$\mathrm{AF}(r, \theta) = \left| \sum_{i=1}^{N_t} w_i e^{j2\pi \cdot \frac{f \Delta r_i}{c}} \right|^2, \tag{2.62}$$

where Δr_i denotes the propagation distance of the antenna minus that of the reference antenna. The location of Bob and Eve is represented by $(r_{\mathrm{B}}, \theta_{\mathrm{B}})$ and $(r_{\mathrm{E}}, \theta_{\mathrm{E}})$, respectively. Then, the secrecy rate can be expressed by

$$R_s = \left[\log\left(1 + \frac{P_{\mathrm{Tx}}|h_{\mathrm{B}}|^2 \mathrm{AF}(r_{\mathrm{B}}, \theta_{\mathrm{B}})}{\sigma^2}\right) \right. \\ \left. - \log\left(1 + \frac{P_{\mathrm{Tx}}|h_{\mathrm{E}}|^2 \mathrm{AF}(r_{\mathrm{E}}, \theta_{\mathrm{E}})}{\sigma^2}\right) \right]^+, \tag{2.63}$$

where h_{B} and h_{E} stand for the complex channel impulse response at Bob and Eve, respectively. σ^2 represents the noise power at the receiver side.

Traditional antenna array structure: For conventional antenna array, the relative distance Δr_i does not depend on the distance r based on the planar-wave approximation. For example, $\Delta r_i = id \sin\theta$ for a linear array and $\Delta r_i = ((i-1) \bmod N_y + 1)d \sin\theta$ for a planar array. If the situation is that Eve has the same propagation angle with Bob, i.e., $\theta_{\mathrm{B}} = \theta_{\mathrm{E}}$, then the equivalent array gain at Bob and Eve is equal. Then, the maximum secrecy rate is given by

$$R_s = \left[\log\left(1 + \frac{P_{\mathrm{Tx}}|h_{\mathrm{B}}|^2 N_{\mathrm{t}}}{\sigma^2}\right) - \log\left(1 + \frac{P_{\mathrm{Tx}}|h_{\mathrm{E}}|^2 N_{\mathrm{t}}}{\sigma^2}\right) \right]^+, \tag{2.64}$$

where the array factor and the phase at the beamformer are given by

$$\begin{aligned} \mathrm{AF}(r_{\mathrm{B}}, \theta_{\mathrm{B}}) &= \mathrm{AF}(r_{\mathrm{E}}, \theta_{\mathrm{E}}) = N_{\mathrm{t}}, \\ \phi_i &= -\frac{2\pi f \Delta r_i}{c}. \end{aligned} \tag{2.65}$$

Based on these results, we draw the conclusion that the traditional antenna array cannot address the near eavesdropping problem due to the same array factor at Bob and Eve.

Widely spaced antenna array structure: As the result of the spherical-wave propagation, we widen the spacing between antennas in an array to ensure that Δr_i is dependent of r. For the planar antenna array in Figure 2.10,

$$\Delta r_i = \sqrt{(r \sin\theta - (i_{\mathrm{row}} - 1)d)^2 + ((i_{\mathrm{col}} - 1)d)^2 + (r\cos\theta)^2} - r, \tag{2.66}$$

where the row antenna index is $i_{\mathrm{row}} = (i-1) \mod N_y + 1$ and the column antenna index is $i_{\mathrm{col}} = (i - i_{\mathrm{row}})/N_y + 1$. Even though the angle of Bob and Eve is the same, their AFs can be different. Therefore, we can beamform matrix $\mathbf{W}$ such that the array factor received by Bob is larger than that received by Eve.

2.5.2 Beamforming design guideline

Now we explore how to design the beamformer $\mathbf{W}$ of the THz WSA. We express the array factor and the channel in the form of a matrix as

$$\mathbf{H}_{\mathrm{i}} = \left[e^{j2\pi \frac{\Delta r_{1,i}}{c}}, \cdots, e^{j2\pi \frac{\Delta r_{1,i}}{c}} \right]^T \cdot |h_{\mathrm{i}}|. \tag{2.67}$$

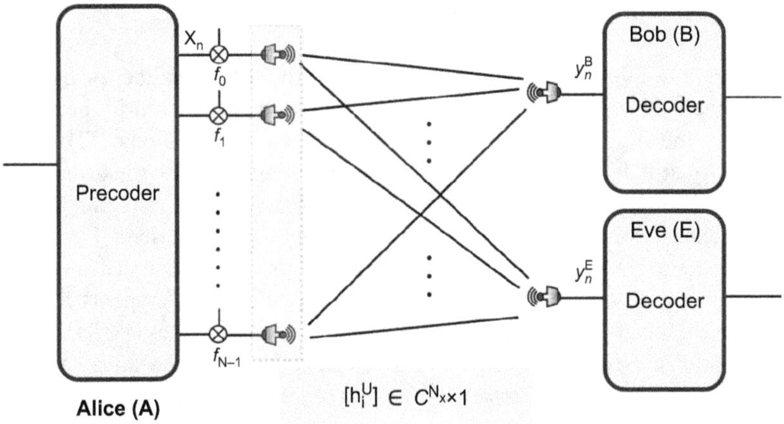

Figure 2.10 THz widely spaced antenna array in the secure communication model

Then, the secrecy rate R_s is simplified as

$$\begin{aligned} R_s &= \left[\log\left(1+\frac{P_{\mathrm{Tx}}\left|\mathbf{W}^{\dagger}\mathbf{H}_{\mathrm{B}}\right|^2}{\sigma^2}\right) - \log\left(1+\frac{P_{\mathrm{Tx}}\left|\mathbf{W}^{\dagger}\mathbf{H}_{\mathrm{E}}\right|^2}{\sigma^2}\right)\right]^+ \\ &= \left\{\log\left[1+\frac{\mathbf{W}^{\dagger}\left(\mathbf{H}_{\mathrm{B}}\mathbf{H}_{\mathrm{B}}^{\dagger}-\mathbf{H}_{\mathrm{E}}\mathbf{H}_{\mathrm{E}}^{\dagger}\right)\mathbf{W}}{\mathbf{W}^{\dagger}\left(\frac{\sigma^2}{P_{\mathrm{Tx}}}\mathbf{I}+\mathbf{H}_{\mathrm{E}}\mathbf{H}_{\mathrm{E}}^{\dagger}\cdot\right)\mathbf{W}}\right]\right\}^+, \end{aligned} \tag{2.68}$$

where $\mathbf{I} \in \mathbb{R}^{N_t \times N_t}$ represents a $N_t \times N_t$ identity matrix. By denoting $\mathbf{A} \triangleq \mathbf{H}_{\mathrm{B}}\mathbf{H}_{\mathrm{B}}^{\dagger} - \mathbf{H}_{\mathrm{E}}\mathbf{H}_{\mathrm{E}}^{\dagger}$ and $\mathbf{B} \triangleq \frac{\sigma^2}{P_{\mathrm{Tx}}}\mathbf{I} + \mathbf{H}_{\mathrm{E}}\mathbf{H}_{\mathrm{E}}^{\dagger}$, we discover that maximizing R_s is indeed equivalent to maximizing the generalized Rayleigh quotient $\lambda_\Sigma \triangleq \frac{\mathbf{W}^{\dagger}\mathbf{A}\mathbf{W}}{\mathbf{W}^{\dagger}\mathbf{B}\mathbf{W}}$. λ_Σ can be further simplified as

$$\lambda_\Sigma = |\langle \mathbf{W}', \mathbf{v}^{(a)}\rangle|^2\lambda_a + |\langle \mathbf{W}', \mathbf{v}^{(b)}\rangle|^2\lambda_b, \tag{2.69}$$

where $\mathbf{W}' = \frac{\mathbf{B}^{1/2}\mathbf{W}}{\|\mathbf{B}^{1/2}\mathbf{W}\|}$, and λ_a and λ_b denote the two non-zero eigenvalues of $\mathbf{B}^{-\frac{1}{2}}\mathbf{A}\mathbf{B}^{-\frac{1}{2}}$ with $\lambda_a \geq \lambda_b$. Let $\mathbf{v}^{(a)}$ and $\mathbf{v}^{(b)}$ be the normalized eigenvectors. Since $|\langle \mathbf{w}', \mathbf{v}^{(a)}\rangle| \leq |\langle \mathbf{v}^{(a)}, \mathbf{v}^{(a)}\rangle| = 1$, when $\mathbf{w}'$ coincides with the eigenvector of the maximum eigenvalue $\mathbf{v}^{(a)}$, the optimal beamformer can be achieved. This is the non-constrained optimal solution as

$$\mathbf{W}^{\mathrm{opt}} = \frac{\mathbf{B}^{-\frac{1}{2}}\mathbf{v}^{(a)}}{\|\mathbf{B}^{-\frac{1}{2}}\mathbf{v}^{(a)}\|}, \quad \lambda_\Sigma = \lambda_a, \quad R_s^{\mathrm{opt}} = \log_2(1+\lambda_a). \tag{2.70}$$

Conventional THz UM-MIMO techniques apply hybrid beamforming technology [34] to overcome the distance limitation of THz communication. Hybrid beamforming puts additional constraints on the beamformer, and further hybrid beamforming design based on the WSA structure is awaited to be explored.

2.6 Conclusion and future directions

In this chapter, we described the fundamental system model, challenges, and technical solutions to THz physical layer security. Despite the good security performance provided by the high directivity of THz wireless communications, THz physical layer security still faces security challenges, including eavesdroppers inside beam sector, eavesdroppers in close proximity and THz covertness problem. In light of these challenges, three THz-specific security schemes were described. First, the DA-APM proposes to transmit signals at sides of molecular absorption peaks to enhance the wireless security and covertness. This scheme effectively combats eavesdroppers inside the beam sector and can even enhance the covertness of THz wireless communications. Second, the SIC-free RAN scheme takes the advantage of the distance-dependent temporal broadening effect to jam Eve in close proximity while not jamming Bob. This method handles the eavesdroppers in close proximity with a lower-complexity receiver AN mechanism. Third, we described the WSA and beamforming scheme to combat near eavesdropping without the aid of other nodes. The size rationality of practical THz WSA is justified.

Despite the fruitful research progress on THz physical layer security, there are still open challenges remaining to be tackled, which are described as follows. First, since the optimal molecular absorption coefficients depend on the propagation distance, it is necessary to adjust the carrier frequency as the distance between Alice and Bob varies. How to efficiently and practically tune the carrier frequency to use proper sub-bands needs investigation. Second, the hybrid beamforming structure is usually adopted in THz communications. However, the hybrid beamformer design for the WSA is a non-convex optimization problem. It is necessary to investigate efficient and low-complexity solutions for the WSA beamformer design. Finally, active security attacks, for instance, eavesdroppers may send the jamming signal to the legitimate receiver while overhearing the secret message, need to be handled for THz physical layer security. In line with this, security mechanisms simultaneously combating active and passive eavesdropping are awaited to be explored as a future direction.

References

[1] Han C, Wei G, Yang N, *et al.* Molecular absorption effect: a double-edged sword of terahertz communications. *IEEE Wireless Communications*. 2023;30(4):140–146.

[2] Gao W, Han C, and Chen Z. DNN-powered SIC-free receiver artificial noise aided terahertz secure communications with randomly distributed eavesdroppers. *IEEE Transactions on Wireless Communications*. 2021;21(1):563–576.

[3] Gao W, Chen Y, Han C, *et al.* Distance-adaptive absorption peak modulation (DA-APM) for terahertz covert communications. *IEEE Transactions on Wireless Communications*. 2021;20(3):2064–2077.

[4] Rappaport TS, Xing Y, Kanhere O, *et al.* Wireless communications and applications above 100 GHz: opportunities and challenges for 6G and beyond. *IEEE Access*. 2019;7:78729–78757.

[5] Yang N, Wang L, Geraci G, *et al.* Safeguarding 5G wireless communication networks using physical layer security. *IEEE Communications Magazine*. 2015;53(4):20–27.
[6] Cao K, Wang B, Ding H, *et al.* Secure transmission designs for NOMA systems against internal and external eavesdropping. *IEEE Transactions on Information Forensics and Security*. 2020;15:2930–2943.
[7] Shafie A, Yang N, Han C, *et al.* Terahertz communications for 6G and beyond wireless networks: challenges, key advancements, and opportunities. arXiv preprint arXiv:220711021. 2022.
[8] Ma J, Shrestha R, Adelberg J, *et al.* Security and eavesdropping in terahertz wireless links. *Nature*. 2018;563(7729):89–93.
[9] Gao W, Lu X, Han C, *et al.* On multiple-antenna techniques for physical-layer range security in the terahertz band. arXiv preprint arXiv:220106253. 2022.
[10] ITU-R. Attenuation by Atmospheric Gases and Related Effects. 2019.
[11] Jornet JM and Akyildiz IF. Channel modeling and capacity analysis for electromagnetic wireless nanonetworks in the terahertz band. *IEEE Transactions on Wireless Communications*. 2011;10(10):3211–3221.
[12] Han C and Chen Y. Propagation modeling for wireless communications in the terahertz band. *IEEE Communications Magazine*. 2018;56(6):96–101.
[13] Han C, Bicen AO, and Akyildiz IF. Multi-ray channel modeling and wideband characterization for wireless communications in the terahertz band. *IEEE Transactions on Wireless Communications*. 2015;14(5):2402–2412.
[14] Wyner AD. The wire-tap channel. *Bell System Technical Journal*. 1975;54(8):1355–1387.
[15] Liu S, Hong Y, and Viterbo E. Practical secrecy using artificial noise. *IEEE Communications Letters*. 2013;17(7):1483–1486.
[16] Lv L, Ding Z, Ni Q, *et al.* Secure MISO-NOMA transmission with artificial noise. *IEEE Transactions on Vehicular Technology*. 2018;67(7):6700–6705.
[17] Chen G, Gong Y, Xiao P, *et al.* Physical layer network security in the full-duplex relay system. *IEEE Transactions on Information Forensics and Security*. 2015;10(3):574–583.
[18] Chen X, Ng DWK, Gerstacker WH, *et al.* A survey on multiple-antenna techniques for physical layer security. *IEEE Communications Surveys & Tutorials*. 2016;19(2):1027–1053.
[19] Hu J, Yan S, Zhou X, *et al.* Covert communication achieved by a greedy relay in wireless networks. *IEEE Transactions on Wireless Communications*. 2018;17(7):4766–4779.
[20] Zheng TX, Wang HM, Ng DWK, *et al.* Multi-antenna covert communications in random wireless networks. *IEEE Transactions on Wireless Communications*. 2019;18(3):1974–1987.
[21] Bash BA, Goeckel D, and Towsley D. Limits of reliable communication with low probability of detection on AWGN channels. *IEEE Journal on Selected Areas in Communications*. 2013;31(9):1921–1930.
[22] Abdelaziz A and Koksal CE. Fundamental limits of covert communication over MIMO AWGN channel. In: *Proceedings of the IEEE CNS*; 2017.

[23] Lee S, Baxley RJ, Weitnauer MA, *et al.* Achieving undetectable communication. *IEEE Journal of Selected Topics in Signal Processing*. 2015;9(7): 1195–1205.

[24] Gao W, Chen Y, Han C, *et al.* Distance-adaptive absorption peak modulation (DA-APM) for terahertz covert communications. *IEEE Transactions on Wireless Communications*. 2020;20(3):2064–2077.

[25] Pitchappa P, Kumar A, Singh R, *et al.* Electromechanically tunable frequency-agile metamaterial bandpass filters for terahertz waves. *Advanced Optical Materials*. 2022;10(2):2101544.

[26] Shafie A, Yang N, Han C, *et al.* Novel Spectrum Allocation Among Multiple Transmission Windows for Terahertz Communication Systems. arXiv preprint arXiv:220702401. 2022.

[27] Zheng TX, Wang HM, Yuan J, *et al.* Physical layer security in wireless ad hoc networks under a hybrid full-/half-duplex receiver deployment strategy. *IEEE Transactions on Wireless Communications*. 2017;16(6):3827–3839.

[28] Zheng G, Krikidis I, Li J, *et al.* Improving physical layer secrecy using full-duplex jamming receivers. *IEEE Transactions on Signal Processing*. 2013;61(20):4962–4974.

[29] Yan S, Zhou X, Yang N, *et al.* Secret channel training to enhance physical layer security with a full-duplex receiver. *IEEE Transactions on Information Forensics and Security*. 2018;13(11):2788–2800.

[30] Ahmed E and Eltawil AM. All-digital self-interference cancellation technique for full-duplex systems. *IEEE Transactions on Wireless Communications*. 2015;14(7):3519–3532.

[31] Gao W, Han C, and Chen Z. DNN-powered SIC-free receiver artificial noise aided terahertz secure communications with randomly distributed eavesdroppers. *IEEE Transactions on Wireless Communications*. 2022;21(1):563–576.

[32] Myers NJ and Heath RW. InFocus: a spatial coding technique to mitigate misfocus in near-field los beamforming. *IEEE Transactions on Wireless Communications*. 2022;21(4):2193–2209.

[33] Chen Y, Yan L, and Han C. Hybrid spherical- and planar-wave modeling and DCNN-powered estimation of terahertz ultra-massive MIMO channels. *IEEE Transactions on Communications*. 2021;69(10):7063–7076.

[34] Yuan H, Yang N, Yang K, *et al.* Hybrid beamforming for terahertz multi-carrier systems over frequency selective fading. *IEEE Transactions on Communications*. 2020;68(10):6186–6199.

Chapter 3
Eavesdropping attacks for terahertz wireless links

Jianjun Ma[1,2], *Yige Qiao*[1,2] *and Houjun Sun*[1,2]

Physical-layer security (PLS) technique protects wireless data transmission against eavesdropping and other security attacks by taking full advantage of link propagation characteristics. It provides an additional level of security, besides the traditional security protocols at the network and transport layers, which is not enough now due to the rapid improvement of computing hashrate and the requirement on low latency. Before the design and application of PLS algorithms, investigations on security threats should be conducted comprehensively. In this chapter, we will present a review of the characteristics of several eavesdropping techniques at the physical layer used in indoor and outdoor terahertz wireless communication systems and hope it can be helpful for future transceiver and protocol designs.*

3.1 Introduction

Terahertz frequency band has been a potential candidate for future wireless networks with high data rates of up to Tbps. Its narrow beam size makes it to be difficult for signal interception in line-of-sight (LoS) links and available for secure data transmission [1,2]. However, when link power is reflected or scattered at some situations, which has been an attractive scheme to achieve wide data coverage, serious scattering happens and eavesdropping threat becomes more serious [3].

Traditional methods of encryption based on a secret key are not enough now, because of the rapid improvement of computing hashrate [4,5]. The common secret key has risks of being deciphered. Besides, encryption and decryption can increase message size and add transmission overheads, which increases time latency and affects the application of terahertz communication in some scenarios with ultralow-latency required (e.g., telemedicine and intelligent transportation) [6]. What's more, these secrecy techniques are mainly applied at the network and higher layers and do not take full advantage of the relevant features of the physical layer [7].

*This chapter has been a part of [28,29].

[1]School of Integrated Circuit and Electronics, Beijing Institute of Technology, China
[2]Beijing Key Laboratory of Millimeter and Terahertz Wave Technology, China

Physical layer is the lowest layer in the Open System Interconnection (OSI) model. Terahertz communication has an extremely high information transmission rate and channel capacity. But it also makes channel dispersion characteristics more complex, and superimposed environmental factors can cause multipath scattering effects. These features make the channel endogenous security elements richer and easier to extract. Physical layer security technology can make full use of the differentiated design of channels to improve the security of information transmission. The implementation of physical-layer security (PLS) technique in terahertz communication systems is essential for secure data transmission [8,9]. Despite the fact that its widespread adoption may be some time away, lots of efforts have been conducted on the characteristics of security threats and PLS techniques. The purpose of this chapter is to review present researches on the performance of terahertz eavesdropping techniques in indoor and outdoor scenarios. We will not show how the equations in this chapter are derived, which could be found in the references.

3.2 Security performance characterization

A typical scheme on terahertz eavesdropping is shown in Figure 3.1. A LoS link setup between Alice and Bob is designed, with scattering objects, such as plates, cups, and cell phones, can be located inside the beam path. The object scatters part of the beam power toward Eve (the eavesdropper) sitting outside of the beam path. Eve's purpose is to choose composition, size, shape, and location of the object to make a successful signal interception and degrade the physical-layer security of the link.

In order to evaluate security performance at physical layer, several parameters, such as blockage, secrecy capacity, backscattering, and outage probability, are necessary and required.

3.2.1 Blockage

Blockage on Bob is used to evaluate the signal attenuation suffered by Bob due to the presence of the object [10]. It can be derived by the following equation:

$$b = 1 - \left(SNR_{Bob}^{\text{object}} / SNR_{Bob}^{\text{no object}} \right). \tag{3.1}$$

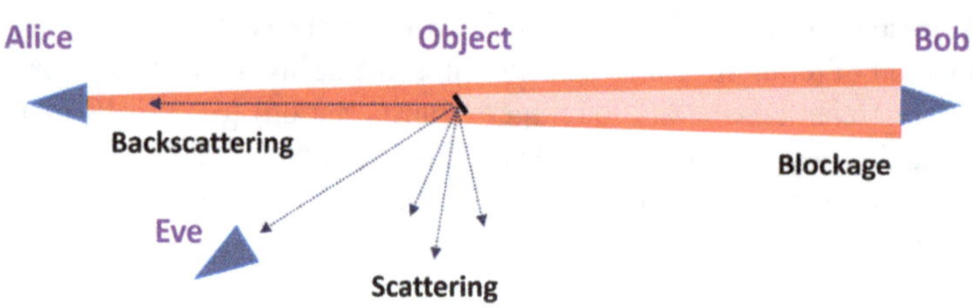

Figure 3.1 Typical scheme on security characterization of a LoS link

Here, *SNR* represents the signal-to-noise ratio with/without the intended object (eavesdropper) inside the LoS beam path. By this parameter, Bob can be able to check whether there is an attack when a threshold value for b is pre-determined. Beyond this value, Bob is certain to be aware of the changes in the characteristics of his LoS link. Usually, it is set to be $b = 0.5$, corresponding to a situation where Eve's object is blocking half of Bob's signal.

3.2.2 Secrecy capacity

Secrecy capacity is usually defined as the maximum data rate from Alice to Bob when perfect secrecy performance is maintained [11]. It is usually expressed as $C_s = [I\,(X;Y) - I(X;Z)]^{+}$with $[x]^{+} = \max\{0, x\}$ determining that the secrecy capacity can never be less than 0. Terms $I\,(X;Y)$ and $I(X;Z)$ represent the mutual information of legitimate (to Bob) and illegitimate (to Eve) links [12], respectively. Parameters X, Y, and Z represent the signals from Alice, to Bob, and to Eve, respectively. As a small modification to this expression, the secrecy capacity parameter is re-defined to characterize empirical limits of Bob and Eve's reception capabilities [10]. For the ease of study, normalized channel capacity is used as a performance indicator of channel security. Under particular modulation and coding methods, it is expressed as

$$C_s = \frac{\log_{10}(1 + SNR_{Bob}) - \log_{10}(1 + SNR_{Eve})}{\log_{10}(1 + SNR_{Bob})} \tag{3.2}$$

with *SNR* representing the signal-to-noise ratio with intended objects inside the beam. It is not easy to define a threshold value for the parameter C_s as Eve's ability to decode a signal depends on several factors, such as modulation scheme and noise level of her receiver. Clearly, a value of $C_s = 1$ indicates that Eve receives no signal and $C_s = 0$ means Eve and Bob receive the same signal. From the point of information theoretic sense and practical implementation [13,14], a value of $C_s = 0.5$ is defined as the arbitrary threshold, below which, successful eavesdropping occurs.

3.2.3 Backscattering parameter

Sometimes, both the blockage and secrecy capacity parameters would not be enough to detect eavesdropping attacks and more strategies are required. Several studies have considered a fact that the presence of scattering object would also backscatter some radiation toward Alice [15,16] and develop a backscattering parameter under the assumption that Alice has previously characterized the link backscattering before any attempted attack by Eve [10]. Combining with other more conventional eavesdropping measures [16–18], the parameter is defined by the following equation:

$$S_{180} = \left|1 - \left(SNR_{Alice}^{\text{no object}} / SNR_{Alice}^{\text{object}}\right)\right|. \tag{3.3}$$

A value of $S_{180} = 0$ means that there is no change in the backscattered signal to Alice when an object is placed inside the beam path by Eve. But a value larger than $S_{180} = 0.5$ (a pre-determined threshold) can be regarded as a warning of a possible security threat. It has been confirmed that the backscattering parameter could be able to enhance the security of terahertz links in some situations [10].

3.2.4 Outage probability

In actual application scenarios, the received signal power would fluctuate spatially and temporally due to the scattering-induced scintillation effect along LoS link paths [19]. Parameters considering random fluctuations and stochastic characterizations should be introduced for security characterization. Secrecy outage probability is a probabilistic indicator and is defined as the probability that instantaneous secrecy capacity falls below a target rate R [20]. This parameter can be expressed by the equation $P_0(R) = P_r\{C_s < R\}$ with target rate $R \geq 0$ always, and represents the probability of secure data transmission. For a LoS link, it can be rewritten as

$$P_o(R) = \int_{C_s \leq R} f_{\mathrm{LOS}}(G_{\mathrm{LOS}}) dG_{\mathrm{LOS}} = \int_0^G f_{\mathrm{LOS}}(G_{\mathrm{LOS}}) dG_{\mathrm{LOS}}, \tag{3.4}$$

with G as the solution of $C_s = R$ and G_{LOS} as the instantaneous LoS link gain. Parameter $f_{\mathrm{LOS}}(G_{\mathrm{LOS}})$ is a probability density function of the instantaneous LoS link gain as the definition in Ref. [21].

3.3 Eavesdropping attacks in indoor terahertz links

Indoor wireless communication is one of the specific applications of interest for the terahertz frequencies. Terahertz links can provide high-speed connection, huge local capacity, and even high directionality inside a room or an office, which has been demonstrated [22]. However, the idea of engineering specialized reflectors/scatters (such as plates and tubes) to facilitate specular non-line-of-sight (NLoS) link has been discussed [23], and a few examples have been demonstrated [24,25].

3.3.1 Attacks in LoS link

By placing a scattering object (eavesdropper) inside the beam path, there would be a direct shadow on Bob's receiver aperture as in Figure 3.1. Measurement results in Figure 3.2(a) show that the insertion of several objects can make blockage $b \leq 0.5$, which means Bob cannot find the existing of eavesdroppers. And with the decrease of the object's size, the blockage is reduced. Oppositely, a larger object intercepts more signal to Eve and secrecy capacity decreases as in Figure 3.2(b). When both parameters $b \leq 0.5$ and $C_s \leq 0.5$, Eve is able to implement a successful eavesdropping attack.

However, there is a disadvantage for Eve—she employs a metal pipe (cylinder shape and composed by metal) as the scattering object. The scattered signal is dispersed in all directions and the power collected by the Eve is relatively low. This leads to the possibility of successful interception being reduced. Flat plates are proven to be able to scatter signal power more selectively with a sacrifice of Eve's freedom of location [10].

Instead of scattering objects, dielectric waveguides can also be used to intercept signals and build an eavesdropping link. A 3-D printed polystyrene (PS) waveguide is fabricated and used in the experiment in Figure 3.3(a). Geometry of the waveguide

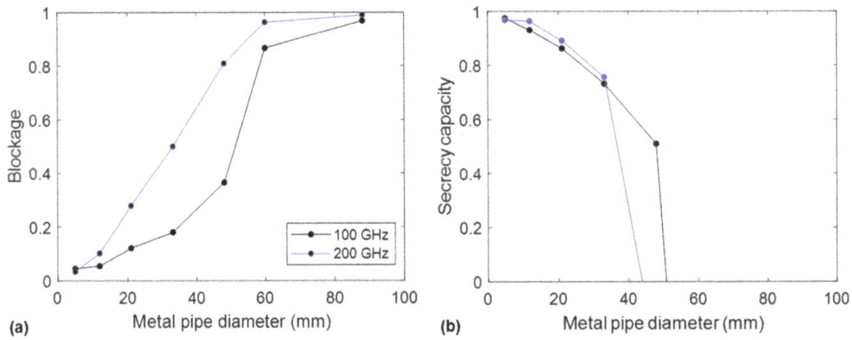

Figure 3.2 Performance of blockage and secrecy capacity due to the insertion of metal pipes inside a LoS terahertz link in Figure 3.1. ©nature 2018. Reprinted with permission from [10].

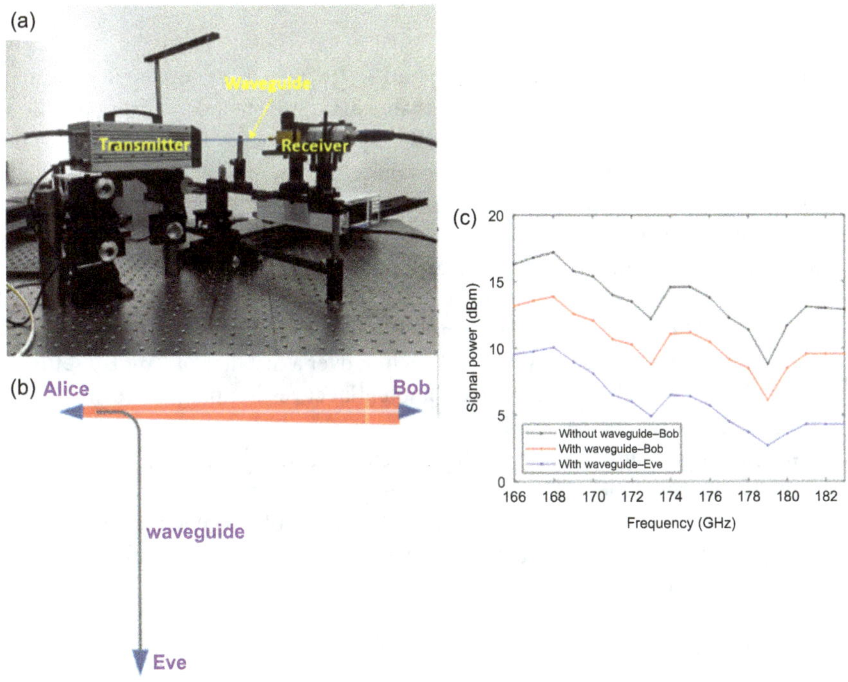

Figure 3.3 Eavesdropping by a PS waveguide inserted in a LoS terahertz link

is generally bent at 90° with its shape detail and geometric design can be found in Ref. [26].

We set up a LoS data transmission link by a single transmitter (Alice) and a single receiver (Bob) as in Figure 3.3(b). The input pot of the waveguide is set inside the LoS (legitimate) beam path, which collects part of the link power and launches it onto

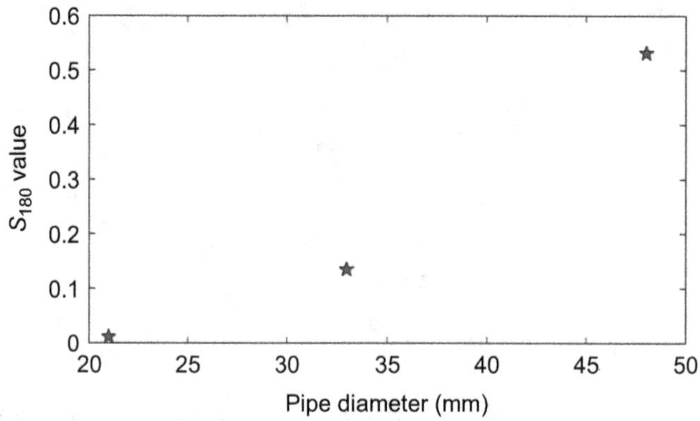

Figure 3.4 Performance of backscattering due to insertion of metal pipes inside a LoS terahertz link

the receiver of Eve. The cross-section (1.6 mm × 1.6 mm) of the waveguide is small enough that there is a small shadow (less than 3 dB) on Bob and the cross profile is perpendicular to the link path to increase the signal coupling. In Figure 3.3(c), the signal power collected by Eve through the waveguide is large enough for decoding and the blockage is small than the threshold $b = 0.5$.

As stated in Section 3.2, Alice can provide an additional level of security by assuming she is able to monitor the backscattering components [16–18]. The insertion of scattering objects, which leads to any change (increase or decrease) in the received signal power of Alice, can be detected by her. Figure 3.4 shows detected backscattering by metal pipes located inside a 100 GHz LoS link over a distance of 2 m. By setting a threshold of $S_{180} = 0.5$ for the detection of potential eavesdropping, there are several attacks, but not all, can be detected.

This parameter can also be effective when a lossless beam splitter is used to replace the scattering object. The splitter is designed large enough to encompass the entire beam generated and can split off a desired fraction of the generated power [10]. Measurements show that it could not detect all the attacks, but makes the beam splitter attacks less devastating.

3.3.2 Attacks in non-LoS link

The method of engineering specialized surfaces to facilitate specular NLoS links has been investigated [23–25] and the impact of surface roughness has also been discussed [3,27]. Rough surfaces could play an important role in scattering link power to more directions with small or without power degradation observed on Bob's side. This gives more chances for Eve to intercept signals from an NLoS link.

Figure 3.5(a) shows an experimental setup used to explore the diffuse-scattering-induced signal eavesdropping due to rough metal surfaces. The picture illustrates the transmitter (Alice) and receiver (Bob or Eve) directed toward a spot on the metal

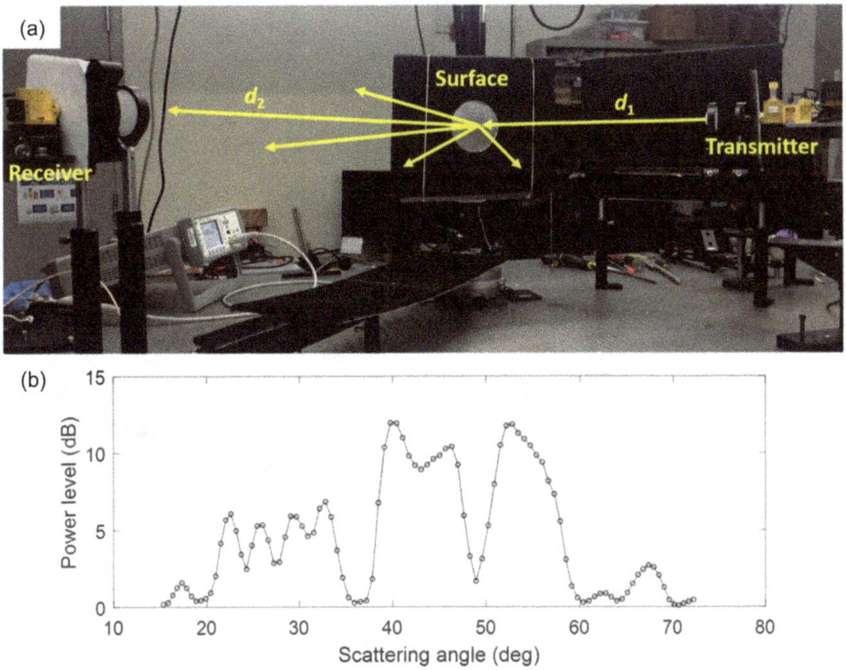

Figure 3.5 Scattering and induced signal eavesdropping in a terahertz wireless link by a rough metal surface

surface. The distance from the transmitter and the receiver antenna to the surface is constant (d_1 = 0.5 m and d_2 = 1 m). The antennas always point to the surface, and a direct LoS free-space calibration was performed at a link distance of 1.5 m (d1 + d_2 = 1.5 m). The receiver is mounted on a movable rotator which permits it to rotate along an arc with its center at the rough surface. By scanning the detector along the arc, the angular distribution of the beam arriving at the receiver horn can be mapped. The detector angle is optimized to ensure accurate pointing, prior to each scan. The incident angle is restricted to 45deg, while the scattering (detection) angle was varied in most cases between 10deg and 80deg with a step of 1deg. Scattering in the backward direction (backscattering) is not measured in this work due to the fact that the transmitter obstructs the view of the receiver.

The scattering pattern is shown in Figure 3.5(b) for vertical input polarization. Comparing with that after reflection by a smooth surface [28], the scattering pattern is broadened with a strong coherent component along the specular direction, which makes a small or no signal variation on Bob's side. The angular broadening is due to the presence of the incoherent component and generates radiation along directions other than the specular direction. This result illustrates that successful data links along diffused scattering directions and the security of this link is significantly reduced.

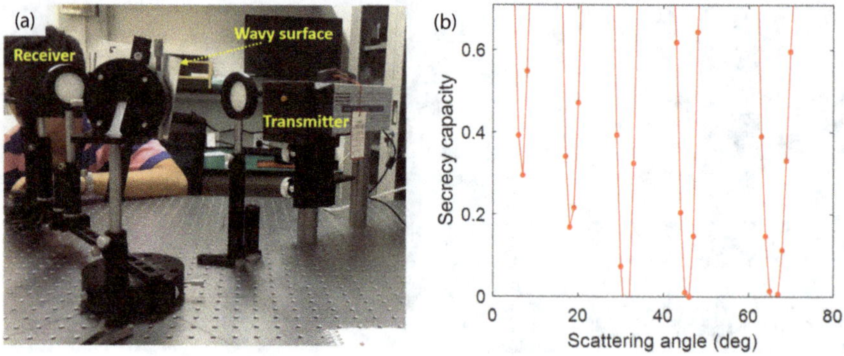

Figure 3.6 Signal eavesdropping attack in a LoS terahertz link caused by a metal wavy surface. The surface can be rotated along its longitudinal axis. ©IEEE 2023. Reprinted with permission from [29].

Due to the randomness of the rough surfaces, the direction of scattered signals is difficult to be controlled [28]. Wavy surfaces, with their height variation following a sinusoidal function $h = A \sin (2\pi x/\Lambda)$, could enable scattered signals toward selective directions. Here, A as the ripple amplitude and Λ as the period.

The idea for investigating multipath scattering is illustrated in Figure 3.6(a). A link between the transmitter (Alice) and the mobile receiver (Bob) is maintained by a piece of metallic wavy surface with the same total transmission path length as the distance from Alice to Bob (56 cm). The radiated signal could be launched onto the surface with an incidence angle of 45Â° and a beam spot size of 3 cm. The beam size is much larger than the surface period (Λ). The surface is parallel to the principle plane and the receiver is mounted on a pivoting rail to vary its angular position (scattering angle). The wavy surface could lead to interference (scattering) patterns, which are collected and analyzed by another identical antenna–lens couple of the eavesdropper (Eve).

Figure 3.6(b) [29] shows that the wavy surface could enable successful signal eavesdropping in a local area network, by employing both the rotation of the surface and the variation of its period Λ. The surface rotation makes sure that there is no power change monitored at Bob. The period could be estimated for the requirement on angular position of possible interception links. This information could be acquired thoroughly using Bragg's Law with experimental confirmation by unmodulated and modulated terahertz links.

3.4 Eavesdropping attacks in outdoor terahertz links

The emphasis of this section is ground-to-ground terahertz wireless links, which suffer significant power loss and degradation due to absorption, reflection, scattering, and

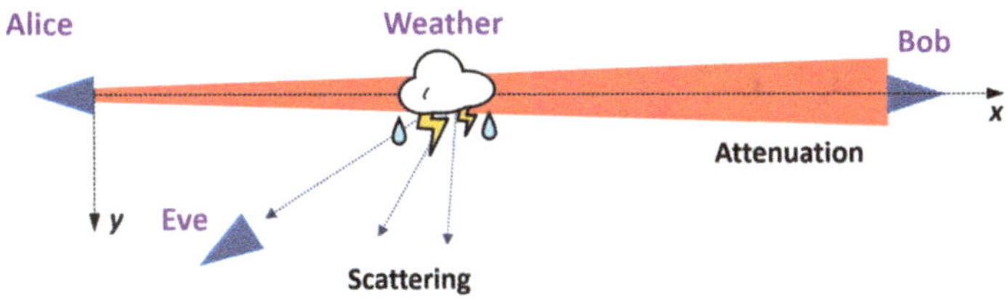

Figure 3.7 Typical scheme on security characterization of an outdoor LoS link in different weathers

spreading losses. Unlike wire communications, the broadcasting nature of wireless links could make it to be vulnerable to eavesdropping attacks. In the case of adverse weather conditions, due to the serious scattering effect by weather particles, it may be necessary to investigate the possibility of signal eavesdropping by an NLoS path scattered away the legitimate (LoS) link path.

Assuming a point-to-point outdoor THz link with the configuration as shown in Figure 3.7, a transmitter (Alice) sends information to a legitimate receiver (Bob) by a LoS link which suffers absorption and scattering effects in rain, snow, or atmospheric turbulence. Meanwhile, an eavesdropper (Eve) exists outside but near the legitimate link path and aims to capture the information through an NLoS link scattered away from the legitimate (LoS) path. The link distance is set to be $d = 1$ km and the locations of Alice and Bob are always fixed, while Eve can change its location and adjust its pointing direction to obtain optimal received signal.

In the presence of rain/snow/turbulence, the LoS link suffers atmospheric attenuation G_A and divergence attenuation $G_D = 4A/\left(\pi d^2\alpha_A^2\right)$ with α_A as the full divergence angle of Alice and A as the effective receiving area of Bob. The atmospheric attenuation including attenuation only by rain/snow/turbulence (with coefficient as α_t) and gaseous attenuation by air (with coefficient as α_g) can be obtained from the previous section. So the atmospheric attenuation coefficient is $\alpha_{atm} = \alpha_t + \alpha_g$ and the atmospheric attenuation can be obtained by $G_A = exp(-\alpha_{atm}d)$.

3.4.1 Deterministic evaluation

In the calculation, the link distance between Alice and Bob is 1 km. The FOV angles of Bob and Eve are set to be 15° and the receiving areas for Bob and Eve are all 1 cm^2. We set temperature $T = 25$°C, pressure $P = 1013$ hPa, and relative humidity RH = 80%. Eve is located at an x-position of 200 m with her y-position varying from 0.

Choosing a carrier frequency of 340 GHz, rainfall rate $Rr = 10$ mm/h, snowfall rate (equivalent rainfall rate) $Rr = 10$ mm/h and atmospheric turbulence strength $C_n^2 = 10^{-10}m^{-2/3}$, the secrecy capacity variation with respect to arbitrary positions of Eve in rain, dry snow, wet snow, and atmospheric turbulence are calculated and shown in Figure 3.8. At positions of secrecy capacity $C_s = 0$, which is called "insecure region." Outside of this region, the secrecy capacity increases dramatically and reaches a maximum capacity. This implies a method to improve link security by keeping Eve away the legitimate link path as far as possible.

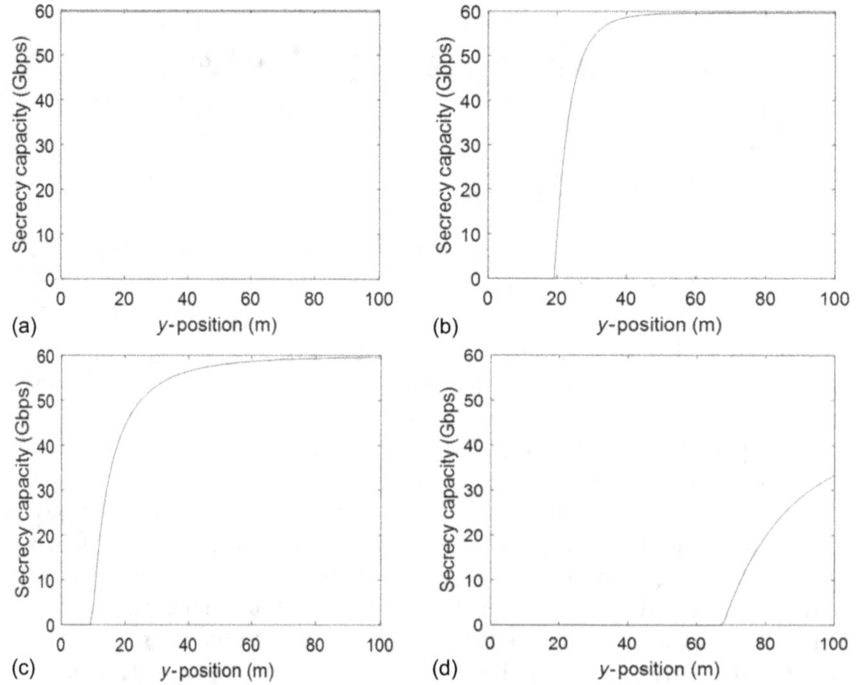

Figure 3.8 Variation of secrecy capacity with respect to the y-position of Eve in (a) rain, (b) dry snow, (c) wet snow, and (d) atmospheric turbulence. (x-position $x = 200\,m$, temperature $T = 30°C$, pressure $P = 1013\,hPa$, humidity $RH = 80\%$, rain/snowfall rate $= 10\,mm/h$, turbulence strength $C_n^2 = 10^{-10} m^{-2/3}$, divergence angle $\alpha_A = 1\,mrad$, $FOV = 10°$).

Two-dimensional (2-D) secrecy capacity distribution can be used to reduce eavesdropping risks by scaling secure and insecure regions. In Figure 3.9, the secrecy capacity distributions with respect to arbitrary positions of Eve are calculated and presented with the same parameter settings as in Figure 3.8. The color bar on the right denotes the safe transmission rate C_s in Gbps, which increases gradually as the color changes from blue to yellow.

In Figure 3.9(a) for rain condition, the whole region is yellow, which means all the region is secure and Eve could never intercept signal if he is located in this region. Thus, the yellow region could be called "secure region." In Figure 3.9(b), a dark blue region appears, which means secure secrecy capacity $C_s = 0$. This region is "insecure region." The insecure region in dry snow (Figure 3.9(b)) is larger than that in wet snow (Figure 3.9(c)) under the same snowfall rate. The snowfall rate is usually represented by an equivalent rainfall rate as in Ref. [30], so, under the same fall rate of 10 mm/h, there should be much more dry snow particles than rain particles dropping down, which would lead to more serious scattering effect in dry snow. This phenomenon

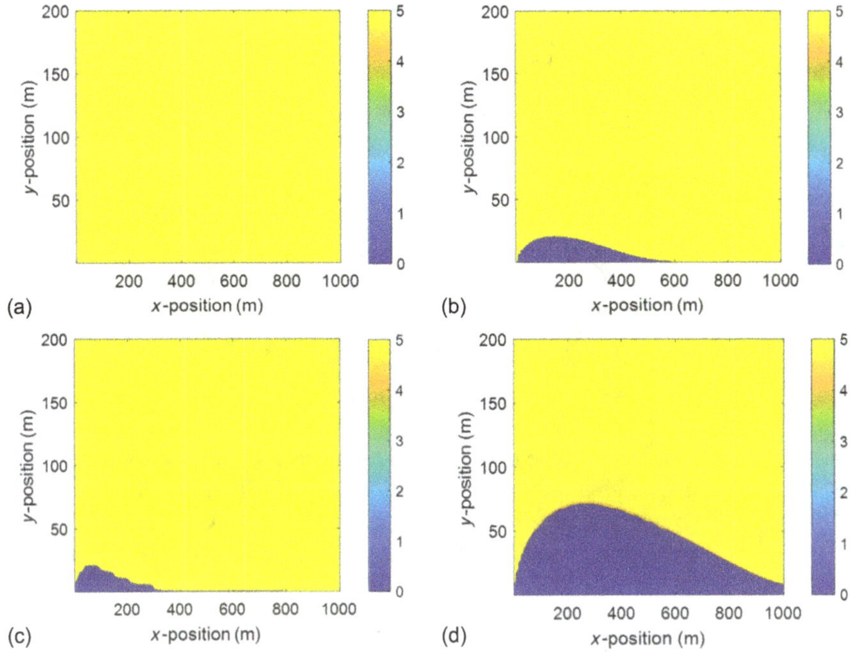

Figure 3.9 Secrecy capacity distribution for 2-D positions of Eve when in (a) rain, (b) dry snow, (c) wet snow, and (d) atmospheric turbulence. (The color bar denotes the safe transmission rates in Gbps). (x-position $x = 200$ m, temperature $T = 30°C$, pressure $P = 1013$ hPa, humidity $RH = 80\%$, rain/snowfall rate 10 mm/h, turbulence strength $C_n^2 = 10^{-10} m^{-2/3}$, divergence angle $\alpha_A = 1$ mrad, FOV = 10°).

is also observed for the 140 and 220 GHz links [31], which indicates that the link secrecy performance in wet snow is better than that in dry snow. However, when we decrease the carrier frequency to 340 GHz, the area of insecure region is enlarged because of the decrease in signal absorption on the LoS link, which leads to a much higher SNR at Bob [31]. This indicates that the link secrecy performance in rain is better than that in wet snow, and the latter one is better than that in dry snow.

3.4.2 Probabilistic evaluation

Since the received power and channel gain should always fluctuate spatially and temporally due to the scattering-induced scintillation effect along the channel path [19], the secrecy capacity should be assumed to be a random variable always. Thus, the outage probability is another evaluating factor for channel secrecy performance at the physical layer, which is different from deterministic secrecy performance analysis above.

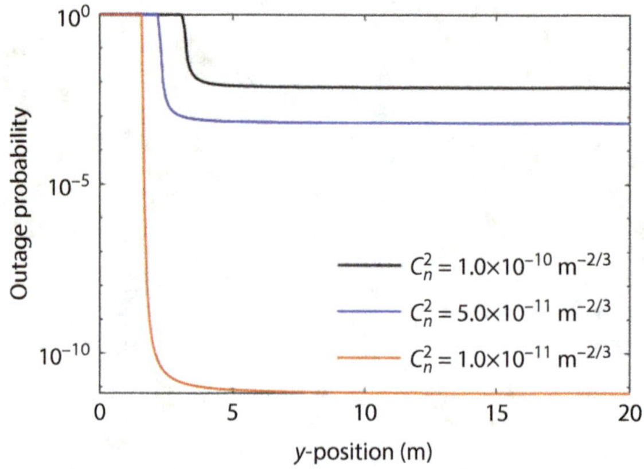

Figure 3.10 Variation of outage probability with respect to the y-position of Eve under different turbulence strength (intended data rate $R = 100$ Gbps, carrier frequency 340 GHz, temperature $T = 30°C$, pressure $P = 1013$ hPa, humidity $RH = 80\%$, divergence angle $\alpha_A = 1$ mrad, FOV angle = 10°)

Figure 3.10 presents predicted outage probability for a 340-GHz link propagating through atmospheric turbulence at different turbulence strengths with the basic link configurations as in reference [32] and the intended data rate $R = 100$ Gbps which is a common achievement for wireless channels with carriers at THz range [33]. It is shown that, due to the turbulence-induced channel degradation, the outage probability increases significantly when the turbulence becomes stronger and the eavesdropping risk becomes more serious.

3.5 Conclusion

We presented a review of characteristics of eavesdropping attacks when scattering objects and dielectric waveguides are located inside a LoS terahertz link; when random rough and wavy surfaces are placed inside a non-LoS terahertz link; and when weather–particle-induced scattering occurs in an outdoor terahertz link setup. Evaluations on the eavesdropping threats are conducted by using parameters related to blockage on Bob, backscattering to Alice, secrecy capacity, and outage probability. An eavesdropper can intercept signal power in a LoS or a non-LoS terahertz wireless links, by choosing appropriate composite, size, and location of the scattering objects or surfaces. For the outdoor link, due to the randomness of the link power fluctuations, both the deterministic and probabilistic evaluations are necessary.

The proliferation of smart devices prompts rapid development of wireless networks, which makes multi-party coordinated interference to be a potential method for secure data transmission. It can employ advanced signal processing technology to conduct artificial noise scrambling, increase the capacity difference between legal and eavesdropping links, and improve physical-layer security in wireless communications. A combination of intelligent reflective surfaces, orbital angular momentum, etc. can also be an efficient approach, even though further progress in theory and facility is still necessary and required.

Acknowledgments

This work was supported in part by the National Science Foundation of China, the US Army Research Office, and the W.M. Keck Foundation. The authors gratefully acknowledge the contributions of members of his research groups at the Beijing Institute of Technology and Brown University, in particular, Wenbo Liu and Peian Li (Figure 3.3), Peian Li, and Jianchen Wang (Figure 3.6).

References

[1] Hussein AF and Elgala H. Wireless link pairing toward secured 6G networks. *Optics Letters*. 2020;45(14):4005–4008.

[2] Yerrapragada AK, Eisman T, and Kelley B. Physical layer security for beyond 5G: Ultra secure low latency communications. *IEEE Open Journal of the Communications Society*. 2021;2:2232–2242.

[3] Jansen C, Priebe S, Moller C, *et al*. Diffuse scattering from rough surfaces in THz communication channels. *IEEE Transactions on Terahertz Science and Technology*. 2011;1(2):462–472.

[4] Wu Y, Khisti A, Xiao C, *et al*. A survey of physical layer security techniques for 5G wireless networks and challenges ahead. *IEEE Journal on Selected Areas in Communications*. 2018;36(4):679–695.

[5] Yang N, Wang L, Geraci G, *et al*. Safeguarding 5G wireless communication networks using physical layer security. *IEEE Communications Magazine*. 2015;53(4):20–27.

[6] Zhang J, Duong TQ, Woods R, *et al*. Securing wireless communications of the Internet of Things from the physical layer, an overview. *Entropy*. 2017;19(8):420.

[7] Zhou X, Song L, and Zhang Y. *Physical layer security in wireless communications*. CRC Press, 2013.

[8] Bloch M and Barros J. *Physical-layer security: from information theory to security engineering*. Cambridge University Press, 2011.

[9] Trappe W. The challenges facing physical layer security. *IEEE Communications Magazine*. 2015;53(6):16–20.

[10] Ma J, Shrestha R, Adelberg J, *et al*. Security and eavesdropping in terahertz wireless links. *Nature*. 2018;563(7729):89–93.

[11] Bustin R, Liu R, Poor HV, *et al.* An MMSE approach to the secrecy capacity of the MIMO Gaussian wiretap channel. *EURASIP Journal on Wireless Communications and Networking*. 2009;2009:1–8.

[12] Wyner AD. Capacity and error exponent for the direct detection photon channel. II. *IEEE Transactions on Information Theory*. 1988;34(6):1462–1471.

[13] Csiszár I and Korner J. Broadcast channels with confidential messages. *IEEE Transactions on Information Theory*. 1978;24(3):339–348.

[14] Ju Y, Wang HM, Zheng TX, *et al.* Safeguarding millimeter wave communications against randomly located eavesdroppers. *IEEE Transactions on Wireless Communications*. 2018;17(4):2675–2689.

[15] Dorney TD, Symes WW, Baraniuk RG, *et al.* Terahertz multistatic reflection imaging. *JOSA A*. 2002;19(7):1432–1442.

[16] Yang N, Wang L, Geraci G, *et al.* Safeguarding 5G wireless communication networks using physical layer security. *IEEE Communications Magazine*. 2015;53(4):20–27.

[17] Zou Y, Zhu J, Wang X, *et al.* A survey on wireless security: technical challenges, recent advances, and future trends. *Proceedings of the IEEE*. 2016; 104(9):1727–1765.

[18] Zhu J, Zou Y, and Zheng B. Physical-layer security and reliability challenges for industrial wireless sensor networks. *IEEE Access*. 2017;5:5313–5320.

[19] Alkholidi A and Altowij K. Effect of clear atmospheric turbulence on quality of free space optical communications in Western Asia. In: *Optical Communications Systems*. *InTech*, 2012;1:41–74.

[20] Bloch M, Barros J, Rodrigues MR, *et al.* Wireless information-theoretic security. *IEEE Transactions on Information Theory*. 2008;54(6): 2515–2534.

[21] Andrews L and Phillips RL. *Laser beam propagation through random media*: 2nd ed. SPIE Press, 2005.

[22] Petrov V, Kokkoniemi J, Moltchanov D, *et al.* Last meter indoor terahertz wireless access: Performance insights and implementation roadmap. *IEEE Communications Magazine*. 2018;56(6):158–165.

[23] Krumbholz N, Gerlach K, Rutz F, *et al.* Omnidirectional terahertz mirrors: a key element for future terahertz communication systems. *Applied Physics Letters*. 2006;88(20):202905.

[24] Piesiewicz R, Jacob M, Koch M, *et al.* Performance analysis of future multigigabit wireless communication systems at THz frequencies with highly directive antennas in realistic indoor environments. *IEEE Journal of Selected Topics in Quantum Electronics*. 2008;14(2):421–430.

[25] Peng B, Rey S, Kürner T. Channel characteristics study for future indoor millimeter and submillimeter wireless communications. In: *2016 10th European Conference on Antennas and Propagation (EuCAP)*. IEEE; 2016. p. 1–5.

[26] Ma J, Weidenbach M, Guo R, *et al.* Communications with THz waves: switching data between two waveguides. *Journal of Infrared, Millimeter, and Terahertz Waves*. 2017;38(11):1316–1320.

[27] Piesiewicz R, Jansen C, Mittleman D, *et al.* Scattering analysis for the modeling of THz communication systems. *IEEE Transactions on Antennas and Propagation*. 2007;55(11):3002–3009.
[28] Ma J, Shrestha R, Zhang W, *et al.* Terahertz wireless links using diffuse scattering from rough surfaces. *IEEE Transactions on Terahertz Science and Technology*. 2019;9(5):463–470.
[29] Li P, Wang J, Zhao L, *et al.* Scattering and eavesdropping in terahertz wireless link by wavy surfaces. *IEEE Transactions on Antennas and Propagation*. 2023;71(4):3590–3597.
[30] Amarasinghe Y, Zhang W, Zhang R, *et al.* Scattering of terahertz waves by snow. *Journal of Infrared, Millimeter, and Terahertz Waves*. 2020;41(2):215–224.
[31] Wang R, Mei Y, Meng X, *et al.* Secrecy performance of terahertz wireless links in rain and snow. *Nano Communication Networks*. 2021;28:100350.
[32] Mei Y, Ma Y, Ma J, *et al.* Eavesdropping risk evaluation on terahertz wireless channels in atmospheric turbulence. *IEEE Access*. 2021;9:101916–101923.
[33] Kürner T, Mittleman D, and Nagatsuma T. *THz Communications: Paving the Way Towards Wireless Tbps*. Springer. 2022.

Chapter 4

Intelligent reflecting surface-aided physical layer security of wireless communications

Tong-Xing Zheng[1], Xin Chen[1], Ying Ju[2], Haoyu Wang[2], Chaowen Liu[3] and Limeng Dong[4]

Existing physical layer security (PLS) techniques are generally channel-adaptive, the achievable secrecy performance of which is severely constrained by the wireless propagation environment. In order to break through the technical bottleneck, intelligent reflecting surface (IRS) provides a new engine for PLS enhancement for the next generation wireless networks by smartly reconfiguring the wireless channels. In this chapter, we collect recent research fruits on the IRS-aided PLS, putting an emphasis on four scenarios, namely, the multi-input multi-output system, the pattern index modulation system, the integrated sensing and communication system, and the artificial intelligence-driven communication system. For each case, we describe the system model, formulate the optimization problem, build the design framework, and provide numerical results. We also present several interesting and challenging open-ended issues for future research works.*

4.1 Introduction

4.1.1 Intelligent reflecting surface-aided physical layer security

Physical layer security (PLS) has been widely recognized as an appealing approach for safeguarding the bit-level secrecy of the wireless communication systems. However, existing PLS techniques are generally channel-adaptive, meaning that the achievable secrecy performance is significantly determined and meanwhile heavily limited by the wireless propagation environment. For example, if the wiretap channel quality is superior to the legitimate channel quality or the two channels are highly correlated, it might not be easy or possible to achieve a positive secrecy rate using conventional

*This chapter has been a part of [1–3].

[1]School of Information and Communications Engineering, Xi'an Jiaotong University, China
[2]School of Telecommunications Engineering, Xidian University, China
[3]School of Communications and Information Engineering, Xi'an University of Posts and Telecommunications, China
[4]School of Electronics and Information, Northwestern Polytechnical University, China

PLS techniques such as transmit beamforming and artificial noise. In order to break through the above technical bottleneck, intelligent reflecting surface (IRS), a software-controlled metasurface consisting of a large number of passive adjustable reflecting elements [4], provides a new engine for boosting PLS for the next-generation wireless networks. By deploying an IRS in the wireless secure transmission system, the wireless propagation environment can be reconfigured flexibly and controllably by adjusting the amplitudes and phases of the signals incident on the reflecting elements of the IRS, and hence more spatial degrees of freedom can be exploited to enhance the PLS performance.

4.1.2 Recent research and evolution of IRS-aided PLS

In the following, we present a brief review of recent research on the IRS-aided PLS under various wireless application scenarios, as well as combined research with some promising technologies for next-generation wireless networks.

4.1.2.1 Point-to-point IRS-PLS systems

The IRS-aided PLS has been extensively studied in point-to-point wireless communication systems considering different optimization objectives of secure transmission, including secrecy rate maximization [1,5–11] and secure transmit power minimization [12–14]. Specifically, the authors in [5] proposed the secrecy rate maximization algorithm for the IRS-aided secure multi-input single-output (MISO) communication system. More complicated scenarios were examined later, e.g., multiple eavesdroppers with imperfect channel state information (CSI) in the mmWave MISO system [6], multiple IRSs and multiple eavesdroppers [7], multi-input multi-output (MIMO) system [1,8,9], and the artificial noise (AN)-aided MIMO secure transmission [10]. Besides, the authors in [11] investigated the issue of maximizing the ergodic secrecy rate with eavesdropper's statistical CSI. As for the secrecy energy efficiency issue, the authors in [12] minimized the transmit power subject to a secrecy rate constraint for the IRS-aided secure MISO system, and this work was extended to the scenario where only the eavesdropper's statistical CSI is available [13,14]. In addition to the secrecy performance optimization, the analysis of secrecy performance in terms of the average secrecy rate and secrecy outage probability of the IRS-aided PLS has been carried out for single-input single-output system, MISO system, vehicle-to-vehicle system, unmanned aerial vehicle (UAV) relay system, and indoor wireless communication system [15–19], respectively.

4.1.2.2 Multi-user and multi-access IRS-PLS systems

Apart from the point-to-point scenarios, the IRS-aided PLS has recently been investigated for the multi-user systems [20–27]. For example, the IRS-aided secure downlink MISO broadcast system was considered in [20], where the base station (BS) wishes to transmit independent data streams to multiple legitimate receivers secretly in the presence of multiple eavesdroppers. The authors in [21] studied the multi-user downlink communication assisted by a self-sustainable IRS which harvests energy from the incident signals while reflecting them, and they examined the problem of maximizing the system sum rate subject to the constraints of the IRS's energy harvesting capability, the secure transmission, and the robustness against the imperfect CSI. The authors in [22] minimized the transmit power for an IRS-aided secure multi-user MISO system by jointly designing the transmit beamforming, the jamming matrix, and the phase shift matrix while satisfying a secrecy rate constraint. A very recent work [23] explored the IRS-aided secure multi-user visible light communication where the sum secrecy rate was maximized by optimizing the association relationship between the IRS units and light emitting diodes.

The combination of IRS and the non-orthogonal multiple access (NOMA) technique has been exploited to improve the PLS performance. For instance, the authors in [24] employed IRS along with NOMA to improve the internal secrecy of the multiple users with heterogeneous secrecy constraints. The AN technique was leveraged to guarantee the security of the IRS-NOMA transmission [25]. The authors in [26] proposed the distributed IRS-assisted NOMA transmission framework against eavesdroppers with imperfect CSI. Considering the case without the eavesdropper's CSI, the legitimate sum rate was maximized via jointly optimizing the AN precoding and IRS reflecting coefficients [27].

We point out that the aforementioned works mainly consider the multi-user MISO system. As for the MIMO case, how to appropriately utilize an IRS to benefit PLS still remains an open problem. Besides, perfect CSI is assumed in most existing works, and it is necessary to design more efficient and robust schemes for the IRS-aided multi-user secure systems with only imperfect CSI available at transmitter.

4.1.2.3 IRS-empowered augmented pattern index modulation for PLS

The philosophy of spatial-domain pattern index modulation (PIM) has attracted considerable research interests in fulfilling the energy-efficient data delivery for 5G/B5G applications [28]. According to the distinct realization mechanisms, current PIM schemes can be mainly classified into three categories, namely, the transmitter-side PIM, the receiver-side PIM, and the joint transceiver PIM, where the indices of the spatial transmission patterns constructed via transmit array element/pattern activation and transmitter preprocessing-based receive array element/pattern activation are individually or jointly exploited to convey information bits.

The undeniable potential of incorporating PIM into an IRS has been exploited to enhance the reliability and spectral efficiency for wireless communication systems, considering the transmitter-side [29], receiver-side [30], and joint transceiver PIM schemes [31]. By applying the principle of PIM to the reflection domain, the indices of reflection patterns, which are constructed by the IRS elements with different combinations of ON/OFF status, can be utilized to embed extra information [32].

It is worth noting that, owing to the high spatial-domain resolution energized by the IRS-assisted beamforming, the augmented PIM (APIM) can assure in-built secrecy for the wireless information delivery at the physical layer. However, as neither the existence statuses nor the locations of the passive eavesdroppers are certified in practical scenarios, we can hardly assert that no direct links exist between the IRS and eavesdroppers. In particular, if an eavesdropper is located near the transmitter without being perceived, the secrecy of the MIMO systems employing any of the existing IRS-empowered PIMs [33] might not be properly pledged, since the baseband amplitude phase modulation (APM) symbols, which can be a sub-part of the PIM-category super-symbols, are always emitted without extra cover. Therefore, dedicated IRS-based PLS enhancing schemes are demanded to assure the security of the PIM-assisted MIMO systems.

4.1.2.4 IRS-enabled secure integrated sensing and communications

The integrated sensing and communications (ISAC) system has been envisioned as a disruptive technology to overcome the challenges of spectrum scarcity, high hardware cost, and the increasing demand for intelligence for the next-generation wireless networks. A dual-functional radar-communication (DFRC) transmitter can realize radar sensing and communications simultaneously using integrated hardware platforms and waveforms, thus facilitating spectrum sharing between radar and communications and lowering the hardware cost. However, there is high-security risk in sharing waveforms when facing a malicious radar target which attempts to intercept the embedded information. Undoubtedly, to strike a good balance between secure communications and radar detection is vital to the design of DFRC systems.

Recently, IRS has been introduced to realize secure DFRC systems. The authors in [34] investigated an IRS-assisted secure DFRC system, where the received signal-to-interference-plus-noise ratio (SINR) of the MIMO radar was maximized by designing the transmit beamforming vectors for useful signals and AN, and the phase shift matrix of the IRS. The work was extended to a multi-cast system with multiple eavesdroppers where the secrecy rate was maximized by optimizing the precoding matrix and the phase shift matrix subject to the SINR constraints of the targets [35]. However, these studies are still in an early stage and there are some interesting and important open problems worth investigation, e.g., considering the target mobility and target location uncertainty, the constant-modulus constraint of waveforms, and more refined design of radar beam patterns.

4.1.2.5 AI-driven IRS-assisted secure communication system

Although the great significance of RIS-PLS has been demonstrated, the majority of existing works have been conducted in relatively simple communication scenarios. For example, the optimization problem therein just refers to single time slot, which might lead to a highly suboptimal solution due to the ignorance of the historical system state and the long-term benefit. Hence, it is generally infeasible to apply the traditional optimization techniques such as semidefinite programming, alternating optimization (AO), and minorization–maximization to achieve an effective secure beamforming/optimization policy in complicated dynamic environments.

Inspired by the recent advances in artificial intelligence (AI), various deep learning (DL) or reinforcement learning (RL) algorithms have been exploited for effectively solving the challenges brought by the randomness, dynamism, mathematical complexity of optimization, as well as the computational cost [36]. Advanced AI algorithms have also been utilized to realize secure communications by jointly optimizing the parameters of transceivers and IRS [37–40]. To be specific, the authors in [37] developed a DL technique to tune the IRS reflections in real-time for effectively solving the secrecy rate maximization problem. The authors in [38] proposed a deep RL (DRL)-based secure transmission method by jointly optimizing the transmit beamforming and the phase-shifting matrix in an IRS-aided multi-user wireless system. The authors in [39] exploited IRS to efficiently resist malicious jamming launched by smart jammers, where the BS, through a win or learn fast policy hill-climbing method, can jointly optimize the anti-jamming power allocation and reflecting beamforming strategy without the knowledge of the jamming model. The authors in [40] proposed a multi-agent DRL-based buffer-aided relay selection scheme for an IRS-assisted secure cooperative network and examined the problems of maximizing the average secrecy rate with the delay constraint and maximizing the throughput with the delay and secrecy constraints.

If a communication optimization problem can be approximated as a convex function, it can be solved by traditional convex optimization methods. However, when the optimization objective is complicated, adopting traditional algorithms might result in a great computational complexity. In particular, it is unrealistic to apply traditional convex optimization methods to complicated continuous decision scenarios. Fortunately, AI algorithms with strong learning capability can constantly learn from and adapt to the dynamic transmission environments, which provides significant support for solving more complex optimization issues in dynamic scenarios. As a vital application scenario in 6G, the security of mobile edge computing (MEC) has been widely concerned, and the deployment of IRS and AI algorithms has the potential in enhancing the PLS of MEC services.

4.1.3 Contributions and organization

Motivated by the above discussion, this chapter collects latest research works on the IRS-aided PLS, focusing on four typical scenarios, namely, the MIMO system, the PIM system, the ISAC system, and the AI-driven communication system. The main contributions of this chapter are summarized as follows:

- The IRS-aided secure MIMO communication system is investigated with and without the eavesdropper's CSI. For the former case, we formulate a secrecy rate maximization problem with respect to the transmit covariance matrix at transmitter and phase shift matrix of IRS, and we propose a barrier-based or convex approximation (SCA)-based AO algorithm to solve it. For the latter case, an AN scheme is proposed to enhance the secrecy rate, where we use the minimal power to transmit useful signals to meet the quality of service (QoS) constraint at the intended user and generate AN using the residual power to jam the eavesdropper.
- A novel PIM structure is proposed to modulate information in both spatial and reflection domains. In the spatial domain, the indices of active transmit and receive patterns are utilized to bear information bits, whilst in the reflection domain, the index of the reflection pattern generated by selectively setting a subset of IRS elements with OFF status is exploited to embed information bits. We also employ an APIM to convey information to further improve the spectral efficiency.

- An IRS-aided secure transmission scheme is proposed for an ISAC system. We maximize the secrecy rate by jointly designing the beamforming vector at BS and the phase shift matrix of IRS while satisfying the radar detection constraint. AO algorithms based on semidefinite relaxation (SDR) and Riemannian conjugate gradient (RCG) are proposed to solve the formulated non-convex problem.
- An AI-driven IRS-aided secure MEC vehicular network is investigated where the BS aims to minimize the service time for the whole MEC process while ensuring communication security for all vehicle users under the potential eavesdropping threats. A deep deterministic policy gradient (DDPG)-based scheme is proposed to realize secure transmission and resource allocation for the considered system.

The remainder of this chapter is organized as follows. The IRS-aided secure MIMO communication system, APIM system, ISAC system, and AI-driven MEC vehicular system are investigated in Sections 4.2–4.5, respectively. Section 4.6 concludes the chapter and presents some open issues.

4.2 IRS-assisted MIMO secure communications

This section investigates an IRS-assisted secure MIMO communication system. We aim to maximize the secrecy rate by designing the optimal active beamforming at transmitter and passive beamforming of IRS.

4.2.1 System model and problem description

As shown in Figure 4.1, we consider an IRS-assisted PLS system including a transmitter Alice, an intended receiver Bob, an eavesdropper Eve, and an IRS with M reflecting elements. Alice, Bob, and Eve are equipped with N_A, N_B, and N_E antennas, respectively. By adjusting the phase shifts of the IRS reflecting elements, the direct signals from Alice can add up with the reflected signals from IRS constructively at

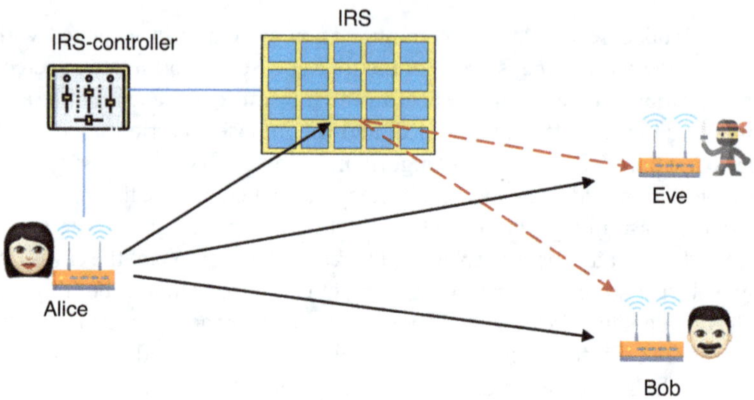

Figure 4.1 Illustration of an IRS-assisted MIMO PLS system

Bob but destructively at Eve thus guaranteeing secure transmission. The received signals at Bob and Eve can uniformly be expressed as

$$\mathbf{y}_X = \mathbf{H}_{AX}\mathbf{s} + \mathbf{H}_{IX}\mathbf{Q}\mathbf{H}_{AI}\mathbf{s} + \mathbf{n}_X,\ X \in \{B, E\}, \tag{4.1}$$

where $\mathbf{s} \in \mathbb{C}^{N_A\times 1}$ is the transmitted signal; $\mathbf{H}_{AB}$, $\mathbf{H}_{IB}$, $\mathbf{H}_{AI}$, $\mathbf{H}_{AE}$, and $\mathbf{H}_{IE}$ denote the channels of Alice-Bob, IRS-Bob, Alice-IRS, Alice-Eve, and IRS-Eve, respectively; $\mathbf{n}_B \sim CN\left(\mathbf{0}, \sigma_B^2\mathbf{I}\right)$ and $\mathbf{n}_E \sim CN\left(\mathbf{0}, \sigma_E^2\mathbf{I}\right)$ denote the received noise at Bob and Eve, respectively; $\mathbf{Q} = \text{diag}\left(q_1, q_2, \ldots, q_M\right)$ denotes the phase shift matrix, and $q_i = e^{j\theta_i}$ with $\theta_i \in [0, 2\pi)$ denoting the phase shift coefficient of the ith IRS element for $i = 1, 2, \ldots, M$. Thus, the achievable rate of Bob and Eve can be uniformly derived as

$$R_X = \log_2\left|\mathbf{I} + \left(\mathbf{H}_{AX} + \mathbf{H}_{IX}\mathbf{Q}\mathbf{H}_{AI}\right)\mathbf{R}\left(\mathbf{H}_{AX} + \mathbf{H}_{IX}\mathbf{Q}\mathbf{H}_{AI}\right)^H/\sigma_X^2\right|, \tag{4.2}$$

where $X \in \{B, E\}, \mathbf{R} = \mathbb{E}\left\{\mathbf{s}\mathbf{s}^H\right\}$ denotes the transmit covariance matrix.

In order to enhance the secure communication capability, we aim to maximize the secrecy rate $R_s \triangleq R_B - R_E$ by jointly designing the active beamforming of transmitter (i.e., transmit covariance matrix $\mathbf{R}$) and the passive beamforming of IRS (i.e., phase shift matrix $\mathbf{Q}$). The optimization problem can be formulated as follows:

$$\mathscr{P}_0 : \max_{\mathbf{R},\mathbf{Q}} \quad R_s\left(\mathbf{R}, \mathbf{Q}\right) = R_B - R_E \tag{4.3a}$$

$$\text{s.t.} \quad \text{Tr}(\mathbf{R}) \leq P,\ \mathbf{R} \succeq \mathbf{0} \tag{4.3b}$$

$$\left|\mathbf{Q}_{i,i}\right| = 1, \forall i = 1, 2, \ldots, M, \tag{4.3c}$$

where P denotes the total transmit power and (4.3c) denotes the unit modulus constraint of the diagonal elements of $\mathbf{Q}$. Due to the non-convex objective function with highly coupled variables and non-convex unit modulus constraint, problem $\mathscr{P}_0$ is a complicated non-convex problem and difficult to be solved directly.

4.2.2 Joint active and passive beamforming design

This section adopts an AO algorithm to solve problem $\mathscr{P}_0$, where the original problem is decomposed into two sub-problems with simpler structure before being solved alternatively. The sub-problem of optimizing $\mathbf{R}$ with fixed $\mathbf{Q}$ is expressed as

$$\mathscr{P}_1 : \max_{\mathbf{R}} R_s\left(\mathbf{R}\right) = \log_2\frac{\left|\mathbf{I} + \mathbf{H}_B\mathbf{R}\mathbf{H}_B^H/\sigma_B^2\right|}{\left|\mathbf{I} + \mathbf{H}_E\mathbf{R}\mathbf{H}_E^H/\sigma_E^2\right|},\ \text{s.t. (4.3b)}, \tag{4.4}$$

where $\mathbf{H}_B \triangleq \mathbf{H}_{AB} + \mathbf{H}_{IB}\mathbf{Q}\mathbf{H}_{AI}$ and $\mathbf{H}_E \triangleq \mathbf{H}_{AE} + \mathbf{H}_{IE}\mathbf{Q}\mathbf{H}_{AI}$. Although problem $\mathscr{P}_1$ is still non-convex, it can be equivalently transformed as [41]

$$\mathscr{P}_2 : \max_{\mathbf{R}} \min_{\mathbf{K}} \quad f\left(\mathbf{R}, \mathbf{K}\right) = \log_2\frac{\left|\mathbf{I} + \mathbf{K}^{-1}\mathbf{H}\mathbf{R}\mathbf{H}^H/\sigma_B^2\right|}{\left|\mathbf{I} + \mathbf{H}_E\mathbf{R}\mathbf{H}_E^H/\sigma_E^2\right|} \tag{4.5a}$$

$$\text{s.t.} \quad \text{(4.3b)}, \mathbf{K} = \begin{bmatrix} \sigma_B^2\mathbf{I} & \mathbf{N}^H \\ \mathbf{N} & \sigma_E^2\mathbf{I} \end{bmatrix}, \mathbf{K} \succeq \mathbf{0}, \tag{4.5b}$$

where $\mathbf{H} = \left[\mathbf{H}_B^H, \mathbf{H}_E^H\right]^H$, $\mathbf{N} = \mathbb{E}\left[\mathbf{n}_E\mathbf{n}_B^H\right]$, and $\mathbf{K} = \mathbb{E}\left[\mathbf{n}\mathbf{n}^H\right]$ with $\mathbf{n} = \left[\mathbf{n}_B^H, \mathbf{n}_E^H\right]^H$. Problem $\mathscr{P}_2$ is a convex–concave problem with a saddle point solution and the

Karush–Kuhn–Tucker (KKT) conditions are sufficient and necessary for optimality. A barrier method in combination with the Newton method and backtracking line search is proposed to solve problem $\mathscr{P}_2$ and is summarized as Algorithm 1 in [1].

The sub-problem of optimizing $\mathbf{Q}$ with fixed $\mathbf{R}$ can be expressed as

$$\mathscr{P}_3 : \max_{\mathbf{Q}} \; R_s(\mathbf{Q}) \;\; \text{s.t. (4.3c).} \tag{4.6}$$

We use an AO algorithm to solve the non-convex problem $\mathscr{P}_3$ by optimizing the ith phase shift q_i while fixing the other $n-1$ phase shifts, and we can obtain a closed-form solution to q_i. The algorithm is summarized as Algorithm 3 in [1].

Considering a more practical scenario with a passive eavesdropper whose channels $\mathbf{H}_{AE}$ and $\mathbf{H}_{IE}$ are not available at Alice, we propose an AN-aided secure transmission scheme, the key idea of which is to satisfy the QoS constraint at Bob using the minimal power and then transmit the AN to degrade Eve with the residual power. The power minimization problem can be described as

$$\mathscr{P}_4 : \min_{\mathbf{R},\mathbf{Q}} \; \text{Tr}(\mathbf{R}) \;\; \text{s.t. (4.3b), (4.3c)}, \log_2 \left|\mathbf{I} + \mathbf{H}_B\mathbf{R}\mathbf{H}_B^H/\sigma_B^2\right| \geq \gamma, \tag{4.7}$$

where γ is the lowest required rate at Bob. The problem is non-convex and we adopt an AO algorithm to solve it. The sub-problem of optimizing $\mathbf{R}$ can be expressed as

$$\mathscr{P}_5 : \min_{\mathbf{R}} \; \text{Tr}(\mathbf{R}) \;\; \text{s.t. (4.3b)}, \log_2 \left|\mathbf{I} + \mathbf{H}_B\mathbf{R}\mathbf{H}_B^H/\sigma_B^2\right| \geq \gamma. \tag{4.8}$$

We prove that the optimal $\mathbf{R}$ to problem $\mathscr{P}_5$ is the solution to the following problem:

$$\mathscr{P}_6 : \max_{\mathbf{R}} \; R(\mathbf{R}) = \log_2 \left|\mathbf{I} + \mathbf{H}_B\mathbf{R}\mathbf{H}_B^H/\sigma_B^2\right| \;\; \text{s.t. (4.3b)}, \text{Tr}(\mathbf{R}) \leq P^*, \tag{4.9}$$

where P^* is the optimal value of $\mathscr{P}_5$. This is a general channel capacity maximization problem and the objective function is non-decreasing with P^*. Thus, binary search can be applied to find P^* such that the optimal value $R(\mathbf{R}^*) = \gamma$.

We then solve the problem of maximizing Bob's achievable rate by designing $\mathbf{Q}$,

$$\mathscr{P}_7 : \max_{\mathbf{Q}} \; \log_2 \left|\mathbf{I} + \mathbf{H}_B\mathbf{R}\mathbf{H}_B^H/\sigma_B^2\right| \;\; \text{s.t. (4.3c).} \tag{4.10}$$

The solution to $\mathscr{P}_4$ is also applied to $\mathscr{P}_7$ via setting $\mathbf{H}_{AE} = \mathbf{0}$ and $\mathbf{H}_{IE} = \mathbf{0}$.

After obtaining the minimal signal power P^*, AN is transmitted isotropically to the null space of Bob's effective channel, i.e., null($\mathbf{W}_B$) with $\mathbf{W}_B = \mathbf{H}_B^H\mathbf{H}_B$, using the residual power $P - P^*$. The AN covariance matrix can be given by

$$\mathbf{R}_{AN} = \frac{P - P^*}{N_A - \text{rank}(\mathbf{W}_B)} \mathbf{U}_{AN}\mathbf{U}_{AN}^H, \tag{4.11}$$

where $\mathbf{U}_{AN}$ is the semi-unitary matrix composed by the eigenvectors corresponding to the zero eigenvalues of $\mathbf{W}_B$. Finally, the secrecy rate can be given by

$$R_s = \gamma - \log_2 \left|\mathbf{I} + \frac{\mathbf{H}_E\mathbf{R}\mathbf{H}_E^H}{\sigma_E^2\mathbf{I} + \mathbf{H}_E\mathbf{R}_{AN}\mathbf{H}_E^H}\right|. \tag{4.12}$$

4.2.3 Numerical results and discussions

In this section, we present numerical results to evaluate the secrecy performance of the IRS-aided MIMO secure communication system.

Figure 4.2 demonstrates the achievable secrecy rate R_s returned by the AO algorithm and two benchmark schemes versus the IRS element number M and the Eve's antenna number N_E, respectively. As can be seen, the AO algorithm significantly outperforms the benchmark schemes. From Figure 4.2(a), it can be observed that R_s increases with M since more reflecting elements bring more spacial degrees of freedom. For the scheme of random phase shifts, increasing M has a slight secrecy performance improvement. It is expected that R_s for the scheme without IRS stays constant as M increases. From Figure 4.2(b), we see that R_s of all schemes has a decreasing trend with N_E. This is inevitable since a larger N_E decreases the spatial degree of freedom for Alice's transmission. In fact, if Eve is equipped with more antennas, an effective solution is to deploy more reflecting elements at IRS so as to enhance the secrecy rate.

Figure 4.3 depicts the actual achievable secrecy rate of the AN-aided scheme without Eve's CSI versus the target QoS threshold γ. We show that the secrecy rate of the AN scheme is larger than that without AN. As long as γ is set properly, Alice could have abundant residual power to transmit AN to degrade Eve thus improving the secrecy rate.

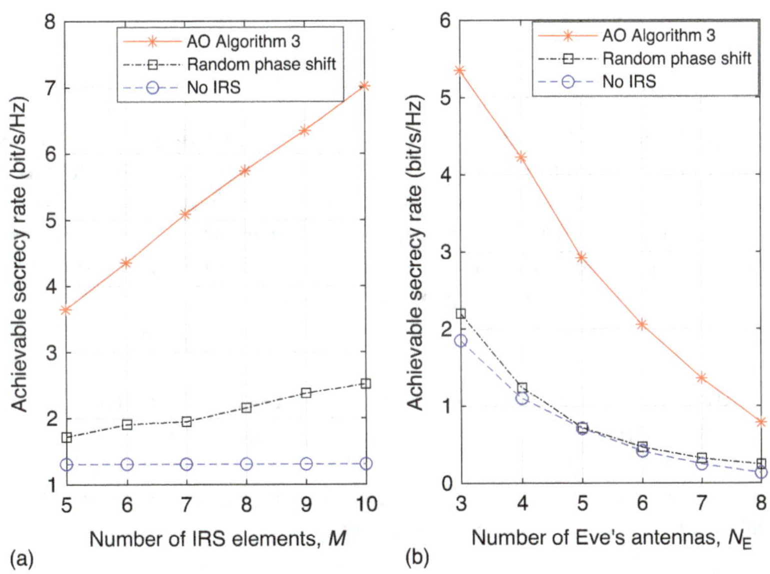

Figure 4.2 Secrecy rate R_s versus IRS element number M and Eve's antenna number N_E, with $N_A = N_B = N_E = 3$ in (a) and $N_A = N_B = 3$, $M = 6$ in (b)

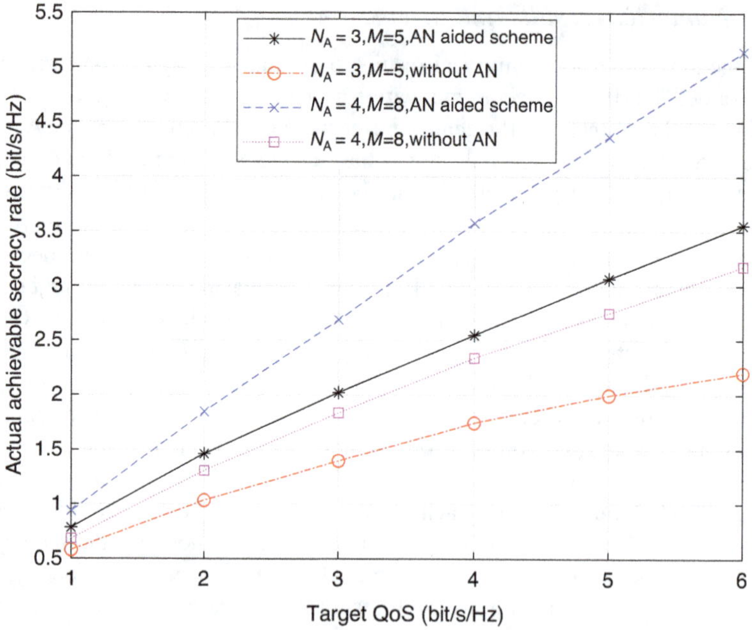

Figure 4.3 *Secrecy rate without Eve's CSI versus target QoS threshold γ with different values of N_A and M with $N_B = N_E = 2$, $P = 35$ dBm*

4.2.4 Summary

This section studies an IRS-aided secure MIMO communication system. The phase shift matrix of IRS and transmit covariance matrix are jointly designed to maximize the system secrecy rate. For the case with global CSI, a barrier method-based AO algorithm is proposed to solve the secrecy rate maximization problem; while an AN-aided scheme is proposed for the case without Eve's CSI. Numerical results validate that IRS can enhance the secrecy rate of the considered MIMO system.

4.3 Enhancing PLS using IRS-empowered pattern index modulations

This section constructs a general framework for the IRS-empowered PIMs and proposes two PLS schemes to enhance the system security performance.

4.3.1 System model and problem description

As depicted in Figure 4.4, we consider an IRS-assisted MIMO transmission system, which consists of a transmitter (Tx), a receiver (Rx), and an IRS, with $A_T = 2^T$

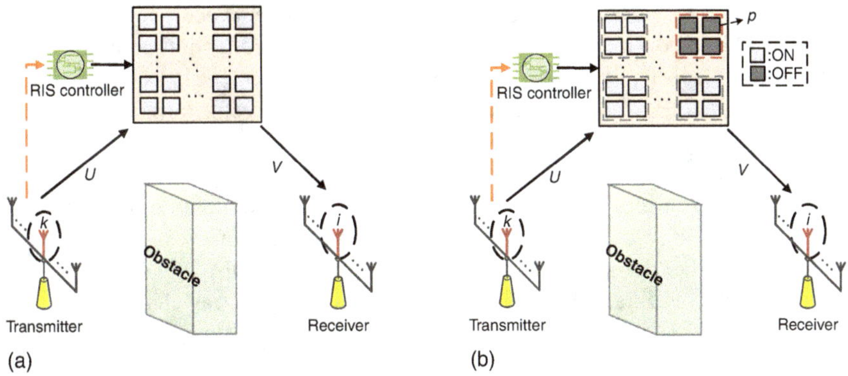

Figure 4.4 Illustration of the IRS-empowered PIM system: (a) IRS-empowered joint transmitter-and-receiver PIM (JPIM) without reflecting elements grouping; (b) IRS-empowered augmented PIM (APIM) with reflecting elements grouping. Copyright IEEE 2022. Reprinted with permission from [3].

Tx array elements, $A_R = 2^R$ Rx array elements, and $M = 2^S$ reconfigurable passive reflecting elements, respectively. We assume that the direct link between Tx and Rx is blocked by the obstacle. We first briefly introduce the IR-enabled JPIM [31], which utilizes both the indices of Tx and Rx patterns, i.e., the Tx and Rx antenna indices, to deliver different portions of information. We then propose a generalized framework of PIM, namely, the APIM, to enhance the system spectral-efficiency by invoking the indices of activated IRS patterns to bear extra portion of information.

4.3.1.1 System model of IRS-empowered JPIM/APIM

As illustrated in Figure 4.4(a), the IRS-empowered JPIM can be recognized as a degraded version of the IRS-assisted joint Tx-and-Rx spatial modulation in [31].

In the JPIM-modulated MIMO, the information bits delivered in every time-slot are separated into two portions of the Tx-/Rx-pattern bits. When the Tx-pattern bits are mapped to the index of $k \in \{1, 2, \ldots, A_T\}$, the single RF chain is connected to the kth Tx array element, which is activated to emit the unmodulated carrier signal. Then, with the Rx-pattern bits being mapped to the index of $i \in \{1, 2, \ldots, A_R\}$, the coefficients of the IRS elements are adjusted accordingly to maximize the observation power of the reflected signal waves at the ith Rx array element, such that the ith Rx pattern is activated. The achievable data rate of IRS-JPIM can be given by

$$DR_J = \underbrace{\lfloor \log_2 (A_T) \rfloor}_{\text{Tx-pattern bits}} + \underbrace{\lfloor \log_2 (A_R) \rfloor}_{\text{Rx-pattern bits}} = T + R. \tag{4.13}$$

Denote $\mathbf{U} \triangleq [\mathbf{u}_1, \mathbf{u}_2, \ldots, \mathbf{u}_{A_T}] \in \mathbb{C}^{M \times A_T}$ and $\mathbf{V} \triangleq [\mathbf{v}_1{}^T, \mathbf{v}_2{}^T, \ldots, \mathbf{v}_{A_R}{}^T]^T \in \mathbb{C}^{A_R \times M}$ as the channels from Tx to IRS and from IRS to Rx, respectively, and denote $\mathbf{q}_w =$

$[\beta_1 e^{jQw_1}, \beta_2 e^{jQw_2}, \ldots, \beta_{A_S} e^{jQw_M}]^T$ as the reflecting-coefficient vector of IRS with $\beta_s \in [0,1]$ and $Qw_s \in (0, 2\pi]$ for $s \in \{1, 2, \ldots, M\}$ being the reflection amplitudes and phase-shifts, respectively. Given that the kth Tx array element is powered on, the observed signal at Rx can be formulated as the following vector form:

$$\mathbf{r} = \sqrt{E_t}\mathbf{VQUe}_k + \mathbf{z} = \sqrt{E_t}\mathbf{VQu_k} + \mathbf{z}, \tag{4.14}$$

where E_t is the transmit power, $\mathbf{Q} = \mathrm{diag}\{\mathbf{q}_w\}$, $\mathbf{e}_k$ is the kth column of the identity matrix, and $\mathbf{z} = [z_1, z_2, \ldots, z_{A_R}]^T$ is the additive zero-mean complex Gaussian noise.

Recalling the definition of $[\mathbf{U}]_{ks} \triangleq \varrho_{sk} e^{-j\vartheta_{sk}}$ and $[\mathbf{V}]_{ks} \triangleq \xi_{is} e^{-j\kappa_{is}}$, we can characterize the instantaneous received power at the i-th Rx array element as

$$E_{r_i} = \left|\sqrt{E_t}\mathbf{v_i Q u_k} + z_i\right|^2 = E_t \left|\sum_{s=1}^{M} \xi_{is}\beta_s \varrho_{sk} e^{j(Qw_s - \kappa_{is} - \vartheta_{sk})}\right|^2 + N_0. \tag{4.15}$$

In order to maximize E_{r_i}, we should guarantee that $\beta_s = 1$ and $Qw_s = \kappa_{is} + \vartheta_{sk}$.

As shown in Figure 4.4(b), the proposed IRS-APIM only turns on a partial of reflecting elements for forming IRS patterns. Hence, the indices of Tx, Rx, and IRS pattern can be jointly employed for conveying information. We employ the reflecting-element grouping approach for constituting the IRS patterns. Particularly, the IRS elements are divided into $A_P = 2^{S_P}$ groups each of which has $A_E = M/A_P \geq 2$ abutting elements with $S_P \leq S - 1$. Given that the kth Tx pattern is activated for transmission, $A_P - 1$ out of A_P subsets are randomly turned on for dedicated reflection of incident signals. Hence, the data rate of IRS-APIM is given by

$$DR_A = DR_J + \underbrace{\lfloor \log_2 (A_P) \rfloor}_{\text{IRS-pattern bits}} = T + R + S_P. \tag{4.16}$$

Denote $p \in \{1, 2, \ldots, A_P\}$ as the index of the activated IRS pattern and that of the reflecting group with an OFF status, and the reflecting coefficients can be given as

$$\mathbf{q}_w^g = \begin{cases} \mathbf{0}_{A_E}, & g = p \\ [\beta_1^g e^{jQw_1^g}, \beta_2^g e^{jQw_2^g}, \ldots, \beta_{A_E}^g e^{jQw_{A_E}^g}]^T, & g \in \mathscr{P} \text{ and } g \neq p \end{cases} \tag{4.17}$$

where $\mathbf{0}_{A_E}$ denotes the A_E-length zero vector, $\beta_e^g \in [0,1]$ and $Qw_e^g \in (0, 2\pi]$ for $e \in \{1, 2, \ldots, A_E\}$ denote the reflecting amplitudes and phase shifts for the gth IRS subset, respectively, and $\mathscr{P} \triangleq \{1, 2, \ldots, A_P\}$. Let $\mathbf{\Pi}_g \in \mathbb{C}^{A_E \times A_T}$ and $\mathbf{\Lambda}_g \in \mathbb{C}^{A_R \times A_E}$ denote the channels from Tx to the gth IRS subset and from the g-th IRS subset to Rx, respectively. Given that the kth Tx pattern and pth IRS pattern are activated, the signal observations at the A_R Rx array elements are

$$\mathbf{r} = \sqrt{E_t}\mathbf{V}\bar{\mathbf{Q}}_p\mathbf{U}\mathbf{e}_k + \mathbf{z} = \sqrt{E_t}\sum_{g=1}^{A_P} \mathbf{\Lambda}_g \mathrm{diag}\{\mathbf{q}_w^g\}\boldsymbol{\pi}_k^g + \mathbf{z}, \tag{4.18}$$

where $\bar{\mathbf{Q}}_p = \mathrm{diag}\{[(\mathbf{q}_w^1)^T, (\mathbf{q}_w^2)^T, \ldots, (\mathbf{q}_w^{A_P})^T]\}$ denotes the reflecting coefficients matrix for the IRS and $\boldsymbol{\pi}_k^g$ denotes the kth column of $\mathbf{\Pi}_g$.

Note that IRS-JPIM can be regarded as a special case of IRS-APIM with $A_P = 1$. With the same configurations, IRS-APIM with $A_P \geq 2$ can significantly outperform

IRS-JPIM in spectral efficiency, which is implied by comparing (4.13) and (4.16). When the reflecting amplitudes within each ON-state IRS subset are set to their maximum, the magnitudes of the reflected waves can be maximized. Therefore, we below consider only the phase shifts for reflection design.

4.3.1.2 Reflection designs for IRS-empowered APIM

EPA-based reflection design

Recalling (4.18), the received signal at the ith Rx antenna can be given as

$$r_i = \sqrt{E_t}\sum_{g=1}^{A_P} \boldsymbol{\lambda}_i^g \mathrm{diag}\{\mathbf{q}_w^g\}\boldsymbol{\pi}_k^g + z_i \overset{(a)}{=} \sqrt{E_t}\sum_{g=1,g\neq p}^{A_P}\sum_{e=1}^{A_E} \lambda_{i,e}^g \pi_{e,k}^g e^{jQw_e^g} + z_i, \quad (4.19)$$

where $\boldsymbol{\lambda}_i^g$ is the ith row of $\boldsymbol{\Lambda}_g$, $\lambda_{i,e}^g$ $(\pi_{e,k}^g)$ is the eth components of $\boldsymbol{\lambda}_i^g$ $(\boldsymbol{\pi}_k^g)$, and (a) holds as the amplitudes of the activated IRS elements are set to the maximum. Let $\lambda_{i,e}^g = \xi_{ie}^g e^{-j\kappa_{ie}^g}$ and $\pi_{e,k}^g = \varrho_{ek}^g e^{-j\vartheta_{ek}^g}$. We use the phase alignment in Section 4.3.1.1 and derive the reflecting phase-shifts as $Qw_e^g = (\kappa_{ie}^g + \vartheta_{ek}^g) \bmod 2\pi$ for $g \neq p$.

GPA-based reflection design

For ease of hardware implementation, we assign common reflecting coefficients in the same IRS group, meaning that $Qw_e^g = Qw^g$ holds except for the case $g = p$. The common phase shift Qw^g can be obtained as $Qw^g = -\arg\left(\sum_{e=1}^{A_E} \xi_{ie}^g \varrho_{ek}^g e^{-j(\kappa_{ie}^g + \vartheta_{ek}^g)}\right)$, $g \neq p$, where $\arg(\mathfrak{C})$ calculates the phase angle of a complex scalar $\mathfrak{C}$.

Remark 1. *One can see that the signal-to-noise ratio (SNR) of the activated Rx array element achieved by EPA is higher than that by GPA, since* $\sum_{e=1}^{A_E} |\lambda_{i,e}^g||\pi_{e,k}^g| > |\boldsymbol{\lambda}_i^g \boldsymbol{\pi}_k^g|^2$ *always holds for* $A_E \geq 2$. *However, the hardware complexity of IRS reflection and control for the GPA-based IRS-APIM can be decreased significantly, since each subset is assigned with one common phase shift. In this sense, EPA can be recognized as the optimal reflection design, while GPA can be regarded as a low-complexity suboptimal reflection design for IRS-APIM.*

4.3.2 PLS-enhancing schemes for IRS-empowered PIMs

Considering that only one Tx antenna is activated and the eavesdropper's location is generally unknown, the conventional additive AN-assisted secrecy scheme cannot be employed to cover every sub-part of the PIM-category super-symbols from being intercepted. To this end, we propose two IRS-based PLS enhancing schemes, namely, random perturbation alignment (RPA) and random vector synthesize (RVS), to carefully mask the sub-symbols for any individual PIM-category super-symbol.

4.3.2.1 IRS-based random perturbation alignment

The basic idea of IRS-based RPA is to add multiplicative random perturbation to emitted signals and then accomplish the phase-rotation reflection for each IRS element, such that the added random phase perturbation can be canceled and the multi-path reflected signals can be superposed in-phase at Rx. However, the inherent disturbance updated at the symbol rate is formed at the eavesdropper, thereby completing the secure transmission of the PIM-category super-symbols. To be specific, the appended multiplicative random scrambling can be expressed as $w = e^{-j\theta_t}$, where θ_t is randomly selected from the continuous set of $(0, 2\pi]$. Note that the EPA and GPA-based RPAs can be well employed here to balance complexity and performance.

4.3.2.2 IRS-based random vector synthesize

Different from RPA mentioned above, IRS-based RVS imposes the random perturbation at IRS and updates with symbol rate, while the randomness is alleviated at Tx with the multi-path signal vector synthesize. With proper design, the legitimate Rx is able to synthesize out the regular modulated constellations. However, without knowing the random reflection coefficients of IRS and the cascaded channel of the legitimate link, the eavesdropper can only achieve highly scrambled signal observations in each PIM-category super-symbol duration, and hence the secrecy of the PIM-category super-symbols transmission can be effectively guaranteed.

We propose two RVS schemes, namely, element-wise vector synthesize (EVS) and group-wise vector synthesize (GVS). For EVS, IRS elements are configured with different reflection coefficients. For GVS, IRS elements are divided into multiple groups with each group sharing one common reflection coefficient. The phase shifts of reflection coefficients for the two schemes are uniformly distributed in $(0, 2\pi]$. The reflected signal to be synthesized at Rx can be given by $v_{sk}^t = e^{-jF_t} s_k^t$, $t \in \{e, g\}$, where e and g stand for EVS and GVS, respectively, s_k^t denotes the baseband APM symbol, and e^{-jF_t} denotes the random perturbation compensation factor. Here, $F_e = \arg \sum_{g=1, g \neq p}^{A_P} \sum_{e=1}^{A_E} \lambda_{i,e}^g \pi_{e,k}^g e^{-j\theta_e^g}$ and $F_g = \arg \sum_{g=1, g \neq p}^{A_P} \sum_{e=1}^{A_E} \lambda_{i,e}^g \pi_{e,k}^g e^{-j\theta_g}$, where θ_e^g and θ_g denote the random perturbation phase selected for the eth element in the gth reflection group and for the gth reflection group, respectively.

4.3.3 *Numerical results and discussions*

This subsection characterizes and compares the 3D constellations observations at Rx and eavesdropper. The system parameters are set as $A_T = 2$, $A_R = 2$, $M = 64$, $A_P = 2$, and SNR $= 5$dB. The index of the desired receive antenna is 1, and we consider QPSK modulation and Rayleigh flat fading channels.

Figures 4.5 and 4.6 show that, due to the use RPA and RVS, the legitimate Rx can achieve regular PIM-category super-symbol constellation while the eavesdropper can only observe scrambled constellation. Hence, the secrecy transmission of IRS-empowered PIMs can be guaranteed for passive eavesdropping scenarios.

4.3.4 *Summary*

We investigate a unified framework for IRS-PIM and propose two enhanced PLS schemes with IRS element-wise and group-wise reflection implementations. Simulation results validate the superiority of the proposed schemes in secure transmission of PIM-category super-symbols even over flat fading channels.

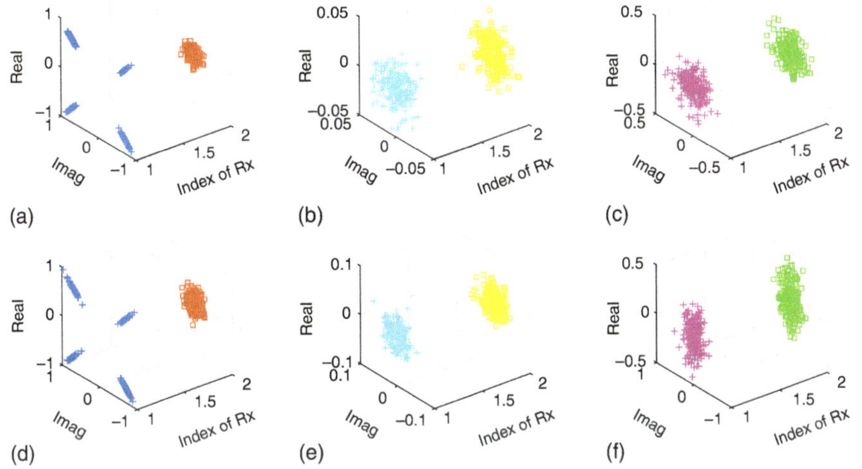

Figure 4.5 *3D constellations for RPA. (a) EPA, legitimate Rx; (b) EPA, eavesdropper without direct link from IRS; (c) EPA, eavesdropper with direct link from IRS; (d), (e), and (f) are for GPA with the same cases of (a), (b), and (c), respectively.*

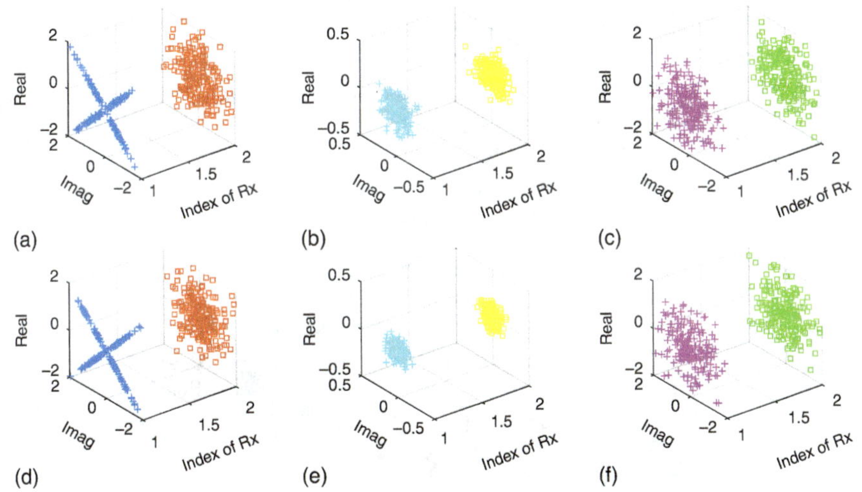

Figure 4.6 *3D constellations for RVS. (a) EVS, legitimate Rx; (b) EVS, eavesdropper without direct link from IRS; (c) EVS, eavesdropper with direct link from IRS; (d), (e), and (f) are for GVS with the same cases of (a), (b), and (c), respectively.*

4.4 IRS-enabled secure integrated sensing and communication systems

This section studies an IRS-enabled secure ISAC system and builds a design framework of secrecy rate maximization subject to the radar detection constraint.

4.4.1 System model and problem description

Consider an IRS-enabled secure ISAC system as shown in Figure 4.7. A DFRC BS (Alice) equipped with an N-antenna uniform linear array transmits the signals for both communication and detection to a single-antenna communication user (Bob) and a single-antenna radar target (Eve). An IRS with M reflecting elements forming a uniform planar array is deployed to secure the communication. The received signals at Bob and Eve can be uniformly expressed as

$$y_X = (\mathbf{h}_{IX}\mathbf{Q}\mathbf{H}_{AI} + \mathbf{h}_{AX})\,\mathbf{w}\sqrt{P}s + n_X,\ X \in \{B, E\}, \tag{4.20}$$

where s denotes the transmitted signal for communication and detection, P denotes the total transmit power, $\mathbf{w} \in \mathbb{C}^{N\times 1}$ denotes the beamforming vector at BS with $\|\mathbf{w}\| = 1$, $\mathbf{H}_{AI}$, $\mathbf{h}_{IB}$, $\mathbf{h}_{AB}$, $\mathbf{h}_{IE}$, and $\mathbf{h}_{AE}$ denote the channels of Alice–IRS, IRS–Bob, Alice–Bob, IRS–Eve, and Alice–Eve, respectively, $n_B \sim CN\left(0, \sigma_B^2\right)$ and $n_E \sim CN\left(0, \sigma_E^2\right)$ denote the received noise at Bob and Eve, respectively; $\mathbf{Q} = \text{diag}(\mathbf{q}) = \text{diag}\left(e^{j\theta_1}, e^{j\theta_2}, \ldots, e^{j\theta_M}\right)$ denotes the phase shift matrix of IRS with $\theta_i \in [0, 2\pi)$ being the phase shift of the ith reflecting element of IRS. Thus, the achievable rate of Bob and Eve $X \in \{B, E\}$ can be uniformly expressed as

$$R_X = \log_2\left(1 + P\left(\mathbf{h}_{IX}\mathbf{Q}\mathbf{H}_{AI} + \mathbf{h}_{AX}\right)\mathbf{w}\mathbf{w}^H\left(\mathbf{h}_{IX}\mathbf{Q}\mathbf{H}_{AI} + \mathbf{h}_{AX}\right)^H/\sigma_X^2\right). \tag{4.21}$$

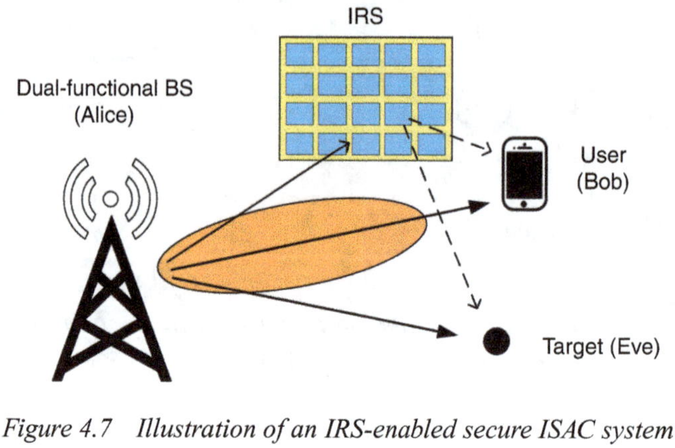

Figure 4.7 Illustration of an IRS-enabled secure ISAC system

We focus on maximizing the secrecy rate $R_s \triangleq R_B - R_E$ subject to the radar detection constraint by jointly optimizing $\mathbf{w}$ and $\mathbf{q}$, which is formulated as below:

$$\mathscr{P}_0 : \max_{\mathbf{w},\mathbf{q}} \quad R_s(\mathbf{w},\mathbf{q}) = R_B - R_E \tag{4.22a}$$

$$\text{s.t.} \quad \|\mathbf{w}\| = 1, \tag{4.22b}$$

$$|\mathbf{q}_i| = 1, \forall i = 1, 2, \ldots, M, \tag{4.22c}$$

$$P_E \geq \gamma, \tag{4.22d}$$

where (4.22b) is the unit norm constraint of $\mathbf{w}$, (4.22c) is the unit modulus constraint of $\mathbf{q}$, and (4.22d) specifies the radar detection constraint, i.e., the target's illumination power P_E should exceed a predetermined threshold γ with $P_X \triangleq P\left(\mathbf{h}_{IX}\mathbf{Q}\mathbf{H}_{AI} + \mathbf{h}_{AX}\right)\mathbf{w}\,\mathbf{w}^H\left(\mathbf{h}_{IX}\mathbf{Q}\mathbf{H}_{AI} + \mathbf{h}_{AX}\right)^H$ for $X \in \{B, E\}$. Problem $\mathscr{P}_0$ is non-convex due to the non-convex objective function and constraints (4.22c) and (4.22d).

4.4.2 Joint active and passive beamforming design

It is easy to observe that problem $\mathscr{P}_0$ can be equivalently expressed as

$$\mathscr{P}_1 : \min_{\mathbf{w},\mathbf{q}} \frac{1 + P_E/\sigma_E^2}{1 + P_B/\sigma_B^2} \quad \text{s.t. (4.22b)} - \text{(4.22d)}. \tag{4.23}$$

We can employ the Dinkelbach method to transform the fractional objective function and then optimize the variables by SDR. However, it is of high computational complexity. Inspired by the fact that the feasible set of $\mathbf{w}$ and $\mathbf{q}$ is a complex hypersphere and an oblique manifold, respectively, the RCG algorithm can be leveraged to handle the problem. To proceed, we first use the penalty function method to incorporate the constraint (4.22d) into the objective function as follows:

$$\min_{\mathbf{w},\mathbf{q}} \frac{1 + P_E/\sigma_E^2}{1 + P_B/\sigma_B^2} + \zeta \max\{0, \gamma - P_E\}, \tag{4.24}$$

where ζ is the penalty factor. When using the RCG algorithm, the objective function should be differentiable and thus we can obtain its gradient. However, the max function in (4.24) is non-smooth and its gradient does not exist. To this end, we approximate it to a smooth function, i.e., $\max\{0, \gamma - P_E\} \leq \epsilon_1 \ln\left(1 + e^{(\gamma - P_E)/\epsilon_1}\right)$, where ϵ_1 is a small positive number. Thus, problem $\mathscr{P}_1$ can be rewritten as

$$\mathscr{P}_2 : \min_{\mathbf{w},\mathbf{q}} f = \frac{1 + P_E/\sigma_E^2}{1 + P_B/\sigma_B^2} + \zeta \epsilon_1 \ln\left(1 + e^{\frac{\gamma - P_E}{\epsilon_1}}\right) \quad \text{s.t. (4.22b) } \textit{and} \text{ (4.22c)} \tag{4.25}$$

Then, $\mathbf{w}$ and $\mathbf{q}$ can be optimized alternatively by RCG. We denote the feasible set of $\mathbf{w}$ and $\mathbf{q}$ by $\mathscr{W} = \{\mathbf{w} \mid \|\mathbf{w}\| = 1\}$ and $\mathscr{Q} = \{\mathbf{q} \mid [\mathbf{q}\mathbf{q}^H]_{i,i} = 1, \forall i = 1, 2, \ldots, M\}$, respectively. With fixed $\mathbf{q}$, the problem of optimizing $\mathbf{w}$ can be expressed as

$$\mathscr{P}_3 : \min_{\mathbf{w}\in\mathscr{W}} f(\mathbf{w}) = \frac{1 + \mathbf{w}^H\mathbf{M}_E\mathbf{w}}{1 + \mathbf{w}^H\mathbf{M}_B\mathbf{w}} + \zeta\epsilon_1 \ln\left(1 + e^{\frac{\gamma - \mathbf{w}^H\mathbf{M}_E'\mathbf{w}}{\epsilon_1}}\right) \quad \text{s.t. (4.22b)}, \tag{4.26}$$

where $\mathbf{M}_X \triangleq P/\sigma_X^2 \left(\mathbf{H}_{AI}^H\mathbf{Q}^H\mathbf{h}_{IX}^H\mathbf{h}_{IX}\mathbf{Q}\mathbf{H}_{AI}+\mathbf{H}_{AI}^H\mathbf{Q}^H\mathbf{h}_{IX}^H\mathbf{h}_{AX}+\mathbf{h}_{AX}^H\mathbf{h}_{IX}\mathbf{Q}\mathbf{H}_{AI}+\mathbf{h}_{AX}^H\mathbf{h}_{AX}\right)$, and $\mathbf{M}_E^{'} = \sigma_E^2\mathbf{M}_E$. Following the RCG algorithm, we have the next point $\mathbf{w}_{k+1}$ as

$$\mathbf{w}_{k+1} = R_{\mathbf{w}_k}\left(s_k\boldsymbol{\eta}_k\right) = \frac{\mathbf{w}_k + s_k\boldsymbol{\eta}_k}{\|\mathbf{w}_k + s_k\boldsymbol{\eta}_k\|}, \tag{4.27}$$

where R is a retraction on $\mathscr{W}$, $s_k \geq 0$ is the step size, and $\boldsymbol{\eta}_k$ is the search direction

$$\boldsymbol{\eta}_k = -\text{grad}f\left(\mathbf{w}_k\right) + \mu_k\mathscr{T}_{\mathbf{w}_{k-1}}\left(\boldsymbol{\eta}_{k-1}\right) \tag{4.28}$$

where $\mu_k \geq 0$ and various choices of μ_k are available like Fletcher and Reeves parameter and Polak–Ribière parameter; grad$f\left(\mathbf{w}_k\right)$ denotes the Riemannian gradient of f at $\mathbf{w}_k$ and is calculated based on the orthogonal projection of the Euclidean gradient $\nabla f\left(\mathbf{w}_k\right)$ onto the tangent space $T_{\mathbf{w}_k}\mathscr{W}$ [42], grad$f\left(\mathbf{w}_k\right) = \nabla f\left(\mathbf{w}_k\right) - \Re\left(\mathbf{w}_k^H\nabla f\left(\mathbf{w}_k\right)\right)\mathbf{w}_k$ and $\mathscr{T}_{\mathbf{w}_{k-1}}\left(\boldsymbol{\eta}_{k-1}\right)$ denotes the vector transport of $\boldsymbol{\eta}_{k-1}$ and is defined as the projection of $\boldsymbol{\eta}_{k-1}$ to $T_{\mathbf{w}_k}\mathscr{W}$, i.e.,

$$\mathscr{T}_{\mathbf{w}_{k-1}}\left(\boldsymbol{\eta}_{k-1}\right) = \boldsymbol{\eta}_{k-1} - \Re\left(\mathbf{w}_k^H\boldsymbol{\eta}_{k-1}\right)\mathbf{w}_k. \tag{4.29}$$

We choose the Polak–Ribière parameter as μ_k to achieve fast convergence [43], i.e.,

$$\mu_k = \frac{\left\langle \text{grad}f\left(\mathbf{w}_k\right), \text{grad}f\left(\mathbf{w}_k\right) - \mathscr{T}_{\mathbf{w}_{k-1}}\left(\text{grad}f\left(\mathbf{w}_{k-1}\right)\right)\right\rangle}{\left\langle \text{grad}f\left(\mathbf{w}_{k-1}\right), \text{grad}f\left(\mathbf{w}_{k-1}\right)\right\rangle}, \tag{4.30}$$

where $\langle \mathbf{x}, \mathbf{y}\rangle = \Re\left(\mathbf{x}^H\mathbf{y}\right)$ and $\mathscr{T}_{\mathbf{w}_{k-1}}\left(\text{grad}f\left(\mathbf{w}_{k-1}\right)\right)$ denotes the vector transport of grad$f\left(\mathbf{w}_{k-1}\right)$. The step size s_k should be chosen to satisfy the Armijo condition

$$f\left(\mathbf{w}_{k+1}\right) \leq f\left(\mathbf{w}_k\right) + c_1 s_k\left\langle \text{grad}f\left(\mathbf{w}_k\right), \boldsymbol{\eta}_k\right\rangle, \tag{4.31}$$

where $0 < c_1 < 1$. When $|\text{grad}f\left(\mathbf{w}_k\right)|$ is less than the predetermined convergence accuracy, $\mathbf{w}_k$ can be outputted as the solution to problem $\mathscr{P}_3$.

When $\mathbf{w}$ is fixed, the problem of optimizing $\mathbf{q}$ can be formulated as

$$\mathscr{P}_4: \min_{\mathbf{q}\in\mathscr{Q}} f\left(\mathbf{q}\right) \text{ s.t. (4.22c)}, \tag{4.32}$$

where $f\left(\mathbf{q}\right) = \frac{1+P\left(\mathbf{q}^H\mathbf{F}_E\mathbf{q}+\mathbf{q}^H f_E+f_E^H\mathbf{q}+g_E\right)/\sigma_E^2}{1+P\left(\mathbf{q}^H\mathbf{F}_B\mathbf{q}+\mathbf{q}^H f_B+f_B^H\mathbf{q}+g_B\right)/\sigma_B^2} + \zeta\epsilon_1\ln\left(1+e^{\frac{\gamma-P\left(\mathbf{q}^H\mathbf{F}_E\mathbf{q}+\mathbf{q}^H f_E+f_E^H\mathbf{q}+g_E\right)}{\epsilon_1}}\right)$,

$\mathbf{F}_X = \left(\mathbf{h}_{IX}^H\mathbf{h}_{IX}\right)\odot\left(\mathbf{H}_{AI}\mathbf{w}\mathbf{w}^H\mathbf{H}_{AI}^H\right)^T$, f_X is the vector constituted by the diagonal elements of matrix $\mathbf{G}_X = \mathbf{h}_{IX}^H\mathbf{h}_{AX}\mathbf{w}\mathbf{w}^H\mathbf{H}_{AI}^H$, and $g_X = \text{Tr}\left(\mathbf{h}_{AX}^H\mathbf{h}_{AX}\mathbf{w}\mathbf{w}^H\right)$, $X \in \{B, E\}$. Given the kth iteration point $\mathbf{q}_k$, the next point $\mathbf{q}_{k+1}$ can be calculated as

$$\mathbf{q}_{k+1} = R_{\mathbf{q}_k}\left(s_k\boldsymbol{\eta}_k\right) = unt\left(\mathbf{q}_k + s_k\boldsymbol{\eta}_k\right), \tag{4.33}$$

where $unt\left(\mathbf{x}\right) \triangleq \left[\frac{x_1}{|x_1|}, \frac{x_2}{|x_2|}, \ldots, \frac{x_M}{|x_M|}\right]^T$. The design of $\mathbf{q}$ is similar to that of $\mathbf{w}$. We first obtain the Riemannian gradient of $f\left(\mathbf{q}\right)$ at point $\mathbf{q}_k$, i.e., grad$f\left(\mathbf{q}_k\right) = \nabla f\left(\mathbf{q}_k\right) - \Re\{\nabla f\left(\mathbf{q}_k\right)\odot\mathbf{q}_k^*\}\odot\mathbf{q}_k$, and then derive the vector transport $\mathscr{T}_{\mathbf{q}_{k-1}}\left(\boldsymbol{\eta}_{k-1}\right) = \boldsymbol{\eta}_{k-1} - \Re\left\{\boldsymbol{\eta}_{k-1}\odot\mathbf{q}_k^*\right\}\odot\mathbf{q}_k$. The values of μ_k and s_k follow from (4.30) and (4.31), respectively. Finally, we obtain the phase shift matrix $\mathbf{Q} = \text{diag}\left(\mathbf{q}\right)$.

4.4.3 Numerical results and discussions

This subsection provides numerical results to validate the performance of the IRS-aided secure ISAC system. Consider a 3D Cartesian coordinate system, where Alice, IRS, Bob, and Eve are located at $(0, 0, 30)$, $\left(-2.5, 2.5\sqrt{3}, 35\right)$, $\left(0, 30, 30 + 10\sqrt{3}\right)$, and $\left(45, 15\sqrt{3}, 0\right)$, respectively, and the IRS is parallel to the $y-z$ plane. Model channel $\mathbf{h}_{XE}, X \in \{A, I\}$, as the array response vector $\mathbf{a}_{XE}^H$ given below

$$[\mathbf{a}_{AE}]_n = e^{j2\pi d_A(n-1)\sin\theta/\lambda}, \quad [\mathbf{a}_{IE}]_m = e^{j2\pi d_I\left(\lfloor\frac{m}{M_y}\rfloor\sin\phi\sin\theta + \left(m-\lfloor\frac{m}{M_y}\rfloor\right)\sin\phi\cos\theta\right)/\lambda},$$

where $n = 1, 2, \ldots, N$, $m = 1, 2, \ldots, M$, λ denotes the wavelength, $d_K = \frac{\lambda}{2}$ denotes the spacing between adjacent elements, M_y denotes the number of IRS elements in each row along y-axis, and $\theta \in [\ 0, 2\pi\ )$ and $\phi \in [\ -\pi/2, \pi/2\ )$ denote the azimuth AoA/AoD and elevation AoA/AoD, respectively. The entries in other channels are complex Gaussian random variables with zero mean and unit variance. We set $N = 4$, $M = 9$, $M_y = 3$, $P = 35$dBm, $\sigma_B^2 = \sigma_E^2 = 0.1$, and $\gamma = 1.5$W.

Figure 4.8 plots the secrecy rate versus the IRS element number M. It is clear that the secrecy rate can be significantly improved with the aid of IRS, and it increases monotonically with the IRS element number, since adding IRS elements brings extra spatial degrees of freedom and more energy can be focused on Bob. In addition, we show that the gap between the secrecy rate of the RCG algorithm and the SDR algorithm is negligible while the former is of low-complexity.

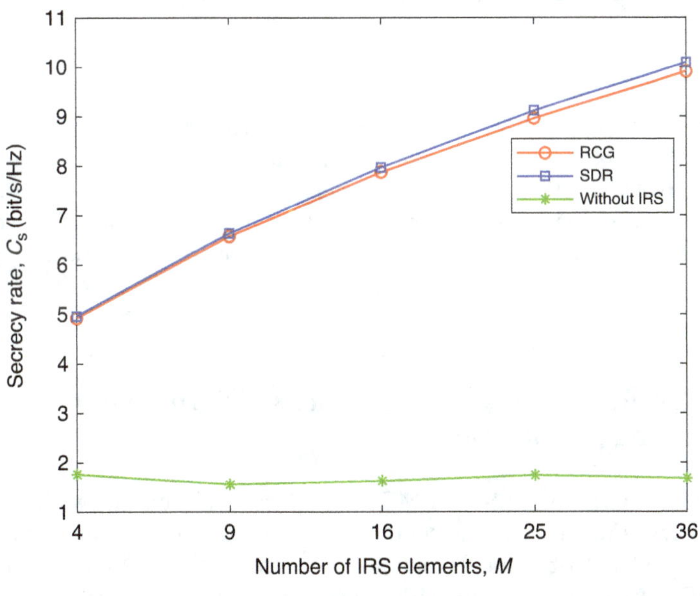

Figure 4.8 Secrecy rate versus IRS element number M

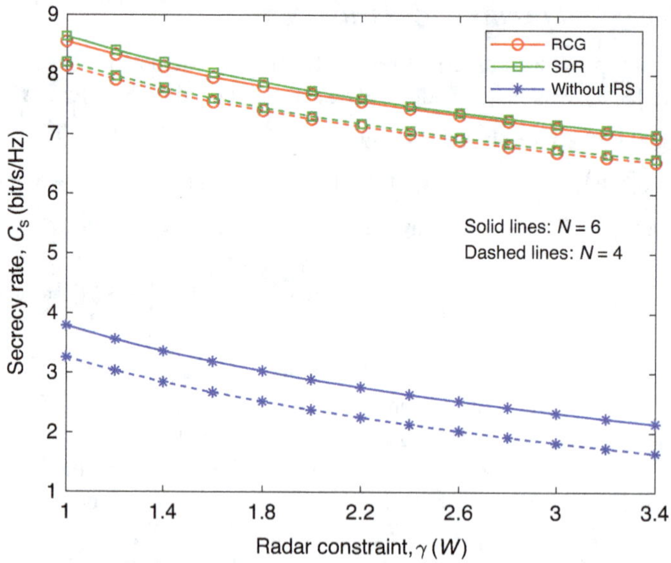

Figure 4.9 Secrecy rate versus radar detection constraint γ

Figure 4.9 plots the secrecy rate versus the radar detection threshold γ. We show that the secrecy rate decreases with γ, as a larger γ will leak more energy to Eve, whereas increasing the transmit antenna number N can improve the secrecy rate.

4.4.4 Summary

This section studies an IRS-aided secure ISAC system, and we propose the SDR-based and RCG-based algorithms to maximize the secrecy rate by optimizing the phase shift matrix of IRS and the beamforming vector at transmitter while satisfying the radar detection constraint. Numerical results demonstrate the significance of IRS in achieving a secure ISAC system and uncover the performance trade-off between secure transmission and radar detection.

4.5 IRS-aided secure communications: a deep reinforcement learning perspective

This section studies the PLS of a MEC-empowered vehicular network, where an IRS is deployed to enhance channel quality while reducing eavesdropping threats. We establish a DDPG-based DRL optimization framework to minimize the maximal secure MEC service latency of the MEC process by jointly designing the IRS elements and computing resource allocation.

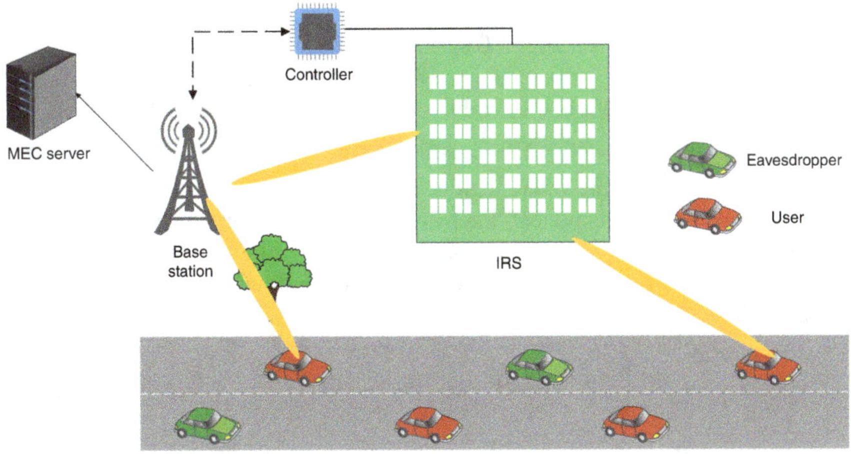

Figure 4.10 Illustration of an IRS-aided MEC vehicular network

4.5.1 System model and problem formulation

4.5.1.1 IRS-aided MEC vehicular network

Consider a MEC-empowered vehicular network illustrated in Figure 4.10, where a BS simultaneously communicates with vehicle users in different sub-bands, as discussed in 3GPP [44]. The resource-constrained vehicles can offload their computation tasks to the MEC BS through the vehicle-to-infrastructure (V2I) links, and the BS can flexibly allocate computing resources for different task requests and feedback on the results. We denote the V2I links by $\mathscr{N} = \{1, 2, \ldots, N\}$. In order to ensure communication reliability, an IRS is deployed to reconfigure the channel environment and alleviate the impact of blocking and multi-path fading.

(1) *Potential wiretapping threats*: The vehicles not served by the BS would be considered as potential eavesdroppers, denoted by the set $\mathscr{E} = \{1, 2, \ldots, E\}$.
(2) *IRS-aided secure transmission*: We aim to stably offload computing tasks to the BS and leverage the IRS to decrease signal leakage to potential eavesdroppers.

4.5.1.2 Communication model

All vehicles are equipped with one omni-directional antenna, and the BS is equipped with a K-antenna uniform linear array (ULA). We also model the IRS as an M-element ULA, where the amplitude reflection coefficient is set to one for all the elements and the phase reflection coefficient of the mth element is presented by $\theta_m = e^{j\phi_m}$ with $\phi_m \in [0, 2\pi)$. Thus, the IRS reflection–coefficient matrix can be defined as $\mathbf{Q} = \text{diag}([\theta_1, \theta_2, \ldots, \theta_M])$. Since there is no in-band interference, BS can utilize maximal-ratio combining (MRC) scheme to design the receiving beamforming matrix for enhancing the signal quality. It can be expressed by $F = [f_1, \ldots, f_N] \in \mathbb{C}^{K \times N}$, where f_n represents the beamforming vector for the n-th V2I link.

(1) *Channel model*: Denote $\mathbf{h}_{n,b} \in \mathbb{C}^{K\times 1}$, $\mathbf{h}_{n,i} \in \mathbb{C}^{M\times 1}$, and $h_{n,e} \in \mathbb{C}$ as the channels from the nth served vehicle to BS, IRS, and the eth eavesdropper, respectively, and denote $\mathbf{h}_{i,e} \in \mathbb{C}^{1\times M}$ and $\mathbf{H}_{i,b} \in \mathbb{C}^{K\times M}$ as the channels from IRS to the eth eavesdropper and BS, respectively. Assuming there is no direct V2I links, the channel gain of the nth V2I link is denoted by $\mathbf{h}_{n,b} = \sqrt{\rho d_{n,b}^{-\alpha_{n,b}}}\mathbf{h}_{n,b}^{*}$, where ρ is the path loss at the reference distance $d_0 = 1$ m, $d_{n,b}$ is the distance, and $\alpha_{n,b}$ is the path-loss exponent. Each element of $\mathbf{h}_{n,b}^{*}$ follows an independent and identically distributed (i.i.d.) complex Gaussian distribution with zero mean and unit variance. The channel gain from IRS to BS follows the Rician distribution given below:

$$\mathbf{H}_{i,b} = \sqrt{\rho d_{i,b}^{-\alpha_{i,b}}}\left(\sqrt{\frac{\kappa_{i,b}}{1+\kappa_{i,b}}}\mathbf{H}_{i,b}^{LoS} + \sqrt{\frac{1}{1+\kappa_{i,b}}}\mathbf{H}_{i,b}^{NLoS}\right), \tag{4.34}$$

where $\kappa_{i,b}$ is the Rician factor related to small-scale fading and $\alpha_{i,b}$ is the path-loss exponent of the IRS-to-BS link. Specifically, the line-of-sight (LoS) component $\mathbf{H}_{i,b}^{LoS} \in \mathbb{C}^{K\times M}$ represents the ULA array response, and each element of the non-LoS (NLoS) component $\mathbf{H}_{i,b}^{NLoS} \in \mathbb{C}^{K\times M}$ follows an i.i.d. complex Gaussian distribution with zero mean and unit variance. The channel gains $\mathbf{h}_{n,i}$ and $\mathbf{h}_{i,e}$ follow the Rician distribution similar to (4.34). We assume that the CSI of all users is known by BS.

(2) *Signal receiving process*: The received signal of the nth V2I link can be given as

$$y_n = \mathbf{f}_n^H\left[\sqrt{P_n^*}(\mathbf{H}_{i,b}\mathbf{Q}\mathbf{h}_{n,i} + \mathbf{h}_{n,b})s_n + \mathbf{n}_n\right], \tag{4.35}$$

where s_n and P_n^* represent the computing task with unit energy and the transmit power of the nth served vehicle, respectively, $\mathbf{n}_n = [n_1, \ldots, n_K]^T$ denotes the noise vector with $n_k \sim N(0, \delta^2)$. Then, the SINR of the nth V2I link is given by

$$\eta_n^* = \frac{P_n^*|\mathbf{f}_n^H(\mathbf{H}_{i,b}\mathbf{Q}\mathbf{h}_{n,i} + \mathbf{h}_{n,b})|^2}{\delta^2|\mathbf{f}_n^H|^2}. \tag{4.36}$$

Similarly, the SINR of eavesdropper e wiretapping the n-th V2I link is given by

$$\eta_{e,n}^* = \frac{P_n^*|\mathbf{h}_{i,e}\mathbf{Q}\mathbf{h}_{n,i} + h_{n,e}|^2}{\delta^2}. \tag{4.37}$$

Accordingly, the achievable rate of the nth V2I link and of eavesdropper e can be denoted by $C_n^* = \log_2(1 + \eta_n^*)$ and $C_{e,n} = \log_2(1 + \eta_{e,n}^*)$, respectively. The secrecy rate of the nth V2I link can be defined as $R_S^n = \max\left\{0, C_n^* - \max_{e\in\mathcal{E}} C_{e,n}\right\}$.

4.5.1.3 Problem formulation

In the MEC network, BS will flexibly allocate computing resource *CoP* to the users. To provide timely and stable service, BS aims to minimize the MEC service time while ensuring communication security for all users. The secure service time of the nth V2I link can be roughly denoted by $t_n^S = \frac{S_n}{R_S^n} + \frac{S_n}{Cop_n}$, where S_n is the task size and Cop_n is the allocated computing resource. Intuitively, the entire MEC service period is determined by the maximum service time of the N V2I links. Thus, the goal of minimizing the overall service time delay can be translated into the following minimum–maximum problem:

$$\mathscr{P}_0: \min_{\mathbf{Q},CoP} \max_{n\in\mathcal{N}} \left\{t_n^S\right\}, \tag{4.38a}$$

$$\text{s.t.} \quad \sum_{n=1}^{N} Cop_n = CoP,\ \forall Cop_n \in (0, CoP) \tag{4.38b}$$

$$R_S^n \geq R_{min}^n,\ \forall n \in \mathcal{N}, \tag{4.38c}$$

$$Tr(FF^H) \leq P_{max} \tag{4.38d}$$

$$|e^{j\phi_m}| = 1,\ \phi_m \in [0, 2\pi),\ \forall m = 1, 2, \ldots, M \tag{4.38e}$$

where R_{min}^n is the target secrecy rate of the nth V2I link. Constraint (4.38b) indicates the limit of computing resource allocated for different computing tasks, Constraint (4.38c) is to satisfy the worst-case secrecy rate, Constraint (4.38d) is a power budget constraint, and Constraint (4.38e) is the unit modulus constraint of IRS elements.

The optimization problem is non-convex and hard to solve. Fortunately, with the strong learning and decision-making capability of DRL, we propose a DRL scheme to solve the above complicated continuous decision problem.

4.5.2 DRL-based scheme

In this section, we propose a DDPG-based DRL scheme to realize secure transmission and resource allocation in the IRS-aided MEC vehicular networks. In DRL, the IRS-aided MEC vehicular network is treated as an environment, and the BS is regarded as a learning agent to interact with the unknown environment and accumulate experience. The DDPG-based scheme is detailed as follows.

4.5.2.1 Algorithm description

The framework of the proposed DDPG algorithm has four sub-networks, as shown in Figure 4.11, including a critic network with parameters ω_q, an actor network with parameters ω_μ, a target critic network with parameters $\omega_{q'}$, and a target actor network with parameters $\omega_{\mu'}$. The DDPG algorithm uses these networks to approximate an action according to its obtained deterministic policy and to approximate the Q value of an action-state pair under the policy. To let the DRL agent learn from its interaction with the environment, the experience replay buffer D is constructed as follows. At time step t, the DDPG agent takes an action a_t under the state s_t, then it obtains a reward s_t and goes to the next state s_{t+1}. Thus, the transition tuple (s_t, a_t, r_t, s_{t+1}) is obtained and is stored in the experience replay D.

4.5.2.2 Implementation issues

The constructions of state, action, reward, and target Q value of the proposed DDPG algorithm are described below.

(1) State and environment: The current state s_t of the BS agent includes:

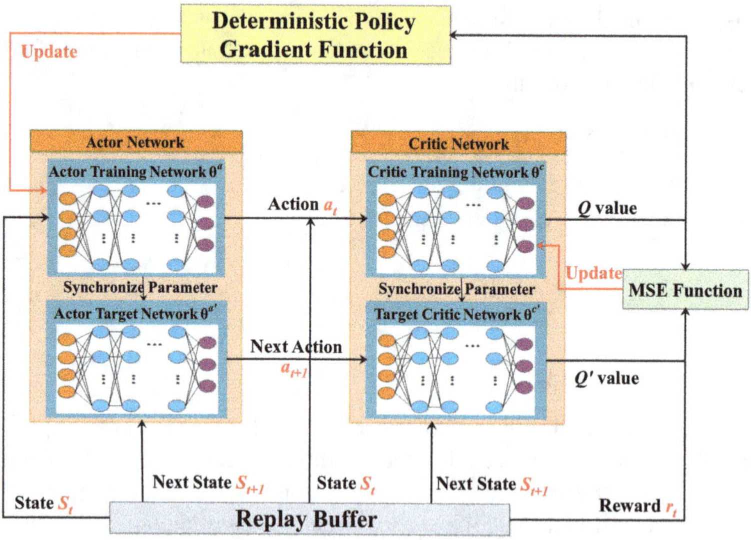

Figure 4.11 Framework of the DDPG-based scheme

- Channel information: The BS can obtain all channel information as current state. Thus, the global channel information G can be expressed by $G = \{\mathbf{h}_{n,i}, \mathbf{h}_{n,b}, \mathbf{h}_{n,e}, \mathbf{H}_{i,b}, h_{i,e}\}$.
- Offloading task size state: We define $\mathscr{K} = \{K_j, j = 1, 2, \ldots, N\}$ to indicate the remained offloading task of each vehicle.
- Rate: Denote R_{sec} and R_{trans} as the secrecy rate and transmission rate, respectively. Then, the state of the BS agent can be denoted by $\mathscr{S}_t = \{G, \mathscr{K}, R_{sec}, R_{trans}\}$.

(2) Action space: The BS agent takes a combination action a_t of $\mathbf{Q}$ and CoP, which aims to minimize the maximal MEC service delay t^S.

(3) Reward function design: In response to MEC delay minimization while maintaining secure communications for users, we adopt the estimated value of $\max_{n \in \mathcal{N}} \{t_n^S\}$ as the reward value for each step. To ensure Constraint (4.38d), we introduce a penalty factor based on the secrecy rate threshold.

4.5.2.3 Learning algorithm

Similar to most DRL algorithms, our proposed DDPG algorithm that consists of two stages, i.e., training and implement. The training process is presented in Algorithm 1. A central controller at the BS is responsible for collecting environment information and making decisions. By reasonably predicting $\mathbf{Q}$ and CoP, the proposed optimization problem can be solved by the DRL scheme.

4.5.2.4 Numerical results and discussions

In this section, we evaluate the performance of MEC vehicular networks via numerical simulations. The origin of the 3D coordinate is set at the beginning of the

Algorithm 1: DDPG-based algorithm

Randomly initialize the parameters of the critic network and actor network with ω_q and ω_μ. Set $\omega_{q'} \leftarrow \omega_q$ and $\omega_{\mu'} \leftarrow \omega_\mu$. Empty the experience replay set D with the size D_B.
for each iteration n_e **do**
  Initialize a noise distribution N for action exploration.
  observe the state s_0.
  for each training episode n_t **do**
    Obtain the action a_t from the actor network according to policy π.
    Obtain the next state s_{t+1} given the action a_t, calculate the reward r_t, and then store the transition tuple (s_t, a_t, r_t, s_{t+1}) in the experience replay D.
    Sample a mini-batch of transition tuples from D with size N_B.
    Update the parameters of the critic network ω_q and ω_μ successively.
    Update the parameters of the target critic network $\omega_{q'}$ and target actor network $\omega_{\mu'}$.
    Update the state $s_t = s_{t+1}$.
  end for
end for

first lane, and BS and the RIS are located at (−10 m, 150 m, 25 m) and (−15 m, 170 m, 15 m), respectively. The path loss exponents of the Vehicle–BS, Vehicle–Eavesdropper, Vehicle–RIS, and BS–RIS links are 3, 3, 2.2, and 2, respectively. Besides, we set $M = 30$, $K = 32$, $N = 6$, and $E = 2$. The carrier frequency is 28 GHz. Several benchmark schemes are adopted for comparison, which is elaborated as follows:

* **Without-RIS.** Vehicle users offload tasks without the deployment of RIS. DDPG is adopted to minimize the maximal MEC time by dynamically allocating MEC resources to users once they complete the task offloading.
* **R-CVX.** RIS element values are randomly generated. After vehicle users complete all task offloading, the MEC server pursues the MEC resource allocation, which is optimized by CVX toolbox (minimizing the maximal computing time is a convex optimization).

BS will face the challenge that the tasks cannot be completely offloaded all the while or it may take a very long time under malicious eavesdropping. We utilize three significant performance metrics: (1) **Average maximal MEC service time.** (2) **Success probability of MEC service.** By setting a threshold of service time $T_{\max}$ or step number $S_{\max}$, if the maximal MEC service time is smaller than it, this MEC service episode is a success, otherwise, a failure. (3) **Secrecy outage probability (SoP) of MEC service.** Security outage event occurs when user's secrecy rate is 0.

Figure 4.12 depicts the average maximal MEC service time and MEC success probability under different task sizes. By comparison, the deployment of IRS can realize dramatic performance enhancement. Malicious eavesdropping significantly

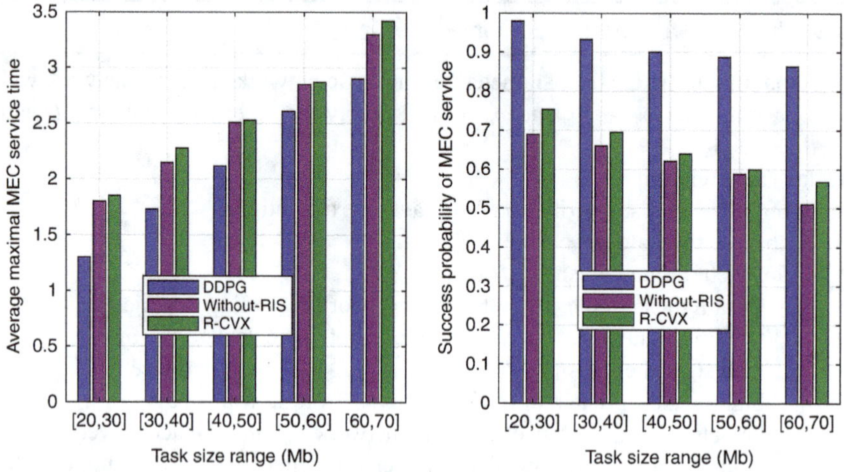

Figure 4.12 MEC service performance under different ranges of task size

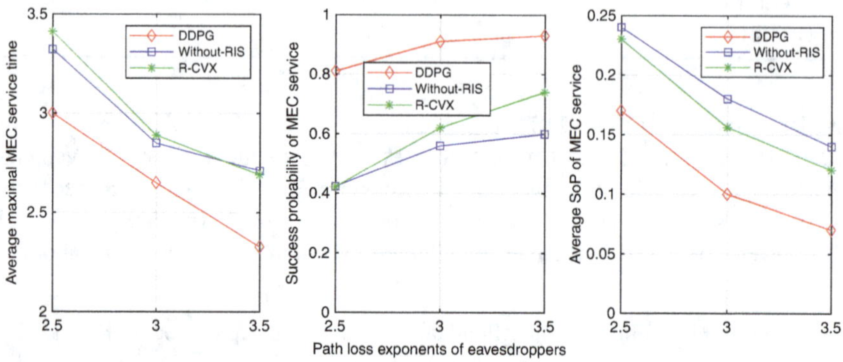

Figure 4.13 MEC service under different capabilities of eavesdropping

limits the achievable secrecy rate of vehicle users, resulting in the tasks cannot be quickly offloaded when the task is large. Fortunately, when adopting the DDPG-based scheme, the MEC success probability can be much higher than those of other two schemes, allowing more users to get secure MEC services successfully. In the left sub-figure, the difference in average MEC service time among the three schemes is not quite large when the task size continuously increases. However, the DDPG-based scheme has realized excellent decisions to optimize the overall performance. It helps many episodes successfully realize the secure MEC service, which cannot implement in the required maximal time under other two schemes. It shows that the DDPG-based scheme can thoroughly learn reasonable action strategies according to the optimization requirements in dynamic communication scenarios.

Figure 4.13 illustrates the system performance under different capabilities of eavesdropping (α_{me}). With the decrease of α_{me}, eavesdroppers can realize higher wiretap channel gain, which seriously deteriorates the overall performance of MEC service. Compared with the other two schemes, the DDPG-based scheme can dramatically reduce the average maximal MEC service time while enhancing the MEC success probability, especially under severe eavesdropping threats ($\alpha_{me} = 2.5$). A higher MEC success probability means that BS can allow more vehicle users to get secure MEC services within the specified service time threshold, which is vital for the robustness of MEC vehicular networks. Besides, the deployment of RIS brings great benefits to secure MEC services. It is because, after sufficient policy learning, DDPG can reasonably design RIS elements based on the current state information to reduce the SoP and balance the secrecy rate of all users, where the maximal task offloading time is reduced. Thus, the overall system performance gets enhanced.

4.5.3 Summary

The physical layer security issue of MEC service in IRS-aided vehicular networks is investigated. To improve the service quality of MEC, i.e., reducing the maximal service delay, a DDPG-based DRL scheme is proposed to realize the joint design of IRS elements, computing resource allocation, and receiving beamforming matrix, which can effectively reduce the service delay while ensuring the security of legitimate communications.

4.6 Conclusions and future works

In this chapter, we explore the potential of IRS to enhance PLS for various wireless networks. Specifically, we investigate the IRS-aided PLS in an MIMO communication system with and without Eve's CSI and proposed several algorithms to enhance the secrecy rate; we study the IRS-empowered APIM, which constructs a unified framework for the IRS-empowered PIMs. Upon implementing the random perturbation injection at transmitter or IRS, we propose the RPA- and RVS-based perturbation cancellation schemes to guarantee the PIM-category super-symbols with comprehensive secrecy; we examine an IRS-enabled secure ISAC system and we propose two algorithms, namely, the SDR algorithm and a low-complexity RCG algorithm, to maximize the secrecy rate while satisfying the radar detection constraint; we focus on using DL optimization methods to realize the secure MEC service in IRS-assisted vehicular networks.

The above endeavors have verified the great potential of IRS in achieving PLS for wireless communication systems. It is worth noting that, as powerful AI algorithms continue to evolve, more advanced and secure transmission mechanisms can be developed for more complicated and dynamic communication environments. With the coming of 6G era, there will be more application scenarios worth exploring where the IRS-aided PLS can be an enabling technology. In the following, we list some interesting and challenging open issues.

1. **IRS-aided secure UAV-MEC networks**: Due to the advantages of high mobility and LoS transmission, UAVs can serve as aerial MEC servers to efficiently

process urgent tasks for ground users. Nevertheless, UAV communications are heavily restrained by the dynamic open wireless propagation environments and are particularly vulnerable to eavesdropping attacks. Introducing an IRS can be beneficial for achieving reliable and secure UAV communications by flexibly and adaptively reconfiguring the wireless propagation environment. Where and how to deploy the IRS and how to jointly design the IRS reflecting coefficients, UAV trajectory, and task scheduling to support secure MEC applications are to be studied.

2. **Massive IRS-aided secure ISAC networks**: In future large-scale heterogeneous wireless networks, the security issue will become an unprecedented severe challenge, and an increasing number of IRSs can be deployed to provide higher-quality secure wireless services through continuous perception and calculation of the dynamic and time-varying wireless communication environment. A rational plan for IRS distribution from a network-wide viewpoint is the foremost problem to settle. Besides, more advanced AI algorithms such as federated learning and transfer learning can be leveraged to intelligently perceive and predict the surrounding environment and then to optimize the parameters of the multiple IRSs to make key observations exchanged and shared securely and reliably over the whole network.
3. **IRS-aided secure symbiotic radio**: IRS-aided symbiotic radio has been recognized as a promising approach for achieving high spectrum efficiency and energy efficiency for the 6G Internet of Things (IoTs). By incorporating the IRS-empowered APIM, the rate of the embedded IoT information can be further boosted. However, the IoT information is prone to eavesdropping since it is directly modulated onto the emitted RF signal. Moreover, the symbiotic relationship between the primary and IoT information makes the primary information readily inferred after the IoT information is deciphered. It is significant to design novel and enhanced IRS-APIM secure symbiotic radio schemes for 6G-IoT networks, balancing well high spectrum efficiency, energy efficiency, and security.

References

[1] Dong L and Wang HM. Enhancing secure MIMO transmission via intelligent reflecting surface. *IEEE Transactions on Wireless Communications*. 2020;19(11):7543–7556.

[2] Chen X, Zheng TX, Wen Y, *et al.* Intelligent reflecting surface aided secure dual-functional radar and communication system. In: *2022 IEEE International Conference on Communications Workshops (ICC Workshops)*; 2022. p. 975–980.

[3] Ma J, Liu C, Zheng TX, *et al.* Augmented pattern index modulation empowered by RIS. In: *ICC 2022 – IEEE International Conference on Communications*; 2022. p. 2858–2863.

[4] Wu Q and Zhang R. Towards smart and reconfigurable environment: intelligent reflecting surface aided wireless network. *IEEE Communications Magazine*. 2020;58(1):106–112.

[5] Shen H, Xu W, Gong S, *et al.* Secrecy rate maximization for intelligent reflecting surface assisted multi-antenna communications. *IEEE Communications Letters*. 2019;23(9):1488–1492.
[6] Lu X, Yang W, Guan X, *et al.* Robust and secure beamforming for intelligent reflecting surface aided mmWave MISO systems. *IEEE Wireless Communications Letters*. 2020;9(12):2068–2072.
[7] Yu X, Xu D, Sun Y, *et al.* Robust and secure wireless communications via intelligent reflecting surfaces. *IEEE Journal on Selected Areas in Communications*. 2020;38(11):2637–2652.
[8] Jiang W, Zhang Y, Wu J, *et al.* Intelligent reflecting surface assisted secure wireless communications with multiple-transmit and multiple-receive antennas. *IEEE Access*. 2020;8:86659–86673.
[9] Chu Z, Hao W, Xiao P, *et al.* Secrecy rate optimization for intelligent reflecting surface assisted MIMO system. *IEEE Transactions on Information Forensics and Security*. 2021;16:1655–1669.
[10] Hong S, Pan C, Ren H, *et al.* Artificial-noise-aided secure MIMO wireless communications via intelligent reflecting surface. *IEEE Transactions on Communications*. 2020;68(12):7851–7866.
[11] Feng B, Wu Y, Zheng M, *et al.* Large intelligent surface aided physical layer security transmission. *IEEE Transactions on Signal Processing*. 2020;68:5276–5291.
[12] Chu Z, Hao W, Xiao P, *et al.* Intelligent reflecting surface aided multi-antenna secure transmission. *IEEE Wireless Communications Letters*. 2020;9(1): 108–112.
[13] Hong S, Pan C, Ren H, *et al.* Robust transmission design for intelligent reflecting surface-aided secure communication systems with imperfect cascaded CSI. *IEEE Transactions on Wireless Communications*. 2021;20(4):2487–2501.
[14] Feng B, Wu Y, and Zheng M. Secure transmission strategy for intelligent reflecting surface enhanced wireless system. In: *2019 11th International Conference on Wireless Communications and Signal Processing (WCSP)*; 2019. p. 1–6.
[15] Yang L, Yang J, Xie W, *et al.* Secrecy performance analysis of RIS-aided wireless communication systems. *IEEE Transactions on Vehicular Technology*. 2020;69(10):12296–12300.
[16] Lai L, Hu J, Chen Y, *et al.* Directional modulation-enabled secure transmission with intelligent reflecting surface. In: *2020 IEEE 3rd International Conference on Information Communication and Signal Processing (ICICSP)*; 2020. p. 450–453.
[17] Makarfi AU, Rabie KM, Kaiwartya O, *et al.* Physical layer security in vehicular networks with reconfigurable intelligent surfaces. In: *2020 IEEE 91st Vehicular Technology Conference (VTC2020-Spring)*; 2020. p. 1–6.
[18] Wang W, Tian H, and Ni W. Secrecy performance analysis of IRS-aided UAV relay system. *IEEE Wireless Communications Letters*. 2021;10(12): 2693–2697.
[19] Tuan VP and Hong IP. Secrecy performance analysis and optimization of intelligent reflecting surface-aided indoor wireless communications. *IEEE Access*. 2020;8:109440–109452.

[20] Chen J, Liang YC, Pei Y, *et al.* Intelligent reflecting surface: a programmable wireless environment for physical layer security. *IEEE Access*. 2019;7: 82599–82612.

[21] Hu S, Wei Z, Cai Y, *et al.* Robust and secure sum-rate maximization for multiuser MISO downlink systems with self-sustainable IRS. *IEEE Transactions on Communications*. 2021;69(10):7032–7049.

[22] Kawai Y and Sugiura S. QoS-constrained energy-efficient beamforming and jamming with intelligent reflecting surface for secure multi-user downlink. *IEEE Transactions on Green Communications and Networking*. 2022;6(1):187–197.

[23] Sun S, Yang F, Song J, *et al.* Optimization on multiuser physical layer security of intelligent reflecting surface-aided VLC. *IEEE Wireless Communications Letters*. 2022;11(7):1344–1348.

[24] Li N, Li M, Liu Y, *et al.* Intelligent reflecting surface assisted NOMA with heterogeneous internal secrecy requirements. *IEEE Wireless Communications Letters*. 2021;10(5):1103–1107.

[25] Zhang Z, Lv L, Wu Q, *et al.* Robust and secure communications in intelligent reflecting surface assisted NOMA networks. *IEEE Communications Letters*. 2021;25(3):739–743.

[26] Zhang Z, Chen J, Wu Q, *et al.* Securing NOMA networks by exploiting intelligent reflecting surface. *IEEE Transactions on Communications*. 2022;70(2):1096–1111.

[27] Wang W, Liu X, Tang J, *et al.* Beamforming and jamming optimization for IRS-aided secure NOMA networks. *IEEE Transactions on Wireless Communications*. 2022;21(3):1557–1569.

[28] Wen M, Zheng B, Kim KJ, *et al.* A survey on spatial modulation in emerging wireless systems: research progresses and applications. *IEEE Journal on Selected Areas in Communications*. 2019;37(9):1949–1972.

[29] Li Q, Wen M, Wang S, *et al.* Space shift keying with reconfigurable intelligent surfaces: phase configuration designs and performance analysis. *IEEE Open Journal of the Communications Society*. 2021;2:322–333.

[30] Basar E. Reconfigurable intelligent surface-based index modulation: a new beyond MIMO paradigm for 6G. *IEEE Transactions on Communications*. 2020;68(5):3187–3196.

[31] Ma T, Xiao Y, Lei X, *et al.* Large intelligent surface assisted wireless communications with spatial modulation and antenna selection. *IEEE Journal on Selected Areas in Communications*. 2020;38(11):2562–2574.

[32] Yan W, Yuan X, and Kuai X. Passive beamforming and information transfer via large intelligent surface. *IEEE Wireless Communications Letters*. 2020;9(4):533–537.

[33] Shu F, Teng Y, Li J, *et al.* Enhanced secrecy rate maximization for directional modulation networks via IRS. *IEEE Transactions on Communications*. 2021;69(12):8388–8401.

[34] Fang S, Chen G, Xu P, *et al.* SINR maximization for RIS-assisted secure dual-function radar communication systems. In: *2021 IEEE Global Communications Conference (GLOBECOM)*; 2021. p. 01–06.

[35] Mishra KV, Chattopadhyay A, Acharjee SS, *et al.* Optm3sec: optimizing multicast Irs-aided multiantenna Dfrc secrecy channel with multiple eavesdroppers. In: *ICASSP 2022–022 IEEE International Conference on Acoustics, Speech and Signal Processing (ICASSP)*; 2022. p. 9037–9041.

[36] Yang H, Alphones A, Xiong Z, *et al.* Artificial-intelligence-enabled intelligent 6G networks. *IEEE Network*. 2020;34(6):272–280.

[37] Song Y, Khandaker MRA, Tariq F, *et al.* Truly intelligent reflecting surface-aided secure communication using deep learning. In: *2021 IEEE 93rd Vehicular Technology Conference (VTC2021-Spring)*; 2021. p. 1–6.

[38] Yang H, Xiong Z, Zhao J, *et al.* Deep reinforcement learning-based intelligent reflecting surface for secure wireless communications. *IEEE Transactions on Wireless Communications*. 2021;20(1):375–388.

[39] Yang H, Xiong Z, Zhao J, *et al.* Intelligent reflecting surface assisted anti-jamming communications based on reinforcement learning. In: *GLOBECOM 2020–2020 IEEE Global Communications Conference*; 2020. p. 1–6.

[40] Huang C, Chen G, and Wong KK. Multi-agent reinforcement learning-based buffer-aided relay selection in IRS-assisted secure cooperative networks. *IEEE Transactions on Information Forensics and Security*. 2021;16:4101–4112.

[41] Khisti A and Wornell GW. Secure transmission with multiple antennas—Part II: the MIMOME wiretap channel. *IEEE Transactions on Information Theory*. 2010;56(11):5515–5532.

[42] Boumal N, Mishra B, Absil P, *et al.* Manopt, a MATLAB toolbox for optimization on manifolds. *Journal of Machine Learning Research*. 2014;15:1455–1459.

[43] Absil PA, Mahony R, and Sepulchre R. Optimization Algorithms on Matrix Manifolds. Princeton University Press, USA. 2008.

[44] Liang L, Ye H, and Li GY. Spectrum sharing in vehicular networks based on multi-agent reinforcement learning. *IEEE Journal on Selected Areas in Communications*. 2019;37(10):2282–2292.

Part II

Authentication using channel features and hardware impairments

Chapter 5
Intelligent physical layer authentication*

Yun Ma[1], *He Fang*[1], *Xianbin Wang*[2] *and Li Xu*[3]

Physical layer authentication technology has been extensively studied for secure communications. The essence of this technology is to make full use of the state characteristics of wireless channels and devices to resist the eavesdropping and spoofing attacks, realizing the protection for legitimate communications. However, the performance of physical layer authentication suffers from the time variations and imperfect estimations of channel-related attributes. The commonly adopted static hypothesis testing for physical layer authentication also faces significant challenges in the time-varying communication environment due to the dynamic propagation and interference conditions.

To circumvent these impediments, this chapter overviews the physical layer authentication technique and its challenges first. Two adaptive multi-dimensional physical layer authentication schemes are proposed based on artificial intelligence (AI) techniques. To be more specific, a non-parametric model based on the back-propagation neural network (BPNN) is developed first to learn and fuse multiple time-varying attributes, and, hence, to improve the reliability of physical layer authentication. Then, a parametric model based on the fuzzy learning is proposed to further decrease the data required and model parameters to be determined in the intelligent authentication process in the cases that some statistical properties of the attributes are available and intermittent availability of some attributes are considered.

5.1 Introduction

The next-generation networks connect billions of things that include sensors, actuators, services, and other Internet-connected objects, enabling the future smart life and connected industries with Internet-of-Things (IoT) applications, as exemplified by smart cities, intelligent transportation, and smart building [2–4]. Wireless communications are extremely vulnerable to interception and spoofing attacks, mainly due to

*This chapter has been a part of [1].

[1]School of Electronic and Information Engineering, Soochow University, China
[2]The Department of Electrical and Computer Engineering, Western University, Canada
[3]The College of Computer and Cyber Security and the Fujian Provincial Key Laboratory of Network Security and Cryptology, Fujian Normal University, China

the open broadcast nature of radio signal propagation, intermittent communications, standardized transmission schemes used, and significant increase of low-cost wireless devices [5,6]. In the context of increasing convergence of wireless infrastructure and vertical industrial applications brought about by IoT, authentication is paramount to secure legitimate communications by confirming the identity of all devices.

Although traditional cryptography methods have been widely used, they may be limited in meeting the expected performance of high reliability, low latency, enhanced mobile broadband, and massive machine-like communications in many emerging scenarios [7–9]. It is mainly because that the time-consuming processes of secret key management, i.e., generation, distribution, update, and revocation of secret keys, and the abundant resources are required for storing the keys [10–12]. In addition, with the significant increase in arithmetic power of devices and the development of quantum computation, the risk of secret key cracking is dramatically increasing. More importantly, traditional authentication mechanisms relying on cryptography and security protocols ignore the inherent physical characteristics of communication devices and users, thus cannot easily detect compromised secret keys.

Physical layer security extends the authentication from the higher protocol layer to the bottom layer, making full use of the uniqueness and randomness of the communication channels and hardware [13–19]. The physical layer attributes provide unique distinguishing characteristics for authenticating the wireless communication devices. The physical layer attributes studied for authentication include the channel impulse response (CIR) [20], received signal strength indicator (RSSI) [21], carrier frequency offset (CFO) [22], in-phase and quadrature imbalance (IQI) [23], and so on. These physical layer attributes can also be used to generate unique combinations for authentication. At the same time, due to the natural refreshing of the physical layer attributes, it provides higher security variation than the conventional scheme.

However, despite these advantages over traditional authentication schemes, physical layer authentication inevitably suffers from several other challenges. As shown in Figure 5.1, imperfect estimation and variation of physical layer attributes are inevitable due to time-varying channels, dynamic interference conditions, device mobility, asymmetric observations of transmitters and receivers, and measurement errors [5,6,24]. Although the time variations of physical layer attributes provide further randomness and natural renew of authentication information to increase the difficulty of being predicted, they also pose challenges for authentication at legitimate devices. The authentication reliability suffers from the imperfect estimation and variation of physical layer attributes, especially in the complex dynamic communication environment.

Moreover, the performance of single-attribute-based physical layer authentication schemes is limited not only by the imperfect estimation of the attributes used but also by the fact that the limited range of the distribution of specific attributes may not be sufficiently wide spread to distinguish the devices all the time. These make it necessary to use a combination of multiple physical layer attributes for authentication to achieve higher reliability. Utilizing multiple physical layer attributes is capable of achieving higher security in the presence of spoofing attacks, but increases the computation overhead imposed on the legitimate users as well. Not only that, due

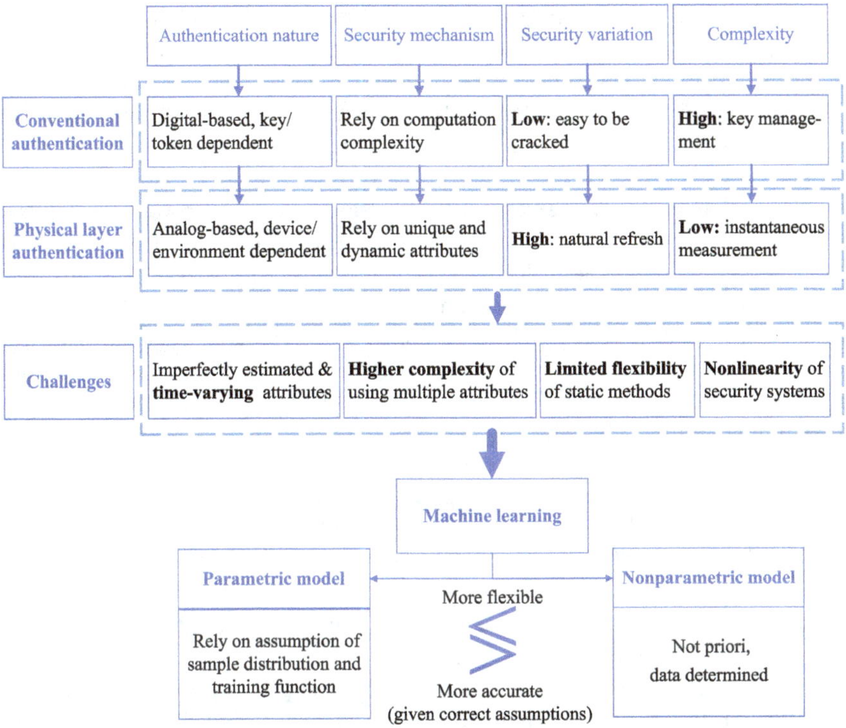

Figure 5.1 Comparison of conventional and physical authentication techniques, challenges of physical layer authentication, and machine learning-based solutions.

to the unknown variations of the physical layer attributes at the design stage, it is very difficult to predesign a static physical layer authentication scheme. Hence, the conception of adaptive multi-dimensional physical layer authentication is extremely helpful for enhancing the security, which can promptly fuse multiple attributes and adapt to the complex time-varying environment.

The timely detection and fusion of multiple time-varying physical layer attributes as well as the adaptation of the authentication process require sophisticated near-instantaneously adaptive processing techniques. To meet the expected performance of the next-generation networks, multiple objectives should be also considered, including high cost-efficiency, high reliability, model independence, continuous protection, and situation awareness. What's more, the flexible designs of security management are extremely helpful in providing automatic authentication in different communication scenarios. The machine-learning techniques [25,26] provide promising methods to design intelligent physical layer authentication schemes for achieving the above objectives. To be more specific, machine-learning techniques are able to learn the dynamic adversarial environment for self-optimization and self-organization, thereafter achieving automatic authentication. More importantly, the machine-learning techniques could help in the detection and fusion of time-varying attributes as well as the adaptation of the authentication process [27,28]. Hence, by exploring the machine-learning techniques, we introduce intelligence to authentication in the complex time-varying environment under unknown network conditions and unpredictable dynamics, supporting radically new applications of next-generation wireless networks.

In this chapter, we utilize the non-parametric learning and parametric learning methods for better design of physical layer authentication to overcome the mentioned challenges. The non-parametric learning methods are not specified a priori, but determined from the available data [29]. Examples include kernel estimator [6], k-nearest neighbors [30], decision trees [31], and so on. This kind of method dynamically learns from the time-varying environment without requiring any assumption about the training model. It provides greater flexibility for intelligent authentication in the complex dynamic environment. However, it requires more training data than the parametric learning methods and may lead to overfitting result. In contrast to non-parametric learning methods, parametric models typically rely on the assumption that the sample distribution is known (i.e., the physical layer attribute estimations) and on a particular form of the training function, such as linear regression and polynomial regression functions [32–34]. Hence, parametric learning methods can achieve high correct rates with less training data when the assumptions are correct, but higher probability of failure when the assumptions do not match.

This chapter first presents the system model and physical layer authentication problem in Section 5.2. Then, a non-parametric model is designed to utilize and fuse multiple physical layer attributes based on the BPNN to improve the authentication reliability in the time-varying communication environment in Section 5.3. To further decrease the training data and parameters to be determined, a fuzzy learning (FL)-based physical layer authentication scheme is developed in Section 5.4. Finally, Section 5.5 concludes this chapter and provides our future works.

In this chapter, matrices and vectors are represented by bold letters and scalars are non-bolded letters, respectively. Whatsmore, $(\cdot)^{\mathrm{T}}$ represents the operations of taking the transpose. A summary of the abbreviations used is given in Table 5.1.

Table 5.1 Table of abbreviations

Abbreviation	**Full name**
FA	False alarm
MD	Misdetection
FL	Fuzzy learning
BP	Back propagation
CFO	Carrier frequency offset
CIR	Channel impulse response
RSSI	Received signal strength indicator
IQI	In-phase and quadrature imbalance

5.2 System model

Figure 5.2 shows an adversarial wireless communication system, where two legitimate devices, i.e., Alice and Bob, communicate with each other in the presence of an eavesdropper, i.e., Eve. Eve intends to intercept and impersonate Alice, and then to send spoofing signals to obtain illegal advantages from the system. Bob's main objective is to uniquely and unambiguously identify the transmitter by the physical layer authentication technique. The basic physical layer authentication aims for supporting this pair of legitimate devices by a reciprocal wireless link, while the device/channel-dependent attributes can be used as a unique security signature.

Different physical layer attributes can characterize the attributes of different devices, environments, and channels, and carry different security information. Using different physical layer attributes for authentication may have different authentication effects. Again, a combination of multiple attributes can be used for improving the authentication performance, since it is more difficult for Eve to simultaneously infer multiple attributes of a large search-space from the received signal. We express N features used for physical layer authentication as $\boldsymbol{H} = (H_1, H_2, \ldots, H_N)^{\mathrm{T}}$, such as channel impulse response (CIR), carrier frequency offset (CFO), received signal strength indicator (RSSI), round trip time (RTT), and in-phase and quadrature imbalance (IQI). Naturally, the various combinations of physical layer attributes provide a high grade of uncertainty for the adversaries and simultaneously improve multidimensional protection for the legitimate devices.

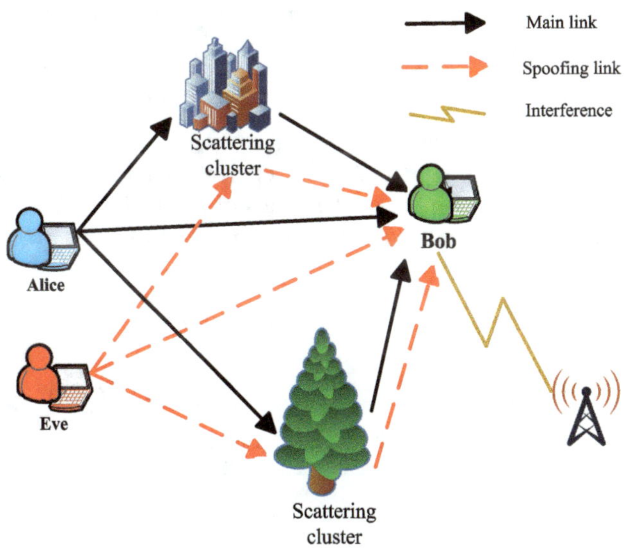

Figure 5.2 System model. Bob authenticates its transmitter based on the physical layer authentication technique

In order to make the experiment closer to the actual situation, we stipulate the following important assumptions:

1. The physical signals transmitted between a pair of legitimate devices rapidly become decorrelated in space, time, and frequency. This implies that it is hard for Eve to observe and predict the channel state between legitimate devices if Eve is at a third location, which is farther than a wavelength away from Alice and Bob.
2. Both the wireless channels and the interferences are time-varying, the devices are moving, thus the wireless environment is dynamically changing. These all lead to time variations of the physical layer attributes.
3. The estimates of the physical layer attributes are imperfect because the legitimate devices suffer from different interferences in a dynamic propagation environment when the devices roam in different locations.

These assumptions characterize a practical scenario, but naturally, it is more difficult for us to deal with these imperfectly estimated time-varying physical layer attributes.

As shown in Figure 5.3, the physical layer authentication comprises two phases, as described below:

Phase I: Alice broadcasts one or more messages to Bob at time t. From the received signal, Bob infers an imperfect estimate of the multiple attributes as

$$\boldsymbol{H}_A^I[t] = (H_{A1}^I[t], H_{A2}^I[t], \ldots, H_{AN}^I[t])^{\mathrm{T}}, \tag{5.1}$$

which are associated with Alice. Meanwhile, Eve overhears the transmission.

Phase II: Either Alice or Eve transmits a message to Bob at time $t+\tau$. Then Bob obtains another imperfect estimate as

$$\boldsymbol{H}^{II}[t+\tau] = (H_1^{II}[t+\tau], H_2^{II}[t+\tau], \ldots, H_N^{II}[t+\tau])^{\mathrm{T}}, \tag{5.2}$$

where τ represents the time interval between the two phases.

Bob should compare $\boldsymbol{H}_A^I[t]$ with $\boldsymbol{H}^{II}[t+\tau]$. If these two estimates are likely to come from the same channel implementation and the same imperfect hardware, the message at time $t+\tau$ is considered to be from Alice.

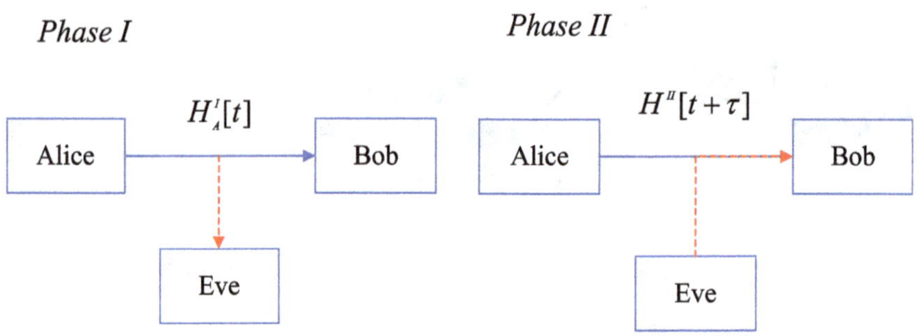

Figure 5.3 System phase diagram of physical layer authentication

An intelligent function $F(\cdot)$ is conceived in this chapter, which is used for fusing N independent physical layer attributes. It can be formulated as

$$\boldsymbol{y} = (y_1, y_2, \ldots, y_M)^{\mathrm{T}} = F(\boldsymbol{H}^{II} - \boldsymbol{H}_A^{I}), \tag{5.3}$$

where $\boldsymbol{y}$ is the output vector of the intelligent fusion of N physical layer attributes. After getting the output $\boldsymbol{y}$ of the physical layer attribute fusion, we design an intelligent authentication function $G(\cdot)$ to preprocess it as follows:

$$G(\boldsymbol{y}) = G(y_1, y_2, \ldots, y_M) = \theta_1 y_1 + \theta_2 y_2 + \cdots + \theta_M y_M + C, \tag{5.4}$$

where θ_m is the coefficient of mth output element of $\boldsymbol{y}$ and C is a bias. The intelligent authentication function $G(\cdot)$ containing θ_m and C is designed to enable adaptive authentication in the dynamic environment.

Then, the physical layer authentication can be expressed as

$$\begin{cases} \Phi_0 : G(\boldsymbol{y}) \leq 0 \\ \Phi_1 : G(\boldsymbol{y}) > 0 \end{cases}, \tag{5.5}$$

where Φ_0 indicates that the signal is from Alice, while Φ_1 indicates that it is from Eve. Due to the variations and imperfect estimates of the physical layer attributes, both false alarms and misdetections are encountered. Therefore, the parameters in $F(\cdot)$ should be promptly updated to achieve low false alarm rate and misdetection rate in the time-varying environment.

The objective of this chapter is to propose intelligent authentication schemes relying on multiple physical layer attributes to increase the authentication reliability and to achieve situation awareness in the complex dynamic environment. The proposed schemes aim at achieving reliable and robust authentication through discovering and learning the complex operating environment. Two physical layer authentication schemes will be proposed for specific communication scenarios, i.e., back propagation neural network-based scheme and fuzzy learning-based scheme, which are nonparametric method and parametric method, respectively.

5.3 BPNN-based intelligent physical layer authentication: a nonparametric method

This section proposes an attribute fusion model based on the BP neural network, which makes full use of multiple time-varying physical layer attributes, thereby improving the performance of authentication schemes in the time-varying environment. A BP neural network-based model is developed to fuse multiple physical layer attributes for authentication without needing to know their statistical characteristics. More importantly, the parameters of the designed authentication model are intelligently updated through the situation awareness after each authentication round to further perform the adaptive authentication for improving the reliability of physical layer authentication scheme in the time-varying environment.

5.3.1 BPNN-based multiple physical layer attribute fusion

The BP neural network has good nonlinear mapping ability, self-learning adaptive ability, and generalization ability [35–37]. The nonlinear mapping ability can realize the fitting from physical layer attributes to Alice and Eve's identity; the self-learning and adaptive ability can update training process intelligently; the good generalization ability increases the practical feasibility of the authentication system. Hence, the BP neural network is utilized for fusing the multiple physical layer attributes.

In the BP neural network-based multiple attribute fusion, the difference between estimates of Phase I and Phase II, i.e., $\boldsymbol{H}_A^I[t]$ and $\boldsymbol{H}^{II}[t+\tau]$, is denoted as $\boldsymbol{h} = (h_1, h_2, \ldots, h_N)^{\mathrm{T}}$. Each element $h_n \in [a_n, b_n]$ is formulated as

$$h_n = H_n^{II}[t+\tau] - H_{An}^I[t], n = 1, 2, \ldots, N. \tag{5.6}$$

The extracted physical layer attributes have different units and large numerical differences. According to the characteristics of the network, if they are directly input into the network, the weights will become extremely large after passing through the accumulator, which makes it difficult for the network to converge. Therefore, it is necessary to normalize the input vector of the network. In the following, the attributes having ranges $[a_n, b_n]$, $n = 1, 2, \ldots, N$, are normalized to $[-1, 1]$ by invoking

$$\widetilde{h}_n = \frac{2}{b_n - a_n}\left(h_n - \frac{a_n + b_n}{2}\right), n = 1, 2, \ldots, N. \tag{5.7}$$

The property estimates are normalized to the range of $[-1, 1]$ by the two equations (5.6) and (5.7), so that the model can handle multiple physical layer properties in the same range. In practical systems, we only need to know the approximate range of variation of the property. This is reasonable because we can always have a detailed understanding of the communication system and environment before designing an authentication system.

Let us assume that a set of observations $(\widetilde{\boldsymbol{h}}_l, \bar{\boldsymbol{y}}_l)_{l=1}^L \in [-1, 1]^N \times \{[1, 0]^{\mathrm{T}}, [0, 1]^{\mathrm{T}}\}$ is given, which is used for training the authentication process, where $\widetilde{\boldsymbol{h}}_l = (\widetilde{h}_{1l}, \widetilde{h}_{2l}, \ldots, \widetilde{h}_{Nl})^{\mathrm{T}}$ is the lth estimate after the normalization, with each element $\widetilde{h}_l$ is defined in (5.7). $\bar{\boldsymbol{y}}_l = [\bar{y}_1, \bar{y}_2]^{\mathrm{T}} \in \{[1, 0]^{\mathrm{T}}, [0, 1]^{\mathrm{T}}\}$ is the one-hot code for Alice and Eve, which can be defined as

$$\bar{\boldsymbol{y}}_l = \begin{cases} [0, 1]^{\mathrm{T}} & \Phi_0 \\ [1, 0]^{\mathrm{T}} & \Phi_1 \end{cases}. \tag{5.8}$$

As shown in Figure 5.4, the normalized estimates $(\widetilde{\boldsymbol{h}}_l)$, $l = 1, 2, \ldots, L$, are considered as the inputs of the BP neural network-based multiple attribute fusion, and $F(\widetilde{\boldsymbol{h}}_l)$ represents the outputs of the BP neural network-based multiple attribute fusion with the corresponding inputs given by $\widetilde{\boldsymbol{h}}_l \in [-1, 1]^N, l = 1, 2, \ldots, L$. Note that for the legitimate users, the training data of a legitimate communication session is relatively straightforward to obtain.

The task is then to infer the underlying mapping function $\widetilde{\boldsymbol{y}} = F(\widetilde{\boldsymbol{h}}_l)$ from the training data set (the samples), i.e., $(\widetilde{\boldsymbol{h}}_l, \bar{\boldsymbol{y}}_l)_{l=1}^L \in [-1, 1]^N \times \{[1, 0]^{\mathrm{T}}, [0, 1]^{\mathrm{T}}\}$. In other words, the task in this section is to represent the authentication system $\widetilde{\boldsymbol{y}} = F(\widetilde{\boldsymbol{h}}_l)$,

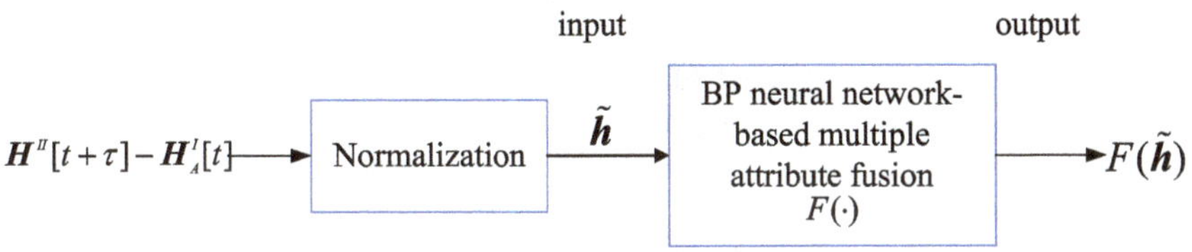

Figure 5.4 The process of the BP neural network-based multiple attribute fusion

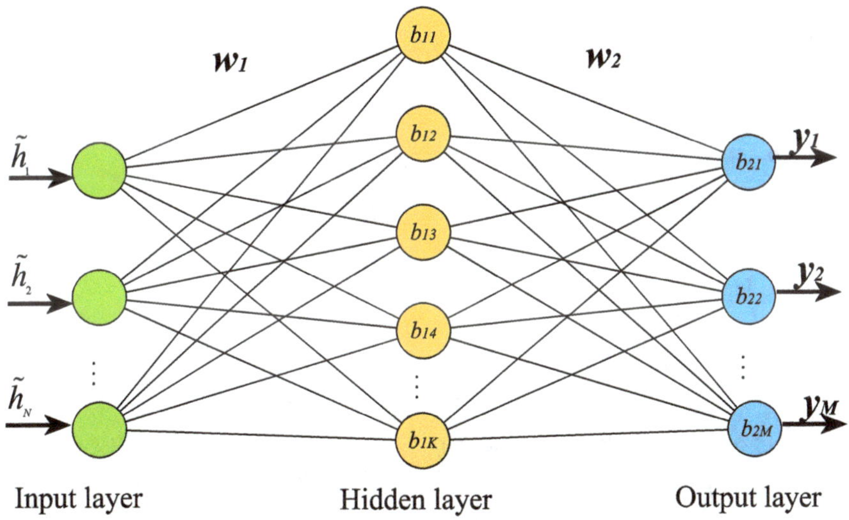

Figure 5.5 BP neural network-based multiple attribute fusion model

and to model the relationship between the estimates of multiple attributes and the corresponding authentication results. After this, once the new normalized estimate $\bar{\boldsymbol{h}}_l = (\bar{h}_1, \bar{h}_2, \ldots, \bar{h}_N)^{\mathrm{T}}$ is obtained, it is possible to verify whether the authenticating transmitter is Alice or Eve.

The designed physical layer attribute fusion model based on BP neural network is shown in Figure 5.5. This model consists of an input layer, a hidden layer, and an output layer. Where $\bar{\boldsymbol{h}}_l = (\bar{h}_1, \bar{h}_2, \ldots, \bar{h}_N)^{\mathrm{T}}$ is the normalized input vector. An inequality of the physical layer attribute fusion model can be shown as:

$$K < \sqrt{N+M} + a, \tag{5.9}$$

where M is the number of output layer nodes, a is a constant between 0 and 10, and K is the number of hidden layer nodes.

If the number of selected hidden layer nodes is low, the model generalization ability will be low as well; while the higher number will lead to overfitting. Hence, the numbers of both hidden layers and hidden layer nodes in the designed model need to be adjusted according to the actual situation and performance requirements.

Given the number of hidden layer nodes, the number of output neurons is determined according to the actual requirements. In the model designed in this section, the input $\widetilde{\boldsymbol{h}}$ corresponds to the output $\boldsymbol{y} = (y_1, y_2, \ldots, y_M)^{\mathrm{T}}$. The output is a fit to the one-hot code of Alice and Eve.

In the one-layer hidden layer BPNN fusion model, the input weight of the k-th node of the hidden layer is designed as $\boldsymbol{w}_{1k} = (w_{1k1}, w_{1k2}, \ldots, w_{1kN})$ and the corresponding bias is b_{1k}. The weights and biases of the hidden layer can be expressed as $\boldsymbol{W}_1 = (\boldsymbol{w}_{11}, \boldsymbol{w}_{12}, \ldots, \boldsymbol{w}_{1K})^{\mathrm{T}}_{K\times N}$ and $\boldsymbol{b}_1 = (b_{11}, b_{12}, \ldots, b_{1K})^{\mathrm{T}}$, respectively. The input weight of the mth node of the output layer is $\boldsymbol{w}_{2m} = (w_{2m1}, w_{2m2}, \ldots, w_{2mK})$ and the corresponding bias is b_{2k}, and the weights and biases of the output layer can be expressed as $\boldsymbol{W}_2 = (\boldsymbol{w}_{21}, \boldsymbol{w}_{22}, \ldots, \boldsymbol{w}_{2M})^{\mathrm{T}}_{M\times K}$ and $\boldsymbol{b}_2 = (b_{21}, b_{22}, \ldots, b_{2M})^{\mathrm{T}}$, respectively.

Then, the sigmod function is chosen as the activation function of the hidden layer of the BP neural network, which is shown as

$$f(x) = \frac{1}{1 + e^{-x}}. \tag{5.10}$$

The kth hidden layer neuron input is

$$d_k = \sum_{i=1}^{N} \widetilde{h}_k w_{1ki}, \tag{5.11}$$

The output of the kth hidden layer neuron is

$$\beta_k = f(d_k + b_{1k}). \tag{5.12}$$

The output layer activation function is chosen as a constant function, which is expressed as

$$g(x) = x. \tag{5.13}$$

Then the input of the mth output layer neuron is formulated as

$$\gamma_m = \sum_{i=1}^{K} \beta_k w_{2mi}, \tag{5.14}$$

The output of the mth output layer neuron is expressed as

$$y_m = \gamma_m + b_{2m}. \tag{5.15}$$

Assuming that the output of the developed model and the desired output corresponding to input $\widetilde{\boldsymbol{h}}$ are $\boldsymbol{y} = (y_1, y_2, \ldots, y_M)^{\mathrm{T}}$ and $\bar{\boldsymbol{y}} = (\bar{y}_1, \bar{y}_2, \ldots, \bar{y}_M)^{\mathrm{T}}$, the mean square error of the multiple attribute fusion can be obtained as

$$E = \frac{1}{M} \sum_{i=1}^{M} (y_i - \bar{y}_i)^2. \tag{5.16}$$

To achieve accurate multiple attribute fusion, we apply the error back propagation algorithm [36,38] to update the weights and biases. After obtaining the mean square error E on the output layer, the weights and biases can be updated via

$$\Delta z_i = -\mu \frac{\partial}{\partial z_i} E, \tag{5.17}$$

where $\mu \in [0, 1]$ is the learning rate. The larger learning rate μ will result in greater modification of the weights and biases, and in faster learning speed of the neural network. However, a large μ will also lead to the oscillation of learning process and converge slower, and even result in divergence of learning when it is larger than 1. When the learning rate μ is too small, the convergence speed will be seriously slow and the weights will hardly converge to stability.

Therefore, it is also critical to select an appropriate learning rate μ. After the learning rate μ is determined, the bias of the weights of the BP neural network is updated via (5.17), and the mean square error is finally reduced to complete the fitting. The proposed BP neural network-based multi-physical layer feature fusion process can be summarized as Algorithm 1.

The estimates of the multiple physical layer attributes H are time-variant, which may lead to a low authentication reliability without adaptation. Therefore, next subsection focuses on proposing adaptive learning procedures for promptly adjusting the authentication system.

Algorithm 1: BP neural network-based physical layer attribute fusion

---

Input: samples $\left(\widetilde{\boldsymbol{h}}_l, \bar{\boldsymbol{y}}_l\right)_{l=1}^{L} \in [-1, 1]^N \times \{[1, 0]^{\mathrm{T}}, [0, 1]^{\mathrm{T}}\}$
Output: trained model weights

Initialization:
give the initial value of the weights W_1, W_2 and biases b_1, b_2
Iteration:
while samples $\left(\widetilde{\boldsymbol{h}}_l, \bar{\boldsymbol{y}}_l\right)_{l=1}^{L} \in [-1, 1]^N \times \{[1, 0]^{\mathrm{T}}, [0, 1]^{\mathrm{T}}\}$ available **do**
Hidden layer output calculation. Obtain the hidden layer output $\boldsymbol{\beta}$ via (5.10)-(5.12).
Output layer calculation. The multiple attribute fusion output y can be obtained via (5.15).
Error calculation. The mean square error of the multiple attribute fusion E is obtained via (5.16).
Weight and bias updates. Update the weights and biases of the multiple attribute fusion model via (5.17).
$l = l + 1$.
end while

5.3.2 Adaptive authentication process based on BPNN

In this section, a learning procedure is proposed for adaptive authentication based on the BPNN, which depends on the observed time-varying samples $(\widetilde{\boldsymbol{h}}_l[t], \bar{\boldsymbol{y}}_l[t])_{l=1}^{L} \in [-1,1]^N \times \{[1,0]^{\mathrm{T}}, [0,1]^{\mathrm{T}}\}$. Explicitly, the proposed learning procedure can be viewed as an intelligent process of learning the time-varying environment for updating the system parameters to achieve a reliable and robust authentication.

After the model training in Section 5.3.1, we obtained a BPNN-based physical layer attribute fusion model, and the next thing we need to do is to perform physical layer authentication and further update the model based on the obtained legal feature to achieve better authentication performance.

According to (5.4) and (5.5), the adaptive authentication process corresponding to the two-output fusion model is formulated as

$$\begin{cases} \Phi_0 : \theta_1 y_1 + \theta_2 y_2 + C \le 0 \\ \Phi_1 : \theta_1 y_1 + \theta_2 y_2 + C > 0 \end{cases}. \tag{5.18}$$

Then, combining (5.18) and the one-hot code used, we can obtain a simple authentication expression as

$$\begin{cases} \Phi_0 : y_1 < y_2 \\ \Phi_1 : y_1 > y_2 \end{cases}. \tag{5.19}$$

The intelligent authentication process can be divided into five steps: attribute extraction, attribute pre-processing, attribute fusion, authentication, and learning sample update, where the first three steps have been processed in Section 5.3.1. In the identity authentication step, the output result of the designed multiple attribute fusion model can be obtained via (5.19). If the authentication result is Alice, this multi-dimensional estimation will be saved as the reference training sample for the next round of authentication. If the authentication result is Eve, this authentication is directly ended and the communication request from this device will be denied. The proposed adaptive physical layer authentication process can be summarized in Algorithm 2.

The flow chart of the BP neural network-based intelligent physical layer authentication scheme is shown in Figure 5.6. The intelligent authentication process requires a recursive execution of the authentication step, since each authentication round uses the difference between the extracted physical layer attribute estimation and the latest. Through the situation-awareness of the time-varying environment, the proposed intelligent physical layer authentication scheme can maintain good authentication accuracy.

5.3.3 Simulation results

In order to evaluate the performance of the proposed BP neural network-based intelligent physical layer authentication scheme, this subsection analyzes the effects of different physical layer features, different learning rates, different number of hidden layer units, different motion states, and different intensity noise on its performance.

Algorithm 2: Adaptive authentication process based on the BPNN

1: For each round of authentication at time t :
2: **Attribute extraction.** Extracts the selected attributes $\boldsymbol{H}[t] = (h_1[t], h_2[t], \ldots, h_N[t])^{\mathrm{T}}$ of the received signal from the transmitter.
3: **Attribute pre-processing.** Obtain the normalized physical layer attribute estimations $\widetilde{\boldsymbol{h}}_l[t] = \left(\widetilde{h_{1l}}[t], \widetilde{h_{2l}}[t], \ldots, \widetilde{h_{Nl}}[t]\right)^{\mathrm{T}}$ via (5.7).
4: **Attribute fusion.** Obtain the fitting result $y[t]$ via (5.10)–(5.15).
5: **Authentication.** Authentication result can be made via (5.19). If the result is Eve, authentication is directly ended and the communication request from this device will be denied.
6: **Learning sample update.** Save the latest estimation $\boldsymbol{H}[t]$ of the transmitter, which has been authenticated as Alice.

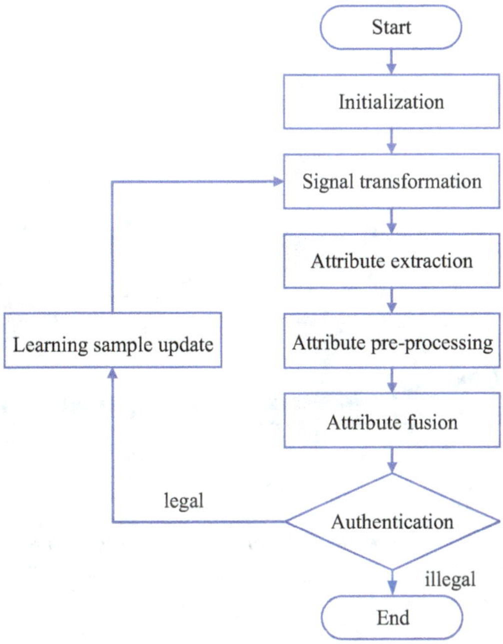

Figure 5.6 Flow chart of the proposed intelligent physical layer authentication scheme based on the BP neural network

The channel model in this section is built using the WINNER II module by MATLAB®. In this section, the proposed scheme is studied in an urban scenario, and the simulation parameters in this case study are given in Table 5.2.

As shown in Figure 5.7, the channel impulse response (CIR) and received signal strength indicator (RSSI) are utilized in the proposed scheme. The learning rate of all

Table 5.2 Simulation parameters of Section 5.3

Parameter	Value	Parameter	Value
Initial position of Alice	[257;274] m	Scenario	Non-line-of-sight
Initial position of Eve	[228;115] m	Center frequency	5.25 GHz
Initial position of Bob	[150;150] m	Sampling number per frame	1600

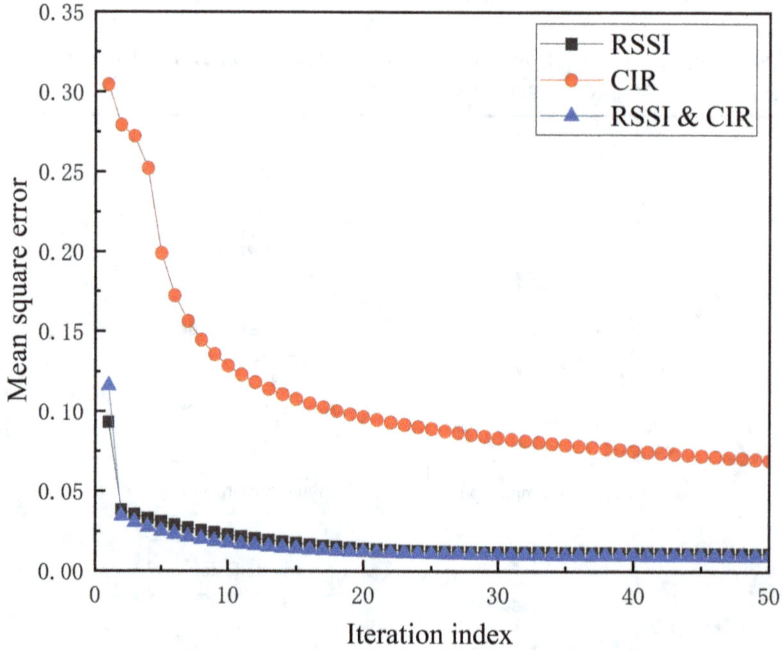

Figure 5.7 Training performance of the proposed BP neural network-based scheme utilizing different physical layer attributes, i.e., CIR, RSSI, and CIR & RSSI

three schemes is set to $\mu = 0.2$. Given 500 CIR and RSSI samples of Alice, the mean square error of learning decreases significantly as the number of iterations increases from 0 to 50. It can be also observed from Figure 5.7 that the intelligent authentication process using CIR & RSSI results in the best training performance compared to that of using CIR or RSSI separately. Moreover, the proposed scheme relying on CIR has the slowest convergence speed and the highest steady-state value of convergence. The mean square error values of the proposed scheme utilizing RSSI and that of utilizing CIR & RSSI decrease dramatically after two iterations. As a conclusion, the

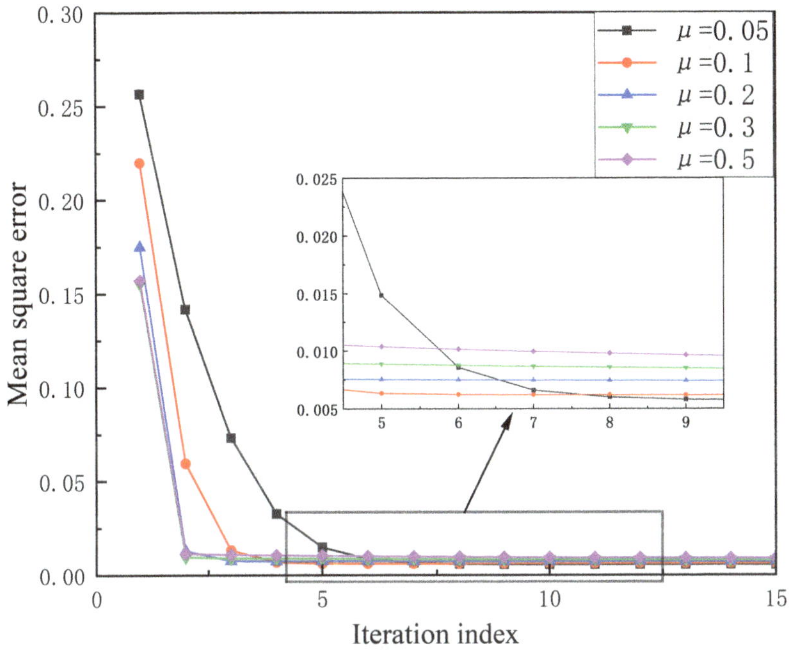

Figure 5.8 Effect of different learning rates ($\mu = 0.05$, $\mu = 0.1$, $\mu = 0.2$, $\mu = 0.3$, $\mu = 0.5$) on the convergence performance of the BP neural network-based scheme

convergence performance of the scheme utilizing RSSI & CIR is better than that of the scheme using CIR alone, and slightly better than that of using RSSI alone.

Figure 5.8 compares the effects of different learning rates on the convergence performance of the proposed scheme relying on RSSI & CIR. We can observe from Figure 5.8 that the higher μ is in the suitable range, the faster the convergence speed is, but the higher the final convergence error value is, where $\mu \in [0, 1]$.

Figure 5.9 characterizes the convergence of the proposed BP neural network-based scheme relying on $\mu = 1.5$. It can be seen that the mean square error always oscillates and fails to converge, which is obviously contrast to the convergence results in Figure 5.8. The normal range of the learning rate μ is [0, 1], and when its value is not in this range, the BP neural network-based scheme cannot converge, which fails the authentication process. In summary, it is helpful to select a suitable learning rate μ to balance the convergence speed and final convergence error in practical scenarios. After a number of simulations, the learning rate μ is finally determined to be 0.2 in this section.

Then, the effects of a number of hidden layer units on the convergence performance and the authentication accuracy of the proposed scheme are studied in Figure 5.10. According to (5.9), the number of hidden layer nodes K should be lower

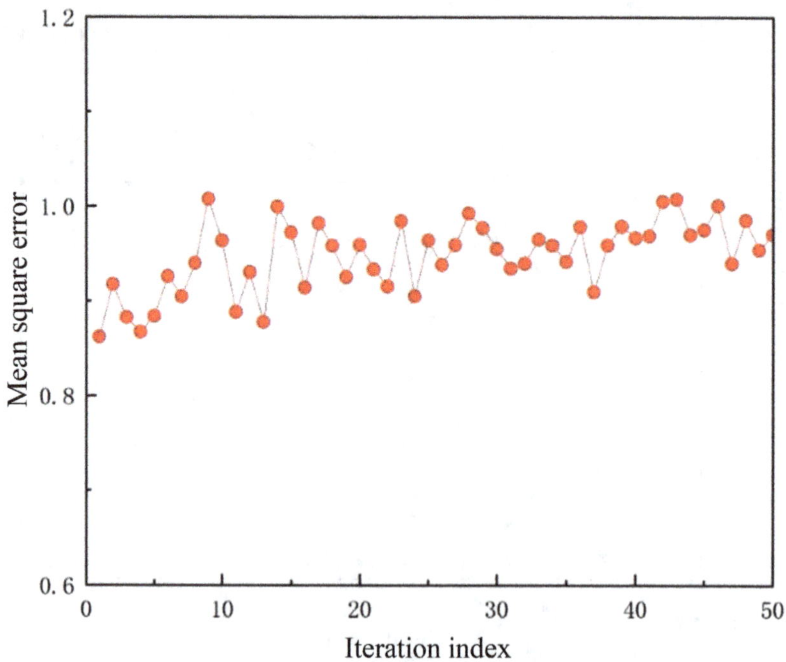

Figure 5.9 Mean square error of the BP neural network-based scheme when the learning rate μ is higher than 1 (μ = 1.5)

than 12. Figure 5.10 sets the number of hidden layer nodes from 5 to 10, and tests the model mean square error of 300 samples after 30 iterations in the proposed scheme utilizing RSSI and CIR & RSSI. The mean square error is tested for the number of nodes of 15 and 20 for comparison.

As we can observe from Figure 5.10, in the proposed scheme utilizing RSSI, the mean square error oscillates within a certain range, i.e., [5,10]. Then, with the increase of number of hidden layer nodes, the mean square error of the proposed scheme utilizing RSSI increases significantly. For another, the mean square error of the proposed scheme utilizing CIR & RSSI is lower than that of utilizing RSSI when the numbers of hidden layer nodes are 5–7. As it increases from 7 to 10, the mean square error increases. When it increases from 15 to 20, the mean square error decreases, but is still higher than that of 5–7.

The authentication accuracy comparison results of the proposed scheme in dynamic and static communication scenarios are given in Figure 5.11. The velocity of both Alice and Eve is set to be 36 km/h in the dynamic scenario. It can be observed that the proposed scheme relying on different attributes performs a bit better in the static scenario. The reason is that the selected physical layer attributes change slower in the static scenario so that the learning performance of the proposed scheme will be

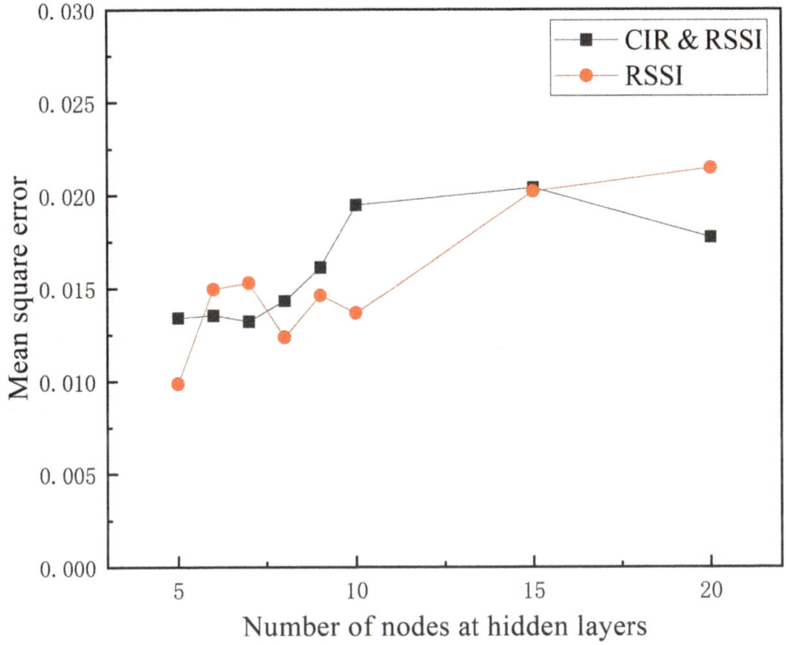

Figure 5.10 *Convergence performance of the proposed BP neural network-based scheme with different numbers of hidden layer nodes*

more accurate. More importantly, we can observe from Figure 5.11 that the authentication accuracy of the proposed BP neural network-based scheme in the dynamic scenario is very closed to that of static scenario. It characterizes that the proposed scheme performs robust in the complex dynamic communication network.

We can also observe from both Figures 5.7 and 5.11, the proposed BP neural network-based scheme using RSSI & CIR has better performance on convergence, authentication accuracy, and robustness. The reason is that more physical layer attributes used for authentication provide multi-dimensional security information and increase the difficulty of being cloned.

Figure 5.12 characterizes the authentication accuracy comparison between the proposed BPNN-based scheme utilizing RSSI & CIR and the support vector machine (SVM)-based scheme. The authentication accuracy of BPNN-based scheme is higher than that of SVM-based scheme when the signal-to-noise ratio (SNR) is higher than −5 dB. In particular, the authentication accuracy of proposed BPNN-based scheme achieves 99% in the case that the SNR is higher than 5 dB, while the SVM-based scheme cannot reach the same authentication accuracy until the SNR is higher than 25 dB. Compared with the traditional SVM-based authentication scheme, the proposed BPNN-based scheme achieves better security performance in the communication environment.

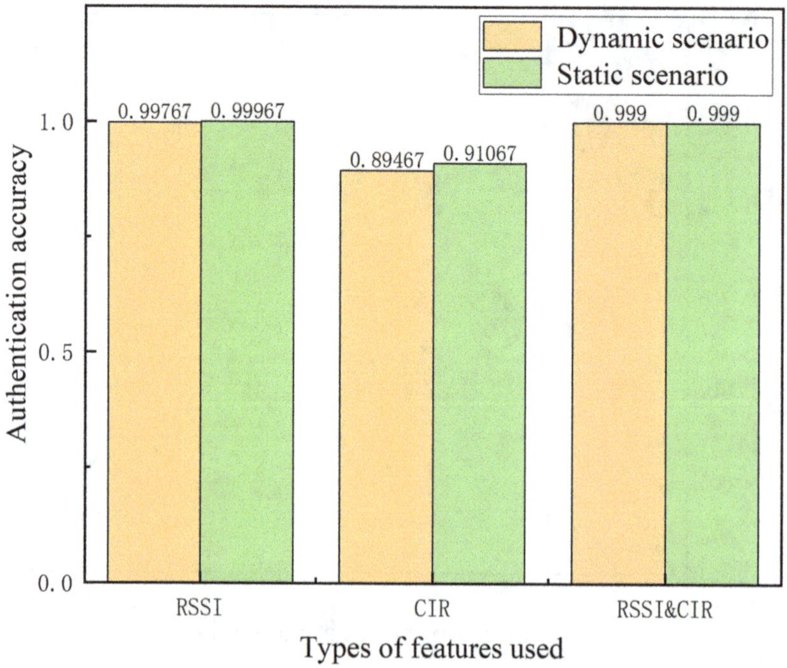

Figure 5.11 Authentication accuracy comparison of the proposed BP neural network-based scheme used in dynamic and static communication scenarios

5.4 FL-based intelligent physical layer authentication: a parametric method

In order to improve the authentication performance when some statistical properties of the attributes are available and intermittent availability of some attributes are considered, a fuzzy learning-based scheme is proposed in this section. Different from the BPNN-based scheme, it is a parametric method, where the particular form of the training function (i.e., multiple attribute fusion function) is designed to be a Gaussian function, so that the parameters to be designed and the training samples required will be much less. As a consequence, this scheme is more suitable for those devices having limited storage resources and for the application scenarios requiring lower computation overhead.

5.4.1 FL-based multiple attribute fusion model

In the designed fuzzy learning-based multiple attribute fusion model, the imperfect estimates of attributes are not numbers, but labels of fuzzy sets, which are classes of objects in which the transition from membership to non-membership is gradual rather than abrupt. In this way, the accuracy requirement of each attribute's estimates can

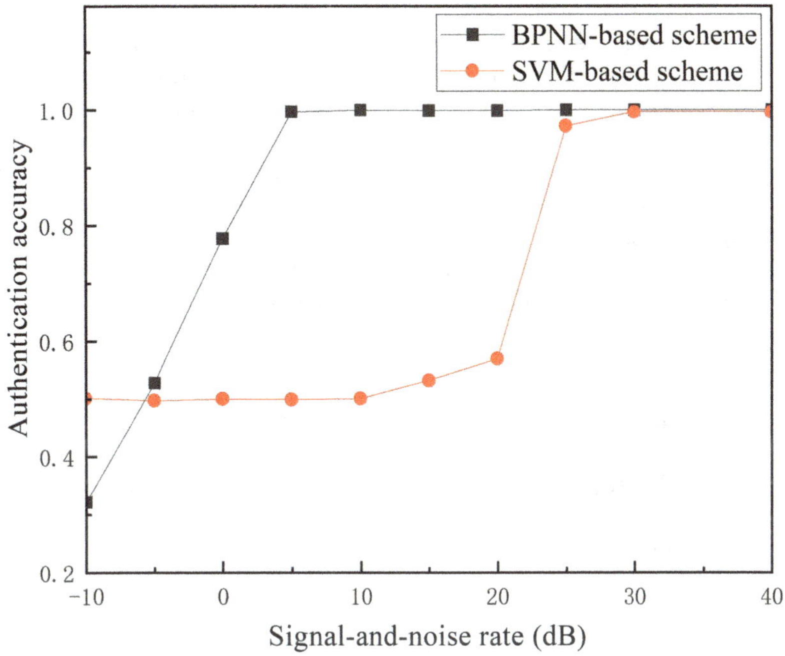

Figure 5.12 Authentication accuracy comparison between the proposed BPNN-based scheme and SVM-based scheme

be relaxed, which leads to high robustness in dealing with the imperfect estimates of attributes. In a nutshell, by using fuzzy theory, we can combine the multiple attributes with imperfect estimates, i.e., $\boldsymbol{H}^{II}[t+\tau]-\boldsymbol{H}_A^{I}[t]$, and then turn these estimates into an accurate authentication result, thus the effects of their imperfectness can be mitigated and the authentication accuracy can be improved.

In our scheme, the fuzzy membership function is designed as the Gaussian function with means μ_n and variances σ_n^2 as

$$u_{[-1,1]}\left(h_n'[t]\right)=\exp\left(\frac{-\left(h_n'[t]-\mu_n\right)^2}{2\sigma_n^2}\right), \tag{5.20}$$

where $h_n'[t]$ is the normalized attribute. For different physical layer attributes, μ_n and σ_n^2, $n=1,2,\ldots,N$, experience different values and control our fuzzy multiple attribute combination model. Without loss of generality, through using product inference engine and center average defuzzifier [39], the fuzzy multiple attribute combination model is designed as

$$\begin{aligned}&\text{Inputs}\rightarrow\text{Outputs}\\ &H^{II}[t+\tau]-H_A^{I}[t]\rightarrow y[t+\tau]=\sum_{n=1}^{N}\alpha_n u_{[-1,1]}\left(h_n'[t+\tau]\right),\end{aligned} \tag{5.21}$$

where α_n are system parameters, which adjust the weights of multiple physical layer attributes. Therefore, the designed fuzzy multiple attribute combination model can well control the different contributions of attributes on the authentication accuracy.

It can also be observed from (5.21) that this fuzzy multiple attribute combination model is a parametric method, and each attribute is independently taken into account. Hence, although some attributes are unavailable at some moments, the remaining attributes still contribute to the authentication and make the model applicable, resulting in a robust approach.

Through the designed fuzzy multiple attribute combination model, we can obtain a real-valued output $y[t]$ from (5.21) at time t. Then we propose a fuzzy authentication process

$$\begin{cases} \Phi_0 : v < y[t] \leq 1 \\ \Phi_1 : 0 < y[t] < v \end{cases}. \tag{5.22}$$

Note that $F(\boldsymbol{H}[t] - \boldsymbol{H}_A[t-\tau]) = 1 - y[t]$ satisfies. The false alarm rate and misdetection rate of the designed fuzzy model at time t can be formulated based on the fuzzy authentication process of (5.22) as

$$P_{\text{FA}}[t] = P(0 < y[t] < v \mid \Phi_0), \tag{5.23}$$

and

$$P_{\text{MD}}[t] = P\left(v \leq y[t] \leq 1 \mid \Phi_1\right). \tag{5.24}$$

In the following, it is assumed that the estimate deviations ΔH, which can be seen as the noises of estimates, are zero-mean complex Gaussian random variables with variances $\boldsymbol{\rho}^2 = \left(\rho_1^2, \rho_2^2, \ldots, \rho_N^2\right)^{\text{T}}$, denoted as $\Delta H \sim \aleph\left(0, \boldsymbol{\varrho}^2\right)$, and are independently distributed. Note that only the noises of attribute estimates are assumed to be the zero-mean complex Gaussian random variables. The multiple physical layer attributes can follow different distributions.

Upon denoting $X_n[t] = \left(h'_n[t] - \widehat{\mu}_n\right)/\rho'_n$, it obeys Gaussian random distribution, which can be expressed as $X_n[t] \sim \aleph\left(\bar{\mu}'_n[t], 1\right)$, where $\rho_n'^2 = 4\left(\rho_{An}^2 + \rho_n^2\right)/(b_n - a_n)^2$ and $\bar{\mu}'_n[t] = 2\left(\bar{H}_n[t+1] - \bar{H}_{An}[t+1] + v_n[t] - \widehat{\mu}_n\right)/\left(\rho'_n\left(b_n - a_n\right)\right)$. Note that the X_n at time instant t is the Gaussian random variable, but X_n at different time instants does not obey Gaussian random distribution, since the mean $\bar{\mu}'_n[t]$ is time-varying. Therefore, each $X_n^2[t]$ obeys the noncentral chi-squared distribution, which can be expressed as $X_n^2[t] \sim \mathscr{X}_n^2\left(1, \lambda_n[t]\right)$. It has two parameters: the number of degrees of freedom, namely for 1, and the noncentrality parameter $\lambda_n[t]$, satisfying $\lambda_n[t] = \bar{\mu}_n'^2[t]$. Using the relationship between the central and noncentral chi-squared distributions, the probability density function (PDF) and the cumulative distribution function (CDF) of variable $X_n^2[t]$ can be given as

$$f_{X_n^2[t]}\left(x; 1, \lambda_n[t]\right) = \frac{1}{2}\exp\left(-\frac{x + \lambda_n[t]}{2}\right)\left(\frac{x}{\lambda_n[t]}\right)^{-\frac{1}{4}} I_{-\frac{1}{2}}\left(\sqrt{\lambda_n[t]x}\right), \tag{5.25}$$

and

$$P_{X_n^2[t]}(x;1,\lambda_n[t]) = \exp\left(-\frac{\lambda_n[t]}{2}\right)\sum_{j=1}^{\infty}\frac{(\lambda_n[t]/2)^j}{j!}Q(x;1+2j), \tag{5.26}$$

respectively. $I_{-\frac{1}{2}}\left(\sqrt{\lambda_n[t]x}\right)$ is a modified Bessel function of the first kind given as

$$I_{-\frac{1}{2}}\left(\sqrt{\lambda_n[t]x}\right) = \left(\frac{\sqrt{\lambda_n[t]x}}{2}\right)^{-\frac{1}{2}}\sum_{j=0}^{\infty}\frac{(\lambda_n[t]x/4)^j}{j!\Gamma(j+1/2)}. \tag{5.27}$$

Moreover, $Q(x;1+2j)$ is the CDF of the central chi-squared distribution with 1 degree of freedom shown as

$$Q(x;1+2j) = \frac{\Upsilon((1+2j)/2, x/2)}{\Gamma((1+2j)/2)}, \tag{5.28}$$

where $\Upsilon(\cdot)$ is the lower incomplete Gamma function, and $\Gamma(\cdot)$ represents the Gamma function. Then, the output of the developed fuzzy multiple attribute combination model y of (5.21) at time t can be rewritten as

$$y[t] = \sum_{n=1}^{N}\hat{\alpha}_n[t]\exp\left(-\frac{\rho_n'^2}{2\hat{\sigma}_n^2[t]}X_n^2[t]\right). \tag{5.29}$$

Lemma 1. *The false alarm rate and misdetection rate of the designed fuzzy multiple attribute combination model can be given, respectively [1], by*

$$P_{\mathrm{FA}}[t] = P(0 < Y[t] < v) = \overbrace{\int\cdots\int_{C_1}}^{N}\prod_{n=1}^{N}f_{\xi_n[t]}(x_n)\,dx_1\cdots dx_N, \tag{5.30}$$

and

$$P_{\mathrm{MD}}[t] = P(v \le Z[t] \le 1) = \overbrace{\int\cdots\int_{C_2}}^{N}\prod_{n=1}^{N}f_{\varsigma_n[t]}(x_n)\,dx_1\cdots dx_N, \tag{5.31}$$

where

$$Y[t] = \sum_{n=1}^{N}\xi_n[t],\ \xi_n[t] = \widehat{\alpha}_n[t]\exp\left(-\frac{\rho_n'^2}{2\widehat{\sigma}_n^2[t]}X^2_{n,\lambda_n^{\mathrm{FA}}[t]}\right),$$
$$\lambda_n^{\mathrm{FA}}[t] = \left(\frac{2\left(v_n[t]-\widehat{\mu}_n[t]\right)}{\rho_n'\left(b_n-a_n\right)}\right)^2. \tag{5.32}$$

$f_{\xi_n[t]}(x_n)$ and $F_{\xi_n[t]}(x_n)$ are the PDF and CDF of variable $\xi_n[t]$, respectively, which are given as

$$\begin{aligned} F_{\xi_n[t]}(x_n) &= 1 - P_{X_n^2[t]}\left(-\frac{2\widehat{\sigma}_n^2[t]}{\rho_n'^2}\ln\left(\frac{x_n}{\widehat{\alpha}_n[t]}\right); 1, \lambda_n^{\mathrm{FA}}[t]\right) \\ &= \int_{-\infty}^{x_n} f_{\xi_n[t]}(\tau)d\tau. \end{aligned} \tag{5.33}$$

Moreover,

$$\begin{aligned} Z[t] &= \sum_{n=1}^{N} \varsigma_n[t],\ \varsigma_n[t] = \widehat{\alpha}_n[t]\exp\left(-\frac{\rho_n'^2}{2\widehat{\sigma}_n^2[t]}X^2_{n,\lambda_n^{\mathrm{MD}}[t]}\right), \\ \lambda_n^{\mathrm{MD}}[t] &= \left(\frac{2\left(\bar{H}_{En}[t] - \bar{H}_{An}[t] + v_n[t] - \widehat{\mu}_n\right)}{\rho_n'(b_n - a_n)}\right)^2. \end{aligned} \tag{5.34}$$

$f_{\varsigma_n[t]}(x_n)$ and $F_{\varsigma_n[t]}(x_n)$ represent the PDF and CDF of variable $\varsigma_n[t]$, which are given as

$$\begin{aligned} F_{\varsigma_n[t]}(x_n) &= 1 - P_{X_n^2[t]}\left(-\frac{2\widehat{\sigma}_n^2[t]}{\rho_n'^2}\ln\left(\frac{x_n}{\widehat{\alpha}_n[t]}\right); 1, \lambda_n^{\mathrm{MD}}[t]\right) \\ &= \int_{-\infty}^{x_n} f_{\varsigma_n[t]}(\tau)d\tau. \end{aligned} \tag{5.35}$$

Furthermore, $\prod$ represents the product of a sequence. C_1 and C_2 are denoted as the domains of multiple integral in (5.30) and (5.31), respectively, which satisfy $0 < Y[t] < v$ and $v \leq Z[t] \leq 1$.

5.4.2 Multi-dimensional adaptive authentication process

In this section, an optimization problem based on the expressions of false alarm rate and misdetection rate and a multi-dimensional adaptive authentication architecture is designed. Then, a hybrid learning algorithm is developed to describe and update the system parameters of the designed fuzzy multiple attribute combination model. The hybrid learning algorithm focuses on updating the system parameters to solve the designed optimization problem, which describes the FA versus MD trade-off, thus reliable authentication in the time-varying environment can be achieved.

As discussed above, the authentication performance of the designed fuzzy multiple attribute combination model can be improved by well deciding and updating the parameters $\hat{\mu}_n, \hat{\sigma}_n^2$ and $\hat{\alpha}_n$ in (5.29). Due to the inevitable trade-off between false alarm and misdetection, the authentication optimization problem is designed as

$$\begin{aligned} &\min_{(\widehat{\mu}_n, \widehat{\sigma}_n^2, \widehat{\alpha}_n, v), n=1,2,\ldots,N} P_{\mathrm{FA}}, \\ &\text{s.t. } P_{\mathrm{MD}} \leq \delta \end{aligned} \tag{5.36}$$

where δ represents the threshold of misdetection rate.

By expressing the output of the multiple attribute fusion as

$$y[t] = \hat{\alpha}_1 \exp\left(\frac{-\left(h'_1[t] - \widehat{\mu}_1\right)^2}{2\widehat{\sigma}_1^2}\right) + \hat{\alpha}_2 \exp\left(\frac{-\left(h'_2[t] - \widehat{\mu}_2\right)^2}{2\widehat{\sigma}_2^2}\right) + \cdots + \hat{\alpha}_N \exp\left(\frac{-\left(h'_N[t] - \widehat{\mu}_N\right)^2}{2\widehat{\sigma}_N^2}\right), \tag{5.37}$$

we can observe that the output is linear in $\hat{\alpha}_n$, while it is nonlinear in $\hat{\mu}_n$ and $\hat{\sigma}_n^2$. Therefore, $\hat{\alpha}_n$ are called as linear parameters, while $\hat{\mu}_n$ and $\hat{\sigma}_n^2$ are named as nonlinear parameters. To determine both the linear and nonlinear parameters of the proposed fuzzy learning-based scheme, a hybrid learning algorithm is developed by combining the gradient descent and the least-square estimator for fast updating the system parameters.

5.4.2.1 Gradient descent for nonlinear parameter update

Given weights $\hat{\alpha}_n, n = 1, 2, \ldots N$, the gradient descent is applied to update the nonlinear parameters $\hat{\mu}_n$ and $\hat{\sigma}_n^2$, thereby to solve the problem of (5.36), which can be expressed as

$$\hat{\mu}_n[k+1] = \hat{\mu}_n[k] - \varepsilon \left.\frac{\partial P_{\text{FA}}}{\partial \widehat{\mu}_n}\right|_k, \tag{5.38}$$

and

$$\widehat{\sigma}_n^2[k+1] = \hat{\sigma}_n^2[k] - \varepsilon \left.\frac{\partial P_{\text{FA}}}{\partial \widehat{\sigma}_n^2}\right|_k, \tag{5.39}$$

where ε represents the step size of Gradient descent. $\partial P_{\text{FA}}/\partial \hat{\mu}_n$ and $\partial P_{\text{FA}}/\partial \hat{\sigma}_n^2$ are the first-order partial derivatives of P_{FA} with respect to $\hat{\mu}_n$ and that with respect to $\hat{\sigma}_n^2$, respectively. Again, given the physical layer attributes adopted and attribute observations of Alice, the closed-from expression of P_{FA} can be obtained. Hence, the gradients in (5.38) and (5.39) can be derived.

5.4.2.2 Least-square estimator for linear parameter update

Given the non-linear parameters $\hat{\mu}_n$ and $\hat{\sigma}_n^2$, the least-square estimator is applied to update the weights $\hat{\alpha}_n$ in (5.37). Consider that a set of Alice's observations $\left\{\left(\boldsymbol{h}'_l, \hat{y}_l\right), l = 1, 2, \ldots, L\right\}$ is given, which is used for training the weights $\widehat{\boldsymbol{\alpha}} = (\hat{\alpha}_1, \hat{\alpha}_2, \ldots, \hat{\alpha}_N)^{\text{T}}$ in the least-square estimator. Then, using the matrix notation, the following equation holds:

$$\widehat{\boldsymbol{y}} = \boldsymbol{y} + \boldsymbol{e} = \boldsymbol{U}\widehat{\boldsymbol{\alpha}} + \boldsymbol{e}, \tag{5.40}$$

where $\hat{\boldsymbol{y}} = (\hat{y}_1, \hat{y}_2, \ldots, \hat{y}_L)^{\mathrm{T}}$ are the observations for training, while $\boldsymbol{y} = (y_1, y_2, \ldots, y_L)^{\mathrm{T}}$ represent the outputs of the fuzzy learning-based scheme with the corresponding inputs $h'_l, l = 1, 2, \ldots, L$. Moreover, $\boldsymbol{e} = \hat{\boldsymbol{y}} - \boldsymbol{y}$ represents the errors between the observations and outputs of the fuzzy learning-based scheme, and

$$\boldsymbol{U} = \begin{bmatrix} u_{[-1,1]}\left(h'_{11}\right) & \cdots & u_{[-1,1]}\left(h'_{N1}\right) \\ \vdots & \ddots & \vdots \\ u_{[-1,1]}\left(h'_{1L}\right) & \cdots & u_{[-1,1]}\left(h'_{NL}\right) \end{bmatrix}_{L\times N}. \tag{5.41}$$

Instead of finding the exact solution for (5.36), the proposed fuzzy learning-based scheme aims at searching for optimal linear parameters, i.e., $\hat{\alpha}_n^*$, for minimizing the sum of squared error, which is given by

$$E(\hat{\boldsymbol{\alpha}}) = \sum_{l=1}^{L} \gamma^{L-l} e_l^2 = \sum_{l=1}^{L} \gamma^{L-l} \left(\hat{y}_l - y_l\right)^2 = (\hat{\boldsymbol{y}} - \boldsymbol{U}\hat{\boldsymbol{\alpha}})^{\mathrm{T}} \gamma (\hat{\boldsymbol{y}} - \boldsymbol{U}\hat{\boldsymbol{\alpha}}), \tag{5.42}$$

where the $\gamma \in [0, 1]$ is forgetting factor to place the heavier emphasis on more recent data and is formulated as

$$\boldsymbol{\gamma} = \begin{bmatrix} \gamma^{L-1} & 0 & \cdots & 0 \\ 0 & \gamma^{L-2} & \cdots & 0 \\ \vdots & \vdots & \ddots & \vdots \\ 0 & 0 & \cdots & \gamma^0 \end{bmatrix}_{L\times L}. \tag{5.43}$$

The utilization of forgetting factor γ in the proposed hybrid learning algorithm can decrease the influence of historical samples in the authentication system gently due to the decorrelation of physical layer attributes in the time-varying environment. According to [40], the closed-form of solution for minimizing $E(\widehat{\boldsymbol{\alpha}})$ of (5.24) can be obtained as

$$\hat{\boldsymbol{\alpha}}^*[L] = \left(\boldsymbol{U}^{\mathrm{T}} \boldsymbol{\gamma} \boldsymbol{U}\right)^{-1} \boldsymbol{U}^{\mathrm{T}} \boldsymbol{\gamma} \hat{\boldsymbol{y}}. \tag{5.44}$$

Hence, given non-linear parameters $\hat{\mu}_n$ and $\hat{\sigma}_n^2$, (5.44) provides an optimal linear parameter $\hat{\alpha}_n^*$ for multi-dimensional adaptive authentication.

If estimate of an attribute, i.e., H_n, cannot be obtained at some moments, $\widehat{\alpha}_n = 0$ is set directly, and update the other linear parameters via (5.44). Since each attribute in the authentication model (5.21) is independently taken into account, the parameter update for each attribute is also independent. This demonstrates the benefit of the proposed solution compared with the nonparametric methods, that is, once some attributes are unavailable at some moments, the remaining attributes will still contribute to the authentication model and make the model work, leading to a robust authentication performance.

5.4.2.3 Hybrid learning algorithm for intelligent authentication process

In order to improve the convergence performance of parameters updating, the gradient descent and least-square estimator presented above is combined, named as hybrid learning. The hybrid learning is implemented in a batch mode, each learning epoch is composed of forward pass and backward pass [40]. Given a threshold v and a set of observations $\{(\boldsymbol{h}'_l, \hat{y}_l), l = 1, 2, \ldots, L\}$, the weights $\hat{\alpha}_n$ are updated according to (5.44) in the forward pass. After the weights $\hat{\alpha}_n$ are identified, the $\hat{\mu}_n$ and $\hat{\sigma}_n^2$ can be updated via (5.38) and (5.39), respectively, in the backward pass. Then, the threshold v is updated by letting $P_{\mathrm{MD}} = \delta$ with the updated $\hat{\mu}_n^*, \hat{\sigma}_n^{*2}, \hat{\alpha}_n^*$. As a conclusion, the proposed hybrid learning algorithm can be summarized in Algorithm 3.

Algorithm 3: Adaptive authentication process based on the hybrid learning

---

Input: N physical layer attributes adopted, Alice's observations,
Output: predict authentication result.

Iteration:
Gradient descent:
obtain $(\hat{\mu}_n^*[t], \hat{\sigma}_n^{*2}[t])$ through updating means $\hat{\mu}_n[k+1]$ and variances $\hat{\sigma}_n^2[k+1]$ via (5.38) and (5.39), respectively, until $|\hat{\mu}_n[k] - \hat{\mu}_n[k-1]| < \varepsilon_1$ & $|\hat{\sigma}_n^2[k] - \hat{\sigma}_n^2[k-1]| < \varepsilon_2$;
update $P_{\mathrm{FA}}(\hat{\mu}_n^*[t], \hat{\sigma}_n^{*2}[t])$ via (5.30);
Least square estimator:
if the estimate of attribute n is unavailable **then**
set $\hat{\alpha}_n[t] = 0$
else
obtain weights $\hat{\alpha}_n[t]$ through (5.44) and update $P_{\mathrm{FA}}(\hat{\alpha}_n^*[t])$ via (5.30);
end if
update threshold v by letting $P_{\mathrm{MD}}(\widehat{\mu_n^*}[t], \hat{\sigma}_n^{*2}[t], \hat{\alpha}_n^*[t]) = \delta$;
Multi-dimensional adaptive authentication: $t = L + 1$
get new physical layer attribute estimate of transmitter $\boldsymbol{H}[t]$;
predict authentication result using $(\hat{\mu}_n^*[L], \hat{\sigma}_n^{*2}[L], \hat{\alpha}_n^*[L])$ via (5.22): Alice/Eve;
update parameters via Step 1 by renewing observation set to $\{(\boldsymbol{h}'_l, \widehat{y}_l), l = 1, \ldots, L+1\}$

5.4.3 Simulation results

In order to show the performance of the proposed fuzzy learning-based intelligent physical layer authentication scheme, case study and simulation results are provided in this section. First, the proposed scheme is validated in a simulated urban scenario by utilizing carrier frequency offset (CFO), channel impulse response (CIR), received signal strength indicator (RSSI), and in-phase and quadrature imbalance (IQI). The parameter design and convergence performance of Algorithm 3 are demonstrated. Then, the fuzzy learning-based scheme is validated in a simulated wireless communication scenario to show its performance. The comparison results between the proposed fuzzy learning-based scheme and some existing schemes are presented, demonstrating the superior permanence of the FL-based scheme.

Table 5.3 Simulation parameters of section 5.4

Parameter	**Value**	**Parameter**	**Value**
Initial position of Alice	[400;600] m	SNR	20 dB
Initial position of Eve	[400;600] m	Range	1000 m × 1000 m
Initial position of Bob	[900;900] m	Velocity of Alice	[8; −9] m/s
Antenna number of Bob	8	Velocity of Eve	[−6.72;2.26] m/s
Center frequency f_c	5 GHz	Sampling rate	20 MHz
Antenna number of Alice/Eve	4	Antenna height	1 m

In this subsection, the proposed scheme is studied in an urban scenario, and the simulation parameters in this case study are given in Table 5.3. Observations of CIR, CFO, RSSI, and IQI of both Alice and Eve can be obtained in this simulation. In the proposed scheme, 12 system parameters should be determined, namely for weights $\hat{\alpha}_n$, means $\hat{\mu}_n$ and widths $\hat{\sigma}_n^2$, $n = 1, 2, 3, 4$, corresponding to CIR, CFO, RSSI, and IQI, respectively.

Considering the utilization of different numbers of physical layer attributes, the authentication performance of the proposed fuzzy learning-based scheme is characterized in Figure 5.13. We can observe from Figure 5.13 that a larger number of attributes used leads to a more obvious improvement in authentication accuracy. The reason is that it is extremely unlikely for an attacker to predict or imitate all the attributes from the received signals. Moreover, reliable and robust authentication can be achieved by utilizing multiple attributes because of more information considered for identifying the transmitter.

We can observe that Figure 5.14 characterizes the robustness of the proposed FL-based scheme. The adaptive authentication process is performed relying on four attributes, during 11th–100th authentication time instant. Then, the IQI is unavailable at the receiver from 101st observation. It is observed from Figure 5.14 that once some of the adopted attributes are unavailable, the remaining attributes still contribute to the authentication model and make the model work, demonstrating the robustness of the proposed FL-based scheme in the time-varying environment.

Figure 5.15 characterizes the comparison results of the FL-based scheme and the optimal weight-based (OW) scheme [41] in different cases, namely for CIR & CFO & RSSI & IQI and CIR & CFO & RSSI. It is observed from Figure 5.15 that with the increasing authentication time, the FA rate values of the proposed FL scheme in both cases keep stable, while that of the OW scheme dramatically increase. The reason is that the optimal weights-based scheme is a static scheme and cannot update the system parameters according to the variations of physical layer attributes, thus cannot adapt to the dynamic environment. These all indicate that the proposed scheme performs better than the weights-based authentication scheme of [41].

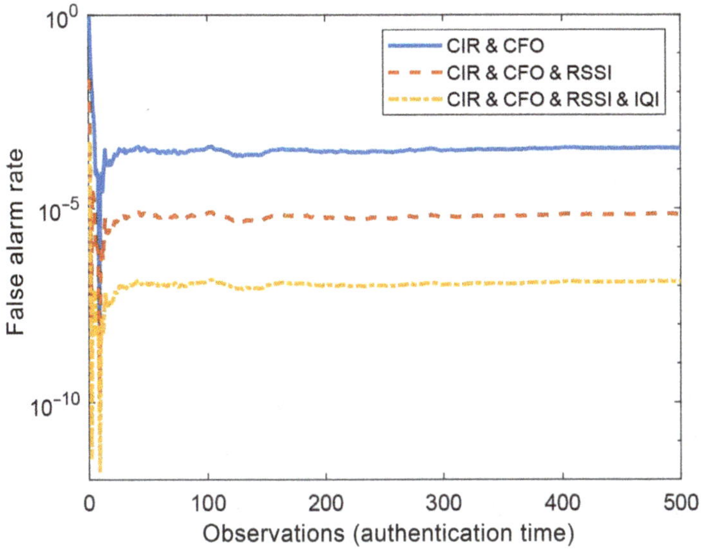

Figure 5.13 False alarm rate of the proposed fuzzy learning-based scheme by utilizing different numbers of attributes. ©IEEE 2020. Reprinted with permission from [1].

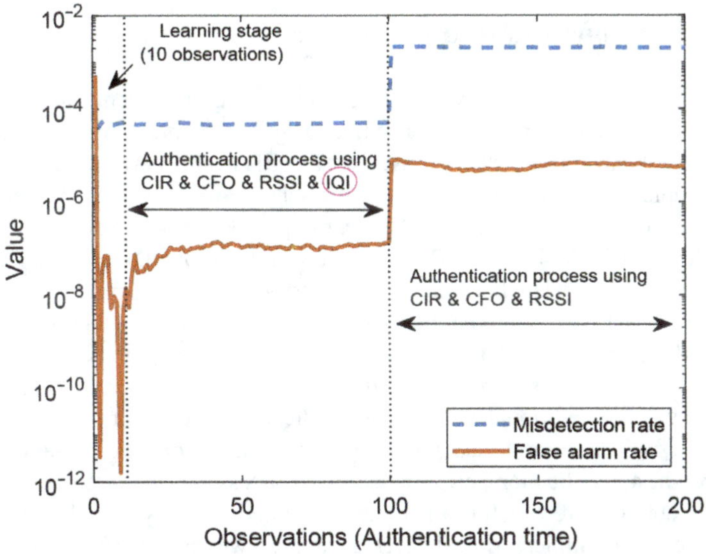

Figure 5.14 Robust performance of the proposed fuzzy learning-based scheme. ©IEEE 2020. Reprinted with permission from [1].

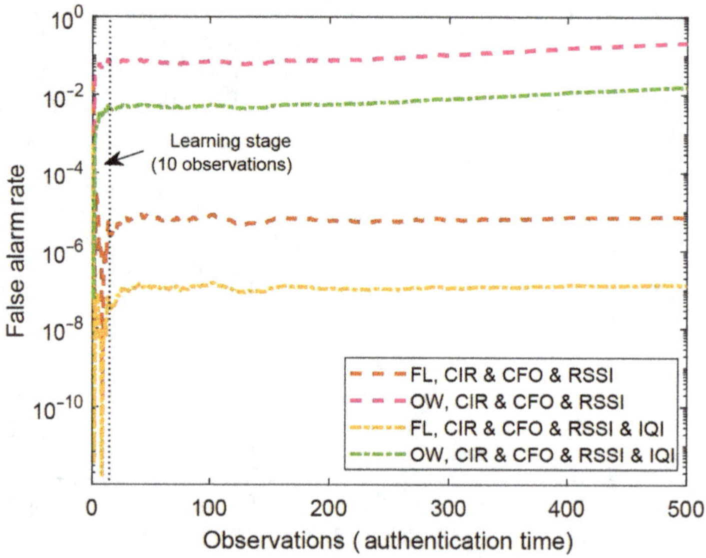

Figure 5.15 Comparison results of the fuzzy learning-based scheme and the optimal weight-based scheme of [41]. ©IEEE 2020. Reprinted with permission from [1].

5.5 Conclusions and future researches

To improve the accuracy and reliability of physical layer authentication, multiple channel/device-related attributes were utilized in this chapter. First of all, a non-parametric authentication scheme based on the BP neural network was presented in the complex dynamic environment. This scheme improved the authentication reliability in dynamic environments by learning and fusing the selected multiple physical layer attributes based on BP neural networks. Through the simulation experiments, the results verified the authentication performance of the proposed BP neural network-based scheme in a dynamic time-varying environment. In particular, when the signal-to-noise ratio is larger than 5 dB, the authentication accuracy is higher than 99%. The proposed scheme achieved higher authentication accuracy at low SNR compared with the traditional SVM-based physical layer authentication scheme.

Then, a fuzzy learning-based physical layer authentication scheme was developed to improve the authentication performance based on less training samples and parameters to be determined, which is a parametric method. A fuzzy model was designed to combine the multiple physical layer attributes, which can efficiently mitigate the imperfectness of estimated attributes. By proposing an adaptive hybrid learning algorithm, i.e., the combination of least-square estimator and gradient method, the system parameters can be updated in real-time to adapt to the time-varying environment. It was demonstrated that the proposed scheme has much better authentication performance than the optimal weights-based scheme.

Both parametric and non-parametric methods come with their own advantages and disadvantages. This chapter proposes an authentication scheme for each of these two methods for reference. Future works will be considered in the following:

1. Using convolutional neural networks for attribute selection and fusion, exploiting their locally connected, weight-sharing features to further improve the accuracy of multiple attribute fusion model and to improve the authentication performance in the fast time-varying environment.
2. Designing a parameter model that can better fit the channel state, exploring the solution of progressive parameter determination, and making full use of channel feature to achieve adaptive authentication.

References

[1] Fang H, Wang X, and Xu L. Fuzzy learning for multi-dimensional adaptive physical layer authentication: a compact and robust approach. *IEEE Transactions on Wireless Communications*. 2020;19(8):5420–5432.

[2] Liu Y, Qin Z, Elkashlan M, *et al.* Non-orthogonal multiple access for 5G and beyond. *Proceedings of the IEEE*. 2017;105(12):2347–2381.

[3] Fang H, Wang X, and Tomasin S. Machine learning for intelligent authentication in 5G and beyond wireless networks. *IEEE Wireless Communications*. 2019;26(5):55–61.

[4] Ye Q and Zhuang W. Distributed and adaptive medium access control for Internet-of-Things-enabled mobile networks. *IEEE Internet of Things Journal*. 2016;4(2):446–460.

[5] Wu X, Yang Z, Ling C, *et al.* Artificial-noise-aided physical layer phase challenge-response authentication for practical OFDM transmission. *IEEE Transactions on Wireless Communications*. 2016;15(10):6611–6625.

[6] Fang H, Wang X, and Hanzo L. Learning-aided physical layer authentication as an intelligent process. *IEEE Transactions on Communications*. 2018;67(3):2260–2273.

[7] Choo KKR, Gritzalis S, and Park JH. Cryptographic solutions for industrial Internet-of-Things: Research challenges and opportunities. *IEEE Transactions on Industrial Informatics*. 2018;14(8):3567–3569.

[8] Lin C, He D, Kumar N, *et al.* Security and privacy for the internet of drones: challenges and solutions. *IEEE Communications Magazine*. 2018;56(1): 64–69.

[9] Halford TR, Courtade TA, Chugg KM, *et al.* Energy-efficient group key agreement for wireless networks. *IEEE Transactions on Wireless Communications*. 2015;14(10):5552–5564.

[10] Li N, Liu D, and Nepal S. Lightweight mutual authentication for IoT and its applications. *IEEE Transactions on Sustainable Computing*. 2017;2(4): 359–370.

[11] Mick T, Tourani R, and Misra S. LASeR: lightweight authentication and secured routing for NDN IoT in smart cities. *IEEE Internet of Things Journal*. 2017;5(2):755–764.

[12] Wang F, Xu Y, Zhang H, *et al.* 2FLIP: a two-factor lightweight privacy-preserving authentication scheme for VANET. *IEEE Transactions on Vehicular Technology*. 2015;65(2):896–911.

[13] Paul LY, Baras JS, and Sadler BM. Physical-layer authentication. *IEEE Transactions on Information Forensics and Security*. 2008;3(1): 38–51.

[14] Paul LY and Sadler BM. MIMO authentication via deliberate fingerprinting at the physical layer. *IEEE Transactions on information forensics and security*. 2011;6(3):606–615.

[15] Aman MN, Chua KC, and Sikdar B. Mutual authentication in IoT systems using physical unclonable functions. *IEEE Internet of Things Journal*. 2017;4(5):1327–1340.

[16] Chen D, Zhang N, Qin Z, *et al.* S2M: a lightweight acoustic fingerprints-based wireless device authentication protocol. *IEEE Internet of Things Journal*. 2016;4(1):88–100.

[17] Choi J. A coding approach with key-channel randomization for physical-layer authentication. *IEEE Transactions on Information Forensics and Security*. 2018;14(1):175–185.

[18] Polak AC, Dolatshahi S, and Goeckel DL. Identifying wireless users via transmitter imperfections. *IEEE Journal on Selected Areas in Communications*. 2011;29(7):1469–1479.

[19] Baracca P, Laurenti N, and Tomasin S. Physical layer authentication over MIMO fading wiretap channels. *IEEE Transactions on Wireless Communications*. 2012;11(7):2564–2573.

[20] Caparra G, Centenaro M, Laurenti N, *et al.* Energy-based anchor node selection for IoT physical layer authentication. In: *2016 IEEE International Conference on Communications (ICC)*; 2016. p. 1–6.

[21] Xiao L, Wan X, and Han Z. PHY-layer authentication with multiple landmarks with reduced overhead. *IEEE Transactions on Wireless Communications*. 2017;17(3):1676–1687.

[22] Hou W, Wang X, Chouinard JY, *et al.* Physical layer authentication for mobile systems with time-varying carrier frequency offsets. *IEEE Transactions on Communications*. 2014;62(5):1658–1667.

[23] Wang X, Hao P, and Hanzo L. Physical-layer authentication for wireless security enhancement: current challenges and future developments. *IEEE Communications Magazine*. 2016;54(6):152–158.

[24] Xiao L, Greenstein LJ, Mandayam NB, *et al.* Using the physical layer for wireless authentication in time-variant channels. *IEEE Transactions on Wireless Communications*. 2008;7(7):2571–2579.

[25] Jiang C, Zhang H, Ren Y, *et al.* Machine learning paradigms for next-generation wireless networks. *IEEE Wireless Communications*. 2016;24(2):98–105.

[26] Challita U, Ferdowsi A, Chen M, *et al.* Machine learning for wireless connectivity and security of cellular-connected UAVs. *IEEE Wireless Communications.* 2019;26(1):28–35.

[27] Restuccia F, D'Oro S, and Melodia T. Securing the Internet of Things in the age of machine learning and software-defined networking. *IEEE Internet of Things Journal.* 2018;5(6):4829–4842.

[28] Xiao L, Wan X, Lu X, *et al.* IoT security techniques based on machine learning: How do IoT devices use AI to enhance security? *IEEE Signal Processing Magazine.* 2018;35(5):41–49.

[29] Xu H, Wang W, and Qian Y. Fusing complete monotonic decision trees. *IEEE Transactions on Knowledge and Data Engineering.* 2017;29(10):2223–2235.

[30] Wang S, Bao Z, Culpepper JS, *et al.* Reverse k nearest neighbor search over trajectories. *IEEE Transactions on Knowledge and Data Engineering.* 2017;30(4):757–771.

[31] Cheng YC and Wang PC. Packet classification using dynamically generated decision trees. *IEEE Transactions on Computers.* 2013;64(2):582–586.

[32] Liu J, Cosman PC, and Rao BD. Robust linear regression via ℓ_0 regularization. *IEEE Transactions on Signal Processing.* 2017;66(3):698–713.

[33] Finlayson GD, Mackiewicz M, and Hurlbert A. Color correction using root-polynomial regression. *IEEE Transactions on Image Processing.* 2015;24(5):1460–1470.

[34] Tan X, Sun C, and Pham TD. Edge-aware filtering with local polynomial approximation and rectangle-based weighting. *IEEE Transactions on Cybernetics.* 2015;46(12):2693–2705.

[35] Wang L. A hydropower project investment estimation model based on chaos BP combined neural network. In: *2022 6th International Conference on Intelligent Computing and Control Systems (ICICCS)*; 2022. p. 1052–1055.

[36] Xie Y, Wang K, and Huan H. BPNN based indoor fingerprinting localization algorithm against environmental fluctuations. *IEEE Sensors Journal.* 2022;22:12.

[37] Guo F, Zhang L, and Liu X. An optimized back propagation neural network based on genetic algorithm for line loss calculation in low-voltage distribution grid. In: *2020 Chinese Automation Congress (CAC)*; 2020. p. 663–667.

[38] Sagar V and Kumar K. A symmetric key cryptography using genetic algorithm and error back propagation neural network. In: *2015 2nd International Conference on Computing for Sustainable Global Development (INDIACom)*; 2015. p. 1386–1391.

[39] Jantzen J. *Foundations of Fuzzy Control: A Practical Approach.* John Wiley & Sons; 2013.

[40] Jang JSR, Sun CT, and Mizutani E. Neuro-fuzzy and soft computing—a computational approach to learning and machine intelligence. *IEEE Transactions on Automatic Control.* 1997;42(10):1482–1484.

[41] Duan X and Wang X. Fast authentication in 5G HetNet through SDN enabled weighted secure-context-information transfer. In: *2016 IEEE International Conference on Communications (ICC)*; 2016. p. 1–6.

Chapter 6

Deep learning-enhanced radio frequency fingerprint identification

Guanxiong Shen[1], Junqing Zhang[1] and Alan Marshall[1]

Radio frequency fingerprint identification (RFFI) is a promising authentication technique for Internet of Things (IoT) devices. The wireless transmitter chain of an IoT device consists of analog components that inevitably deviate from nominal values during the manufacturing process. These imperfections slightly distort the transmitted waveform, and an RFFI system can be equipped at the receiver to uniquely identify the device by analyzing the characteristics of the received signal. The deep learning technique is frequently used in RFFI for its effective feature extraction capabilities. After sufficient training, the neural network is capable of predicting the device identity by analyzing the input signal. In this chapter, we first introduce the basic components of an RFFI system and then use LoRa as a case study to demonstrate implementation details. The channel-independent spectrogram is proposed as the neural network input to combat channel effects. Then we design CNN and LSTM networks to process the channel-independent spectrogram. Experiments are conducted involving five LoPy4 devices and a USRP N210 software-defined radio (SDR). The results demonstrate that the designed LoRa-RFFI system is resilient to location changes, and the classification accuracy is consistently above 90%.

6.1 Introduction

The population of Internet of Things (IoT) devices has been blooming exponentially in recent years. Some IoT applications such as smart homes and smart medical carry sensitive and private data. Therefore, their security has received much attention. The authentication of IoT devices is an essential module to prevent an attack from accessing a user's private data. Conventional authentication schemes rely on software addresses, e.g., MAC addresses, and cryptographic algorithms. However, the software address is prone to impersonation without much effort. For example, the MAC address can be easily modified to be exactly the same as the device the attacker is trying to spoof. In addition to this, cryptographic algorithms usually require secure key

[1]Department of Electrical Engineering and Electronics, University of Liverpool, UK

distribution and storage mechanisms, which is hard to achieve in low-cost IoT end nodes. A lightweight and efficient IoT authentication scheme is needed for energy-constraint IoT applications.

Radio frequency fingerprint identification (RFFI) is an authentication technique suitable for IoT applications. The transmitter chain of a wireless IoT node consists of numerous analog components, e.g., digital-to-analog converter (DAC), oscillator, mixer, antenna [1–3]. These components are low-cost and inevitably deviate from nominal values during the manufacturing process, making the transmitted waveform slightly distorted from the standard one. An IoT network secured by the RFFI system is shown in Figure 6.1, the receiver/gateway is equipped with an RFFI system that analyzes the characteristics of the received signals to predict from which IoT end node the signal is sent. The hardware characteristics of a wireless IoT end node are hard to mimic without much effort, which increases the system's security. RFFI technique does not require any hardware modifications to the IoT end nodes or updates to the existing communication protocol. Therefore, it is a lightweight and low-cost authentication technique applicable to IoT applications.

The state-of-the-art RFFI systems are mostly based on deep learning because of their excellent feature extraction and classification ability. The receiver equips a neural network whose input is the received wireless signal and output is the inferred device identity. Various neural network models are designed in previous studies: convolutional neural network (CNN) [2,4–23], recurrent neural network (RNN)/long-short term memory (LSTM) network [7,10,17,24,24,25], and the transformer [24]. Although these deep learning models can classify the received signals with high accuracy, there are still issues that were not considered in previous studies.

The RFFI performance can degrade due to the change in environment. First, the transmitted signal propagates over the air before reaching the receiver, which affects the characteristics of the received signal. The RFFI system trained with the signals sent from one location may not be able to correctly classify the signal sent from another since this violates the fundamental independent and identically distributed (i.i.d.) assumption in deep learning. Hence, the mitigation scheme needs to be designed to combat channel effects. In addition to this, the temperature variation can cause the transceiver hardware characteristics to drift, reducing the stability of the RFFI system. Experiments have shown that oscillators are sensitive to temperature changes [26],

Figure 6.1 An IoT network secured by the RFFI system

and the frequency offset in the received signal can change over time [7,8]. The effects of frequency offset drift need to be investigated and solutions need to be proposed to combat performance degradation.

In this chapter, we introduce the basic modules in deep learning-based RFFI systems. Then we employ LoRa as a case study to show the implementation details. The channel-independent spectrogram and data augmentation are used to combat channel effects, making the RFFI system robust to location changes. The frequency offset compensation is also applied to the received signals to prevent performance degradation caused by temperature variations. The LoRa-RFFI system is experimentally evaluated using five LoPy4 devices and a USRP N210 software-defined radio (SDR). The RFFI system is experimentally shown to be robust to location variations. It consistently maintained more than 90% accuracy on datasets collected at five different locations. Experiments show that frequency offset compensation is necessary for RFFI systems to effectively prevent performance degradation due to unavoidable temperature changes.

The rest of this chapter is organized as follows. Sections 6.2 and 6.3 introduce related work and system overview, respectively. Section 6.4 presents the basic components of the RFFI system, and Section 6.5 takes LoRa-RFFI as a case study for a more detailed introduction. Experimental evaluation of the LoRa-RFFI system is provided in Section 6.6. Section 6.7 finally concludes the work.

6.2 Related work

RFFI is a lightweight technique for authenticating wireless devices. An RFFI system analyzes the characteristics of the received physical layer signal for unique identification. Traditional RFFI approaches rely on handcrafted features, that is, the researchers design a feature extraction algorithm with expertise and knowledge of wireless communication, and use it to analyze the received signals. The traditional features include signal statistical features, frequency spectrum [1,27], power amplifier characteristics [28], frequency offset [29–34], IQ offset [32,35], etc. However, the hardware impairments are interrelated and it is difficult to estimate each feature accurately. Deep learning is recently used in RFFI for its excellent performance and automatic feature extraction capability. It learns a feature extraction algorithm from a large number of labeled packets, which outperforms the manually designed one. After sufficient training, the neural network can correctly predict the device label from the received physical layer signal.

The received signals are IQ samples which are complex numbers that cannot be processed by most neural networks. Moreover, some features are difficult to extract in the time domain. Therefore, the IQ samples need to be first converted to a signal representation that is compatible with neural networks. Designing the signal representation that makes the feature more obvious is a popular research topic in RFFI. Signal representations that are often utilized include error signal [19], frequency spectrum [4], differential constellation trace figure [5], channel-independent spectrogram [9], and decomposed signals [36]. In addition, decomposing IQ samples into

I and Q components is also an effective signal representation that can be used as input to a neural network.

Numerous neural network types have been studied in previous research, including CNN [7,8], LSTM/GRU network, and transformer. The design of the neural network should take into account the characteristics of the signal representation. For example, CNN can be used to process IQ samples and frequency spectrum, while LSTM/GRU networks can be used to process IQ samples and time–frequency domain spectrograms. It is experimentally shown that LSTM is not effective on frequency spectrum classification since there is no time-related information.

The change of wireless channels can significantly degrade the performance of a deep learning-based RFFI system, which has been one of the major challenges in the development of RFFI. Designing a channel-agnostic signal representation as the neural network input is a possible solution to the channel problem. For example, Li *et al.* design the amplitude of quotient feature as the neural network input, which is not affected by channel variations [37]. In addition, emulating more channel conditions during training can also improve the system's robustness to channel variations, which is known as data augmentation [9,11,14,17]. Moreover, intentionally adding impairments to the transmitter chain can enhance the RF fingerprint, making the RFFI system more robust to the channel changes [38,39].

6.3 System overview and problem statement

6.3.1 System overview

The overview of an RFFI system is shown in Figure 6.2. The training stage produces a well-trained classification neural network, which is then used in the working stage to predict the device identity. In the training stage, we first collect signals from all the IoT devices operating in the communication network. The collected signals are augmented for channel robustness and then converted to appropriate signal representations as the input to the neural network. Afterward, we define the architecture of the neural network and update its parameters using the training data.

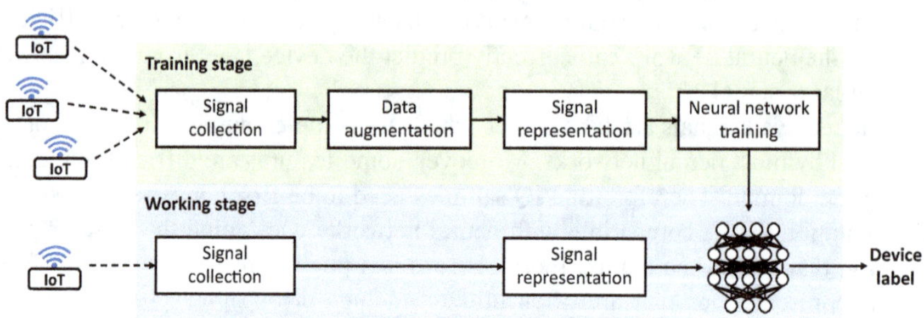

Figure 6.2 Overview of an RFFI system

In the working stage, the collected signal is directly converted to the signal representation and then fed into the neural network. The output is a list of probabilities over all the predictable devices. A common approach is to select the label with the highest probability as the final prediction.

6.3.2 Problem statement

There are D IoT end nodes operating in the network and the end node d emitted an RF signal $F^d(x(t))$ to the air. $F^d(\cdot)$ is the hardware distortion of device d and $x(t)$ is the ideal waveform. In the proposed RFFI system, we only focus on the preamble part, therefore, $x(t)$ is the same for all the transmitters. The RF signal is propagated through the wireless channel and finally captured by the receiver. The received signal $r(t)$ is mathematically expressed as

$$r(t) = h(t) * F^d(x(t)) + n(t), \tag{6.1}$$

where $h(t)$ is the channel impulse response (CIR) and $n(t)$ is the additive white Gaussian noise (AWGN). $*$ denotes convolution operation. The analog-to-digital converter (ADC) digitalizes $r(t)$ with a sampling rate of T_s, given as

$$r(t) \longrightarrow r[nT_s], \tag{6.2}$$

where $r[nT_s]$ is the digitalized signal which is a vector composed of complex numbers. $r[nT_s]$ is also called the IQ samples. We denote $r[nT_s]$ as $\mathbf{r}$ in the rest of the paper for simplicity.

The objective of an RFFI system is to correctly predict the device label d by analyzing the received signal $\mathbf{r}$.

6.4 System design

6.4.1 Signal processing

The captured signal $\mathbf{r}$ needs to be processed to satisfy the fundamental requirements for RFFI. This includes preamble extraction, frequency offset compensation, and normalization.

- **Preamble extraction**: Deep learning-based RFFI systems typically only employ the preamble portion for identification in order to prevent the neural network from learning the payload or header information. After capturing a packet, we only keep the preamble and discard the rest. Note that the preamble is available in most practical communication systems for packet receiving. Using the preamble part for RFFI does not require any dedicated changes to the packet format.
- **Frequency offset compensation**: Frequency offset compensation is crucial in RFFI systems. The frequency offset is a discriminative feature of the transmitter chain. However, it is easy for the attacker to manipulate the frequency offset thus making the RFFI system vulnerable [40]. Furthermore, prior research demonstrates that frequency offset drifts with temperature changes, making the RFFI system unstable over time [7,8].

- **Normalization**: The magnitudes of the received signals are affected by the varying channel conditions. Normalization is to scale the signals to equal magnitudes so that channel variations will not have an impact on system performance.

The processed signal is denoted $\mathbf{r}'$.

6.4.2 Training data collection and data augmentation

To train a classification neural network, a large number of labeled signals should be collected from all IoT end nodes. The collected training dataset $\mathcal{X}^{train}$ is given as

$$\mathcal{X}^{train} = \left\{(\mathbf{r}'_m, \mathbf{p}_m)\right\}_{m=1}^{M_{train}}, \tag{6.3}$$

where $\mathbf{r}'_m$ is the mth training signal, $\mathbf{p}_m$ is the one-hot encoding result of the transmitter label d. M_{train} is the number of training signals.

Before the training process, $\mathcal{X}^{train}$ needs to be augmented to increase the system robustness to channel variations, which is called data augmentation. Its principle is to simulate various channel conditions during the training process so that the trained neural network can generalize well in the inference stage. Thanks to the fast-developed wireless communication technology, well-designed wireless channel models can be employed during data augmentation. The tapped delay line (TDL) models are widely used in RFFI for their simplicity. The power delay profile (PDP), root mean square (RMS) delay spread, and Doppler spectrum need to be determined before building a TDL channel model. The PDP and RMS delay spread characterize the multipath richness while the Doppler spectrum indicates how the channel varies over time. Note that the Doppler effects can be neglected for some wireless technologies such as WiFi because the wireless channel can be assumed constant during a packet duration. Besides, additive white Gaussian noise (AWGN) is added to the training signals to increase the system performance in low SNR scenarios.

The data augmentation is performed on all the collected training signals. For each training signal $\mathbf{r}'_m$, we first randomly select a set of parameters to synthesize a wireless channel $\mathbf{h}^s_m$, and then pass $\mathbf{r}'_m$ through it. After this, a synthesized AWGN $\mathbf{n}^s$ is added to the signal. The augmentation is mathematically expressed as

$$\mathbf{y}_m = \mathbf{h}^s_m * \mathbf{r}'_m + \mathbf{n}^s_m, \tag{6.4}$$

where $\mathbf{y}_m$ is the mth training signal after augmentation. Note that $\mathbf{h}^s{}_m$ and $\mathbf{n}^s{}_m$ are different for each training signal $\mathbf{r}'{}_m$. This procedure is repeated until all the signals are augmented, and the training dataset after augmentation becomes

$$\mathcal{X}^{train} = \{(\mathbf{y}_m, \mathbf{p}_m)\}_{m=1}^{M_{train}}. \tag{6.5}$$

6.4.3 Signal representation

The received IQ samples are complex numbers that cannot be processed by most neural networks. Therefore, they need to be first transformed into appropriate signal representations. A well-designed signal representation can reveal the RF fingerprints and improve the identification performance. There are a number of signal representations proposed in previous studies. For some examples, error signal [19], frequency spectrum [4], differential constellation trace figure [5], channel-independent spectrogram [9], and decomposed signals [36]. Note that the IQ samples can also serve as neural network inputs after some processing [8,10,38]. The characteristics of the modulation technique should be considered when designing the signal representation. After signal representation, the training dataset becomes

$$\mathscr{X}^{train} = \{(\mathbf{C}_m, \mathbf{p}_m)\}_{m=1}^{M_{train}}, \tag{6.6}$$

where $\mathbf{C}_m$ is the signal representation converted from the mth training signal.

6.4.4 Neural network

The neural network is used to extract discriminative features from the input signal representations and then use them for classification. First, we need to determine the architecture of the neural network and initialize its parameters. The neural network is denoted as $f(\cdot;\Theta)$ and Θ is its parameters. The properties of the signal representations should be considered when designing the neural network architecture. A loss function $\mathscr{L}$ is defined after the initialization of the neural network, which is usually cross-entropy in classification problems, given as

$$\mathscr{L}(\hat{\mathbf{p}}, \mathbf{p}) = -\sum_{d=1}^{D} p_d \log(\hat{p}_d), \tag{6.7}$$

where $\hat{\mathbf{p}} = f(\mathbf{C};\Theta)$ is the output of the neural network. Then the training dataset $\mathscr{X}^{train}$ is used to optimize the parameters of the neural network, the optimization process is mathematically expressed as

$$\Theta = \underset{\Theta}{\operatorname{argmin}} \sum_{(S,\mathbf{p})\in \mathscr{X}^{train}} \mathscr{L}(f(\mathbf{C};\Theta), \mathbf{p}). \tag{6.8}$$

In the inference stage, the newly received signal can be processed and converted to the signal representation $\mathbf{C}^{infer}$. Then the well-trained neural network can make a prediction, given as

$$\hat{\mathbf{p}} = f(S^{infer};\Theta), \tag{6.9}$$

The element index with the highest probability in $\hat{\mathbf{p}}$ can be simply selected as the predicted device label.

6.5 Case study: LoRa-RFFI system

6.5.1 LoRa physical layer primer

LoRa is a proprietary modulation technique developed from chirp spread spectrum (CSS) technology, which falls under non-cellular low-power wide area network

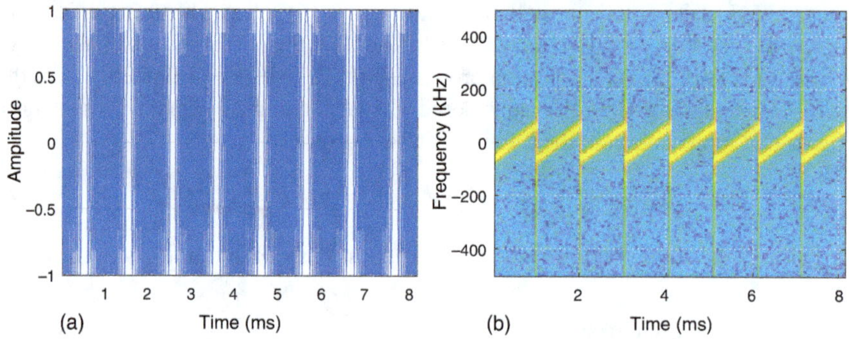

Figure 6.3 The preamble part in a LoRa packet. (a) Time domain waveform (I-component). (b) Time–frequency domain spectrogram.

(LPWAN) technologies. A LoRa symbol modulates information into its initial frequency. An unmodulated baseband LoRa symbol (upchirp) $s[n]$ is mathematically given as

$$s[n] = Ae^{j(-\pi Bn + \pi \frac{B}{T}n^2)}, \tag{6.10}$$

where A and B are signal amplitude and bandwidth, respectively. T is the LoRa symbol duration which is jointly determined by spreading factor SF and bandwidth B, given as

$$T = \frac{2^{SF}}{B}. \tag{6.11}$$

The instantaneous frequency of $s(t)$ is defined

$$f[n] = -\frac{B}{2} + \frac{B}{T}n. \tag{6.12}$$

It can be observed that $f(t)$ of a LoRa symbol varies linearly over time. Due to the frequency-varying nature, LoRa signals are often analyzed and represented in the time–frequency domain, i.e. spectrogram. The short-time Fourier transform (STFT) is a widely used time–frequency analyzing algorithm that converts a time domain signal into a spectrogram.

There are usually eight repeated unmodulated LoRa symbols at the beginning of a LoRa packet, named preamble and denoted as $r[n]$. The time domain preamble and the corresponding spectrogram are shown in Figure 6.3.

6.5.2 Signal processing

- **Preamble extraction**: The preamble part of a LoRa packet consists of eight identical unmodulated chirps. After capturing a LoRa packet, we extract its preamble part to prevent the neural network from learning the payload/header information.
- **Frequency offset compensation**: As the two reasons discussed in Section 6.4, the frequency offset of the received signal needs to be compensated for system

security and stability. The compensation process consists of two steps, namely coarse compensation and fine compensation.

As shown in (6.12), the instantaneous frequency of an unmodulated LoRa symbol increases linearly from $-\frac{B}{2}$ to $\frac{B}{2}$ in ideal conditions. In the presence of frequency offset, (6.12) can be re-written as

$$f[n] = -\frac{B}{2} + \frac{B}{T}n + \Delta f, \tag{6.13}$$

where Δf is the frequency offset between the transmitter and receiver. Thanks to the linearity of $f[n]$, Δf can be coarsely estimated by taking the average of $f[n]$, expressed as

$$\Delta\hat{f}_{coarse} = \frac{1}{L}\sum_{n=0}^{L-1} f[n], \tag{6.14}$$

where $L = \frac{T}{T_s}$ is the length of $f[n]$. Then we compensate the signal by multiplying $\Delta\hat{f}_{coarse}$, given as

$$r_{coarse}[n] = r[n]e^{-j2\pi\,\Delta\hat{f}_{coarse} n T_s}, \tag{6.15}$$

where $r_{coarse}[n]$ is the signal after coarse frequency offset compensation.

There is a residual frequency offset in $r_{coarse}[n]$ since the coarse compensation is not sufficiently accurate. We further estimate the residual frequency offset in $r_{coarse}[n]$ by

$$\Delta\hat{f}_{fine} = -\frac{1}{2\pi T_s L}\angle\left(\sum_{n=0}^{L-1} r_{coarse}[n]r^*_{coarse}[n+L]\right), \tag{6.16}$$

where $\angle(\cdot)$ returns the angle. Then the finely compensated signal $r_{fine}[n]$ can be calculated by

$$r_{fine}[n] = r_{coarse}[n]e^{-j2\pi\,\Delta\hat{f}_{fine} n T_s}. \tag{6.17}$$

- **Normalization**: The signal needs to be normalized to prevent the neural network from learning the received signal strength. This is achieved by dividing the root mean square, given as

$$r'[n] = \frac{r[n]}{RMS(r[n])}, \tag{6.18}$$

where $RMS(\cdot)$ returns the root mean square value.

6.5.3 Data augmentation

The channel simulator is designed with the exponential PDP that is denoted as $P(k)$ and given as

$$P(k) = \frac{1}{\tau_d}e^{kT_s/\tau_d}, k = 1, 2, ..., k_{\max}, \tag{6.19}$$

where τ_d is the RMS delay spread and T_s is the receiver sampling rate. k is the path index and k_{max} is the index of the last non-negligible path. The PDP, $P(k)$, is normalized. Each discrete path is implemented with Rician distribution, and we use the Jakes Doppler spectrum for all the paths, given as

$$J(f) = \frac{1}{\pi f_p \sqrt{1 - (f/f_p)^2}}, \tag{6.20}$$

where f_p is the maximum Doppler shift. $J(f)$ characterizes how the path changes over time. For each training signal, we select an RMS delay spread (τ_d) between 10 and 300 ns, a maximum Doppler frequency f_p between 0 and 10 Hz, a Rician K-factor between 0 and 10, and an SNR between 20 and 80 dB to synthesize a wireless channel $\mathbf{h}_m^s$. Then (6.4) is used for augmentation.

The training signal after augmentation is visualized in Figure 6.4. Figure 6.4(a) is the signal $\mathbf{r}'_\mathbf{m}$ in (6.4) collected in a static line-of-sight (LOS) scenario, which is nearly the same as the ideal time-domain signal shown in Figure 6.3(a). When the channel impulse response $\mathbf{h}_\mathbf{m}^\mathbf{s}$ is generated with a maximum Doppler frequency f_p of 10 Hz and an RMS delay spread τ_d of 10 ns, the augmentation result y_m is shown in Figure 6.4. It can be observed that the 10 Hz Doppler effect caused the signal amplitude to vary over time. In contrast, when the maximum Doppler frequency f_p

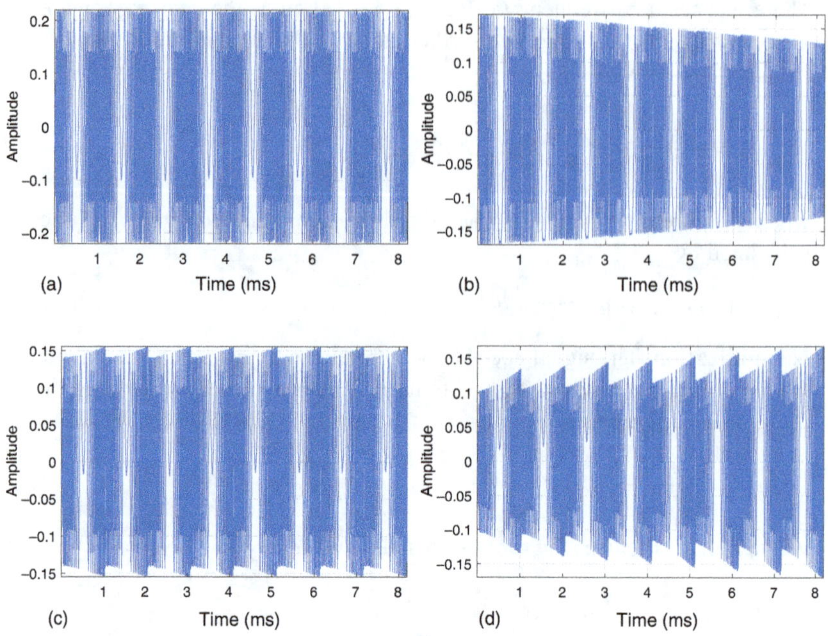

Figure 6.4 Training signal after data augmentation. (a) Collected packet in LOS static environment. (b) Augmentation result with 10 Hz f_p and 10 ns τ_d. (c) Augmentation result with 0 Hz f_p and 300 ns τ_d. (d) Augmentation result with 10 Hz f_p and 300 ns τ_d.

is as small as 0 Hz and the RMS delay spread τ_d is 300 ns, the augmented signal in Figure 6.4(c) shows a "sawtooth" shape and the signal amplitude does not change. When both Doppler and multipath effects are evident (10 Hz f_p and 300 ns τ_d), the augmentation results in Figure 6.4(d) show both "sawtooth" shape and its amplitude varies. The waveform well matches the LoRa signal collected in the moving scenario. In summary, the "sawtooth" shape of the signal is caused by the multipath effect and the amplitude variation is caused by the Doppler effect.

6.5.4 Signal representation

As discussed in Section 6.5.1, the frequency of a LoRa signal varies linearly over time thus spectrogram is often used for representation. However, the spectrogram generated by STFT is affected by the wireless channel, which may degrade the RFFI performance. Therefore, we use channel-independent spectrogram as the signal representation fed into the neural network.

We first perform discrete STFT on the received time-domain signal, generating a complex matrix $\mathbf{S}$ with each element given as

$$S_{k,m} = \sum_{n=0}^{N-1} y[n]W[n - mZ]e^{-j2\pi \frac{k}{N}n}, \tag{6.21}$$

where $k = 1, 2, \ldots, K$, and $m = 1, 2, \ldots, M$ are the column and the row index, respectively. $W[n]$ is the window function and Z is the hop size. The rectangular window is used in this paper. The column number of $\mathbf{S}$, K is determined by the length of the received signal $y[n]$, and the row number M is equal to the length of $W[n]$.

The spectrogram S can be rearranged as

$$\begin{aligned}\mathbf{S} &= \begin{bmatrix} H_{1,1}F(X_{1,1}) & H_{1,2}F(X_{1,2}) & \cdots & H_{1,M}F(X_{1,M}) \\ H_{2,1}F(X_{2,1}) & H_{2,2}F(X_{2,2}) & \cdots & H_{2,M}\ F(X_{2,M}) \\ \vdots & \vdots & \cdots & \vdots \\ H_{K,1}F(X_{K,1}) & H_{K,2}F(X_{K,2}) & \cdots & H_{K,M}F(X_{K,M}) \end{bmatrix} \\ &= [\mathbf{H}_1 \circ F(\mathbf{X}_1) \quad \mathbf{H}_2 \circ F(\mathbf{X}_2) \ \cdots \ \mathbf{H}_M \circ F(\mathbf{X}_M)],\end{aligned} \tag{6.22}$$

where $\circ$ denotes element-wise product. $X_m = [X_{1,m}, X_{2,m}, ..., X_{k,m}]^T$ is the ideal signal frequency spectrum of the mth signal segment. $H_m = [H_{1,m}, H_{2,m}, ..., H_{k,m}]^T$ is the corresponding channel frequency response.

Then we calculate the amplitude of $\mathbf{S}$ since the phase information is easily affected by the phase noise, sampling offset, frequency offset, etc., given as

$$|\mathbf{S}| = [\,|\mathbf{H}_1| \circ |F(\mathbf{X}_1)| \quad |\mathbf{H}_2| \circ |F(\mathbf{X}_2)| \ \cdots \ |\mathbf{H}_M| \circ |F(\mathbf{X}_M)|\,], \tag{6.23}$$

where $|\cdot|$ returns the absolute value. After that, we derive the channel-independent spectrogram $\mathbf{C}$ by dividing the adjacent two columns

$$\mathbf{C} = \left[\frac{|\mathbf{H}_2| \circ |F(\mathbf{X}_2)|}{|\mathbf{H}_1| \circ |F(\mathbf{X}_1)|} \quad \frac{|\mathbf{H}_3| \circ |F(\mathbf{X}_3)|}{|\mathbf{H}_2| \circ |F(\mathbf{X}_2)|} \quad \cdots \quad \frac{|\mathbf{H}_M| \circ |F(\mathbf{X}_M)|}{|\mathbf{H}_{M-1}| \circ |F(\mathbf{X}_{M-1})|} \right]. \tag{6.24}$$

In low-speed scenarios, it is reasonable to assume that the channel frequency response $\mathbf{H}$ nearly remains the same in a short time period, that is $\mathbf{H}_m \approx \mathbf{H}_{m+1}$. Then (6.24) can be simplified as

$$\mathbf{C} = \left[\frac{|F(\mathbf{X}_2)|}{|F(\mathbf{X}_1)|} \quad \frac{|F(\mathbf{X}_3)|}{|F(\mathbf{X}_2)|} \quad \cdots \quad \frac{|F(\mathbf{X}_M)|}{|F(\mathbf{X}_{M-1})|} \right]. \tag{6.25}$$

where it can be observed that C is not related to channel frequency response H and thus channel independently. We finally express C in dB scale, given as

$$\mathbf{C}_{dB} = 10 \log_{10} (|\mathbf{C}|^2). \tag{6.26}$$

$\mathbf{C}_{dB}$ is the signal representation that acts as the input to the neural network. It is a real-valued matrix that does not contain channel information.

6.5.5 Neural network

The characteristics of the signal representation, specifically the channel-independent spectrogram introduced in Section 6.5.4, should be taken into account when designing neural networks. We designed a CNN and an LSTM network to process the channel-independent spectrograms.

6.5.5.1 CNN

The channel-independent spectrogram can be considered a 2-dimensional gray-scale image. Therefore, the popular CNNs that are good at processing images can be used to classify channel-independent spectrograms.

The designed CNN, which is referred to as the ResNet architecture, is shown in Figure 6.5. It consists of 10 convolutional layers, a max pooling layer, an average pooling layer, and a dense layer with softmax activation. There are also skip connections that can enhance the feature extraction capability. Batch normalization is used after each convolutional layer. The designed CNN has 920,350 parameters in total.

6.5.5.2 LSTM network

The channel-independent spectrogram is also a time-varying frequency spectrum by considering each column as a timestep. There are time correlations among columns

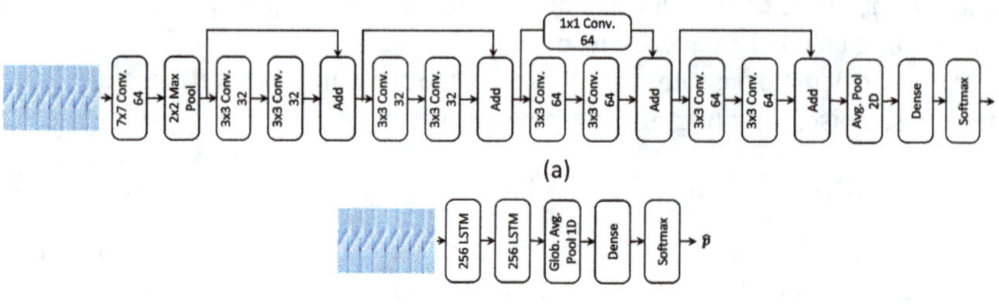

Figure 6.5 Architecture of the neural networks

and thus sequence models are applicable. The LSTM network is a widely used sequence model and is taken as an example.

The architecture of the designed LSTM network is shown in Figure 6.5. It consists of two LSTM layers of 256 units, a global average pooling 1D layer and a dense layer with softmax activation. The LSTM network has learnable 902,686 parameters.

6.6 Experimental evaluation

This section conducts an experimental evaluation of the proposed RFFI system. We collect data to train the CNN and LSTM networks and assess their performance. We also add AWGN to the test signals to emulate different SNR conditions.

6.6.1 Experimental settings

There are five LoPy4 devices and a USRP N210 SDR involved in the experiment. The LoPy4 devices act as end nodes in a LoRaWAN network, and the USRP N210 SDR acts as the LoRa gateway. The goal of the RFFI system is to correctly classify the five LoPy4 devices by analyzing the physical layer signals collected by the USRP N210 SDR. The USRP N210 SDR and LoPy4 devices are situated roughly one meter apart during signal collection. The data collection is carried out one after another and the LoPy4 devices are placed at the same location. The bandwidth and spreading factor of all the LoPy4 devices are set to 125 kHz and 7, respectively. The sampling rate of the USRP N210 SDR is 1 MHz.

The neural networks are trained with the same hyperparameters. The ADAM optimizer is used for training with an initial learning rate of 0.001 and a batch size of 32. Ten percent of the training data is used for validation. Once the validation loss stops decreasing for 10 epochs, the learning rate is reduced to 0.2 times the current value. When the validation loss stops reducing after 20 epochs, the training is terminated.

The LoRa data collection program is based on MATLAB® and the deep learning models are built with the Tensorflow library. The training is done on a PC with an NVIDIA GeForce GTX 1660 GPU.

6.6.2 Effect of locations

We collect six datasets, i.e., datasets A–F, at different locations in an office building. The floor plan is shown in Figure 6.6. Dataset A contains 2000 LoRa packets from each LoPy4 device, while each of the remaining datasets consists of 200 packets from each LoPy4. Locations A–C are in line-of-sight (LOS) conditions and locations D–F are in non-line-of-sight (NLOS) conditions with stronger multipath effects.

Dataset A is used to train the neural networks and datasets B–F are used for testing. The classification results are shown in Figure 6.7. It can be observed that the accuracy of both the LSTM and CNN networks is above 90% at different locations. The LSTM network performs marginally better than the CNN network, which may be because it is designed to have fewer parameters and is easier to train. The results

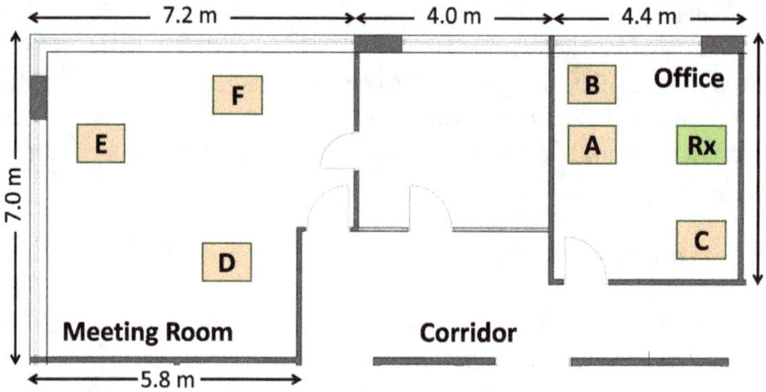

Figure 6.6 Floor plan

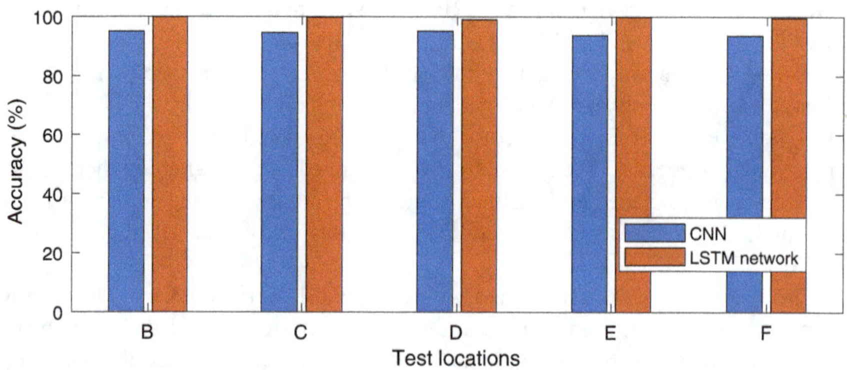

Figure 6.7 Effect of locations. Both CNN and LSTM networks are trained with the dataset collected at location A.

demonstrate that using channel-independent spectrograms as neural network inputs can effectively mitigate channel effects and achieve location-invariant RFFI.

6.6.3 Performance in different SNR scenarios

We add artificial noise to dataset B to emulate different SNR conditions. Dataset B is collected in the LOS scenario and the estimated SNRs of the collected packets are above 40 dB. Therefore, it can be considered as a dataset consisting of noiseless signals, and adding noise to it can synthesize datasets collected under different SNR conditions.

The classification results are shown in Figure 6.8. It can be observed that both the CNN and LSTM networks perform excellently in high SNR scenarios above 40 dB, and the classification accuracy exceeds 90%. In low SNR scenarios under 20 dB, the system nearly does not work as the classification accuracy is close to random guesses.

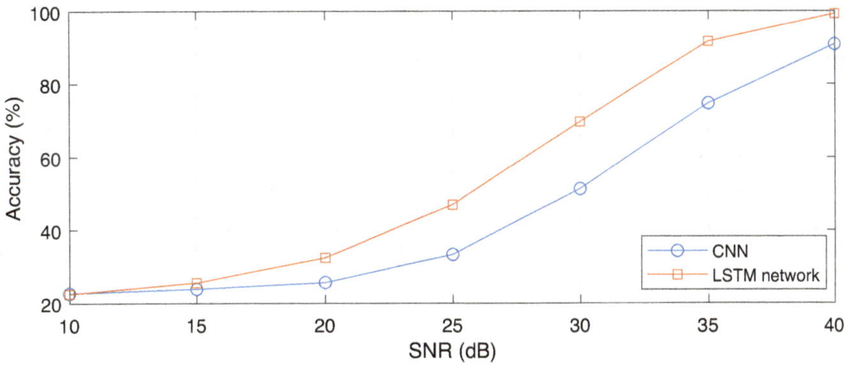

Figure 6.8 Performance under different SNR conditions

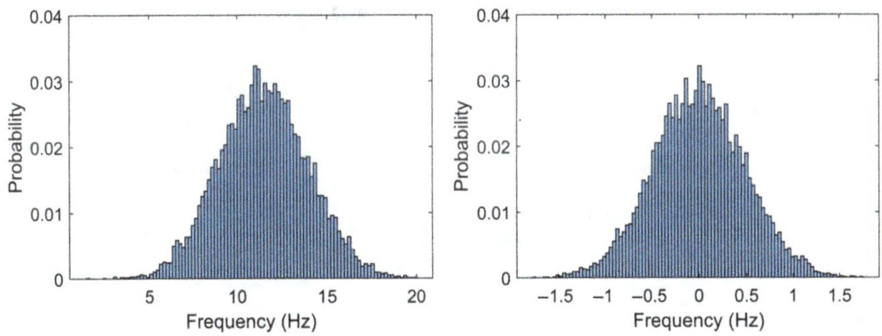

Figure 6.9 Performance of the frequency offset compensation algorithm. (a) Residual frequency offset after coarse compensation. (b) Residual frequency offset after fine compensation.

6.6.4 Performance of the frequency offset compensation algorithm

Simulation is used to evaluate the performance of the proposed coarse and fine frequency offset compensation algorithms. We generate eight consecutive unmodulated LoRa symbols as the preamble part in a LoRa packet and add a random frequency offset between −10,000 and +10,000 Hz. Then we use the proposed algorithms to estimate and compensate for the frequency offset. The simulation is repeated 10,000 times and the residual frequency offset after coarse and fine compensation are shown in Figure 6.10 as histograms. As can be observed that the residual frequency offset after coarse compensation roughly ranges from 5 Hz to 20 Hz. The fine compensation can further reduce the frequency offset to less than 1.5 Hz. The SNR is set to 20 Hz during simulation.

6.6.5 *Effect of frequency offset compensation on RFFI performance*

The oscillators in the transceivers are sensitive to temperature variations thus the frequency offset in the received signal can change over time. This reduces the performance of the RFFI system, as the frequency offsets in the training data may be different from those in the test data.

We used dataset A to train the neural network, and dataset B to evaluate and compare the system performance with and without frequency offset compensation. Comparing the results in Figure 6.10(a) and (b), we can observe that not using the frequency offset compensation reduces the classification accuracy of CNN by about 20%, namely from 95% to 73.60%. For the results of the LSTM network shown in Figure 6.10(c) and (d), the accuracy also drops from 99.90% to 95.00% when frequency offset compensation is not applied. The results show that the frequency

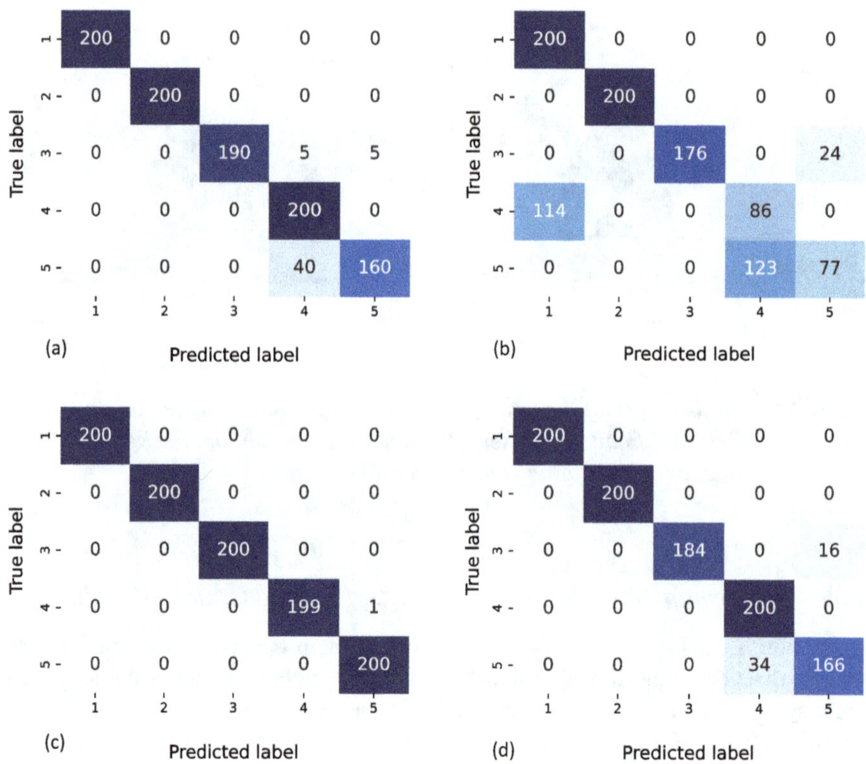

Figure 6.10 Classification results at location B. (a) CNN with frequency offset compensation, 95.00%. (b) CNN without frequency offset compensation, 73.60%. (c) LSTM network with frequency offset compensation, 99.90%. (d) LSTM network without frequency offset compensation, 95.00%.

offset in the received signal is not stable, and compensating for it can effectively improve system performance.

6.7 Conclusion

In this chapter, we introduce the essential modules in an RFFI system and take LoRa-RFFI as a case study. We propose channel-independent spectrogram to mitigate the channel effects in LoRa-RFFI, alleviating the multipath effect while preserving the hardware distortions. As channel-independent spectrograms can be regarded as either 2D images or time-varying frequency spectrum, both CNN and LSTM networks are designed to process them. Frequency offset compensation is used to prevent performance degradation caused by temperature changes. Experiments involving five COTS LoPy4 devices and a USRP N210 SDR are conducted in an office building. The LoPy4 devices are placed at different locations to evaluate the robustness of the system to channel variations. The experimental results show that the RFFI system is robust to location changes and the classification accuracy is always above 90%. It is also experimentally found that frequency compensation is necessary for RFFI to prevent performance degradation. The accuracy can be improved by about 20% after frequency offset compensation is applied.

References

[1] W. Wang, Z. Sun, S. Piao, B. Zhu, and K. Ren, "Wireless physical-layer identification: Modeling and validation," *IEEE Trans. Inf. Forens. Secur.*, vol. 11, no. 9, pp. 2091–2106, 2016.

[2] J. Zhang, R. Woods, M. Sandell, M. Valkama, A. Marshall, and J. Cavallaro, "Radio frequency fingerprint identification for narrowband systems, modelling and classification," *IEEE Trans. Inf. Forens. Secur.*, vol. 16, pp. 3974–3987, 2021.

[3] Q. Xu, R. Zheng, W. Saad, and Z. Han, "Device fingerprinting in wireless networks: challenges and opportunities," *IEEE Commun. Surveys Tuts.*, vol. 18, no. 1, pp. 94–104, 2015.

[4] P. Robyns, E. Marin, W. Lamotte, P. Quax, D. Singelée, and B. Preneel, "Physical-layer fingerprinting of LoRa devices using supervised and zero-shot learning," in *Proceedings of the ACM Conference on Security Privacy Wireless Mobile Networks (WiSec)*, 2017, pp. 58–63.

[5] L. Peng, J. Zhang, M. Liu, and A. Hu, "Deep learning based RF fingerprint identification using differential constellation trace figure," *IEEE Trans. Veh. Technol.*, vol. 69, no. 1, pp. 1091–1095, 2019.

[6] A. Al-Shawabka, F. Restuccia, S. D'Oro, *et al.*, "Exposing the fingerprint: Dissecting the impact of the wireless channel on radio fingerprinting," in *Proceedings of the IEEE International Conference on Computer Communications (INFOCOM)*, July 2020, pp. 646–655.

[7] G. Shen, J. Zhang, A. Marshall, L. Peng, and X. Wang, "Radio frequency fingerprint identification for LoRa using deep learning," *IEEE J. Sel. Areas Commun.*, vol. 39, no. 8, pp. 2604–2616, 2021.

[8] G. Shen, J. Zhang, A. Marshall, L. Peng, and X. Wang, "Radio frequency fingerprint identification for LoRa using spectrogram and CNN," in *Proceedings of the IEEE International Conference on Computer Communications (INFOCOM)*, Virtual Conference, May 2021, pp. 1–10.

[9] G. Shen, J. Zhang, A. Marshall, and J. R. Cavallaro, "Towards scalable and channel-robust radio frequency fingerprint identification for LoRa," *IEEE Trans. Inf. Forens. Secur.*, vol. 17, pp. 774–787, 2022.

[10] D. Roy, T. Mukherjee, M. Chatterjee, E. Blasch, and E. Pasiliao, "RFAL: adversarial learning for RF transmitter identification and classification," *IEEE Trans. Cogn. Commun. Netw.*, vol. 6, no. 2, pp. 783–801, 2019.

[11] M. Cekic, S. Gopalakrishnan, and U. Madhow, "Wireless fingerprinting via deep learning: the impact of confounding factors," in *Proceedings of the Asilomar Conference on Signals, Systems, and Computers*, 2021, pp. 677–684.

[12] J. Yu, A. Hu, G. Li, and L. Peng, "A robust RF fingerprinting approach using multisampling convolutional neural network," *IEEE Internet Things J.*, vol. 6, no. 4, pp. 6786–6799, 2019.

[13] T. Jian, Y. Gong, Z. Zhan, *et al.*, "Radio frequency fingerprinting on the edge," *IEEE Trans. Mobile Comput.*, vol. 21, no. 11, pp. 4078–4093, 2021.

[14] N. Soltani, K. Sankhe, J. Dy, S. Ioannidis, and K. Chowdhury, "More is better: Data augmentation for channel-resilient RF fingerprinting," *IEEE Commun. Mag.*, vol. 58, no. 10, pp. 66–72, 2020.

[15] N. Soltani, G. Reus-Muns, B. Salehihikouei, J. Dy, S. Ioannidis, and K. Chowdhury, "RF fingerprinting unmanned aerial vehicles with non-standard transmitter waveforms," *IEEE Trans. Veh. Technol.*, vol. 69, no. 12, pp. 15518–15531, 2020.

[16] Y. Qian, J. Qi, X. Kuai, G. Han, H. Sun, and S. Hong, "Specific emitter identification based on multi-level sparse representation in automatic identification system," *IEEE Trans. Inf. Forensics Security*, vol. 16, pp. 2872–2884, 2021.

[17] A. Al-Shawabka, P. Pietraski, S. B Pattar, F. Restuccia, and T. Melodia, "DeepLoRa: fingerprinting LoRa devices at scale through deep learning and data augmentation," in *Proceedings of the ACM International Symposium on Mobile Ad Hoc Networking and Computing (MobiHoc)*, Shanghai, China, July 2021.

[18] M. Piva, G. Maselli, and F. Restuccia, "The tags are alright: Robust large-scale RFID clone detection through federated data-augmented radio fingerprinting," in *Proceedings of the ACM International Symposium on Mobile Ad Hoc Networking and Computing (MobiHoc)*, Shanghai, China, Jul. 2021.

[19] K. Merchant, S. Revay, G. Stantchev, and B. Nousain, "Deep learning for RF device fingerprinting in cognitive communication networks," *IEEE J. Sel. Topics Signal Process.*, vol. 12, no. 1, pp. 160–167, 2018.

[20] A. Elmaghbub and B. Hamdaoui, "LoRa device fingerprinting in the wild: disclosing RF data-driven fingerprint sensitivity to deployment variability," *IEEE Access*, vol. 9, pp. 142 893–142 909, 2021.
[21] S. Hanna, S. Karunaratne, and D. Cabric, "Wisig: a large-scale WiFi signal dataset for receiver and channel agnostic RF fingerprinting," *IEEE Access*, vol. 10, pp. 22808–22818, 2022.
[22] R. Xie, W. Xu, Y. Chen, *et al.*, "A generalizable model-and-data driven approach for open-set RFF authentication," *IEEE Trans. Inf. Forens. Secur.*, vol. 16, pp. 4435–4450, 2021.
[23] S. Rajendran and Z. Sun, "RF impairment model-based IoT physical-layer identification for enhanced domain generalization," *IEEE Trans. Inf. Forens. Secur.*, vol. 17, pp. 1285–1299, 2022.
[24] G. Shen, J. Zhang, A. Marshall, M. Valkama, and J. Cavallaro, "Towards length-versatile and noise-robust radio frequency fingerprint identification," *arXiv preprint arXiv:2207.03001*, 2022.
[25] R. Das, A. Gadre, S. Zhang, S. Kumar, and J. M. Moura, "A deep learning approach to IoT authentication," in *Proceedings of the IEEE International Conference on Communications (ICC)*, 2018, pp. 1–6.
[26] S. D. Andrews, "Extensions to Radio Frequency Fingerprinting," Ph.D. dissertation, Virginia Tech, 2019.
[27] B. Danev, T. S. Heydt-Benjamin, and S. Capkun, "Physical-layer identification of RFID devices," in *Proceedings of the USENIX Security Symposium*, 2009, pp. 199–214.
[28] A. C. Polak, S. Dolatshahi, and D. L. Goeckel, "Identifying wireless users via transmitter imperfections," *IEEE J. Sel. Areas Commun.*, vol. 29, no. 7, pp. 1469–1479, 2011.
[29] P. Liu, P. Yang, W.-Z. Song, Y. Yan, and X.-Y. Li, "Real-time identification of rogue WiFi connections using environment-independent physical features," in *Proceedings of the IEEE International Conference on Computer and Communications (INFOCOM)*, Paris, France, April 2019, pp. 190–198.
[30] K. Joo, W. Choi, and D. H. Lee, "Hold the door! fingerprinting your car key to prevent keyless entry car theft," in *Proceedings of the Network and Distributed System Security Symposium (NDSS)*, Virtual Conference, February 2020.
[31] J. Hua, H. Sun, Z. Shen, Z. Qian, and S. Zhong, "Accurate and efficient wireless device fingerprinting using channel state information," in *Proceedings of the IEEE International Conference on Computer and Communications (INFOCOM)*, Honolulu, HI, USA, April 2018, pp. 1700–1708.
[32] Y. Shi and M. A. Jensen, "Improved radiometric identification of wireless devices using MIMO transmission," *IEEE Trans. Inf. Forens. Secur.*, vol. 6, no. 4, pp. 1346–1354, 2011.
[33] A. C. Polak and D. L. Goeckel, "Wireless device identification based on RF oscillator imperfections," *IEEE Trans. Inf. Forens. Secur.*, vol. 10, no. 12, pp. 2492–2501, 2015.

[34] L. Peng, A. Hu, J. Zhang, Y. Jiang, J. Yu, and Y. Yan, "Design of a hybrid RF fingerprint extraction and device classification scheme," *IEEE Internet of Things J.*, vol. 6, no. 1, pp. 349–360, 2018.

[35] V. Brik, S. Banerjee, M. Gruteser, and S. Oh, "Wireless device identification with radiometric signatures," in *Proceedings of the International Conference on Mobile Computer Network (MobiCom)*, San Francisco, CA, USA, September 2008, pp. 116–127.

[36] B. He and F. Wang, "Cooperative specific emitter identification via multiple distorted receivers," *IEEE Trans. Inf. Forens. Secur.*, vol. 15, pp. 3791–3806, 2020.

[37] G. Li, J. Yu, Y. Xing, and A. Hu, "Location-invariant physical layer identification approach for WiFi devices," *IEEE Access*, vol. 7, pp. 106974–106986, 2019.

[38] K. Sankhe, M. Belgiovine, F. Zhou, S. Riyaz, S. Ioannidis, and K. Chowdhury, "ORACLE: Optimized radio classification through convolutional neural networks," in *Proceedings of the IEEE International Conference on Computing Communications (INFOCOM)*, Paris, France, 2019, pp. 370–378.

[39] K. Sankhe, M. Belgiovine, F. Zhou, *et al.*, "No radio left behind: Radio fingerprinting through deep learning of physical-layer hardware impairments," *IEEE Trans. Cogn. Commun. Netw.*, vol. 6, no. 1, pp. 165–178, 2019.

[40] K. Merchant and B. Nousain, "Enhanced RF fingerprinting for IoT devices with recurrent neural networks," in *Proceedings of the IEEE Military Communications Conference (MILCOM)*, 2019, pp. 590–597.

Chapter 7
Radio frequency fingerprint-based wireless device identification

Ming Liu[1], Hui Luo[1], Xin Wang[1], Linning Peng[2] and Hua Fu[2]

With the rapid evolution of the wireless networking technologies, various wireless network services emerged and penetrated into all aspects of life. Numerous wireless networking applications, such as smart cities, industrial Internet, and Internet of Vehicles, are developed and deployed around the world. More and more wireless devices are connected to the Internet, forming an enormous Internet of Things (IoT). Especially, in the era of fifth-generation wireless communication (5G), IoT becomes one of the three natively supported scenarios, which will consequently promote the development of IoT-based applications [1]. The forthcoming sixth-generation wireless communication (6G) will bring stronger communication capabilities with more features and will lead to new wireless applications [2]. One can foresee that the wireless networking technology, together with artificial intelligence, big data, cloud computing, will form the cornerstone of the future smart life. While providing extraordinary convenience to connect people and "things," the wireless networks are facing more and more security challenges from adversaries. In this chapter, a security measure that exploits the physical layer characteristics of the transceivers is introduced. The related research progress is reviewed and a representative technique is presented. Some potential research directions are highlighted.

7.1 Security risks in wireless networks

In a wireless network, user devices and the network transmit information through wireless channels in a broadcast manner. The communication medium is open, allowing any device to transmit to, and receive from the surrounding area. This offers great

[1]Engineering Research Center of Network Management Technology for High Speed Railway of Ministry of Education, Collaborative Innovation Center of Railway Traffic Safety, School of Computer and Information Technology, Beijing Jiaotao University, China
[2]School of Cyber Science and Engineering, Southeast University, China

flexibility for network deployment and ubiquitous connectivity for users. In the meantime, the wireless communication links also contain potential information security risks. The wireless signal can be intercepted and detected by the malicious users. The information carried by the signal could be decoded and deciphered if the encryption mechanism is cracked or if the secret key is stolen, resulting in possible information leakage. More seriously, the digital identities of the wireless devices could be tampered with and disguised by deliberate attackers. If the attackers can impersonate the identities of legitimate wireless devices, the identification and authentication mechanisms of the wireless network will be compromised and the attackers may gain access to the network leading to more serious security consequences. Obviously, security has become one of the main threats faced in the development of wireless networks.

Notably, besides the risks resulted from the openness of communication medium, security protocols themselves may also contain some flaws that can be exploited by adversaries to launch cyber attacks. Take the mobile telephone network as an example. The Global System for Mobile Communications (GSM) standard only performs one-way authentication. It does not require the user equipment (UE) to authenticate the base stations (BSes) [3]. Therefore, a malicious user can take advantage of this weakness to disguise itself as a BS. With the false base station problem, a UE faces the challenges of both passive and active attacks [4].

When performing passive attacks, the false base stations can steal the identities of legitimate users, or locate their positions, based on the received wireless signal. The passive attacks do not interfere the legitimate communication but could lead to identity leakage and privacy loss. In active attacks, the attackers can carry out more serious attacks by sending fake signaling and information with the identity of BS. The active attackers can perform either Man-In-The-Middle (MITM), or Man-On-The-Side (MOTS) type of attacks [5]. In MITM attacks, the attacker connects to both legitimate BS and UE. It can replay forged signaling and messages to legitimate UEs and can literally hijack the communication link to grab the control of UEs. As such, the MITM attacker can cheat and manipulate the UEs to affect their own communication or do harm to the network on a larger scale. As for MOTS type of attacks, the attacker does not need to take over the control of UEs as in the MITM cases. It injects messages into the legitimate communications in order to acquire information from UE or deliver information to UE. For instance, the attacker can steal the IMSI of UE by sending an identity request message. More commonly, the MOTS attacker can be forced to send phishing short messages to UE and potentially cause financial loss on the mobile network subscribers.

It is worth noting that the new-generation network will co-exist with the previous ones for a long time. The attackers can attack the new-generation network and force the user devices to fall back to the legacy network [6]. So the security breach existing in the legacy network can be exploited to achieve the attacks to the mobile network. The security threat can last for decades during the whole life-span of the wireless standard. The security "blind spots" exist in not only the legacy wireless communication system but the "newer generation" ones, as well [7,8]. New attack methods can be devised, bringing more severe security risks.

The Internet-of-Things (IoT) is the most active sector of the Internet applications. It is estimated that the number of connected IoT devices reaches 14.4 billion worldwide, and it is expected that this number will increase to 27 billion by 2025 [9]. Along with the global popularization, the security risks associated with IoT applications raise more and more attention in recent years. On the one hand, wireless devices are often directly exposed to the open wireless signal transmission environment. Most IoT devices are unattended and generally lack anti-virus software and network firewall protection, making them more vulnerable to direct attacks by brute force cracking. On the other hand, once the authentication mechanism of IoT devices is cracked, the massive IoT devices can be hijacked by malicious users and could become a source of cyber attacks to affect the Internet infrastructure and services. For instance, the US domain name service (DNS) company Dyn was paralyzed by a large-scale distributed denial-of-service (DDoS) attack with historically high traffic volume. It was later revealed that this attack originated from massive IoT devices infected by "Mirai" botnet [10].

Moreover, due to the cost, size, and energy consumption restrictions, IoT devices usually have very limited hardware resources. Its data storage, information processing, and reliable communication capabilities are obviously weaker than those of traditional wireless communication equipment. These factors make it difficult to employ sophisticated security mechanisms. The signaling interaction in traditional wireless networks becomes too cumbersome and complicated for the authentication of IoT devices. Therefore, novel lightweight security methods are highly desired for the IoT applications.

In fact, not only limited to wireless networks, the current network security measures are facing more and more challenges. Traditionally, the authenticity of user identity and the confidentiality of network communication are protected by high-layer security protocols with some cryptographic mechanisms. For instance, the users' identities should be authenticated when users' devices are initially connected to the network. For now, the authentication mainly relies on the secure credentials, such as security certificate, password, or security questions, that are associated with the user's identity. However, these kinds of cryptographic mechanisms are subject to security risks, such as weak passwords, key leakage, and permission bypass. With the continuous development of computing technology and decryption techniques, the traditional cryptographic protocol-based high-layer authentication mechanism may encounter more and more challenges from malicious users with higher computation capability. The emerging quantum computers have a huge impact on the security of asymmetric keys and can greatly reduce the complexity of brute force cracking of symmetric key algorithms. We may need to review the existing security measures that rely solely on encryption in a short future.

In general, as can be seen from the above facts, the network security mechanisms that solely rely on encryption protocols have become less reliable. Securing the authentication of the identities of wireless devices at the network access stage and preventing the intrusion of fake devices into the network are critical prerequisites for the secure communication. *It is necessary to investigate a more reliable device identification mechanism that is less susceptible to tampering, impersonation and permission bypassing.* Therefore, it is important for the future wireless network infrastructure to incorporate various processes including encryption, multi-factor authentication to secure data and services. A viable solution is to develop the *inherent features in the wireless communication that are uniquely determined by communication participants with high stability*, as the basis for identifying the communication devices.

7.2 Radio frequency fingerprint

As discussed in previous sections, the high-layer encryption-based network security mechanism may face more severe challenges in the coming future. In recent years, the device identification methods leveraging the hardware characteristics have drawn more and more attention. It is commonly referred to as radio frequency (RF) fingerprint (RFF) [11,12]. The basic idea is to extract, from the received RF signal, distinctive signal features that are associated with the specific transmitter. As the signal features are affected by the hardware imperfections of the transmitter, they can be leveraged as the unique "signature" to verify the identity of the transmitter. It can be used as an identification method that is independent of the traditional authentication mechanisms.

7.2.1 Properties of RF fingerprint

Like human biometric measures such as fingerprints, irises, and palmprints, RFF should preserve some important properties to achieve practical implementation [13].

Universality: A fingerprint should be extracted from all considered devices. It means that the fingerprint features are not sporadic and limited to a portion of devices. The fingerprint can be observed in the signal of all devices in common application scenarios.

Uniqueness: Different devices should have distinct RFF features. The features of different devices can be differentiated in one or more dimensions.

Permanence: The representation of a fingerprint should be relatively stable over time. Note that this does not necessarily mean that the features are absolutely static. In fact, the characteristics may vary, but the manner of the variation (distribution of characteristics) should remain unchanged.

Collectability: The features of a fingerprint should be able to be captured with available (affordable) equipment. In other words, the features should be observed by the devices that need to perform the identification.

Robustness: A fingerprint should function properly under the internal and external influences. The internal influences are the factors that directly apply to the hardware of the transmitter and affect its operation characteristics. The common internal influences include ambient temperature, voltage, and power levels. The external influences are the factors that affect the signal propagation conditions and have an effect on the signal characteristics. The external influences mainly are the physical channel conditions like surrounding materials that may change the electromagnetic (EM) wave propagation, as well as the electrical impacts like RF interference that may directly alter the signal distribution.

Data-(in)dependency: A device's fingerprint is a collection of physical-layer signal features that are uniquely associated with the specific device. In theory, a fingerprint should be only determined by physical layer factors and be independent of any *information* that is carried by the signal. That is, a fingerprint should be irrelevant to the pattern of baseband data bits, which means that the extracted fingerprint should remain the same with randomly changing data.

Note that not all the RFF extraction methods proposed in the literature can fulfill all the properties listed above. In particular, the robustness and data-independency properties are still challenging issues that attract much research attention recently. The unsolved difficulties and possible solutions will be elaborated in Section 7.5.

7.2.2 Source of RF fingerprint

Figure 7.1 shows the RF front-end framework of a typical wireless digital communication system, consisting of the transmitter, wireless channel, and receiver. The information-bearing baseband signal is modulated by the quadrature modulator and is then up-converted to RF band and power amplified before sending over-the-air. At the receiver side, the reverse process is carried out to recover the transmitted information. The whole transmission and reception process can be expressed as the complex-valued baseband equivalent transmission model [14].

The main sources of RFF are the hardware imperfections of the electronic components of the transmitter RF circuits [11]. The manufacturing process of electronic components inevitably contains some "errors" in terms of parameter inaccuracies. In other words, the actual parameters of every component may deviate from their nominal values. In fact, the production of electronic devices can accept components with parameters varying within a small range, commonly referred to as component tolerance, without hindering the proper function of the devices [15]. Thus, the actual component parameters are diverse for the different devices. Moreover, the assembly process further introduces some variations in the distributed components. This will cause inherent hardware defects in the RF circuit of the transmitter. These imperfect factors form a combination of characteristics that reflect the hardware, and thus can be used as the basis of the identification. Three main sources of hardware impairments, namely carrier frequency offset (CFO) [16], I/Q imbalance [17], and non-linearity [18], will be discussed in this section.

7.2.2.1 Carrier frequency offset

In most communication systems, the transmitter and the receiver are not synchronized by a common clock reference. Instead, they use their own local oscillators (LOs) to generate a frequency reference signal for up-/down-conversion. Since the LOs used at the two sides cannot offer exactly the same reference signal, the actual frequencies generated at each side may deviate from the nominal frequency. This yields a CFO between the transmitter and the receiver. The CFO causes signal distortion at the

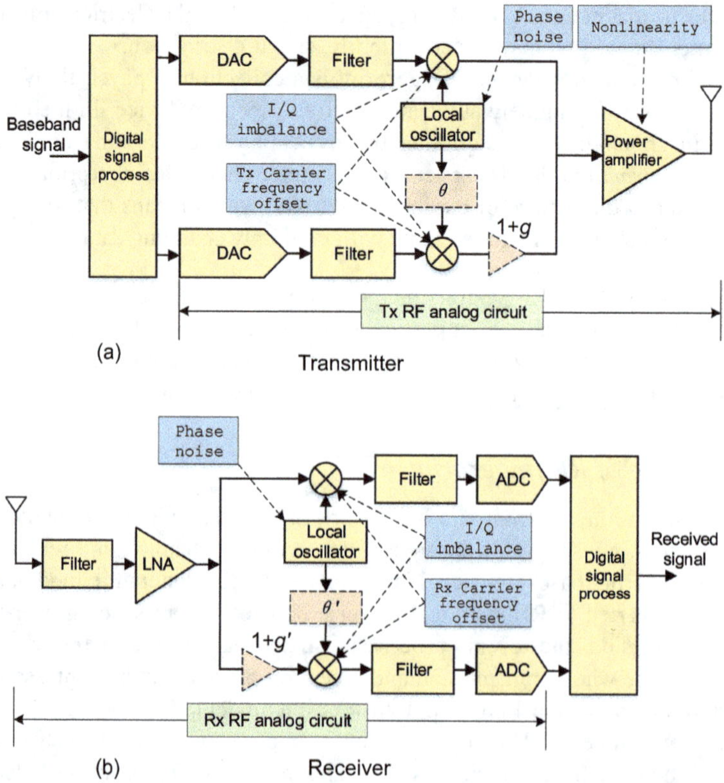

Figure 7.1 RF transmission chain using quadrature modulation structure. Various prominent hardware impairments are highlighted. The dashed blocks are the virtual blocks caused by the imperfection of quadrature modulator.

receiver side, which can be observed and leveraged as an indicator to determine the identity of transmitter.

Concretely, the real-valued passband signal of the transmitter is written as [14]

$$\begin{aligned} s(t) &= \mathfrak{Re}[s_b(t) \cdot e^{j2\pi f_c t}] \\ &= \mathfrak{Re}[s_b(t)] \cos(2\pi f_c t) - \mathfrak{Im}[s_b(t)] \sin(2\pi f_c t), \end{aligned} \tag{7.1}$$

where $s_b(t)$ is the so-called complex baseband equivalent signal; f_c is the carrier frequency of the transmitter; $\mathfrak{Re}[\,\cdot\,]$ and $\mathfrak{Im}[\,\cdot\,]$ represent the real and imaginary parts of a complex number, respectively.

After passing the passband propagation channel, the signal received by the receiver can be written as

$$r(t) = \sum_{l=1}^{L} \rho_l(t) s(t - \tau_l) + w(t), \tag{7.2}$$

where $\rho_l(t)$ is the real-valued passband channel coefficient of the lth path; L is the number of propagation paths; τ_l is the corresponding delay; $w(t)$ is the additive white Gaussian noise. Consequently, the receiver demodulates the received signal with a frequency f_c' generated by its local oscillator. Assume that there exists a CFO between transmitter and receiver sides, i.e., $f_o \triangleq f_c - f_c'$. For the ease of derivation in the following sections, we denote a nominal carrier frequency f, such that $f_c = f + \frac{f_o}{2}$ and $f_c' = f - \frac{f_o}{2}$. Note that f does not necessarily need to be the accurate carrier frequency, but is only introduced to facilitate the following derivation.

The real part of the baseband signal is obtained by demodulating $s(t)$ with $\cos(2\pi f_c' t)$ followed by a low-pass filtering at the baseband. Accordingly, the imaginary part of the baseband signal is obtained by multiplying $-\sin(2\pi f_c' t)$ followed by a low-pass filtering. Combining the real and imaginary parts, the complex-valued baseband signal is expressed as

$$\begin{aligned} y(t) &= \mathscr{F}[r(t) \cdot e^{-j2\pi f_c' t}] \\ &= \mathscr{F}\left[\sum_{l=1}^{L} \rho_l(t) s(t-\tau_l) e^{-j2\pi f_c' t}\right] + w'(t) \\ &= e^{j2\pi f_o t} \sum_{l=1}^{L} h_l(t) s_b(t-\tau_l) + w'(t), \end{aligned} \tag{7.3}$$

where $\mathscr{F}[\,\cdot\,]$ is the low-pass filtering function; $h_l(t) \triangleq \rho_l(t) e^{-j2\pi f_c \tau_l}$ is the baseband channel coefficient [14]; $w'(t)$ is the low-pass filtered noise. It can be seen that the transmitted signal $s_b(t)$ is affected by channel uncertainties, including multipath fading and noise, as well as CFO. If there is no CFO between transmitter and receiver, i.e., $f_c = f_c'$, the received signal (7.3) turns to the classic baseband signal with a tapped delay line channel model.

7.2.2.2 I/Q imbalance

The hardware asymmetries between the in-phase (I) and quadrature (Q) branches of the quadrature modulator yield different gains and phase shifts to the I/Q signal components. The origin of I/Q imbalance is demonstrated by the dashed blocks in Figure 7.1. The two branches of the quadrature modulator and demodulator have certain imbalances including a gain mismatch g and a phase mismatch θ in one branch, as shown in Figure 7.1. Instead of multiplying by a pure complex frequency $e^{j2\pi f_c t}$ in (7.1), the passband signal after quadrature modulation is

$$s(t) = \mathfrak{Re}[s_b(t) \cdot x_t], \tag{7.4}$$

where

$$\begin{aligned} x_t &= \cos(2\pi f_c t) + j(1+g)\sin(2\pi f_c t + \theta) \\ &= K_1 e^{j2\pi f_c t} + K_2 e^{-j2\pi f_c t}, \end{aligned} \tag{7.5}$$

is the up-conversion signal, and the coefficients are

$$K_1 = \frac{1}{2}(1 + (1+g)e^{j\theta}), \tag{7.6}$$

$$K_2 = \frac{1}{2}(1 - (1+g)e^{-j\theta}). \tag{7.7}$$

Reversely, the down-conversion signal for the quadrature de-modulator is

$$\begin{aligned} x_r &= \cos(2\pi f_c' t) - j(1+g)\sin(2\pi f_c' t + \theta') \\ &= K_1' e^{-j2\pi f_c' t} + K_2' e^{j2\pi f_c' t}, \end{aligned} \tag{7.8}$$

where the coefficients are

$$K_1' = \frac{1}{2}(1 + (1+g')e^{-j\theta'}), \tag{7.9}$$

$$K_2' = \frac{1}{2}(1 - (1+g')e^{j\theta'}), \tag{7.10}$$

with θ' and g' being the phase and gain mismatches at the receiver side, respectively. Replacing the pure complex frequency $e^{-j2\pi f_c' t}$ in (7.3) by (7.8) and simplifying the multipath channel fading to flat fading, the received baseband signal is

$$\begin{aligned} y(t) &= \mathscr{F}[r(t) \cdot x_r] \\ &= s_b(t)(K_1 K_1' e^{j2\pi f_o t} + K_2 K_2' e^{-j2\pi f_o t}) \\ &+ s_b^*(t)(K_2^* K_1' e^{j2\pi f_o t} + K_1^* K_2' e^{-j2\pi f_o t}). \end{aligned} \tag{7.11}$$

In the case where the I/Q imbalance is negligible, i.e., $g = 0, \theta = 0, g' = 0$, and $\theta' = 0$, the coefficients become $K_1 = 1, K_2 = 0, K_1' = 1$, and $K_2' = 0$. The received signal converts to the form with only CFO as given in (7.3).

7.2.2.3 Power amplifier non-linearity

After quadrature modulation and up-conversion, the RF passband signal goes through the power amplifier which elevates the signal power to a certain level. Ideally, the amplification process should be a linear increase in signal amplitude. However, in reality, the power amplifier exhibits nonlinear behavior due to the limit of electronics. As a consequence, the amplified RF signal contains nonlinear distortion. This nonlinearity factor gives a clue of the characteristics of the power amplifier, and, therefore, can be leveraged to determine the identity of the transmitter.

More concretely, the output signal of the amplifier can be expressed as [19]

$$p(t) = h_{PA}(s(t), \mathbf{a}_{tx}) * h_{BP}(t), \tag{7.12}$$

where $h_{PA}(\cdot, \mathbf{a}_{tx})$ is the nonlinear response of the power amplifier at the transmitter; $\mathbf{a}_{tx}$ is the complex power series; $*$ is the convolution operation; and $h_{BP}(t)$ is the response of bandpass filter whose frequency response is

$$H_{BP}(t) = \begin{cases} 1, & |f| < W_c \\ 0, & |f| > W_c \end{cases}. \tag{7.13}$$

The passband filter is assumed to be linear and constant within the carrier bandwidth W_c. The nonlinear response of the amplifier can be approximately expressed by the Taylor series [20], or Volterra series [21].

The nonlinearity in power amplifier leads to complicated signal distortion such as harmonic distortion, gain compression, cross modulation, and intermodulation. These factors consequently affect the characteristics of RF signal, which in turn can be leveraged to determine the identity of the transmitter.

7.2.3 RF fingerprint-based device identification

RFF exploits physical layer signal features to determine the transmitter that sends the signal. Therefore, it can be adopted as a complementary security measure to enhance the traditional encryption-based security mechanisms. Figure 7.2 illustrates the basic RFF-based device identification process. The RF signal sent by the transmitters is collected by an "identifier." Consequently, necessary signal processing is carried out to extract RFF features which are employed as the basis of the fingerprint. The extracted fingerprints are then stored in the database and will be used to identify the new RF signal. Two types of device identification tasks, namely *known device verification* and *unknown device detection*, are usually considered when trying to apply RFF in the authentication procedure of wireless networks.

Known device verification refers to the tasks that determine the signal-under-test belongs to which known device in the database. It is a many-to-many matching problem by nature and can be solved by traditional *classification models*. In other words, this type of task decides "who is who?"

Unknown device detection refers to the tasks that evaluate whether the signal-under-test belongs to a particular device. It is a one-vs-rest binary classification problem [22]. This type of task decides "who is real?"

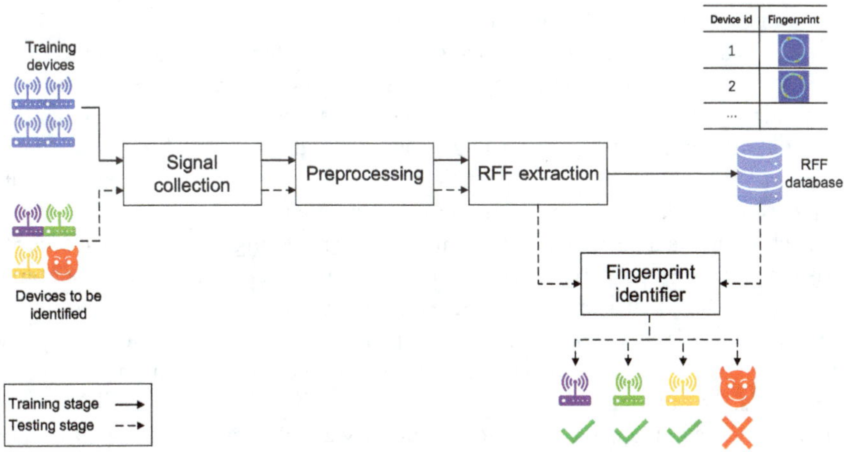

Figure 7.2 RFF based device identification method

It is worth noting that both verification and detection modes can be used to evaluate the legitimacy of transmitter. The identity acquired from both modes offers a complementary means to guarantee that network access is from a genuine device. More concretely, an RFF identification-empowered network is able to extract a "physical layer identity" of the user device and can make a cross-validation with its digital identity extracted from higher layers. When the two identities coincide with each other, the user device is recognized to be safe, and the network access request is granted. Otherwise, when the two identities do not match, some abnormal situation may incur in the current communication session. The network access is temporally suspended, and follow-up abnormal processing will be triggered to further verify the legitimacy of the user device.

Note that the application scenarios of these two modes are slightly different from each other. The *known device verification* mode is essentially a multi-class classification, which makes it suitable for identifying one or more devices from a set consisting of multiple known devices. It is natural to use known device verification mode in the cases where the identification should be done simultaneously for multiple target devices. A typical example is the user authentication in the cellular network where the base station needs to determine whether the access request is from a legitimate user equipment, or not. In such cases, this known device verification mode reflects a scenario where the base station can simultaneously have two identities of each device, one obtained from the higher layer protocol of the system and another extracted from the physical layer signal waveform. If the two identities of the same device do not match each other, the base station may need to pay more attention to re-identify the device under test. This mode aims to perform fast identity verification and is suitable for the lightweight security check or for the low risk application scenarios.

As for the *unknown device detection* mode, it aims at checking whether a signal is sent from a known, legitimate device, or not. Therefore, it is more suitable to be used in the "single user" communication scenarios. This implies that the communicating devices are exchanging information in a point-to-point manner. An example of this type is the base station authentication carried out by the user device in the cellular network. In such case, the user device only needs to identify the base station with which it is currently communicating. It only needs to evaluate whether the received signal fulfills the characteristics of the target base station. Any signal with unfamiliar characteristics will probably indicate a potential attack from the fake base station. Therefore, this unknown device detection mode is suitable for the scenarios with higher security risks or with higher security requirements.

A challenge that needs to be dealt with when considering the RFF-based security measures is how to properly discover the unknown devices. This challenge comes from the sad fact that it might not be easy to have the exact information on the characteristics of the attacker. As a consequence, we cannot leverage the a priori information of the attacker when designing and tuning our identification methods. In other words, the data samples of the attacker may not be exhaustively available at the training stage as in the naive classification tasks where the training data are assumed to be drawn from the same label and feature spaces as the testing data. Therefore, complete information on both classes and features should be available at the training stage which helps the classifier/identifier get familiar with the queries in testing time. However, if the samples from the unknown classes appear during testing, they will not be correctly recognized and will be misclassified into a closest known class. It is not straightforward to classify unknown devices with various possible characteristics into an "abnormal" category.

A possible solution to this problem is to adopt the so-called *open-set recognition* techniques [23]. The term "open-set" indicates that the knowledge is incomplete at the classifier training time. The target is not only to accurately classify the samples from the known classes but also to properly handle the queries from *unknown classes* after the classifier/identifier being trained. Unlike the traditional classification problem where the whole feature space is *partitioned* into different classes, for the open-set recognition problem, the classification boundaries are *shrunk* to the areas that the samples of the known classes crowded. This will effectively reduce the misclassification and can reject the samples from unknown classes in the test. A complete overview of the open-set recognition techniques on a broader scope can be found in [23] and the references therein. A few recent research works investigated the unknown device detection using the open-set recognition [24–26]. For instance, in [24], it was proposed to detect the unseen devices through a classification confidence evaluation. The unseen devices can be found when the classification confidence of the current signal samples is lower than the confidence threshold obtained from the known devices available in the training. In [25], several open-set recognition approaches, including one-vs-all binary classification, discriminating classifier, OpenMax, and auto-encoder, were studied for the unknown wireless device detection. The RFF features were directly extracted by the neural networks from the IQ samples of Wi-Fi preamble signal. The unknown device detection performance was evaluated using real-world RF samples and showed encouraging detection accuracy. In [26], a neural synchronization technique and a hypersphere representation were proposed for the RFF extraction. The RFF identification and unknown device detection were achieved by comparing the similarity of the extracted RFF and the RFFs of known devices.

7.2.4 State-of-the-art

Inherent differences introduced in the manufacture of wireless devices can have an impact on the physical layer signal. In the transmitter, modulator, power amplifiers, RF oscillators and other components, all cause changes in wireless signals. Reversely, these signal features can uniquely reflect the identity of the transmitter and are recognized as the fingerprint of the device, which is referred to as *RF fingerprint*, or *RF-DNA* [27]. Sometimes, this technique is also called as *physical layer identification* [13], or *specific emitter identification* [28], depending on its particular application scenario.

Signal feature extraction and identification methods have been extensively researched in recent years. Different identification methods are suitable for different features, and the achievable identification efficiency also varies. This section reviews the research progress of RFF extraction and identification techniques. According to the state of the identifiable signal, the RFF extraction technology is divided into the

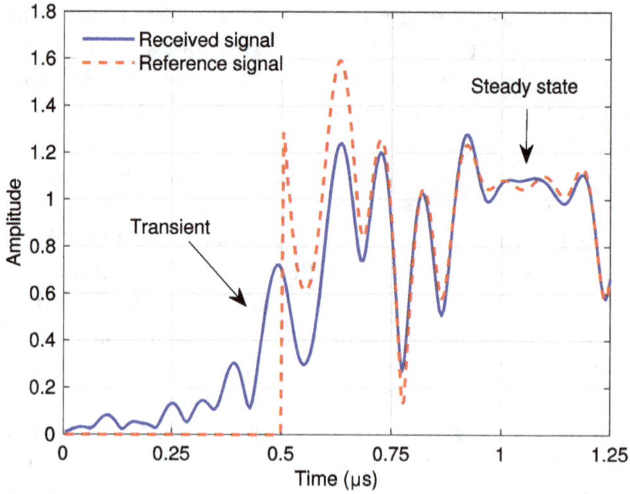

Figure 7.3 Instantaneous amplitude of RF signal of LTE terminal

methods based on the *transient signal features* and methods based on the *steady-state signal features*.

Figure 7.3 is an example of an RF waveform of a long-term evolution (LTE) terminal [29]. It can be seen that there is a short period of "build-up time" after the transmitter being turned on, where comprehensive elements and parts start building their working status. This is the so-called transient signal phase. Once the device achieves to its normal working status, it will send signals with stable characteristics. The device is said to be in a steady state. It has been demonstrated that both transient and steady-state signal carry unique features that can be leveraged to determine the transmitter's identity. Various feature extraction techniques have been investigated to construct the fingerprint. A list of features that have been researched in the literature is given in Table 7.1.

7.2.4.1 Transient state fingerprint

Early research on RFF extraction and recognition technology mainly focused on transient signals. The transient signal refers to the signal variation stage after the transmitter being turned on and before its electronic characteristics reaching a stable status. As this part of the signal does not carry any data information, it is only related to the hardware characteristics of the transmitter and is independent of data. The fingerprint feature can be extracted from signal parameter, or via some signal transform, or from their combination.

In [30], authors proposed to construct fingerprint using signal features, such as transient length, variance of the normalized amplitude, number of peaks of the carrier signal, the first discrete wavelet transform (DWT) coefficient, and difference between

Table 7.1 Overview of RFF feature extraction methods

Signal parameter features	Transient		Amplitude [30]
			Phase [31]
			Power [30]
			Length [30]
			Different between mean & max values [30]
			Multifractal trajectory [32]
			Singular spectrum analysis [33]
			Nature measure [34]
			Signal statistics[1] [35]
			Frequency hopping features [36]
	Steady		Error vector magnitude [37]
			Carrier frequency offset [37,38]
			Phase error [37,39]
			I/Q imbalance [37,40]
			DC offset [40]
			Amplifier nonlinearity [18,20,41,42]
			Constellation heat map [43]
			Signal statistics[1] [44,45]
Transform domain features	Wavelet	Transient	Wavelet coefficients [31,46]
	Frequency	Transient	Spectrum profile [47,48]
		Steady	Spectrum of preamble [49,50]
	Time-Frequency	Steady	Spectrogram [51–54], Gabor-Wigner transform [55]
	Bispectrum	Steady	Bispectrum [51,56,57]
	Hilbert-Huang	Steady	Hilbert-Huang transform [51,58,59]
Implicit signal features	Raw I/Q samples	Steady	Based on known symbols [60,61], based on random symbols [41,62,63]

[1]Signal statistics include variance, skewness, and kurtosis of instantaneous amplitude, frequency, and phase.

the mean and maximum values of the transient. In [32], multifractal trajectory of turn-on transients was considered to identify transmitters. In [33], singular spectrum analysis (SSA) was adopted to extract features for transient signal. In [34], the nature measure of the one-dimensional component of the received signal was extracted as fingerprints. In [35], variational mode decomposition (VMD) was used to obtain higher order statistical (HOS) features of transient signals. Specifically, variance, skewness, and kurtosis of instantaneous amplitude, frequency, and phase are extracted as fingerprints.

Besides, different kinds of transforms were considered to extract the transient signal features. Dated back in the 1990s, wavelet features were adopted for the transmitter identification purposes [46,64]. For instance, multi-resolution wavelet analysis was introduced in [46] to characterize the features contained in transient signal, and wavelet coefficients were used as fingerprint. The wavelet-based methods can analyze non-stationary signal effectively, but the obtained wavelet features are usually high-dimensional and the feature selection operation is needed in general. The spectrum features were also considered to characterize the transient signal. Fast Fourier transform (FFT) was applied to convert transient signal to the frequency domain which was consequently used to identify RFID cards in [48].

Signal parameter features and transform domain features were also combined together to achieve complementary identification performance. In [31], RFF that combines features including amplitude, phase, in-phase data, quadrature data, power, and DWT coefficients was proposed to detect intrusion for Wi-Fi system. Recently, an RFF based on the energy spectrum of transient signal was developed for Wi-Fi in [47] which shows a reasonable classification performance at low SNR region.

The main limitation of transient signal-based RFF is the difficulty of signal collection. In general, the duration of transient signal is very short, and high-end instrument is usually needed for high-speed signal acquisition, which increases the application cost. In addition, it is necessary to accurately detect the starting point of the signal to capture the transient signal part [65,66], which might affect the quality of the fingerprint. Moreover, as the transient state does not last long, it is not easy to collect a large number of signal samples, which makes the channel fading mitigation and noise suppression very challenging.

7.2.4.2 Steady-state fingerprint

Compared to the transient signal, the steady-state signal represents a much longer signal duration. It is therefore easier to capture a large amount of signal samples in practice. In recent years, many RFF-related researches have focused on the steady-state signal. The steady-state feature extraction methods also target on signal parameter and transform domain features.

For the steady-state signal, the signal parameter features are commonly reflected by the modulation errors such as error vector magnitude (EVM) [37], carrier frequency offset (CFO) [37,38], phase error [37–39], I/Q imbalance [37,38,40], DC offset [40], and SYNC correlation [37,38]. The nonlinearity of power amplifier was considered as the RFF feature in [18,20,41,42].

Various transforms, such as Fourier transform, bispectrum, Wigner transform and Hilbert–Huang transform (HHT), were also investigated to enhance the RFF features originated from different hardware impairments. Spectrum features of the preamble were adopted to construct RFF for Universal Mobile Telecommunications Service (UMTS) devices [49]. The power spectral density of OFDM signals was used as the fingerprint in [50]. The bispectrum was employed to extract nonlinearity features in [56]. Time–frequency domain features were acquired using short-time Fourier Transform (STFT) [51–54]. The obtained spectrogram was used to identify LoRa devices [52,53]. It was then applied with deep metric learning that enables the RFF extractor to obtain fingerprints from previously unseen devices. Gabor–Wigner transform was considered to extract time–frequency RFF features in [27] and the references therein.

In [58], HHT in combination with variational mode decomposition (VMD) was used to extract temporal features like entropy and moments, as well as spectral features like flatness, brightness, and roll-off. In [59], RFF extraction methods based on empirical mode decomposition (EMD) and HHT were proposed. The entropy, first- and second-order moments were selected as the identification features. In [51], multiple transforms including HHT, STFT, and bispectrum were used together to extract fusion fingerprints, which is expected to possess the advantages of each individual feature extraction method.

Recently, some researchers investigated directly to extract features from the raw signal samples to avoid the potential information loss due to the signal processing and transform used in feature extraction [41,60–63]. In [63], it was shown that the RFF can be extracted from the raw IQ samples by a convolutional neural network (CNN) and can be applied to identify a large number of wireless devices under practical conditions like channel variation, noise, etc. A so-called partial equalization technique was considered to mitigate the influence of the propagation channel. The work in [60] exploited the repetitive signal patterns in the ZigBee signal, which are static over time and known to the receiver, as the fingerprint. The Gaussian probabilistic linear discriminant analysis technique was used to determine whether a signal pattern under test matches a legitimate device's characteristics. In [61], the received preamble signal was down-sampled with multiple rates to offer different time resolutions reflecting from short-term to long-term time domain features. The complex-valued multi-sampling signal was then input into a convolutional neural network to extract features and make identification. These two methods chose the known signal as the region of interest for the fingerprint extraction. As the preamble signal is known to the receiver, the neural network-based identifier is not affected by the randomness of the data. However, it does not satisfy the data-independency property of RFF. In contrast, in [41,62], it was proposed to use complex-valued CNN to directly extract RFF from raw data samples. The network is trained in an end-to-end manner, and it does not require to know the data when performing the identification. Yet, there is still a potential problem that the extracted identity contains channel effect and the identification is location-dependent.

7.3 Case study: RFF extraction of Bluetooth devices

As seen from the literature, various RFF extraction and identification measures have been investigated in the past. This section will present an RFF extraction method for Bluetooth Low Energy (BLE) devices [67]. This case study serves as an intuitive example to help readers understand the entire process from signal process to RFF extraction. Bluetooth is a popular short-range communication standard that enables the inter-connection of IoT devices, personal digital devices, medical devices, and industrial devices [68]. However, it was shown recently that the Bluetooth standard contains certain vulnerabilities that enable an attacker to impersonate a legitimate Bluetooth device and to establish a secure connection with a victim [69]. Therefore, it makes sense to investigate advanced security measures to avoid impersonation attack for Bluetooth standard.

Bluetooth standard adopts the Gaussian Frequency-Shift Keying (GFSK) modulation as the basic physical layer transmission scheme. GFSK is a typical narrow-band waveform with good spectral efficiency, smooth phase transition, and constant envelope, and is suitable for energy-efficient, low-cost applications [70].

The received GFSK signal can be written in discrete-time form as

$$r(n) = \sqrt{\frac{2E_b}{T}} h(n) e^{j(2\pi f_o nT + \theta(n) + \theta_0)} + w(n) \tag{7.14}$$

where $h(n)$ is the channel fading coefficient; f_o is the difference of the carrier frequencies between transmitter and receiver; $\theta(n)$ is the continuous varying phase that conveys information; θ_0 is a random initial phase; $w(n)$ is the additive white Gaussian noise.

Many RFF extraction methods [16] can be considered to capture the unique features from the received signal. For instance, the RFF features can be extracted either from different signal parameters or by employing various transfer operations. A general overview of the potential RFF extraction techniques can be found in Section 7.2.4.

In this section, we take the differential constellation trace figure (DCTF) method as an example to show how the RFF features are extracted and stabilized by proper signal processing measures from the original RF signal samples where the signal feature is imperceptible and may be varying. DCTF can be built via a differential operation of the received baseband signal [16]. In particular, when applying differential operation to signal samples separated by τ sampling periods, the result is

$$d(n, \tau) = r(n) \cdot r(n + \tau)^*, \tag{7.15}$$

where $(\cdot)^*$ is the conjugate operation of a complex variable. Without loss of generality, one can assume that the wireless channel does not significantly vary within a coherence time of λ samples, i.e. $h_l(n) \approx h_l(n + \lambda)$. When $\tau < \lambda$, the differential result in (7.15) becomes

$$d(n, \tau) = \alpha \cdot e^{j(\theta(n) - \theta(n+\tau) - 2\pi f_o \tau T)} + w, \tag{7.16}$$

where $\alpha = \frac{2E_b}{T} h(n) h(n + \tau)^*$ is the correlation of channel coefficients at time n and $n + \tau$, and it turns to a constant $\frac{2E_b}{T} |h(n)|^2$ when differential separation τ is not large compared to the channel coherence time λ.

It is shown in (7.16) that DCTF contains the contribution components from the channel coefficients α, the differential separation τ, and CFO f_o. Among them, the first two kinds of components converge to a few regions related to the differential results of all combinations of constellation points. The third component introduces a phase rotation determined by the CFO between transmitter and receiver and the differential separation.

The phase rotation can be decomposed into two aspects:

1. *Information baring phases* $\theta(n)$ and $\theta(n+\tau)$ are determined by the information bits priori to time n and $n+\tau$, respectively. According to the phase transition characteristics of GFSK modulation, the phase changes $\pi/2$ or $-\pi/2$, depending on the information bit, when the modulation index equals to 0.5 as defined in BLE specification. As a result, the value of $(\theta(n) - \theta(n+\tau))$ equals to $\{-\frac{\pi}{2}, \frac{\pi}{2}\}$ when τ is odd, or equals to $\{0, \pi\}$ when τ is even.
2. *CFO between transmitter and receiver* f_o yields a cumulative phase rotation of $2\pi f_o \tau T$ over time τT. Therefore, the overall phase of the differential results concentrate in two diametrically opposite positions if plotted on the complex plane. The angular offset that these positions deviate from real or imaginary axis is determined by f_o for given τ and T. Since each BLE device transmits at a carrier frequency slightly different from the nominal value, the carrier frequency offset f_o can be used to identify the specific device. Hence, the phase rotation $2\pi f_o \tau T$ in the differential results offers a direct way to demonstrate the device identity.

As such, the differential results contain the CFO information that can uniquely demonstrate the identity of the transmitter. Hence, it can be exploited to construct the RFF.

Performing differential operation of (7.16) for a certain length (typically, several hundreds to thousands) of received signal, and plotting the results $d(n,\tau)$ in the complex plane, it yields the DCTF. Obviously, this is an accumulation of differential results over time, which essentially visualizes the statistical distribution. In other words, the concentration of the points in DCTF actually represents the statistical average of the differential results. Therefore, the high-intensity areas in DCTF present the stable components of the differential results, i.e. those related to signal constellation, or caused by inherent hardware imperfections like CFO.

Some DCTF examples of two BLE devices with several differential intervals (τ) are illustrated in Figure 7.4. The signal was captured in three BLE advertising channels, namely channel numbers 37, 38, and 39, respectively. The carrier frequencies of the three advertising channels are 2402 MHz, 2026 MHz, and 2480 MHz, locating in the low, medium, and high parts of the BLE spectrum, respectively. The advertising channel is used for the transmission and reception of various advertising packets and is the first medium for establishing the wireless connection between BLE devices. Therefore, it makes sense to evaluate the possibility of extracting RFF from the signal in advertising channel. Three differential intervals are considered to generate DCTFs. BLE signal is emitted from wireless modules that compile with the BLE specification. The BLE signal is captured in an indoor scenario by a software-defined radio (SDR) platform which will be described in detail in Section 7.4.3. Both the BLE devices and the signal sampling platform stay in stationary status. From Figure 7.4, we can have following observations:

(1) DCTF remains stable in different sampling occasions. This can be seen by the similarity among the DCTF examples generated from the same channel and differential interval. It means that the DCTFs of the same device possess stable features that can be retrieved as RFF.

(2) DCTFs of BLE devices show significant similarities in different channels. The three advertising channels are located in the low, medium, and high parts of BLE spectrum. Especially, channel 37 and channel 39 locate in the lowest and highest channels of the spectrum. Similar features can be observed in the DCTFs generated from different channels. It means that DCTF is not sensitive to the variation of channel response of different channels and can maintain feature similarities among different channels across the BLE spectrum band. This provides great flexibility and convenience for deploying DCTF-based RFF recognition in real application scenarios.

(3) DCTFs of different devices are distinct. Obvious differences can be observed by comparing the DCTFs of the two BLE devices. It means that different BLE devices generate distinct DCTF features which can be leveraged as RFF to identify BLE devices.

In general, the above observations provide the intuitive evidence that DCTF can serve as the fingerprint to identify wireless devices. It will be shown in the following section that the fingerprints extracted from BLE signals can be applied to identify the transmitter in real communication scenarios.

7.4 Deep learning-based device identification

In this section, the known device verification and unknown device detection tasks will be discussed. Their performance will be evaluated with signal samples collected in the real-world communication scenario.

As discussed in Section 7.2.3, the known device verification task is to determine which legitimate transmitter the signal-under-test is sent from. This task is to decide a known class that best matches the signal's characteristics and is commonly modeled as a traditional classifier like CNN or support vector machine (SVM). It is suitable to be adopted in the scenarios where the RFF identification mechanism is used to double-check whether the identity derived from physical layer signal coincides with the identity provided in higher-layer authentication mechanism. In contrast, the unknown device detection task aims to determine whether a device-under-test belongs to the set of legitimate devices. This task is to determine whether a signal sample is from any signal distribution that is recognized by the identifier/detector. It can be modeled as an open-set recognition problem. The unknown device detection is suitable to be employed to discover whether there exists any intrusion from unknown (rogue) devices.

The rest of this section will discuss deep learning-based solutions to the two identification tasks mentioned above. Without loss of generality, the DCTF-based RFF extraction is considered in the following discussion.

7.4.1 CNN-based known device verification

As presented in Section 7.3, DCTF visualizes the modulation and RFF features on two-dimensional (2D) images. The visual features can easily be extracted and analyzed

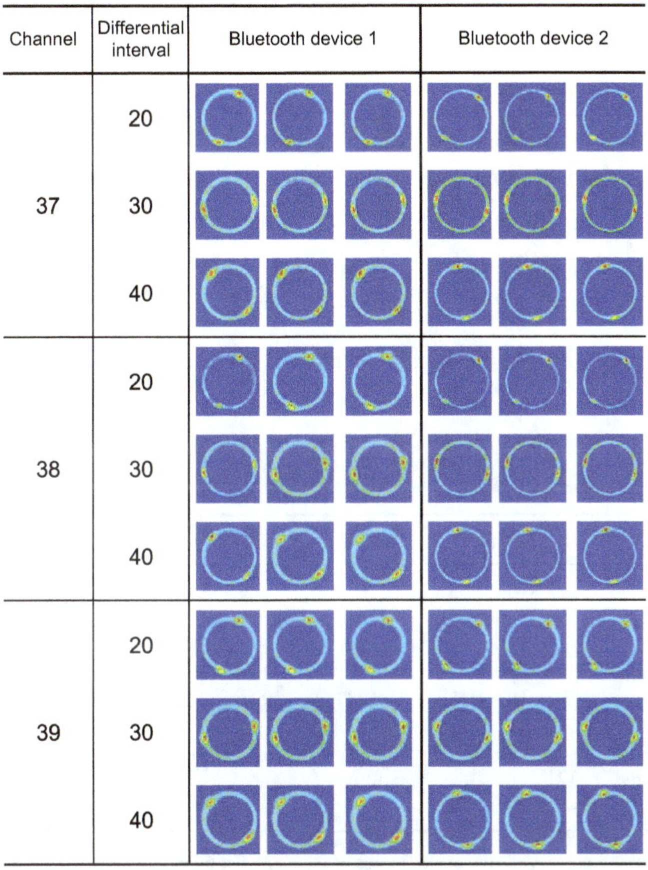

Figure 7.4 DCTF of two BLE devices with various differential intervals. The intensity of the pixel indicates the density of the points in a certain area on the complex signal plane (red: high, blue: low).

by the computer vision-based pattern recognition techniques. CNN is a feedforward neural network that employs convolution kernels to capture local and low-level features in 2D area. Through concatenation of multiple convolution layers, the local and low-level features are synthesized into global and high-level (semantic) features. This structure makes CNN possess the shift invariant properties and strong feature extraction and representation capability. CNN has been shown to be a very powerful tool for the classification jobs in various applications [56,71]. Due to these merits, CNN is employed to analyze RFF features and verify the identities of known devices.

A CNN with three 2D convolutional layers followed by two fully connected layers is designed in this section. The network structure is depicted in Figure 7.5, and the network details are given in Table 7.2. The dimension of the input DCTF is set to

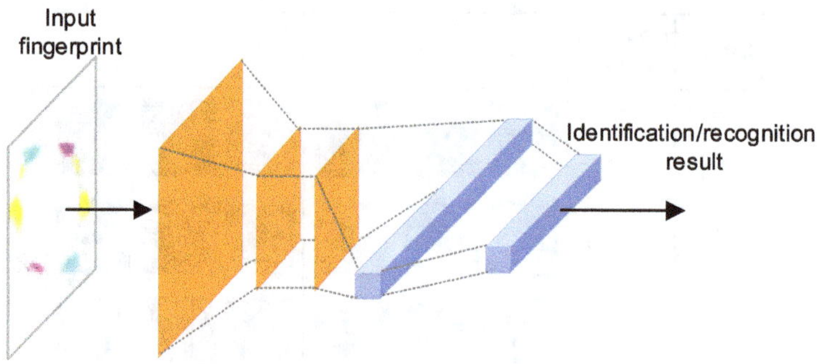

Figure 7.5 RF fingerprint classification by CNN network

Table 7.2 CNN network structure for classification

Layer	Dimension	Activation	Options
Input	$[64 \times 64] \times 3$[1]	–	Zero-center normalization
2D convolution	$[5 \times 5] \times 16$[2]	ReLU	Stride 1, padding 1, batch normalization
Max-pooling	$[3 \times 3]$	–	Stride 2
2D convolution	$[3 \times 3] \times 32$	ReLU	Stride 1, padding 1, batch normalization
2D convolution	$[3 \times 3] \times 64$	ReLU	Stride 1, padding 1
Fully connected	64	–	–
Dropout	–	–	Rate 0.5
Fully connected	32[3]	Softmax	–
Classification	–	–	–

[1]The input is three channels of gray-scale figures of the size 64×64 pixels.
[2]$[5 \times 5] \times 16$ means that this layer contains 16 convolutional kernels of the size 5×5 pixels.
[3]The size of last fully connected layer is determined by the number of classes to be classified.

64×64 pixels, which was shown to be a good trade-off between the classification accuracy and computational complexity [16]. The network can be tuned to adapt to other DCTF size with minor effort, offering a good flexibility to different requirements and/or constraints associated with specific applications. The features extracted by the three consecutive 2D convolutional layers are synthesized by the two fully-connected layers. The use of activation layer brings some non-linearity when representing the feature. The classification results are finally given by a classification layer. With this network, the identity information will be output for each input RFF.

7.4.2 Anomaly detection-based unknown device detection

The major challenge of unknown (rogue) device detection is that the signal of unknown device is commonly unavailable or unlabeled in the training stage. The missing of negative samples makes it impossible to take the unknown devices' signal samples into account in the training process. Recent research progress in generative adversarial network (GAN) offers an effective way to handle the problem of data imbalance [72], or class data missing [73]. This provides a potential way to detect the unknown class without specific a priori knowledge about it.

By analogy with [73], the unknown device detection can be modeled as an anomaly detection problem. The legitimate devices' signal is treated as the "normal" case, while the unknown devices' signal being seen as "abnormal" case. The so-called *GANomaly* model is trained solely by the legitimated devices' data (known and labeled). Through the training process, the GAN model can accurately characterize the feature distribution of the legitimate devices' signal. When the signal-under-test comes from the legitimate device, the trained GAN will be able to recognize this is a "normal" signal. However, when the signal from unknown device is input into the trained GAN, it will give an "abnormal" sign, because the signal seems to be unusual compared to the distribution it has learned.

Figure 7.6 presents the GANomaly network structure which can determine the legitimate devices from the mixture of legitimate and rogue devices. The overall network consists of two main parts: *generator G* and *discriminator D*. The former is a bow-tie autoencoder that aims to learn the representation of the input data samples (RFF of legitimate devices). The latter is a standard discriminator with convolutional network structure as in the Deep Convolutional GAN (DCGAN) [74].

In more detail, the generator contains an encoder network G_E and a decoder network G_D. The encoder network G_E gradually compresses the input fingerprint r into a "bottleneck" vector z which is supposed to carry the best representation of the input fingerprint within the smallest dimension and is treated as the feature of the input for the distinguishing job. The decoder network G_D, in the contrary, aims to reconstruct a best recovery of the input $\hat{r}$ from the bottleneck vector z. Another encoder

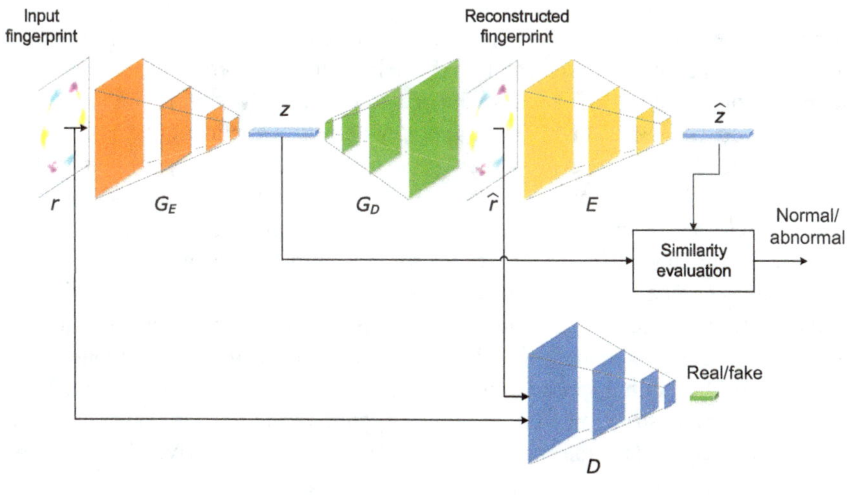

Figure 7.6 Device RF fingerprint identification through GANomaly network

network E which has the same structure as G_E but different parametrization is used to compress the reconstructed input $\hat{r}$ in order to generate its feature representation $\hat{z}$. This second encoder network learns to minimize the distance between $\hat{z}$ and z. More importantly, the detection of rogue device is achieved with this minimization. The discriminator network is to determine whether its input is a real sample or a fake one generated by the generator network. The training process of the whole network can be seen as a two-player zero-sum game. The generator wants to fool the discriminator by improving the similarity of its output. In the meantime, the discriminator is trained to distinguish whether its input is real or fake. Through this adversarial game both networks are getting better and better.

The rationale behind GANomaly is that the network's parameters are only tuned for generating normal samples. Yet, some abnormal features cannot be properly represented due to the lack of samples during the training process. As a consequence, the generated fingerprint will show some perceptible dissimilarity to the discriminator, and thus can be detected when comparing $\hat{z}$ and z. Throughout the iterative training process, the generator actually generates different realizations of the feature outside the distribution of the legitimate devices' features. The discriminator helps to update the enclosure that characterizes the legitimate devices. Eventually, the model captures the training data distribution within both fingerprint and latent vector space. At inference time, a larger distance metric from this learned data distribution is indicative of an abnormal one from that distribution – a rogue device. That is the reason why the network can be trained only using the normal samples, but can be used to detect the abnormal ones. Moreover the better we train the network using the legitimate devices' features the easier the network can detect the signal from rogue devices. In the literature, the anomaly detection can also be realized using autoencoder, where the anomaly is determined by directly evaluating the reconstruction loss [75]. In contrast, GANomaly judges the anomaly by evaluating the latent representations of the input and the reconstructed samples, which leads to advantages in terms of detection robustness. Moreover, GANomaly involves a discriminator network that helps the training of the encoder and decoder networks, which yields a good latent representation of the input samples and can be very helpful in the anomaly detection. This is the main motivation that the more sophisticated GANomaly is more preferable than the autoencoder-based methods in the anomaly detection applications.

Each encoder/decoder element network consists of five 2D convolutional layers with Leaky ReLU activation. The dimension of the convolutional layers of encoder network decreases by half compared with the previous layer, reducing from 64×64 pixels of the first layer to 4×4 of the last one. The number of channels, on the other hand, doubles compared with the previous layer, increasing from 3 of the first layer to 48 of the last one. Batch normalization is performed in the three hidden convolutional layers. The decoder element networks have a reverse organization compared with the encoder. The 2D transposed convolutional layers replace the convolutional one to realize the reconstruction from the latent vector. In the testing phase, with each RFF-under-test, the decision on legitimate or rogue devices is given by the output GANomaly network.

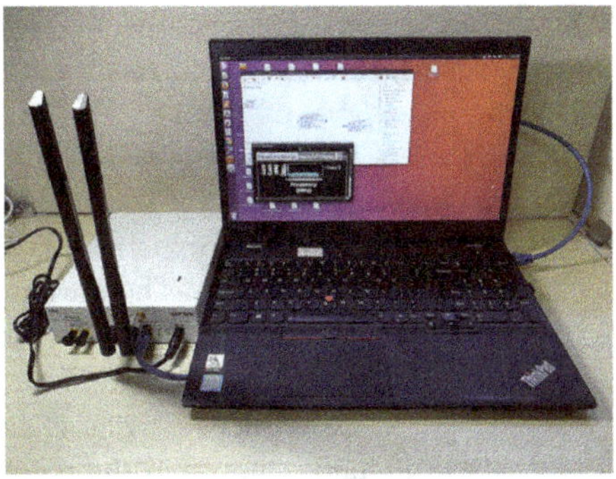

(a) RF sampling platform.

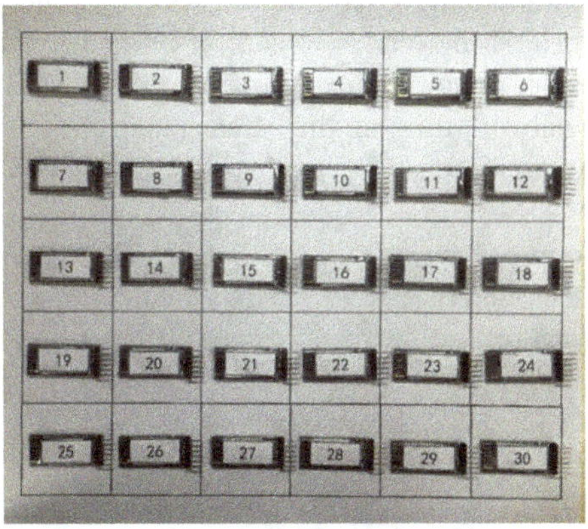

(b) Bluetooth transceivers.

Figure 7.7 Hardware-based experimental platform: (a) USRP-based RF sampling platform; (b) Bluetooth transceivers

7.4.3 Performance evaluation and analysis

RF signal was collected by the SDR platform Universal Software Radio Peripheral (USRP) N210 (with a UBX-40 RF daughterboard) as shown in Figure 7.7(a). BLE transceivers (TI CC2541) served as the target devices for the identification, as shown in Figure 7.7(b). The advertising signal transmitted over the BLE's advertisement channels is captured. The sampling rate of SDR was chosen as 10 MSamples/s, achieving an oversampling rate of 10. Two hundred signal traces were collected for each BLE device. Each trace contains roughly 18 advertising frames. The SNR of collected signal is greater than 40 dB, ranging from 41 dB to 48 dB. For the ease of presentation, these parts of data are referred to as "40 dB data" in the sequel. Artificial noise is added to the collected data (using MATLAB® routine *awgn*) in order to augment the volume of data traces. The noise level is adjusted so that the noisy data achieves eight SNR levels, ranging from 0 dB to 35 dB with steps of 5 dB.

The noisy data together with the original collected data constitutes the data set of 1000 traces per BLE device. Finally, one DCTF is generated from each data trace. The dimension of DCTF is chosen as 64 × 64 pixels.

The networks are trained in a server equipped with NVIDIA Titan Xp GPU. The training of neural networks was performed on the PyTorch (version 0.4.1 with Python 3.5.2) platform. In the known device verification experiments, 32 BLE devices were involved in the classification accuracy evaluation. The training data set consists of 80% samples of all devices, and the testing data set contains the rest 20% samples. The number of testing samples belonging to different classes is roughly balanced.

In the unknown device detection experiments, 10 BLE devices were involved. One of the ten BLE devices was randomly chosen as the rogue device, while other nine devices were treated as the legitimate ones. The training data set consists of 90% samples of all legitimate devices. The test data set contains the whole data of the rogue device as well as the rest 10% data of all legitimate devices. Note that, the number of samples from legitimate and rogue devices is roughly balanced in the testing dataset. In the network training the batch size is 64. The network weights are optimized using the Adam method with the learning rate of 0.0002, and momentums $\beta_1 = 0.5$, $\beta_2 = 0.999$.

The known device verification accuracy using DCTF and CNN is shown in Figure 7.8. It can be seen that the DCTF-CNN scheme offers reliable known device identification (classification) performance under different SNR levels. At high SNR

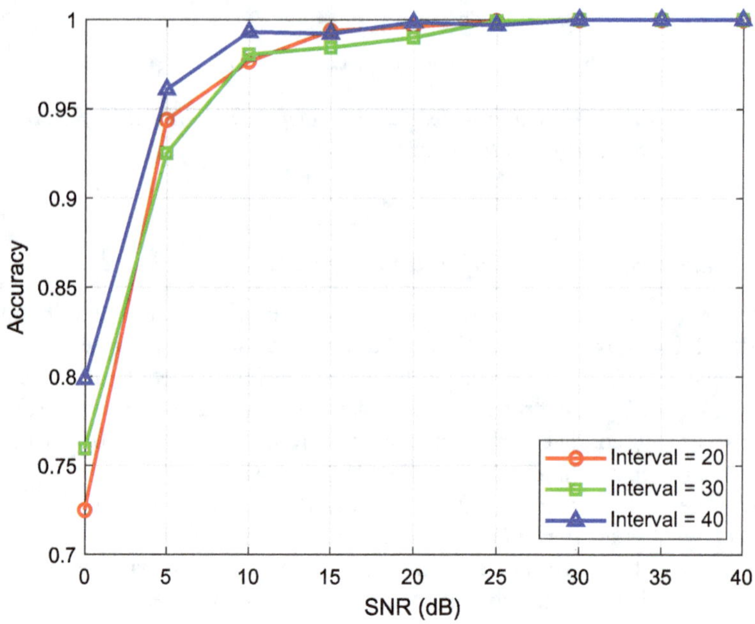

Figure 7.8 Known device verification performance using CNN

region, the accuracy can reach as high as 97%. At low SNR level like 5 dB, the accuracy is still higher than 90%, which shows good robustness with bad signal quality. The performance is similar when the differential interval is chosen as 20, 30, or 40 for the DCTF construction. This suggests that the classification accuracy is not sensitive to the differential operation as long as the channel does not vary significantly within the differential interval.

A comparison of the known device verification accuracy using different RFF extraction methods and classification methods is given in Figure 7.9. DCTF, bispectrum (noted as "BISP"), and short-time Fourier transform (noted as "STFT") are considered as the RFF feature extraction methods. Both CNN and k-nearest neighbor (noted as "KNN") are used as the classification measures. Through the comparison, it can be found that DCTF performs well in combination with both classic KNN-based classifier and CNN-based classifier. In contrast, the BISP- and STFT-based RFF extraction methods show significant performance degradation compared with DCTF. This can be understood by the fact that CFO is a relatively obvious RFF factor in BLE devices, which yields more distinct features in DCTF than in other methods.

The unknown device detection accuracy is presented in Figure 7.10 where the horizontal axis represents the index of the device that is selected as the rogue device. The accuracy shown in the figure is the ratio of the number of samples that are correctly determined whether it is from a legitimate device or a rogue one over the number of all testing samples. It can be seen from the figure that the GANomaly-based anomaly detection method can accurately detect the legitimacy of a device. The detection

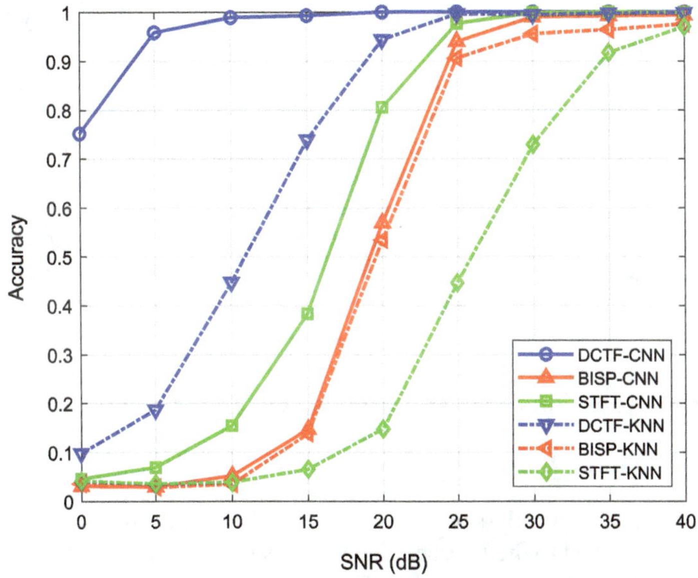

Figure 7.9 Performance comparison of different RFF extraction methods in known device verification scenario

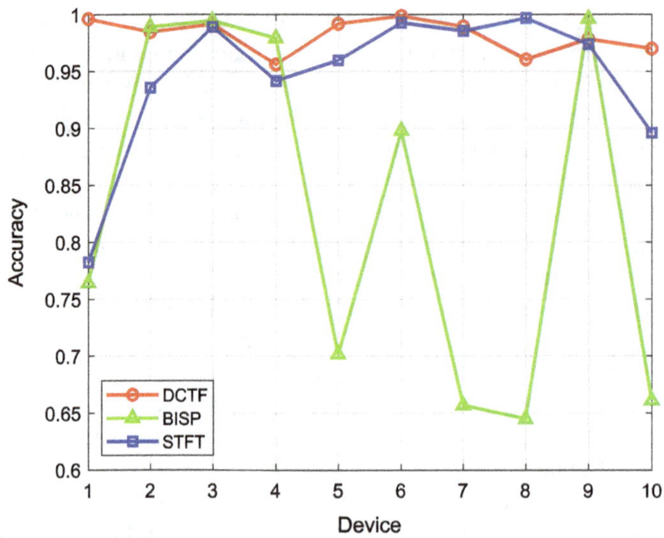

Figure 7.10 Unknown device detection performance using GANomaly network

accuracy is above 92% for all 10 rogue devices and can reach as high as 99% for some device. In comparison with other RFF extraction methods, namely bispectrum and short-time Fourier transform, DCTF offers more reliable and stable anomaly detection performance. In contrast, bispectrum-based RFF can provide high anomaly detection accuracies for some devices like 3 and 4. However, its performance is not satisfactory for the cases where the other eight devices are chosen as the unknown devices. Similar conclusion can be drawn with short-time Fourier transform-based RFF. Although performing well when detecting device 8, its detection accuracies are not as good as the DCTF-based ones. These results give an intuitive evidence that the anomaly detection technique can be used to identify the fingerprints of the devices that are unknown in the training phase. This makes it possible to discover the adversaries that are unknown to the RFF identifier, which is very meaningful in the real communication scenarios.

7.5 Open questions and potential research directions

7.5.1 High capacity fingerprint for massive devices

With the popularity of wireless networking and wireless applications, more and more devices are connected to the Internet. The wireless networks are required to serve a large number of devices in a given area. The massive machine-type communications (mMTC) scenario of current 5G supports up to 1 million online devices per square kilometer (km^2). The future 6G mobile network is believed to boost this number by an order of magnitude. 6G is currently slated to support a connection of as high as 10^7 devices/km^2 [76]. This suggests that the RFF-based identification method should be able to distinguish a specific device from a set of many devices. Obviously, it poses great challenges to both the RFF extraction method and the fingerprint identification/verification method. Therefore, it is important to investigate the RFF extraction and identification method that can be applied in the network scenarios with massive devices. However, it is unclear whether RFFs can achieve effective identification of a large number of wireless devices and what is the theoretical limit of this technique. The existing identification methods usually only rely on one or a few features of the RF signal as the basis for device differentiation. Existing work such as [77] analyzed the "capacity" that RFF-based identification can achieve. However, analysis of the fingerprint with extensive impairment factors and real-world impact factors has not been sufficiently conducted, yet. In fact, the impairment of each component in the signal processing path will cause specific signal features. The features associated with different hardware impairments possess different properties in terms of their levels of distinction and robustness against interference. In other words, features from different "dimensions" may contain complementary advantages, and may avoid the insufficiency of single signal feature. It makes sense to extend the feature dimensions from which the RFF is constructed. Hence, it will be interesting to investigate the origin and the correlation of different types of signal features. With this basis, it will be helpful to study how to construct a more distinctive fingerprint by synthesizing features from different signal dimensions.

7.5.2 Channel-independent fingerprint extraction

The wireless channel causes signal distortion and covers up the RFF characteristics. The resulting fingerprint is affected by the channel response and becomes "location-dependent" [11]. At present, there have been researches related to RFF of narrowband wireless devices. As the symbol duration of the narrowband signal is relatively long compared with the channel delay spread, the influence of multipath propagation is represented as flat fading which is, in general, not difficult to deal with. For instance, in the DCTF method, the flat fading effect is converted into a real-valued scale factor after performing the delayed differential operation on the received signal. The scale factor can be compensated by proper signal normalization. Note that this power normalization preserves the fingerprint characteristics (i.e., CFO features) while eliminating the channel influence.

However, in the broadband wireless communication system, the symbol duration is shorter compared with the channel delay spread, and the influence of multipath effect is more significant. The channel does not uniformly influence the signal components. Instead, it will cause a frequency selective fading and destroy the fingerprint features. In fact, the RFFs extracted by many existing methods contain both hardware impairments and channel effects. Whereas, this kind of fingerprints can only be applied in the static channel scenarios and is channel (location) dependent. So, the channel effect must be removed from the signal before extracting fingerprint in order to achieve a reliable and channel-independent identification capability.

The traditional channel compensation techniques proposed for the signal equalization cannot be directly applied here. This is because the channel response obtained from channel estimation is actually the equivalent baseband channel which includes both wireless channel and RF up-/down-convert process [14]. It is difficult to separate the distortion caused by the RF component impairments from the distortion caused by wireless channel fading. Hence, the fingerprint features may varnish when the signal is restored by the traditional equalization process using the baseband channel estimation. Therefore, it will be interesting to investigate how to extract reliable RFF that is irrespective of channel effect, especially for the broadband communication devices.

7.5.3 Security of RF fingerprint

Like all other authentication mechanisms, RFF itself also has the security risks that it could be compromised or even bypassed. It was shown in [78] that the modulation-based and transient-based identification methods can be impersonated by feature replay and signal replay. In the feature replay attack, the radio signal characteristics of the attacker device are modified to match all or part of the features of the legitimate transmitter which will be used by the RFF identification system. In the signal replay attack, the RF signal of legitimate device is recorded by high-end instruments, so that the RFF features can be captured and could be replayed to thwart the RFF identification mechanism.

The potential limitation of the replay attacks lies in the fact that the attacker itself commonly contains imperfections of some kinds, and it will inevitably involve its own fingerprint in signal reception and transmission. The attacker ought to use high-precision hardware to guarantee the quality (fidelity) of the replay. It should calibrate its transmission and reception response to wipe out its trace of impersonation. These requirements make the replay attack not easy to be launched in reality. It was also pointed out in [79] that a strong adversary who is aware of the fingerprinting methods could impose artificial distortion to the signal in order to disguise its fingerprint. However, accurately observing and delicately manipulating signal features to match the legitimate user's fingerprint are still challenging for an attacker with imperfect hardware.

To circumvent the requirement of hardware calibration, Ref. [80] first proposed a GAN-based signal spoofing attack framework. A collaborative attacker was adopted as the discriminator to help the attacker adjust its feature to match the legitimate user's device property (i.e., fingerprint) and communication channel property (i.e., channel response). This work provides a viable idea to impersonate RFF with low requirements on the hardware capability. However, the RFF feature that this method learns is a composite of "real RFF" that comes from hardware imperfection and the channel response. Therefore, the impersonation performance is severely degraded once the channel condition varies. A policy gradient-based reinforcement learning method was proposed to penetrate the RFF-based authentication mechanism in [81]. The attacker gradually updates its signal impersonator in the reinforcement learning process based on the authentication decision feedback from the legitimate user. Experiments show that this attack method can fool the deep learning-based physical layer authentication measures of wireless communication system. Whereas, this attack method can only achieve untargeted attacks. In fact, it learns the adversarial samples of the authentication network, which may not necessarily carry meaningful information. Recently, a GAN-based colluding RFF impersonation attack framework was proposed in [82]. The method can effectively impersonate the legitimate user's RFF, but not the location-specific features, under dynamic channel conditions. The attack framework can be applied under white-box attack or gray-box attack assumptions, which pushes forward the research on RFF attack to more practical scenarios.

Despite these efforts, the research on the RFF attack and its prevention is still in its infant stage. There are still some challenges that need to be properly handled. An obvious difficulty when forging an RFF is the varying channel conditions and random data. The randomness will make the training process of the spoofing method very unstable. The attack may need to update its way of spoofing to be synchronized with the changes. How to devise an effective RFF impersonation learning and updating strategy is an interesting topic with profound impact on the future research. With this basis, we need to investigate the method to design a secure RFF that can resist impersonation attack. Exploring new counterfeit-proofing RFF features or adding artificial distortion to enhance the randomness of the fingerprint could be possible ways to prevent RFF impersonation attacks in the future.

Acknowledgments

M. Liu, H. Luo, and X. Wang were supported by the National Natural Science Foundation of China under Grant 61971029, National Key R&D Program of China (No.2021YFE0110500), and the Fundamental Research Funds for the Central Universities (2019JBM021, 2022JBXT001). L. Peng and H. Fu were supported by the National Natural Science Foundation of China under Grant 62001106 and 62171120, Natural Science Foundation of Jiangsu Province under Grant BK20200350, National Key R&D Program of China (2022YFB4300300, 2020YFE0200600), Guangdong Key R&D Program under Grant 2020B0303010001, Jiangsu Provincial Key Laboratory of Network and Information Security No. BM2003201, and the Purple Mountain Laboratories for Network and Communication Security. The authors would like to express their gratitude to Miss Lanlan Wei, Dr. Wenyuan Xue, Mr. Lei Zhang, Miss Xiaoyi Han, Miss Nian Liu, Miss Yawen Zheng, Mr. Tianzhuang Zhang, Mr. Yuxuan Xu, Miss Manqi Li, Mr. Ziyi Fang, and Mr. Haoyu Yi who made contributions to the RFF research project.

References

[1] Navarro-Ortiz J, Romero-Diaz P, Sendra S, *et al.* A survey on 5G usage scenarios and traffic models. *IEEE Communications Surveys & Tutorials*. 2020;22(2):905–929.

[2] Gui G, Liu M, Tang F, *et al.* 6G: opening new horizons for integration of comfort, security, and intelligence. *IEEE Wireless Communications*. 2020;27(5):126–132.

[3] Li Z, Wang W, Wilson C, *et al.* FBS-radar: uncovering fake base stations at scale in the wild. In: *Proceedings of Network and Distributed System Security (NDSS) Symposium*; 2017. p. 1–15.

[4] 3GPP TR 33.809. Study on 5G security enhancement against false base stations (FBS) [Release 18]; 2022.

[5] Wan T. False Base Station or IMSI Catcher: What You Need to Know; 2019. Available from: https://www.cablelabs.com/blog/help-an-innovator-win-10g-challenge.

[6] Huang L. LTE redirection attack – forcing targeted LTE cellphone into unsafe network. In: *Hack in the Box Security Conference*; 2016.

[7] Rupprecht D, Kohls K, Holz T, *et al.* Breaking LTE on layer two. In: *Proceedings of IEEE Symposium on Security and Privacy (S&P)*; 2019. p. 1121–1136.

[8] Yang H, Bae S, Son M, *et al.* Hiding in plain signal: physical signal overshadowing attack on LTE. In: *Proceedings of USENIX Security Symposium (USENIX Security)*; 2019. p. 55–72.

[9] Hasan M. State of IoT 2022: Number of Connected IoT Devices Growing 18% to 14.4 Billion Globally; 2022. Available from: https://iot-analytics.com/number-connected-iot-devices/.

[10] Gund Y, Tiwari R, Beek C. Mirai, the IoT Botnet [McAfee Labs Threats Report]; 2017. Available from: https://www.mcafee.com/blogs/other-blogs/mcafee-labs/mcafee-labs-threats-report-explores-threat-intelligence-sharing-mirai-iot-botnet/.

[11] Xu Q, Zheng R, Saad W, *et al.* Device fingerprinting in wireless networks: challenges and opportunities. *IEEE Communications Surveys & Tutorials*. 2016;18(1):94–104.

[12] Mucchi L, Jayousi S, Caputo S, *et al.* Physical-layer security in 6G networks. *IEEE Open Journal of the Communications Society*. 2021;2:1901–1914.

[13] Danev B, Zanetti D, and Capkun S. On physical-layer identification of wireless devices. *ACM Computing Surveys*. 2012;45(1):1–29.

[14] Tse D and Viswanath P. *Fundamentals of Wireless Communication*. Cambridge University Press; 2005.

[15] Ludwig R and Bogdanov G. *RF Circuit Design: Theory and Applications*. Pearson International Edition. Prentice-Hall; 2009.

[16] Peng L, Zhang J, Liu M, *et al.* Deep learning based RF fingerprint identification using differential constellation trace figure. *IEEE Transactions on Vehicular Technology*. 2020;69(1):1091–1095.

[17] Ramasubramanian M, Banerjee C, Roy D, *et al.* Exploiting spatio-temporal properties of I/Q signal data using 3D convolution for RF transmitter identification. *IEEE Journal of Radio Frequency Identification*. 2021;5(2):113–127.

[18] Yuan Y, Huang Z, Wang F, *et al.* Radio specific emitter identification based on nonlinear characteristics of signal. In: *Proceedings of IEEE International*

Black Sea Conference on Communications and Networking (BlackSeaCom); 2015. p. 77–81.

[19] Wang W, Sun Z, Piao S, *et al.* Wireless physical-layer identification: modeling and validation. *IEEE Transactions on Information Forensics and Security*. 2016;11(9):2091–2106.

[20] He B and Wang F. Cooperative specific emitter identification via multiple distorted receivers. *IEEE Transactions on Information Forensics and Security*. 2020;15:3791–3806.

[21] Wan Y, Dodd TJ, and Harrison RF. Modeling of power amplifier nonlinearities using Volterra series. *IFAC Proceedings Volumes*. 2005;38(1): 785–790.

[22] Bishop CM. *Pattern Recognition and Machine Learning*. Berlin, Heidelberg: Springer-Verlag; 2006.

[23] Geng C, Huang SJ, and Chen S. Recent advances in open set recognition: a survey. *IEEE Transactions on Pattern Analysis and Machine Intelligence*. 2021;43(10):3614–3631.

[24] Gritsenko A, Wang Z, Jian T, *et al.* Finding a 'New' needle in the haystack: unseen radio detection in large populations using deep learning. In: *2019 IEEE International Symposium on Dynamic Spectrum Access Networks (DySPAN)*; 2019. p. 1–10.

[25] Hanna S, Karunaratne S, and Cabric D. Open set wireless transmitter authorization: deep learning approaches and dataset considerations. *IEEE Transactions on Cognitive Communications and Networking*. 2021;7(1):59–72.

[26] Xie R, Xu W, Chen Y, *et al.* A generalizable model-and-data driven approach for open-set RFF authentication. *IEEE Transactions on Information Forensics and Security*. 2021;16:4435–4450.

[27] Reising DR, Temple MA, and Jackson JA. Authorized and rogue device discrimination using dimensionally reduced RF-DNA fingerprints. *IEEE Transactions on Information Forensics and Security*. 2015;10(6): 1180–1192.

[28] Talbot KI, Duley PR, and Hyatt MH. Specific emitter identification and verification. *Technology Review Journal*. 2003;113:113–130.

[29] Yin P, Peng L, Zhang J, *et al.* LTE device identification based on RF fingerprint with multi-channel convolutional neural network. In: *Proceedings of IEEE Global Communications Conference (GLOBECOM)*; 2021. p. 1–6.

[30] Rasmussen KB and Capkun S. Implications of radio fingerprinting on the security of sensor networks. In: *Proceedings of International Conference on Security and Privacy in Communications Networks (SecureComm)*; 2007. p. 331–340.

[31] Hall J, Barbeau M, and Kranakis E. Radio frequency fingerprinting for intrusion detection in wireless networks. *IEEE Transactions on Defendable and Secure Computing*. 2005;12:1–35.

[32] Shaw D and Kinsner W. Multifractal modelling of radio transmitter transients for classification. In: *Proceedings of IEEE Communications, Power and Computing (WESCANEX)*; 1997. p. 306–312.

[33] Sun D, Li Y, Xu Y, *et al.* A novel method for specific emitter identification based on singular spectrum analysis. In: *Proceedings of IEEE Wireless Communications and Networking Conference (WCNC)*; 2017. p. 1–6.

[34] Jia Y, Zhu S, and Gan L. Specific emitter identification based on the natural measure. *Entropy*. 2017;19(3):117.

[35] Aghnaiya A, Dalveren Y, and Kara A. On the performance of variational mode decomposition-based radio frequency fingerprinting of Bluetooth devices. *Sensors*. 2020;20(6):1704.

[36] Ali AM, Uzundurukan E, and Kara A. Improvements on transient signal detection for RF fingerprinting. In: *Proceedings of IEEE Signal Processing and Communications Applications Conference (SIU)*; 2017. p. 1–4.

[37] Shi Y and Jensen MA. Improved radiometric identification of wireless devices using MIMO transmission. *IEEE Transactions on Information Forensics and Security*. 2011;6(4):1346–1354.

[38] Brik V, Banerjee S, Gruteser M, *et al.* Wireless device identification with radiometric signatures. In: *Proceedings of ACM International Conference on Mobile Computing and Networking*; 2008. p. 116–127.

[39] Knox DA and Kunz T. Wireless fingerprints inside a wireless sensor network. *ACM Transactions on Sensor Networks (TOSN)*. 2015;11(2):1–30.

[40] Sankhe K, Belgiovine M, Zhou F, *et al.* No radio left behind: radio fingerprinting through deep learning of physical-layer hardware impairments. *IEEE Transactions on Cognitive Communications and Networking*. 2019;6(1): 165–178.

[41] Wang Y, Gui G, Gacanin H, *et al.* An efficient specific emitter identification method based on complex-valued neural networks and network compression. *IEEE Journal on Selected Areas in Communications*. 2021;39(8): 2305–2317.

[42] Polak AC, Dolatshahi S, and Goeckel DL. Identifying wireless users via transmitter imperfections. *IEEE Journal on Selected Areas in Communications*. 2011;29(7):1469–1479.

[43] Peng Y, Liu P, Wang Y, *et al.* Radio frequency fingerprint identification based on slice integration cooperation and heat constellation trace figure. *IEEE Wireless Communications Letters*. 2022;11(3):543–547.

[44] Jia Y, Ma J, and Gan L. Combined optimization of feature reduction and classification for radiometric identification. *IEEE Signal Processing Letters*. 2017;24(5):584–588.

[45] Patel HJ, Temple MA, and Baldwin RO. Improving ZigBee device network authentication using ensemble decision tree classifiers with radio frequency distinct native attribute fingerprinting. *IEEE Transactions on Reliability*. 2015;64(1):221–233.

[46] Toonstra J and Kinsner W. Transient analysis and genetic algorithms for classification. In: *Proceedings of IEEE Communications, Power, and Computing (WESCANEX)*. vol. 2; 1995. p. 432–437.

[47] Köse M, Taşcioğlu S, and Telatar Z. RF fingerprinting of IoT devices based on transient energy spectrum. *IEEE Access*. 2019;7:18715–18726.

[48] Danev B, Heydt-Benjamin TS, and Capkun S. Physical-layer identification of RFID devices. In: *Proceedings of USENIX Security Symposium*; 2009. p. 199–214.

[49] Kennedy IO, Scanlon P, Mullany FJ, *et al.* Radio transmitter fingerprinting: a steady state frequency domain approach. In: *Proceedings of IEEE Vehicular Technology Conference (VTC)*; 2008. p. 1–5.

[50] Suski II WC, Temple MA, Mendenhall MJ, *et al.* Using spectral fingerprints to improve wireless network security. In: *Proceedings of IEEE Global Telecommunications Conference (GLOBECOM)*; 2008. p. 1–5.

[51] Liu M, Wang J, Zhao N, *et al.* Radio frequency fingerprint collaborative intelligent identification using incremental learning. *IEEE Transactions on Network Science and Engineering*. 2021; p. 1–11.

[52] Shen G, Zhang J, Marshall A, *et al.* Radio frequency fingerprint identification for LoRa using spectrogram and CNN. In: *Proceedings of IEEE Conference on Computer Communications (INFOCOM)*; 2021. p. 1–10.

[53] Shen G, Zhang J, Marshall A, *et al.* Radio frequency fingerprint identification for LoRa using deep learning. *IEEE Journal on Selected Areas in Communications*. 2021;39(8):2604–2616.

[54] Shen G, Zhang J, Marshall A, *et al.* Towards scalable and channel-robust radio frequency fingerprint identification for LoRa. *IEEE Transactions on Information Forensics and Security*. 2022;17:774–787.

[55] Reising DR, Temple MA, and Oxley ME. Gabor-based RF-DNA fingerprinting for classifying 802.16 e WiMAX mobile subscribers. In: *Proceedings of International Conference on Computing, Networking and Communications (ICNC)*; 2012. p. 7–13.

[56] Ding L, Wang S, Wang F, *et al.* Specific emitter identification via convolutional neural networks. *IEEE Communications Letters*. 2018;22(12): 2591–2594.

[57] Gong J, Xu X, and Lei Y. Unsupervised specific emitter identification method using radio-frequency fingerprint embedded InfoGAN. *IEEE Transactions on Information Forensics and Security*. 2020;15:2898–2913.

[58] Satija U, Trivedi N, Biswal G, *et al.* Specific emitter identification based on variational mode decomposition and spectral features in single hop and relaying scenarios. *IEEE Transactions on Information Forensics and Security*. 2018;14(3):581–591.

[59] Zhang J, Wang F, Dobre OA, *et al.* Specific emitter identification via Hilbert–Huang transform in single-hop and relaying scenarios. *IEEE Transactions on Information Forensics and Security*. 2016;11(6): 1192–1205.

[60] Zhou X, Hu A, Li G, *et al.* A robust radio-frequency fingerprint extraction scheme for practical device recognition. *IEEE Internet of Things Journal*. 2021;8(14):11276–11289.

[61] Yu J, Hu A, Li G, *et al.* A robust RF fingerprinting approach using multisampling convolutional neural network. *IEEE Internet of Things Journal*. 2019;6(4):6786–6799.

[62] Riyaz S, Sankhe K, Ioannidis S, *et al.* Deep learning convolutional neural networks for radio identification. *IEEE Communications Magazine*. 2018;56(9):146–152.

[63] Jian T, Rendon BC, Ojuba E, *et al.* Deep learning for RF fingerprinting: a massive experimental study. *IEEE Internet of Things Magazine*. 2020;3(1): 50–57.

[64] Choe HC, Poole CE, Andrea MY, *et al.* SPIE. Novel identification of intercepted signals from unknown radio transmitters. *Wavelet Applications II*. 1995;2491:504–517.

[65] Klein R, Temple MA, Mendenhall MJ, *et al.* Sensitivity analysis of burst detection and RF fingerprinting classification performance. In: *Proceedings of IEEE International Conference on Communications (ICC)*; 2009. p. 1–5.

[66] Lakafosis V, Traille A, Lee H, *et al.* RF fingerprinting physical objects for anticounterfeiting applications. *IEEE Transactions on Microwave Theory and Techniques*. 2010;59(2):504–514.

[67] SIG B. Bluetooth Core Specifications Version 4.2; 2014.

[68] Collotta M, Pau G, Talty T, *et al.* Bluetooth 5: a concrete step forward toward the IoT. *IEEE Communications Magazine*. 2018;56(7):125–131.

[69] Antonioli D, Tippenhauer NO, and Rasmussen K. BIAS: Bluetooth impersonation attacks. In: *Proceedings of IEEE Symposium on Security and Privacy (S&P)*; 2020. p. 549–562.

[70] Liu H, Venkatesan V, Nilsen C, *et al.* Performance of frequency hopped noncoherent GFSK in correlated Rayleigh fading channels. In: *Proceedings of IEEE International Conference on Communications (ICC)*. vol. 4; 2003. p. 2779–2783.

[71] Merchant K, Revay S, Stantchev G, *et al.* Deep learning for RF device fingerprinting in cognitive communication networks. *IEEE Journal of Selected Topics in Signal Processing*. 2018;12(1):160–167.

[72] Wang C, Yu Z, Zheng H, *et al.* CGAN-plankton: towards large-scale imbalanced class generation and fine-grained classification. In: *Proceedings of IEEE International Conference on Image Processing (ICIP)*; 2017. p. 855–859.

[73] Akcay S, Atapour-Abarghouei A, and Breckon TP. GANomaly: semi-supervised anomaly detection via adversarial training. In: *Proceedings of Asian Conference on Computer Vision*; 2018. p. 622–637.

[74] Radford A, Metz L, and Chintala S. Unsupervised representation learning with deep convolutional generative adversarial networks. In: *Proceedings of International Conference on Learning Representations (ICLR)*; 2016. p. 1–16.

[75] Wu Z, Peng L, Zhang J, *et al.* Authorized and rogue LTE terminal identification using wavelet coefficient graph with auto-encoder. In: *Proceedings of IEEE 96th Vehicular Technology Conference (VTC2022-Fall)*; 2022. p. 1–5.

[76] Tataria H, Shafi M, Molisch AF, *et al.* 6G wireless systems: vision, requirements, challenges, insights, and opportunities. *Proceedings of the IEEE*. 2021;109(7):1166–1199.

[77] Wang W, Sun Z, Ren K, *et al.* User capacity of wireless physical-layer identification. *IEEE Access*. 2017;5:3353–3368.

[78] Danev B, Luecken H, Capkun S, *et al.* Attacks on physical-layer identification. In: *Proceedings of ACM Conference on Wireless Network Security*; 2010. p. 89–98.

[79] Polak AC and Goeckel DL. Identification of wireless devices of users who actively fake their RF fingerprints with artificial data distortion. *IEEE Transactions on Wireless Communications*. 2015;14(11):5889–5899.

[80] Shi Y, Davaslioglu K, and Sagduyu YE. Generative adversarial network for wireless signal spoofing. In: *Proceedings of the ACM Workshop on Wireless Security and Machine Learning (WiseML)*; 2019. p. 55–60.

[81] Karunaratne S, Krijestorac E, and Cabric D. Penetrating RF Fingerprinting-based authentication with a generative adversarial attack. In: *Proceedings of IEEE International Conference on Communications (ICC)*; 2021. p. 1–6.

[82] Xu Y, Liu M, Peng L, *et al.* Colluding RF fingerprint impersonation attack based on generative adversarial network. In: *Proceedings of IEEE International Conference on Communications (ICC)*; 2022. p. 1–6.

Part III

Key generation from wireless channels

Chapter 8
Secret key generation from wireless channels

Magnus Sandell[1]

8.1 Introduction

Wireless communication networks are rapidly expanding, especially in the Internet of Things (IoT) and machine-to-machine (MTM) applications. Although this offers flexibility, mobility and simplicity, there are security challenges which have become a critical concern. While some of these risks are similar to those in wired networks, others are completely new to wireless networks. The most significant one is that the communication medium is open to intruders, who may passively or actively attack the network. An example of the former is eavesdropping (listening to messages communicated by legitimate devices) and examples of the latter are spoofing (introducing malicious messages) and jamming (disturbing the legitimate communication channel). Other security challenges to wireless networks include resource constraints, such as battery life and computational ability. Devices may have limited transmission power or use weak cryptographic mechanisms to save power, making them susceptible to powerful adversaries. Self-configuring heterogeneous networks may have weak lower-layer links that could be a potential breach for the whole system. A direct consequence of these risks is the loss of data confidentiality (providing information to someone not authorised to access it), data integrity (modifying some legitimate information) or denial of service (exhausting resources needed to provide service) attacks. Unsecure wireless networks are also susceptible to attacks that consume network bandwidth or generally degrade network performance.

The tasks of authentication, confidentiality and privacy in communication systems are usually controlled by the upper layers of the communication stack. This is handled by variations of public- and private-key cryptosystems, which are usually based on mathematical operations, such as the factorisation into prime factors, which are deemed to be hard to crack by an attacker with limited computational power. The security provided by these systems is referred to as *computational security*. While it has shown its benefits over recent years, the dawn of quantum computing may exploit some limitations in this approach. Furthermore the development of new mobile *ad hoc*

[1]Toshiba Europe Ltd., Bristol Research and Innovation Laboratory, UK

networks comprising varying degrees of capabilities may be an inefficient way to provide security. Simple IoT devices with limited power and/or computational resources may not be able to handle complex operations such as advanced encryption.

In recent years, research has been directed towards the physical layer (PHY) of communication systems. In particular, the reciprocity of wireless propagation channels has been studied as a way of generating symmetric secret keys and has attracted considerable attention. Reciprocal channels refer to the case where the 'forward' channel (from devices A to B) is the same as the 'backward' channel (from devices B to A). This is usually the case in time-division duplex (TDD) system which transmits in the same frequency band at different times. If these two transmissions (forward and backward) take place within the coherence interval, they are virtually the same. By coherence interval, we mean the time during which the propagation characteristics have not changed significantly. This varies with the fading rate of the channel, e.g., an indoor channel may have a long coherence interval, while an outdoor channel for a mobile device may have a much shorter one. However, a communication system can be designed to allow both channels to be measured very quickly (milliseconds or less), thereby ensuring that the channel is reciprocal. Furthermore, an eavesdropper who is only a few wavelengths away from the legitimate devices will experience an almost uncorrelated channel, thereby providing security from a passive attacker.

This means that crypto keys generated from the random wireless channel can provide *information-theoretical security* as opposed to the computational security. If keys generated in this fashion can be made to be uniformly distributed, they can be used for one-time pad encryption which is impossible to crack. There are two major advantages with keys generated from the random wireless channel: (1) the keys are already shared between the two legitimate users, which overcomes the problem of key distribution, (2) these keys can be replenished dynamically as the wireless channel changes over time. Hence, this PHY approach can not only provide information-theoretic secure keys but also information-theoretic secure encryption.

8.1.1 Secret key rate

Important theoretical work on secret key generation was done in the early 1990s [1,2]. The problem of sharing a secret key was investigated from a probabilistic point of view, where a passive eavesdropper is assumed to observe correlated versions of the legitimate communication. This key-generation model is shown in Figure 8.1 where the legitimate users Alice and Bob want to establish a secure cryptographic key in the presence of the eavesdropper Eve. Since the channel between Alice and Bob is wireless, it must be assumed that everything they transmit can be overheard by Eve. However, the propagation channels Alice–Eve and Bob–Eve are different from that of Alice–Bob as Eve is assumed to be at least a few wavelengths away from either legitimate user.

The devices Alice and Bob have the N correlated observations $X^{(N)} = (X_1, \ldots, X_N)$ and $Y^{(N)} = (Y_1, \ldots, Y_N)$, respectively. Alice and Bob will exchange a common message s over a public channel, which can be overheard by Eve in its entirety. Below we distinguish between three types of public channels.

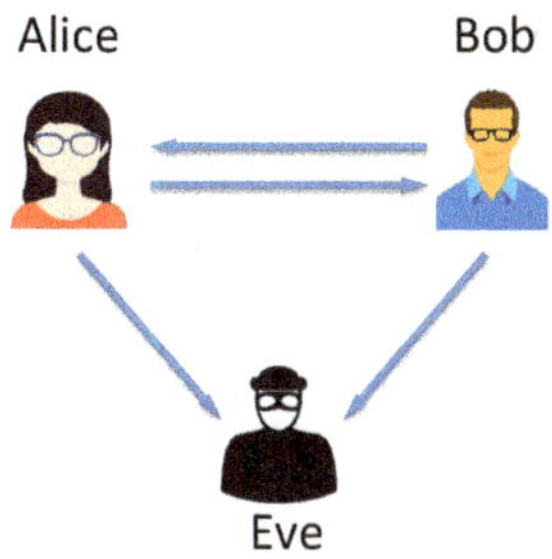

Figure 8.1 Secret communication system. Legitimate users Alice and Bob use a (public) wireless channel that an adversary Eve can eavesdrop.

8.1.1.1 Unlimited public discussion

Besides observing the sequences $X^{(N)}$ and $Y^{(N)}$, Alice and Bob can communicate over a noiseless public channel with unlimited capacity. It is assumed that the eavesdropper Eve has perfect access to this public channel. The objective is for Alice and Bob to agree on a common random key while only allowing Eve a negligible amount of information about it. If there exist key-generation functions $K_A = f_A(X^{(N)}, s)$ and $K_B = f_B(Y^{(N)}, s)$ which satisfy

$$\Pr\{K_A \neq K_B\} < \varepsilon \tag{8.1}$$

$$\frac{1}{N} I(s; K_A) < \varepsilon \tag{8.2}$$

$$\frac{1}{N} H(K_A) > R - \varepsilon \tag{8.3}$$

$$\frac{1}{N} \log_2 |\mathscr{K}| < \frac{1}{N} H(K_A) + \varepsilon \tag{8.4}$$

for any $\varepsilon > 0$ and sufficiently large N, then R is the achievable rate [3]. The first condition, (8.1), means that Alice and Bob can produce the same key with high probability. The second one, (8.2), means that the mutual information between the public message and the secret key is limited. This ensures that the information leaked to an eavesdropper can be made arbitrary small. The achievable rate R in (8.3) is bounded by the entropy per observation, while (8.4) dictates that the generated key is (almost) uniformly distributed where $\mathscr{K}$ is the alphabet of the key. The secret key capacity in this case is given by the mutual information between X and Y

$$C = I(X; Y) \tag{8.5}$$

and represents the largest achievable key rate. This can be achieved by only one transmission from Alice.

8.1.1.2 Constrained rate public discussion

If the public channel does not have unlimited capacity, the secret key capacity is reduced accordingly. If it has a rate constraint R_0, the secret key capacity can be computed by introducing an auxiliary random variable U which is subject to a Markov chain relationship

$$C = \max_U I(U;Y) \tag{8.6}$$

$$\text{s.t.} \begin{cases} U \to X \to Y \\ I(U;X) - I(U;Y) \leq R_0 \end{cases}. \tag{8.7}$$

Again this can be achieved by transmitting from Alice only.

8.1.1.3 Side information at Eve

In certain cases, it might be that Eve can observe a sequence $Z^{(N)} = (Z_1, \ldots, Z_N)$ which is correlated with $X^{(N)}$ and $Y^{(N)}$. In this case, the security condition in (8.2) should be replaced by

$$\frac{1}{N} I\left(s, Z^{(N)}; K_A\right) < \varepsilon. \tag{8.8}$$

Although the secret key capacity has not been characterised, it is possible to achieve the rate

$$R = [I(U;Y) - I(U;Z)]^+ \tag{8.9}$$

$$\text{s.t.} \begin{cases} U \to X \to Y \\ I(U;X) - I(U;Y) \leq R_0 \end{cases}. \tag{8.10}$$

where $[x]^+ = \max\{x, 0\}$ and U is again an auxiliary random variable subject to the Markov chain relationship. This rate can be achieved by transmitting from Alice only.

8.1.2 Key generation

The previous section discussed the information-theoretic limits of secret key generation. They serve as upper bounds on the key-generation rate and can be useful to study different types of communication channels. However, in this chapter, we are more concerned with practical schemes to generate secret keys from wireless channels. Hence, instead of capacity, we will be discussing *key-disagreement probability* (KDP) and *key-generation rate* (KGR). The former is defined as

$$KDP = \Pr\{K_A \neq K_B\} \tag{8.11}$$

i.e., the probability that Alice and Bob will generate different keys. The KGR is defined as the number of bits in the secret key that can be generated per second.

Generally speaking, there are five stages in the secret key generation from wireless channels:

1. **Channel sounding.** Obtain information about the channel.
2. **Feature extraction.** Process the observations.
3. **Channel quantisation.** Generate a key from the features.

4. **Information reconciliation.** Reduce key mismatch.
5. **Privacy amplification.** Enhance key randomness and wipe off leaked information.

The first three items are sometimes combined as *advantage distillation* [2], i.e., creating a random variable about which Alice and Bob have more information than Eve. However, as we are specifically concerned with generating secret keys from the wireless channel, we divide it into three specific tasks. These five subtasks will be discussed in detail in the next sections.

8.2 Channel sounding

8.2.1 Channel probing

In order for Alice and Bob to estimate their channels, it is necessary to exchange some packets. Typically, a simple protocol as shown in Figure 8.2 is used. In most wireless systems, coherent data detection is used which requires some form of training. Hence, it is possible to use these packets to estimate the channel for secret key generation without changing the packet structure. This also reduces overhead since ordinary data communication can be achieved at the same time. However, it is important that Bob transmits as soon as possible after receiving the packet from Alice as the propagation channel must remain as unchanged as possible. Furthermore, if more advanced techniques are used, such as precoding (see next subsection), a new packet format might be needed. Since the only objective of the channel probing is to estimate the channel, the packet does not need any payload which means it can be made quite small. Even if some processing is required, it might be possible to do this within the boundaries of an existing standard by, e.g., providing the required information in the payload. This means that minimal changes to the communication stack are needed and that off-the-shelf components can be used to build a secret key-generation system.

8.2.2 Static environments

In a static environment, the channel may change very slowly in time. This will reduce the KGR since the secret key generation depends on randomness. One way to mitigate this is for Alice and Bob to transmit randomly generated signals which do not need to be exchanged (they can be generated locally at Alice and Bob). If the frequency

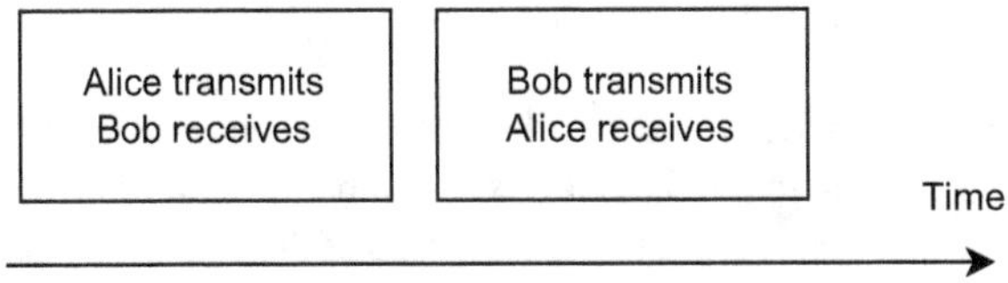

Figure 8.2 A simple protocol for estimating the channel in both directions

response in an orthogonal frequency-division multiplexing (OFDM) system is H_k on the kth subcarrier, Alice can transmit a random signal A_k such that Bob receives

$$R_k^{(B)} = H_k A_k + V_k^{(B)} \tag{8.12}$$

where $V_k^{(B)}$ is the additive noise. Bob then transmits his own random signal B_k such that Alice receives

$$R_k^{(A)} = H_k B_k + V_k^{(A)} \tag{8.13}$$

where we have assumed that the channel H_k is reciprocal, i.e., the same in both directions. Since Alice and Bob know their own signals A_k and B_k, respectively, they can form the composite channel

$$A_k R_k^{(A)} = A_k H_k B_k + A_k V_k^{(A)} \tag{8.14}$$

$$B_k R_k^{(B)} = A_k H_k B_k + B_k V_k^{(B)}. \tag{8.15}$$

Hence, both Alice and Bob will have access to the (noisy) channel $A_k H_k B_k$, which can then be quantised to generate a secret key. This has been investigated in [4,5] (A_k,B_k were random complex variables), in [6] (random QAM symbols were used) and in [7] (random FIR filters). However, it should be noted that although the key-generation rate may increase, the protection against eavesdropping is still based on the propagation channels being different. If an eavesdropper knows the channel between Alice and Bob, it can work out the random sequences and glean information about the generated secret key.

Other approaches to introduce artificial fading is to use electronically steerable parasitic array of radiators (ESPAR). This is a type of antenna which has a tunable radiation pattern and, by varying this, the received signal strength will fluctuate and can, hence, be used for secret key generation in slowly fading channels [8]. A related approach is to apply random beamforming to produce time-varying channel responses [9]. More recently, intelligent reflecting surfaces have also been studied in secret key-generation applications [10] to achieve similar effects, as well as adding protection against eavesdroppers.

8.2.3 Precoding

There may be several reasons why Alice and Bob will not generate the same secret key. It might be due to poor SNR, channels are not reciprocal (see Section 8.2.4), ineffective channel quantisation (see Section 8.4), etc. The latter can be improved by applying some form of precoding. For instance, once Bob has estimated the channel, he does not have to send the same sounding packet as Alice. Instead, he can use the knowledge of the channel to provide a signal for Alice which is better suited for quantisation.

This principle is shown in Figure 8.3 for an abstract channel quantisation scheme. Generally speaking, the signal to be quantised can be thought of as residing in one of the hexagons (quantisation regions) and all signals in the same hexagon are quantised to the same value (+ in Figure 8.3). Hence, a quantisation error will occur (Alice and Bob generate different keys) when their respective signals are in different quantisation

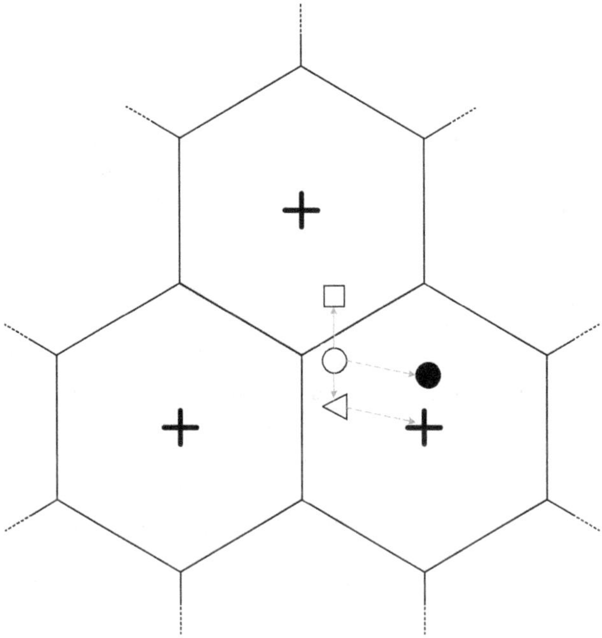

Figure 8.3 Signal quantisation and the effects of precoding [11]. If the noiseless channel (○) is close to a decision boundary, the noisy estimates of Alice (□) and Bob (△) may fall in different regions. With precoding, the noiseless precoded channel for Alice (•) is closer to the centre of the quantisation region (+).

regions due to the noise, nonreciprocity, etc. It is easy to see that this is most likely to happen when the noiseless signal is close to a quantisation boundary (○ in Figure 8.3). A small amount of noise/distortion can take the channel into different quantisation regions.

To mitigate this, Bob can precode his transmitted signal to move it closer to the centre of the quantisation region. For instance, if his estimate is $\widehat{H}_k^{(B)}$ and he has quantised the channel as $H_k^{(Q)}$, he can transmit $\alpha \frac{H_k^{(Q)}}{\widehat{H}_k^{(B)}}$ which Alice will receive as $\alpha \frac{H_k^{(Q)}}{\widehat{H}_k^{(B)}} H_k$, where α is an appropriate scaling factor to ensure the correct transmit power and we have ignored the noise for clarity. If $\widehat{H}_k^{(B)} \approx H_k$, then Alice will have $aH_k^{(Q)}$; the factor α is irrelevant as we will see in Section 8.4. Hence, Alice's channel should be in (or near) the centre of the quantisation region and will be less affected by noise since it is far away from the quantisation boundary.

Precoding to reduce KDP has been applied using phase rotations in [12]. That secret key-generation scheme is based on phase-shift keying (PSK) modulation and by offsetting the transmitted signal with the phase of the channel, a more robust quantisation can be achieved. A similar approach was used in [13] for an OFDM system, where the phase response on each subcarrier was used for precoding. It is worth mentioning that precoding may reveal some information about the Alice–Bob channel to Eve if it is not done properly. Precoding is related to decode-and-forward and the security implications compared to amplify-and-forward (which is the case of no precoding) are discussed in [14].

8.2.4 Reciprocity

One important aspect to consider when using the propagation channel to generate secret keys is how identical the channels are that Alice and Bob see. Reciprocity is essential for Alice and Bob to generate the same key but in practice there might be differences between the two observed channels. Apart from additive noise, the channels might be observed at slightly different times and due to fading, the propagation channel has changed slightly. Methods to combat this include fractional interpolation [15], exploiting correlation [16] and sending information about the distance from quantised values over a public channel [13].

Another issue is sampling offsets as Alice and Bob may not be perfectly time-synchronised due to oscillator differences. These offsets will mean that the channels are sampled at slightly different time instances and will produce slightly different channel responses. The impact of this can be mitigated by, e.g., applying robust transformations [17], considering the channel correlation where timing offsets have a small effect [16], including it in the key reconciliation system based on error-correcting codes [18] or incorporating the phase offset in the hypothesis testing [19].

8.2.5 Frequency-division duplex

So far we have discussed reciprocal channels that require TDD, i.e., transmission in the same frequency band at different times. Many systems today employ frequency-division duplex (FDD) where the transmission takes place in different frequency bands, i.e., the Alice-to-Bob channel has a different carrier frequency/channel band than the Bob-to-Alice channel. In this case, the propagation channels are not the same and not even the signal strength (see Section 8.3.1) might be possible to use due to small-scale fading. One way to overcome this problem is to use amplify-and-forward in two phases. Alice and Bob exchange sounding packets as usual to estimate their channels, $H_k^{(AB)}$ and $H_k^{(BA)}$, respectively, which now are assumed to be different. Alice then transmits the channel she estimated, $H_k^{(BA)}$, so Bob can estimate the combined channel $H_k^{(AB)}H_k^{(BA)}$. Bob then transmits his estimated channel, $H_k^{(AB)}$, such that Alice can estimate $H_k^{(BA)}H_k^{(AB)}$. In this way, both Alice and Bob will have access to the same channel which they then quantise. This principle has been used in [20] and in a related form in [21]. Other approaches for FDD include using the angle and delay of the transmitted signals together with the Chinese remainder theorem [22].

Another approach is to assume that the channel propagation environment is similar for both channels. This means that the scattering and reflecting objects that determine the propagation channel are the same. As the two channels $H_k^{(AB)}$ and $H_k^{(BA)}$ are in slightly different frequency bands, they will have different realisations. Although the propagation paths might be the same, they will have different amplitude and phase for different frequency bands; hence, they will add up to create different impulse responses. However, it can be expected that there is some relationship between the two channels, even though it is very difficult to express this mathematically. This was investigated in [23] for massive MIMO and these ideas have been applied to secret key generation in FDD systems. In [24], a deep learning approach was taken where a neural network was trained to extract channel features for one frequency band using observations from another. It was shown that this can be used for secret key generation in FDD systems.

8.3 Feature extraction

8.3.1 RSSI

The most straightforward way of extracting a feature from the channel observations is to see how the received power varies with time. Most commercially available chip sets provide the receive signal strength indicator (RSSI) as defined by the corresponding standard. This varies over time due to fading but will affect Alice and Bob in the same way if the channel is reciprocal. Its simplicity also makes it suitable for low-cost devices that have limited computational and energy resources.

The drawback of using only the power of the estimated channel is that some information is lost. This was investigated in [25] for a narrowband channel, where it was found that the secret key capacity scales as $\frac{1}{2}\log_2(\gamma/2) - 0.69$ for large SNR γ when only the envelop of the signal is used; if the full channel information is used (the complex values), the capacity scales as $\log_2(\gamma/2)$. Hence, there is a loss associated with discarding the phase of the channel and only using the magnitude. In fact, it was shown in [18] that the phase contains more information than the magnitude. Hence, it is worth considering features of the channel observations that do not only depend on the envelop.

Another metric closely related to the received signal strength is the instantaneous uncoded BER. With a fading channel and a sufficiently large data packet, it can be expected that the uncoded BER depends directly on the received power of the packet. Hence, the fluctuations in BER should follow the channel fading and be the same at both Alice and Bob. This was explored in [26] where it was shown that a 128-bit key could be generated in 12.42 s if the Doppler rate was 10 Hz. The time can be reduced by using smaller measurement intervals; however, this makes the generated bits correlated and, hence, reduces security. The fundamental limit is set by the Doppler rate as uncorrelated bits cannot be generated quicker if the channel is not fading fast enough.

One drawback with using RSSI or related features is that it might take a long time to collect enough observations to be able to generate uncorrelated bits. This depends heavily on the fading rate of the channel and, in static or slow-changing environments, other features can be explored.

8.3.2 Frequency response

Many wireless communication systems today operate in wideband channels, such as WiFi and 4G/5G. To handle the multipath channel and facilitate simple equalisation, OFDM systems are often used. This modulation technique splits the available frequency band into a number of independent narrowband channels which can be estimated and equalised separately. For secret key generation, this means that the available channel to quantise is a vector of channel samples. It is worth noting that these are highly correlated, so when they are converted into a secret key, care must be taken to ensure that the key is uniformly distributed. Using the frequency response of the channel as the channel feature to be quantised has been investigated in, e.g., [6,13,16,18,27–37].

8.3.3 Spatial domain

Another approach to extract feature vectors is to use the spatial domain. Multiple-input multiple-output (MIMO) channels can be decomposed using a singular value decomposition (SVD) [38] to obtain the matrices for Bob

$$\mathbf{H}_k^{(AB)} = \mathbf{U}_k^{(AB)} \boldsymbol{\Sigma}_k^{(AB)} \mathbf{V}_k^{(AB)H} \tag{8.16}$$

$$\mathbf{H}_k^{(AB)} \in \mathbb{C}^{N_R^{(B)} \times N_T^{(A)}} \tag{8.17}$$

$$\mathbf{U}_k^{(AB)} \in \mathbb{C}^{N_R^{(B)} \times N_R^{(B)}} \tag{8.18}$$

$$\boldsymbol{\Sigma}_k^{(AB)} \in \mathbb{C}^{N_R^{(B)} \times N_T^{(A)}} \tag{8.19}$$

$$\mathbf{V}_k^{(AB)} \in \mathbb{C}^{N_T^{(A)} \times N_T^{(A)}} \tag{8.20}$$

where the number of transmit antennas at Alice and receive antennas at Bob are denoted by $N_T^{(A)}$ and $N_R^{(B)}$, respectively. Alice can similarly obtain

$$\mathbf{H}_k^{(BA)} = \mathbf{U}_k^{(BA)} \boldsymbol{\Sigma}_k^{(BA)} \mathbf{V}_k^{(BA)H} \tag{8.21}$$

$$\mathbf{H}_k^{(BA)} \in \mathbb{C}^{N_R^{(A)} \times N_T^{(B)}} \tag{8.22}$$

$$\mathbf{U}_k^{(BA)} \in \mathbb{C}^{N_R^{(A)} \times N_R^{(A)}} \tag{8.23}$$

$$\boldsymbol{\Sigma}_k^{(BA)} \in \mathbb{C}^{N_R^{(A)} \times N_T^{(B)}} \tag{8.24}$$

$$\mathbf{V}_k^{(BA)} \in \mathbb{C}^{N_T^{(B)} \times N_T^{(B)}} \tag{8.25}$$

where the number of transmit antennas at Bob and receive antennas at Alice are denoted by $N_T^{(B)}$ and $N_R^{(A)}$, respectively. Note that if Alice and Bob are using all their antennas for both transmission and reception, we have $N_T^{(A)} = N_R^{(A)}$ and $N_T^{(B)} = N_R^{(B)}$. Secret key generation is simplified if all numbers of antennas are the same, although this is not necessary. After the channel sounding, both Alice and Bob can decompose the MIMO channel as described above and use the results for secret key generation.

In [39], secret keys are generated by first separating the complex SVD matrices into magnitude and phase. These are then quantised by comparing the elements with the average of the whole matrix; in this way, a large number of bits can be generated. However, care needs to be taken since correlated bits are generated and the SVD matrices are only unique up to complex rotation factor. MIMO is combined with OFDM in [40], where subcarriers are activated/deactivated to produce secret keys. Preprocessing using principle component analysis has also been studied in [30].

8.3.4 Decorrelation

Some of the approaches above can generate a large number of bits from a small set of observations. However, these observations may be dependent and thus produce keys with correlated bits. Hence, an important technique is the decorrelation of the observations to only extract statistically independent information from them. If a vector with N observations has a joint-Gaussian distribution with covariance matrix $\mathbf{R}$, then the mutual information per element between Alice's and Bob's observations, $\mathbf{r}_a$ and $\mathbf{r}_b$, respectively, is [41]

$$I_{AB} = I\left(\mathbf{r}_a; \mathbf{r}_b\right) = \frac{1}{N}\sum_{n=1}^{N}\log_2\left(\frac{1}{1-\left(\frac{\lambda_n/\sigma^2}{1+\lambda_n/\sigma^2}\right)^2}\right). \tag{8.26}$$

Here λ_n are the eigenvalues of $\mathbf{R}$ and σ^2 is the variance of the Gaussian observation noise affecting $\mathbf{r}_a$ and $\mathbf{r}_b$. This quantity is an upper bound to the secret key rate [2] and is the maximum key length in bits that Alice and Bob can achieve from their common random channel. Note that the vectors $\mathbf{r}_a$ and $\mathbf{r}_b$ can take different forms as we have seen above; e.g., it can be the N subcarriers in an OFDM system, the signal power over N time intervals, the channel response from N antennas, etc.

One interpretation of (8.26) is that the observation vector can be transformed into N parallel channels, where channel n has the secret rate bound

$$-\log_2\left(1-\left(\frac{\lambda_n/\sigma^2}{1+\lambda_n/\sigma^2}\right)^2\right). \tag{8.27}$$

In fact, the mutual information in (8.26) can be derived by applying the eigenvalue decomposition of $\mathbf{R}$ and use the eigenvectors to produce uncorrelated observations. Element n in the decorrelated vector will then have the SNR λ_n/σ^2, which produces the corresponding secret key rate.

In [15], measurements of the channel are taken over time and interpolated to create observations taken at the same time instances for both Alice and Bob. As these channel observations are then by definition correlated, they are decorrelated by using the Karhunen–Loeve transform [42]. Using an estimate of the correlation matrix of the observations, a vector of uncorrelated samples can be obtained. Another technique is to adjust the probing rate to obtain as uncorrelated observations as possible [16], which has the advantage of reducing processing and data size.

8.3.5 Other processing

Other general types of preprocessing can be done once the channel has been estimated. For instance, neighbouring subcarriers in an OFDM system can be averaged [33]. As they are highly correlated, this will produce a smoothed estimate with substantial noise reduction. Furthermore, the subcarriers may not be suitable to be quantised directly as they will produce correlated bits. If the elements in the feature vector are complex, another decision to make is if they should be split into real/imaginary parts [6] or magnitude/phase [39]. For all such considerations, even though the KGR could be high, further processing to ensure independent and uniformly distributed bits might be necessary. In general, there will also be a trade-off between KGR and KDP; although it is easy to generate a large number of bits (high KGR), their quality might be low (high KDP). Conversely, very reliable bits might be generated but at a too low rate to be of practical interest.

8.4 Channel quantisation

A key step in generating secret keys is the quantisation. From the discussion above, a feature vector is extracted (impulse response, frequency response, received signal strength, etc.) after channel sounding. This should be converted into a bit string which will form the secret key. In this subsection, we will discuss a few different ways this can be done.

8.4.1 Scalar quantisation

A straightforward approach to quantise the feature vector is to work with each element separately. These can then be quantised by dividing the range of samples into quantisation regions. The range can be determined from the smallest and largest samples or large-scale statistics can be used. Furthermore, it can be uniformly or nonuniformly quantised; in the latter case, an assumed distribution, e.g., Gaussian, is used. If there are more than two quantisation regions, i.e., more than one bit is produced, they can be grey-coded to minimise the bit-error rate.

This type of *scalar quantisation* will produce a secret key with Nq bits, where N is the number of elements in the feature vector and q is the number of bits per element. Although this can produce a large secret key, some papers have found that $q = 1$, i.e., only one bit per sample, is a good value (note the discussion of the trade-off between KGR and KDP in the previous subsection). Even with reciprocal channels, the noisy estimates may produce different bits for Alice and Bob if too many bits per sample is produced. An example of scalar quantisation is shown in Figure 8.4(a).

It is easy to see that most quantisation errors (Alice produces different bits to Bob) occur when the noiseless sample is close to the quantisation boundary. Even a small amount of noise can take the observations into different regions and, hence, produce

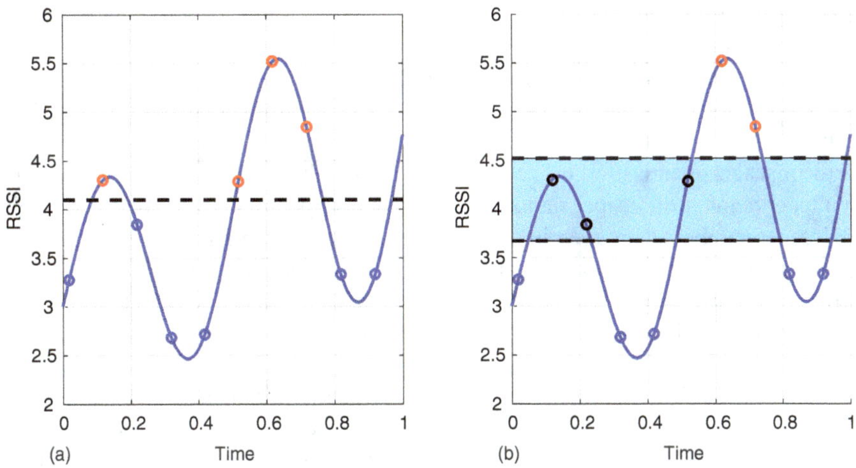

Figure 8.4 Scalar quantisation of an extracted feature vector: (a) uniform quantisation and (b) uniform quantisation with a guard band. The circles represent the observations with blue quantised as 0, red as 1 and black not quantised.

different bits. One way to mitigate this is to introduce guard or censor intervals; this means that samples that fall within these regions near the quantisation boundaries are marked as not quantised. This is illustrated in Figure 8.4(b). The probability that Alice and Bob will generate different keys can then be substantially reduced. A variation of this approach was used in [43,44], where excursions (sequences of consecutive samples which are in the same quantisation region) are detected. The location of these are then shared over a public channel and Alice and Bob will use the excursions that they both share.

However, the drawback with guard intervals and/or excursions is that they may produce a variable number of bits. This may impact the KGR negatively and also increase signalling overhead, key-generation processing, etc., due to its nondeterministic nature.

8.4.2 Vector quantisation

An alternative to scalar quantisation is *vector quantisation* [45]. This works by grouping multidimensional signals together that are close in some sense and represent them by a single value, known as the centroid point. The quantisation function Ω of a vector $\mathbf{x} \in \mathbb{C}^{N\times 1}$ then becomes

$$\Omega(\mathbf{x}) = \mathbf{c}_i \tag{8.28}$$

$$i = \arg\min_j \|\mathbf{x} - \mathbf{c}_j\| \tag{8.29}$$

where $\mathbf{c}_j \in \mathbb{C}^{N\times 1}$ are the centroids (+ in Figure 8.3), or codewords, of the vector quantiser. The norm $\|\|$ is often taken to be the Euclidean distance. The advantage of vector quantisation is that it considers the whole vector $\mathbf{x}$ rather than its elements separately which is the case for scalar quantisation. If the vector $\mathbf{x}$ is, e.g., the frequency response of the channel, it can handle the correlation between the samples. In such a case, scalar quantisation would produce highly correlated bits which is undesirable for secret key generation. It is possible to decorrelated all samples and perform scalar quantisation on each sample separately [15]; however, this requires an integer number of bits per sample, where vector quantisation can effectively apply noninteger numbers. However, in general, if all samples are highly uncorrelated, e.g., channel impulse response taps, vector quantisation does not add much advantage in terms of performance. Vector quantisation has been investigated in [46] for secret key generation, where the design of multiple codebooks was also considered; for a given channel estimate, a codebook with a small distance to its nearest centroid can be chosen and if the other device is told which one, the KDP can be reduced.

The main drawback of vector quantisation is its complexity. If Q bits are to be generated, a codebook must consist of 2^Q codewords. The quantisation then requires the computation of the distance from the feature vector to all these codewords. This computation can be substantial for large Q and devices with computational/power limitations. Techniques to reduce the complexity include tree-structured vector quantisation [47]. This is based on the idea to sequentially dividing the vector space into regions to reduce the overall distortion (quantisation error). At a given stage in the design procedure, the region with the largest distortion is split into two and thereby reducing the overall distortion. This continues until the total distortion is small enough or the number of regions has reached a predetermined value. One example of this is the binary search tree proposed in [48]; at each level in the tree, a binary decision is made which branch to follow. It uses a balanced tree, i.e., all leaves are at the same level, thus quantisation always produces a fixed number of bits. An example is shown in Figure 8.5, where at each level, the subregion where the input vector lies is found. This decision can be made by finding in which hyperplane the input vector is, i.e.,

$$\|\mathbf{x} - \mathbf{c}_1\|^2 < \|\mathbf{x} - \mathbf{c}_2\|^2 \Leftrightarrow \mathrm{Re}\left\{\mathbf{x}^H \mathbf{a}\right\} \leq b \tag{8.30}$$

$$\mathbf{a} = \|\mathbf{c}_2 - \mathbf{c}_1\|^2, \; b = \|\mathbf{c}_2\|^2 - \|\mathbf{c}_1\|^2 \tag{8.31}$$

where $\mathbf{c}_1$ and $\mathbf{c}_2$ are the centroids for the two subregions being compared. If there are Q levels in the decision tree, this quantisation will produce Q bits since each comparison produces one bit. The hyperplane $P_{i,j}, j = 1, \ldots, 2^{i-1}$ on level $i = 1, \ldots, Q$ can be precomputed and stored; note that they are all generally different.

8.4.3 Statistical properties

The previous subsections have discussed how to generate bits from a wireless channel. However, apart from the number of bits generated (KGR) and their reliability (KDP), it is also important to consider their quality. For instance, they should be uniformly distributed to provide as large entropy as possible but also independent to avoid giving an eavesdropper any information about them. In fact, there are many statistical properties that should be satisfied before the secret key can be considered to be 'random'. The National Institute for Standards and Technology (NIST) in the United

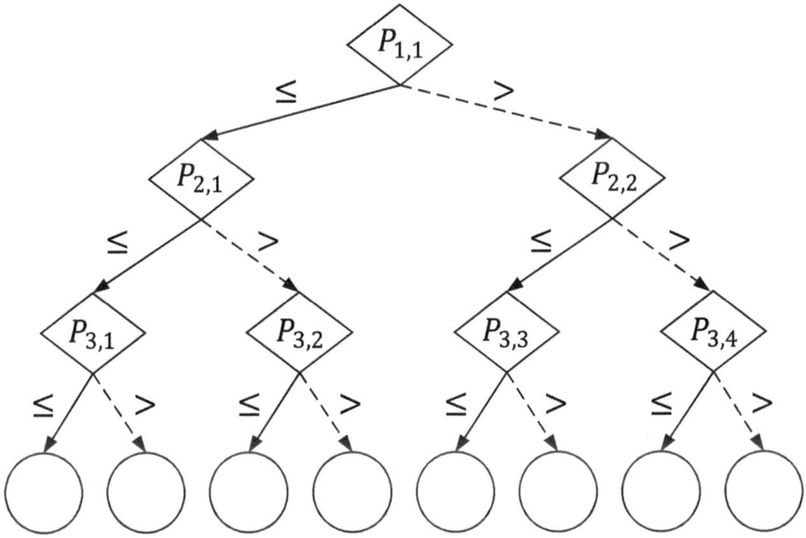

Figure 8.5 *Binary decision tree for tree-structured vector quantisation [11]. At each level, a decision is made, in which subregion, the input vector lies. Hyperplane j on level i is denoted* $P_{i,j}$.

States has quantified this and published a report [49] which is widely used to measure the randomness of bit strings. There are 15 tests and these include frequency tests (overall and within blocks), run tests (consecutive bit values), spectral test (discrete Fourier transform), approximate entropy test and random excursion tests (how the cumulative sum behaves). These can be used to evaluate how good the generated secret keys are and how suitable they are for cryptographic applications.

8.5 Information reconciliation

As shown in the previous section, the secret key generation is not completely error free. In a noisy channel, there is always a chance that Alice and Bob will generate different keys. As this could have catastrophic consequences, it is necessary to perform some kind of *information reconciliation*. This aims at correcting any discrepancies between the two keys. Since the keys cannot be compared in a secure way, some form of public discussion is needed. This can be achieved by using a *secure sketch*, which was introduced in [50]. This is a procedure that produces public information about its input w that does not reveal w, and yet allows exact recovery of w given another value that is close to w. Thus, it can be used to reliably reproduce noisy inputs without revealing vital information about them to eavesdroppers.

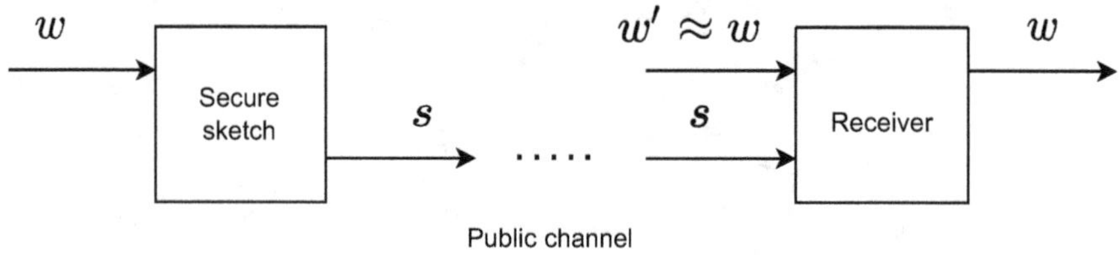

Figure 8.6 Secure sketch: from the input w, a secure sketch s is produced. This has the property that the entropy of w is high, even given s. The string w can be recovered from a noisy observation w′ (close to w) using s.

A secure sketch is illustrated in Figure 8.6 and its formal definition is [50].

Definition 1. *A ($\mathcal{M}$, m, m′, t)-secure sketch scheme consists of a sketch function* `SSGen` *and a recovery function* `SSRep` *such that the following properties hold:*

1. *The input to the sketch function is $w \in \mathcal{M}$, where $\mathcal{M}$ is a metric space, and the output is $s =$ `SSGen` (w).*
2. *The recovery function takes s and a noisy observation $w' \in \mathcal{M}$ and returns $w =$ `SSRep`(s, w'). This is guaranteed to work as long as the distance between w and w' is less than a certain threshold t.*
3. *For any random variable W over $\mathcal{M}$ with min-entropy $H_\infty(W) \geq m$, an adversary observing S has an average min-entropy $H_\infty(W|S) \geq m'$.*

The min-entropy of a random variable X is defined as [50]

$$H_\infty(X) = -\log_2\left(\max_x \Pr(X = x)\right) \tag{8.32}$$

which can be viewed as the 'worst-case' entropy. The average min-entropy of X conditioned on Y is defined as $H_\infty(X|Y) = -\log_2 \mathbb{E}_{y\leftarrow Y}(\max_x \Pr(X = x|Y = y))$, which can be viewed as the average case for Y and the worst case for X. For the secure sketch, this means that for any W with $H_\infty(W) \geq m$, the probability of guessing W after observing S is at most $2^{-m'}$, where $m' \leq H_\infty(W|S)$; hence, the entropy loss due to releasing S is $m - m'$. Note that min-entropy is a more conservative notion than Shannon entropy and, therefore, often preferred within cryptology [51].

In most cases, the secure sketch is constructed using error-correcting codes. A common approach is the code offset, denoted fuzzy commitment in [52]. If the secret keys generated by Alice and Bob have length n, a linear code (n, k) can be used, where $k < n$ is the number of information bits of the code. The secure sketch in [50] randomly selects one of the 2^k codewords of length n at Alice and adds it modulo-2 to her secret key. In this instance, the secret key would be w and c is the codeword. The secure sketch is then obtained as

$$s_i = c_i \oplus w_i, \; i = 1, \ldots, n \tag{8.33}$$

which can be sent to Bob over a public (and error-free) channel. The recovery function at Bob then adds his secret key w' to produce

$$y_i = s_i \oplus w'_i = c_i \oplus w_i \oplus w'_i = c_i \oplus e_i,\ i = 1, \ldots, n \tag{8.34}$$

where $e_i = w_i \oplus w'_i$ can be interpreted as an error vector. If Alice and Bob's secret keys are similar (but not identical), e_i will contain only a few nonzero elements which the error-correcting code can correct. Hence, Bob can recover c_i and derive Alice's secret key as $s_i \oplus c_i = w_i$.

So for a binary code, we can set $\mathcal{M} = \{0, 1\}^n$ and use an (n, k) error-correcting code with minimum codeword distance $2t + 1$ which can correct t errors. The loss in entropy for this scheme is bounded by $n - k$ since the codewords have length n but only dimension k. Hence, this would constitute a secure sketch scheme ($\{0, 1\}^n$, m, $m - (n - k)$, t).

As mentioned above, most information reconciliation schemes are based on error-correcting codes. A brief summary of some popular schemes is given in Table 8.1 where it is noted how they exploit the error-correcting capability. The *syndrome* can be used as the secure sketch [53], whereby the modulo-2 difference between Alice's and Bob's syndrome can be computed. The error patterns are obtained with the error-correcting code which then corrects any errors in the secret key. In the *code offset* method, a random codeword is chosen and the secure sketch is the sum of the secret key and the codeword. Differences between the secret keys of Alice and Bob can be removed with the help of the error-correcting code and Table 8.1 shows a few variations of this idea [51]. It is possible to use the codeword [52], the corrected secret key [50] or the information word corresponding to the code word [54]. It is

Table 8.1 A summary of some common information reconciliation schemes [51]. The functions SSGen *and* SSRep *produce and recover the secure sketch, respectively, whereas H, $\mathcal{C}$ and $\|$ denote the parity check matrix, set of codewords and concatenation, respectively.*

$s \leftarrow \texttt{SSGen}(w)$	$y \leftarrow \texttt{SSRep}(w', s)$	Method	Ref
$s \leftarrow wH^T$	$w'H^T \oplus s = eH^T$ determine e' $y = \widehat{w} \leftarrow w' \oplus e'$	Syndrome	[53]
Random $c \in \mathcal{C}$ $s \leftarrow w \oplus c$	$x \leftarrow w' \oplus s = w \oplus e$ $y = \widehat{c} \leftarrow \texttt{Correct}(x)$	Code offset	[52]
	$x \leftarrow w' \oplus s = w \oplus e$ $y = \widehat{w} \leftarrow s \oplus \texttt{Correct}(x)$		[50]
	$x \leftarrow w' \oplus s = w \oplus e$ $y = \widehat{m} \leftarrow \texttt{Decode}(x)$		[54]
$s \leftarrow w_{1:k}A \oplus w_{k+1:n}$	$x \leftarrow \texttt{Correct}(w' \oplus (0\| s))$ $y = \widehat{w} \leftarrow x \oplus (0\| s)$	Systematic	[55]
	$y = \widehat{w}_{1:k} \leftarrow \texttt{Decode}(w' \oplus (0\| s))$		[56]
$s \leftarrow j$ such that $w \in \mathcal{C}_j$	$y = \widehat{m} \leftarrow \texttt{Decode}_{\mathcal{C}_j}(w')$	Multicode	[1]

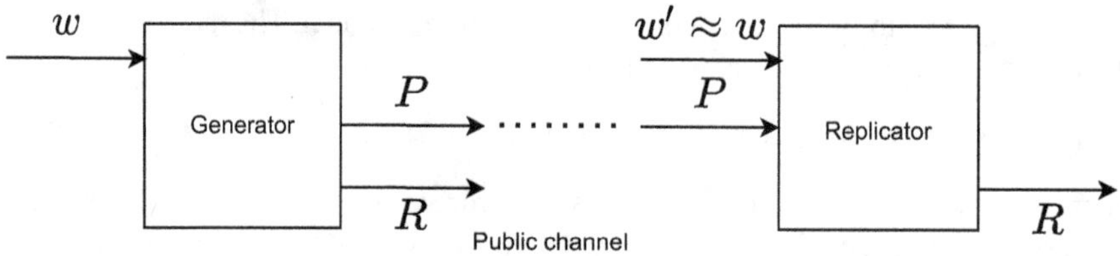

Figure 8.7 Fuzzy extractor: the generator extracts a random string R from the input w in a noise-tolerant way. It also output a nonsecret string P, such that R remains uniformly random even given P. The string R can be extracted using a noisy observation w′ and the public P.

also possible to use a *systematic* generator matrix for the error-correcting code [55,56] or using *multiple codes* that cover mutually disjoint sets of codewords [1]. It should be noted that it was shown in [51] that all schemes described above have the same min-entropy loss and are largely equivalent.

A related technique is that of *fuzzy extractors*, which can be created using secure sketches. For this construction, see Figure 8.7, the generator takes the random string w as input and outputs a public string P and a secret string R. The latter should have the property of being almost uniformly distributed even given P. The replicator can then reproduce the string R using a noisy w' (close to w) and the public string P. As an example, we can consider the scheme in [52], see Table 8.1, where the public string is $s = w \oplus c$ and the secret string is c. From the noisy w' and the public string s, the replicator can output the secret string c. A fuzzy extractor addresses nonuniformity of the secret, which a secure sketch does not. Furthermore, fuzzy extractors can be viewed as combining information reconciliation with privacy amplification, which will be discussed in the next section.

8.6 Privacy amplification

At this stage, Alice and Bob have ensured that they have generated secret keys which are identical with a high probability. However, a passive eavesdropper Eve may have gathered some information about the key. This could have happened during the advantage distillation phase if, e.g., Eve has a propagation channel that is correlated with the one between Alice and Bob (she places herself physically close to either Alice or Bob, or both if she has another collaborating device). Another way that Eve can glean information is from the information reconciliation process. The message s on the public channel can easily be overheard by Eve and while it is designed to reveal very little information about the secret key, some will still be available for Eve. For instance, if an error-correcting code is used, it does not have the full entropy due the redundancy, which can be gathered by Eve.

The general formulation considers the case where Alice and Bob are sharing a secret S but Eve has some information about it. The problem of transforming the secret S into a virtual secret S' which Eve has no information about is called *privacy amplification* and was introduced in [57]. It was shown that if Eve has deterministic information about the secret, e.g., knows a few bits of it, universal hashing is a good technique (more about that later). More specifically it was shown that S' must be shorter than S and that the difference in length must be equal to the information Eve has plus a security parameter. It was later generalised in [58] to probabilistic information about S. From Eve's point of view, the length of S' can in this case be roughly equal to the collision (or second-order Rényi) entropy of S [59]. This is defined as $-\log_2 \Pr(X = Y) = -\log_2 \sum_i p_i^2$, where X and Y are independent and identically distributed random variables with $p_i = \Pr(X = x_i)$. It is worth noting that collision entropy is upper bounded by the Shannon entropy [58].

The amplification of secrecy is realised with the help of another auxiliary random source, which is public and called a random seed. The random hash functions used for this purpose are often called extractors. If there are no computational assumptions (Eve is modelled as being computationally unbounded), it implies that all the entropy in S' comes from S. Thus, one of the most important requirements of such a protocol is that the output length of S' must be as long as possible given the entropy of S. The difference between the entropy of S and the length of S' is called the entropy loss. If Eve is passive, i.e., merely listens in on the messages between Alice and Bob but cannot modify them, strong extractors [60] provide an optimal solution to this problem. Alice simply sends an extractor seed to Bob, they both extract S' from S using the seed.

A key concept in privacy amplification is the hash function, which is defined as:

Definition 2. *$\mathcal{H} = \{h : \{0,1\}^n \rightarrow \{0,1\}^m, n > m\}$ is a universal family of hashing functions if for all $x_1 \neq x_2 \in \{0,1\}^n$*

$$\Pr_{h \in \mathcal{H}} \{h(x_1) = h(x_2)\} \leq 2^{-m} \tag{8.35}$$

when h is sampled uniformly from $\mathcal{H}$.

A simple example of a universal hashing family can be defined by using multiplication with random $m \times n$ matrices over $GF(2)$, i.e., modulo-2 arithmetic. If M is a $m \times n$ matrix with elements in $\{0,1\}$ drawn uniformly and independently, then $y = Mx$ produces a vector in $\{0,1\}^m$. So if the two vectors $y_1 = Mx_1$ and $y_2 = Mx_2$ should be equal, we need $Mx_1 = Mx_2 \Rightarrow M(x_1 + x_2) = 0$, i.e., $x_1 + x_2$ needs to be in the null space of M. Since we are working in $GF(2)$, the matrix multiplication $M(x_1 + x_2)$ is the same as adding the columns of M corresponding to the nonzero elements of $x_1 + x_2$. Since all elements of M are independent and equiprobable, this produces a uniformly distributed vector in $\{0,1\}^m$. Hence, the probability of y_1 and y_2 being equal is 2^{-m}. The property in (8.35) is called *collision resistance*.

A common way of building hash functions is the Merkle–Damgård construction [61,62]. This is a method which builds collision-resistant cryptographic hash functions from collision-resistant one-way compression functions. The latter are functions that transform two fixed-length inputs into a fixed length output. The term one-way relates

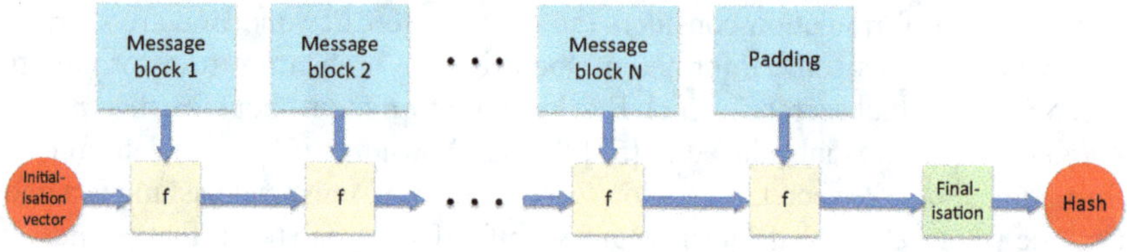

Figure 8.8 The Merkle–Damgård method for building collision-resistant cryptographic hash functions where f denotes a one-way compression function

to the difficulty of finding the inputs from the output and compression refers to the size reduction. The input to the Merkle–Damgård hash functions can have variable length and its key property is that if an appropriate padding scheme is used and the compression function is collision-resistant, than the hash function will also be collision-resistant. The construction is illustrated in Figure 8.8. To harden the hash further, the last result is then sometimes fed through a finalisation function, which can compress a bigger internal state (the last result) into a smaller output hash size or to guarantee a better mixing.

The Merkle–Damgård construction has been used in many popular hash algorithms such as MD5 (Message Digest), SHA-1 (Secure Hash Algorithm) and SHA-2; the latter are published by NIST as a US Federal Information Processing Standard. However, the early hash functions from the 1990s have been caught up by the advance of computational power. In 2008, the CMU Software Engineering Institute* concluded that MD5 was essentially cryptographically broken and unsuitable for further use; the weaknesses of MD5 have been exploited in the field, sometimes in computer malware. More recent hash functions are constructed with larger block sizes and/or different approaches such as Keccak [63]. Although they can be broken in theory, even the use of present-day quantum computer cannot do that in practice [64]. Apart from collision resistance, other important properties of a hash function are speed, avalanche effect (a small change in the input should result in a larger change in the output) and determinism (should always produce the same output from the same input).

Hash functions can be combined with the generated secret key (after information reconciliation) to provide privacy amplification. A scheme is described in [6] where the binary sequence generated by Alice and Bob is split into two equal parts, $\mathbf{q}_1$ and $\mathbf{q}_2$. The first part is used to select a hash function from a family of universal hash functions, whereas the second part of the binary sequence is used as its input. The family of universal hash functions is created by finding p which is the largest prime $< 2^m$, where m is the output size of the hash function. The sequences $\mathbf{q}_1$ and $\mathbf{q}_2$ are then split into l parts with size m (padding can be used if the length of $\mathbf{q}_1$ or $\mathbf{q}_2$ is not divisible by m), denoted by $q_{i,l',m'}$ with $i = 1, 2,\ l' = 1, \ldots, l,\ m' = 1, \ldots, m$. If $\widetilde{q}_{i,l'}$

*https://www.sei.cmu.edu/

denotes the decimal number corresponding to the binary sequence $q_{i,l',m'}$, the hash function is computed as

$$h_{\mathbf{q}_1}(\mathbf{q}_2) = \sum_{l'=1}^{l} \widetilde{q}_{1,l'} \widetilde{q}_{2,l'} \bmod p. \tag{8.36}$$

The output of this hash function is the final secret key and both Alice and Bob perform this hashing to reduce any leaked information to Eve. To verify consistency, i.e., to ensure that Alice and Bob are using the same key, they can hash their keys using the same principle to produce checksums [6].

Universal hash functions are good randomness extractors, as stated by the leftover hash lemma (LHL) [65]. Despite its numerous applications, LHL-based extractors suffer from two limitations: large entropy loss and large seed length. To extract v bits from distribution X with min-entropy m which are ε-close to uniform, one must set $v \leq m - 2\log(1/\varepsilon)$, meaning that the entropy loss is $m - v \geq 2\log(1/\varepsilon)$; for many applications, such entropy loss is too large. The seed length n of the universal hash function required by the LHL must be at least $n \geq \min(\text{len}(S) - v, v + 2\log(1/\varepsilon)) - O(1)$ and must grow with the number of extracted bits [66].

It might be worth adding that the previous discussion has focussed on a passive eavesdropper. The attacker may be active, i.e., it can change the messages that Alice and Bob exchange. In this case, the optimal entropy loss cannot be achieved by strong extractors. However, with the use of the so-called nonmalleable† extractors, it was nonconstructively shown in [67] that there exists a protocol with optimal entropy loss and optimal round complexity (number of messages exchanged by Alice and Bob). Explicit nonmalleable extractors have since been constructed in, e.g., [68].

8.7 Multidevice networks

In the previous sections, we have assumed that there is a direct link between the two legitimate users Alice and Bob. However, in some circumstances, this link might not exist or be too weak to provide secret keys. In that case, relays can be used, which bridge the poor connection between Alice and Bob. Another reason to use a relay may be that the Alice–Bob link is almost static and thus provides a very low key-generation rate. In that case, a relay can be used to aid the secret key generation. If the relay is moving, e.g., by being a mobile user in the network or even an aerial base station, its channel to both Alice and Bob is time-varying. By including the relay in the secret key-generation process, it can increase both range and key-generation rate.

A simple relaying scheme was studied in [69] where it is assumed that Alice and Bob are within communication range and the relay helps to generate a higher secret key rate by collaboration. The relay works in an amplify-and-forward mode, i.e., it simply forwards the signal with processing it. In this scenario, lower and

†Loosely speaking, nonmalleability is if a message has been tampered with, it should be either the original message or something completely unrelated.

upper bounds on the secrecy rate are derived. These are shown to be very close under practical communication scenarios, meaning that they are tight. The authors in [69] propose a practical secret key agreement protocol and consider the trade-off between security and protocol efficiency in the joint design of advantage distillation, information reconciliation, and privacy amplification. The protocol parameters are optimised to achieve the tight bound on the secret key rate.

In [70], group secret key generation, i.e., when multiple keys are needed, was studied by considering star and chain topologies. The collaborative key extraction via the star topology is designed for scenarios when multiple wireless devices are within each other's communication range, e.g., people travelling together, whereas the approach via the chain topology deals with scenarios when not all wireless devices under consideration are within each other's communication range but they are interconnected. The secret key generation was based on the received signal strength.

If Alice and Bob are not within communication distance, the relays can be used as range extenders. In [71], a multihop wireless OFDM system was considered with multiple relays. A two-hop communication system is illustrated in Figure 8.9 with two cooperating relays, Carol and David. In the first phase of the secret generation protocol, all connected nodes generate a secret key based on the propagation channel between them. Alice then generates a dummy message M_A and sends it encrypted with $K_{AC} \oplus K_{AD}$ over a public channel which a potential eavesdropper has access to. The relay Carol will then receive $M_A \oplus K_{AC} \oplus K_{AD}$ and adds $K_{BC} \oplus K_{AC}$ from her own secret key generations; this produces $M_A \oplus K_{BC} \oplus K_{AD}$ which is sent to Bob. Upon

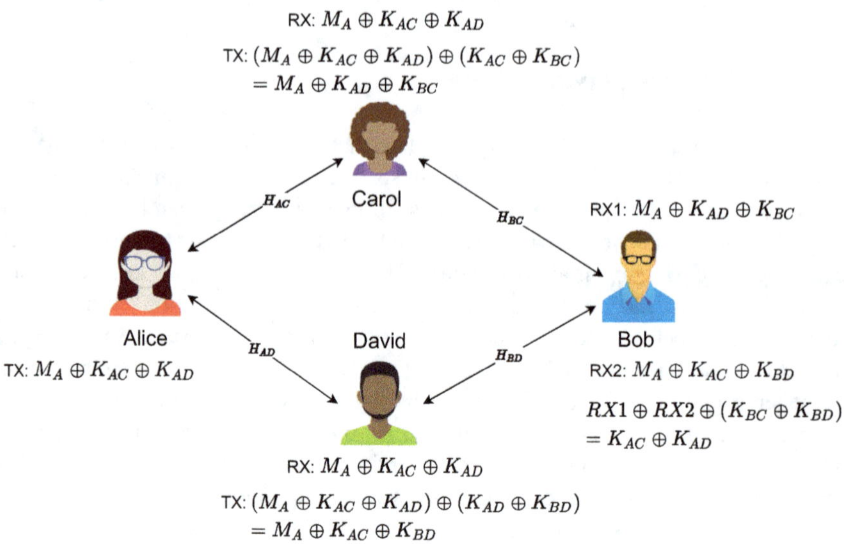

Figure 8.9 Secret key generation in a multihop system with untrusted relays Carol and David [71]. The keys K_{XY} are first generated from the channels H_{XY}, where $X \in \{A, B\}$ and $Y \in \{C, D\}$.

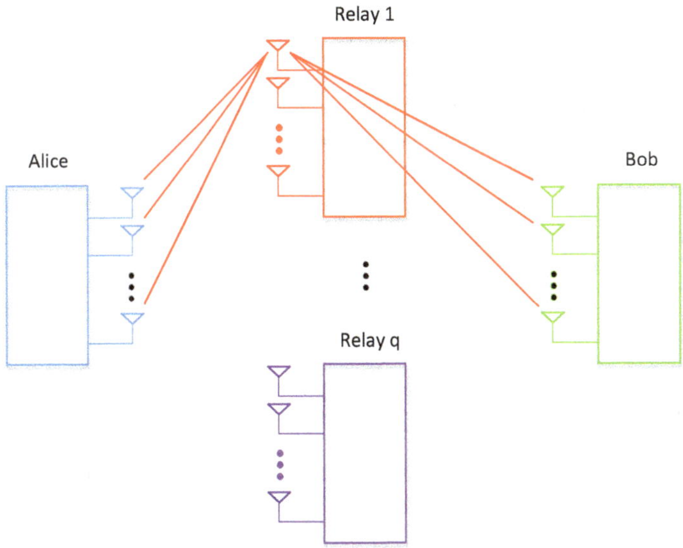

Figure 8.10 Multiantenna and multirelay secret key generation [72], showing the first time slot of the first phase; the first antenna on the first relay broadcasts a pilot to Alice and Bob

reception, he adds $K_{BC} \oplus K_{BD}$ to get $M_A \oplus K_{AD} \oplus K_{BD}$. The same procedure is done by the relay David, i.e., he receives $M_A \oplus K_{AC} \oplus K_{AD}$ from Alice, adds $K_{AD} \oplus K_{BD}$ to get $M_A \oplus K_{AC} \oplus K_{BD}$ which is sent to Bob. He can then add $K_{BC} \oplus K_{BD}$ to get $M_A \oplus K_{AC} \oplus K_{BD}$. If Bob adds his two messages together, he gets $K_{AC} \oplus K_{AD}$. These two individual keys are also available at Alice, so it can be used as the final secret key. Note that the neither relay has knowledge about the final key $K_{AC} \oplus K_{AD}$, Carol only knows K_{AC} and David only knows K_{AD}. Hence, this scheme can be used when the relays are collaborating but untrusted ("nice but curious"). It can also be extended to systems with more than two hops using the same principle.

The case of untrusted relays was studied in [72], where a system with multiple relays equipped with multiple antennas was considered, see Figure 8.10. The secret key protocol consisted of three phases: (1) the relays transmit sounding packets for Alice and Bob to estimate their channels to the relays; (2) Alice and Bob simultaneously transmit sounding packets to the relays; (3) the relays amplify and forward their received signal to Alice and Bob. When Alice and Bob simultaneously transmit their sounding packets in phase 2, the relays can only estimate the combination of them. This avoids the relays learning about the individual Alice-relays and Bob-relays channels, which the secret key is based on. Hence, untrusted relays can be used in such a system as they will not have access to the generated secret key. Three types of colluding relays were considered: (1) all relays work together; (2) only a subgroup of relays are colluding; (3) none of the relays are colluding. The selection of relays, as well as antennas at Alice and Bob, can be optimised to maximise the secret key rate.

Another approach for untrusted relays was taken in [73] where Alice and Bob are assumed to have intermittent ability to generate secret keys from a direct link. The untrusted relay node participates in the transmission only when the Alice–Bob link is not secure. In this case, secrecy can be guaranteed if the legitimate nodes have exchanged enough secret-key bits without being eavesdropped on by the relay node. The case of limited size key queues at both Alice and Bob was considered and the stored key bits were used to secure data transmissions whenever needed. The secret-key buffer was modelled as a queueing system and its Markov chain was analysed, where the achievable rates were derived.

8.8 Summary

As we have mentioned, wireless communications open itself up to security threats due its open nature (eavesdropping, jamming, etc.). On the other hand, it also provides a means to enhance the security with its unique propagation channel between legitimate users. This can be exploited to generate secret keys which are only available to the intended users. These keys can be used for cryptographic protection and offer additional security on the physical layer which complements that on other layers in the OSI model. The main advantages are distribution and dynamism; they solve the distribution problem by being created directly at the nodes that need them and they can be replenished due to the changing propagation environment. This makes them suitable to be used in one-time pads (or variations thereof), which are known to be theoretically uncrackable.

In this chapter, we have described the three main steps in generating secret keys from wireless channels: advantage distillation, information reconciliation and privacy amplification. The first one consists of exploiting a random variable that the legitimate users have more knowledge about than an eavesdropper. In our case, it is the wireless channel and we discussed how to convert it into a bit string. This involves channel sounding, feature extraction and quantisation. An overview of the details and practical schemes were presented, illustrating the approaches that can be taken. Although advantage distillation, information reconciliation and privacy amplification can be considered as separate tasks and optimised individually, it is important to remember that they are connected. For instance, a large number of bits may be generated in the advantage distillation but if they are unreliable, they need to be corrected in the information reconciliation phase. This will affect the overall key-generation rate and also impact the leaked information that needs to be wiped by the privacy amplification. Hence, a good design of secret key generation should include the interaction of these tasks and consider the whole system.

References

[1] R. Ahlswede and I. Csiszar, "Common randomness in information theory and cryptography. I. Secret sharing," *IEEE Trans. Inf. Theory*, vol. 39, no. 4, pp. 1121–1132, 1993.

[2] U. Maurer, "Secret key agreement by public discussion from common information," *IEEE Trans. Inf. Theory*, vol. 39, pp. 733–742, May 1993.

[3] X. S. Zhou, L. Song, and Y. Zhang, *Physical Layer Security in Wireless Communications*. Boca Raton, USA: CRC Press, 2014.

[4] S. Zhang, L. Jin, Y. Lou, and Z. Zhong, "Secret key generation based on two-way randomness for TDD-SISO system," *China Communications*, vol. 15, no. 7, pp. 202–216, 2018.

[5] Y. Huang, L. Jin, H. Wei, Z. Zhong, and S. Zhang, "Fast secret key generation based on dynamic private pilot from static wireless channels," *China Communications*, vol. 15, no. 11, pp. 171–183, 2018.

[6] N. Aldaghri and H. Mahdavifar, "Physical layer secret key generation in static environments," *IEEE Trans. Inf. Forensics Security*, vol. 15, pp. 2692–2705, 2020.

[7] G. Li, H. Yang, J. Zhang, H. Liu, and A. Hu, "Fast and secure key generation with channel obfuscation in slowly varying environments," in *IEEE INFOCOM 2022 – IEEE Conference on Computer Communications*, pp. 1–10, 2022.

[8] T. Aono, K. Higuchi, T. Ohira, B. Komiyama, and H. Sasaoka, "Wireless secret key generation exploiting reactance-domain scalar response of multipath fading channels," *IEEE Trans. Antennas Propag.*, vol. 53, no. 11, pp. 3776–3784, 2005.

[9] M. G. Madiseh, S. W. Neville, and M. L. McGuire, "Applying beamforming to address temporal correlation in wireless channel characterization-based secret key generation," *IEEE Trans. Inf. Forensics Security*, vol. 7, no. 4, pp. 1278–1287, 2012.

[10] X. Lu, J. Lei, Y. Shi, and W. Li, "Intelligent reflecting surface assisted secret key generation," *IEEE Signal Process. Lett.*, vol. 28, pp. 1036–1040, 2021.

[11] M. Sandell, "Secret key generation with multiantenna relays," *IEEE Access*, vol. 11, pp. 8387–8396, 2023.

[12] S. Althunibat, V. Sucasas, and J. Rodriguez, "A physical-layer security scheme by phase-based adaptive modulation," *IEEE Trans. Veh. Technol.*, vol. 66, no. 11, pp. 9931–9942, 2017.

[13] Y. El Hajj Shehadeh, O. Alfandi, and D. Hogrefe, "Towards robust key extraction from multipath wireless channels," *Journ. Commun. Netw.*, vol. 14, no. 4, pp. 385–395, 2012.

[14] L. J. Rodriguez, N. H. Tran, T. Q. Duong, T. Le-Ngoc, M. Elkashlan, and S. Shetty, "Physical layer security in wireless cooperative relay networks: state of the art and beyond," *IEEE Commun. Mag.*, vol. 53, no. 12, pp. 32–39, 2015.

[15] N. Patwari, J. Croft, S. Jana, and S. K. Kasera, "High-rate uncorrelated bit extraction for shared secret key generation from channel measurements," *IEEE Trans. Mobile Comput.*, vol. 9, no. 1, pp. 17–30, 2010.

[16] J. Zhang, A. Marshall, R. Woods, and T. Q. Duong, "Efficient key generation by exploiting randomness from channel responses of individual OFDM subcarriers," *IEEE Trans. Commun.*, vol. 64, pp. 2578–2588, June 2016.

[17] M. Sandell, "Robust secret key generation for frequency-selective channels," in *IEEE Veh. Techn. Conf. Spring*, (Helsinki, Finland), pp. 1–5, June 2022.

[18] Y. Liu, S. C. Draper, and A. M. Sayeed, "Exploiting channel diversity in secret key generation from multipath fading randomness," *IEEE Trans. Inf. Forensics Security*, vol. 7, pp. 1484–1497, July 2012.
[19] L. Xiao, L. Greenstein, N. Mandayam, and W. Trappe, "Fingerprints in the ether: Using the physical layer for wireless authentication," in *IEEE Intern. Conf. Commun.*, (Glasgow, UK), pp. 4646–4651, June 2007.
[20] A. M. Allam, "Channel-based secret key establishment for FDD wireless communication systems," *Commun. Appl. Electr.*, vol. 7, pp. 27–31, November 2017.
[21] X. Wu, Y. Peng, C. Hu, H. Zhao, and L. Shu, "A secret key generation method based on CSI in OFDM-FDD system," in *Conf. Global. Commun.*, (Atlanta, USA), pp. 1297–1302, December 2013.
[22] W. Wang, H. Jiang, X. Xia, P. Mu, and Q. Yin, "A wireless secret key generation method based on Chinese remainder theorem in FDD systems," *Sci. China Inf. Sci.*, vol. 55, no. 7, pp. 1605–1616, 2012.
[23] M. Alrabeiah and A. Alkhateeb, "Deep learning for TDD and FDD massive MIMO: Mapping channels in space and frequency," in *Asilomar Conf. Signals Syst. Comput.*, (Pacific Grove, USA), pp. 1465–1470, November 2019.
[24] X. Zhang, G. Li, J. Zhang, A. Hu, Z. Hou, and B. Xiao, "Deep learning-based physical-layer secret key generation for FDD systems," *IEEE Internet Things J.*, vol. 9, pp. 6081–6094, April 2022.
[25] F. Rottenberg, T.-H. Nguyen, J.-M. Dricot, F. Horlin, and J. Louveaux, "CSI-based versus RSS-based secret-key generation under correlated eavesdropping," *IEEE Trans. Commun.*, vol. 69, no. 3, pp. 1868–1881, 2021.
[26] T. Kitano, A. Kitaura, H. Iwai, and H. Sasaoka, "A private key agreement scheme based on fluctuations of BER in wireless communications," in *Intern. Conf. Adv. Commun. Techn.*, vol. 3, (Pyeongchang, South Korea), pp. 1495–1499, February 2007.
[27] A. Badawy, T. Elfouly, T. Khattab, C. F. Chiasserini, A. Mohamed, and D. Trinchero, "Robust secret key extraction from channel secondary random process," *Wireless Commun. Mob. Comp.*, vol. 16, no. 11, pp. 1389–1400, 2016.
[28] M. Jacovic, M. Kraus, G. Mainland, and K. R. Dandekar, "Evaluation of physical layer secret key generation for IoT devices," in *Wireless Microw. Techn. Conf.*, (Cocoa Beach, USA), pp. 1–6, April 2019.
[29] G. Li, A. Hu, L. Peng, and C. Sun, "The optimal preprocessing approach for secret key generation from OFDM channel measurements," in *Conf. Global. Commun.*, (Washington, USA), pp. 1–6, December 2016.
[30] G. Li, A. Hu, J. Zhang, L. Peng, C. Sun, and D. Cao, "High-agreement uncorrelated secret key generation based on principal component analysis preprocessing," *IEEE Trans. Commun.*, vol. 66, pp. 3022–3034, July 2018.
[31] H. Liu, Y. Wang, J. Yang, and Y. Chen, "Fast and practical secret key extraction by exploiting channel response," in *IEEE Int.Conf. Comp. Commun.*, (Turin, Italy), pp. 3048–3056, April 2013.

[32] M. F. Marzban, R. Chabaan, N. Al-Dhahir, and A. El Shafie, "Securing OFDM-based wireless links using temporal artificial-noise injection," in *Cons. Commun. Netw. Conf.*, (Las Vegas, USA), pp. 1–6, January 2018.

[33] A. Sayeed and A. Perrig, "Secure wireless communications: secret keys through multipath," in *Intern. Conf. Acoust. Speech Sig. Proc.*, (Las Vegas, USA), pp. 3013–3016, March 2008.

[34] C.-Y. Wu, P.-C. Lan, P.-C. Yeh, C.-H. Lee, and C.-M. Cheng, "Practical physical layer security schemes for MIMO-OFDM systems using precoding matrix indices," *IEEE J. Sel. Areas Commun.*, vol. 31, no. 9, pp. 1687–1700, 2013.

[35] W. Xi, M. Duan, X. Bai, K. Zhao, L. Mo, and J. Zhao, "Keep: Secure and efficient communication for distributed IoT devices," *IEEE Internet of Things J.*, vol. 8, no. 16, pp. 12758–12770, 2021.

[36] J. Zhang, A. Marshall, R. Woods, and T. Q. Duong, "Design of an OFDM physical layer encryption scheme," *IEEE Trans. Veh. Technol.*, vol. 66, no. 3, pp. 2114–2127, 2017.

[37] J. Zhang, R. Woods, T. Q. Duong, A. Marshall, Y. Ding, Y. Huang, and Q. Xu, "Experimental study on key generation for physical layer security in wireless communications," *IEEE Access*, vol. 4, pp. 4464–4477, 2016.

[38] G. Strang, *Linear algebra and its applications*. Harcourt Brace Jovanovich, 3rd ed., 1988.

[39] H. M. Furqan, J. M. Hamamreh, and H. Arslan, "Secret key generation using channel quantization with SVD for reciprocal MIMO channels," in *Intern. Symp. Wireless Commun. Sys.*, (Poznan, Poland), pp. 597–602, Sept. 2016.

[40] J. Tang, H. Song, A. Xu, Y. Jiang, H. Wen, Y. Zhang, and K. Qin, "Secret sharing simultaneously in internet of things," in *Intern. Conf. Pow. Intel. Comp. Syst.*, (Shenyang, China), pp. 815–819, July 2020.

[41] C. Chen and M. A. Jensen, "Secret key establishment using temporally and spatially correlated wireless channel coefficients," *IEEE Trans. Mobile Comput.*, vol. 10, no. 2, pp. 205–215, 2011.

[42] N. Jayant and P. Noll, *Digital coding of waveforms: Principles and applications to speech and video*. Englewood Cliffs: Prentice-Hall, 1984.

[43] S. Mathur, W. Trappe, N. Mandayam, C. Ye, and A. Reznik, "Radio-telepathy: Extracting a secret key from an unauthenticated wireless channel," in *ACM Int. Conf. Mobile Comput. Network (MobiCom)*, (San Francisco, USA), pp. 128–139, Sept. 2008.

[44] C. Ye, S. Mathur, A. Reznik, Y. Shah, W. Trappe, and N. B. Mandayam, "Information-theoretically secret key generation for fading wireless channels," *IEEE Trans. Inf. Forensics Security*, vol. 5, pp. 240–254, June 2010.

[45] Y. Linde, A. Buzo, and R. Gray, "An algorithm for vector quantizer design," *IEEE Trans. Commun.*, vol. 28, pp. 84–95, January 1980.

[46] Y.-W. P. Hong, L.-M. Huang, and H.-T. Li, "Vector quantization and clustered key mapping for channel-based secret key generation," *IEEE Trans. Inf. Forensics Security*, vol. 12, pp. 1170–1181, May 2017.

[47] J. Lin and J. A. Storer, "Design and performance of tree-structured vector quantizers," *Info. Proc. Manag.*, vol. 30, pp. 851–862, November–December 1994.

[48] A. Buzo, A. Gray, R. Gray, and J. Markel, "Speech coding based upon vector quantization," *IEEE Trans. Acoust., Speech, Signal Process.*, vol. 28, pp. 562–574, October 1980.

[49] A. Rukhin, J. Soto, J. Nechvatal, M. Smid, and E. Barker, "A statistical testsuite for random and pseudorandom number generators for cryptographic applications," Tech. Rep. 800-22, Nat. Inst. Standards Technol., Gaithersburg, MD, USA, April 2010.

[50] Y. Dodis, R. Ostrovsky, L. Reyzin, and A. Smith, "Fuzzy extractors: How to generate strong keys from biometrics and other noisy data," *SIAM Journ. Comp.*, vol. 38, no. 1, pp. 97–139, 2008.

[51] J. Delvaux, D. Gu, I. Verbauwhede, M. Hiller, and M.-D. M. Yu, "Secure sketch metamorphosis: Tight unified bounds," *Intern. Ass. Crypt. Res.*, 2015.

[52] A. Juels and M. Wattenberg, "A fuzzy commitment scheme," *ACM Conf. Comp. Commun. Sec.*, vol. 1, November 1999.

[53] C. H. Bennett, G. Brassard, C. Crépeau, and M.-H. Skubiszewska, "Practical quantum oblivious transfer," in *Ann. Intern. Crypto. Conf.*, (Santa Barbara, USA), pp. 351–366, August 1991.

[54] P. Tuyls, A. H. M. Akkermans, T. A. M. Kevenaar, G.-J. Schrijen, A. M. Bazen, and R. N. J. Veldhuis, "Practical biometric authentication with template protection," in *Int. Conf. Audio Video-Based Biom. Person Auth.*, (Hilton Rye Town, USA), pp. 436–446, July 2005.

[55] M. Yu, "Turn FPGAs into "key" players in the cryptographics field." Electronic Design Magazine, July 2009.

[56] H. Kang, Y. Hori, T. Katashita, M. Hagiwara, and K. Iwamura, "Cryptographic key generation from PUF data using efficient fuzzy extractors," in *Intern. Conf. Adv. Commun. Techn.*, (Pyeongchang, South Korea), February 2014.

[57] C. H. Bennett, G. Brassard, and J.-M. Robert, "Privacy amplification by public discussion," *SIAM Journ. Comp.*, vol. 17, no. 2, pp. 210–229, 1988.

[58] C. H. Bennett, G. Brassard, C. Crkpeau, and U. M. Maurer, "Generalized privacy amplification," *IEEE Trans. Inf. Theory*, vol. 41, pp. 1915–1923, November 1995.

[59] U. Maurer and S. Wolf, "Secret-key agreement over unauthenticated public channels - Part III: Privacy amplification," *IEEE Trans. Inf. Theory*, vol. 49, no. 4, pp. 839–851, 2003.

[60] N. Nisan and D. Zuckerman, "Randomness is linear in space," *Journ. Comp. Syst. Sci.*, vol. 52, no. 1, pp. 43–53, 1996.

[61] R. C. Merkle, "One way hash functions and DES," in *Ann. Intern. Crypto. Conf.*, (Santa Barbara, USA), pp. 428–446, Springer, August 1989.

[62] I. B. Damgård, "A design principle for hash functions," in *Ann. Intern. Crypto. Conf.*, (Santa Barbara, USA), pp. 416–427, August 1989.

[63] G. Bertoni, J. Daemen, M. Peeters, and G. V. Assche, "The making of KECCAK," *Cryptologia*, vol. 38, no. 1, pp. 26–60, 2014.

[64] J. Guo, G. Liu, L. Song, and Y. Tu, "Exploring SAT for cryptanalysis:(quantum) collision attacks against 6-round SHA-3," *Cryptology ePrint Archive*, 2022. https://eprint.iacr.org/2022/184.pdf.

[65] R. Impagliazzo, L. A. Levin, and M. Luby, "Pseudo-random generation from one-way functions," in *ACM Symp. Theory Comp.*, STOC '89, (New York, NY, USA), pp. 12–24, Association for Computing Machinery, 1989.

[66] B. Barak, Y. Dodis, H. Krawczyk, O. Pereira, K. Pietrzak, F.-X. Standaert, and Y. Yu, "Leftover hash lemma, revisited," in *Ann. Intern. Crypto. Conf.*, (Santa Barbara, USA), pp. 1–20, August 2011.

[67] Y. Dodis and D. Wichs, "Non-malleable extractors and symmetric key cryptography from weak secrets," in *ACM Symp. Theory Comp.*, (Bethesda, USA), pp. 601–610, May 2009.

[68] X. Li, "Non-malleable extractors, two-source extractors and privacy amplification," in *Symp. Found. Comp. Sci.*, (New Brunswick, USA), pp. 688–697, October 2012.

[69] N. Wang, N. Zhang, and T. A. Gulliver, "Cooperative key agreement for wireless networking: Key rates and practical protocol design," *IEEE Trans. Inf. Forensics Security*, vol. 9, pp. 272–284, February 2014.

[70] H. Liu, J. Yang, Y. Wang, Y. Chen, and C. E. Koksal, "Group secret key generation via received signal strength: Protocols, achievable rates, and implementation," *IEEE Trans. Mobile Comput.*, vol. 13, no. 12, pp. 2820–2835, 2014.

[71] F. He, W. Wang, X. Xu, L. Zhou, and H. Man, "A network coding method for channel signatures based key distribution," in *IEEE Intern. Conf. Commun.*, (Ottawa, Canada), pp. 5036–5041, June 2012.

[72] C. D. T. Thai, J. Lee, and T. Q. S. Quek, "Physical-layer secret key generation with colluding untrusted relays," *IEEE Trans. Wireless Commun.*, vol. 15, pp. 1517–1530, February 2016.

[73] A. El Shafie, A. Sultan, A. Mabrouk, K. Tourki, and N. Al-Dhahir, "Secret-key-aided scheme for securing untrusted DF relaying networks," *IEEE Trans. Veh. Technol.*, vol. 67, pp. 6727–6731, July 2018.

Chapter 9

Impact of reconfigurable intelligent surfaces on physical layer key generation in NextG wireless networks

Guyue Li[1], Lei Hu[1], Long Jiao[2], Kai Zeng[2], Chen Sun[3] and Aiqun Hu[3]

Physical layer key generation (PKG) is a promising means to provide on-the-fly shared secret keys by exploiting the intrinsic randomness of the radio channel. However, the performance of PKG is highly dependent on the wireless propagation environment. Due to its feature of controlling the wireless environment, reconfigurable intelligent surface (RIS) is appealing to be applied in PKG. In this chapter, we explore the effects of RIS on PKG from both constructive and destructive perspectives. For the constructive aspect, we have identified static and wave-blockage environments as two RIS-empowered-PKG applications in future wireless systems. Furthermore, we also point out that RIS could be utilized by an attacker to perform new jamming and leakage attacks and provide countermeasures, respectively. At last, possible directions for future research have also been given.*

9.1 Introduction

The tremendous growth in connectivity and the ubiquity of wireless communications has substantially increased the awareness of security and privacy. However, the heterogeneous, dynamic, and decentralized architecture of these networks leads to difficulties for cryptographic key management and distribution, especially in resource-constrained large-scale mobile networks. Meanwhile, physical layer key generation (PKG) inaugurates a novel idea to generate symmetric secret keys from estimates of the wireless channel between them, which is developed based on the information-theoretic source model studied in [5].

*This chapter has been a part of [1–4].

[1]School of Cyber Science and Engineering, Southeast University, China
[2]Department of Electrical and Computer Engineering, George Mason University, USA
[3]School of Information Science and Engineering, Southeast University, China

PKG is based on the reciprocity feature of the physical propagation channel, and its security is given by the following two aspects. First, the unpredictable movement of users or surrounding objects brings in channel fluctuations, which contribute to continuously updated key material that makes key cracking difficult. Second, due to spatial diversity, a third party can hardly obtain any information about the key, as long as it is not within the vicinity of legitimate users. Based on these principles, PKG is a practically feasible concept with a lot of current research focusing on fabricating relevant hardware prototypes and implementing it on Commercial Off-The-Shelf (COTS) devices.

Nevertheless, despite its attractive advantages, PKG also has limitations for some harsh propagation scenarios as shown in Figures 9.1 and 9.9. For example, in a factory automation scenario where equipment is densely deployed, radio waves may be blocked by obstacles, leading to an interruption of communication as well as the secret key generation. Further, this wave-blockage phenomenon is even more common at higher frequencies which are a key feature of beyond fifth-generation (B5G) communications [6]. Another example is the static environment case which is a common situation in wireless sensor networks (WSN). Since the randomness source for PKG is the wireless channel, for each coherence time interval, it provides essentially only one random symbol. Hence, when the wireless channel varies slowly, the generated key rate can be very low and secret keys cannot be renewed frequently. Previous research works addressed this challenge by evading to other diversity domains, e.g., by utilizing multiple antenna systems or by employing relays [5]. However, the former imposes higher requirements on the utilized hardware while the latter has a trustworthy problem. Besides, the key rate has limited growth unless the antenna or the relay node keeps moving all the time. These vivid examples highlight that PKG still faces challenges and thus there is a strong need to facilitate PKG in the respective scenarios.

Recently, reconfigurable intelligent surface (RIS) has emerged as a disruptive wireless communication technology, which is capable of creating controllable propagation environments by adjusting the discrete phase shifts of its reflecting elements [7–10]. It has been extensively studied in the wireless communication field to mitigate blockage vulnerability and enhance the coverage capability [6,11]. Due to its feature of controlling the wireless environment [12,13], RIS may be a suitable solution for the synthesis of dynamic channels with high entropy, avoiding the mobility requirement for the relay nodes. Furthermore, RIS is viewed as a planar array of nearly-passive reflecting elements, which makes attackers difficult to estimate the instantaneous channel information and derive the secret key. In this context, RIS is not subject to the issue of trustworthiness that affects relay nodes [14]. These advantages of RIS make it a perfect helper to assist Alice and Bob in forming an RIS-induced fluctuating channel, which serves as the common randomness for generating secret keys.

In this chapter, we investigate the impact of RIS on PKG in NextG wireless networks. The impact is twofold [1]: first, RIS is capable to construct desired channel in harsh propagation scenarios and thereby tackle the challenge of PKG; second, RIS may help adversarial eavesdropping. As an example, when an RIS is controlled by a malicious party, it could break the key agreement of legitimate parties. These facts motivate our study of both constructive and destructive impacts of RIS in the following sections.

9.2 New changes in fundamentals when PKG meets RIS

9.2.1 Fundamentals of PKG

In the model of a typical point-to-point PKG, there are generally three parties, where Alice and Bob are two legitimate parties who treat the wireless channel as a shared randomness source to extract a symmetric key, while Eve is a passive or active attacker who intends to eavesdrop on the key or to prevent Alice and Bob from agreeing on the same key. The general process of PKG comprises the following four phases [15,16]:

- *Channel sounding and randomness extraction:* In a time division duplex (TDD) system, Alice and Bob exchange pilots alternatively to obtain bidirectional estimates of the channel between them and extract appropriate channel features, e.g., received signal strength (RSS) and channel state information (CSI) [17].
- *Quantization:* Channel features are mitigated from disadvantageous stochastic properties (with respect to entropy as well as to reciprocity robustness) and converted into binary bit sequences, which are referred to as raw keys.
- *Information reconciliation:* Possible disagreements between the two sequences are corrected via error-detection protocols or error-correcting codes.
- *Privacy amplification:* Alice and Bob distill the key and compress the key to its real entropy by discarding leaked information to an attacker in the previous phases.

As observed from the process, the feasibility of the PKG method is strongly associated with the wireless channel between Alice and Bob.

9.2.2 RIS-involved channel model

When an RIS participates in the process of PKG, either constructively or destructively, appropriate channel models must be found. Here, we illustrate the RIS-involved channel model using an example, in which each user is equipped with a single antenna. On the basis of traditional channel models, the RIS introduces an additional channel between Alice and Bob. The RIS-induced channel conventionally includes the channel coefficients vector from Alice to the RIS, $\mathbf{h}_{ar}$, a diagonal matrix that models the RIS' signal reflection, $\mathbf{\Phi}_1$, and the channel coefficients vector from the RIS to Bob, $\mathbf{h}_{rb}$. Specifically, the RIS receives the pilot signal from Alice and then reflects the signal while applying amplitude and phase changes adjusted by the RIS control layer. As a result, the Alice–RIS–Bob link, $\tilde{h}_{ab}$, is represented by a multiplicative channel model, which is added coherently with the direct link, h_{ab}, to form the new channel model as

$$\tilde{h}_{ab} = \mathbf{h}_{ar}^T \mathbf{\Phi}_1 \mathbf{h}_{rb} + h_{ab}. \tag{9.1}$$

Similarly, the reverse channel from Bob to Alice, $\tilde{h}_{ba}$, satisfies that

$$\tilde{h}_{ba} = \mathbf{h}_{br}^T \mathbf{\Phi}_2 \mathbf{h}_{ra} + h_{ba}, \tag{9.2}$$

where $\mathbf{h}_{br}$, $\mathbf{h}_{ra}$, and h_{ba}, respectively, denote the channels from Bob to the RIS, from the RIS to Alice, and from Bob to Alice, and $\mathbf{\Phi}_2$ denotes the reflection matrix of the RIS in the reverse channel. The channels observed by Eve also follow the above model by substituting the direct channel with the channel from Alice/Bob to Eve and substituting the channel responses from the RIS to Bob/Alice in the RIS-induced channel with those from the RIS to Eve.

The reflection coefficients of RIS elements in $\mathbf{\Phi}_1$ and $\mathbf{\Phi}_2$ are often programmed to achieve purposeful manipulation of wireless channels. Ideally, these reflection coefficients, including phase shifts and amplitudes, are assumed to be continuously tuneable to achieve the optimal performance, which provides a theoretical bound for RIS-assisted systems. However, considering the hardware cost and implementation complexity, some works assume the set of reflection coefficients to be discrete with finite amplitude or phase shift levels. For example, in a two-level phase shift control, the reflection coefficient of an element may be controlled by a PIN diode [6], selecting from $\{1, -1\}$, i.e., by switching biasing "On" and "Off."

9.2.3 Implications of RIS-involved channel

Since the fundamental principles of PKG, i.e., channel reciprocity, spatial diversity, and temporal fluctuation, depend on the channel model, an interesting question is whether these principles will still be met under the new RIS-involved channel model.

- *Channel reciprocity*: The reciprocity between $\tilde{h}_{ab}$ and $\tilde{h}_{ba}$ is not merely dependent on the reciprocity of the physical channels. It also depends on the agreement of the diagonal matrices $\mathbf{\Phi}_1$ and $\mathbf{\Phi}_2$ in the bidirectional channel probing phases. Besides, for the channel features, the non-reciprocity effect of noise is changed as the received signal strength can be either boosted or attenuated through the RIS.
- *Spatial diversity:* The correlation between channel features of Bob and Eve is also changed by RIS. According to the central limit theorem, when the number of reflecting elements is large, the RIS-induced channels between Bob and Eve asymptotically satisfy the independence as long as the RIS reflecting channels of Eve and legitimate parties are independent [3].
- *Temporal variation:* RIS introduces an additional degree of freedom, which is capable to boost the randomness. Through a dynamical configuration of RIS, the channels of $\tilde{h}_{ab}$ and $\tilde{h}_{ba}$ vary along time, even for environments where the physical channel would not fluctuate strongly.

Accordingly, when the RIS is controlled by legitimate users, it is possible for them to meet these principles, even in some harsh scenarios. Conversely, RIS can also be exploited by Eve to disrupt the secret key generation between Alice and Bob by breaking these principles.

9.3 RIS-assisted PKG: potential improvements

Since an RIS can artificially adjust the phase shifts and induce randomness in the wireless propagation environment, it has been considered for improving the key-generation performance in the field of PKG [4,18–25].

In [18–21], the authors studied RISs with random configurations. The authors of [18], in particular, studied RIS-assisted PKG schemes in scenarios with a single-user and multiple nearby eavesdroppers. They derived an expression of the maximum achievable key-generation rate (KGR) under the assumption of a large number of RIS elements. Also, the authors designed an optimal time slot allocation algorithm for both the key-generation and the ciphertext transmission phases. A similar idea has been extended in a recent work [20], where RIS-assisted key generation was investigated in quasi-static environments and its SKR is found to be determined by the number of RIS elements, the correlation coefficient, the pilot length, and the quality of the reflecting channel. In addition, the authors of [19] implemented a practical RIS-assisted PKG system with commodity Wi-Fi transceivers. The experimental measurements demonstrated the effectiveness of the RIS-assisted PKG scheme in low-entropy environments. However, in [21], the authors found that collaborative eavesdroppers can estimate the legitimate Alice–Bob channel and erode their secret key rates in low-entropy channels lacking small-scale fading.

To further improve the secret key rate, the authors of [22–24] studied algorithms for optimizing the configuration of RISs. In [23], in particular, the authors optimized the switching states of the RIS units to maximize the secret key rate. A limited number of RIS units are turned on, which correspond to the largest variances of the RIS channels. The calculation of the secret key rate and the parameter design were greatly simplified in [23] because the RIS-induced channel of the eavesdropper was assumed to be independent from the channels of the legitimate users. The authors of [24] considered a PKG scenario with an access point and a mobile user in the presence of multiple eavesdroppers, whose observed channels are correlated with those of the legitimate users. By assuming that the eavesdroppers are curious users with known channel statistics, the reflection coefficients of the RIS elements were optimized to maximize a lower bound of the secret key rate.

In this section, we analyze the constructive effects of RIS by considering two case studies where RIS plays a "must-have" role.

9.3.1 Case I: static environments

As shown in Figure 9.1, in static environments, the limited randomness found in the direct channel renders key-generation challenging, while RIS-induced channel variations can now provide the basis for PKG. Thus, with the help of RIS in PKG, the entropy of the secret key can be enhanced [4,18,19].

Here, we introduce the system model and the experimental results of an RIS-assisted PKG system in static environments.

9.3.1.1 System model

We consider an RIS-assisted PKG system where two legitimate parties, Alice and Bob, seek to establish a shared cryptographic key. Alice and Bob deploy a standard TDD wireless communication protocol, e.g., IEEE 802.11n Wi-Fi with orthogonal frequency-division multiplexing (OFDM), and obtain CSI as an essential part of their communication. Using their respective CSI data, both parties implement a PKG

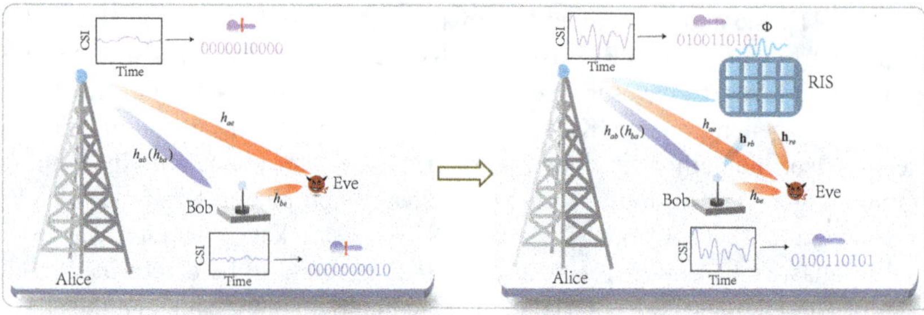

Figure 9.1 RIS-assisted PKG in static environments

procedure. Furthermore, we assume that a passive RIS is within reach of Alice and/or Bob and thus can partially control the wireless propagation channel. The passive eavesdropper Eve can capture messages sent by Alice and Bob and also obtains CSI, representing the effective channels between Eve and the legitimate parties Alice and Bob, respectively. Eve is aware of the key-generation procedure of Alice and Bob and thus can derive her own key material. All parties operate in a normally static indoor environment, i.e., the parties do not move, nor do objects in the environment. Hence, the environment does not contribute to temporal variation and thus exhibits a limited amount of randomness. In this case, the passive RIS can be used to enhance the channel temporal variation.

The RIS-assisted PKG scheme follows a standard architecture which includes channel probing, randomness extraction, quantization, information reconciliation, and privacy amplification stages. The key difference of this scheme is the channel probing which is performed in conjunction with a channel randomization step from the RIS. During the channel probing stage, Alice and Bob exchange packets in a ping-pong manner. At the receiver side, Alice and Bob use a standard least-squares (LS) channel estimator to obtain CSI. We assume that the complex OFDM baseband signal transmitted by Alice is generated by taking the inverse discrete Fourier transform of the complex modulated data symbols $s_\ell[n]$ of all $L, \ell = 0, 1, \cdots, L-1$, subcarriers of the nth OFDM symbol. In the time domain, a cyclic prefix, which is assumed to be longer than the channel's maximum delay spread, is then prepended to each OFDM symbol. Then, after time and frequency synchronization, removal of the cyclic prefix and discrete Fourier transform, the received baseband signals of Bob and Alice corresponding to the ℓth subcarrier of the nth OFDM symbol in the frequency domain are, respectively, given by

$$y_\ell^{ul}[n] = \tilde{h}_{ba,\ell}[n]s_\ell[n] + z_\ell^{ul}[n], \tag{9.3}$$

$$y_\ell^{dl}[n] = \tilde{h}_{ab,\ell}[n]s_\ell[n] + z_\ell^{dl}[n]. \tag{9.4}$$

where $\tilde{h}_{ab} = \mathbf{h}_{ar,\ell}^T \mathbf{\Phi} \mathbf{h}_{rb,\ell} + h_{ab,\ell}$ and $\tilde{h}_{ab,\ell}$ are the complex channel gains of the link between Alice and Bob for the ℓth subcarrier, and $z_\ell^{ul}[n] \sim \mathscr{CN}\left(0, \sigma^2\right)$ is the additive white Gaussian noise (AWGN). The reflection coefficients for the M-element RIS can be expressed as $\mathbf{\Phi} = \mathrm{diag}\left\{e^{j\theta_1[n]}, e^{j\theta_2[n]}, \cdots, e^{j\theta_M[n]}\right\}$, where $\theta_i[n], i \in \{1, \cdots, M\}$

denotes the discrete phase shift introduced by the nth reflecting element. To facilitate the hardware design of RIS, the phase shift $\theta_i[n]$ is generated based on an independent random variable (RV) uniformly distributed in a set of discrete values $\left\{0, \frac{2\pi}{2^B}, \cdots, \frac{(2^B-1)2\pi}{2^B}\right\}$, where B is the number of the quantization bits for phase shifts.

After receiving the signals as (9.3) and (9.4), Alice and Bob use a standard LS channel estimator to obtain CSI, which is expressed as

$$\hat{h}_\ell^{ul}[n] = y_\ell^{ul}[n]/s_\ell^{ul}[n] = h_{ba,\ell}[n] + z_k^{ul}[n]/s_\ell^{ul}[n], \tag{9.5}$$

$$\hat{h}_\ell^{dl}[n] = y_\ell^{dl}[n]/s_\ell^{dl}[n] = h_{ab,\ell}[n] + z_k^{dl}[n]/s_\ell^{dl}[n],. \tag{9.6}$$

Note that the RIS is assumed to be the only source of channel variation. As the RIS switches channel states, Alice and Bob effectively observe a static channel during one RIS configuration period. Thus, averaging over L consecutive samples obtained per RIS configuration can help Alice and Bob to reduce noise components.

After the quantization, information reconciliation, and privacy amplification stages, the key materials $\hat{h}_\ell^{ul}[n]$ and $\hat{h}_\ell^{dl}[n]$ are finally converted into key bits $\mathscr{K}_A$ and $\mathscr{K}_B$, at Alice and Bob, respectively.

Next, we will present some promising experimental results in [1] to show the performance of an RIS-assisted PKG system in static environments.

9.3.1.2 Experimental results

In the field trials, the parties Alice and Bob as well as an RIS are placed in a long-term static basement environment. The commodity Wi-Fi devices, implementing OFDM communication at Alice and Bob, in conjunction with an RIS prototype to realize a PKG system as outlined in Section 9.2 are used. For the channel probing, an application written in C language is used to control ath9k-based peripheral component interconnect express (PCIe) network interface cards (NICs) implementing IEEE 802.11n Wi-Fi in a 2×2 MIMO configuration. The RIS consists of two modules with a total of 128 elements, each having 1-bit phase control, i.e., an impinging wave is either reflected with phase shift 0 or π. The elements are switched via a microcontroller using serial communication. To generate the desired channel variation, the RIS applies random surface configurations at a fixed update rate. In parallel, Alice and Bob exchange Wi-Fi packets to obtain CSI in the channel probing phase. The authors in [1] use Wi-Fi channel 60 at 5.3 GHz which is close to the RIS' optimum operation frequency at 5.37 GHz. Alice and Bob are 3 m and 1.5 m apart from the RIS, with line-of-sight for the direct and the RIS channels.

The authors in [1] consider uncoordinated RIS operation where Alice and Bob adapt their channel probing to the speed at which the channel varies, as is the case with conventional PKG implementations (Figure 9.2). Due to the switched RIS behavior, the corresponding channel component changes instantaneously and thus Alice and Bob may *oversample* the channel in between, i.e., spend L channel measurements per RIS configuration. Exemplary for the instantaneous RIS-induced channel variation and oversampling ($L = 4$), we plot the CSI magnitude of Alice and Bob for a single subcarrier $k = 25$. After block averaging over L samples to improve SNR, both parties translate their respective channel profiles to bit strings using single-bit CDF

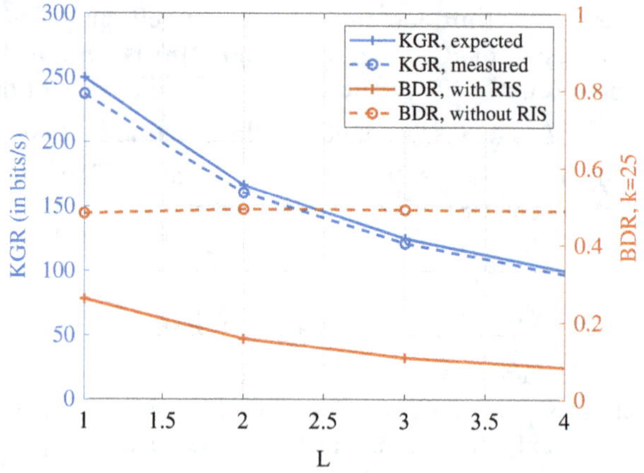

Figure 9.2 The KGR and BDR in single subcarrier

quantization [26]. In the implementation of [1], one bidirectional packet exchange for channel probing takes $T_p \approx 2$ ms and we consider an RIS configuration duration of $T_{RIS} = L\,T_p$. Accounting for an additional RIS update time $T_u \approx 2$ ms, we can estimate the effective KGR after quantization with $(L\,T_p + T_u)^{-1}$. This is consistent with the observation that the RIS here dictates the channel coherence time, which in turn allows to formulate an upper bound on the KGR. Figure 9.2 shows the expected and experimentally achieved KGR and bit disagreement ratio (BDR) results for an exemplary OFDM subcarrier $k = 25$ with varying values of L with and without the RIS. For the latter, as the fluctuation of channel measurements is caused by noise, its BDR of approximately 50%, which highlights the infeasibility of PKG in static environments.

Next, the randomness of the key material generated by the RIS-assisted PKG procedure is assessed via applying the statistical test suite for random number generators provided by the American NIST [27]. It is examined with a bit sequence of length 600000 bits that was obtained at the output of the quantization stage without further processing. All applicable tests were passed, indicating that the randomized RIS-induced channel variations are suitable for cryptographic applications.

In another experiment, an eavesdropper Eve is placed at 1 m distance to Alice. Here, the security of the key material is linked to the correlation of the channels from the RIS to Alice and Eve, respectively. The mutual information in the raw channel sequences of 100000 random RIS configurations observed by Alice, Bob, and Eve for a single OFDM subcarrier are estimated. To estimate the empirical mutual information, a k-nearest neighbor estimation [28] is utilized. The average mutual information across OFDM subcarriers of Alice' and Bob's and Alice' and Eve's observations were 0.69 and 0.04 bit/observation, respectively. This shows that Eve is at a significant disadvantage compared to the legitimate parties. For a detailed description of the experimental results, refer to [1,19].

These experimental results highlight the RIS as a promising extension to PKG, enabling key generation in static environments. As the RIS is a part of the environment, operating directly at the physical layer and external to user terminals, all users within reach—not just Alice and Bob—can potentially benefit from the provided channel randomization effect. As demonstrated, existing devices, wireless standards, and PKG procedures (cf. Section 9.2) can be reused with little extra overhead. work, e. g., on optimized surface configurations with SNR enhancement and reduction of information leakage to eavesdroppers.

9.3.1.3 RIS configuration optimization to enrich the scattering environment

The existing work mentioned above does not cover the discussion over the optimal RIS configuration strategies to get the highest KGR.[†] To find the optimal RIS configuration strategies, the following questions have to be answered: (a) which distribution that the RIS reflective channels should follow? (b) how to specify the weight of RIS phase shifter? To answer these questions, a novel RIS configuration strategy has been proposed in [4]. The authors evaluate the efficiency of key generation under various RIS configuration strategies from two perspectives: (a) reducing the redundancy among reconfigured RIS channels; (b) increasing consistency among the binary bits quantized from the reconfigured RIS channels' samples. A channel-probing scheme has been proposed and the key factors of key generation, affected by RIS configuration strategies, have been discussed in detail. In [4], the authors consider an RIS-assisted wireless system with two single-antenna users, Alice and Bob. They are communicating with each other in a TDD mode. To enhance the communication throughput, an RIS equipped with M passive reflecting elements is deployed in this system to enhance the KGR. The authors assume the RIS is controlled by Alice through a reliable and low delay channel. The wireless channels between Alice–Bob and Alice–RIS–Bob are named as the direct and RIS channels, respectively. For each channel, the narrowband and quasi-static block-fading channel model is considered. In [4], the authors indicate the KGR is limited by the high temporal correlation among channel samples collected from the static/slow fading wireless environments. Based on this observation, the authors propose a new paradigm to reconfigure the channel propagation environment for key generation. As shown in Figure 9.3, Alice has access to RIS. Right before the channel probing, the RIS channels are reconfigured based on the real-time CSI observed at the Alice. In Figure 9.4, the detailed channel-probing and RIS reconfiguration strategies are illustrated. For instance, after dividing a single fading block into $\tau_1 + \tau_2$ slots, the channel probing is conducted within the first τ_1 slots to get the estimation of RIS reflective channels. In the following τ_2 slots, Alice will perform RIS channel reconfiguration based on the RIS reflective channels estimated in the first τ_1 slots. We assume that for each RIS channel reconfiguration, τ_4 slots are used by Alice/Bob to estimate the reconfigured wireless channels. That is, for a single RIS channel reconfiguration including the bi-directional channel probing, $2 \times \tau_4$ slots are needed. In this case, within a fading block, the RIS will be reconfigured by

[†]This chapter has been a part of [4], with DOI as https://doi.org/10.1145/3468218.3469042.

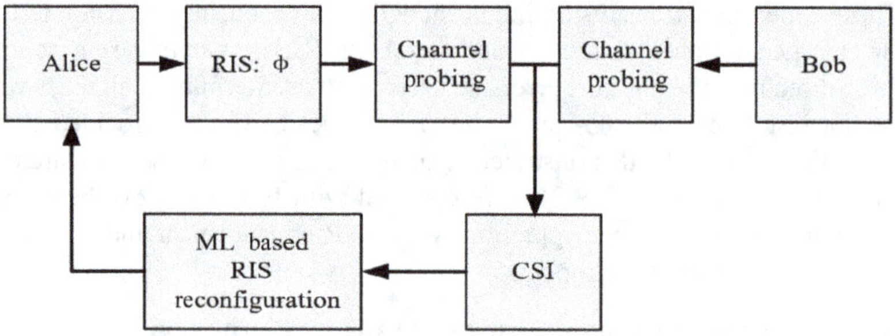

Figure 9.3 RIS channel reconfiguration for PHY key generation

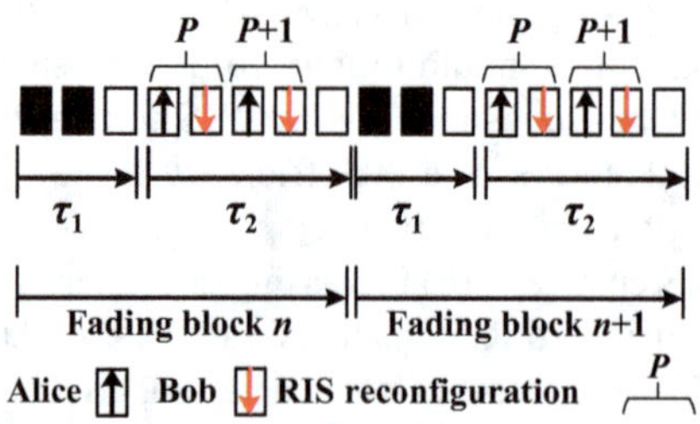

Figure 9.4 Channel probings and RIS channels reconfiguration. ©ACM 2021. Reprinted with permission from [4].

$\tau_3 = \tau_2/(2 \times \tau_4)$ times. In this step, after reconfiguring the scattering environment within a single block, Alice and Bob are able to observe newly reconfigured wireless channels or common randomness for $\tau_2/(2 \times \tau_4)$ times.

Methods to reconfigure RIS channels

For the RIS channels observed at Alice/Bob, the channels of each element can be estimated separately or Alice/Bob can directly estimate the combined RIS channels. There are two methods to process reconfigured channels: separating the RIS reflective channels by tuning each RIS element individually (separated reconfigured channels), or combining the RIS reflective channels with phase shifting arrays (combined reconfigured channels). To have a better comparison of these two methods, they are listed in [4] below:

$$1: \sum_{1}^{M} \phi_{m,1}\, g_m, \ldots, \sum_{1}^{M} \phi_{m,\tau_3}\, g_m \mid \mathbf{\Phi}, \tag{9.7}$$

$$2: \phi_{m,1}\, g_m, \ldots, \phi_{m,\tau_3}\, g_m \mid [\phi_{m,1}, \ldots, \phi_{m,\tau_3}] \tag{9.8}$$

As observed in (9.7), $\mathbf{\Phi}$ indicates the weight of RIS phase shifters. In (9.7), the combined reconfigured channels are represented, whereas the separate reconfigured channels are defined in (9.8). In (9.7), the channel-probing signals are transmitted over M reflective paths and superposed at the Rx with the fading coefficients $\sum_1^M \phi_{m,i}\, g_m$. For the separate reconfigured channels in (9.8), at a single time instant i, signals reflected over mth single RIS phase shift are affected with the fading coefficients $\phi_{m,i}\, g_m$. According to the triangular inequality, the upper limit of the received power for a combined reconfigured channel is represented as $\sum_1^M |\phi_{m,i}\, g_m|^2 p$ and that of the separate reconfigured channels is represented as $|\phi_{m,i}\, g_m|^2 p$. Based on this observation, the first method has a higher upper limit, which thus supports a wider quantization range.

Considering the content discussed, the authors in [4] utilize the combined reconfigured channels as the common randomness. After the radio environment reconfiguration, the common randomness turns into $\sum_1^M \phi_{m,t}\, g_m,\ \ t \in \{1, ..., \tau_3\}$, which requires a careful design of RIS phase shifting array $\phi_{\mathbf{t}},\ \mathbf{t} \in \{\mathbf{1}, ..., \tau_3\}$. The efficiency regarding different RIS designs will be discussed later.

Based on $\hat{\mathbf{g}}_{m,a}$, which is the estimated CSI at Alice, Alice has to reconfigure RIS channels τ_3 times and $\tau_3 = \tau_2/(2 \times \tau_4)$. After performing bi-directional channel probing, a sequence of distinct observations $[\tilde{\mathbf{g}}_{m,a}^{(1)}, \tilde{\mathbf{g}}_{m,a}^{(2)}, ..., \tilde{\mathbf{g}}_{m,a}^{(\tau_3)}]$ are collected at Alice. Until now, we have introduced the reconfigured RIS channels $[\tilde{\mathbf{g}}_{m,a}^{(1)}, \tilde{\mathbf{g}}_{m,a}^{(2)}, ..., \tilde{\mathbf{g}}_{m,a}^{(\tau_3)}]$, which is consisted of the estimation for RIS reflecting channel at Alice as $\hat{\mathbf{g}}_{m,a}$ and the RIS weights $[\boldsymbol{\phi}^{(1)}, ..., \boldsymbol{\phi}^{(\tau_3)}]$.

Measuring the redundancy among temporally correlated channel observations

There might be a temporal correlation existed among quantized bits from reconfigured RIS channels $[\tilde{\mathbf{g}}_{m,a}^{(1)}, \tilde{\mathbf{g}}_{m,a}^{(2)}, ..., \tilde{\mathbf{g}}_{m,a}^{(\tau_3)}]$. RIS phase shifting array $[\boldsymbol{\phi}^{(1)}, ..., \boldsymbol{\phi}^{(\tau_3)}]$ has to be configured in a way that the number of key bits (with the low redundancy) extracted from the current random source is maximized. We have to point out that the amplitude of RIS reflective channels is non-Gaussian. Based on mutual information, it is hard to evaluate the dependence among $[\tilde{\mathbf{g}}_{m,a}^{(1)}, \tilde{\mathbf{g}}_{m,a}^{(2)}, ..., \tilde{\mathbf{g}}_{m,a}^{(\tau_3)}]$. In order to measure the temporal correlation among quantized bits, entropy rate can be served as a metric.

The entropy rate H, or the per-symbol entropy, of random variable X is

$$H = H(X) = \lim_{N\to\infty} \frac{1}{N} H(X_1, ..., X_N), \tag{9.9}$$

whenever the limit exists. For the samples of a stochastic random process $(X_1; X_2; ...; X_N)$, the entropy rate H(X) is the limit of the joint entropy of N members of the process X divided by the length of observations N, as N tends to infinity. In fact, the knowledge of the probability distribution involved in its calculation requires, in principle, an extensive sampling that usually cannot be performed. In contrast, the complexity as originally formulated by Lempel and Ziv (LZ76) is a property of individual sequences that can be used to estimate the entropy rate. If we denote the estimation of entropy rate based on LZ76 as $C_{\mathrm{LZ}}(X_N)$ [29], for an ergodic and stationary sequence X_N, the entropy rate can be expressed as

$$H = \lim_{N\to\infty} C_{\mathrm{LZ}}(X_N). \tag{9.10}$$

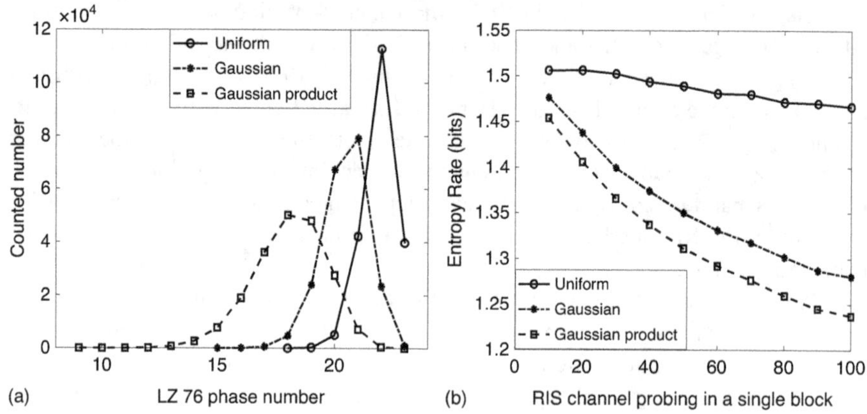

Figure 9.5 Entropy rate analysis of different pdfs. © ACM 2021. Reprinted with permission from [4]. (a) Occurrence of LZ 76 phase number. (b) Entropy rate versus probing numbers.

In [4], the authors compared the entropy rate of samples with the various distribution enabled by reconfiguring RIS. The authors compare the uniform distribution, Gaussian distribution, and Gaussian product distribution. The authors find that reconfigured RIS channels under a uniform distribution have the maximum entropy rate or the least redundancy among samples. Due to this property, the reconfigured RIS channels are going to be drawn uniformly among the quantization ranges.

Evaluating the BDR among reconfigured RIS channels' samples

Besides the redundancy among the samples of reconfigured RIS channels, the second factor that has a great impact on the efficiency of PHY key generation is the BDR between the binary bits quantized at Alice and Bob sides. To achieve a high consistency between the quantized bits at both sides, where the strict channel reciprocity holds, decreasing the effect of additive noise on the common randomness or a low quantization level is usually adopted.

To reduce the impact of noise on samples, Alice and Bob can either increase the transmission power or reduce the probability that channel samples cross the boundaries of quantization ranges. For a given quantization level L, every channel sample locates in a quantization range $l \in \{1, 2, \ldots, L\}$. To decrease the BDR, reconfigured RIS channel measurements are designed to be located around the middle of the quantization range to decrease the probability to cross the boundaries of quantization levels. An optimal algorithm has been developed, i.e., (9) in [4] for this problem.

Machine learning-based adaptive quantization level prediction

To obtain the binary bit sequences with high consistency at Alice and Bob, one of the major measures is to set appropriate quantization levels. In general, the lower the quantization level, the smaller the BDR, but the KGR also decreases. In existing

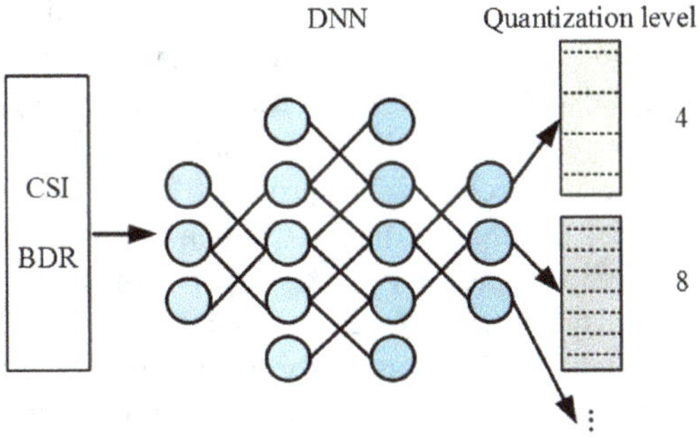

Figure 9.6 Adaptive quantization level prediction. ©ACM 2021. Reprinted with permission from [4].

works, to maintain a pre-set BDR, the extracted channels' samples are usually quantized in a unified quantization level after specifying the maximum and minimum of channel samples. As a result, an adaptive quantization level that fits for samples with different amplitudes is desired. For a given CSI sample of RIS channels, the sequential samples of RIS reconfigured channels are highly correlated with the current CSI samples. Machine learning (ML) has the potential to evaluate the mapping between a certain quantization level and the BDR based on the collected dataset.

To maximize the KGR while maintaining a low BDR, in [4], the authors propose an adaptive quantization level prediction scheme based on the ML techniques. For instance, while tuning the weights of RIS phase shifters to get a sequence of reconfigured channels within a fading block, the dynamic range of reconfigured channels is restricted by the value of the combined RIS channels at current block. At the same time, under a certain BDR threshold, the dynamic range of RIS reconfigured channels can directly affect the number of quantization levels. The authors developed an ML algorithm to learn the mapping between the CSI of the current RIS combined channels and the maximum of quantization level that can be applied under a defined BDR. In this way, the proposed scheme can adaptively specify the quantization level based on the CSI of current RIS combined channels.

The feedforward neural network (FNN) is considered in [4], which is one of the common deep neural networks (DNNs). It consists of three types of layers, namely, input, hidden, and output layers. In feedforward neural networks, information moves in only one direction: from the input nodes, through the hidden nodes, and to the output nodes. Back propagation (BP) is a well-known method for training multilayer feedforward networks based on the error back-propagation algorithm. In fitting a neural network, BP computes the gradient of the loss function with respect to the weights of the network for a single input–output example. The neural network used in the proposed scheme has two hidden layers with 30 and 100 neural cells, respectively, where the neural transfer functions are "tansig" and "purelin."

This scheme aims to learn a function that maps given channel observations, i.e., CSI, to the quantization levels based on input–output pairs where a certain BDR threshold is satisfied. For each channel observation, each labeled example (pair) consists of the quantization level and its CSI (a vector) and an indicator that represents if the BDR threshold is satisfied (1 standing for true and 0 for false). Note that for each instance of CSI, all of the possible quantization levels are given and the corresponding outputs (true or false) are also included. The proposed algorithm at first analyzes the training data and produces an inferred function, which can be used for mapping new channel observations. For instance, after the convergence of DNN trainings, given the unseen CSI and its current quantization level, trained DNN models are able to correctly determine the optimal quantization levels for this instance. In other words, the trained models are expected to generalize from the training data to unseen situations.

Simulation results

We assume an RIS, equipped with $M = 5$ elements, is deployed between Alice and Bob. Here, we focus on the RIS channel reconfiguration with a reconfiguration factor $P = 10$. The distance between Alice and RIS is set to be 50 m. The path loss model and the corresponding path loss factors for the directive channels and RIS reflective channels are as those defined in [30]. The channel bandwidth is assumed to be 1 MHz, and the power spectrum density of the AWGN at both sides is −169 dBm/Hz. For the correlation matrix, we adopt the exponential correlation matrix model. The activation functions of the three layers are chosen as "tansig" for input and hidden layers and "purelin" for output layer. The learning rate is set to 0.01. The default value for maximum validation failures is set as 6. The default value of momentum constant is 0.9. The performance goal is evaluated based on the MSE between targets and predictions, which is set to 10^{-3}. For each training process, the maximum number of epochs to train is selected to be 1,000.

In Figure 9.7, the performance of the proposed ML prediction model for quantization level prediction is given. From this figure, we can clearly observe that the

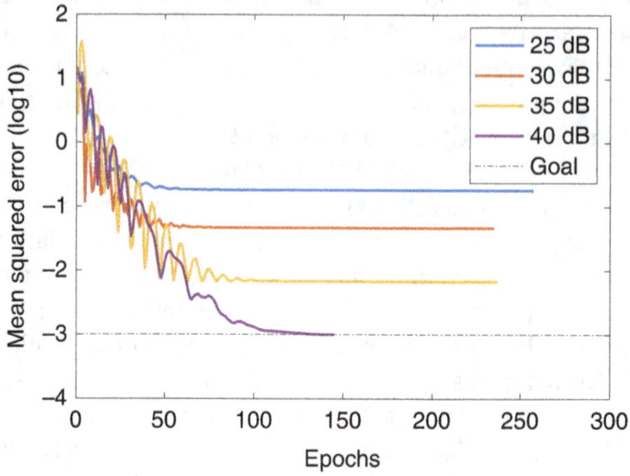

Figure 9.7 Training performance. ©ACM 2021. Reprinted with permission from [4].

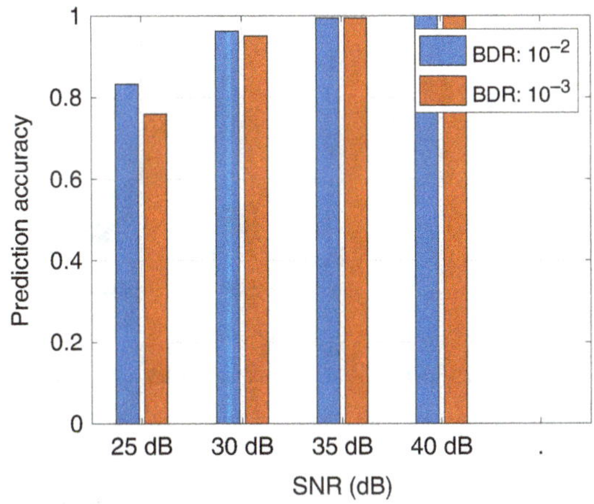

Figure 9.8 Prediction accuracy versus SNR. ©ACM 2021. Reprinted with permission from [4].

convergence level of the model training is directly impacted by SNRs. A higher SNR enables a lower convergence MSE level. Besides, we can see that the convergence speed of model training is determined by the SNRs. For instance, higher SNRs need more training epochs. This effect can also be reflected in the dynamic of gradient levels. The proposed quantization level prediction scheme based on DNN can achieve a high prediction accuracy under various BDR thresholds. In Figure 9.8, the plot indicating the prediction accuracy under different SNRs and BDR thresholds is given. Based on Figure 9.8, we can observe that the prediction accuracy approaches 1 as increasing of SNRs. When SNR = 35 dB and BDR threshold is set to be 10^{-2}, we can get a prediction accuracy of 98.2%. If we set a stricter BDR threshold (10^{-3}) for a lower BDR tolerance level, the prediction accuracy decreases. We can observe that in lower SNR regime (i.e., SNR = 25 dB), the prediction accuracy decreases at a stricter BDR threshold (i.e., 10^{-3}). However, in higher SNR regime (i.e., 35 dB and 40 dB), the prediction accuracy of quantization levels is nearly not affected by the BDR threshold as the accuracy approaches to 1.

9.3.2 Case II: wave-blockage environments

In wave-blockage environments, the direct link is blocked, leading to low SNR at the receiver. Consequently, CSI is submerged in noise and, thus, results in a low KGR, which is the upper bound on the number of bits per channel observation that Alice and Bob can generate, about which Eve cannot obtain any useful information based on her own observation [31]. To facilitate PKG in wave-blockage environments, RIS shapes an RIS-induced fluctuating channel, which serves as the common randomness for generating secret keys. It has been shown in the literature that the user's SNR can be significantly increased by designing the configuration of an RIS appropriately [6].

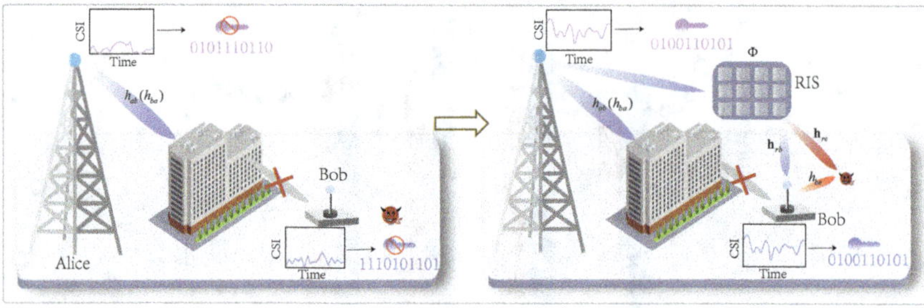

Figure 9.9 RIS-assisted PKG in wave-blockage environments

As an extension, in multi-user systems where some remote users are in worse channel conditions, the channel reciprocity of these users can be improved by the deployment of RISs.

In the former case, we have demonstrated the constructive effect of an RIS with random phase shifts on a single-user PKG system in a static environment. On this basis, we further report promising theoretical results of an RIS-assisted multi-user PKG system with different RIS configuration algorithms in wave-blockage environments.

9.3.2.1 System model

Consider an narrow-band RIS-assisted multiuser key-generation system, which comprises a wireless AP (Alice), an RIS, K legitimate user terminals (UTs). All parties, including Alice and the UTs, are assumed to be equipped with a single antenna. Alice intends to generate secret keys $\kappa = \{\kappa_1, \kappa_2, \cdots, \kappa_K\}$ with the UTs, from the wireless channels between them. A TDD protocol is assumed. The considered setup is different from existing works on group key generation [32,33], as the keys generated between Alice and the multiple UTs are not the same. We assume that the direct wireless channels between Alice and the UTs are blocked, therefore, an RIS with M elements is deployed to enable the key generation (Figure 9.9). The phase shifts of the RIS are programmed and reconfigured via a controller.

The procedure of RIS-assisted multiuser key generation (RMK) between Alice and the UT k encompasses five main steps.

Step 1: The RIS phase shift matrix, $\mathbf{\Phi}$, is configured according to statistical CSI, i.e., the cross-correlation matrix $\mathbf{R}_{kk'}$. The optimization of $\mathbf{\Phi}$ is discussed in [3].

Step 2: Alice broadcasts the pilot signal, s_{dl}, and the received signal at the UT k is given by

$$y_{dl,k} = h_{ak}s_{dl} + n'_{dl,k}, \tag{9.11}$$

where h_{ak} is the RIS-induced downlink channel from Alice to the kth UT, which is modeled as

$$h_{ak} = \mathbf{h}_{rk}^T \mathbf{\Phi} \mathbf{h}_{ar}, \tag{9.12}$$

where $\mathbf{h}_{ar} \in \mathbb{C}^{M\times 1}$ and $\mathbf{h}_{rk} \in \mathbb{C}^{M\times 1}$ denote the channel from Alice to the RIS and the channel from the RIS to the kth UT, respectively. The channel gains are assumed to be zero-mean complex Gaussian random variables and the channel from Alice to the RIS is independent of those from the RIS to the UTs. The diagonal matrix $\mathbf{\Phi} = \text{diag}(e^{j\phi_1}, e^{j\phi_2}, \cdots, e^{j\phi_M})$ collects the phase shifts of the RIS, where $\phi_m \in [-\pi, \pi]$ represents the phase shift of the mth element of the RIS. Similar to [34], the phase shifts are assumed to be continuously adjustable. $n'_{dl,k}$ is the AWGN at the UT k whose mean is zero and whose variance is σ^2.

Step 3: The UT k estimates the downlink CSI as

$$z_k^{dl} = y_k^{dl} s_{dl}^* = h_{ak} + n_{dl,k} = \mathbf{h}_{rk}^T \mathbf{\Phi} \mathbf{h}_{ar} + n_{dl,k}, \tag{9.13}$$

where the pilot signal satisfies $s_{dl}s_{dl}^* = 1$, and the equivalent noise is $n_{dl,k} = n'_{dl,k}s_{dl}^*$.

Step 4: The UTs send pilot signals, $\mathbf{s}_{ul,k} \in \mathbb{C}^{1\times K}, k \in \{1, 2, \cdots, K\}$ to Alice simultaneously. The received signal is

$$\mathbf{y}_{ul} = \sum_{k=1}^{K} h_{ka}\mathbf{s}_{ul,k} + \mathbf{n}_{ul}, \tag{9.14}$$

where $\mathbf{n}_{ul}$ is the AWGN at Alice, whose entries have zero mean and variance is equal to σ^2. The RIS-induced uplink channel from the kth UT to Alice is modeled as

$$h_{ka} = \mathbf{h}_{ra}^T \mathbf{\Phi} \mathbf{h}_{kr}, \tag{9.15}$$

where $\mathbf{h}_{kr}$ and $\mathbf{h}_{ra}$ represent the channel from the kth UT to the RIS and the channel from the RIS to Alice, respectively.

Step 5: Alice estimates the uplink CSI of the UT k as

$$\begin{aligned} z_k^{ul} &= \mathbf{y}_{ul}(\mathbf{s}_{ul,k})^H \\ &= h_{ka}\mathbf{s}_{ul,k}(\mathbf{s}_{ul,k})^H + h_{k'a}\sum_{k'\neq k}^{K} \mathbf{s}_{ul,k'}(\mathbf{s}_{ul,k})^H + n_{ul,k}, \end{aligned} \tag{9.16}$$

where $n_{ul,k} \triangleq \mathbf{n}_{ul}(\mathbf{s}_{ul,k})^H$. To distinguish each UT, the pilot signals of different UTs are designed to satisfy the orthogonality condition

$$\mathbf{s}_{ul,k}(\mathbf{s}_{ul,k'})^H = \begin{cases} 1, & k = k' \\ 0, & k \neq k'. \end{cases} \tag{9.17}$$

Therefore, the CSI estimate in (9.16) reduces to

$$z_k^{ul} = h_{ka} + n_{ul,k} = \mathbf{h}_{ra}^T \mathbf{\Phi} \mathbf{h}_{kr} + n_{ul,k}. \tag{9.18}$$

Since s_{dl} is normalized to one and $\mathbf{S}_{ul} = [\mathbf{s}_{ul,1}^T, \mathbf{s}_{ul,2}^T, \cdots, \mathbf{s}_{ul,K}^T]$ is a unitary matrix, the noises $n_{dl,k}$ and $n_{ul,k}$ are also Gaussian distributed, because the distribution of circularly symmetric complex Gaussian random variables is isotropic and thereby unitarily invariant.

As customary in TDD systems, we assume that channel reciprocity holds, i.e., $\mathbf{h}_{ra} = \mathbf{h}_{ar}$ and $\mathbf{h}_{kr} = \mathbf{h}_{rk}$ during one key generation [35]. Therefore, the channels between Alice and the UT k are equal as well, i.e., $h_{ka} = h_{ak}$, which can be used to generate a pair of symmetric keys.

After channel probing, Alice and the UT k quantize their channel estimates, z_k^{ul} and z_k^{dl}, into binary sequences, $\mathbf{q}_k^{ul}$ and $\mathbf{q}_k^{dl}$, respectively. Since these binary sequences are not exactly the same due to the impact of noise, they are referred to as initial keys. Then, Alice and the UT k obtain, from the initial keys, a consistent secret key κ_k through the steps of information reconciliation and privacy amplification.

9.3.2.2 Secret key rate

When Alice communicates with a UT, however, the other UTs are regarded as potential non-colluding curious users. Under the TDD mode, each UT cannot transmit and receive signals at the same time, therefore, the UT only has the channel observation in the downlink. Under these assumptions, the secret key rate is the minimum mutual information given the observation of the other UTs and Eve. The number of secure bits for the RIS-induced link from Alice to the UT k is the conditional mutual information expressed as [36]

$$I_k = \min_{k' \neq k} I(z_k^{dl}; z_k^{ul} | z_{k'}^{dl}). \tag{9.19}$$

For notational simplicity, but without loss of generality, we normalize the variance of the noise to $\sigma^2 = 1$.

Based on the assumptions above, the secret key rate in different eavesdropping conditions can be calculated as follows.

Theorem 1. *When the channels of the different UTs are statistically independent among each other, the secret key rate of the UT k is*

$$I(z_k^{dl}; z_k^{ul}) = 2\log\left(1 + \mathbf{v}^H \mathbf{R}_k \mathbf{v}\right) - \log(1 + 2\mathbf{v}^H \mathbf{R}_k \mathbf{v}), \tag{9.20}$$

where $\mathbf{v} = [e^{j\phi_1}, e^{j\phi_2} \cdots, e^{j\phi_M}]^T$, *the channel variance of the UT k is*

$$R_k = \mathbb{E}\{(\mathbf{h}_{rk}^T \boldsymbol{\Phi} \mathbf{h}_{ar})(\mathbf{h}_{rk}^T \boldsymbol{\Phi} \mathbf{h}_{ar})^*\} \triangleq \mathbf{v}^H \mathbf{R}_k \mathbf{v}, \tag{9.21}$$

and $\mathbf{R}_k = \mathbb{E}\{\boldsymbol{\Lambda}_k \mathbf{h}_{ar}^* \mathbf{h}_{ar}^T \boldsymbol{\Lambda}_k^H\}$.

From Theorem 1, we can observe that the secret key rate only depends on the phase vector $\mathbf{v}$ and the correlation matrix $\mathbf{R}_k$. By optimizing the phase shifts at RIS, the secret key rate could be maximized.

Theorem 2. *When the channels among the different UTs are correlated, the secret key rate* $I(z_k^{dl}; z_k^{ul} | z_{k'}^{dl})$ *can be formulated as in (9.22)*

$$I(z_k^{dl}; z_k^{ul} | z_{k'}^{dl}) = \log \frac{\left(\left(1 + \mathbf{v}^H \mathbf{R}_k \mathbf{v}\right)\left(1 + \mathbf{v}^H \mathbf{R}_{k'} \mathbf{v}\right) - \left|\mathbf{v}^H \mathbf{R}_{kk'} \mathbf{v}\right|^2\right)^2}{\left(1 + \mathbf{v}^H \mathbf{R}_{k'} \mathbf{v}\right)\left((1 + \mathbf{v}^H \mathbf{R}_{k'} \mathbf{v})(1 + 2\mathbf{v}^H \mathbf{R}_k \mathbf{v}) - 2\left|\mathbf{v}^H \mathbf{R}_{kk'} \mathbf{v}\right|^2\right)}, \tag{9.22}$$

where $R_{kk'}$ *is the cross-correlation between* $\mathbf{h}_{ak}$ *and* $\mathbf{h}_{k'a}$, *i.e., the channels corresponding to the UT k and the UT k'.* $R_{kk'}$ *can be expressed as*

$$R_{kk'} = \mathbb{E}\{(\mathbf{h}_{rk}^T \boldsymbol{\Phi} \mathbf{h}_{ar})(\mathbf{h}_{rk'}^T \boldsymbol{\Phi} \mathbf{h}_{ar})^*\} \triangleq \mathbf{v}^H \mathbf{R}_{kk'} \mathbf{v}, \tag{9.23}$$

and the cross-correlation matrix is $\mathbf{R}_{kk'} = \mathbb{E}\{\mathbf{\Lambda}_{k'}\mathbf{h}_{ar}^{*}\mathbf{h}_{ar}^{T}\mathbf{\Lambda}_{k}^{H}\}$ *with* $\mathbf{\Lambda}_k = \mathrm{diag}(\mathbf{h}_{rk}^{H})$, *where* diag(·) *extracts each element of the input vector as the corresponding diagonal element of the output diagonal matrix.*

In the correlated eavesdropping case, information leakage could be caused among the UTs. By optimizing the RIS configuration, we can prevent the information leakage and enhances the sum secret key rate in multiuser systems.

9.3.2.3 Simulation results

We consider an RIS-assisted multiuser key-generation system. The direct wireless channels between Alice and the UTs are blocked, therefore, an RIS is deployed to enable the key generation. Eve is assumed to be a passive eavesdropper, who is located at least several wavelengths away from the legitimate parties. In the work of [3], it has been proved that when the number of antenna elements of the RIS is large, the RIS-induced channels of Eve are independent of those of Alice and the UTs, therefore, they do not affect the secret key rate. However, when Alice communicates with a UT, the other UTs are regarded as potential non-colluding curious users. The channel from Alice to the RIS and the channel from the RIS to each UT are zero-mean complex Gaussian random variables.

In the simulation, we evaluate the feasibility of RIS-assisted multi-user PKG systems through the metrics of sum secret key rate, which is defined as the sum of secret key rates between Alice and all UTs. The secret key rate between Alice and one UT is expressed as the minimum mutual information given the observation of the other UTs. For traditional PKG systems without RIS, the sum secret key rate is zero, as the direct wireless channels between Alice and the UTs are blocked. For RIS-assisted multi-user PKG systems, we compare the sum secret key rates of them under three different configuration algorithms of RIS: (1) random phase shifts, (2) on-off switching states, and (3) optimized phase shifts. In the second algorithm [23], the reflection coefficients are in a two-level amplitude control and the secret key rate is maximized by turning on a limited number of RIS units, which correspond to the largest variances of the RIS channels. In the last algorithm, we solve the optimal phase shift ranging from $-\pi$ to π to maximize the sum secret key rate.

Figure 9.10 reports the sum secret key rates obtained by RIS-assisted multi-user PKG systems under these RIS configuration algorithms over independent and correlated channels.‡ When the reflection channels among the different UTs are correlated, the RIS-induced channels are correlated as well, which causes information leakage among the UTs. It is observed that the spatial correlation among the channels of the UTs reduces the sum secret key rate for all of the considered RIS configuration algorithms. The performance loss is roughly 5 dB for the optimal solution, 2 dB and 4 dB for the random configuration algorithm and the on-off switching algorithm, respectively. For both independent and correlated channels, the optimal solution provides the best sum key rate, while the on–off switching algorithm has lower sum secret key rate than others due to its limited controllable states of RIS units. The

‡The correlated channels are realized by putting two independent channels through a correlation matrix.

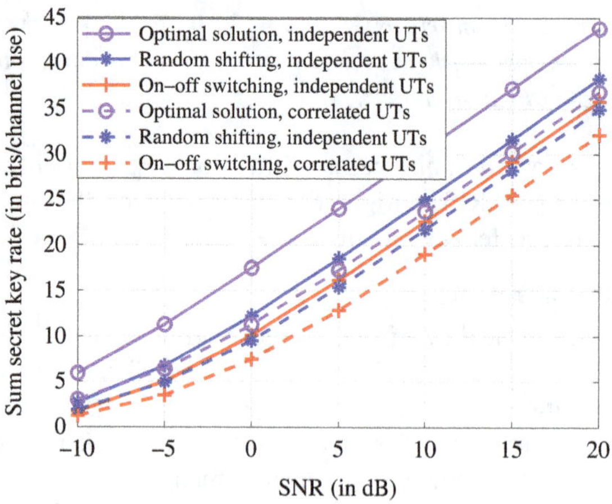

Figure 9.10 The comparison of sum secret key rates for different RIS configurations under assumptions of independent and correlated channels among UTs. The number of RIS elements is 16 and the correlation coefficient is set to be 0.5 for the simulation of correlated channels.

performance gap between the optimal solution and the on–off switching algorithm is 7 dB and 4 dB for independent and correlated channels, respectively. It should also be noted that although the optimal solution provides the best sum secret key rate, it has higher computation complexity than the other two configuration algorithms. The other two configurations are also appealing for practical usage in systems with limited computing capacity.

9.4 RIS-based attacks and countermeasures

Despite constructive effects as shown in Section 9.3, RIS can take destructive as well as constructive roles, decided by who owns the control right. In [37], it has been shown that when an RIS is controlled by a malicious party, a number of novel attacks against PKG become possible. The potential destructive aspects of RIS are largely overlooked in existing works. Based on the attacker's goal, we divide RIS-based attacks into two categories and discuss along with countermeasures in this section.

9.4.1 RIS jamming (RISJ) attack

9.4.1.1 Attack description

RISJ attack is an active attack which aims at disrupting the key establishment process between Alice and Bob by adjusting the RIS reflection matrices. According to the channel model in Section 9.2.2, the RIS-involved channel is the superposition of the direct link and the RIS-induced link, therefore, the channel reciprocity will be affected

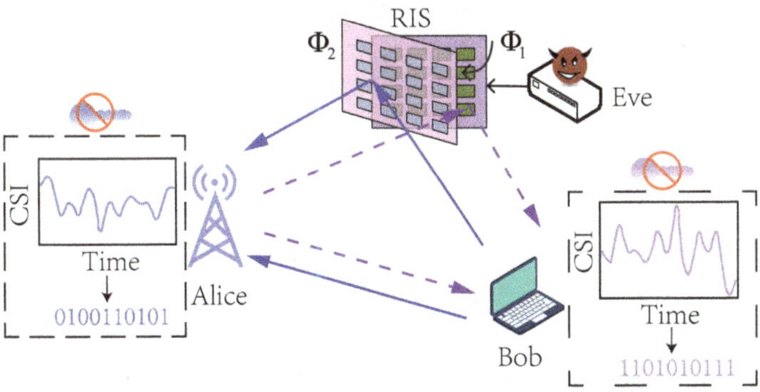

Figure 9.11 System model of the RISJ attack

by either of them. This can be taken advantage of by Eve to launch a jamming-like attack through RIS. Specifically, Eve can change the RIS reflection matrices in the Alice–RIS–Bob link and the Bob–RIS–Alice link, i.e., $\mathbf{\Phi}_1 \neq \mathbf{\Phi}_2$, in the process of a bidirectional channel probing. For example, when the RIS applies random surface configurations at an update rate higher than the channel sampling rate,[§] the observed CSI at Alice and Bob will most likely to be different.

Figure 9.11 shows an RISJ attack in PKG system. Considering an OFDM system, a single-antenna AP Alice and a single-antenna user Bob exploit the wireless channels in TDD mode to generate keys. A malicious attacker Eve employs an M elements RIS to destroy the PKG process. Specifically, Eve randomly and rapidly changes the reflection phase of each element and the time interval of the change is much smaller than the channel coherence time. During the channel probing stage, the pilot symbol transmitted by Alice is denoted as $\mathbf{X} = \text{diag}(x_1, x_2, \cdots, x_L)$, where L is the number of subcarriers. The received signal at Bob side after removing cyclic prefix is represented as

$$\mathbf{y}_b = \mathbf{X}\mathbf{h}_d + \mathbf{X}\sum_{m=1}^{M} e^{j\phi_{b,m}}\mathbf{h}_{ar,m} \odot \mathbf{h}_{rb,m} + \mathbf{n}_b \tag{9.24}$$

where $\mathbf{h}_d \in \mathbb{C}^{L\times 1}$ denotes the direct channel from Alice to Bob, $\mathbf{n}_b \sim \mathcal{CN}\left(\mathbf{0}, \sigma_n^2\mathbf{I}_L\right)$ is the complex additive white Gaussian noise. $\odot$ denotes the Hadamard product, $\mathbf{h}_{ar,m} \in \mathbb{C}^{L\times 1}$ and $\mathbf{h}_{rb,m} \in \mathbb{C}^{L\times 1}$ are the links from Alice to mth sub-surface of RIS and from the mth sub-surface to Bob, respectively. $\mathbf{h}_r\boldsymbol{\theta}_b$ is the equivalent cascaded channel matrix and $\mathbf{H}_r = [\mathbf{h}_{r,1}, \mathbf{h}_{r,2}, \cdots, \mathbf{h}_{r,M}]$, where $\mathbf{h}_{r,m} \triangleq \mathbf{h}_{ar,m} \odot \mathbf{h}_{rb,m}$, $m = 1, 2, \cdots, M$.

[§]For PIN diode-based RIS implementations, the switching frequency can be up to 5 MHz, which corresponds to the switching time of 0.2 μs [6]. This is much smaller than the typical channel coherence times in the order milliseconds.

The RIS reflection vector is $\mathbf{v}_b = [e^{j\phi_{b,1}}, e^{j\phi_{b,2}}, \cdots, e^{j\phi_{b,M}}]^T$. Then, Bob performs the LS estimation and the estimated channel is given by

$$\hat{\mathbf{h}}_b = \mathbf{h}_d + \sum_{m=1}^{M} e^{j\phi_{b,m}} \mathbf{h}_{ar,m} \odot \mathbf{h}_{rb,m} + \tilde{\mathbf{n}}_b, \tag{9.25}$$

$$= \mathbf{h}_d + \mathbf{H}_r \mathbf{v}_b + \tilde{\mathbf{n}}_b, \tag{9.26}$$

where the noise component $\tilde{\mathbf{n}}_b = \mathbf{X}^{-1}\mathbf{n}_b$ and $\tilde{\mathbf{n}}_b \sim \mathcal{CN}\left(\mathbf{0}, \sigma_n^2 \mathbf{X}^{-1}\right)$. For the purpose of exposition, we assume $x_1 = x_2 = \cdots = x_L = 1$.

Similarly, the channel estimation at Alice in RISJ can be expressed as

$$\hat{\mathbf{h}}_a = \mathbf{h}_d + \mathbf{H}_r \mathbf{v}_a + \tilde{\mathbf{n}}_a, \tag{9.27}$$

where the reflection phase shift $\phi_{u,m}, u \in \{a, b\}$ follows an independent and identically distributed (i.i.d.) uniform distribution over $[0, 2\pi)$. As can be observed from (9.26) and (9.27), the channel estimation of Alice and Bob is not relevant when $\mathbf{v}_a \neq \mathbf{v}_b$.

Then, to evaluate the impact of RISJ on channel reciprocity, we could calculate Pearson's cross-correlation coefficient of the channel estimations at Alice and Bob. Specifically, we assume the ℓth element of $\hat{\mathbf{h}}_u$ is expressed as $\widehat{h}_{u,\ell} = h_{d,\ell} + \sum_{m=1}^{M} \theta_{u,m} h_{r,m}^{\ell} + \tilde{n}_{u,\ell}, u \in \{a, b\}$. Hence, the correlation coefficient of the ℓth channel estimation in the frequency domain is

$$\rho_\ell = \frac{\mathbb{E}\{(\widehat{h}_{a,\ell} - \mu_{a,\ell})(\widehat{h}_{b,\ell} - \mu_{b,\ell})^*\}}{\sqrt{\mathbb{E}\{|\widehat{h}_{a,\ell} - \mu_{a,\ell}|^2\}\mathbb{E}\{|\widehat{h}_{b,\ell} - \mu_{b,\ell}|^2\}}} \tag{9.28}$$

$$= \frac{\sigma_{d,\ell}^2}{\sigma_{d,\ell}^2 + \xi_\ell + \sigma_{n,\ell}^2} \tag{9.29}$$

where $\sigma_{d,\ell}^2$ and $\sigma_{n,\ell}^2$ denote the variance of direct channel and noise in the ℓth subcarrier, respectively. Also, $\xi_\ell = \sum_{m=1}^{M} \sigma_{r,m,\ell}^2$ is the sum of the cascaded channel variances. It is noted that since the variance $\xi_\ell > 0$, RISJ will inevitably lead to a decrease in the CFR correlation coefficient. As the ξ_ℓ increases, the correlation coefficient tends to zero, resulting in low channel reciprocity.

Different from traditional jamming attacks, the RISJ attack reflects a jamming signal instead of transmitting it. Due to this fact, it is difficult to resist this attack by countermeasures based on location detection, since the location of Eve will be hardly exposed. Therefore, new ideas are desired for Alice and Bob to establish secret keys against the RISJ attack.

9.4.1.2 Countermeasures

One possible countermeasure is to exploit the fact that Eve can only change the RIS-induced channel, while the direct link, if exists, will not be affected by the RIS. Thus, Alice and Bob could separate them to distinguish the contaminated channel and still generate secret keys from the remaining uncontaminated channel. This idea, referred to as countermeasure based on channel separation (CCS), can be implemented in

PKG systems with high multipath resolution. For example, in a wideband OFDM system, these channels are separated through an inverse discrete Fourier transform (IDFT) and the RIS-induced channel can be detected from its high variance [37].

Specifically, the process of path separation can be described as

$$\mathbf{h}_u^t = \frac{1}{L}\mathbf{F}_L^H(\mathbf{h}_d + \mathbf{h}_r\boldsymbol{\theta}_u + \tilde{\mathbf{n}}_u) \tag{9.30}$$

$$= \mathbf{h}_d^t + \mathbf{h}_{r,u}^t + \bar{\mathbf{n}}_u^t \tag{9.31}$$

where $\mathbf{F}_L$ refers to the L-order discrete Fourier transform (DFT) matrice. $\mathbf{h}_d^t$ and $\mathbf{h}_{r,u}^t$ are the direct link and the attacked link, respectively. The direct link has multiple paths and the attacked link has one path. Then, we distinguish which paths are available for key generation and which path is attacked by Eve. We propose to detect RISJ by sounding the channel multiple times in a coherence time, which can also reduce noise on unattacked paths. Specifically, according to the time chart of proposed key generation based on IEEE 802.11a DATA frame, a round of channel probing can be carried out every 2.8 ms, providing the possibility of attack detection. In order to minimize the impact of noise, the detection method is shown in (9.32)

$$\frac{\sum_{q=1}^{Q} h_{i,q}h_{i,q}^*}{Q} - \frac{1}{L_d+1}\sum_{\ell=1}^{L_d+1}\frac{\sum_{q=1}^{Q} h_{\ell,q}h_{\ell,q}^*}{Q} \underset{\mathcal{H}_1}{\overset{\mathcal{H}_0}{\gtrless}} \alpha \tag{9.32}$$

where L_d is the number of paths of the direct link, $h_{i,q}$ is the qth round channel probing on ith path, and Q is the number of probing. $\mathcal{H}_0$ indicates the ith path is attacked by RISJ and $\mathcal{H}_1$ indicates that the RISJ is absent on ith path. Note that α depends on the variance of the attacked path and is set to be an empirical value. With the increase of the number of probes, the estimation of the variance will be more accurate. However, the channel probing will result in the increase of pilot overhead and the number of channel probing rounds is limited in the coherence time. Therefore, a tradeoff should be made between the number of channel probing and the success rate. Finally, Alice and Bob will discard the attacked path, and the remaining paths possess high channel reciprocity.

After IDFT is performed, different paths will have different SNRs. To make full use of high SNR paths and reduce the BDR of low SNR paths, we use a flexible quantization method to maximize the generated key bits. Specifically, the SNR in the frequency domain is defined as

$$\text{SNR} = \frac{\mathbb{E}\{(\mathbf{h}_d + \mathbf{H}_r\mathbf{v}_u)^H(\mathbf{h}_d + \mathbf{H}_r\mathbf{v}_u)\}}{\mathbb{E}\{\tilde{\mathbf{n}}_u^H\tilde{\mathbf{n}}_u\}}, \tag{9.33}$$

$$= \frac{\mathbb{E}\{\mathbf{h}_d^H\mathbf{h}_d\} + M\mathbb{E}\{\mathbf{h}_r^H\mathbf{h}_r\}}{L\sigma_n^2}. \tag{9.34}$$

After the RISJ attack detection, the path through RIS is removed. Then, the SNR of the ith path is [38]

$$\mathrm{SNR}_i = \frac{\mathbf{f}_i/L\mathbb{E}\{\mathbf{H}_d\mathbf{H}_d^H\}\mathbf{f}_i^H/L}{\sigma_n^2/L} \tag{9.35}$$

$$= \frac{\mathbf{f}_i/L\mathbf{F}_L\mathbb{E}\{\mathbf{h}_d\mathbf{h}_d^H\}\mathbf{F}_L^H\mathbf{f}_i^H/L}{\sigma_n^2/L} \tag{9.36}$$

$$= \frac{L\mathbb{E}\{|\bar{h}_i|^2\}}{\sigma_n^2}, \quad i \neq k \tag{9.37}$$

where $\mathbf{f}_i$ is the ith row of the DFT matrix $\mathbf{F}$ and $\bar{h}_i$ is the ith element of $\mathbf{h}_d^{\prime}$. In the practical system, in order to ensure the bits consistency, the BDR needs to be below 10^{-3}. To meet this requirement, we set quantization levels according to the SNR of each path. Particularly, each path will use higher quantization bits to increase the key rate when it is lower than a given KER.

9.4.1.3 Simulation results

Figure 9.12 shows the probability of successful detection. As the variance of noise decreases, the probability increases. In addition, the success rate increases with the increase of the ratio of the variance of the attacked path to the variance of the total paths, γ. Also, as the number of probes increases, a higher detection probability can be obtained. In the simulation setup, assume the speed $v = 1m/s$, then the Doppler spread D_s is about 6.7 Hz. Thus, the coherence time is $T_c = \frac{1}{2D_s} = 75$ ms. Since multiple probing rounds increase the pilot overhead, we set the round of probing $Q = 16$ to achieve a tradeoff between the success rate and the overhead.

Since as the number of RIS reflection elements increases, the subpath of the attacked path increases, increasing the variance of the attacked path. Figure 9.13 compares the BDR of the CSI versus different γ after the RISJ attack, where the quantization bit $m = 1$. We can find that as the SNR increases, the BDR corresponding

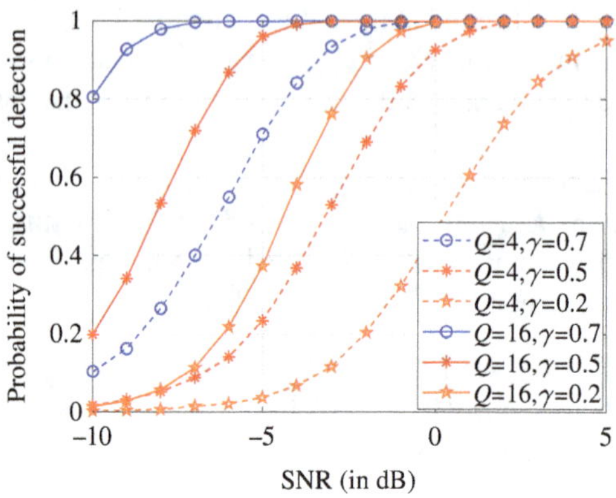

Figure 9.12 Probability of successful detection versus SNR and probing times

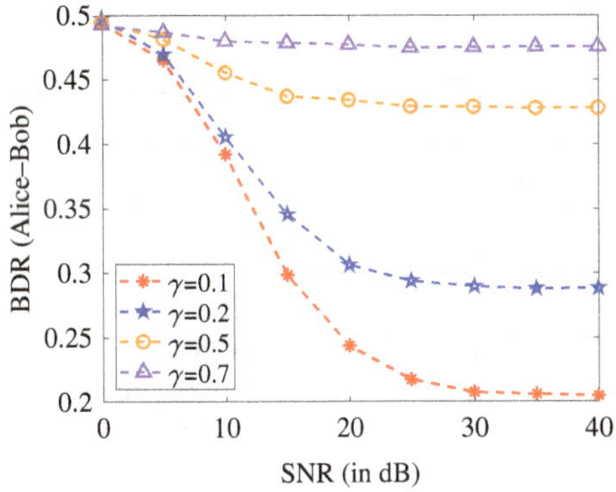

Figure 9.13 The BDR of Alice and Bob versus SNR under the RISJ attack and CCS-protected PKG method in an OFDM system

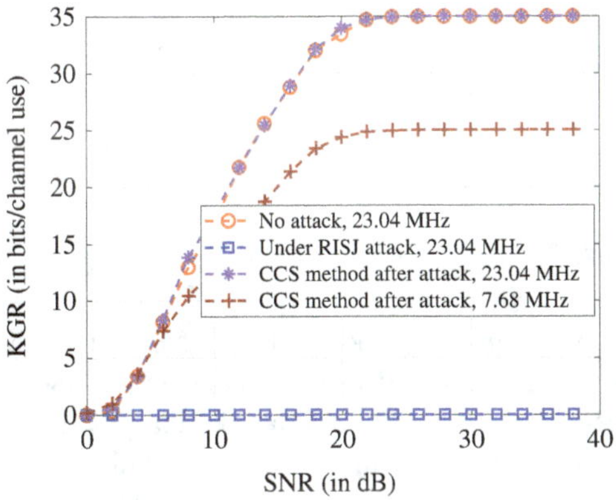

Figure 9.14 KGR under the RISJ attack and the CCS method

to different γ gradually decreases and approaches a constant value. However, we note that the BDR increases with the increase of γ. Specifically, when γ is 0.2, the BDR tends to 0.29 with the increase of SNR. In this case, existing information reconciliation methods cannot correct the generated key bits and this round of key generation fails.

Finally, Figure 9.14 presents the practical key rate after channel probing, CDF-based quantization, and low-density parity-check code (LDPC) bit-flipping-based information reconciliation. We notice that the BDR of conventional CFR-based key-generation method is high, resulting in low KGR. In contrast, under the CCS method, with the increase of SNR, the number of quantization bits of each path increases, leading to the increase of total key bits generated. We also note that in the CCS method, the KGR in high bandwidth is higher than that in low bandwidth. This is because in low bandwidth, such as 7.68 MHz bandwidth, some of the reciprocal paths and the attacked path are overlapped in the same tap. Therefore, these reciprocal paths are discarded and the KGR after the attack is lower than that before the attack. Under high bandwidth of 23.04 MHz, the path resolution is high enough that the attacked path can be removed directly. In this case, the near-optimal key rate can be achieved.

9.4.2 RIS leakage (RISL) attack

9.4.2.1 Attack description

In an RISL attack, Eve is more interested in obtaining the extracted secret keys between Alice and Bob, but not preventing them from agreeing on the same key. Given the fact that the RIS-induced channel, as a component of the channel for key generation, is under control by Eve, he/she is likely to obtain partial or complete information of the secret key. Notably, RISL attack sounds like the man-in-the-middle (MitM) attacks [39], where Eve infers part of the secret key generated between Alice and Bob by injecting her own packet upon observing highly correlated channel measurements from Alice and Bob (Figure 9.15). Different from MitM, RISL has a much broader scope as it does not need to transmit any packets and does not need to wait for an appropriate time to attack.

Next, we introduce two types of RISL attacks. The first RISL type intends to cause desired or predictable changes in the channel measurements by changing the RIS reflection matrices in a planned way. This RISL attack is similar to the well-known predictable channel attack, but one clear difference between them lies in the manipulation object. Compared with controlling the movements of some intermediate object, manipulating RIS reflection matrices is more stealthy and easier to implement. For example, when Eve switches the "On" and "Off" states of all elements on the RIS alternatingly, the channel gain observed at Alice and Bob is likely to be boosted and attenuated regularly with time. Specifically, we consider a RSS-based key-generation system, where Alice and Bob exchange pilots in adjacent slots to obtain a similar RSS for key generation. In this case, the RISL attack could be performed as follows. First, during the odd time slot, all elements are switched "On" mode and the channel between Alice and Bob is expressed as

$$\tilde{h}_{ba}^{odd} = \mathbf{h}_{br}^T \mathbf{h}_{ra} + h_{ba}, \tag{9.38}$$

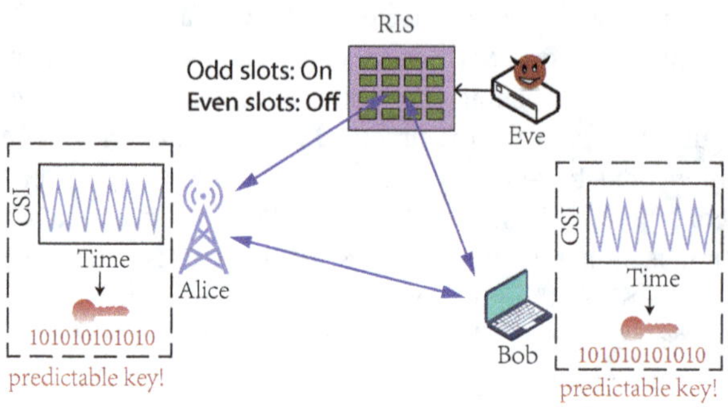

Figure 9.15 System model of the RISL attack

Then, during the even time slots, Eve switches all elements "Off" mode. Hence, the channel between Alice and Bob is

$$\tilde{h}_{ba}^{even} = -\mathbf{h}_{br}^{T}\mathbf{h}_{ra} + h_{ba}, \tag{9.39}$$

These channel values would lead to an alternating sequence of 0 and 1 in raw keys. Such predictable keys are easy to break.

In the second type of RISL attack, Eve speculates on the legitimate channel measurement by calculating the gain of the RIS-induced channel from the multiplicative channel model. In this attack, Eve needs to accurately estimate the channel gains from Alice and Bob to RIS, which might impose a high requirement on Eve. Since Eve can hardly obtain the channel information of the direct link, these RISL attacks would be more effective in some harsh propagation environments, e.g., static environments and wave-blockage environments.

9.4.2.2 Countermeasures

The effect of RISL is limited in fast-varying and rich multipath scattering environments; therefore, we focus on countermeasures in static and wave-blockage environments, respectively. In a static environment, RISL attacks can be prevented by increasing the fluctuation of the direct link. With the help of dynamic private pilots or artificial noise, Alice and Bob could construct an artificial fast-fading channel [16]. Since the channel fluctuation is not completely controlled by the RIS, it is difficult for Eve to obtain the secret keys, either by predictable changes or by channel speculation.

Specifically, the dynamic private pilots method to alleviate the RISL attack can be described as the following steps.

Step 1: Alice generates a random coefficient w, which follows a complex Gaussian distribution with zero mean and unit variance. Then, Alice sends out the probe signal which is the product of pilot s and w. The received signals at Bob can be expressed as

$$y_B = \left(\mathbf{h}_{ar}^{T}\mathbf{\Phi}\mathbf{h}_{rb} + h_{ab}\right) ws + n_B, \tag{9.40}$$

where n_B is the AWGN at Bob with variance σ^2. Then, Bob estimates the channel by performing the LS estimation as

$$h_B = \left(\mathbf{h}_{ar}^{T}\mathbf{\Phi}\mathbf{h}_{rb} + h_{ab}\right) w + \bar{n}_B, \tag{9.41}$$

where the estimation noise $\bar{n}_B = n_B/s$.

Step 2: Bob sends out the pilot s. The received signal at Alice is given by

$$\tilde{y}_A = \left(\mathbf{h}_{ar}^{T}\mathbf{\Phi}\mathbf{h}_{rb} + h_{ab}\right) s + n_A, \tag{9.42}$$

where n_A is the AWGN at Alice with variance σ^2. After the LS estimation, Alice obtain the channel as

$$\tilde{h}_A = \left(\mathbf{h}_{ar}^{T}\mathbf{\Phi}\mathbf{h}_{rb} + h_{ab}\right) + \tilde{n}_A, \tag{9.43}$$

where the estimation noise $\tilde{n}_A = n_A/s$.

Step 3: To obtain a similar combined channel gains, Alice multiplies the channel estimation $\tilde{h}_A$ by her generated AR coefficient w as

$$h_A = w\tilde{h}_A = w\left(\mathbf{h}_{ar}^T \mathbf{\Phi}\mathbf{h}_{rb} + h_{ab}\right) + \bar{n}_A, \tag{9.44}$$

where the noise $\bar{n}_A = w\tilde{n}_A$. Then, Alice and Bob quantize the h_A in (9.44) and h_B in (9.41) to generate raw keys.

It is noted that the reciprocal component, $w\left(\mathbf{h}_{ar}^T \mathbf{\Phi}\mathbf{h}_{rb} + h_{ab}\right)$, is dependent on the RIS and the random coefficient w. Hence, the generated keys are difficult for Eve to predict.

Figures 9.16 and 9.17 give the simulation results of the RISL attack and the countermeasure based on dynamic private pilots (CDPP). In Figure 9.16, we show the RSS waveform at Alice within 40 channel probing. It is observed that under RISL attack the RSS observed at Alice is boosted and attenuated regularly, contributing to a low-entropy key. In contrast, under the CDPP-based countermeasure, the RSS changes irregularly and is difficult to predict. Furthermore, Figure 9.17 illustrates the BDR performance between Alice and Eve, where single-threshold quantization method based on RSS is employed. From the results, the BDR of Alice and Eve is observed to reduce with the increase of γ and SNR. Specifically, in the case of $\gamma = 0.2$ and SNR = 10 dB, the bits of Alice are completely obtained by Eve. In our proposed CDPP method, under different γ and SNR, the BDR is always 0.5, which makes legitimate user-generated keys unpredictable by Eve. In a wave-blockage environment, the issue of RISL attack becomes similar to the problem of untrusted relay, which could be solved by protocols based on private randomness [40]. Another possible way to mitigate the impact of RISL attack is to increase the number of helper RIS devices in the system. When multiple helper RIS are deployed, the number of multipaths between Alice and Bob will increase, resulting in a decrease in the proportion of leaked key information.

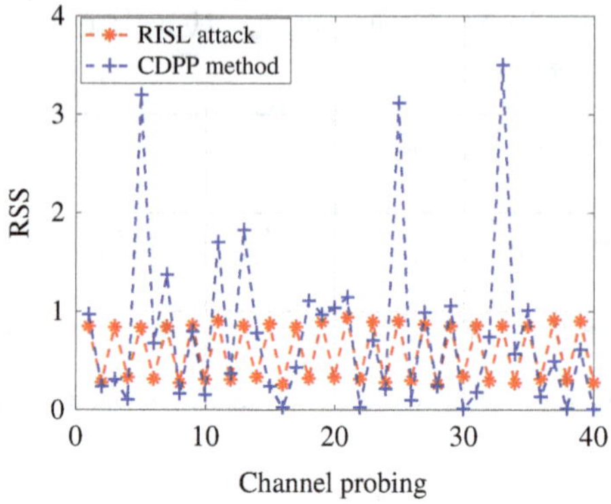

Figure 9.16 The waveform at Alice under RISL attack and CDPP-protected method in static environments

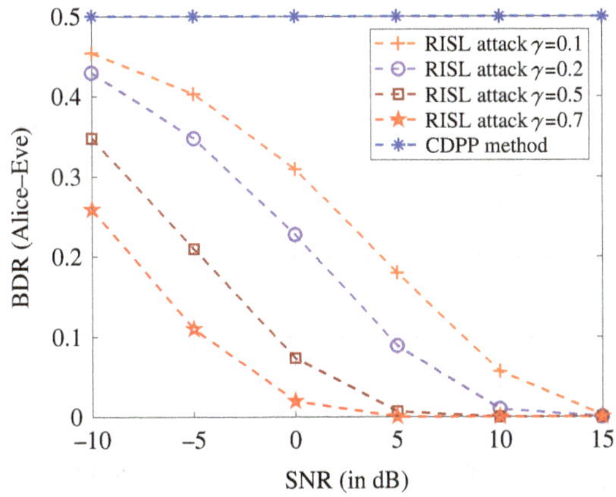

Figure 9.17 The BDR of Alice and Eve under RISL attack and CDPP-protected PKG method in static environments

9.5 Conclusion and future directions

In this chapter, we have reviewed PKG in view of the recently emerging RIS. Based on our observations, we outline future directions.

- *The optimization of surface configurations in a secure and practical manner*: The current RIS literature pursues feedback of channel estimations. However, this is perpendicular to treating the channel response as a shared secret. Therefore, more work is required on how to optimally configure RIS for PKG without the optimization itself giving away useful information to eavesdroppers. Another open question remains on how to find strongly quantized RIS configurations in real-time to accomplish the remaining goals of SNR improvement and leakage reduction.
- *The effects of RIS on other key-generation processes and in frequency division duplex (FDD) systems*: The available literature mainly studies the role of RIS in the channel-sounding stage of PKG. However, the influence of RIS on other key-generation processes needs to be explored. For example, the RIS could help reduce information leakage during information reconciliation and quantization schemes may be optimized towards RIS-induced channel variations. Also, RIS-aided PKG in FDD systems is one of the promising topics in the future.
- *The optimal deployment of multiple RISs*: Compared with a single RIS, multiple RISs can greatly improve the time-variability of the channel. However, there is no literature considering the optimal deployment of RISs in PKG. Also, when a large number of RISs are controlled by the BS, jointly optimizing the reflection coefficients of multiple RISs at the same time would trigger an explosion of computational overhead, resulting in an intolerable key-generation delay.

- *The detection and defense of RIS-based attacks*: Future work should investigate a general scenario where a malicious RIS and a friendly RIS exist concurrently. How can nodes distinguish between a friendly or a malicious RIS? It appears imperative to link the PKG process more to classical cryptography, allowing to authenticate nodes. However, it is yet to be explored, how this may be expanded to the physical layer to authenticate RIS signals.

Acknowledgments

This work was supported in part by the National Natural Science Foundation of China under Grant 62171121, in part by the Natural Science Foundation of Jiangsu Province under Grant BK20211160 and in part by the US NSF under Grant no. 2131507 and the Commonwealth Cyber Initiative (CCI) and its Northern Virginia (NOVA) Node, an investment in the advancement of cyber R&D, innovation, and workforce development.

We thank Paul Staat, Prof. Harald Elders-Boll, Christian Zenger, Prof. Christof Paar, Prof. Wei Xu, and Prof. Marco Di Renzo for their efforts. We also thank Markus Heinrichs and Prof. Rainer Kronberger at the High Frequency Laboratory, TH Cologne—University of Applied Sciences, Cologne, Germany, for providing the RIS prototype devices.

References

[1] Li G, Hu L, Staat P, *et al.* Reconfigurable Intelligent Surface for Physical Layer Key Generation: Constructive or Destructive? *IEEE Wirel. Commun.* 2022;29(4):146–153.

[2] Hu L, Li G, Luo H, *et al.* On the RIS Manipulating Attack and Its Countermeasures in Physical-layer Key Generation. In: *IEEE 94th Vehicular Technology Conference (VTC2021-Fall)*. Virtual; 2021. p. 1–5.

[3] Li G, Sun C, Xu W, *et al.* On Maximizing the Sum Secret Key Rate for Reconfigurable Intelligent Surface Assisted Multiuser Systems. *IEEE Trans. Inf. Forensics Secur.* 2022;17:211–225.

[4] Jiao L, Sun G, Le J, *et al.* Machine Learning-Assisted Wireless PHY Key Generation with Reconfigurable Intelligent Surfaces. In: *Proceedings of the 3rd ACM Workshop on Wireless Security and Machine Learning. WiseML'21*. New York, NY; 2021. p. 61–66. Available from: https://doi.org/10.1145/3468218.3469042.

[5] Zhang J, Li G, Marshall A, *et al.* A New Frontier for IoT Security Emerging From Three Decades of Key Generation Relying on Wireless Channels. *IEEE Access*. 2020;8:138406–138446.

[6] Pan C, Ren H, Wang K, *et al.* Reconfigurable Intelligent Surfaces for 6G Systems: Principles, Applications, and Research Directions. *IEEE Commun. Mag.* 2021;59(6):14–20.

[7] Di Renzo M, Debbah M, Phan-Huy DT, *et al.* Smart Radio Environments Empowered by Reconfigurable AI Meta-Surfaces: An Idea Whose Time Has Come. *EURASIP J. Wirel. Commun. Netw.* 2019;129(2019):1–20.

[8] Di Renzo M, Zappone A, Debbah M, *et al.* Smart Radio Environments Empowered by Reconfigurable Intelligent Surfaces: How It Works, State of Research, and The Road Ahead. *IEEE J. Sel. Areas Commun.* 2020;38(11):2450–2525.

[9] Yan W, Yuan X, He ZQ, *et al.* Passive Beamforming and Information Transfer Design for Reconfigurable Intelligent Surfaces Aided Multiuser MIMO Systems. *IEEE J. Sel. Areas Commun.* 2020;38(8):1793–1808.

[10] Dong L, Wang HM, Xiao H. Secure Cognitive Radio Communication via Intelligent Reflecting Surface. *IEEE Trans. Commun.* 2021;69(7): 4678–4690.

[11] Liu Y, Liu X, Mu X, *et al.* Reconfigurable Intelligent Surfaces: Principles and Opportunities. *IEEE Commun. Surv. Tutor.* 2021;23(3):1546–1577.

[12] Xu P, Chen G, Pan G, *et al.* Ergodic Secrecy Rate of RIS-Assisted Communication Systems in the Presence of Discrete Phase Shifts and Multiple Eavesdroppers. *IEEE Wirel. Commun. Lett.* 2021;10(3):629–633.

[13] Yang L, Yang J, Xie W, *et al.* Secrecy Performance Analysis of RIS-Aided Wireless Communication Systems. *IEEE Trans. Veh. Technol.* 2020;69(10): 12296–12300.

[14] Di Renzo M, Ntontin K, Song J, *et al.* Reconfigurable Intelligent Surfaces Vs. Relaying: Differences, Similarities, and Performance Comparison. *IEEE Open J. Commun. Soc.* 2020;1:798–807.

[15] Jiao L, Wang N, Wang P, *et al.* Physical Layer Key Generation in 5G Wireless Networks. *IEEE Wirel. Commun. Mag.* 2019;26(5):48–54.

[16] Zeng K. Physical Layer Key Generation in Wireless Networks: Challenges and Opportunities. *IEEE Commun. Mag.* 2015;53(6):33–39.

[17] Li G, Hu A, Zhang J, *et al.* High-Agreement Uncorrelated Secret Key Generation Based on Principal Component Analysis Preprocessing. *IEEE Trans. Commun.* 2018;66(7):3022–3034.

[18] Ji Z, Yeoh PL, Chen G, *et al.* Random Shifting Intelligent Reflecting Surface for OTP Encrypted Data Transmission. *IEEE Wirel. Commun. Lett.* 2021;10(6):1192–1196.

[19] Staat P, Elders-Boll H, Heinrichs M, *et al.* Intelligent Reflecting Surface-Assisted Wireless Key Generation for Low-Entropy Environments. In: *Proc. Annual International Symposium on Personal, Indoor and Mobile Radio Communications (PIMRC)*. Virtual; 2021. p. 1–7.

[20] Lu T, Chen L, Zhang J, *et al.* Reconfigurable Intelligent Surface Assisted Secret Key Generation in Quasi-Static Environments. *IEEE Commun. Lett.* 2022;26(2):244–248.

[21] Wei Z, Guo W, Li B. A Multi-Eavesdropper Scheme Against RIS Secured LoS-Dominated Channel. *IEEE Commun. Lett.* 2022;26(6):1221–1225.

[22] Liu Y, Wang M, Xu J, *et al.* Boosting Secret Key Generation for IRS-Assisted Symbiotic Radio Communications. In: *2021 IEEE 93rd Vehicular Technology Conference (VTC2021-Spring)*. 2021. p. 1–6.

[23] Lu X, Lei J, Shi Y, *et al.* Intelligent Reflecting Surface Assisted Secret Key Generation. *IEEE Signal Process. Lett.* 2021;28:1036–1040.

[24] Ji Z, Yeoh PL, Zhang D, *et al.* Secret Key Generation for Intelligent Reflecting Surface Assisted Wireless Communication Networks. *IEEE Trans. Veh. Technol.* 2021;70(1):1030–1034.

[25] Chen L, Cao K, Lu T, *et al.* A One-Time Pad Encryption Scheme Based on Efficient Physical-Layer Secret Key Generation for Intelligent IoT System. *China Commun.* 2022;19(7):185–196.

[26] Zenger C, Zimmer J, Paar C. Security Analysis of Quantization Schemes for Channel-Based Key Extraction. *ICST Trans. Secur. Saf.* 2015; 2(6):e5.

[27] Rukhin A, Soto J, Nechvatal J, *et al.* A Statistical Test Suite for Random and Pseudorandom Number Generators for Cryptographic Applications. DTIC Document; 2001.

[28] Kraskov A, Stögbauer H, Grassberger P. Estimating Mutual Information. *Phys. Rev. E* 2004;69(6):066138.

[29] Wei Y, Zeng K, Mohapatra P. Adaptive Wireless Channel Probing for Shared Key Generation Based on PID Controller. *IEEE Trans. Mob. Comput.* 2012;12(9):1842–1852.

[30] Wang Z, Liu L, Cui S. Channel Estimation for Intelligent Reflecting Surface Assisted Multiuser Communications: Framework, Algorithms, and Analysis. *IEEE Trans. Wirel. Commun.* 2020;19(10):6607–6620.

[31] Patwari N, Croft J, Jana S, *et al.* High-Rate Uncorrelated Bit Extraction for Shared Secret Key Generation from Channel Measurements. *IEEE Trans. Mob. Comput.* 2010;9(1):17–30.

[32] Li G, Zhang Z, Zhang J, *et al.* Encrypting Wireless Communications on the Fly Using One-Time Pad and Key Generation. *IEEE Internet of Things J.* 2021;8(1):357–369.

[33] Liu H, Yang J, Wang Y, *et al.* Group Secret Key Generation via Received Signal Strength: Protocols, Achievable Rates, and Implementation. *IEEE Trans. Mob. Comput.* 2014;13(12):2820–2835.

[34] Wu Q, Zhang R. Intelligent Reflecting Surface Enhanced Wireless Network: Joint Active and Passive Beamforming Design. In: *Proc. IEEE Global Communications Conference (GLOBECOM)*. Abu Dhabi, United Arab Emirates; 2018. p. 1–6.

[35] Tang W, Chen X, Chen MZ, *et al.* On Channel Reciprocity in Reconfigurable Intelligent Surface Assisted Wireless Networks. *IEEE Wirel. Commun. Mag.* 2021;28(6):94–101.

[36] Li G, Sun C, Jorswieck EA, *et al.* Sum Secret Key Rate Maximization for TDD Multi-User Massive MIMO Wireless Networks. *IEEE Trans. Inf. Forensics Secur.* 2021;16:968–982.

[37] Hu L, Li G, Luo H, *et al.* On the RIS Manipulating Attack and Its Countermeasures in Physical-Layer Key Generation. In: *Proc. IEEE Veh. Technol. Conf. (VTC fall)*. Virtual; 2021. p. 1–5.

[38] Li G, Hu A, Zhang J, *et al.* High-Agreement Uncorrelated Secret Key Generation Based on Principal Component Analysis Preprocessing. *IEEE Trans. Commun.* 2018;66(7):3022–3034.

[39] Pan Y, Xu Z, Li M, *et al.* Man-in-the-Middle Attack Resistant Secret Key Generation via Channel Randomization. *MobiHoc'21*. New York, NY: Association for Computing Machinery; 2021. p. 231–240.

[40] Aldaghri N, Mahdavifar H. Physical Layer Secret Key Generation in Static Environments. *IEEE Trans. Inf. Forensics Secur.* 2020;15:2692–2705.

Chapter 10

Physical layer key generation for LoRa-enabled networks

Huanqi Yang[1], *Zehua Sun*[1] *and Weitao Xu*[1]

10.1 Introduction

The past decades have witnessed tremendous growth of diverse Internet of Things (IoT) applications, such as smart building [1], smart industry [2], and smart farming [3]. However, the vast demand for various IoT applications raises new networking security challenges because of the broadcast nature of wireless communications and open-source mediums, which have encouraged the development of cryptographic key management. Conventional public key infrastructure (PKI) may not be a proper option for resource-constrained systems because it relies on a third party and requires a high computational complexity. Pre-shared keys (PSK) also have their drawbacks in that they cannot be deployed at scale and have no revocation capability [4].

Physical layer key generation, which takes advantage of the reciprocity and randomness of wireless fading channels, has recently received much interest. Specifically, three principles mainly govern physical layer key generation: channel reciprocity, temporal variation, and spatial de-correlation [5]. Channel reciprocity refers to the fact that the channel characteristics of two communicating nodes will be nearly identical. The channel parameters will be similar when two devices are measured by sending a series of probing packets to each other within the channel coherence time.* Temporal variation refers to the channel variation due to environmental changes such as position changes of IoT devices. Due to the multi-path effect, devices in various places (i.e., located with more than half a wavelength from each other) will have different radio channels. Despite that many research efforts have been made on legacy communication technologies such as ZigBee [6], Wi-Fi [5,7,8] and 5G [9], we find that the majority of these works concentrate on either temporal or spatial de-correlation. Channel reciprocity has received little attention because the transmission speeds of legacy communication standards are in the tens of Kbps or Mbps range. As the two end devices in such high bandwidth communication technologies may easily exchange packets for channel probing within the channel coherence

*Coherence time is the period of time during which the radio channel can be regarded as stable.

[1]Department of Computer Science, City University of Hong Kong, Hong Kong

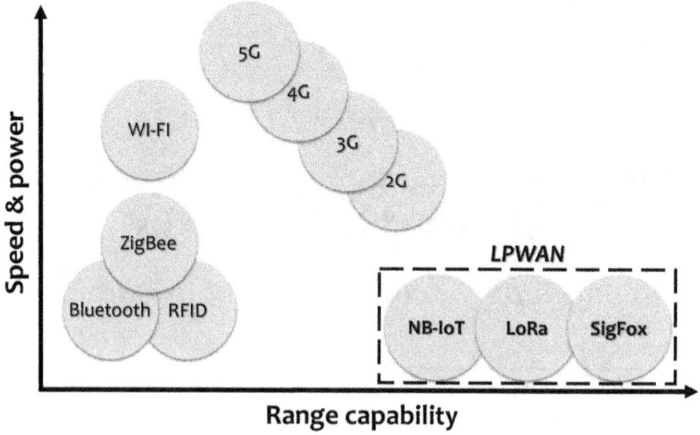

Figure 10.1 LPWAN versus legacy technologies

time, the requirement of channel reciprocity can be effortlessly reached despite in fast fading channels. However, this is not the case with low-power wide-area networks (LPWAN), which may communicate at speeds of tens of bits per second. Because of the extremely low bandwidth, packets take a long time to reach their destination (hundreds of milliseconds rather than a few milliseconds as with high bandwidth technologies). Furthermore, the communication range of legacy communication technologies and LPWAN differ significantly. For example, Zigbee only supports a short communication range of fewer than 100 m. In contrast, Long Range (LoRa), one of the most representative LPWAN technologies, has a communication range of up to ten kilometers. Figure 10.1 shows a more detailed comparison of various technologies. Because the data rate of LoRa is very low (e.g., 183 bps), the two communicating nodes may not be able to exchange the probing packets within channel coherence time, posing significant challenges for LoRa network to generate secret keys. Thus, secret key agreement for LoRa-based networks is critical and challenging. In this chapter, we discuss the principle, process, and existing methods of key generation in LoRa-enabled networks to inspire better work.

The remainder of this chapter is laid out as follows. In Section 10.2, we begin with a primer on LoRa and then introduce the principles, evaluation metrics, and procedure of the wireless key generation. In Section 10.3, we present several state-of-the-art key generation schemes for LoRa-enabled networks. In Section 10.4, we conclude the chapter.

10.2 Background

10.2.1 Primer on LoRa

Although LoRa has been thoroughly discussed in several studies [10,11], we present a brief overview of LoRa, as well as the LoRa PHY layer and LoRaWAN MAC layer.

10.2.1.1 LoRa overview

LoRa adopts chirp spread spectrum (CSS) modulation technique, and LoRaWAN is a data link layer specification developed on top of LoRa. LoRa runs in the sub-GHz ISM radio band, which not only varies depending on deployment locations but also follows duty cycle limitations [11]. LoRa outperforms short-range wireless standards and cellular technologies in terms of long-range transmission and low power consumption. It supports a wide coverage range of up to 15 km (rural areas) and 5 km (urban areas), with a data rate from 0.3 to 37.5 Kbps [12]. Due to its relatively low energy consumption, the LoRa device's battery life can be up to 10 years. Other characteristics, such as cheap device costs, gateway concurrent reception capacity, and modulation attribute robustness against fading, multipath, and Doppler effect compared to other wireless signals [13], make LoRa appealing. Due to the implementation complexity and numerous disturbances, real-world deployed LoRa networks often do not achieve the optimum performance that LoRa offers.

10.2.1.2 LoRa PHY layer

The LoRa PHY layer uses a spread spectrum modulation approach based on CSS with an integrated forward error correction (FEC) mechanism [14], which provides significant processing advantages for link budget improvements and multipath and interference resistance. Within the pre-defined usable narrow bandwidth of LoRa modulation, the frequency of chirps increases or decreases linearly with time. A base up-chirp is defined as a frequency that linearly sweeps from $-BW/2$ to $BW/2$ at a rate of k, as denoted by $C(t) = e^{j2\pi\left(\frac{k}{2}t-\frac{BW}{2}\right)t}$ [15]. The payload is represented by multiple chirps with a shifted initial frequency f_{sym} in LoRa modulation, so the signal of a symbol is denoted as $S\left(t,f_{sym}\right) = C(t) \cdot e^{j2\pi f_{sym}t}$. In terms of demodulation, the receiver performs a "de-chirp" operation, in which each received symbol is multiplied by a base down-chirp, denoted as $S\left(t,f_{sym}\right) \cdot C^{-1}(t) = e^{j2\pi f_{sym}t}$, resulting in a single frequency tone. The receiver then performs a fast Fourier transform (FFT) operation on the multiplication result, with one FFT peak corresponding to the original frequency f_{sym}. A LoRa packet structure is shown in Figure 10.2, which includes an 8-base up-chirp preamble, a 2-base up-chirp synchronization word, a 2.25-base down-chirp Start of Frame Delimiter (SFD), and payload data followed by Cyclic Redundancy Check (CRC).

The following parameters are used to configure a LoRa device: spreading factor (SF), bandwidth (BW), code rate (CR), carrier frequency (CF), and transmission power (TP). The number of symbols per bit of payload is represented by SF, ranging from 7 to 12. A higher applied SF results in a lower data rate, longer propagation time, and higher energy consumption, but it also results in better sensitivity and a longer transmission distance. As a result, devices further away from the gateway demand a larger SF due to the more significant link budget requirement, resulting in increased processing gain but at the expense of a reduced data rate. Furthermore, because SFs are orthogonal (basically quasi-orthogonal [14]), signals modulated with distinct SF can be transmitted in the same channel. 125, 250, and 500 kHz are the most common BW. The CR, which indicates the FEC code's rate, can be adjusted

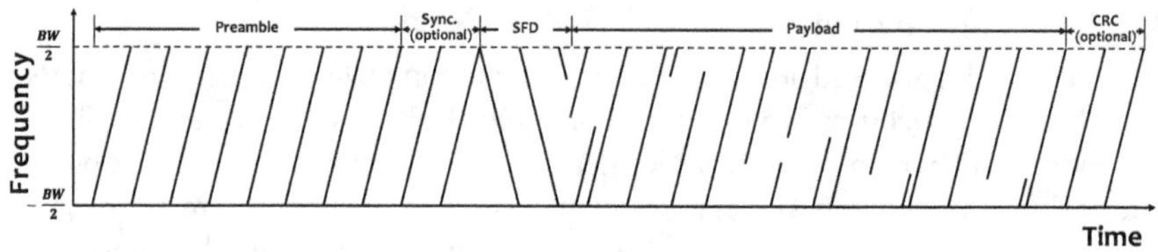

Figure 10.2 LoRa PHY packet structure

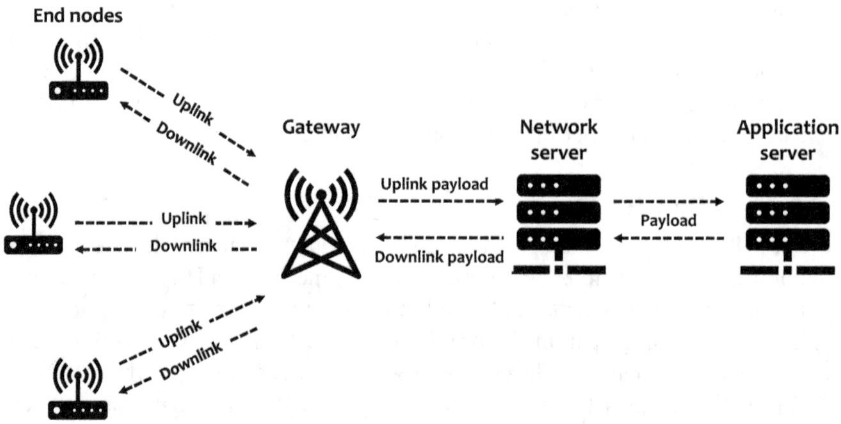

Figure 10.3 LoRa network architecture

to 4/5, 4/6, 4/7, or 4/8, with a higher number providing better protection at the cost of increased broadcast time. The LoRa modulation bit rate R_b can therefore be determined using the formula $R_b = SF \times \frac{BW}{2^{SF}} \times CR$. Furthermore, the LoRa band defines separate multiple frequency channels in different regions, with CF being the center frequency of these channels. The TP on a LoRa radio is limited by the hardware, which typically ranges from 2 to 20 dBm [16]. Tuning the above settings can result in a trade-off between communication range and power consumption, which motivates a variety of configuration strategies.

10.2.1.3 LoRaWAN MAC layer

Due to the fact that LoRa only defines the lower PHY layer of the communication stack, several upper network protocols are available, with LoRaWAN being the most popular. The LoRaWAN MAC protocol defines the LoRa network architecture and communication protocol in both directions. The LoRaWAN network architecture is a typical star topology, as shown in Figure 10.3, with end devices, gateways, and servers as the main components. End devices (also known as nodes, clients, or points) with various sensors are used as data sources in the scenarios. Gateways (also known as base stations) serve as a connection point between end devices and network servers.

Network servers serve as the LoRa network controller, managing devices and applications, while application servers serve as the user interface. In addition, LoRaWAN defines three types of end-device operation classes: A(ll), B(eacon), and C(ontinue). All devices must implement Class A, while Classes B and C are optional. The most energy-efficient of these three classes is Class A, which creates two receive downlink windows shortly after an uplink transmission from devices. End devices add planned downlink ping slots in class B, which are triggered by synchronization beacons broadcast by gateways, in addition to the two receive slots. Class C is the most energy-intensive since end devices continue to listen even when they are not transmitting (i.e., half-duplex). Choosing the appropriate operating class can help users save energy and satisfy the reaction time demands of various application scenarios.

LoRaWAN also provides security, channel access, and adaptive data rate (ADR). The LoRaWAN MAC layer uses the ALOHA MAC protocol, which allows end devices to communicate as soon as they wake up, without waiting for channel discovery or time synchronization. Furthermore, ADR is an important LoRaWAN method that allows end devices to be dynamically set with varying data rates based on network conditions. However, because LoRaWAN does not provide a specification of the ADR algorithm, some research has been conducted [17]. To complete data encryption, message integrity checking, and node authentication, LoRaWAN uses Advanced Encryption Standard (AES) techniques and two 128-bit unique session keys (i.e., NwkSKey and AppSKey).

10.2.2 Primer on wireless key generation

10.2.2.1 System model

As shown in Figure 10.4, we suppose there are two devices in a LoRa network, namely Alice and Bob, that intend to agree on the same key to secure their communication. They are both equipped with LoRa communication modules, with no prior sharing of secrets.

We suppose the existence of an attacker Eve, who tries to intercept the communication with the aim of generating the same key. Theoretically, if Eve is farther away than $\lambda/2$ from legitimate nodes Alice or Bob, she will collect entirely different channel measurements because of the spatial de-correlation property of the wireless channel, where λ is the wavelength of the LoRa signal [7]. In reality, this situation can be easily satisfied. For example, if attacker Eve is less than 17 cm away from the two legitimate nodes, she can be easily discovered (λ = 34 cm for 868 MHz LoRa).

10.2.2.2 Channel measurement for LoRa key generation

LoRa transceivers support to obtain RSSI, which is the most common feature for key generation, especially under the considerations of practical implementation. Two types of RSSIs, raw instantaneous RSSI and averaged RSSI per packet, are provided by the commonly used SX127x LoRa transceivers [18] and are referred to as $RSSI_r$ and $RSSI_p$ in this article. As shown in Figure 10.6, the differences between $RSSI_r$ and $RSSI_p$ are as follows.

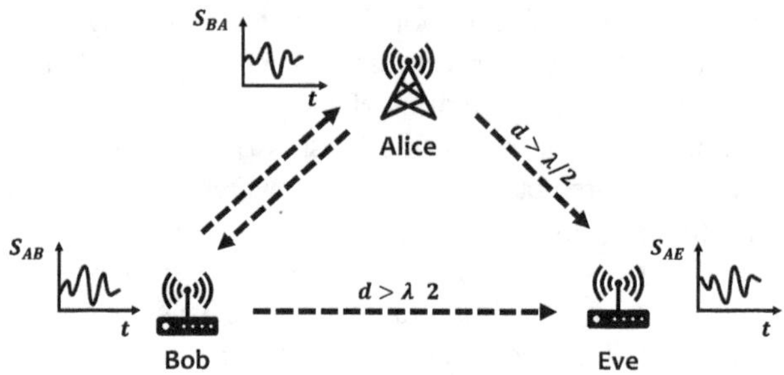

Figure 10.4 Wireless key generation model

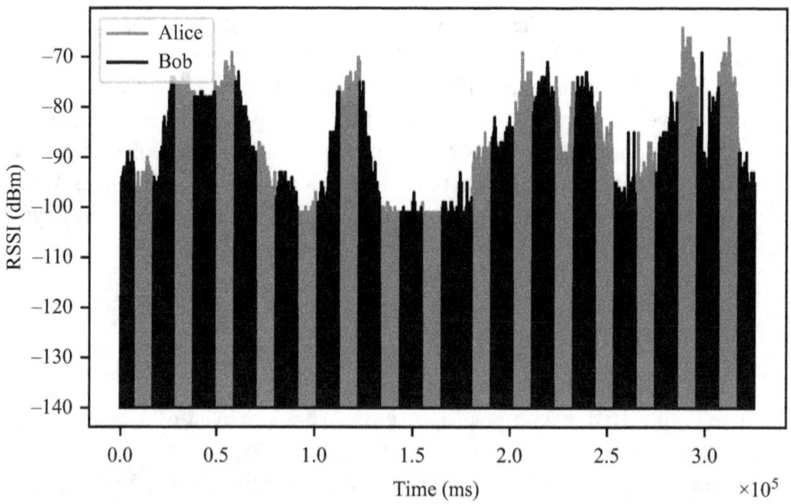

Figure 10.5 $RSSI_r$ of Alice and Bob

First, the RSSIs change substantially during packet transmission and reception due to the low data rate of the LoRa signal, as demonstrated in Figure 10.5. We can see, for example, that the initial half of the samples differ from the second half of the samples. $RSSI_r$ of both Alice and Bob are affected by this phenomenon. As a result, their average RSSI, $RSSI_p$, is unable to represent channel changes during packet reception. However, we observe that the end of Alice's $RSSI_r$ is nearly identical to the start of Bob's $RSSI_r$. This is due to the fact that the sampling timestamps are close for Alice and Bob (within coherence time in the LoRa channel). The findings above encourage us to employ $RSSI_r$ as a channel parameter to generate secret keys to improve the correlation between the channel measurements of legitimate parities.

Second, the quantity of $RSSI_r$ samples that can be measured in a single LoRa packet is considerably larger than $RSSI_p$, implying that $RSSI_r$ can be used to speed

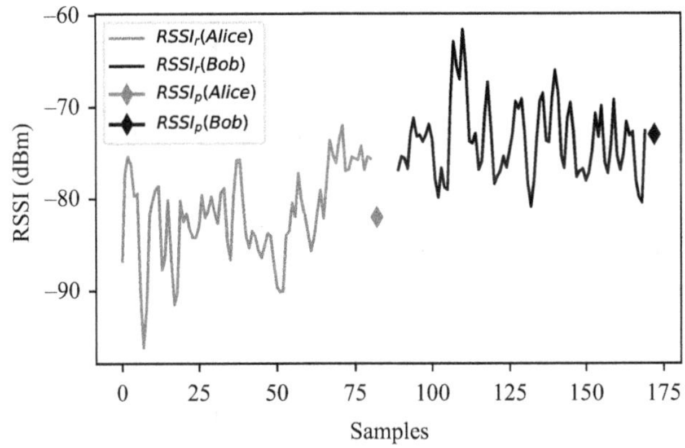

Figure 10.6 $RSSI_p$ versus $RSSI_r$

up key generation. However, because there are still some variances, as illustrated in Figure 10.6, $RSSI_r$ cannot be directly used to produce keys.

10.2.2.3 Protocol

As demonstrated in Figure 10.7, key generation typically comprises four stages: channel probing or measurement, quantization, information reconciliation, and privacy amplification. To keep generality, Alice is chosen as the process's initiator.

- **Channel probing**: Alice and Bob alternately probe and measure the randomness of the public wireless channel. This step necessarily involves bidirectional communication between Alice and Bob. Specifically, one side (say Alice) initializes the probing process by sending a pilot signal to another side (say Bob). Upon receiving the pilot signal, Bob measures the channel variations using a specific type of channel feature. The channel features that are supported by LoRa devices for secret key generation are discussed in Section 10.2.2.2. After that, Bob sends a response signal to Alice. Upon receiving the signal, Alice measures the same channel feature to obtain channel variations. Then Alice and Bob repeat this probing process until they obtain enough channel measurements, X_A and X_B, respectively.

  Because the LoRa physical layer communications operate in half-duplex mode, Alice and Bob cannot measure the selected channel parameter at the same time. For half-duplex hardware, the measuring time difference between Alice and Bob $\Delta t_{AB} = |t_{i,A} - t_{i,B}| > 0$ always holds. Moreover, hardware noises, which exist independently on different platforms, also have an impact on channel measurements. As a result, the non-simultaneous measurements and noises influence the cross-correlation connection of the measured channel parameter between the legitimate parties, i.e., S'_{BA} and S'_{AB}. Signal pre-processing methods, such as interpolation [19–21] and filtering [22–27], can help to attenuate these effects.

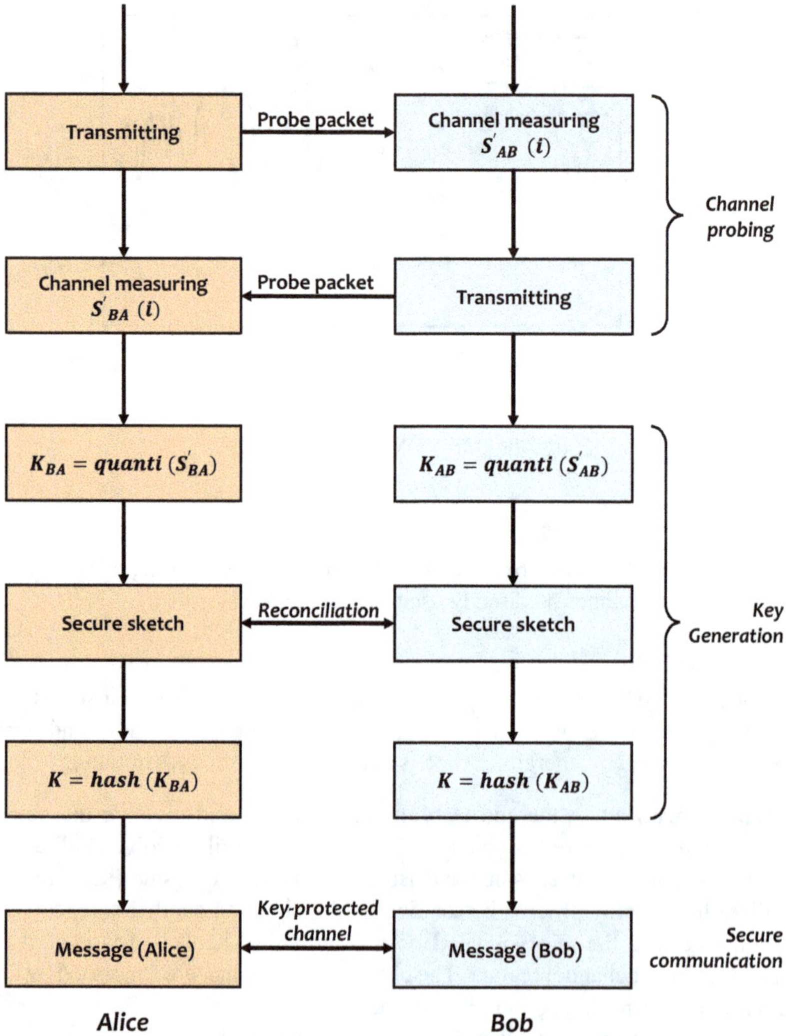

Figure 10.7 *Key generation procedure*

- **Quantization**: The quantization stage converts measurements from analog wireless channels to binary sequences. As cryptographic algorithms require binary input, the collected channel measurements should be quantized into binary sequence. This can be done by comparing channel measurements with thresholds and assign binary sequences. For instance, mean value-based quantization method uses the mean value as a threshold to do quantization.

  Two components of a quantizer are the quantization level and thresholds. The signal-to-noise ratio (SNR) of the channel typically determines the amount of bits that each measurement is quantized to. Thresholds are reference values that can be calculated from the mean and standard deviation or the cumulative distribution function (CDF) [5,28,29].

- **Information reconciliation**: Because of channel variation, hardware asymmetry, and noise, the keys of the legitimate nodes contain some mismatches. Thus, information reconciliation is used to enable the legitimate nodes to correct the mismatches and agree on the same keys.

  Secure sketch [30] is a well-known error correction method that has been widely used in physical layer key agreement systems to correct the mismatches between Alice and Bob [20,31,32]. Alice chooses a codeword c from the ECC C, and utilizes to be XOR-ed to calculate the syndrome s, and sends it to Bob. Bob works out the codeword and decodes it. Bob then uses the XOR operation to get the new K'_{AB}. When the hamming distance between c and c'' is smaller than the ECC's correction capacity, $c' = c$, Bob can receive the same keys as Alice. The cyclic redundancy check (CRC) can verify the agreement. The adopted ECCs or protocols determine the correcting capacity.
- **Privacy amplification**: The information reconciliation stage relies on the information exchanged via the public channel between Alice and Bob, which provides information about the keys to eavesdroppers. To prevent information leakage, privacy amplification is used [33]. Extractor [34] or universal hashing algorithms like the leftover hash lemma [5,35], cryptographic hash functions [36,37], and Merkle-Damgard hash function [38] can be used to amplify privacy.

10.2.2.4 Metrics

The secret keys utilized in cryptography applications such as authentication and encryption have criteria with respect to the key-generation rate (KGR), key agreement rate (KAR), and key randomness for practical situations, which are explained in depth below.

- **KGR**: KGR denotes the number of bits that a key generation system can generate per unit time (e.g., per second). KGR represents the speed of the key agreement scheme to generate secret keys. Because cryptographic methods require a specific length of binary bits, a high KGR is desirable for key agreement systems. For example, AES, one of the cryptographic algorithms, requires a minimum sequence length of 128.
- **Key agreement rate**: KAR denotes the percentage of the same bits over all the generated bits. A considerable KAR can significantly increase the efficiency of a wireless key generation protocol.
- **Key randomness**: In a cryptographic key-generation protocol, key randomness is critical because a key with a low level of randomness can be easily hijacked by attackers. The popular NIST test suite [39] is used to evaluate the randomness of the generated keys in the wireless key generation system. The NIST test suite includes 15 tests for evaluating various randomness features. Each test returns a p-value, indicating whether or not the key passed the NIST test. If the p-value is less than 0.01, it is assumed that the key passed the associated test and has a strong randomness.

10.3 Key-generation schemes for LoRa-enabled networks

In this section, we introduce several LoRa key generation methods with respect to different environments, configurations, and channel parameters.

10.3.1 LoRa key-generation scheme in various environments

The first complete LoRa-based secret key-generation protocol, LoRa-Key, was proposed by Xu *et al.* [19]. They conducted comprehensive experiments in both indoor and outdoor settings with static and mobile modes such as walking, biking, and driving. To boost the KGR, they used a variety of signal-processing algorithms. They also devised a reconciliation framework based on compressive sensing to improve the KAR. The details of LoRa-Key are as follows.

Signal processing: As shown in Figure 10.8, signal processing in LoRa-Key consists of two phases: filtering and interpolation. Filtering is used to remove anomalous data points and noise from the RSSI values. Based on the filtered RSSI sequence, interpolation is used to fill in missing data.

- Filtering: After the channel sampling process, they used outlier identification [40] and the Savitzky–Golay filter [41] to reduce the inconsistencies that were caused by ambient noise.
- Interpolation: They utilized a linear interpolation technique to fill in missing data points based on filtered RSSI readings to reduce the influence of asymmetry measurements produced by transceiver half-duplex operation. Compared to other interpolation methods, linear interpolation consumes less memory and computes faster.

Key generation: As shown in Figure 10.8, the key-generation step in LoRa-Key is composed of three basic steps: quantization, reconciliation, and privacy amplification. During the quantization step, Alice and Bob convert RSSI samples into binary bits. During the reconciliation stage, Alice and Bob exchange error-correcting messages via the public channel, allowing them to agree on the same bits. The openly shared messages, on the other hand, provide attacker Eve with some knowledge about the

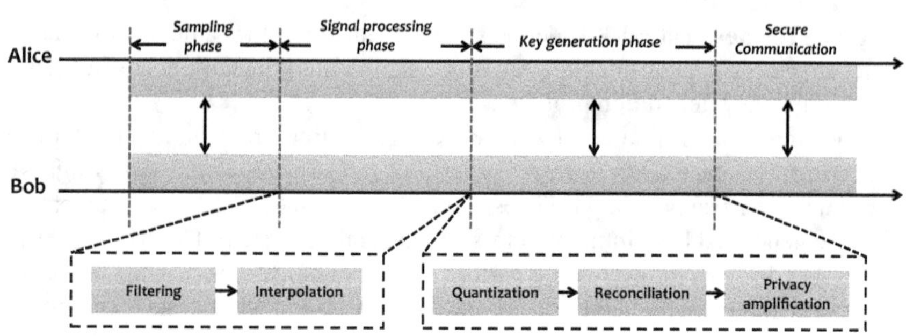

Figure 10.8 System flowchart

bits. To remedy this problem, Alice and Bob use privacy amplification to reduce the leaked information given to Eve.

- Quantization: After the signal processing step, RSSI values are transformed into bit strings using the multilevel quantization approach [42]. More specifically, they used a sliding window to segment the RSSI values, then checked the upper and lower bounds and used the bound to calculate the quantization levels. Each level in the quantization is assigned to an n-bit code (i.e., $n = \log_2 m$, where m represents the number of levels). Following that, the secret bits are retrieved from the result based on their quantization levels. When $n = 2$, the binary key is shown in the equation below:

$$bk_i = \begin{cases} 00 & RSSI_i \geq \text{upperbound}/2 \\ 01 & \text{mean} \leq RSSI_i < \text{upperbound}/2 \\ 10 & \text{lowerbound}/2 \leq RSSI_i < \text{mean} \\ 11 & RSSI_i < \text{lowerbound}/2 \end{cases}$$

- Reconciliation: To lower the bit mismatch rate, they designed a novel CS-based reconciliation mechanism in LoRa-Key. The idea behind CS-based reconciliation is to recover mismatched keys by using a shared matrix. The CS-based method has three advantages: initially, it transmits compressed vectors rather than the initial keys, which minimizes transmission overhead. Second, the heavy calculation is only required on one side, say Alice (e.g., a LoRa gateway), which typically has a powerful CPU and power supply, easing the pressure on resource-constrained end devices. Finally, compressed vectors are exchanged instead of the shared key; Eve, even if it captures the transmitted messages, is unable to recover the initial key. The CS-based method surpasses traditional error-correction code-based reconciliation methods, according to previous research [19,43].
- Privacy amplification: Privacy amplification enables legitimate parities to derive a shorter but almost private key from a shared random variable about which Eve has only partial knowledge. LoRa-Key employs one of the commonly used hash functions, SHA. After privacy amplification, symmetric-key algorithms like AES-128 can employ the secret key to provide secure communication between Alice and Bob. Only the first 128 bits of the final key are used if it is longer than 128 bits.

10.3.2 LoRa key generation scheme for COTS LoRaWAN

Ruotsalainen *et al.* [44] studied the key agreement scheme for commercially available LoRa. They conducted thorough experiments with various spreading factor and bandwidth setups and found that the sampling delay caused by bidirectional transmissions between legitimate parities will be affected by the setup of spreading factor and bandwidth. They also considered the receive delay in LoRaWAN protocol, as shown in Figure 10.10, to generate keys and conducted experiments in both indoor and outdoor settings. They designed a six-element electrically steerable parasitic array of radiator (ESPAR) antenna [6], as shown in Figure 10.11. A set amount of reactive elements

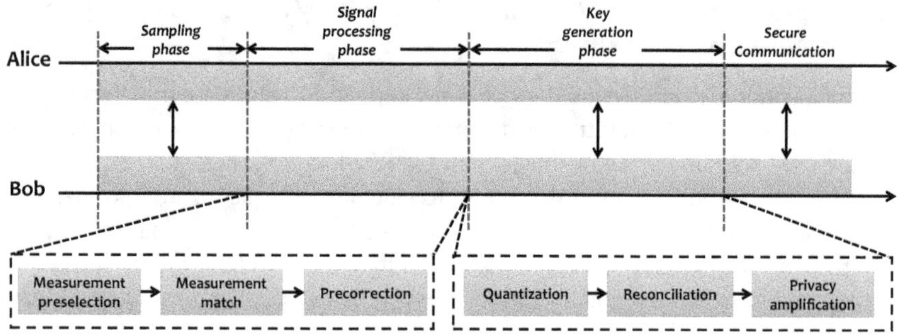

Figure 10.9 System flowchart

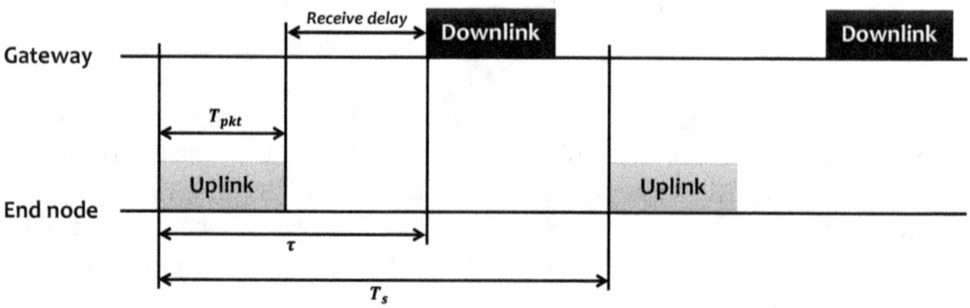

Figure 10.10 Time slots in LoRaWAN

can be used to tune the antenna radiation pattern in this type of antenna design. This kind of design increases the variance in measured RSSI traces. As a result, the reconfigurable antenna can be installed on the LoRaWAN network's gateway and/or end devices to improve the performance of the system. They showed that their technique could operate when there is not much channel variation. The details of their method are as follows.

Signal processing: In this work, signal-processing stage is composed of three phases: (a) measurement preselection, (b) measurement match, and (c) precorrection.

- Measurement preselection: Measurement preselection is the first phase of signal processing. In this work, the gateway assesses the data and selects those samples where ESPAR antenna reconfigurations mostly cause the changes in RSSI values. The sample sets are first partitioned into smaller subsets. Afterwards, they performed the Kolmogorov–Smirnov (KS) test, which is run for each adjacent subset, to test the equality of the probability distributions of two sets. If these two sets pass the test, they will be chosen for further processing since they appear to follow the same distribution.
- Measurement match: Because the end devices have only a limited number of ISM bands available, the CSS modulation used in LoRaWAN results in longer

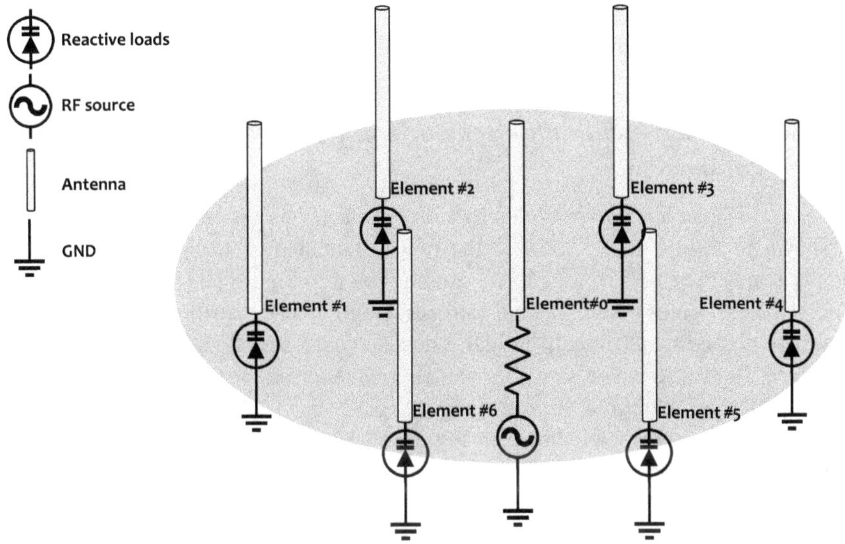

Figure 10.11 ESPAR antenna

packet lengths. And the long packet lengths means more packet airtime, which increases the probability of packet collisions [45]. Therefore, it is not possible to conduct prolonged channel probing for each uplink/downlink packet. And a measurement match between the end device and the gateway must be established before the key-generation process. They proposed the following two steps to solve this problem. (1) The gateway provides the sequence number of the very first and very last uplink packet pre-selected for measurement to the end device. (2) The end device checks the downlink packet numbers and determines subsets of continuous samples, for example, a set of measurements without packet drops, that are transmitted back to the gateway. Afterwards, the preselection results and available samples are delivered to both the end device and the gateway.

- Precorrection: They proposed to use discrete cosine transform (DCT) to rectify the synchronized measurements, which is done blockwise for the measured values. Following that, those frequency components of DCT that produce significant errors in quantization can be deleted as a precorrection step [46]. This step assists in overcoming bit mismatches caused by low SNR signal conditions.

Key generation: During the key-generation process, there are three fundamental phases: (a) quantization, (b) reconciliation, and (c) privacy amplification.

- Quantization: As introduced in Section 10.3.1, multilevel quantization is also applied as the quantization method.
- Reconciliation: The secure sketch approach with BCH encoding was chosen since it is appropriate for low-resource LoRaWAN devices, as stated in Section 10.2.2.3.

- Privacy amplification: The privacy amplification is performed using the SHA256 hash function, which results in the final secret key.

10.3.3 LoRa key-generation scheme using different channel parameters

Gao *et al.* [47] used *RSSIr* provided by LoRa transceivers for secret key generation. As shown in Figure 10.5, $RSSI_r$ is the raw instantaneous strength indicator. The comparison between $RSSI_r$ and $RSSI_p$ can be shown in Figure 10.6. When compared to $RSSI_p$, $RSSI_r$ gives better channel estimation and whose distributions provided by legitimate parties are extremely similar. They also used a random waypoint model to determine the best window size for balancing channel reciprocity and entropy. The workflow of their method is shown in Figure 10.12.

Signal processing: As shown in Figure 10.12, signal processing in this work consists of two phases: (a) modeling and (b) curve fitting.

- Modeling: After gathering sufficient $RSSI_r$ data, the authors used curve fitting to fit the data to a Rician fading model, observing that measurements of Alice and Bob follow the same Rician distribution, while the distribution of Eve follows a different Rician distribution. Based on the above finding, they designed a Rician distribution model-based key-generation method, by estimating the parameters in the Rician distribution. The empirical cumulative distribution function (CDF) and probability density function (PDF) are constructed from $RSSI_r$ in order to estimate the parameters in the Rician distribution.
- Curve fitting: After collecting the distribution of the $RSSI_r$ in a certain period, it is discovered that the Rician distribution parameters from Alice and Bob are significantly closer than the eavesdropper Eve. Following that, they performed polynomial curve fitting to the sets of normalized parameters to ensure that comparable trends are converted to similar keys after quantization. The resulting coefficient set will be utilized as the shared secret for key generation.

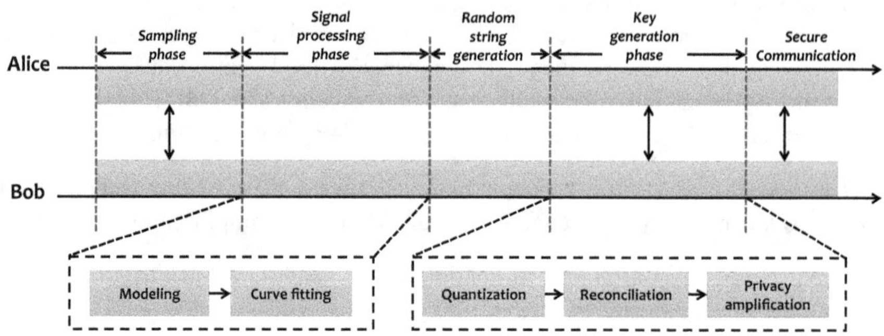

Figure 10.12 System flowchart

Key generation: The key-generation process is mainly comprised of three parts: (a) quantization, (b) reconciliation, and (c) privacy amplification. The details of the process are as follows.

- Quantization: Following the curve fitting, Alice will construct a set of random indices and send them to Bob. The set of random indices is then combined with the result (the coefficient set) from the previous stage by Alice and Bob. After that, Alice and Bob can produce the initial keys using a multilevel quantization approach as described in Section 10.3.1.
- Reconciliation: They also applied a CS-based reconciliation approach to reduce the key mismatches caused by noise, as stated in Section 10.3.1.
- Privacy amplification: Reconciliation increases reliability, but it also exposes secrets, as some information is exposed to the public and can be intercepted by attacker Eve. To solve the above problem, they chose the hash (e.g., SHA-256) function as a privacy amplification method.

10.3.4 LoRa key-generation scheme for long-range and highly mobile scenarios

Vehicle–Key [48] proposed a novel channel parameter adjacent register RSSI (arRSSI) by employing the mean value of adjacent $RSSI_r$ as a new channel measurement for key generation. As shown in Figure 10.5, $RSSI_r$ is the raw instantaneous strength indicator, and the adjacent parts of $RSSI_r$ for Alice and Bob are similar. The zoom-in view of comparison between $RSSI_r$ and $RSSI_p$ can be shown in Figure 10.6. To further improve the channel reciprocity, they also proposed a novel Bi-directional LSTM (BiLSTM)-based model to achieve channel prediction and quantization as the same time. The proposed model addresses the low correlation problem of LoRa channel measurements and hence increases KAR significantly. They also proposed an effective autoencoder-based reconciliation method to achieve high capability of error correction for the mismatches between the keys generated by Alice and Bob. The workflow of their method is shown in Figure 10.13.

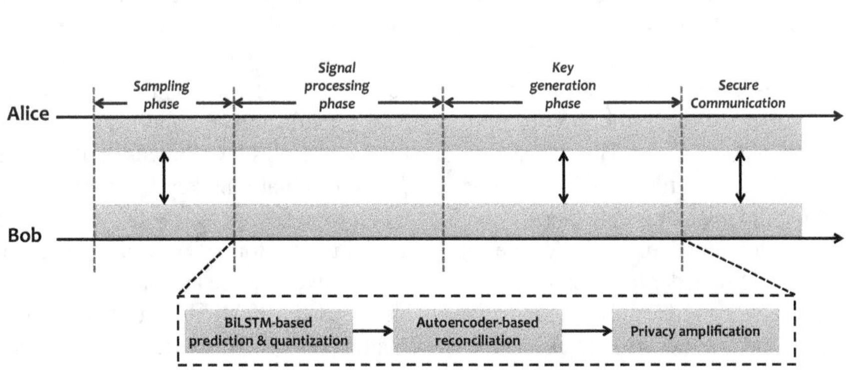

Figure 10.13 System flowchart

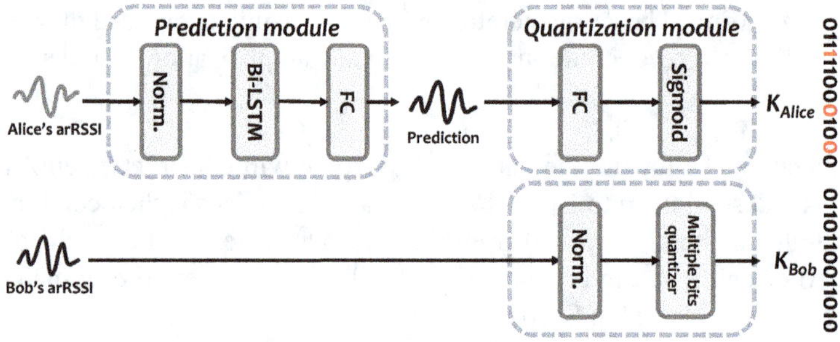

Figure 10.14 BiLSTM-based prediction and quantization model. © IEEE 2021. Reprinted with permission from [48].

Signal processing and key generation: As shown in Figure 10.13, signal processing and key generation in Vehicle–Key consists of the following three phases: (a) BiLSTM-based prediction and quantization, (b) autoencoder-based reconciliation, and (c) privacy amplification.

- BiLSTM-based prediction and quantization: Vehicle–Key reveals that the primary challenge for LoRa-enable long-range Internet of vehicle key generation is the channel measurements observed by Alice and Bob are not close to each other because they cannot be measured within channel coherence time. To address, they proposed a BiLSTM-based model to achieve prediction and quantization as the same time. As shown in Figure 10.14, the proposed BiLSTM-based model consists of a prediction sub-module and a quantization sub-module. The prediction sub-module is designed to predict the channel measurements of one side (i.e., Bob) based on the measurement from another side (i.e., Alice). The quantization sub-module converts the predicted arRSSI values into binary bits. The input of the model is the arRSSIs observed by one LoRa communication node and the output is the generated binary bits.
- Autoencoder-based reconciliation: As shown in Figure 10.15, they proposed a novel autoencoder model with a two-input structure for reconciliation. After the BiLSTM-based prediction and quantization phase, the generated keys of Alice and Bob are passed through an adapted Bloom filter [49] to protect the secret keys against reverse engineering attacks. In the following, the results from the Bloom filter are passed through a pre-trained neural network and compressed into a low-dimensional space. The subtraction of Alice's and Bob's low-dimensional vectors will then be utilized as the decoder's input to calculate the key mismatches. To find the mismatches between Alice and Bob, Alice first performs the pre-trained encoder. She then computes the subtraction of the vectors of Alice and Bob, and uses the decoder to compute the mismatches between Alice and Bob.
- Privacy amplification: Reconciliation increases the KAR, but it also exposes secrets to attackers, because some information is exposed to the public and can be intercepted by attacker Eve. To solve the above problem, they chose the hash (e.g., SHA-256) function as the privacy amplification method.

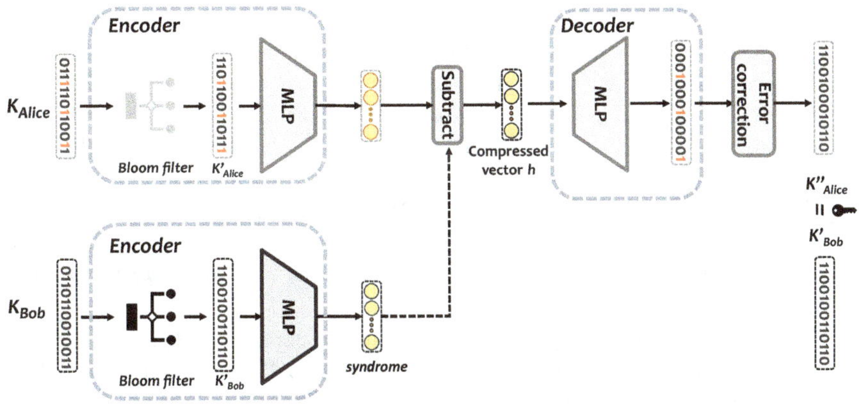

Figure 10.15 *Autoencoder-based reconciliation. © IEEE 2021. Reprinted with permission from [48].*

10.4 Conclusion

This chapter provides an overview of LoRa-enabled wireless key generation as well as representative works of research on LoRa wireless secret key agreement. We begin by discussing the LoRa PHY and MAC layers. After that, we discuss the principles, evaluation metrics, procedure, and channel parameters for LoRa-enabled key generation. We then present several relevant studies on LoRa network physical layer key generation, including LoRa key generation schemes in various environments, LoRa key generation schemes for COTS LoRaWAN, LoRa key generation schemes utilizing various channel parameters, and LoRa key generation schemes for long-range and highly mobile scenarios. Since LoRa networks are often used for large-scale sensing deployments and the current method of key generation for LoRa networks is designed for two legitimate nodes, group key protocols between a large number of LoRa-enabled devices remain an unsolved challenge [50]. Now that LoRa has been identified as a key technology for LPWAN and physical layer security on LoRa is very important, it is envisioned that the research on secret key generation for LoRa networks will attract more attention from academia and industry.

References

[1] Xu W, Kim JY, Huang W, *et al.* EMIoT: a LoRa-enabled smart building solution based on emergency lights. In: *Proceedings of the 7th ACM International Conference on Systems for Energy-Efficient Buildings, Cities, and Transportation*; 2020. p. 330–331.

[2] de Castro Tomé M, Nardelli PH, and Alves H. Long-range low-power wireless networks and sampling strategies in electricity metering. *IEEE Transactions on Industrial Electronics*. 2018;66(2):1629–1637.

[3] Caruso A, Chessa S, Escolar S, *et al.* Collection of data with drones in precision agriculture: analytical model and LoRa case study. *IEEE Internet of Things Journal.* 2021;8(22):16692–16704.

[4] Kim JY, Holz R, Hu W, *et al.* Automated analysis of secure internet of things protocols. In: *Proceedings of the 33rd Annual Computer Security Applications Conference*; 2017. p. 238–249.

[5] Jana S, Premnath SN, Clark M, *et al.* On the effectiveness of secret key extraction from wireless signal strength in real environments. In: *Proceedings of the 15th Annual International Conference on Mobile Computing and Networking*; 2009. p. 321–332.

[6] Aono T, Higuchi K, Ohira T, *et al.* Wireless secret key generation exploiting reactance-domain scalar response of multipath fading channels. *IEEE Transactions on Antennas and Propagation.* 2005;53(11):3776–3784.

[7] Mathur S, Trappe W, Mandayam N, *et al.* Radio-telepathy: extracting a secret key from an unauthenticated wireless channel. In: *Proceedings of the 14th ACM International Conference on Mobile Computing and Networking*; 2008. p. 128–139.

[8] Xi W, Qian C, Han J, *et al.* Instant and robust authentication and key agreement among mobile devices. In: *Proceedings of the 2016 ACM SIGSAC Conference on Computer and Communications Security*; 2016. p. 616–627.

[9] Jiao L, Wang N, Wang P, *et al.* Physical layer key generation in 5G wireless networks. *IEEE Wireless Communications.* 2019;26(5):48–54.

[10] Li C and Cao Z. LoRa networking techniques for large-scale and long-term IoT: a down-to-top survey. *ACM Computing Surveys (CSUR).* 2022;55(3):1–36.

[11] Sun Z, Yang H, Liu K, *et al.* Recent advances in LoRa: a comprehensive survey. In: *ACM TOSN*; 2022.

[12] Centenaro M, Vangelista L, Zanella A, *et al.* Long-range communications in unlicensed bands: the rising stars in the IoT and smart city scenarios. *IEEE Wireless Communications.* 2016;23(5):60–67.

[13] Liando JC, Gamage A, Tengourtius AW, *et al.* Known and unknown facts of LoRa: experiences from a large-scale measurement study. *ACM Transactions on Sensor Networks (TOSN).* 2019;15(2):1–35.

[14] Reynders B and Pollin S. Chirp spread spectrum as a modulation technique for long range communication. In: *2016 Symposium on Communications and Vehicular Technologies (SCVT).* IEEE; 2016. p. 1–5.

[15] Hou N, Xia X, and Zheng Y. Jamming of LoRa PHY and countermeasure. In: *IEEE INFOCOM 2021—IEEE Conference on Computer Communications.* IEEE; 2021. p. 1–10.

[16] Bor M and Roedig U. LoRa transmission parameter selection. In: *2017 13th International Conference on Distributed Computing in Sensor Systems (DCOSS).* IEEE; 2017. p. 27–34.

[17] Kufakunesu R, Hancke GP, and Abu-Mahfouz AM. A survey on adaptive data rate optimization in LoRaWAN: recent solutions and major challenges. *Sensors.* 2020;20(18):5044.

[18] Semtech SX1272. https://www.semtech.com/products/wireless-rf/lora-core/sx1272 Accessed April 4, 2022. [EB/OL].

[19] Xu W, Jha S, and Hu W. LoRa-key: Secure key generation system for LoRa-based network. *IEEE Internet Things Journal*. 2018;6(4):6404–6416.

[20] Liu H, Yang J, Wang Y, *et al.* Collaborative secret key extraction leveraging received signal strength in mobile wireless networks. In: *2012 Proceedings IEEE Infocom*. IEEE; 2012. p. 927–935.

[21] Patwari N, Croft J, Jana S, *et al.* High-rate uncorrelated bit extraction for shared secret key generation from channel measurements. *IEEE Transactions on Mobile Computing*. 2009;9(1):17–30.

[22] Zhang J, Marshall A, Woods R, *et al.* Efficient key generation by exploiting randomness from channel responses of individual OFDM subcarriers. *IEEE Transactions on Communications*. 2016;64(6):2578–2588.

[23] Liu H, Wang Y, Yang J, *et al.* Fast and practical secret key extraction by exploiting channel response. In: *2013 Proceedings IEEE INFOCOM*. IEEE; 2013. p. 3048–3056.

[24] Azimi-Sadjadi B, Kiayias A, Mercado A, *et al.* Robust key generation from signal envelopes in wireless networks. In: *CCS*. ACM; 2007. p. 401–410.

[25] Ali ST, Sivaraman V, and Ostry D. Eliminating reconciliation cost in secret key generation for body-worn health monitoring devices. *IEEE Transactions on Mobile Computing*. 2013;13(12):2763–2776.

[26] Zhu X, Xu F, Novak E, *et al.* Extracting secret key from wireless link dynamics in vehicular environments. In: *2013 Proceedings IEEE INFOCOM*. IEEE; 2013. p. 2283–2291.

[27] Zhang J, Woods R, Marshall A, *et al.* An effective key generation system using improved channel reciprocity. In: *2015 IEEE International Conference on Acoustics, Speech and Signal Processing (ICASSP)*. IEEE; 2015. p. 1727–1731.

[28] Guillaume R, Mueller A, Zenger CT, *et al.* Fair comparison and evaluation of quantization schemes for phy-based key generation. In: *OFDM 2014; 18th International OFDM Workshop 2014 (InOWo'14)*. VDE; 2014. p. 1–5.

[29] Zenger C, Zimmer J, and Paar C. Security analysis of quantization schemes for channel-based key extraction. *EAI Endorsed Transactions on Security and Safety*. 2015;2(6):e5.

[30] Dodis Y, Ostrovsky R, Reyzin L, *et al.* Fuzzy extractors: how to generate strong keys from biometrics and other noisy data. *SIAM Journal on Computing*. 2008;38(1):97–139.

[31] Liu H, Yang J, Wang Y, *et al.* Group secret key generation via received signal strength: protocols, achievable rates, and implementation. *IEEE Transactions on Mobile Computing*. 2014;13(12):2820–2835.

[32] Mathur S, Miller R, Varshavsky A, *et al.* Proximate: proximity-based secure pairing using ambient wireless signals. In: *Proceedings of the 9th International Conference on Mobile Systems, Applications, and Services*; 2011. p. 211–224.

[33] Bennett CH, Brassard G, Crépeau C, *et al.* Generalized privacy amplification. *IEEE Transactions on Information Theory*. 1995;41(6):1915–1923.

[34] Wang Q, Su H, Ren K, *et al.* Fast and scalable secret key generation exploiting channel phase randomness in wireless networks. In: *2011 Proceedings IEEE INFOCOM*. IEEE; 2011. p. 1422–1430.

[35] Premnath SN, Jana S, Croft J, *et al.* Secret key extraction from wireless signal strength in real environments. *TMC*. 2013;12(5):917–930.
[36] Zhang J, Kasera SK, and Patwari N. Mobility assisted secret key generation using wireless link signatures. In: *INFOCOM*. IEEE; 2010. p. 1–5.
[37] Ambekar A, Hassan M, and Schotten HD. Improving channel reciprocity for effective key management systems. In: *2012 International Symposium on Signals, Systems, and Electronics (ISSSE)*. IEEE; 2012. p. 1–4.
[38] Wei Y, Zeng K, and Mohapatra P. Adaptive wireless channel probing for shared key generation based on PID controller. *IEEE Transactions on Mobile Computing*. 2012;12(9):1842–1852.
[39] Rukhin A, Soto J, Nechvatal J, *et al.* A statistical test suite for random and pseudorandom number generators for cryptographic applications. DTIC Document; 2001.
[40] Wang G, Chen Y, and Zheng X. Gaussian field consensus: a robust non-parametric matching method for outlier rejection. *Pattern Recognition*. 2018;74:305–316.
[41] Press WH and Teukolsky SA. Savitzky–Golay smoothing filters. *Computers in Physics*. 1990;4(6):669–672.
[42] Zeng K, Wu D, Chan A, *et al.* Exploiting multiple-antenna diversity for shared secret key generation in wireless networks. In: *INFOCOM*. IEEE; 2010. p. 1–9.
[43] Lin Q, Xu W, Liu J, *et al.* H2B: heartbeat-based secret key generation using piezo vibration sensors. In: *Proceedings of the 18th International Conference on Information Processing in Sensor Networks*; 2019. p. 265–276.
[44] Ruotsalainen H, Zhang J, and Grebeniuk S. Experimental investigation on wireless key generation for low-power wide-area networks. *IEEE Internet of Things Journal*. 2019;7(3):1745–1755.
[45] Adelantado F, Vilajosana X, Tuset-Peiro P, *et al.* Understanding the limits of LoRaWAN. *IEEE Communications Magazine*. 2017;55(9):34–40.
[46] Margelis G, Fafoutis X, Oikonomou G, *et al.* Physical layer secret-key generation with discreet cosine transform for the Internet of Things. In: *2017 IEEE International Conference on Communications (ICC)*. IEEE; 2017. p. 1–6.
[47] Gao J, Xu W, Kanhere S, *et al.* A novel model-based security scheme for LoRa key generation. In: *Proceedings of the 20th International Conference on Information Processing in Sensor Networks (co-located with CPS-IoT Week 2021)*; 2021. p. 47–61.
[48] Yang H, Liu H, Luo C, *et al.* Vehicle-key: a secret key establishment scheme for LoRa-enabled IoV communications. In: *IEEE ICDCS*; 2022.
[49] Xu W, Li Z, Xue W, *et al.* InaudibleKey: generic inaudible acoustic signal based key agreement protocol for mobile devices. In: *IPSN*; 2021.
[50] Xu W, Zhang J, Huang S, *et al.* Key generation for Internet of Things: a contemporary survey. *ACM Computing Surveys (CSUR)*. 2021;54(1):1–37.

Chapter 11

Robust secret key generation from stochastic fading in the presence of passive and active attackers

Arsenia Chorti[1,2]

11.1 Introduction and chapter organization

Sixth generations (6G) systems will be required to meet diverse constraints in an integrated ground–air–space global network. In particular, meeting overly aggressive latency constraints and operating in massive connectivity regimes with low energy footprint and low computational effort while providing explicit security guarantees can be challenging [1]. In addition, the extensive introduction of artificial intelligence (AI) and machine learning (ML) and the rapid advances in quantum computing are further developments that will increase the attack surface of 6G systems [2,3]. More importantly, the massive deployment of low-end Internet of Things (IoT) nodes, often produced following non-homogeneous production processes and with expected lifespans exceeding 10 years, poses pressing questions concerning future security architectures.

In this setting, quality of security (QoSec) is envisioned as a flexible security framework for future networks with diverse requirements. By mirroring the differentiated services (DiffServ) networking paradigm, various security levels could be conceptualized, moving away from static security controls, captured currently in zero-trust security architectures [4]. In parallel, the integration of communications and sensing, along with embedded (on-device) AI, can provide the foundations for building autonomous and adaptive security controls, orchestrated by a vertical security plane in coordination with a vertical semantic plane, dubbed as context-aware smart security [5,6]. It is in this framework that we envision the incorporation of physical layer security (PLS) schemes in 6G security protocols, introducing security controls at all layers for the first time [4]. This exciting prospect does not come, however, without challenges.

Under this prism, in this chapter, we turn our attention to one of the most mature PLS schemes that of distilling symmetric secret keys at two remote locations by

[1] ETIS UMR 8051/CY Paris University, ENSEA, CNRS, France
[2] Barkhausen Institut gGmbH, Germany

observing the same random process. While in the future the source of shared randomness will likely include sensing signals, at present the most widely considered approach is based on distilling shared randomness from stochastic fading in wireless channel realizations. The aim of this chapter is thus to provide a comprehensive view of key requirements that need to be met to render secret key generation (SKG) robust in the presence of both passive and active attackers. While some of the presented materials appear so far only in preprints, the bulk of it has appeared in a multitude of IEEE papers within the last five years, notably in [7–10], among others; I am immensely thankful to all of my contributing co-authors.

With respect to the chapter organization, the material is presented in the following order.

- General directions for the incorporation of PLS and SKG in particular in future networks, starting with 6G, are discussed in Section 11.2.
- In Section 11.3, the performance of SKG and achievable rates are discussed in the short blocklength, pertinent for Internet of Things (IoT) applications with low computational and delay-constrained devices.
- Section 11.4 sets the scene for achieving robustness in SKG in the presence of closely located passive attackers (eavesdroppers), via pre-processing techniques to remove spatial correlations and dependencies from the observed channel state information (CSI) vectors in experimental datasets.
- Section 11.5 explores SKG in man-in-the-middle (MiM) attacks at the physical layer (PHY) and proposes respective countermeasures.
- In Section 11.6, a performance analysis is provided for SKG in denial of service (DoS) attacks at the PHY in the form of reactive jamming.
- In Section 11.7, we move further to present recently proposed hybrid SKG-cryptosystems in order to provide security guarantees against active attacks at upper layers as well.
- Finally, Section 11.8 concludes this chapter and discusses future perspectives.

The present chapter was put together with the hope of providing useful and comprehensive guidelines on how PLS and SKG, in particular, can provide explicit security guarantees and aspires to contributing to the community's ongoing effort to bring PLS to life, for the first in 6G.

11.2 General directions for incorporating PLS in 6G

Despite intense research interest in PLS for more than two decades, its incorporation in actual security products remains largely elusive, with a few exceptions in RF fingerprinting (physec.de) and multi-factor authentication (silencelaboratories.com). The hurdles to be overcome concern primarily two key issues: the difficulty in providing explicit security guarantees with PLS irrespective of the link quality and the potential degradation in terms of achievable rates. For incorporating PLS in 6G, both of these issues need to be addressed.

With this in mind, first we will provide a draft roadmap for the inclusion of PLS, and, SKG in particular, in 6G. Our goal is to pave the way for developing practical PLS schemes with explicit security guarantees, taking into account inconsistencies between idealized channel models—used in information-theoretic proofs—and actual transmission conditions. To this end, we draw inspiration from the methodology used by the cryptographic community. We revisit how idealized semantic security primitives, such as pseudorandom number generators and functions, are used to build real cryptosystems. We use these ideas as guidelines for the proposed roadmap. Furthermore, we explicitly propose using hybrid SKG-crypto schemes, an approach that allows addressing jointly two key issues. First, to relax the pressure for high secrecy transmission rates, and, second, to shield SKG schemes from active attacks, e.g., by ensuring the integrity of exchanged messages or side information with the use of message authentication codes and cryptographic hashing [11].

11.2.1 Roadmap for incorporating PLS in 6G

We aim to provide a methodology that will overcome existing hurdles in translating PLS theory to practical systems. In this direction, we begin the discussion from the fact that the study of reliable communications involves a typical bottom-up approach; building on idealized source and channel models, powerful theorems on the fundamental limits of achievable rates have been built. Such results guide the design of actual wireless systems operating over actual transmission channels, given that in practice (a few) connection outages and errors are tolerated. However, the requirements are different in information security, in which the guarantees to be met need to be tight and explicit.

In cryptographic proofs, such explicit guarantees are expressed through the concept of the "adversarial advantage," which needs to be negligible (i.e., strictly upper-bounded in polynomial time, reflecting limitations in computational resources) [12]. Related proofs of semantic security (chosen-plaintext attack—CPA security, chosen ciphertext attack—CCA security, etc.) are built around *games* played between legitimate and adversarial parties, when the latter possess specific capabilities, ranging from launching passive attacks (eavesdropping) to launching active attacks (MiM, DoS). In the cryptographic proofs, several idealized mathematical constructions are used, such as pseudorandom number generators (PRG), pseudorandom functions (PRF) and permutations (PRP), and, random oracles (RO), as abstractions of the fundamental cryptographic primitives [12].

Consequently, practical cryptosystems are built to *resemble*, as closely as possible, the above-mentioned idealized mathematical structures. In particular, secure stream ciphers should resemble PRGs, while block ciphers should be built to resemble PRFs or PRPs. As an example of the latter, non-linearities are a core element of their design, aiming to induce maximal confusion [13], e.g., with the use of S-boxes. Their design typically involves selected functions from the class of bent functions that are subsequently modified to withstand linear and differential cryptanalysis [14]. As practical crypto primitives are designed to resemble the respective idealized mathematical structures, practical cryptography lingers toward semantic security.

Turning our attention to PLS, information-theoretic proofs guaranteeing perfect secrecy or distributed key distillation without leakage are similarly built using idealized channel models. Importantly, it has been shown that it is possible to prove semantic security for wiretap channels [15], while recently, fundamental results have been published regarding the finite blocklength [16] along with related approaches for analysis [17] and code design [10,18]. However, it remains challenging to translate the theoretical results to practical systems for various reasons, including that actual transmission channels might not match idealized channel models. As an example, although the impact of correlations on the secrecy capacity has been extensively explored [19–21], when accounting for more complex phenomena, captured typically through compound or arbitrarily varying channel models, translating channel secrecy capacity expressions [22,23] to practical code designs becomes tedious. More importantly, the adequacy of practical channel models—proposed by 3GPP—for information-theoretic security is still unclear. We note in passing that, for mmWave and sub-THz bands, existing 3GPP models have been shown to have limitations [24,25].

We, therefore, need to shed light on how PLS can provide explicit security guarantees in non-idealized transmission conditions. A first step to resolve related inconsistencies that will be tangible in 6G systems is the site-specific, online learning and characterization of the channel [26], which can allow constructing secrecy maps [17] from actual channel estimates. It should be noted, however, that there are inherent limitations in the online evaluation of secrecy outages directly from field data, e.g., during the network operation, as the dataset size limits the precision of the estimation of mutual information from measured CSI vectors [27].

To bypass this issue, we propose a further step—to engineer the transmission, given the channel model learned, so that the end-to-end transmission resembles transmission over an idealized channel, for which the security guarantees are explicit. A general approach in this direction could exploit dimensionality reduction as a generic tool to carve out a memoryless/block fading channel from a non-ideal, site-specific channel with correlations across different dimensions. In Section 11.4, we will provide relevant results for the case of SKG.

A final point that should be addressed concerns the potential impact of the proposed pre-processing approaches on the achievable secrecy or secret key rates. Indeed, in numerous measurement campaigns for SKG [28], or secrecy outage probabilities in sub-THz bands [29], it has been shown that in real-world scenarios, high entropy (for the extraction of keys) or a consistent signal-to-noise ratio (SNR) advantage (for wiretap coding) cannot always be guaranteed, e.g., for target rates of 0.5 b/s/Hz. Due to this negative result, there has been considerable skepticism concerning the feasibility of practical PLS systems. In order to address this issue, we first need to identify the target secrecy or secret key rate. We propose to use PLS mainly for two security operations: (i) for the distillation (using SKG) or distribution (in wiretap channels) of secret keys in hybrid PLS—cryptosystems [8,11]; (ii) for localization or RF fingerprint-based authentication [30,31].

Consequently, the target rate would be dictated by cryptographic considerations (mode of operation) or networking parameters (e.g., re-keying due to handovers in high mobility environments). An example of the former is that in the record protocol of the transport layer security (TLS) protocol version 1.3, the suggested cipher suite is the `TLS_AES_256_GCM_SHA384 [GCM]`, requiring essentially 96 bytes of key material (32-byte keys, 48 byte MAC secrets, and 2×8 byte IVs), while the largest record that can be transmitted amounts to 2^{14} bytes [32]. Assuming SKG is used to generate and distribute these keys, then key rates (over multiple subcarriers and antennas) as low as 0.006 bits/sec could be comfortably targeted.* This toy example showcases that in hybrid SKG-cryptosystems, the target rates can be very low, alleviating concerns related to diminishing throughput.

Before getting into details of actual implementations and robustness to attacks, we provide a review of essential background concepts around SKG.

11.2.2 Robust SKG

The baseline SKG system model with two legitimate nodes, denoted by Alice and Bob and a single adversary, denoted by Eve, are depicted in Figure 11.1. Typically, the SKG process consists of three phases [33,34], as depicted in Figure 11.2. In the first phase, referred to as *shared randomness distillation*, Alice and Bob observe dependent random variables denoted by Y_A, Y_B while an eavesdropper, referred to as Eve, observes Y_E. In wireless channels, a readily available source of shared randomness is the multipath fading due to the reciprocity of the wireless medium during the channel's coherence time [35]. To showcase reciprocity, in Figure 11.3, we depict RSS measurements using two nodes, each equipped with an ESP32 low-power system on a microcontroller chip with integrated WiFi [8]. The measurement parameters are summarized in Table 11.1.

In the next two phases, known as *information reconciliation* and *privacy amplification*, side information s_A is exchanged between Alice and Bob, generated by corresponding encoders f_A, f_B [36,37]. Any information leakage due to the public exchange of side information is addressed through hashing at the privacy amplification. At the end of the SKG process, a common key $K \in \mathcal{K}$ is extracted at Alice and Bob such that, for any $\varepsilon > 0$, the following statements hold [38]:

$$Pr\,(K = f_A\,(Y_A, s_A) = f_B\,(Y_B, s_A)) \geq 1 - \varepsilon, \tag{11.1}$$

$$I(K; s_A) \leq \varepsilon, \tag{11.2}$$

$$H(K) \geq \log|\mathcal{K}| - \varepsilon, \tag{11.3}$$

where $H(K)$ denotes the entropy of the key K and $I(K; s_A)$ denotes the mutual information between K and s_A.

The first inequality demonstrates that the SKG process can be made error free; (11.2) ensures that the exchange of side information through public discussion does not leak any information to eavesdroppers; while (11.3) establishes that the generated keys attain maximum entropy (i.e., are uniform). Under the three conditions, an upper bound on the rate for the generation of secret keys is given

*Target key rates can be much lower if we consider key re-use over multiple records or zero-round-trip time operation that can incorporate SKG [11].

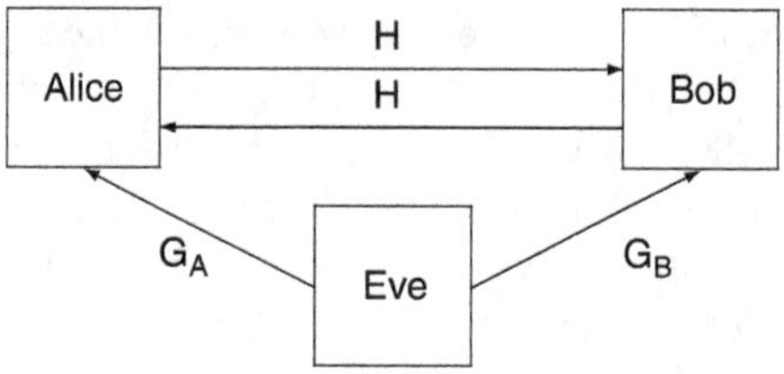

Figure 11.1 SKG system model with two legitimate nodes, referred to as Alice and Bob, in the presence of an adversary, referred to as Eve

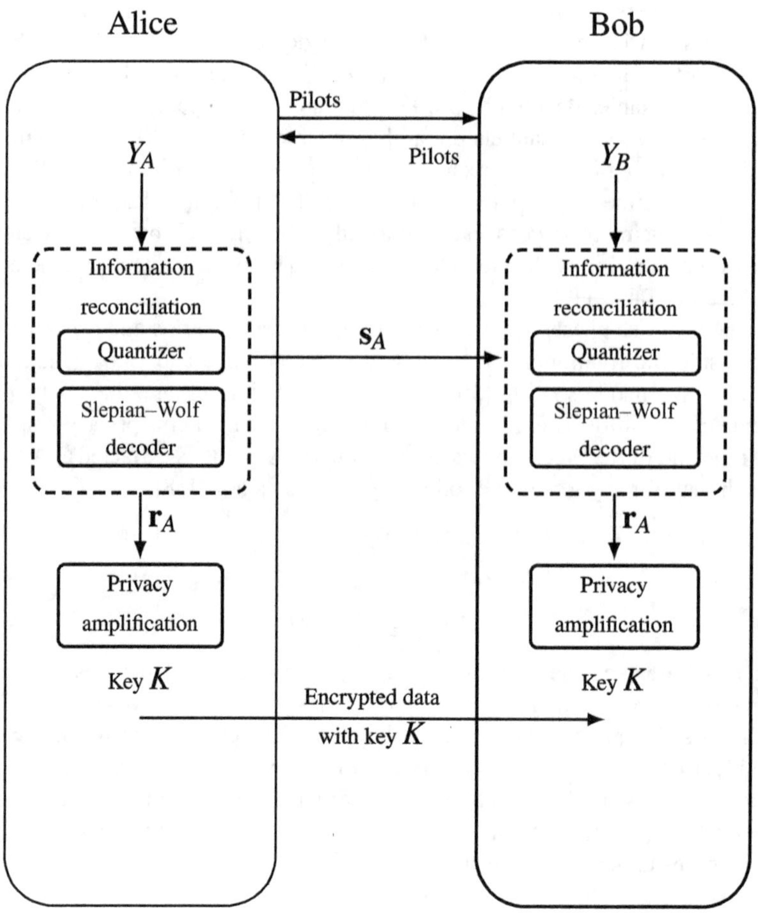

Figure 11.2 Secret key generation between Alice and Bob. Thanks to reciprocity, the quantizer outputs can be expressed as $\boldsymbol{r}_A = \boldsymbol{d} \oplus \boldsymbol{e}_A, \boldsymbol{r}_B = \boldsymbol{d} \oplus \boldsymbol{e}_B$.

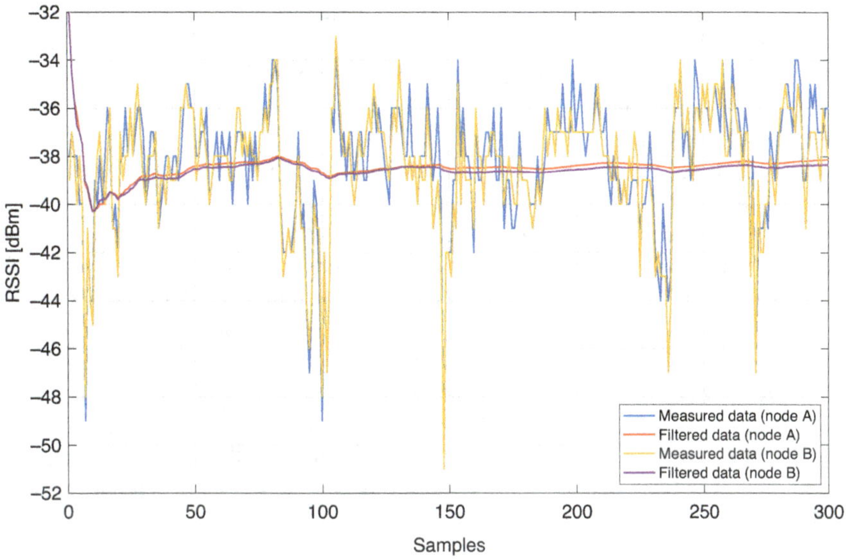

Figure 11.3 Measured RSS data and filtered data using Kalman filter at a distance of 9 m. The filter is initialized using $K_1 = 0.5$ and $Y_0 = -32$ dBm.

Table 11.1 Experimental setup and parameters

Scenario	**Office space**
Transmit power	19.5 dBm
Carrier frequency	2.4 GHz
Measurement period	500 ms
Estimated interference level	≈ -80 dBm
Measured distances	$d \in \{1, 3, 6, 9\}$
Presence of line of sight	80%

by $\min\{I(Y_A; Y_B), I(Y_A; Y_B|Y_E)\}$ [33,39]. Assuming rich multipath environments, the decorrelation properties of the wireless channel over short distances can be exploited to ensure that Eve's observation Y_E is uncorrelated with Y_A and Y_B [35,38,40,41]; in this case, the SKG capacity is given by [39, Section II]

$$C_K = I(Y_A; Y_B). \tag{11.4}$$

However, this condition is rarely met in real-life scenarios. In particular, correlations and dependencies in four domains, *space (location), time, frequency, and antenna* among Alice's, Bob's, and the Eve's observations have to be taken explicitly into account. While subsampling in the time, frequency and antenna domains can constitute simple approaches to re-create a memoryless channel so that the observations between Alice and Bob are independent from the observations of Eve along these domains, correlations and dependencies in space need, on the other hand, to be taken explicitly into account.

In multiple studies, it has been demonstrated that large-scale fading components are dominated by deterministic fading and are strongly dependent on the location and the distance between users, which makes them predictable for eavesdroppers. Therefore, for (11.4) to hold, we need to perform pre-processing to disentangle predictable components of the CSI (deterministic fading) from purely random but reciprocal (stochastic fading). This latter is largely captured in small-scale fading components; appropriate pre-processing enables us to obtain straightforward evaluations of the achievable secret key rate in a given channel.

In the literature, SKG has been extensively analyzed in Rayleigh fading [38,40]. Assuming for convenience that Alice and Bob exchange unit probe signals to excite the fading channel and that appropriate pre-processing has already taken place, they obtain respective observations Y_A and Y_B with

$$Y_A = H + N_A, \; Y_B = H + N_B, \tag{11.5}$$

where H denotes the fading coefficient in the link between the legitimate nodes, modeled as a zero mean Gaussian random variable $H \sim \mathcal{N}(0, \sigma_H^2)$, and N_A and N_B model the effect of additive white Gaussian noise (AWGN) and denote independent and identically distributed (i.i.d.) Gaussian random variables $N_A \sim \mathcal{N}(0, \sigma_A^2)$, $N_B \sim \mathcal{N}(0, \sigma_B^2)$. Using this notation, the SKG capacity has been expressed as [38]:

$$C_K = I(Y_A; Y_B) = \frac{1}{2} \log_2 \left(1 + \frac{\sigma_H^2}{\sigma_A^2 + \sigma_B^2 + \frac{\sigma_A^2 \sigma_B^2}{\sigma_H^2}} \right). \tag{11.6}$$

These rates are achievable in the asymptotic regime. However, as SKG is a prime candidate for IoT systems with constrained devices and ultra-low latencies, it is pertinent to explore the SKG rates that are achievable in short blocklengths. To this end, in the following section, we review some recent implementations of SKG reconciliation in the finite blocklength, before delving into SKG robustness against passive and active attacks.

11.3 SKG reconciliation rates in short blocklengths

In this section, we explore practical Slepian Wolf decoder designs for reconciliation in the short blocklength. In our system model, we assume Alice and Bob have quantized observations $\mathbf{r}_A, \mathbf{r}_B$, respectively, of length n obtained by quantizing their local measurements of a common source of randomness, represented by Y_A and Y_B, respectively. At present, we assume that a passive adversary, referred to as Eve, is unable to obtain any information about the observed sequences, i.e., we assume that the generated binary sequences are i.i.d., i.e., bits at all positions $i = 1, \ldots, n$ are equiprobably 0 and 1:

$$Pr\left(\mathbf{r}_A[i] = 0\right) = Pr\left(\mathbf{r}_B[i] = 0\right) = 0.5. \tag{11.7}$$

Also the corresponding bits of the sequences $\mathbf{r}_A, \mathbf{r}_B$ are different with probability p,

$$Pr\left(\mathbf{r}_A[i] \neq \mathbf{r}_B[i]\right) = p, i = 1, \ldots, n. \tag{11.8}$$

Alice and Bob attempt to agree on a secret key K of length k that must be drawn uniformly from a length n binary tuple $\mathscr{K} = \{0, 1\}^k$. To this end, Alice sends her syndrome $\mathbf{s}_A$, of length $n - k$, through a public channel in order to help Bob to obtain an estimate $\hat{\mathbf{r}}_A$ of her sequence $\mathbf{r}_A$. Additionally, Bob and Eve can listen to the public channel without error.

The code rate is defined as $R = \frac{k}{n}$ and the frame error rate (FER) is defined as the probability that Bob's estimation of $\mathbf{r}_A$ is erroneous $Pr(\mathbf{r}_A \neq \hat{\mathbf{r}}_A)$. In this set-up, Bob first estimates the sequence $\hat{\mathbf{r}}_A$ and then, Alice and Bob extract the common key K independently using privacy amplification. Motivated by the promising performance of LDPC [42], polar [43] and BCH codes [44] with list decoding in short blocklengths for forward error correction, we explore the performance of these three families of SW decoders against an upper bound on SKG rates reported in [45].

Here, we focus the FER performance of the aforementioned codes in the SKG setting for very short blocklengths of 128 and 512 bits. We also compare the FER performance of codes with the information theoretical finite length upper bound reported in [45], which translates to a lower bound in the FER.

For the first instance, we consider a blocklength of 128 bits. We implement a $(n, k) = (128, 75)$ polar code with 11 bits CRC and a $(127, 64, 10)$ BCH code. Also, we pick regular $(3, 6)$ LDPC codes with length 128 bits.† The maximum number of iterations for SP algorithm is set to 50. In Figure 11.4, the FER performances of half rate codes with 128 bits blocklength are depicted (i.e., the key length after privacy amplification is 64) and compared to the upper bound reported in [45]. Note that the upper bound becomes loose for very short blocklengths while its accuracy improves as the blocklength increases. That is the reason we observe comparatively a large gap between the upper bound and the codes' FER performances for $n = 128$ [47].

As it is depicted in Figure 11.4, the polar code with list size 128, second-order BCH list decoding code, and LDPC iterative list decoding for $t = 2$ perform best and show almost similar performance at FER 10^{-3}. Moreover, the results show that while traditional polar codes perform poorly in short blocklengths, their performance can be significantly improved by equipping them with list decoding. For example, the polar code with list size 128 provides around two orders lower FER compared to the classical polar code at $H(r_A[i]|r_B[i]) = 0.1944$. However, this improvement comes at the cost of 128 times more decoding complexity. We also observe that LDPC iterative list decoding outperforms the non-iterative one. Furthermore, list decoding also improves the performance of the BCH and LDPC codes, but the improvement

† We note that the variable and check nodes degree distributions can be optimized to improve the performance of LDPC codes [46]; however, the degree optimization of the LDPC codes is out of the scope of this work.

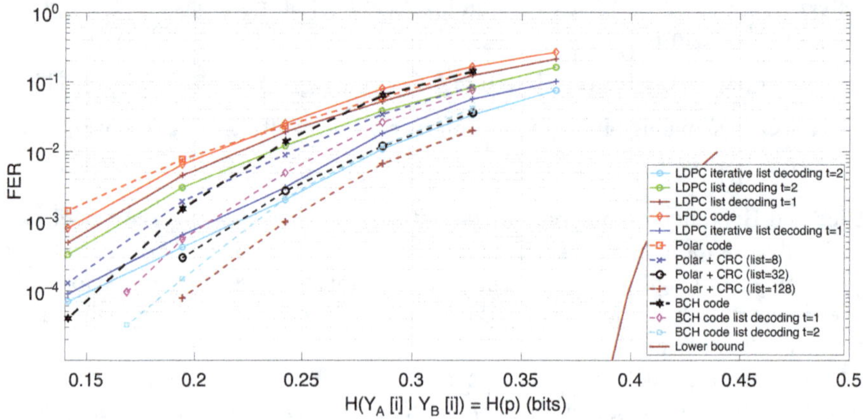

Figure 11.4 Comparison of FER performance of codes with the lower bound in [45] for n = 128

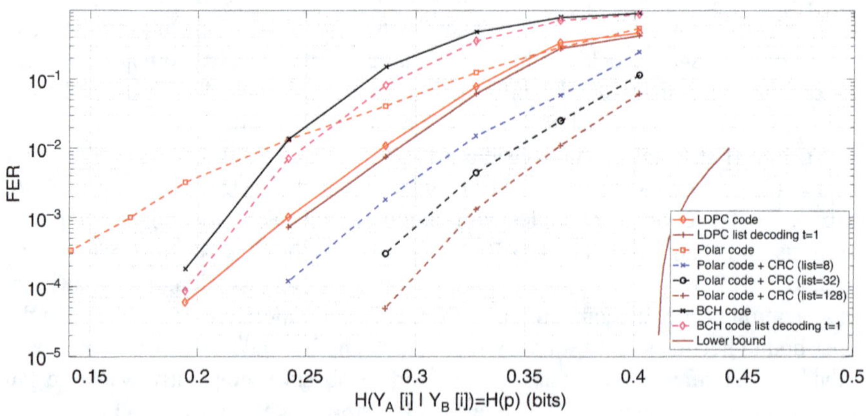

Figure 11.5 Comparison of FER performance of codes with the lower bound in [45] for n = 512

with respect to the additional complexity is not as notable as in the case of the polar code.

Furthermore, to generate keys that can be used with standard block ciphers, e.g., AES-128 or AES-256, the SKG process is assumed to run with blocklength 512 and rate half. As it is demonstrated in Figure 11.5, we observe a significant improvement in the performance of polar codes by using list decoding. On the other hand, we do not observe a noticeable improvement by using list decoding size $t = 1$ for LDPC and BCH code for this blocklength. The strength of list decoding with a fixed size as the blocklength increases diminishes. We do not generate higher list decoding order for LDPC and BCH codes due to their high decoding complexity. Also, at blocklength $n = 512$, the gap between the FER of the polar code with list size 128 and the lower bound is less than the case $n = 128$. We posit that one of the reasons is that the lower bound at this length is tighter than the first instance.

After this brief presentation of finite blocklength considerations on SKG rates, we move on at examining robustness against various types of attacks, beginning with eavesdropping.

11.4 Robust SKG against eavesdropping by a nearby attacker

In this section, we focus on rendering SKG robust to eavesdropping by nearby attackers, i.e., we focus on removing via pre-processing spatial correlations and dependencies at Eves in the vicinity of legitimate users. Without any pre-processing of the observed CSI, an attacker can, in principle, distill highly correlated sequences with these seen by the legitimate users and thus substantially decrease the effective size of the key space.‡ These correlations can indeed be removed through privacy amplification. However, this approach (i.e., without any pre-processing of the channel coefficients) entails the use of much larger quantizers and heavier hashing, potentially leading to less energy-efficient solutions and hinders the straightforward evaluation of the achievable SKG rate.

Another gain of the proposed CSI pre-processing is that it can facilitate relating SKG to the concept of a cryptographic PRG at the advantage distillation phase [48].

Definition 1. *Pseudorandomness of generator G [48]: the ensemble $\{G(U_n)\}_{n\in N}$ is pseudorandom, iff for any probabilistic polynomial-time algorithm A, for any positive polynomial p and for all sufficiently large n,*

$$|Pr(A(G(U_n); 1^{l(n)})) = 1 - Pr(A(U_{l(n)}; 1^{l(n)}) = 1)| < \frac{1}{p(n)}. \tag{11.9}$$

In the definition of the PRG given above, the probabilistic polynomial-time algorithm A can be seen as a polynomial-time *statistical test*. If an algorithm A exists such that the above condition does not hold, we say the generator G fails the test A. From a pseudorandom generator, we expect an observer who knows i bits of the output to be unable to predict the $(i+1)$th bit with a probability non-negligibly greater than $\frac{1}{2}$. This property is called (next-bit) unpredictability.

Various polynomial time statistical tests of randomness have been introduced and form the NIST suite. There also exist non-polynomial time tests, the most famous among them being the spectral test, which measures the correlation between overlapping n-tuples in the output of a PRG. However, these tests do not account for

‡Although any correlations in the time, frequency, space, or antenna domains between the reconciled sequences at Alice, Bob, and potential Eves should be explicitly taken into account in the evaluation of the conditional min-entropy to estimate the target privacy amplification rate, such measures are omitted in many published works.

dependencies and cross-correlations between legitimate and adversarial nodes that can be present in the case of SKG. To address this vital issue, in the following, we alternatively propose a novel metric *to infer* predictability/unpredictability, by measuring spatial dependencies between observed CSI vectors.[§]

11.4.1 Proposed statistical independence metric

In the PLS literature, to the best of our knowledge without exception, the Pearson cross-correlation coefficient (CC) has been used as an indirect measure of unpredictability. However, decorrelation does not suffice to prove unpredictability as there might be non-linear dependencies and/or the underlying distributions might not be Gaussian. On the contrary, independence of the observations at legitimate and adversarial entities suffices to infer unpredictability and allows alignment of the design of SKG with the definition of a PRG, right at the advantage distillation step; hence, we explicitly introduce spatial independence of SKG seeds as a proposed design criterion for semantically secure key distillation from wireless fading coefficients.

In this direction, we propose to measure dependencies between observed CSI vectors through the use of the test statistic of a recently introduced kernel-based statistical test of independence, dubbed as the d-parameter Hilbert–Schmidt independence criterion (d-HSIC) [49]. The test applies a positive-definite kernel on N-dimensional random variables (RVs) and maps its distribution into the reproducing kernel Hilbert space to determine whether the N-dimensions are independent. More precisely, let $\boldsymbol{Y} = (\boldsymbol{y}^1, \ldots, \boldsymbol{y}^N)$ be an $M \times N$ matrix of the observation vectors $\tilde{\boldsymbol{y}}^i = [y_1^i, \ldots, y_M^i]^T$ for $i \in \{1, \ldots, N\}$. The null hypothesis indicates that the vectors $\boldsymbol{y}^i$, $i \in \{1, \ldots, N\}$, are mutually independent and therefore their joint probability density function can be expressed as the product of the marginals, $H_0 : F_{y^1,\ldots y^N} = F_{y^1} \cdots F_{y^N}$, whereas the alternative, H_A : (not H_0), denotes that $\boldsymbol{Y}$ consists of at least two dependent vectors. An estimator (test statistic) $dHSIC$ of the statistical functional was defined as follows [49, Def 2.6]:

$$
\begin{aligned}
dHSIC(\boldsymbol{Y}) &= \frac{1}{M^2} \sum_{i,j=1}^{M} \prod_{l=1}^{N} \left(\mathbf{1}_{M \times M} \circ K_{ij}^l\right) \\
&\quad + \frac{1}{M^{2N}} \prod_{l=1}^{N} \sum_{i,j=1}^{M} K_{ij}^l - \frac{2}{M^{N+1}} \sum_{i,j=1}^{M} \prod_{l=1}^{N} \left(\mathbf{1}_{M \times 1} \circ K_{ij}^l\right),
\end{aligned}
\tag{11.10}
$$

where the operator $\circ$ denotes the Hadamard product and $\mathbf{1}_{M\times M}$ is an $M \times M$ matrix of ones. Also, $\mathbf{K}^l = \left(\mathbf{K}_{ij}^l\right) = \left(k^l(x_i, x_j)\right) \in \mathbb{R}^{M\times M}$ is the Gram matrix of the positive semi-definite Gaussian kernel k^l, defined $\forall x_i, x_j \in \mathbb{R}$ by, $k^l = \exp\left(-\frac{\|x_i - x_j\|^2}{\sigma^2}\right)$, with bandwidth $\sigma = \sqrt{\frac{\mathrm{med}\left(\|x_i - x_j\|^2\right)}{2}}$ and $\mathrm{med}(\cdot)$ is the median heuristic.

[§]The proposed methodology can be straightforwardly extended to incorporate time, frequency, and antenna domains.

According to [49, Theorem 3.1], with respect to the hypothesis test at hand, the critical value (for a specific significance level α) can be obtained by large number of Monte-Carlo realizations of $dHSIC'(\boldsymbol{Y})$ and re-sampling without replacement.

Based on the *dHSIC* test statistic and the corresponding critical value, we propose here a normalized metric for measuring the level of dependence, expressed as

$$\overline{\Delta} = \frac{dHSIC(\tilde{\boldsymbol{Y}})}{CV_\alpha} 1_{\{dHSIC(\boldsymbol{Y}) > CV_\alpha\}}. \tag{11.11}$$

$\overline{\Delta}$ is close to unity in case of low dependence between the variables and grows without bound in case of high dependence. Finally, we note that during the pre-processing, lowering dependencies between observations at different spatial locations should be balanced with preserving, as much as possible, the reciprocity between the observations of Alice and Bob, discussed next.

11.4.2 Reciprocity and mismatch probability

On the one hand, we introduce pre-processing to lower dependencies between legitimate and adversarial observations. On the other hand, this pre-processing will inevitably impact the reciprocity of the SKG seeds. In this work, we use a one-bit quantizer about the median point along the time dimension to check the reciprocity of the SKG seeds in the uplink (Alice to Bob) and the downlink (Bob to Alice). The mismatch probability (MP) between Alice and Bob is given by the ratio of the number of bits in disagreement to the total number of bits. Mismatches can be corrected during the information reconciliation step; however, the higher the MP, the lower the rate of the reconciliation decoder, while reconciliation becomes impractical beyond a certain point for short blocklengths.

11.4.3 Proposed power domain pre-processing

We present a power domain approach for disentangling predictable CSI components (primarily due to large-scale fading) from unpredictable ones (primarily due to small-scale fading) using principal component analysis (PCA), based on the conjecture that predictable components will be dominant in power. To validate the proposed methodology, we perform experiments on a real dataset and described next.

11.4.3.1 Nokia Stuttgart experimental dataset

A massive multiple-input multiple-output (MIMO) channel measurement campaign was conducted on the Nokia campus in Stuttgart, Germany. The campaign area consisted of multiple roads with high buildings (15 m high approximately), acting as reflectors and blockers for the radio wave propagation. The transmit antenna array was placed on the roof-top of one of these buildings. The geometry of 64-element transmit-array was such that there were 4 rows with 16 single-polarization patch antennas, with a horizontal spacing of $\lambda/2$, and vertical spacing of λ.

The transmit antenna array transmitted 64 time–frequency orthogonal pilot signals at 2.18 GHz carrier frequency, using orthogonal frequency division multiplexing

(OFDM) waveforms according to the 10 MHz LTE numerology (i.e., 600 subcarriers with 15 kHz spacing). The pilot signals have been arranged so that the sounding on 50 separate subbands (each consisting of 12 consecutive subcarriers) required 0.5 ms. Within that pilot burst period, the propagation channel was assumed to be time-invariant. The pilot bursts were sent continuously with a periodicity of 0.5 ms.

The receiver user equipment (UE) was mounted on a mobile cart and consisted of a single monopole antenna mounted at 1.5 m height, a Rohde and Schwarz TSMW receiver and a Rohde and Schwarz IQR hard disc recorder, which continuously captured the received base-band signal. Both the transmit array and the receiver were frequency synchronized via GPS. During the measurements, the receiver cart moved along several routes at walking speed (3.6 kmph), which corresponded to a spatial channel sampling distance of less than 0.5 mm. Post-processing was used to extract, for each pilot burst and subband, the 64-dimensional CSI vector.

In this section, we used datasets on tracks 6 and 12, depicted in Figure 11.6, that are parallel at a vertical distance of 1 m. In detail, we assumed that Bob is the base station and Alice walks along track-6, while Eve performs an "on the shoulder attack" and walks in parallel to Alice on track-12, i.e., legitimate and adversarial nodes are at all times 1 m away. In order to remove frequency domain (within the coherence bandwidth) and antenna domain correlations, we have downsampled the dataset. In detail, we kept the measurements from every 10th subcarrier and every 4th antenna, i.e., we used sub-sampling factors of 5 and 6, respectively.

Furthermore, as the Nokia dataset consists of only uplink data, we used alternate consecutive measurements to approximate downlink data and have further downsampled the data in the time domain, keeping every 5th channel sample. In detail, starting from sample index 1, we labeled odd index samples as uplink and even indexed samples as downlink. As a result, for each Alice and Eve, we used CSI vectors of length $M = 800$, concatenating real and imaginary parts. Before presenting the details of the pre-processing, we discuss the statistical analysis of the two datasets.

11.4.4 Pre-processing using PCA

Let $\boldsymbol{Y}_u = [\boldsymbol{y}_{1u}, \cdots, \boldsymbol{y}_{Nu}]$ denote the observed CSI matrix at Bob (aggregating the CSI vectors from all Alices) and $\boldsymbol{U}$ the $M \times M$ matrix whose rows are the eigenvectors of the matrix $\mathrm{Cov}(\mathbf{Y}_u)$ sorted in decreasing order. In many scenarios, e.g., Rician and generally line of sight settings, it is plausible to assume that the first few PCs correspond to the dominant large-scale fading terms, while the rest of the PCs correspond to small-scale fading terms and noise. We want to identify a region of PCs with indices $\{\tilde{D}_1, \ldots, \tilde{D}_2\}$, corresponding to components $\tilde{\boldsymbol{Y}}_u = [\tilde{\boldsymbol{y}}_{1u}, \cdots, \tilde{\boldsymbol{y}}_{Nu}]$ for $u \in \{a, b\}$ over which low dependence and correlation between Alices and Eves is achieved while keeping the MP below a threshold.

Note that the PCs beyond $\tilde{D}_2 + 1$ are dominated by noise and should be neglected (denoising). To efficiently disentangle, the CSI matrix into predictable and unpredictable parts, the pair $\{\tilde{D}_1, \tilde{D}_2\}$ is chosen such that $\overline{\Delta}$ and the MP are kept as low as possible for the range $\{\tilde{D}_1, \ldots, \tilde{D}_2\}$. In the following, we will also discuss the trade-off between minimizing $\overline{\Delta}$ (as a measure of independence) and the MP.

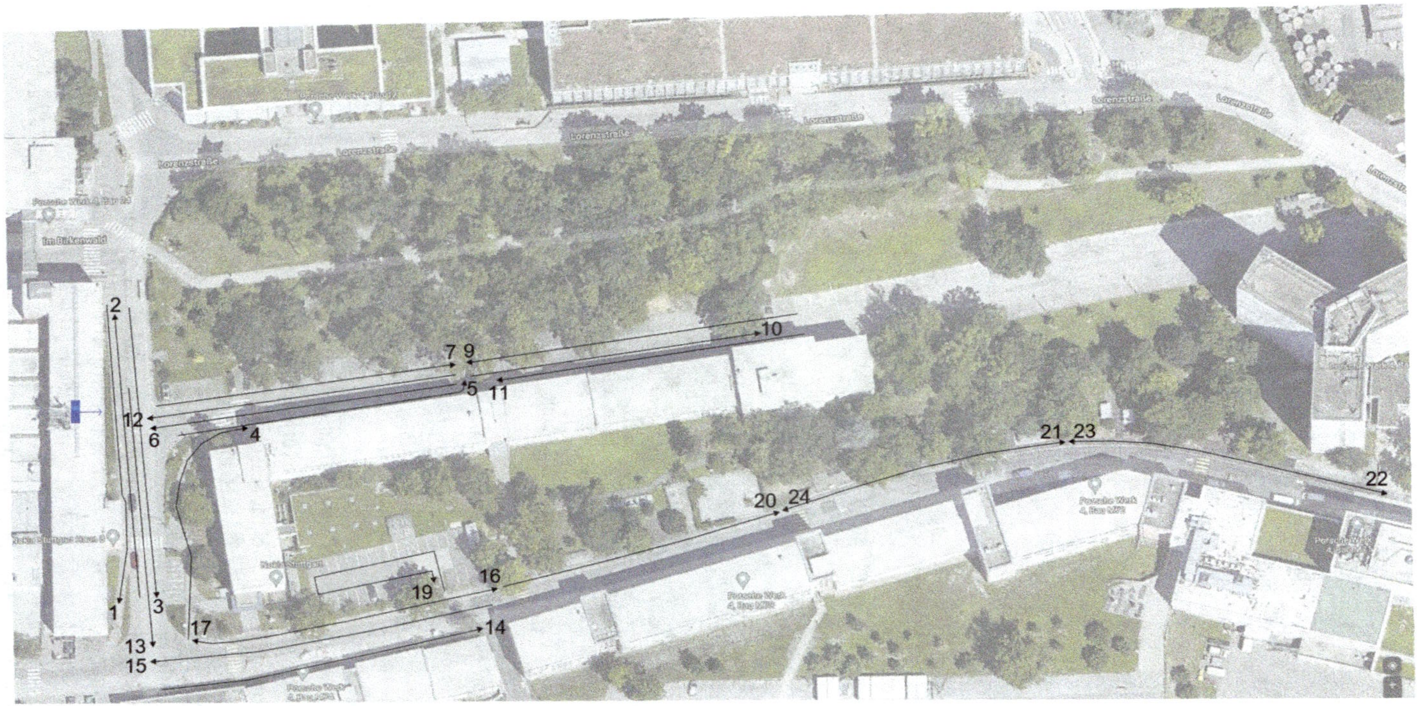

Figure 11.6 *Pictorial representation of the Nokia campus in Stuttgart, Germany, where the dataset was recorded. The blue bar on the left side corresponds to the location of the massive MIMO transmit antenna, placed on a rooftop, while the arrow points the antenna boresight. The black dashed lines depict the measurement tracks, along which the user equipment (UE) cart moved. The tracks are numbered from 1 to 24; in this work, we utilized the datasets of tracks 6 and 12, which are parallel and at 1 m distance from each other. Source: [50].*

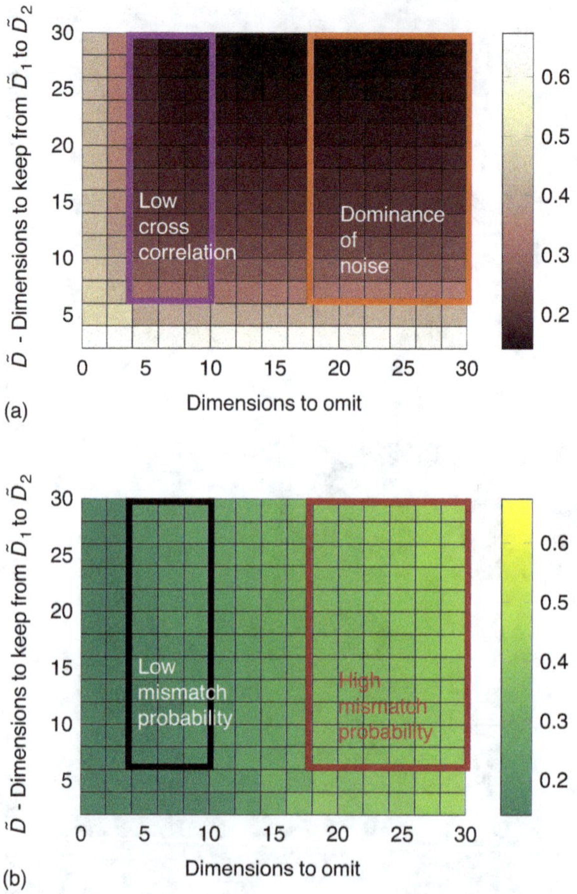

Figure 11.7 Trade-off between CC and MP for SNR = 20 dB for the Nokia dataset. (a) Average CC (Original = 0.38). (b) Average MP.

We present results for the Nokia dataset at an SNR = 20 dB in Figure 11.7. With no pre-processing, the average CC is approximately 0.38. However, with a sufficient number of dimensions retained, an increase in the number of dimensions omitted decreases the CC. Specifically, for $\tilde{D}_1 = 6$ and $\tilde{D}_2 = 30$, we observe a drop in the CC to 0.15, with no significant increase in MP. We posit that this is the regime in which the predominant large-scale predictable components have been removed and the small-scale fading components have been retained. Importantly, a trend similar to CC is observed in the average $\overline{\Delta}$ in Figure 11.8, especially for $\tilde{D}_1 \geq 4$ after which value the dependence level collapses. Based on these results, the proposed pre-processing can efficiently address passive attacks on SKG. However, a number of possible active attacks need also be addressed to achieve robust SKG.

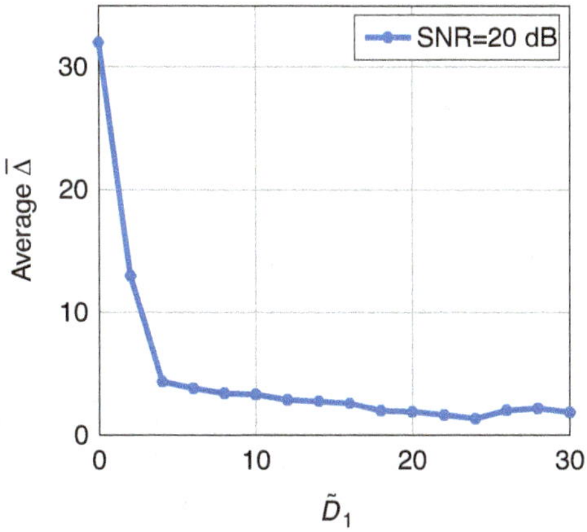

Figure 11.8 Evolution of $\overline{\Delta}$ with $\tilde{D}_1$ for $\tilde{D}_2 = 30$ for the Nokia dataset

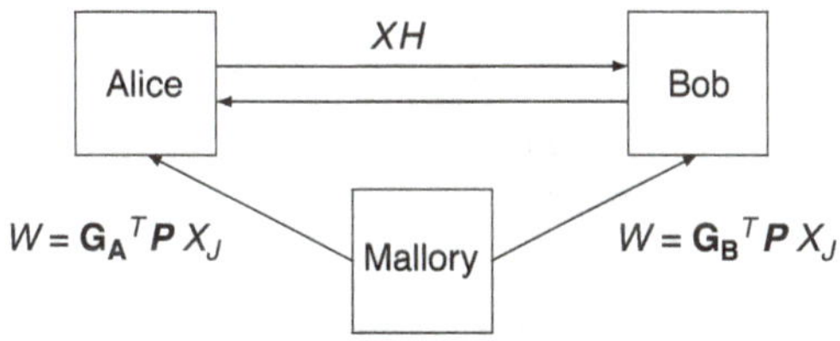

Figure 11.9 Alice and Bob have single transmit and receive antennas and exchange pilot signals X over a Rayleigh fading channel with realization H. A MiM, Mallory, with multiple transmit antennas can inject a suitably pre-coded signal $\boldsymbol{P}X_J$, such that the received signal at both Alice and Bob coincide $W = \boldsymbol{G}_A^T\boldsymbol{P} = \boldsymbol{G}_B^T\boldsymbol{P}$.

11.5 Man-in-the-middle attacks on SKG and countermeasures

MiM in the form of injection attacks constitutes one of the most critical limitations in SKG systems that extract secret keys from received signal strength (RSS) measurements [51–53]. Recently, various possible approaches for injection attacks have been published: in [51], the attacker controlled the movement of objects in an indoor wireless network, thus generating predictable changes in the RSS (e.g., by obstructing, or not, a line-of-sight). In [52], whenever similar channel envelope measurements in the links to the legitimate nodes were observed, the MiM spoofed the SKG process by injecting a strong signal. In the following, we will prove that—even when full CSI is used to extract the keys—it suffices that the adversary has one additional antenna with respect to the legitimate users to be able to mount an injection MiM attack.

To capture the main components of MiM attacks in SKG systems, we employ the system model depicted in Figure 11.9, comprising three nodes: a legitimate transmitter, its intended receiver, and a MiM, referred to as Alice, Bob, and Mallory, respectively. Alice and Bob are assumed to have a single antenna each for simplicity, while Mallory has two transmit antennas.‖ The fading channel realization in the link Alice–Bob is denoted by the complex circularly symmetric Gaussian random variable $H \sim \mathcal{CN}(0, \sigma_H^2)$. To obtain estimates of H, Alice and Bob exchange pilot signals X with $\mathbb{E}[|X|^2] \leq P$. Furthermore, following [52], we assume Mallory has perfect knowledge of the channel vectors in the multiple input single output (MISO) links Mallory–Alice and Mallory–Bob. The channel coefficients are assumed to be i.i.d., i.e., $\mathbf{G_A} = [G_{A1}, G_{A2}]^T, \mathbf{G_B} = [G_{B1}, G_{B2}]^T$ with $(G_{A1}, G_{A2}, G_{B1}, G_{B2}) \sim \mathcal{CN}(\mathbf{0}, \sigma_J^2/2\ \mathbf{I}_4)$; this assumption is realistic since Mallory can estimate the channel vectors while Alice and Bob exchange pilot signals, as long as the channel's coherence time is respected (a plausible scenario in slow fading, low mobility environments).

To mount the attack, Mallory transmits a signal X_J, suitably precoded as $\mathbf{P}X_J$. The precoding matrix $\mathbf{P} = [P_1, P_2]^T$ is chosen such that the same signal is "injected" at both Alice and Bob, i.e.,

$$\mathbf{G_A}^T \mathbf{P} X_J = \mathbf{G_B}^T \mathbf{P} X_J \Rightarrow P_1 = \frac{G_{B2} - G_{A2}}{G_{A1} - G_{B1}} P_2,$$

where, due to the i.i.d. assumption and to the continuous distribution of the channels, $G_{A1} \neq G_{B1}$ almost surely. As a result, Mallory can select a suitable precoding matrix (among infinite possibilities). Assuming a total power constraint $\mathbb{E}[|\mathbf{P}X_J|^2] \leq \Gamma$ for Mallory's transmission, P_2 should be chosen as

$$P_2 \leq \frac{\sqrt{\Gamma}}{\left|\frac{G_{B2}-G_{A2}}{G_{A1}-G_{B1}}\right|}. \tag{11.12}$$

This shows that it is possible to generalize the injection attack presented in [52], in which an attacker injected a strong signal whenever the RSS in the Mallory–Alice and Mallory–Bob links were similar. More importantly, the presented injection attack accounts not only for the RSS but for the full CSI, i.e., it includes the signal phase.

The observations at Alice and Bob, denoted by Y_A and Y_B, are given as

$$Y_A = XH + W + N_A \tag{11.13}$$

$$Y_B = XH + W + N_B, \tag{11.14}$$

where $W = \mathbf{G_A}^T \mathbf{P} X_J = \mathbf{G_B}^T \mathbf{P} X_J$ denotes the observed injected signal at Alice and Bob which is identical at both due the precoding matrix $\mathbf{P}$; and, N_A, N_B denote zero-mean unit variance i.i.d. complex circularly symmetric Gaussian random noise variables, i.e., $N_A, N_B \sim \mathcal{CN}(0, 1)$. The secret key rate controlled by Mallory is upper bounded by [54]

$$L \leq I(Y_A, Y_B; W). \tag{11.15}$$

Identifying the optimal injection signal W corresponds to finding the capacity achieving input signal of the *two-look Gaussian channel* in (11.13)–(11.14). This signal is

‖It is straightforward to see that the scenario can easily be generalized to a multi-antenna setting in which Mallory has one more antenna than Alice and Bob.

known to be Gaussian [55]; hence, a good choice for X_J is to be constant, so that, the overall injected signal is an optimal complex zero-mean circularly symmetric Gaussian signal, $W \sim \mathcal{CN}(0, \sigma_J^2 \Gamma)$.

A countermeasure to injection attacks can be built by randomizing the pilot sequence exchanged between Alice and Bob [53,54]. Here, we propose to randomize the pilots by drawing them from a (scaled) QPSK modulation, as follows: instead of transmitting the same probing signal X, Alice and Bob transmit independent, random probe signals X_A and X_B, respectively, drawn from i.i.d. zero-mean discrete uniform distributions $\mathcal{U}(\{\pm r \pm jr\})$, where $j = \sqrt{-1}, r = \sqrt{P/2}$, so that, $\mathbb{E}[X_A] = \mathbb{E}[X_B] = 0, \mathbb{E}\left[|X_A|^2\right] = \mathbb{E}\left[|X_B|^2\right] = P$, and $\mathbb{E}[X_A X_B] = 0$, i.e., the pilots are randomly chosen QPSK signals. Alice's observation Y_A is modified accordingly as

$$Y_A = X_B H + W + N_A, \quad Y_B = X_A H + W + N_B. \tag{11.16}$$

To establish shared randomness in spite of the pilot randomization, Alice and Bob post-multiply Y_A and Y_B by their randomized pilots, obtaining local observations $\tilde{Y}_A$ and $\tilde{Y}_B$ (unobservable by Mallory), expressed as:

$$\tilde{Y}_A = X_A Y_A = X_A X_B H + X_A W + X_A N_A, \tag{11.17}$$

$$\tilde{Y}_B = X_B Y_B = X_A X_B H + X_B W + X_B N_B. \tag{11.18}$$

Lemma 1. *The source of shared randomness, when the pilots are randomized QPSK symbols, is a circularly symmetric zero mean Gaussian random variable,* $X_A X_B H \sim C\mathcal{N}(0, P^2 \sigma_H^2)$.

For the proof, refer to [7].

Furthermore, due to the fact that X_A and X_B are independent and have zero mean, the variables $X_A W$ and $X_B W$ are uncorrelated, circularly symmetric zero mean Gaussian random variables, and, therefore independent, while the same holds for $X_A N_A, X_B N_B$, i.e., $(X_A W, X_B W) \sim \mathcal{CN}(\mathbf{0}, \sigma_J^2 P \Gamma \mathbf{I}_2)$ and $(X_A N_A, X_B N_B) \sim \mathcal{CN}(\mathbf{0}, P\mathbf{I}_2)$. Alice and Bob extract the common key from the modified source of common randomness $X_A X_B H$ as opposed to XH. On the other hand, since $X_A W, X_B W, X_A N_A, X_B N_B$ are i.i.d. complex circularly symmetric Gaussian random variables, the proposed scheme reduces injection attacks to uncorrelated jamming attacks, i.e., using Lemma 1 we get that

$$L \leq I\left(\tilde{Y}_A, \tilde{Y}_B; W\right) = 0. \tag{11.19}$$

11.6 Analysis of denial of service attacks on SKG

Building on the results of the previous section, we next examine in detail the scenario in which Mallory acts as a reactive jammer. Reactive jamming is a stealthy jamming approach in which the jammer first senses the spectrum and jams only when she detects an ongoing transmission. Due to the effectiveness and difficulty to be detected, reactive jammers are considered as the most harmful [56,57]. Furthermore, as OFDM is used in many actual systems, in our analysis, we assume a BF-AWGN channel. In this context, we assume that Alice and Bob perform SKG over a BF-AWGN channel with N parallel blocks (referred to as subcarriers for clarity). The notation introduced in the previous section is extended with the introduction of a carrier index $i \in \{1, \ldots, N\}$, i.e., X_{Ai}, X_{Bi} denote the randomized pilots on the ith subcarrier, H_i denotes the channel coefficient in the link Alice–Bob, W_i the signal injected by Mallory on the ith subcarrier and $N_{A,i}, N_{B,i}$ noise variables. As a reactive jammer, Mallory senses the spectrum and jams a specific subcarrier only when the power on it exceeds a certain threshold p_{th}. Two scenarios are considered: (i) when p_{th} is fixed (determined in essence by the carrier sensing capability of Mallory's receiver); (ii) when p_{th} is variable (its choice forms part of her strategy).

We can reformulate the expressions of Alice's and Bob's local observations on the ith SKG subcarrier as follows:

$$\tilde{Y}_{A,i} = X_{Ai}X_{Bi}H_i + X_{Ai}W_i + X_{Ai}N_{A,i} \tag{11.20}$$

$$\tilde{Y}_{B,i} = X_{Ai}X_{Bi}H_i + X_{Bi}W_i + X_{Bi}N_{B,i} \tag{11.21}$$

for $i = 1, \ldots, N$ with $H_i \sim \mathcal{CN}(0, \sigma_H^2)$, $W_i \sim \mathcal{CN}(0, \sigma_J^2 \gamma_i)$, $N_{A,i} \sim \mathcal{CN}(0, 1)$, $N_{B,i} \sim \mathcal{CN}(0, 1)$. In this work, we assume that Alice and Bob use the same power p on all pilots, in agreement with common practice during the advantage distillation phase. Based on this assumption, we have that $\mathbb{E}[|X_{Ai}|^2] = \mathbb{E}[|X_{Bi}|^2] = p$ with $p \in [0, P]$.

At the same time, we let Mallory choose the power allocation vector to maximize the impact of her attack. The power Mallory uses on the ith subcarrier is denoted by γ_i, so that $\mathbb{E}[|W_i|^2] = \sigma_J^2\gamma_i$. Denoting the average available power for jamming by Γ and the power allocation of the jammer by $\underline{\gamma} = (\gamma_1, \ldots, \gamma_N)$, we assume the following short-term power constraint:

$$\underline{\gamma} \in \mathbb{R}_+^N, \quad \sum_{i=1}^{N} \gamma_i \leq N\Gamma. \tag{11.22}$$

Assuming that H_i is uncorrelated with $H_{A,i}, H_{B,i}$, $i = 1, \ldots, N$ and that the pilot randomization approach proposed in the previous section is employed, the SKG rate $R(p, \gamma_i) = I\left(\tilde{Z}_{A,i}; \tilde{Z}_{B,i}\right)$ on the ith subcarrier, can be expressed as a function of p and $\gamma_i, i = 1, \ldots, N$ as [58]:

$$R(p, \gamma_i) = \log_2\left(1 + \frac{p\sigma_H^2}{2(1 + \gamma_i\sigma_J^2) + \frac{(1+\gamma_i\sigma_J^2)^2}{p\sigma_H^2}}\right). \tag{11.23}$$

Note that the rate in (11.23) is independent of the instantaneous realizations of the fading coefficients; instead, the variations of the channel gains expressed through the variances σ_H^2, σ_J^2 determine the rate of the secret keys that can be extracted from the wireless medium. The overall SKG sum-rate can then be simply expressed as follows:

$$C_K(p, \underline{\gamma}) = \sum_{i=1}^{N} R(p, \gamma_i). \tag{11.24}$$

11.6.1 Optimal power allocation strategies

Alice and Bob's common objective is to maximize $C_K(p, \underline{\gamma})$ with respect to (w.r.t.) p, while Mallory wants to minimize $C_K(p, \underline{\gamma})$ w.r.t. $\underline{\gamma}$. Given the opposed objectives, a non-cooperative zero-sum game can be formulated to study the strategic interaction between the legitimate users and the jammer: $\mathcal{G} = (\{L, J\}, \{\mathcal{A}_L, \mathcal{A}_J(p)\}, C_K(p, \underline{\gamma}))$. The game $\mathcal{G}$ has three components. First, there are two players: player L representing the legitimate users (Alice and Bob are considered to act as a single player) and player J representing the jammer (Mallory). Second, player L has a set of possible actions $\mathcal{A}_L = [0, P]$ while player J's set of actions is

$$\mathcal{A}_J(p) = \begin{cases} \{(0, \ldots, 0)\}, & \text{if } p \le p_{\text{th}}, \\ \left\{\underline{\gamma} \in \mathbb{R}_+^N \mid \sum_{i=1}^N \gamma_i \le N\Gamma\right\}, & \text{if } p > p_{\text{th}}. \end{cases} \tag{11.25}$$

At last, $C_K(p, \underline{\gamma})$ denotes the payoff function of player L.

Due to the fact that Mallory first observes the transmit power of the legitimate users on the subcarriers and then decides which strategy to choose (a consequence of player J being a reactive jammer), we study a hierarchical game in which player L is the leader and player J is the follower. In this hierarchical game, the solution is the Stackelberg equilibrium (SE)—rather than Nash—defined as a strategy profile $(p^{\text{SE}}, \underline{\gamma}^{\text{SE}})$ where player L chooses his optimal strategy first, by anticipating the strategic reaction of player J (i.e., its best response). This can be rigorously written as:

$$p^{\text{SE}} \triangleq \arg\max_{p \in \mathcal{A}_L} \sum_{i=1}^N R(p, \underline{\gamma}^*(p)), \text{ and } \underline{\gamma}^{\text{SE}} \triangleq \underline{\gamma}^*(p^{\text{SE}}), \tag{11.26}$$

where $\underline{\gamma}^*(p)$ denotes the jammer's best response (BR) function to any strategy $p \in \mathcal{A}_L$ chosen by player L, defined as follows:

$$\underline{\gamma}^*(p) \triangleq \arg\min_{\underline{\gamma} \in \mathcal{A}_J(p)} \sum_{i=1}^N R(p, \underline{\gamma}). \tag{11.27}$$

We also denote by $\gamma_i^*(p)$ the ith component of $\underline{\gamma}^*(p)$.

11.6.2 Stackelberg equilibrium with fixed sensing threshold p_{th}

In the following, we evaluate the SE of the game $\mathcal{G}$ assuming that the threshold p_{th} is predefined and fixed, e.g., determined by Mallory's equipment quality. The case $P \le p_{\text{th}}$ is trivial as $\underline{\gamma}^{\text{SE}} = (0, \ldots, 0)$, whereas the legitimate users will optimally use the maximum available power so that $(p^{\text{SE}} = P)$. Indeed, because of the badly chosen threshold or low sensing capabilities of Mallory, the legitimate transmission will never be detected on any of the subcarriers and hence will not be jammed. In the following, we assume that: $P > p_{\text{th}}$.

Lemma 2. *The BR of the jammer for any $p \in \mathcal{A}_L$ chosen by the leader defined in (11.27) is the uniform power allocation, such that:*

$$\underline{\gamma}^*(p) \triangleq \begin{cases} (\Gamma, \ldots, \Gamma), & \text{if } p > p_{th}, \\ (0, \ldots, 0), & \text{if } p \le p_{th}. \end{cases} \tag{11.28}$$

For the proof, refer to [7].

In light of this result, the SKG sum rate can have two forms:

$$C_K(p, \underline{\gamma}^*(p)) = \begin{cases} NR(p, \Gamma), & \text{if } p > p_{\text{th}}, \\ NR(p, 0), & \text{if } p \le p_{\text{th}}, \end{cases} \tag{11.29}$$

which simplifies the players' options. Next, we address the question of how Alice and Bob should choose their power p optimally by considering the actions available to the players in the game at the key points i.e. at P and p_{th}.

Theorem 1. *Depending on the available power P for W-SKG, player L will either transmit at P or p_{th} on all subcarriers. The SE point of the game is unique when $P \neq p_{th}(\sigma_J^2\Gamma + 1)$ and is given by*

$$(p^{SE}, \underline{\gamma}^{SE}) = \begin{cases} \{(p_{th}, (0, \ldots, 0))\}, & \text{if } P < p_{th}(\sigma_J^2\Gamma + 1), \\ \{(P, (\Gamma, \ldots, \Gamma))\}, & \text{if } P > p_{th}(\sigma_J^2\Gamma + 1). \end{cases} \tag{11.30}$$

When $P = p_{th}(\sigma_J^2\Gamma + 1)$, the game $\mathcal{G}$ has two SEs:
$(p^{SE}, \underline{\gamma}^{SE}) \in \{(p_{th}, (0, \ldots, 0)), (P, (\Gamma, \ldots, \Gamma))\}$.

For the proof, refer to [7].

Some numerical results are presented in Figure 11.10 for a total number of SKG subcarriers $N = 10$ (pertinent to narrowband IoT applications), $p_{\text{th}} = 2$, $\Gamma = 4$, and $\sigma^2 = \sigma_J^2 = 1$. The top figure compares the achievable rates of the SE strategy and of two alternative strategies consisting in transmitting with fixed $p = P$ or $p = p_{\text{th}}$. The bottom figure depicts the following quantities:

$$F = \frac{C_K(p^{\text{SE}}, \underline{\gamma}^{\text{SE}}) - C_K(P, (\Gamma, \ldots, \Gamma))}{C_K^{SE}}, \tag{11.31}$$

$$D = \frac{C_K(p^{\text{SE}}, \underline{\gamma}^{\text{SE}}) - C_K(p_{\text{th}}, (0, \ldots, 0))}{C_K^{SE}}, \tag{11.32}$$

where F and D represent the jammer's gain (or legitimate users' loss) if player L deviates from the SE point (indeed, if player L transmits at $P > p_{\text{th}}$, the jammer will jam at $\gamma_i^*(P) = \Gamma$; and if player L transmits at p_{th} the jammer will not detect it and will remain silent). Both figures show that deviating from the SE point can decrease the achievable sum-rates by up to 85%.

11.7 Novel hybrid SKG-cryptosystems

Apart from MiM and DoS attacks at the PHY, it is important to investigate robustness of hybrid SKG-cryptosystems against tampering, which is discussed in Section 11.7.1, through a novel hybrid SKG-crypto authenticated encryption (AE) scheme, depicted in Figure 11.11. Furthermore, in Section 11.7.2, we propose the replacement of Diffie Hellman by SKG for the generation of resumption keys in a zero-round-trip-time (0-RTT) resumption protocol for low latency applications.

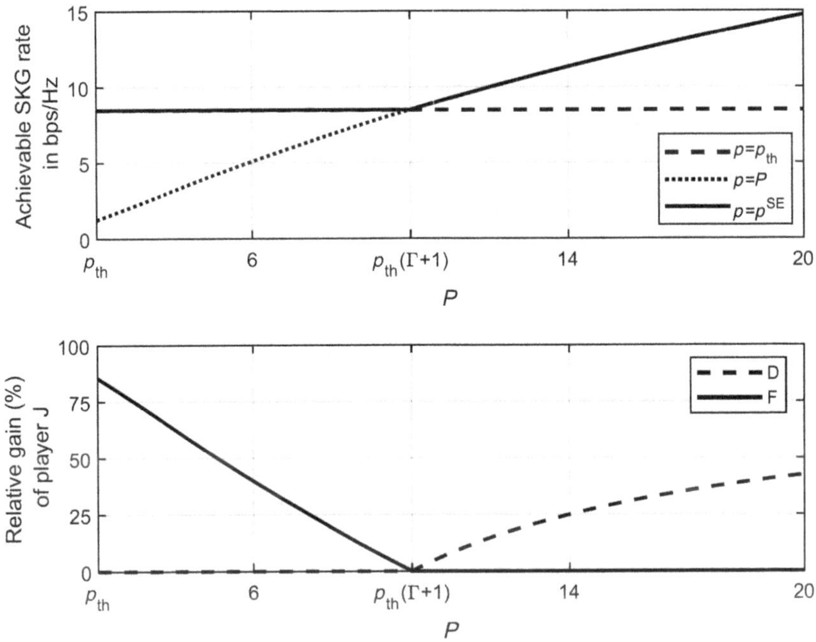

Figure 11.10 *UP: SE policy compared to always transmitting with either full power or with p_{th}. DOWN: functions D and F versus P. In both sub-figures, $p_{th} = 2$, $\Gamma = 4$, $N = 10$, $\sigma_H^2 = \sigma_J^2 = 1$.*

11.7.1 Authenticated encryption using SKG

To develop a hybrid crypto system that can withstand tampering attacks, SKG can be introduced in standard AE schemes in conjunction with standard block ciphers in counter mode (to reduce latency), e.g., AES GCM. As a sketch of such a primitive, let us assume a system with three parties: Alice who wishes to transmit a secret message **m** to Bob with confidentiality and integrity, and Eve, that can act as a passive and active attacker. The following algorithms are employed:

- The SKG scheme denoted by $\mathsf{G} : \mathscr{H} \to \mathscr{K} \times \mathscr{S}$, accepting as inputs N-dimensional vectors of complex numbers (the fading coefficients), and generating as outputs N binary vectors of sizes n and $n - k$, respectively, $n, k \in \mathbb{N}$, (in the key and the syndrome spaces), i.e.,

$$\mathsf{G}(\mathbf{H}) = (\mathbf{K}, \mathbf{S}_A), \tag{11.33}$$

where $\mathbf{K} \in \mathscr{K}$ denotes the key obtained from $\mathbf{H}$ after privacy amplification and $\mathbf{S}_A = [\mathbf{S}_{A,1} \| \ldots \| \mathbf{S}_{A,N}] \in \mathscr{S}$ is the concatenation of Alice's syndromes.

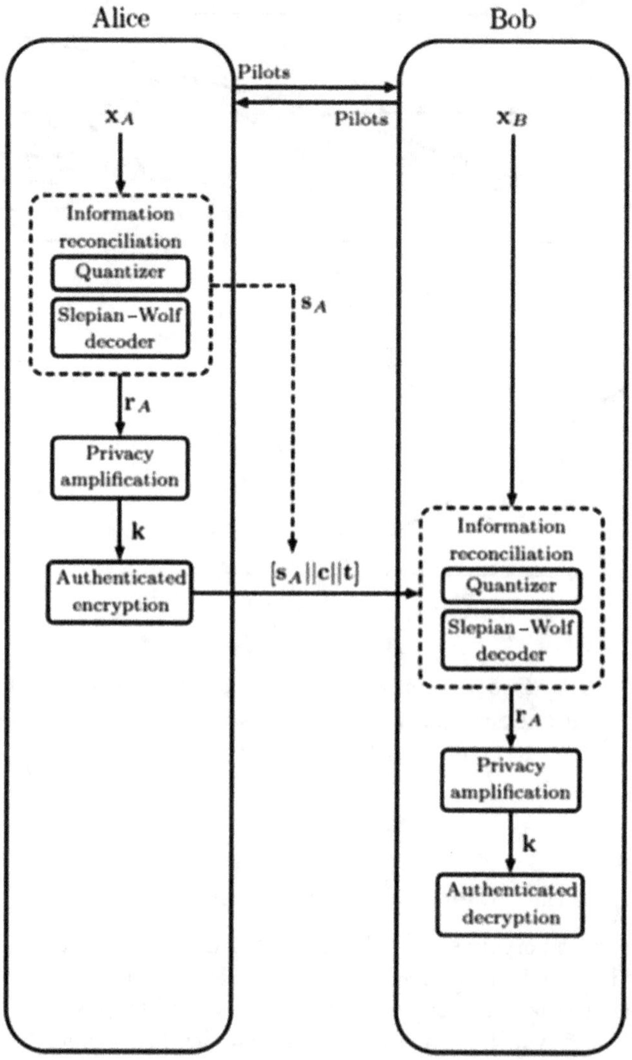

Figure 11.11 Hybrid SKG AE scheme

- A symmetric encryption algorithm, e.g., AES GCM, denoted by $\texttt{Es} : \mathscr{K} \times \mathscr{M} \to \mathscr{C}$ where $\mathscr{C}$ denotes the ciphertext space with the corresponding decryption $\texttt{Ds} : \mathscr{K} \times \mathscr{C} \to \mathscr{M}$, such that

$$\texttt{Es}(\mathbf{K}, \mathbf{m}) = \mathbf{c}, \tag{11.34}$$

$$\texttt{Ds}(\mathbf{K}, \mathbf{c}) = \mathbf{m}, \tag{11.35}$$

for $\mathbf{K} \in \mathscr{K}$, $\mathbf{m} \in \mathscr{M}$, $\mathbf{c} \in \mathscr{C}$.

- A pair of message authentication code (MAC) algorithms, e.g., in HMAC mode, denoted by $\texttt{Sign}: \mathscr{K} \times \mathscr{M} \rightarrow \mathscr{T}$, with a corresponding verification algorithm $\texttt{Ver}: \mathscr{K} \times \mathscr{M} \times \mathscr{T} \rightarrow (yes, no)$, such that

$$\texttt{Sign}(\mathbf{K}, \mathbf{m}) = \mathbf{t}, \tag{11.36}$$

$$\texttt{Ver}(\mathbf{K}, \mathbf{m}, \mathbf{t}) = \begin{cases} \boldsymbol{yes}, & \text{if integrity verified} \\ \boldsymbol{no}, & \text{if integrity not verified} \end{cases} \tag{11.37}$$

A hybrid crypto-PLS system for AE SKG can be built as follows:

1. The SKG procedure is launched between Alice and Bob generating a key and a syndrome $\texttt{G}(\mathbf{H}) = (\mathbf{K}, \mathbf{S}_A)$.
2. Alice breaks her key into two parts $\mathbf{K} = \{\mathbf{K}_e, \mathbf{K}_i\}$ and uses the first to encrypt the message as $\mathbf{c} = \texttt{Es}(\mathbf{K}_e, \mathbf{m})$. Subsequently, using the second part of the key, she signs the ciphertext using the signing algorithm $\mathbf{t} = \texttt{Sign}(\mathbf{K}_i, \mathbf{c})$ and transmits to Bob the extended ciphertext $[\mathbf{S}_A \| \mathbf{c} \| \mathbf{t}]$.
3. Bob checks first the integrity of the received ciphertext as follows: from $\mathbf{S}_A$ and his own observation he evaluates $\mathbf{K} = \{\mathbf{K}_e, \mathbf{K}_i\}$ and computes $\texttt{Ver}(\mathbf{K}_i, \mathbf{c}, \mathbf{t})$. The integrity test will fail if any part of the extended ciphertext was modified, including the syndrome (that is sent as plaintext); for example, if the syndrome was modified during the transmission, then Bob would not have evaluated the correct key and the integrity test would have failed.
4. If the integrity test is successful then Bob decrypts $\mathbf{m} = \texttt{Ds}(\mathbf{K}_e, \mathbf{c})$.

11.7.2 SKG for zero-round-trip time protocols (0-RTT)

Typically, authentication of session keys is required at the start of each communication cycle and at each key renegotiation, inducing non-negligible latency due to the related handshakes. Accounting for recent advances in terms of resumption protocols, we present a solution that is inspired by the 0-RTT authentication mode introduced in TLS version 1.3 version [59]. The use of 0-RTT obviates the need for performing a handshake for every re-authentication through the use of a resumption secret $\mathbf{R}_s$, thus reducing latency. Another strong motivation for using this mechanism is that it is forward secure, i.e., compromising the system at one instance does not carry over to future instances [60]. We first briefly describe the TLS 0-RTT mechanism before describing a similarly inspired 0-RTT mechanism applied to the information reconciliation phase of the standard SKG scheme.

The TLS 1.3 0-RTT handshake works as follows: in the very first connection between client and server, a regular TLS handshake is used. During this step the server sends to the client a look-up identifier $\mathbf{K}_l$ for a corresponding entry in session caches or it sends a session ticket. Then both parties derive a resumption secret $\mathbf{R}_s$ using their shared key and the parameters of the session. Finally, the client stores the resumption secret $\mathbf{R}_s$ and uses it when reconnecting to the same server which also retrieves it during the re-connection.

If session tickets are used the server encrypts, the resumption secret using long-term symmetric encryption key, called a session ticket encryption key (STEK), resulting in a session ticket. The session ticket is then stored by the client and included in subsequent connections, allowing the server to retrieve the resumption secret. Using this approach the same STEK is used for many sessions and clients. On the one hand, this property highly reduces the required storage of the server, however, on the other hand, it makes it vulnerable to replay attacks and not forward secure. Due to these vulnerabilities, in this work, we focus on the session cache mechanism described next.

When using session caches, the server stores all resumption secrets and issues a unique look-up identifier $\mathbf{K}_l$ for each client. When a client tries to reconnect to that server it includes its look-up identifier $\mathbf{K}_l$ in the 0-RTT message, which allows the server to retrieve the resumption secret $\mathbf{R}_s$. Storing a unique resumption secret $\mathbf{R}_s$ for each client requires server storage for each client but it provides forward security and resilience against replay attacks, when combined with a key generation mechanisms such as Diffie–Hellman (or the SKG used in this chapter) which are important goals for security protocols [60]. In the proposed physical layer 0-RTT, given that a node identifier state would be required for link-layer purposes, the session cache places little comparative load and thus is the mechanism proposed here for (re-)authentication.

The physical layer resumption protocol modifies the information reconciliation phase following initial authentication to provide a re-authentication mechanism between Alice and Bob. At the first establishment of communication we assume initial authentication is established. During that, Alice sends to Bob a look-up identifier $\mathbf{K}_l$. Then, both derive a resumption secret $\mathbf{R}_s$ that is identified by $\mathbf{K}_l$. Note, $\mathbf{R}_s$ and the session key have the same length k. Then the following steps are executed:

1. Advantage distillation is carried out as discussed in Sections 11.2 and 11.4, during which both parties obtain observation vectors $\mathbf{r}_A$ and $\mathbf{r}_B$ (channel indices are dropped here for simplicity).
2. During the information reconciliation phase both Alice and Bob exclusive-or the resumption secret $\mathbf{R}_s$ with their observations $\mathbf{r}_A$ and $\mathbf{r}_B$, obtaining syndromes $\mathbf{S}'_A$ and $\mathbf{S}'_B$ with which both parties can carry out reconciliation to obtain the same shared value which is now $\mathbf{c} \oplus \mathbf{R}_s$.
3. The privacy amplification step is carried out, but now on the key space $\mathbf{c} \oplus \mathbf{R}_s$ to produce the final shared key $\mathbf{K}'$ that is a result of both the shared wireless randomness and the resumption secret.

Note that the key $\mathbf{K}'$ can only be obtained if both the physical layer generated key and the resumption key are valid and this method can be shown to be forward secure [60].

11.8 Concluding remarks and future perspectives

In this chapter, we have considered various practical aspects that should be taken into account when designing SKG systems for real applications. Starting from a short blocklength performance comparison, we have observed that although list decoding for LDPC and BCH codes improves performance at very short blocklengths

($\sim$100 bits), the improvement diminishes with increasing blocklength. On the other hand, we have observed that polar codes with list decoding provide a significant improvement for medium size blocklength as well as short blocklength.

With respect to robustness against eavesdropping attacks, we showcased how dimensionality reduction can be a generic tool for disentangling predictable from unpredictable components of observed CSI vectors, in order to perform SKG resilient to eavesdropping. We discussed in detail the trade-off between correlations and dependencies at different locations (that can be generalized to include time, frequency and antenna domains) and reciprocity, or, the lack of mismatch between the uplink and downlink for the unpredictable components used for SKG. Overall, PCA has proven to be a sufficiently straightforward approach for the disentanglement of large from small-scale fading terms in observed CSI and can be incorporated as a pre-processing tool to allow for the simultaneous use of CSI for key distillation. Alternative approaches can be devised in the time domain, e.g., using Kalman filters [8].

In terms of resilience to active attacks, we have demonstrated that an active attacker can launch an injection attack as long as it possesses on extra antenna than the legitimate users. This limitation can be overcome with the introduction of randomization techniques, here in the form of pilot randomization, reducing an injection attack to a less harmful jamming attack. Furthermore, reactive jamming attacks were analyzed in SKG systems and optimal power allocation policies were investigated in BF-AWGN channels. It was shown that an intelligent reactive jammer should optimally jam with equal power on the whole spectrum. Countermeasures against jamming attacks were not reviewed in the present chapter but can include standard techniques based on frequency hopping or more advanced approaches, e.g., using energy harvesting, both investigated in [58].

Finally, we have presented novel hybrid SKG-cryptosystems that provide robustness against tampering attacks by the use of AE and also discussed the possible incorporation of SKG in 0-RTT protocols for faster authentication leveraging resumption secret keys. This discussion showcases the emergence of SKG as a mature technology that can command the confidence of the scientific community, the industry, and end users. Numerous potential applications are conceivable.

Looking at the future, we note that some of the recent, increasingly sophisticated attacks on the wireless edge, e.g., jamming or false base stations, can be implemented with a price tag as low as 1k$ using low-cost software-defined radios. Importantly, in emerging cyberphysical systems, static security controls based on cryptography can only provide answers on one facet for multi-faced trust and trustworthiness evaluation. In the evolution toward truly autonomous cyberphysical systems, new means of reaching situational awareness by harvesting and interpreting the “context” of the communication and actuation will be reached [5]. Incorporating context awareness in security controls is projected to allow handling more efficiently aspects related to identifying the risk or threat level and the required security level, particularly for applications such as autonomous vehicles, platooning, and eHealth.

In this framework, incorporating security controls from the palette of physical layer security (PLS) can be particularly attractive due to their low computational complexity (relevant implementations are based on standard encoders) and their inherent ability to adapt to the transmission medium properties. Notably, the incorporation of SKG leveraging enhanced context awareness, e.g., with respect to the transmission conditions and the presence of adversaries, can be particularly attractive for massive connectivity regimes and aggressive latency constraints.

References

[1] Y. Zou, J. Zhu, X. Wang, and L. Hanzo, "A survey on wireless security: Technical challenges, recent advances, and future trends," *Proc. IEEE*, vol. 104, no. 9, pp. 1727–1765, 2016.

[2] V. Mavroeidis, K. Vishi, M. D. Zych, and A. Jøsang, "The impact of quantum computing on present cryptography," *Int. J. Adv. Comput. Sci. Appl.*, vol. 9, no. 3, pp. 405–414, 2018.

[3] L. Chen, L. Chen, S. Jordan, *et al.*, *Report on Post-Quantum Cryptography*. US Department of Commerce, National Institute of Standards and Technology, 2016, vol. 12.

[4] IEEE INGR Security, "Security and Privacy, International Network Generations Roadmap (INGR) – 2021 Edition."

[5] A. Chorti, A. N. Barreto, S. Kopsel, *et al.*, "Context-aware security for 6G wireless, the role of physical layer security," *IEEE Communications Standards Magazine*, vol. Special Issue on Emerging Security Technologies for 6G Security and Privacy.

[6] A. N. Barreto, S. Köpsell, A. Chorti, *et al.*, "Towards intelligent context-aware 6G security," *arXiv:2112.09411*.

[7] M. Mitev, A. Chorti, E. V. Belmega, and M. Reed, "Man-in-the-middle and denial of service attacks in wireless secret key generation," in *2019 IEEE Global Communications Conference (GLOBECOM)*, 2019, pp. 1–6.

[8] M. Mitev, M. Shakiba-Herfeh, A. Chorti, M. Reed, and S. Baghaee, "A physical layer, zero-round-trip-time, multifactor authentication protocol," *IEEE Access*, vol. 10, pp. 74555–74571, 2022.

[9] M. Srinivasan, S. Skaperas, M. Mitev, *et al.*, "Smart channel state information pre-processing for authentication and symmetric key distillation," *IEEE Trans. Mach. Learn. Commun. Netw.*, vol. early access, pp. 1–1, 2023.

[10] M. Shakiba-Herfeh and A. Chorti, "Comparison of short blocklength Slepian–Wolf coding for key reconciliation," in *2021 IEEE Statistical Signal Processing Workshop (SSP)*, 2021, pp. 111–115.

[11] M. Mitev, A. Chorti, M. Reed, and L. Musavian, "Authenticated secret key generation in delay-constrained wireless systems," *Eurasip J. Wireless Commun. Netw.*, vol. 122, pp. 1–29, 2020.

[12] D. Boneh and V. Shoup, *A Graduate Course in Applied Cryptography*, Stanford, 2017.

[13] C. E. Shannon, *A Mathematical Theory of Cryptography*, September 1, 1945.

[14] K. Nyberg, "Perfect nonlinear s-boxes," in *Advances in Cryptology – EUROCRYPT '91*, Brighton, UK, 1991, p. 378–386.

[15] A. V. M. Bellare, S. Tessaro, "Semantic security for the wiretap channel," in *CRYPTO*. Springer, 2012.

[16] W. Yang, R. F. Schaefer, and H. V. Poor, "Wiretap channels: nonasymptotic fundamental limits," *IEEE Trans. Inf. Theory*, vol. 65, no. 7, pp. 4069–4093, 2019.

[17] Z. Utkovski, M. Frey, P. Agostini, I. Bjelakovic, and S. Stanczak, "Semantic security based secrecy maps for vehicular communications," in *WSA 2021; 25th International ITG Workshop on Smart Antennas*, 2021, pp. 1–5.

[18] M. Shakiba-Herfeh, L. Luzzi, and A. Chorti, "Finite blocklength secrecy analysis of Polar and Reed-Muller codes in BEC semi-deterministic wiretap channels," in *2021 IEEE Information Theory Workshop (ITW)*, 2021, pp. 1–6.

[19] H. Jeon, N. Kim, J. Choi, H. Lee, and J. Ha, "Bounds on secrecy capacity over correlated ergodic fading channels at high SNR," *IEEE Trans. Inf. Theory*, vol. 57, no. 4, pp. 1975–1983, 2011.

[20] J. Si, Z. Li, J. Cheng, and C. Zhong, "Secrecy performance of multi-antenna wiretap channels with diversity combining over correlated Rayleigh fading channels," *IEEE Trans. Wirel. Commun.*, vol. 18, no. 1, pp. 444–458, 2019.

[21] M. L. Akter, M. I. Rumman, M. K. Kundu, M. Z. I. Sarkar, and A. S. M. Badrudduza, "Security analysis in correlated Nakagami-m fading channels with diversity combining," in *2019 3rd International Conference on Electrical, Computer Telecommunication Engineering (ICECTE)*, 2019, pp. 69–72.

[22] J. Nötzel, M. Wiese, and H. Boche, "The arbitrarily varying wiretap channel—secret randomness, stability, and super-activation," *IEEE Trans. Inf. Theory*, vol. 62, no. 6, pp. 3504–3531, 2016.

[23] A. S. Mansour, H. Boche, and R. F. Schaefer, "List decoding for arbitrarily varying wiretap channels," in *2016 IEEE Conference on Communications and Network Security (CNS)*, 2016, pp. 611–615.

[24] S. Ju and T. S. Rappaport, "Sub-terahertz spatial statistical mimo channel model for urban microcells at 142 GHz," in *2021 IEEE Global Communications Conference (GLOBECOM)*, 2021, pp. 1–6.

[25] Y. Xing, T. S. Rappaport, and A. Ghosh, "Millimeter wave and sub-thz indoor radio propagation channel measurements, models, and comparisons in an office environment," *IEEE Commun. Lett.*, vol. 25, no. 10, pp. 3151–3155, 2021.

[26] W. Xia, S. Rangan, M. Mezzavilla, A. Lozano, G. Geraci, V. Semkin, and G. Loianno, "Millimeter wave channel modeling via generative neural networks," in *2020 IEEE Globecom Workshops (GC Wkshps)*, 2020, pp. 1–6.

[27] D. McAllester and K. Stratos, "Formal limitations on the measurement of mutual information," in *Proceedings of the Twenty-Third International Conference on Artificial Intelligence and Statistics*, ser. Proceedings of Machine Learning Research, S. Chiappa and R. Calandra, Eds., vol. 108. PMLR, 26–28 Aug 2020, pp. 875–884.

[28] P. Walther and T. Strufe, "Blind twins: Siamese networks for non-interactive information reconciliation," in *2020 IEEE 31st Annual International*

Symposium on Personal, Indoor and Mobile Radio Communications, 2020, pp. 1–7.

[29] J. Ma, R. Shrestha, J. Adelberg, and *et al.*, "Security and eavesdropping in terahertz wireless links," *Nature*, vol. 563, pp. 89–93, 2018.

[30] E. Jorswieck, S. Tomasin, and A. Sezgin, "Broadcasting into the uncertainty: Authentication and confidentiality by physical-layer processing," *Proc. IEEE*, vol. 103, no. 10, pp. 1702–1724, 2015.

[31] W. Njima, M. Chafii, A. Chorti, R. M. Shubair, and H. V. Poor, "Indoor localization using data augmentation via selective generative adversarial networks," *IEEE Access*, vol. 9, pp. 98337–98347, 2021.

[32] RFC 8446, "The transport layer security (TLS) protocol version 1.3," 2018.

[33] U. Maurer, "Secret key agreement by public discussion based on common information," *IEEE Trans. Inf. Theory*, vol. 39, no. 5, pp. 733–742, 1993.

[34] C. Bennett, G. Brassard, C. Crépeau, and U. Maurer, "Generalized privacy amplification," *IEEE Trans. Inf. Theory*, vol. 50, no. 2, pp. 394–400, 1995.

[35] A. Mukherjee, S.A.A., Fakoorian, H. Jing, and A. Swindlehurst, "Principles of physical layer security in multiuser wireless networks: a survey," *IEEE Commun. Surv. Tuts.*, vol. 16, no. 3, pp. 1550–1573, Third Quarter 2014.

[36] C. Bennet, G. Brassard, and J. Robert, "Privacy amplification by public discussion," *J. Comput.*, vol. 17, no. 2, pp. 210–229, 1998.

[37] G. Brassard and L. Salvail, "Secret-key reconciliation by public discussion," in *Advances in Cryptology – EUROCRYPT*, 1994, pp. 410–423.

[38] C. Ye, S. Mathur, A. Reznik, Y. Shah, W. Trappe, and N. Mandayam, "Information-theoretically secret key generation for fading wireless channels," *IEEE Trans. Inf. Forens. Secur.*, vol. 5, no. 2, pp. 240–154, 2010.

[39] R. Ahlswede and I. Csiszár, "Common randomness in information theory and cryptography – Part I: secret sharing," *IEEE Trans. Inf. Theory*, vol. 39, no. 7, pp. 1121–1132, 1993.

[40] C. Ye, A. Reznik, and Y. Shah, "Extracting secrecy from jointly Gaussian random variables," in *Proceedings of the International Symposium on Information Theory (ISIT)*, Seatle, US, Jul. 2006, pp. 2593–2597.

[41] T.-H. Chou, S. Draper, and A. M. Sayeed, "Key generation using external source excitation: Capacity, reliability and secrecy exponent," *IEEE Trans. Inf. Theory*, vol. 58, no. 4, pp. 2455–2474, 2012.

[42] W. Zhou and M. Lentmaier, "Improving short-length LDPC codes with a CRC and iterative ordered statistic decoding: (invited paper)," in *53rd Annual Conference on Information Sciences and Systems (CISS)*, 2019, pp. 1–6.

[43] I. Tal and A. Vardy, "List decoding of polar codes," *IEEE Trans. Inf. Theory*, vol. 61, no. 5, pp. 2213–2226, 2015.

[44] M. P. C. Fossorier and Shu Lin, "Soft-decision decoding of linear block codes based on ordered statistics," *IEEE Trans. Inf. Theory*, vol. 41, no. 5, pp. 1379–1396, 1995.

[45] H. Tyagi and S. Watanabe, "Converses for secret key agreement and secure computing," *IEEE Trans. Inf. Theory*, vol. 61, no. 9, pp. 4809–4827, 2015.

[46] M. Shakiba-Herfeh, A. K. Tanc, and T. M. Duman, "LDPC code design for fading interference channels," *IEEE Trans. Veh. Technol.*, vol. 68, no. 3, pp. 2374–2385, 2019.

[47] H. Hentilä *et al.*, "On polar coding for finite blocklength secret key generation over wireless channels," in *Proceedings of the IEEE International Conference on Acoustics, Speech and Signal Processing (ICASSP)*, 2020, pp. 5265–5269.

[48] A. Rock, "Pseudorandom number generators for cryptographic applications," Diplomarbeit zur Erlangung des Magistergrades an der Naturwissenschaftlichen Fakult at der Paris-Lodron-Universit at Salzburg, 2005.

[49] N. Pfister, B. Buhlmann, and J. P. Scholkopf, "Kernel-based tests for joint independence," *J. R. Stat. Soc.: Ser. B (Stat. Methodol.)*, vol. 80, no. 1, pp. 5–31, 2018.

[50] M. K. Shehzad, L. Rose, S. Wesemann, and M. Assaad, "ML-based massive MIMO channel prediction: Does it work on real-world data?" *IEEE Wirel. Commun. Lett.*, vol. 11, no. 4, pp. 811–815, 2022.

[51] S. Jana, S. N. Premnath, M. Clark, S. K. Kasera, N. Patwari, and S. V. Krishnamurthy, "On the effectiveness of secret key extraction from wireless signal strength in real environments," in *Proceedings of the 15th Annual International Conference on Mobile Computing Network.* ACM, 2009, pp. 321–332.

[52] S. Eberz, M. Strohmeier, M. Wilhelm, and I. Martinovic, "A practical man-in-the-middle attack on signal-based key generation protocols," in *Computer Security – ESORICS 2012*, ser. Lecture Notes in Computer Science, 2012, vol. 7459, pp. 235–252.

[53] J. Rong and Z. Kai, "Physical layer key agreement under signal injection attacks," in *IEEE Conference on CNS*, 2015, pp. 254–262.

[54] A. Chorti, "A study of injection and jamming attacks in wireless secret sharing systems," *Lecture Notes on Electrical Engineering*, Springer, pp. 1–14, Jan. 2018.

[55] T. M. Cover and J. A. Thomas, *Elements of Information Theory*. Hoboken, NJ: John Wiley and Sons, Inc., 2006.

[56] S. Fang, Y. Liu, and P. Ning , "Wireless communications under broadband reactive jamming attacks," *IEEE Trans. Depend. Sec. Comput.*, vol. 13, no. 3, pp. 394–408, 2016.

[57] M. Spuhler, D. Giustiniano, V. Lenders, M. Wilhelm, and J. B. Schmitt, "Detection of reactive jamming in DSSS-based wireless communications," in *IEEE Transactions on Wireless Communications*, vol. 13, no. 3, pp. 1593–1603, 2014.

[58] E. V. Belmega and A. Chorti, "Protecting secret key generation systems against jamming: energy harvesting and channel hopping approaches," *IEEE Trans. Inf. Forens. Secur.*, vol. 12, no. 11, pp. 2611–2626, 2017.

[59] "The Transport Layer Security (TLS) Protocol Version 1.3. RFC 8446 (2018)," Rescorla, E., available online https://rfc-editor.org/rfc/rfc8446.txt.

[60] N. Aviram, K. Gellert, and T. Jager, "Session resumption protocols and efficient forward security for TLS 1.3 0-RTT," Cryptology ePrint Archive, Report 2019/228, 2019.

Chapter 12

Physical-layer key generation for multi-user massive MIMO wireless communications*

Chen Sun[1], *Guyue Li*[2], *Hongyi Luo*[2] *and Aiqun Hu*[3]

This chapter presents a multi-user secret key generation in massive multiple-input multiple-output (MIMO) wireless networks. Physical-layer key generation (PKG) based on channel reciprocity has emerged as a new technique to establish secret keys between devices. However, in future wireless communications, massive antennas will be employed to support multiple users, which bring new challenges to PKG, e.g., low key agreements caused by long-time delay or strong inter-user interference. By harnessing the sparse characteristics of the beam domain channel model, we propose a general channel dimensionality reduction (CDR)-based secret key generation scheme and optimize the parameters of this scheme. The security of this key generation scheme under different user cases of non-overlapping and overlapping beams is then discussed. Some new directions that generalize and improve the basic scenarios for PKG are described. These new directions include multi-cell multi-user secret key generation, PKG in millimeter wave wireless communications, and PKG in frequency division duplexing systems.

12.1 Introduction

The past decade has experienced a tremendous growth in mobile user terminals, as well as the transmission data rate. Various wireless communication applications have come into our lives, followed by serious threats to data security. Due to the openness of wireless communications, transmission signals can be received by all user terminals, leading to the risk of information leakage. Data security is traditionally guaranteed by high-layer encryption, and its critical issue lies in the secret key sharing among users. Unfortunately, existing key distribution methods based on cryptography techniques face challenges in future wireless networks [3,4]. First, with the tremendous growth of wireless devices, the key distribution among massive devices becomes complicated

*This chapter has been a part of [1,2].

[1]National Mobile Communications Research Laboratory, Southeast University, China
[2]School of Cyber Science and Engineering, Southeast University, China
[3]School of Information Science and Engineering, Southeast University, China

and even impractical for networks such as wireless sensor networks and mobile ad hoc networks. Second, the distributed key usually does not update for a long time, which may be compromised by emerging quantum computers in the future.

Physical-layer key generation (PKG) has provided a new paradigm for information-theoretic secure key sharing. The theoretical research of Maurer showed that two communication users are possible to generate a pair of symmetric keys with a positive secret key rate, so long as they share a correlated source [5]. Fortunately, the wireless channel itself was found as a good correlated source due to its reciprocity, time-varying, and spatial decorrelation properties. Many issues of PKG for point-to-point communications with small-scale antennas have been thoroughly studied [6–9] and some experimental platforms for WiFi [10,11], Zigbee [12], Bluetooth [13], and LoRa [14–16] have been set up to verify the feasibility and effectiveness of PKG. The traditional PKG system considers the secret key generation between two communication ends, named Alice and Bob. In general, the detailed PKG process contains four stages: channel probing, quantization, information reconciliation, and privacy amplification. (1) Alice and Bob estimate the channel information by transmitting pilot signals. (2) After obtaining correlated observations, analog channel measurements are quantized into digital bits. (3) Alice and Bob utilize information reconciliation to correct the key disagreement caused by the calibration errors, the temporal variation, and the noise. (4) Privacy amplification is used to eliminate any potential information leakage to eavesdroppers.

In future wireless networks, one BS needs to simultaneously communicate with multiple user terminals (UTs). However, PKG for multi-user multiple-input multiple-output (MIMO)/massive MIMO communications has not been actively addressed [17]. According to the recent trends in wireless communications, the number of antennas will continue growing up to support the huge increase in serving users. The sixth-generation (6G) wireless communication systems will employ massive/ultra-massive MIMO to support extremely high throughput and multiple UTs [6,18], which brings new challenges to PKG as follows:

- *Long-time delay*: Most existing PKG research works in the time-division duplex (TDD) mode, where two users, Alice and Bob, probe their channel by sending pilot signals to each other alternately. When the time delay is within the channel coherence time, Alice and Bob are able to achieve channel state information (CSI) estimations with high correlation, which establishes a basis for symmetric key generation. However, in a multi-user system with orthogonal pilots, it is challenging to realize a short-time delay. To distinguish different users, the pilot length scales linearly with the number of antennas. When the number of antennas or UTs is large, the time delay is likely to exceed the coherence time, failing the secret key generation.
- *Strong inter-user interference:* If we use non-orthogonal pilot signals for different users (or pilot signals shared among UTs), it can cause severe inter-user interference, leading to two negative effects on the secret key generation. First, the interference reduces the reciprocity of the channel measurements, which makes communication ends difficult to obtain symmetric keys. Second, the interference

results in some correlation between two different users, which may cause secret key leakage.

All the above challenges constitute the motivation for exploring a new secret key generation approach for multi-user massive MIMO wireless communications. Considering the multi-user secret key generation, this chapter is organized as follows. Section 12.2 introduces the system model for massive MIMO key generation. Section 12.3 proposes a general channel dimensionality reduction (CDR)-based multi-user secret key generation scheme. Section 12.4 designs the detailed parameters of CDR-based key generation, including the transmit and receive directions and the power allocation. Security analysis is provided in Section 12.5. Section 12.6 carries out numerical results and is followed by the conclusion.

12.2 System model for massive MIMO key generation

In a narrow-band star topology network, a BS simultaneously communicates with K UTs, as shown in Figure 12.1, where the BS is equipped with M antennas and the kth UT is equipped with N_k antennas. To protect the private information of each UT, the BS generates secret keys $\kappa = \{\kappa_1, \kappa_2, \ldots, \kappa_K\}$ with each UT according to their channel observations, where κ_k is the secret key between the BS and the kth UT.

We consider passive eavesdropping, where an eavesdropper wants to eavesdrop on the secret key in the set of κ based on her own channel observations and all the information exchanged over the public channel. Usually, Eve is assumed to be located at least a half wavelength away from the BS and UTs, otherwise Eve is easy to be seen, e.g., a half wavelength of 2.4 GHz is 6.25 cm. Under this assumption, Eve's channel observations become independent of those at the BS and UTs. Besides Eve, we also consider the potential unintended hearing from other UTs. As wireless communication is open, all the UTs can receive signals. Some UTs, which are treated as the curious user, may be interested in other UTs' signals.

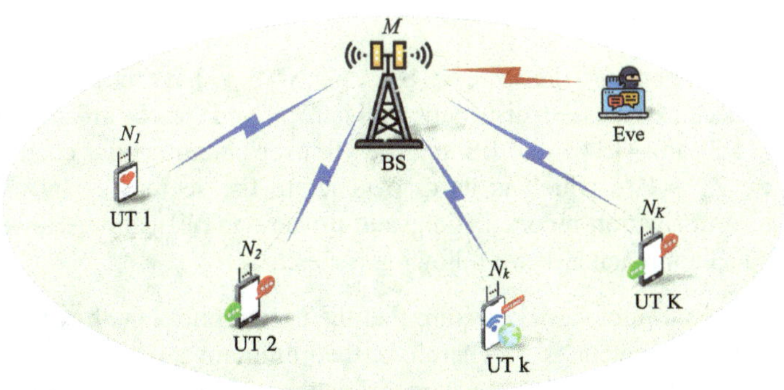

Figure 12.1 System model for multi-user secret key generation

12.2.1 Traditional secret key generation model

In the traditional secret key generation, the BS and UTs transmit orthogonal pilot signals to estimate CSI. Mathematically, in the downlink, the BS broadcasts the pilot signal, $\mathbf{S}^d \in \mathbb{C}^{M\times M}$, satisfying $\mathbf{S}^d(\mathbf{S}^d)^H = \mathbf{I}_M$. Then, the k-th UT receives signal as

$$\mathbf{Y}_k^d = \mathbf{H}_k^d\mathbf{S}^d + \mathbf{N}_k^d, \tag{12.1}$$

where $\mathbf{H}_k^d \in \mathbb{C}^{N_k\times M}$ is the downlink channel of the k-th UT, and $\mathbf{N}_k^d \in \mathbb{C}^{N_k\times M}$ is the Gaussian noise at UT k. By the least square (LS) estimation, UT k estimates this downlink CSI as

$$\mathbf{Z}_k^d = \mathbf{Y}_k^d(\mathbf{S}^d)^H = \mathbf{H}_k^d + \mathbf{N}_k^d(\mathbf{S}^d)^H. \tag{12.2}$$

In the uplink, all UTs transmit pilot signals, $\mathbf{S}_k^u \in \mathbb{C}^{N_k\times N}, k \in \{1, 2, \ldots, K\}$, to the BS at the same time and frequency slot, where $N = \sum_k N_k$ is the total number of UT antennas. The received signal at the BS is given by

$$\mathbf{Y}^u = \sum_{k=1}^{K} \mathbf{H}_k^u\mathbf{S}_k^u + \mathbf{N}^u, \tag{12.3}$$

where $\mathbf{H}_k^u$ is the uplink channel of the kth UT and $\mathbf{N}^u \in \mathbb{C}^{N_k\times N}$ is the complex Gaussian noise. The original uplink and downlink channels are assumed to be reciprocal, i.e., $\mathbf{H}_k^u = (\mathbf{H}_k^d)^T$. The estimated uplink channel of the kth user can be expressed as

$$\mathbf{Z}_k^u = \mathbf{H}_k^u\mathbf{S}_k^u(\mathbf{S}_k^u)^H + \mathbf{H}_k^d \sum_{k'\neq k}^{K} \mathbf{S}_{k'}^u(\mathbf{S}_k^u)^H + \mathbf{N}^u(\mathbf{S}_k^u)^H. \tag{12.4}$$

To eliminate the interference, the pilot signals of different UTs are required to be orthogonal, i.e.,

$$\mathbf{S}_i^u(\mathbf{S}_j^u)^H = \begin{cases} \mathbf{I}_{N_i}, & i = j \\ \mathbf{0}_{N_i\times N_j}, & i \neq j. \end{cases} \tag{12.5}$$

Therefore, (12.4) can be further reduced to

$$\mathbf{Z}_k^u = \mathbf{H}_k^u + \mathbf{N}^u(\mathbf{S}_k^u)^H. \tag{12.6}$$

When $\mathbf{S}^d$ and $\mathbf{S}_k^u$ are unitary, the noise $\mathbf{N}_k^d(\mathbf{S}^d)^H$ and $\mathbf{N}^u(\mathbf{S}_k^u)^H$ have the same distribution as $\mathbf{N}_k^d$ and $\mathbf{N}^u$ because Gaussian distribution is isotropic and thereby unitarily invariant.

From (12.2) and (12.6), the BS and the kth user obtain similar channel observations, i.e., $\mathbf{Z}_k^u \approx \mathbf{Z}_k^d$, which is used to generate the secret key by employing quantization, information reconciliation, and privacy amplification. However, this traditional PKG approach has two following issues:

1. The orthogonal pilot signals require that the uplink pilot length scales with the total number of antennas N. Therefore, the minimum time of channel probing is $T = (M + N)\Delta T + T_{Switch}$, where ΔT is the symbol transmission time, and T_{Switch} is the switching time from downlink to uplink. To keep the high correlation of uplink and downlink channels, the channel probing time should be smaller than

the channel coherence time, which becomes very challenging with large M, K and N_k.

2. The elements of $\mathbf{z}_k^u$ and $\mathbf{z}_k^d$ have high auto-correlation, due to the spatial correlation of the channel, which may lead to long 0s or 1s after quantization. Some preprocessing approaches, e.g., PCA [19], can be used to reduce the auto-correlation, which suffer from high computational complexity with a large number of antennas.

To sum up, the core problems are how to reduce the length of pilots and the high dimension of the channel matrix in the multi-user massive MIMO system. Fortunately, literature and field measurements have shown that the massive MIMO channel model reveals some new characteristics [20,21], which play an important role in multi-user secret key generation.

12.2.2 Massive MIMO channel model

We consider a geometry-based channel model with N_P paths. Then, the $N_k \times M$ physical MIMO channel matrix in the downlink associated with the pth path of the kth UT can be expressed as [22]

$$\mathbf{H}_{k,p}^d = \alpha_{k,p}\mathbf{a}_{UT,k}\left(\theta_{k,p}\right)\mathbf{a}_{BS}^H\left(\varphi_{k,p}\right), \tag{12.7}$$

where $\alpha_{k,p}$ is the complex gain of the pth path, $\mathbf{a}_{UT,k}(\theta_{k,p})$ is the UT antenna array response vector with the angle of arrival (AoA) $\theta_{k,p}$, and $\mathbf{a}_{BS}(\varphi_{k,p})$ is the BS antenna array response vector corresponding to the angle of departure (AoD) $\varphi_{k,p}$. Specifically, under the uniform linear array (ULA) setup [23–25], these vectors are given by

$$\mathbf{a}_{UT,k}\left(\theta_{k,p}\right) = \frac{1}{\sqrt{N_k}}\left[1, e^{-j\frac{2\pi}{\lambda}d\sin\left(\theta_{k,p}\right)}, \ldots, e^{-j(N_k-1)\frac{2\pi}{\lambda}d\sin\left(\theta_{k,p}\right)}\right]^T \tag{12.8}$$

$$\mathbf{a}_{BS}\left(\varphi_{k,p}\right) = \frac{1}{\sqrt{M}}\left[1, e^{-j\frac{2\pi}{\lambda}d\sin\left(\varphi_{k,p}\right)}, \ldots, e^{-j(M-1)\frac{2\pi}{\lambda}d\sin\left(\varphi_{k,p}\right)}\right]^T, \tag{12.9}$$

where λ is the wavelength, and d is the distance between the adjacent antennas. The antenna can be also configured as a uniform planar array [26,27] or a dual-polarized uniform planar array [28,29]. For a narrow-band channel model, the channel response of the kth UT can be expressed as

$$\mathbf{H}_k^d = \sum_{p=1}^{N_P}\mathbf{H}_{k,p}^d. \tag{12.10}$$

To reveal the spatial characteristics, we employ two sampling matrices to separate signals of different directions. Define the angle sampling matrices at the BS and UT k as

$$\begin{aligned}\mathbf{A}_{UT,k} &= \left[\mathbf{a}_{UT,k}\left(\theta_1\right), \mathbf{a}_{UT,k}\left(\theta_2\right), \ldots, \mathbf{a}_{UT,k}\left(\theta_{N_k}\right)\right] \in \mathbb{C}^{N_k\times N_k}\\ \mathbf{A}_{BS} &= \left[\mathbf{a}_{BS}\left(\varphi_1\right), \mathbf{a}_{BS}\left(\varphi_2\right), \ldots, \mathbf{a}_{BS}\left(\varphi_M\right)\right] \in \mathbb{C}^{M\times M},\end{aligned} \tag{12.11}$$

where φ_m and θ_n are the mth and nth sample angles at the BS and the UT, satisfying that $\sin\left(\varphi_m\right) = 2m/M - 1$ and $\sin\left(\theta_n\right) = 2n/N_k - 1$ [30]. With a half wavelength antenna spacing, i.e., $d = \lambda/2$, the matrices $\mathbf{A}_{UT,k}$ and $\mathbf{A}_{BS}$ become the unitary discrete Fourier transform (DFT) matrix, given by [31]

$$[\mathbf{A}_{UT,k}]_{n_1,n_2} = \frac{1}{\sqrt{N_k}} \exp\left(-j2\pi(n_1 - 1)(n_2 - N_k/2)/N_k\right) \tag{12.12}$$

$$[\mathbf{A}_{BS}]_{m_1,m_2} = \frac{1}{\sqrt{M}} \exp\left(-j2\pi(m_1 - 1)(m_2 - M/2)/M\right). \tag{12.13}$$

Then, we can sample the channel matrix as

$$\tilde{\mathbf{H}}_k^d = \mathbf{A}_{UT,k}^H \mathbf{H}_k^d \mathbf{A}_{BS}, \tag{12.14}$$

where different elements of $\tilde{\mathbf{H}}_k^d$ present the channel gains from different transmitting sampling directions to different receiving sampling directions. Each direction is called one beam and the channel matrix $\tilde{\mathbf{H}}_k^d$ is called the beam domain channel matrix.

For a better understanding of the beam domain channel model, we will present some results [21] when the number of antennas grows to infinity.

Proposition 1. *When the number of antennas grows to infinity, the beam domain channel $\tilde{\mathbf{H}}_k^d$ tends to $\mathbf{G}_k^d$, i.e., for arbitrary n and m,*

$$\lim_{M,N_k \to \infty} [\tilde{\mathbf{H}}_k^d - \mathbf{G}_k^d]_{n,m} = 0, \tag{12.15}$$

where $\mathbf{G}_k^d \in \mathbb{C}^{N_k \times M}$ is given by

$$[\mathbf{G}_k^d]_{n,m} = \sum_{p=1}^{N_P} \alpha_{k,p} \delta(\theta_{k,p} - \arcsin(2n/N_k - 1)) \delta(\varphi_{k,p} - \arcsin(2m/M - 1)). \tag{12.16}$$

Remark 1. *Proposition 1 presents that with a large number of antennas, the beam domain channel tends to matrix $\mathbf{G}_k^d$. From the definition of $\mathbf{G}_k^d$ in (12.16), we can find two observations:*

- *The beam domain channel can separate signals of different directions. From (12.16), each element contains at most one path p simultaneously satisfying $\theta_{k,p} = \arcsin(2n/N_k - 1)$ and $\varphi_{k,p} = \arcsin(2m/M - 1)$. One element corresponds to the channel gains of one AoD and AoA pair $(\varphi_{k,p}, \theta_{k,p})$. In this case, different channel gains of different AoAs and AoDs are distributed in different positions of the beam domain channel matrix.*
- *Matrix $\mathbf{G}_k^d$ is a sparse matrix. As there are N_P propagation paths, the number of non-zero entries in matrix $\mathbf{G}_k^d$ is N_P, which is much less than the number of antennas M and N_k. Thus, $\mathbf{G}_k^d$ is very sparse. In practice, with a large (but finite) number of antennas at the BS and the UT, $\mathbf{G}_k^d$ is also a good approximation of the beam domain channel matrix $\tilde{\mathbf{H}}_{k,p}^d$, which indicates that $\tilde{\mathbf{H}}_{k,p}^d$ contains N_P dominant elements.*

Define the beam domain channel covariance matrices at the BS and kth UT respectively as

$$\begin{aligned}\tilde{\mathbf{R}}_{BS,k} &= \mathbb{E}\{(\tilde{\mathbf{H}}_k^d)^H \tilde{\mathbf{H}}_k^d\} = \mathbf{A}_{BS}^H \mathbf{R}_{BS,k} \mathbf{A}_{BS} \\ \tilde{\mathbf{R}}_{UT,k} &= \mathbb{E}\{\tilde{\mathbf{H}}_k^d (\tilde{\mathbf{H}}_k^d)^H\} = \mathbf{A}_{UT,k}^H \mathbf{R}_{UT,k} \mathbf{A}_{UT,k},\end{aligned} \tag{12.17}$$

where $\mathbf{R}_{BS,k} = \mathbb{E}\{(\mathbf{H}_k^d)^H \mathbf{H}_k^d\}$ and $\mathbf{R}_{UT,k} = \mathbb{E}\{\mathbf{H}_k^d (\mathbf{H}_k^d)^H\}$ are covariance matrices of the physical channel. As the number of antennas increases, $\tilde{\mathbf{R}}_{BS,k}$ and $\tilde{\mathbf{R}}_{UT,k}$ become diagonal matrices with the diagonal elements given by

$$\begin{aligned}&\lim_{M\to\infty} \quad [\tilde{\mathbf{R}}_{BS,k}]_{m,m} - \sum_{p=1}^{N_P} |\alpha_{k,p}|^2 \delta(\varphi_{k,p} - \arcsin(2m/M - 1)) = 0, \\ &\lim_{N_k\to\infty} \quad [\tilde{\mathbf{R}}_{UT,k}]_{n,n} - \sum_{p=1}^{N_P} |\alpha_{k,p}|^2 \delta(\theta_{k,p} - \arcsin(2n/N_k - 1)) = 0.\end{aligned} \tag{12.18}$$

This result is coincident with Proposition 1. As different elements of the beam domain channel matrix present channel gains of different paths, these elements are uncorrelated due to the uncorrelated scattering assumption and the diagonal elements of the covariance matrices represent the channel gains of the corresponding transmit or receive angles.

Furthermore, as in the secret key generation, the BS and UTs vectorize the channel measurements, here we also calculate the full correlation of the vectorized channel matrix as

$$\begin{aligned}\mathbf{R}_k &= \mathbb{E}\{\mathrm{vec}(\mathbf{H}_k^d)\mathrm{vec}(\mathbf{H}_k^d)^H\} \\ &= \mathbb{E}\{(\mathbf{A}_{BS}^* \otimes \mathbf{A}_{UT,k})\mathrm{vec}(\tilde{\mathbf{H}}_k^d)\mathrm{vec}(\tilde{\mathbf{H}}_k^d)^H (\mathbf{A}_{BS}^* \otimes \mathbf{A}_{UT,k})^H\}.\end{aligned} \tag{12.19}$$

Noting that as the elements of the beam domain channel matrix $\tilde{\mathbf{H}}_k^d$ become uncorrelated, the matrix $\mathbb{E}\{\mathrm{vec}(\tilde{\mathbf{H}}_k^d)\mathrm{vec}(\tilde{\mathbf{H}}_k^d)^H\}$ converges to a diagonal matrix, denoted as $\mathbf{\Lambda}_k$. Then, the full correlation matrix $\mathbf{R}_k$ can be decomposed as

$$\mathbf{R}_k = (\mathbf{A}_{BS}^* \otimes \mathbf{A}_{UT})\mathbf{\Lambda}_k(\mathbf{A}_{BS}^* \otimes \mathbf{A}_{UT})^H, \tag{12.20}$$

where the $(mN_k + n)$th diagonal element in $\mathbf{\Lambda}_k$ is given by

$$\begin{aligned}[\mathbf{\Lambda}_k]_{mN_k+n,mN_k+n} = &\sum_{p=1}^{N_P} |\alpha_{k,p}|^2 \delta(\theta_{k,p} - \arcsin(2n/N_k - 1)) \\ &\times \delta(\varphi_{k,p} - \arcsin(2m/M - 1)).\end{aligned} \tag{12.21}$$

The matrix $\mathbf{\Lambda}_k$ also represents the channel spatial characteristics in the beam domain. When different scatters correspond to different AoAs and AoDs, there are N_P dominant entries in $\mathbf{\Lambda}_k$, each corresponding to one non-zero channel gain.

In summary, the above analysis presents that the representation of the beam domain channel can bring several benefits. First, the beam domain channel matrix becomes sparse, which indicates that only a few (much less than the number of antennas) elements contain the most channel information. This property can reduce the dimension of channel estimation. Second, the elements of the beam domain channel matrix become mutually independent. This can reduce the redundancy of the channel observation. Third, the beam domain sampling matrix at the BS becomes identical, independent of UTs. One sampling matrix can separate signals of different UTs from different directions. Such property is desirable for multi-user secret key generation.

12.3 General CDR-based key-generation scheme

12.3.1 Key-generation scheme based on CDR

To take full advantage of the spatial properties of the beam domain channel, we propose channel dimensionality reduction (CDR)-based multi-user secret key generation. At the beginning, the BS and UTs transmit sounding signals to acquire the statistical CSI. Then, they design precoding and receiving matrices based on the statistical CSI, in order to reduce the dimension of channel estimation. By using the designed precoding and receiving matrices, the BS and UTs obtain an equivalent channel parameters with high correlation. After quantization, information reconciliation and privacy amplification procedures are used to generate consistent and private secret keys, similar to point-to-point secret key generation.

Specifically, in the downlink, as shown in Figure 12.2, let $\mathbf{P} = [\mathbf{P}_1, \mathbf{P}_2, \cdots, \mathbf{P}_K]$ and $\mathbf{C}^H = [\mathbf{C}_1^H, \mathbf{C}_2^H, \cdots, \mathbf{C}_K^H]$ denote the precoding and receiving matrices. The BS transmits pilot signal $\mathbf{S}_k^d \in \mathbb{C}^{M_e \times T_D}$ through linear precoding matrix $\mathbf{P}_k$ to UT k within T_D time slots. The pilot signal satisfies $\mathbf{S}_k^d(\mathbf{S}_k^d)^H = \mathbf{I}$. At the receiver, UT k multiplies the received signal by the receiving matrix $\mathbf{C}_k^H$ and obtains

$$\mathbf{Y}_k^d = \mathbf{C}_k^H \mathbf{H}_k^d \mathbf{P}_k \mathbf{S}_k^d + \mathbf{C}_k^H \mathbf{H}_k^d \sum_{k' \neq k} \mathbf{P}_{k'} \mathbf{S}_{k'}^d + \mathbf{C}_k^H \mathbf{N}_k. \tag{12.22}$$

By the LS estimation, UT k estimates the downlink CSI as

$$\mathbf{Z}_k^d = \mathbf{C}_k^H \mathbf{H}_k^d \mathbf{P}_k + \mathbf{C}_k^H \mathbf{H}_k^d \sum_{k' \neq k} \mathbf{P}_{k'} \mathbf{S}_{k'}^d \left(\mathbf{S}_k^d\right)^H + \mathbf{C}_k^H \mathbf{N}_k \left(\mathbf{S}_k^d\right)^H. \tag{12.23}$$

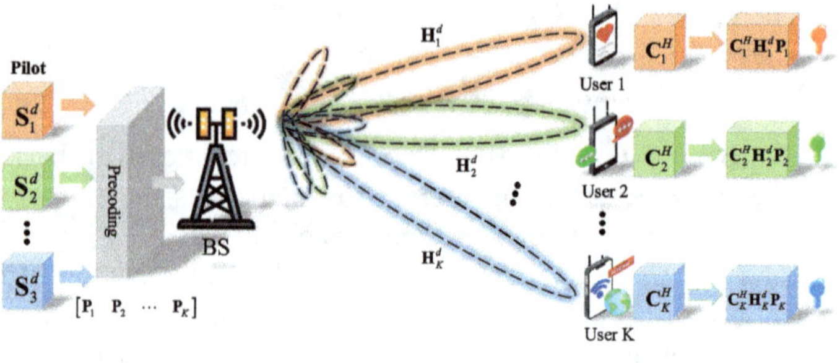

Figure 12.2 Downlink channel probing in secret key-generation scheme

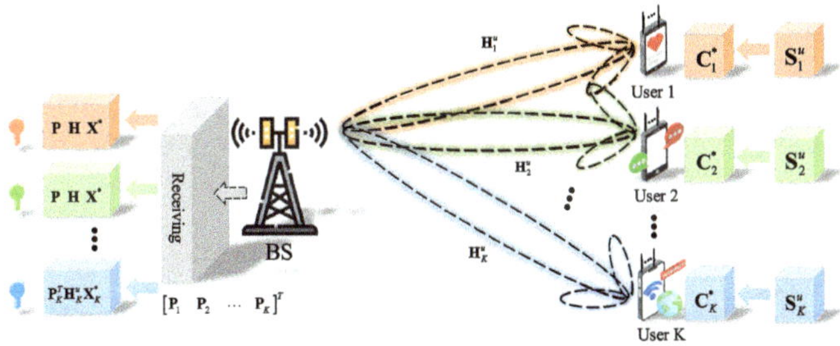

Figure 12.3 Uplink channel probing in secret key-generation scheme

In the uplink, the kth UT transmits pilot signals $\mathbf{S}_k^u \in \mathbb{C}^{N_e \times T_U}$ within T_U time slots through the matrix $\mathbf{C}_k^*$ (Figure 12.3). The pilot signals satisfy $\mathbf{S}_k^u (\mathbf{S}_k^u)^H = \mathbf{I}$. The BS multiplies the received signals by the receiving matrix $\mathbf{P}_k^T$ and obtains the kth UT signal as

$$\mathbf{Y}_k^u = \mathbf{P}_k^T \mathbf{H}_k^u \mathbf{C}_k^* \mathbf{S}_k^u + \mathbf{P}_k^T \sum_{k' \neq k} \mathbf{H}_{k'}^u \mathbf{C}_{k'}^* \mathbf{S}_{k'}^u + \mathbf{P}_k^T \mathbf{N}, \tag{12.24}$$

where $\mathbf{Y}_k^u \in \mathbb{C}^{M_e \times T_U}$ is the received signals of time length T_U. After the LS estimation, the estimated effective channel of the kth UT can be expressed as

$$\mathbf{Z}_k^{UL} = \mathbf{P}_k^T \mathbf{H}_k^{UL} \mathbf{C}_k^* + \mathbf{P}_k^T \sum_{k' \neq k} \mathbf{H}_{k'}^{UL} \mathbf{C}_{k'}^* \mathbf{S}_{k'}^{UL} \left(\mathbf{S}_k^{UL}\right)^H + \mathbf{P}_k^T \mathbf{N} \left(\mathbf{S}_k^{UL}\right)^H . \tag{12.25}$$

Remark 2. *By employing precoding and receiving matrices, the BS and the kth UT obtain the equivalent channel parameter $\mathbf{C}_k^H \mathbf{H}_k^d \mathbf{P}_k$. Note that the dimension of the equivalent channel is $N_e \times M_e$, where M_e and N_e are the reduced dimensions at the BS and UT. As M_e and N_e are approximately equal to the number of paths N_P, which is far smaller than M and N_k, the precoding and receiving matrices can significantly reduce the channel estimation dimension. In addition, different elements of the beam domain present the channel gains of different directions and the channel gains of one UT are concentrated within a few directions. This indicates that the beam domain channel matrix can separate signals of different UTs, which has the potential to reduce the inter-user interference.*

After obtaining the equivalent channel parameters, the BS and UTs vectorize them to generate the secret key. For the simplicity of notation, denote the downlink channel $\mathbf{H}_k^d$ as $\mathbf{H}_k$, and the uplink channel is $\mathbf{H}_k^u = (\mathbf{H}_k)^T$. Then, the vectorized channel matrices can be expressed as

$$\mathbf{z}_k^d = \text{vec}(\mathbf{Z}_k^d) = \sum_{k'} \left((\mathbf{P}_{k'} \mathbf{S}_{k'k}^d)^T \otimes \mathbf{C}_k^H \right) \text{vec}(\mathbf{H}_k) + \text{vec}(\mathbf{C}_k^H \mathbf{N}_k) \tag{12.26}$$

$$\mathbf{z}_k^u = \text{vec}((\mathbf{Z}_k^u)^T) = \sum_{k'} \left(\mathbf{P}_k^T \otimes \mathbf{S}_{k'k}^u \mathbf{C}_{k'} \right) \text{vec}(\mathbf{H}_{k'}) + \text{vec}(\mathbf{N}^T \mathbf{P}_k), \tag{12.27}$$

where $\mathbf{S}_{k'k}^d = \mathbf{S}_{k'}^d (\mathbf{S}_k^d)^H$ (or $\mathbf{S}_{k'k}^u = (\mathbf{S}_{k'}^u (\mathbf{S}_k^u)^H)^T$) represents the covariance matrix of the downlink (or uplink) pilot signals.

Remark 3. *It is worth noting that the interference terms of the uplink and downlink channels are not reciprocal. Specifically, in the downlink transmission, the interference received at the kth UT is the summation of the transmitted signals of the k′th UT propagating through the channel of the kth UT* $\mathbf{H}_k$*, while in the uplink transmission, the interference received at the BS for the kth UT is the summation of the transmitted signals of the k′th UT propagating through the channel of different UTs* $\mathbf{H}_{k'}$*. Thus, the interference will also reduce the agreement.*

12.3.2 Secret key rate

According to the vectorized channel matrices, we will analyze the performance of the CDR-based secret key-generation scheme. We assume potential non-colluding curious users, which indicates that UTs may be interested in the other UTs' signals. Under the TDD operation, UTs cannot simultaneously transmit and receive signals. Each UT only has the channel observation in the downlink transmission. Thus, the key rate is the minimum mutual information given other UT's observations given by [32]

$$I_k = \min_{i \neq k} I(\mathbf{z}_k^d; \mathbf{z}_k^u | \mathbf{z}_i^d). \tag{12.28}$$

Remark 4. *As UTs only receive signals in the downlink transmission, the secret key rate is the minimum mutual information between* $\mathbf{z}_k^d$ *and* $\mathbf{z}_k^u$ *on the condition of* $\mathbf{z}_i^d$*. Moreover, we assume that UTs are far away from the BS, and thus, the channels of different UTs are independent from each other.*

When the channels of different UTs are independent, the secret key rate degrades to

$$I_k = I(\mathbf{z}_k^d; \mathbf{z}_k^u), \tag{12.29}$$

which can be calculated as the following result [1].

Theorem 1. *When the channels of different UTs become independent, the secret key rate of the kth UT is given by*

$$I(\mathbf{z}_k^d;\mathbf{z}_k^u) = -\log\det\Big(\mathbf{I} - \mathbf{V}_{kk}\big(\sum_{k'}\mathbf{V}_{kk'}^H\mathbf{V}_{kk'} + (\tilde{\mathbf{P}}_k^T\tilde{\mathbf{P}}_k^* \otimes \mathbf{I}_{T_U})\big)^{-1}\mathbf{V}_{kk}^H \times\mathbf{V}_k\left(\mathbf{V}_k^H\mathbf{V}_k + \mathbf{I}_{T_D}\otimes\tilde{\mathbf{C}}_k^H\tilde{\mathbf{C}}_k\right)^{-1}\mathbf{V}_k^H\Big), \tag{12.30}$$

where $\mathbf{V}_k$ *and* $\mathbf{V}_{kk'}$ *are, respectively, given by*

$$\begin{aligned}\mathbf{V}_k &= \mathbf{\Lambda}_k^{1/2}\big(\sum_{k'}(\tilde{\mathbf{P}}_{k'}\mathbf{S}_{k'k}^d)^T\otimes\tilde{\mathbf{C}}_k^H\big)^H\\ \mathbf{V}_{kk'} &= \mathbf{\Lambda}_{k'}^{1/2}(\tilde{\mathbf{P}}_k^T\otimes\mathbf{S}_{k'k}^u\tilde{\mathbf{C}}_{k'}^H)^H,\end{aligned} \tag{12.31}$$

$\tilde{\mathbf{P}}_k = \mathbf{A}_{BS}^H\mathbf{P}_k$ *and* $\tilde{\mathbf{C}}_k = \mathbf{A}_{UT,k}^H\mathbf{C}_k$ *are the precoding and receiving matrices in the beam domain.*

Remark 5. *The secret key rate* (12.30) *is complicated, depending not only on the precoding and receiving vectors but also on the covariance matrices of pilot signals. The pilot signals of each UT satisfy* $\mathbf{S}_{kk}^d = \mathbf{I}$ *and* $\mathbf{S}_{kk}^u = \mathbf{I}$ *to estimate the effective CSI [33]. With orthogonal pilot signals among UTs, i.e.,* $\mathbf{S}_{k'k}^d = \mathbf{0}$ *and* $\mathbf{S}_{k'k}^u = \mathbf{0}$*, which can eliminate the inter-user interference, the secret key rate can be simplified as*

$$\begin{aligned}I(\mathbf{z}_k^d;\mathbf{z}_k^u) = &-\log\det\big(\mathbf{I} - \mathbf{\Lambda}_k^{1/2}(\tilde{\mathbf{P}}_k^*\otimes\tilde{\mathbf{C}}_k)\\ &\times\big(\mathbf{I} + (\tilde{\mathbf{P}}_k^T\otimes\tilde{\mathbf{C}}_k^H)\mathbf{\Lambda}_k(\tilde{\mathbf{P}}_k^*\otimes\tilde{\mathbf{C}}_k)\big)^{-1}(\tilde{\mathbf{P}}_k^T\otimes\tilde{\mathbf{C}}_k^H)\mathbf{\Lambda}_k\\ &\times(\tilde{\mathbf{P}}_k^*\otimes\tilde{\mathbf{C}}_k)\big(\mathbf{I} + (\tilde{\mathbf{P}}_k^T\otimes\tilde{\mathbf{C}}_k^H)\mathbf{\Lambda}_k(\tilde{\mathbf{P}}_k^*\otimes\tilde{\mathbf{C}}_k)\big)^{-1}\times(\tilde{\mathbf{P}}_k^T\otimes\tilde{\mathbf{C}}_k^H)\mathbf{\Lambda}_k^{1/2}\big),\end{aligned} \tag{12.32}$$

where the secret key rate is only decided by the precoding and receiving matrices. However, the pilot overhead of the orthogonal pilots scales with the number of UTs, becoming extremely large in multi-user communication systems. Thus, it is impractical to use orthogonal pilots among UTs due to the short coherent time. Alternatively, pilot signals can be reused among UTs, i.e., $\mathbf{S}_{k'k}^d = \mathbf{I}$ *and* $\mathbf{S}_{k'k}^u = \mathbf{I}$*. The reused pilot signals cause inter-user interference and precoding and receiving matrices will dominate the secret key rate of all UTs.*

12.4 Optimization design of CDR-based key generation

From the above analysis, precoding and receiving matrices play an important role in the multi-user secret key generation, especially under the pilot reuse case. In this section, we will design the CDR-based secret key-generation scheme to maximize the secret key rate.

Consider the precoding and receiving matrices design maximizing the sum secret key rate, which can be generally formulated as

$$\begin{aligned}
\max_{\tilde{\mathbf{P}}_k,\tilde{\mathbf{C}}_k} \quad & R_{\text{sum}} = \sum_k I(\mathbf{z}_k^d;\mathbf{z}_k^u) \\
\text{s.t.} \quad & \tilde{\mathbf{P}}_k^H \tilde{\mathbf{P}}_k \le P \\
& \tilde{\mathbf{C}}_k^H \tilde{\mathbf{C}}_k = \mathbf{I},
\end{aligned} \tag{12.33}$$

where P is the total power constraint at the BS. The BS designs the precoding matrix to mitigate inter-user interference under the total power constraint, while each UT designs the receiving matrix to receive signals of the corresponding directions. In general, the precoding matrix $\tilde{\mathbf{P}}_k$ can be decomposed as

$$\tilde{\mathbf{P}}_k = \mathbf{\Phi}_k \mathbf{D}_k, \tag{12.34}$$

where $\mathbf{\Phi}_k \in \mathbb{C}^{M\times M_e}$ presents the transmitted directions and $\mathbf{D}_k \in \mathbb{C}^{M_e\times M_e}$ is a diagonal matrix with the transmitted power in each direction. Next, we will detailedly design the direction matrices $\mathbf{\Phi}_k$, $\tilde{\mathbf{C}}_k$ and the power allocation $\mathbf{D}_k$, respectively.

12.4.1 Design of precoding and receiving directions

First, we consider the transmit and receive direction design under the equal power allocation, i.e., $\mathbf{D}_k = \mathbf{I}$, which can be formulated as

$$\begin{aligned}
\max_{\tilde{\mathbf{P}}_k,\tilde{\mathbf{C}}_k} \quad & R_{\text{sum}} = \sum_k I(\mathbf{z}_k^d;\mathbf{z}_k^u) \\
\text{s.t.} \quad & \mathbf{\Phi}_k^H \mathbf{\Phi}_k = \mathbf{I} \\
& \tilde{\mathbf{C}}_k^H \tilde{\mathbf{C}}_k = \mathbf{I},
\end{aligned} \tag{12.35}$$

where the secret key rate can be expressed as

$$\begin{aligned}
I(\mathbf{z}_k^d;\mathbf{z}_k^u) = & -\log\det\Bigg(\mathbf{I} - \mathbf{\Lambda}_k^{1/2}\left(\mathbf{\Phi}_k^T \otimes \tilde{\mathbf{C}}_k^H\right)^H \\
& \times \left(\mathbf{I} + \sum_{k'}\left(\mathbf{\Phi}_k^T \otimes \tilde{\mathbf{C}}_{k'}^H\right)\mathbf{\Lambda}_{k'}\left(\mathbf{\Phi}_k^T \otimes \tilde{\mathbf{C}}_{k'}^H\right)^H\right)^{-1} \\
& \times \left(\mathbf{\Phi}_k^T \otimes \tilde{\mathbf{C}}_k^H\right)\mathbf{\Lambda}_k\left(\sum_{k'}(\mathbf{\Phi}_{k'})^* \otimes \tilde{\mathbf{C}}_k\right) \\
& \times \left(\mathbf{I} + \left(\sum_{k'}(\mathbf{\Phi}_{k'})^T \otimes \tilde{\mathbf{C}}_k^H\right)\mathbf{\Lambda}_k\left(\sum_{k'}(\mathbf{\Phi}_{k'})^* \otimes \tilde{\mathbf{C}}_k\right)\right)^{-1} \\
& \times \left(\sum_{k'}(\mathbf{\Phi}_{k'})^T \otimes \tilde{\mathbf{C}}_k^H\right)\mathbf{\Lambda}_k^{1/2}\Bigg).
\end{aligned} \tag{12.36}$$

As the optimization matrices $\mathbf{\Phi}_k$ and $\tilde{\mathbf{C}}_k$ are involved in both inside and outside the matrix inversion operation of the objective function (12.36), problem (12.35) is usually non-convex and difficult to solve globally. By utilizing the interference neutralization approach [34] to mitigate the interference, i.e., for arbitrary matrix $\tilde{\mathbf{C}}_{k'}$ ($k' \neq k$), the precoding matrix $\mathbf{\Phi}_k$ satisfies

$$(\mathbf{\Phi}_k^T \otimes \tilde{\mathbf{C}}_{k'}^H)\mathbf{\Lambda}_{k'} = \mathbf{0}, \quad k' \neq k, \tag{12.37}$$

which indicates that the precoding matrix $\mathbf{\Phi}_k$ can eliminate the inter-user interference. This constraint can be satisfied when the precoding beams of one UT are different from channel beams of other UTs, i.e.,

$$\mathbf{\Phi}_k^H \tilde{\mathbf{R}}_{BS,k'} = \mathbf{0}, \quad k' \neq k. \tag{12.38}$$

Under this constraint, problem (12.35) maximizing the sum secret key rate can be degraded to the following subproblems maximizing the secret key rate of each user

$$\begin{aligned} \min_{\mathbf{\Phi}_k, \tilde{\mathbf{C}}_k} \quad & \log\det \big(\mathbf{I} - \mathbf{\Lambda}_k^{1/2})\big)^{-1} \big(\mathbf{\Phi}_k^T \otimes \tilde{\mathbf{C}}_k^H\big)\mathbf{\Lambda}_k\big(\mathbf{\Phi}_k^* \otimes \tilde{\mathbf{C}}_k\big) \\ & \times \big(\mathbf{I} + \big(\mathbf{\Phi}_k^T \otimes \tilde{\mathbf{C}}_k^H\big)\mathbf{\Lambda}_k\big(\mathbf{\Phi}_k^* \otimes \tilde{\mathbf{C}}_k\big)\big)^{-1} \big(\mathbf{\Phi}_k^T \otimes \tilde{\mathbf{C}}_k^H\big)\mathbf{\Lambda}_k^{1/2}\big) \\ \text{s.t.} \quad & \mathbf{\Phi}_k^H \mathbf{\Phi}_k = \mathbf{I} \\ & \tilde{\mathbf{C}}_k^H \tilde{\mathbf{C}}_k = \mathbf{I}. \end{aligned} \tag{12.39}$$

Remark 6. *The secret key rate in* (12.39) *is equal to that using orthogonal pilots in* (12.32). *This indicates that when UTs can be separated in the beam domain by different beams, reusing the identical pilot signals among UTs can approach the performance with orthogonal pilot signals. Both schemes can mitigate inter-user interference. The difference is that the orthogonal pilot scheme uses orthogonal pilot signals to separate different UTs, which requires a large pilot overhead, while the interference neutralization scheme designs the precoding matrices to eliminate the interference, which is independent of pilot signals among UTs.*

Define $\mathbf{U}_k = \big(\mathbf{\Phi}_k^* \otimes \tilde{\mathbf{C}}_k\big)$, and problem (12.39) can be rewritten as

$$\begin{aligned} \min_{\mathbf{U}_k} \quad & \log\det \left(\mathbf{I} - \left(\mathbf{\Lambda}_k^{1/2}\mathbf{U}_k \left(\mathbf{I} + \mathbf{U}_k^H \mathbf{\Lambda}_k \mathbf{U}_k\right)^{-1} \mathbf{U}_k^H \mathbf{\Lambda}_k^{1/2}\right)^2\right) \\ \text{s.t.} \quad & \mathbf{U}_k^H \mathbf{U}_k = \mathbf{I}. \end{aligned} \tag{12.40}$$

The solution of problem (12.40) is obtained as follows:

Theorem 2. *The optimal* $\mathbf{U}_k$ *maximizing the secret key rate is*

$$\mathbf{U}_k = \left[\mathbf{e}_{\eta_1} \ \mathbf{e}_{\eta_2} \ \cdots \ \mathbf{e}_{\eta_{M_e N_e}}\right], \tag{12.41}$$

where $\mathbf{e}_i = [0, 0, \cdots, 0, 1, 0, \cdots, 0]$ *is a unit vector with the ith unit element and* η_i *is the index of the ith sorted eigenvalue of matrix* $\mathbf{\Lambda}_k(\mathbf{I} + \mathbf{\Lambda}_k)^{-1}$. *The optimal rate is*

$$R_k = -\sum_{i=1}^{M_e N_e} \log \lambda_i\big(\mathbf{I} - \mathbf{\Lambda}_k^2(\mathbf{I} + \mathbf{\Lambda}_k)^{-2}\big). \tag{12.42}$$

Remark 7. *From Theorem 2, the optimal* $\mathbf{U}_k$ *maximizing the secret key rate consists of the unit vectors corresponding to the sorted diagonal elements in* $\mathbf{\Lambda}_k$*. However, in general,* $\mathbf{U}_k$ *with the structure of* $\mathbf{\Phi}_k^* \otimes \tilde{\mathbf{C}}_k$ *may not satisfy the optimal condition* (12.41)*. Fortunately, by employing the channel properties and Theorem 2, we can construct the precoding and receiving vectors* $\mathbf{\Phi}_k$ *and* $\tilde{\mathbf{C}}_k$*.*

As $\mathbf{U}_k$ consists of vectors $\mathbf{e}_i$, the beam domain precoding and receiving matrices $\mathbf{\Phi}_k$ and $\tilde{\mathbf{C}}_k$ only have elements of 0 and 1, which present the selected beams for each UT. Recalling Proposition 1, with a large number of antennas, different elements of the beam domain channel matrix represent the channel gains of different AoDs and AoAs, and the most channel gains are concentrated within a few elements. Specifically, with N_P propagation paths, the BS can select the strongest N_P beams, i.e., the precoding matrix $\mathbf{\Phi}_k$ is given by

$$\mathbf{\Phi}_k = \left[\mathbf{e}_{\eta_{t,k,1}} \ \mathbf{e}_{\eta_{t,k,2}} \ \cdots \ \mathbf{e}_{\eta_{t,k,N_P}}\right] \tag{12.43}$$

where $\eta_{t,k,1}$ is the index of the sorted eigenvalue of matrix $\mathbf{R}_{BS,k}$. Similarly, UT k selects the strongest N_P receiving directions, i.e., the receiving matrix $\tilde{\mathbf{C}}_k$ is given by

$$\tilde{\mathbf{C}}_k = \left[\mathbf{e}_{\eta_{r,k,1}} \ \mathbf{e}_{\eta_{r,k,2}} \ \cdots \ \mathbf{e}_{\eta_{r,k,N_P}}\right] \tag{12.44}$$

where $\eta_{r,k,1}$ is the index of the sorted eigenvalue of matrix $\mathbf{R}_{UT,k}$. By choosing $M_e = N_e = N_P$, we can construct $\mathbf{U}_k = \mathbf{\Phi}_k^* \otimes \tilde{\mathbf{C}}_k$ to obtain the N_P^2 elements in $\mathbf{\Lambda}_k$. As the number of paths N_P is relatively small, this approach can significantly reduce the pilot overhead and fits well with the massive MIMO channel model and precoding [35].

Remark 8. *Equations* (12.43) *and* (12.44) *reveal that the optimal precoding and receiving directions for users with non-overlapping beams are corresponding to the strongest beams. When the channel beams of different users are overlapping, the overlapping beams cause inter-user interference, as well as information leakage, which will be discussed in Section 12.5. Thus, the BS first allocates non-overlapping beam set* $\mathscr{A}_k = \{\mathbf{e}_{k_1}, \mathbf{e}_{k_2}, \ldots\}$ *for each UT, satisfying*

$$\mathbf{e}_{k_i} \tilde{\mathbf{R}}_{BS,k'} = \mathbf{0}. \tag{12.45}$$

Then, the BS selects the strongest N_P *beams from the non-overlapping beam set.*

12.4.2 Design of transmitted power allocation

In the above subsection, we have shown that in the beam domain, the optimal transmit and receive directions are to select corresponding beams for each UT. Based on the above result, we now consider the beam scheduling and power allocation on each beam.

As we focus on the power allocation at the BS side, for simplicity, we consider the single antenna at each UT. From (12.34) and (12.43), we find that the precoding matrix is to select transmitting beams and allocate power on the beams. To jointly present the beam scheduling and power allocation, we define an augmented diagonal power allocation matrix $\tilde{\mathbf{D}}_k \in \mathbb{C}^{M \times M}$ (M is the number of BS antennas), in which the indices of non-zero diagonal elements present the beam allocated to UT k and the values present the transmitted power. According to (12.26) and setting $\mathbf{C}_k = 1$, $\mathbf{S}^d_{k'k} = \mathbf{S}^d_{k'k} = \delta(d - d')\mathbf{I}$, $\mathbf{P}_k = \mathbf{A}_{BS}\tilde{\mathbf{D}}_k$, and $\mathbf{h}_k = \tilde{\mathbf{h}}_k\mathbf{A}^H_{BS}$, we can rewrite the downlink beam domain received signal as

$$\tilde{\mathbf{y}}^y_k = \tilde{\mathbf{D}}_k\tilde{\mathbf{h}}^T_k + \sum_{k' \neq k} \tilde{\mathbf{D}}_{k'}\tilde{\mathbf{h}}^T_k + \mathbf{n}_k, \tag{12.46}$$

and the uplink received signal is the summation of all UTs as

$$\tilde{\mathbf{y}}^u = \sum_{k'} \tilde{\mathbf{h}}^T_{k'} + \mathbf{n}, \tag{12.47}$$

where $\tilde{\mathbf{h}}_k$ is the beam domain channel of UT k, $\mathbf{n}_k$, and $\mathbf{n}$ are complex Gaussian noise at UT k and the BS, respectively. Based on the received signals, we can have the following secret key rate [2].

Theorem 3. *The secret key rate is upper bounded by*

$$\begin{aligned} I_k = {} & \log\det\left(\mathbf{I} + \sum_{k'} \mathbf{\Lambda}_{k'}\right) \\ & - \log\det\left(\mathbf{I} + \sum_{k' \neq k} \mathbf{\Lambda}_{k'} + \mathbf{\Lambda}^{1/2}_k(\mathbf{I} + \mathbf{\Lambda}^{1/2}_k\mathbf{Q}\mathbf{\Lambda}^{1/2}_k)^{-1}\mathbf{\Lambda}^{1/2}_k\right), \end{aligned} \tag{12.48}$$

where $\mathbf{Q} = \tilde{\mathbf{D}}^H\tilde{\mathbf{D}}$ *is a diagonal matrix,* $\tilde{\mathbf{D}} = \sum_k \tilde{\mathbf{D}}_k$*, and* $\mathbf{\Lambda}_k$ *is the channel covariance matrix* $\mathbf{\Lambda}_k = \mathbb{E}\{\tilde{\mathbf{h}}^T_k\tilde{\mathbf{h}}^*_k\}$.

Theorem 3 reveals that the upper bound of the secret key rate I_k depends only on the sum of $\tilde{\mathbf{D}}_k$ (or $\mathbf{Q}$), independent of the individual $\tilde{\mathbf{D}}_k$. Next, we first consider the total power allocation optimization problem maximizing the upper bound of the sum secret key rate as

$$\begin{aligned} \max_{\mathbf{Q}} \quad & I = \sum_k I_k \\ \text{s.t.} \quad & \text{tr}(\mathbf{Q}) \leq P \\ & \mathbf{Q} \succeq \mathbf{0}, \end{aligned} \tag{12.49}$$

in which we optimize the total power allocation matrix $\mathbf{Q}$. Recalling (12.48), the first term $\log\det(\mathbf{I} + \sum_{k'} \mathbf{\Lambda}_{k'})$ only depends on the channel correlation matrices, independent of $\mathbf{Q}$. Thus, maximizing the upper bound is equal to minimizing the second term in (12.48). Define function $f_k(\mathbf{Q})$ as

$$f_k(\mathbf{Q}) = \log\det\left(\mathbf{I} + \sum_{k' \neq k} \mathbf{\Lambda}_{k'} + \mathbf{\Lambda}^{1/2}_k(\mathbf{I} + \mathbf{\Lambda}^{1/2}_k\mathbf{Q}\mathbf{\Lambda}^{1/2}_k)^{-1}\mathbf{\Lambda}^{1/2}_k\right). \tag{12.50}$$

Then, problem (12.49) is equal to

$$\begin{aligned} \min_{\mathbf{Q}} \quad & \sum_k f_k(\mathbf{Q}) \\ \text{s.t.} \quad & \text{tr}(\mathbf{Q}) \leq P \\ & \mathbf{Q} \succeq 0. \end{aligned} \tag{12.51}$$

Before solving problem (12.51), the convexity of the problem is shown as follows.

Theorem 4. *Problem* (12.51) *is a convex optimization problem.*

Remark 9. *Due to the convexity of problem* (12.51), *we can obtain the global optimal solution of* $\mathbf{Q}$, *which can be obtained by the MATLAB® CVX software [36]. Besides, problem* (12.51) *considers the power allocation on the total beams, which does not show the beams for each UT. As mentioned in the above subsection, we need to allocate non-overlapping beams for different UTs. Then, we will consider the beam scheduling after power allocation.*

The optimal solution $\mathbf{Q}$ only has some non-zero elements on its diagonal, whose indices present the selected beams of UTs. The values are the allocated power on each beam. Suppose that there are N non-zero diagonal elements in $\mathbf{Q}$, whose indices and values are denoted as $\{b_1, b_2, \ldots, b_N\}$ and $\{[\mathbf{Q}]_{b_1,b_1}, [\mathbf{Q}]_{b_2,b_2}, \ldots, [\mathbf{Q}]_{b_N,b_N}\}$, respectively. With these notations, we can construct a set of vectors $\mathscr{B} = \{\mathbf{s}_1, \mathbf{s}_2, \ldots, \mathbf{s}_N\}$, where $\mathbf{s}_i = [0, \ldots, \sqrt{[\mathbf{Q}]_{b_i,b_i}}, 0, \ldots, 0]$ corresponds to the b_ith non-zero element in $\mathbf{Q}$. Each vector $\mathbf{s}_i$ represents one beam and the set $\mathscr{B}$ contains all the selected beams.

After solving problem (12.51), the BS has the optimal beam set $\mathscr{B}$ for total UTs. Then, we consider the beam selection for each UT from the total beam set, which can be expressed as

$$[\mathbf{D}_k]_{i,:} \in \mathscr{B}, \tag{12.52}$$

where $[\mathbf{D}_k]_{i,:}$ is the ith row of matrix $\mathbf{D}_k$. This constraint indicates that each row of $\mathbf{D}_k$ is constructed by the vector in $\mathscr{B}$ to guarantee the beams for each UT selected from the optimal beam set.

Then, we consider the beam scheduling for each UT, given by

$$\max_{\mathbf{D}_k} \quad \sum_k R_k \tag{12.53a}$$

$$\text{s.t.} \quad [\mathbf{D}_k]_{i,:} \in \mathscr{B} \tag{12.53b}$$

$$[\mathbf{D}_k]_{i,:}([\mathbf{D}_{k'}]_{i,:})^T = 0 \tag{12.53c}$$

$$\text{rank}(\mathbf{D}_k) \leq M_e. \tag{12.53d}$$

Constraint (12.53c) presents that beams of different UTs are non-overlapping and constraint (12.53d) presents that the number of allocated beams for each UT is less than M_e. To tackle this problem, we propose an efficient beam-scheduling algorithm to obtain a sub-optimal solution, as shown in Algorithm 1.

Algorithm 1: Beam scheduling algorithm.

Initialize $\mathbf{D}_k^{(0)} = \mathbf{0}$, $R^{(0)} = 0$, $i = 1$, and $B_k = 0$.
repeat
  Set $R^{(i)} = R^{(i-1)}$.
  for $b = 1$ to N **do**
    for $k = 1$ to K **do**
      if $B_k < M_e$ **then**
        Set $[\mathbf{D}_k]_{B_k+1,:} = \mathbf{s}_b$ and calculate

$$R^{(b,k)} = \sum_k R_k.$$

      end if
    end for
    Select the UT with the maximum rate

$$k^* = \arg\max R^{(b,k)}.$$

    if $R^{(b,k^*)} > R^{(i)}$ **then**
      Set $B_{k^*} = B_{k^*} + 1$ and $I^{(i)} = I^{(b,k^*)}$.
      Update $[\mathbf{D}_{k^*}]_{B_{k^*},:} = \mathbf{s}_b$.
    end if
  end for
  Update $i = i + 1$.
until $\left|I^{(i)} - I^{(i-1)}\right| \leq \varepsilon$.

Note that the allocated beams for secret key generation are not exactly the same as the channel beams. The channel beams present the channel gains in different directions, which reflect the physical channel conditions. The channel beams of different UTs may be overlapping when the channels have similar angles. The allocated beams are the BS selected to generate the secret key with different UTs, which are non-overlapping among UTs. After the beam scheduling, the BS employs non-overlapping allocated beams to generate the secret keys with different UTs. When the number of UTs increase and the BS does not have enough beams, UTs may employ different time slots or frequencies to generate secret keys with the BS.

In Algorithm 1, there are three iterations. The outer iteration allocates the precoding vectors to UTs. As the sum key rate is monotonically increasing, Algorithm 1 converges. Simulation results will illustrate the quick convergence of Algorithm 1.

The inner iteration allocates the selected beams to the UT with the maximum sum key rate. The complexity of the inner iteration is $O(NK)$. Moreover, to satisfy the rank constraint, a variable B_k is used to record the number of allocated beams for UT k. When UT k has been allocated M_e beams, i.e., $B_k = M_e$, the BS will not allocate beams to UT k. Therefore, a sub-optimal solution is obtained by satisfying the constraints.

12.5 Security analysis

In the CDR-based secret key-generation scheme design, the channels of different UTs are assumed to be independent. This assumption can be easily satisfied when the channel beams of different UTs are non-overlapping. In reality, this condition may not hold strictly, when two UTs are close to each other and suffer some identical propagation paths. In this case, the part of the two UTs' beams are overlapping, i.e.,

$$\tilde{\mathbf{R}}_{BS,k}\tilde{\mathbf{R}}_{BS,k'} \neq \mathbf{0}, \quad k' \neq k. \tag{12.54}$$

Since the channels of the overlapping beams between $\tilde{\mathbf{R}}_{BS,k}$ and $\tilde{\mathbf{R}}_{BS,k'}$ may cause correlation, the independent assumption does not hold anymore. Thus, information leakage should be considered for the overlapping case design.

As different elements in the beam domain channel matrix present channel gains of different directions, which are statistically independent, the secret key rate can be decomposed as the summation of the key rate of each beam. Thus, for simplicity, we will analyze the information leakage on one overlapping beam as an example.

Consider one beam (beam b) is occupied by two UTs (UT 1 and UT 2). If there are more than two UTs on one beam, the information leakage only depends on the UT with the highest correlation. Denote the channel gains from beam b to UT 1 (or UT 2) as h_1 (or h_2), which has the unit attenuation power, i.e., $\mathbb{E}\{h_1^2\} = \mathbb{E}\{h_2^2\} = 1$. Define the information leakage ratio as $\gamma = (R_h - R_1)/R_h$, where R_h is the secret key rate with the independent channel assumption and R_1 is the key rate when the channels are correlated. The secret key rate of independent channel is always larger than that of correlated case, i.e., $R_h \geq R_1$, and the information leakage ratio is between 0 and 1.

Theorem 5. *The information leakage ratio can be expressed as*

$$\gamma = 1 - \frac{\log \frac{((1+\sigma^2)^2-\rho^2)^2}{(1+\sigma^2)(\sigma^6+3\sigma^4-2\sigma^2\rho^2+2\sigma^2)}}{\log \frac{(1+\sigma^2)^2}{\sigma^2(2+\sigma^2)}}, \tag{12.55}$$

where σ^2 is the noise variance and ρ is the cross-channel correlation defined as

$$\rho = \frac{\mathbb{E}\{h_1 h_2\}}{\mathbb{E}\{h_1^2\}\mathbb{E}\{h_2^2\}} = \mathbb{E}\{h_1 h_2\}. \tag{12.56}$$

Remark 10. *From* (12.55), *we have the closed-form expression of the information leakage ratio, which is decided by the correlation ρ and the noise variance σ^2. To* *better reveal the characteristics of the information leakage ratio, consider a special case when ρ is 1 or -1. Then, the information leakage can be calculated as*

$$\gamma = \frac{\log \frac{3+2\sigma^2}{\sigma^2(6+11\sigma^2+6\sigma^4+\sigma^6)}}{\log \frac{(1+\sigma^2)^2}{\sigma^2(2+\sigma^2)}}. \tag{12.57}$$

As the SNR increases (or σ^2 decreases), the information leakage ratio becomes

$$\lim_{\sigma^2 \to 0} \gamma = 1, \tag{12.58}$$

which indicates that the secret key rate goes to zero and vanishes.

The information leakage is dominated by the channel correlation, which highly depends on the overlapping channel beams. When the channel correlation is high (tends to 1), the information leakage ratio becomes 1, which indicates that the information is almost leaked and the correlated overlapping beams provide little secret key rate. To avoid the information leakage, the BS should allocate *non-overlapping transmitting beams* to different UTs, i.e., the precoding vector $\tilde{\mathbf{P}}_k$ satisfies

$$\tilde{\mathbf{P}}_k \tilde{\mathbf{R}}_{BS,k'} = \mathbf{0}, \quad k' \neq k. \tag{12.59}$$

The physical channel beams may be overlapping due to the similar propagation angles, which cause severe information leakage. The BS allocates non-overlapping beams to different UTs. After beam allocation, the constraint (12.37) is satisfied, and the channel matrices of the allocated beams for different UTs are independent.

12.6 Numerical result

In this section, we provide some numerical results to illustrate the performance of secret key generation in multi-user massive MIMO wireless communication systems. As the number of BS antennas significantly affects the multi-user performance, we consider that the BS is equipped with a large number of antennas. The detailed simulation parameters are summarized in Table 12.1.

Figure 12.4 provides an example of the beam domain channel gains, which is the diagonal elements in $\tilde{\mathbf{R}}_{BS,k}$. In the figure, the number of the dominant beams of one UT is much smaller than the number of antennas. This reveals the sparse of the beam domain channel. Moreover, when the angles of different UTs are different, the BS can separate UTs in the beam domain, which provides the potential for multi-user secret key generation.

Figure 12.5 compares the secret key rate for different numbers of UTs. The secret key rate grows up as the number of UTs increases. Specifically, the key rate of 4 UTs approaches twice that of 2 UTs. As the number of UTs further increases, the increase of the key rate becomes smaller, due to the interference or leakage among UTs.

In multi-user secret key generation, the bottleneck is the pilot overhead. To jointly reveal the secret key rate and the pilot overhead, we define the unit secret key rate as

$$R_{\text{unit}} = R_{\text{sum}}/T, \tag{12.60}$$

Table 12.1 Simulation parameters

Simulation parameters	Setting value
Number of BS antennas	$M = 128$
Number of UT antennas	$N_k = 4$
Number of UTs	$K = 6$
UT positions	Uniformly distributed
Channel model	3GPP SCM
Antenna configuration	ULA
Antenna spacing	0.5λ
Number of paths	$N_P = 6$
Scenario	Suburban macro
Radius of cell	500 m

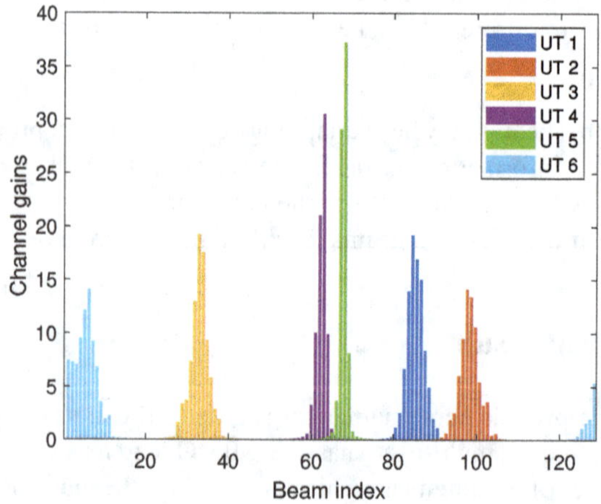

Figure 12.4 The channel gains distribution in the beam domain

where T is the pilot overhead, scaled with the dimension of the effective channel M_e and N_e. We set $N_e = N_k = 4$ equal to the number of UT antennas. In Figure 12.6, we present the unit secret key rate comparison for different M_e with the traditional orthogonal pilot scheme. From the result, one can observe that the unit secret key rates of reused pilot schemes are much higher than that of the orthogonal pilot, as the pilot overhead of the orthogonal scheme is extremely large. The rate of the reused pilot scheme with $M_e = 6$ is slightly better than that with $M_e = 4$.

To show the agreement performance of the secret key, we evaluate the bit disagreement ratio (BDR), which is defined as the ratio of the number of the disagreement bits to the number of total bits of the initial secret key. Figure 12.7 compares the BDR performance of a legitimate user and a curious user. In the figure, the curve of "single

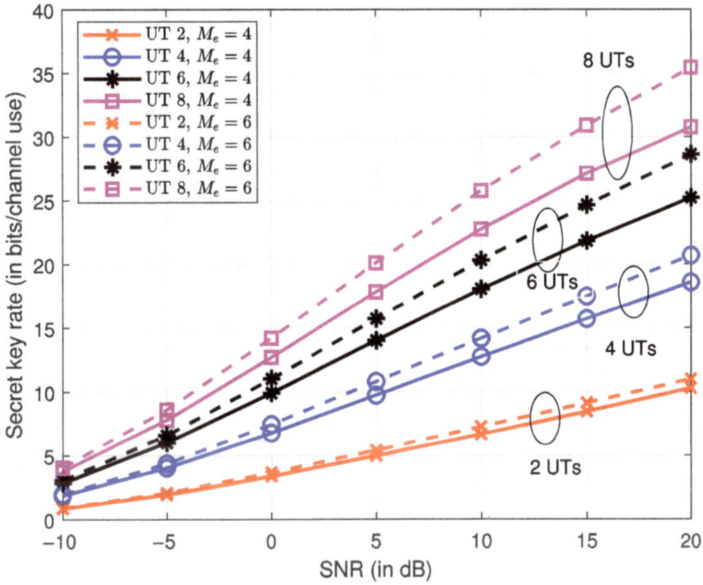

Figure 12.5 Secret key rate comparison for different numbers of UTs. ©IEEE 2020. Reprinted with permission from [1].

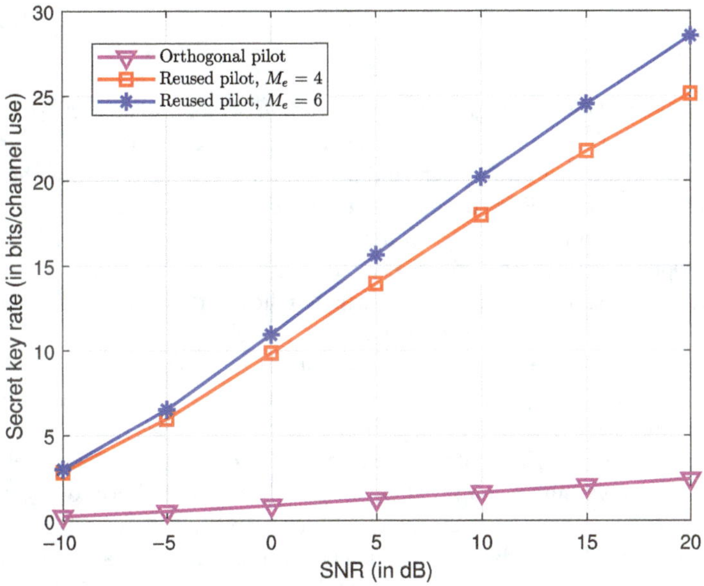

Figure 12.6 Unit secret key rate comparison for orthogonal and reused pilot. ©IEEE 2020. Reprinted with permission from [1].

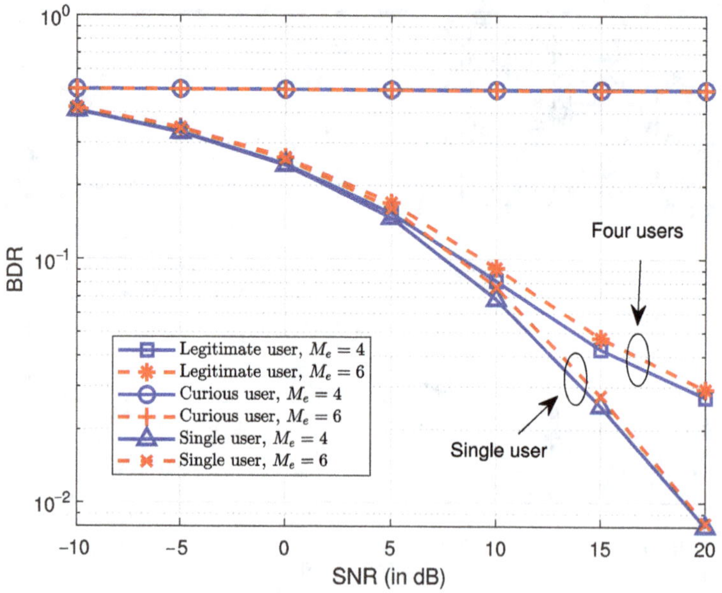

Figure 12.7 BDR comparison for legitimate and curious UTs. ©IEEE 2020. Reprinted with permission from [1].

user" presents the BDR of single UT case, the curve of "legitimate user" presents the BDR between the BS and the legitimate user, and the curve of "curious user" presents the BDR at the curious user. We can find that for the single user case, without inter-user interference, the BDR performance is the lowest. Our proposed multi-user secret key-generation scheme can approach the single-user performance, which indicates that the proposed scheme can effectively reduce the interference, while the BDR of "curious user" remains high (about 0.5), which indicates that the curious user cannot obtain other UTs' secret key based on his own channel observations.

Furthermore, we present the secret key rate of the beam-scheduling algorithm with that of the uniform power allocation, as shown in Figure 12.8. In the uniform power allocation scheme, the BS only selects the strongest M_e beams for each UT. From the results, we can find that the secret key rate of our proposed beam-scheduling algorithm is always higher than that of the uniform power allocation. This result indicates that if the BS only selects the beams for each UT without considering their overlapping beams, the overlapping channel beams may cause inter-user interference, decreasing the performance. We also present the upper bound I, which is the globally optimal result of the problem and cannot be achieved. Our proposed beam scheduling scheme can approach the upper bound in the low and middle SNR regions.

Figure 12.9 shows the convergence of the outer iteration in Algorithm 1 with the SNR set to be -10 dB, 0 dB, and 10 dB, respectively. In the figure, all the rate of

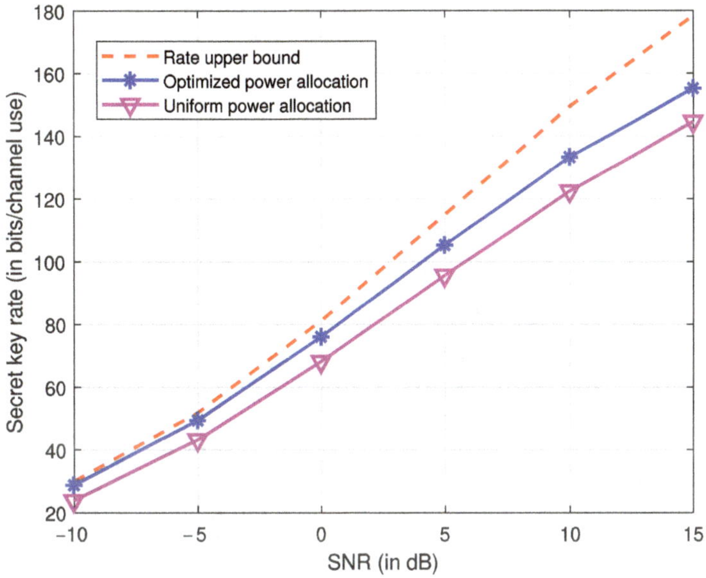

Figure 12.8 Secret key rate comparison of different power allocation schemes. ©IEEE 2020. Reprinted with permission from [2].

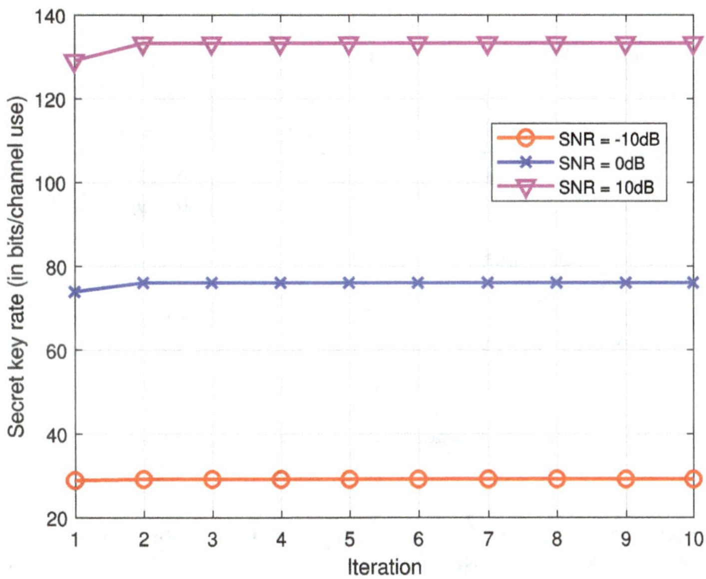

Figure 12.9 Convergence of the proposed beam scheduling algorithm. ©IEEE 2020. Reprinted with permission from [2].

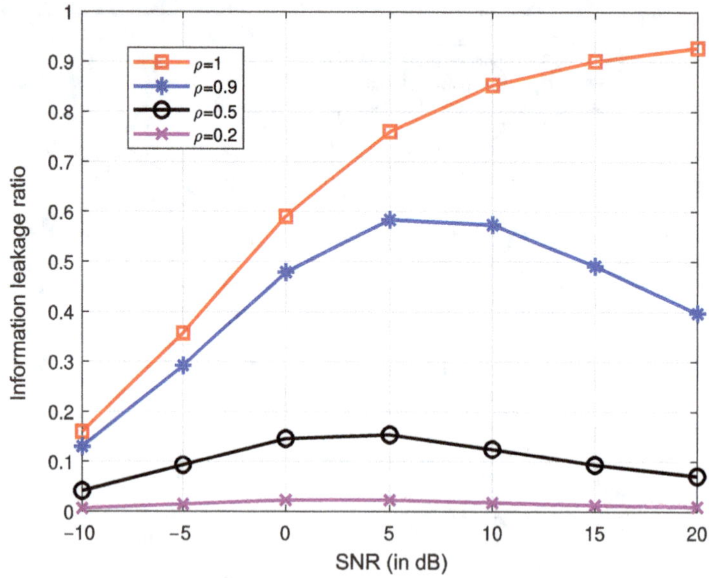

Figure 12.10 Information leakage ratio with different correlation coefficients. ©IEEE 2020. Reprinted with permission from [1].

Table 12.2 NIST random test result

	Pass ratio	***p*-Value**
Approximate entropy	0.9199	0.4233
Runs	0.9262	0.4590
Ranking	0.9128	0.3810
Longest runs of ones	0.9717	0.3701
Frequency	0.9926	0.5254
FFT	0.9983	0.5800
Block frequency	0.9949	0.5466
Cumulative sums	0.9974	0.4997
Serial	0.9255	0.4474, 0.4903
Total pass ratio	0.7759	

different SNRs converge quickly, even within only a few steps. This result confirms the quick convergence of Algorithm 1.

Figure 12.10 compares the information leakage ratio with different correlation coefficient. UTs may obtain similar channel observations when the channels are correlated, which causes information leakage. When the correlation coefficient is 1, which is the highest correlation, the information leakage ratio increases with the SNR and approaches 1 at the high SNR regimes. For other correlation coefficient cases, the information leakage ratio increases in the low and middle SNRs and decreases in the high SNRs, as the BS can extract the difference of different UTs' channel when the BS have a precise channel estimation in the high SNRs. Also the information leakage increases with the channel correlation coefficient.

Finally, we perform the National Institute of Standards and Technology (NIST) random test to show the randomness of the generated keys before privacy amplification. We test 10000 trials, each for 256 bits. The pass ratio and the averaged p-values are summarized in Table 12.2. From the results, for each test term, the pass ratio is high than 90% and the total pass ratio is about 78%. Therefore, the generated key has statistical indistinguishability.

12.7 Conclusions

This chapter provided a fundamental design and analysis of the multi-user secret key generation in massive MIMO wireless networks. We provided a beam domain channel model, representing the channel gains of different directions. Then, we derived the secret key rate and designed precoding and receiving matrices to maximize the sum secret key rate. Furthermore, we provided a security analysis by considering the channel correlation between UTs. When the channels of different UTs are correlated, the BS employs the several strongest non-overlapping beams of each UT to avoid information leakage. Some further research directions are provided as follows.

- Multi-cell multi-user secret key generation. In this chapter, we consider multi-user wireless communications with only one BS. Cellular networks usually contain multiple BSs serving a large number of UTs. In this scenario, inter-user interference as well as inter-cell interference will have a severe effect on the PKG. How to reduce the interference to generate symmetric secret keys needs to be further investigated.
- PKG in millimeter wave wireless communications. In the future 6G wireless networks, to achieve higher data rate, higher frequency resources, such as millimeter wave or even THz, and more antennas will be employed. Higher frequency and ultra-massive antennas bring new channel characteristics, such as larger attenuation, fewer propagation paths, and near-field effect, which bring new challenges and opportunities for PKG.
- PKG in frequency division duplexing (FDD) systems. Existing studies of PKG focus on TDD systems in which uplink and downlink channel observations have a good similarity. FDD operation is also widely used in wireless communications. In FDD systems, the uplink and downlink channel observations are quite different, which cannot be directly used to generate secret keys. However, as the carriers of uplink and downlink are close, there are some reciprocal channel parameters in FDD systems. More work is needed to be explored on extracting these reciprocal channel parameters to facilitate PKG in FDD systems.

Acknowledgments

This work was supported by the National Natural Science Foundation of China under Grant 62171121, 62271145, in part by the Natural Science Foundation of Jiangsu Province under Grant BK20211160.

We thank Prof. Eduard Jorswieck and Dr. Junqing Zhang for their efforts.

References

[1] Li G, Sun C, Jorswieck E, *et al.* Sum secret key rate maximization for TDD multi-user massive MIMO wireless networks. *IEEE Trans Inf Forens Secur*. 2020;16:968–982.

[2] Sun C and Li G. Power allocation and beam scheduling for multi-user massive MIMO secret key generation. *IEEE Access*. 2020;8:164580–164592.

[3] Zhang J, Duong TQ, Woods R, *et al.* Securing wireless communications of the Internet of Things from the physical layer, an overview. *Entropy*. 2017;19(8):420.

[4] Chi C, Lu R, Petzoldt A, *et al.* Securing the Internet of Things in a quantum world. *IEEE Commun Mag*. 2017;55(2):116–120.

[5] Maurer UM. Secret key agreement by public discussion from common information. *IEEE Trans Inf Theory*. 1993;39(3):733–742.

[6] Li G, Sun C, Zhang J, *et al.* Physical layer key generation in 5G and beyond wireless communications: challenges and opportunities. *Entropy*. 2019;21(5):497.

[7] Chen C and Jensen MA. Secret key establishment using temporally and spatially correlated wireless channel coefficients. *IEEE Trans Mobile Comput*. 2010;10(2):205–215.

[8] Wallace JW and Sharma RK. Automatic secret keys from reciprocal MIMO wireless channels: measurement and analysis. *IEEE Trans Inf Forens Secur*. 2010;5(3):381–392.

[9] Li G, Hu L, Staat P, *et al.* Reconfigurable intelligent surface for physical layer key generation: constructive or destructive? *IEEE Wireless Commun*. 2022;29(4):146–153.

[10] Mathur S, Trappe W, Mandayam N, *et al.* Radio-telepathy: extracting a secret key from an unauthenticated wireless channel. In: *ACM International Conference on Mobile Computing and Networking*. San Francisco, CA, USA; 2008. p. 128–139.

[11] Jana S, Premnath SN, Clark M, *et al.* On the effectiveness of secret key extraction from wireless signal strength in real environments. In: *Proceedings of the 15th Annual International Conference on Mobile Computer and Network*, Beijing, China: ACM; 2009. p. 321–332.

[12] Aono T, Higuchi K, Ohira T, *et al.* Wireless secret key generation exploiting reactance-domain scalar response of multipath fading channels. *IEEE Trans Antennas Propag*. 2005;53(11):3776–3784.

[13] Premnath SN, Gowda PL, Kasera SK, *et al.* Secret key extraction using Bluetooth wireless signal strength measurements. In: *Eleventh Annual IEEE International Conference on Sensing, Communication, and Networking (SECON)*; 2014. p. 293–301.

[14] Xu W, Jha S, and Hu W. LoRa-Key: secure key generation system for LoRa-based network. *IEEE Internet Things J.* 2018;6(4):6404–6416.

[15] Ruotsalainen H and Grebeniuk S. Towards wireless secret key agreement with LoRa physical layer. In: *Proceedings of the International Conference on Availability, Reliability and Security*. Hamburg, Germany; 2018. p. 1–6.

[16] Zhang J, Marshall A, and Hanzo L. Channel-envelope differencing eliminates secret key correlation: LoRa-based key generation in low power wide area networks. *IEEE Trans Veh Technol.* 2018;67(12):12462–12466.

[17] Jiao L, Tang J, Zeng K. Physical layer key generation using virtual AoA and AoD of mmWave massive MIMO channel. In: *Proceedings of the IEEE Conference on Communication and Network Security (CNS)*; 2018. p. 1–9.

[18] Busari SA, Huq KMS, Mumtaz S, *et al.* Millimeter-wave massive MIMO communication for future wireless systems: a survey. *Commun Surveys Tuts.* 2017;20(2):836–869.

[19] Li G, Hu A, Zhang J, *et al.* High-agreement uncorrelated secret key generation based on principal component analysis preprocessing. *IEEE Trans Commun.* 2018;66(7):3022–3034.

[20] Rusek F, Persson D, Lau BK, *et al.* Scaling up MIMO: opportunities and challenges with very large arrays. *IEEE Signal Process Mag.* 2013;30(1): 40–60.

[21] Sun C, Gao X, Jin S, *et al.* Beam division multiple access transmission for massive MIMO communications. *IEEE Trans Commun.* 2015;63(6): 2170–2184.

[22] Tse D and Viswanath P. *Fundamentals of Wireless Communication*. New York, NY: Cambridge University Press; 2005.

[23] Yin H, Gesbert D, Filippou M, *et al.* A coordinated approach to channel estimation in large-scale multiple-antenna systems. *IEEE J Sel Areas Commun.* 2013;31(2):264–273.

[24] Adhikary A, Nam J, Ahn JY, *et al.* Joint spatial division and multiplexing—the large-scale array regime. *IEEE Trans Inf Theory*. 2013;59(10):6441–6463.

[25] You L, Gao X, Xia XG, *et al.* Pilot reuse for massive MIMO transmission over spatially correlated Rayleigh fading channels. *IEEE Trans Wireless Commun.* 2015;14(6):3352–3366.

[26] Yu H, You L, Wang W, *et al.* Active channel sparsification for uplink massive MIMO with uniform planar array. *IEEE Trans Wireless Commun.* 2021;20(9):6018–6032.

[27] Han Y, Liu Q, Wen CK, *et al.* FDD massive MIMO based on efficient downlink channel reconstruction. *IEEE Trans Commun.* 2019;67(6):4020–4034.

[28] Khalilsarai MB, Yang T, Haghighatshoar S, *et al.* Dual-polarized FDD massive MIMO: a comprehensive framework. *IEEE Trans Wireless Commun.* 2021;21(2):840–854.

[29] Yu H, Yi X, and Caire G. Downlink precoding for DP-UPA FDD massive MIMO via multi-dimensional active channel sparsification. *IEEE Trans Wireless Commun*. 2022;21(8):6698–6713.

[30] Sayeed AM. Deconstructing multiantenna fading channels. *IEEE Trans Signal Process*. 2002;50(10):2563–2579.

[31] You L, Gao X, Li GY, *et al.* BDMA for millimeter-wave/terahertz massive MIMO transmission with per-beam synchronization. *IEEE J Sel Areas Commun*. 2017;35(7):1550–1563.

[32] Liang Y, Poor HV, and (Shitz) SS. Information theoretic security. *Found Trends Commun Inf Theory*. 2009;5(4-5):355–580.

[33] Hassibi B and Hochwald BM. How much training is needed in multiple-antenna wireless links? *IEEE Trans Inf Theory*. 2003;49(4):951–963.

[34] Ho Z, Jorswieck E, and Engelmann S. Information leakage neutralization for the multi-antenna non-regenerative relay-assisted multi-carrier interference channel. *IEEE J Sel Areas Commun*. 2013;31(9):1672–1686.

[35] Ayach OE, Rajagopal S, Abu-Surra S, *et al.* Spatially sparse precoding in millimeter wave MIMO systems. *IEEE Trans Wireless Commun*. 2014;13(3):1499–1513.

[36] Grant M and Boyd S. *CVX: Matlab Software for Disciplined Convex Programming*, version 2.1; 2014. http://cvxr.com/cvx.

Part IV

Applications of physical layer security

Chapter 13
Physical layer security for UAV wireless networks

Chenxi Liu[1], Rui Ma[1], Jemin Lee[2] and Tony Q.S. Quek[3]

Unmanned aerial vehicle (UAV) wireless communication has been considered to be a promising solution to provide ubiquitous coverage and enhanced capacity performance in the sixth-generation (6G) wireless networks. Against this background, safeguarding the UAV networks using physical layer security (PLS) techniques has been attracting significant research attention. In this chapter,* we provide a comprehensive overview of PLS techniques in the UAV networks. Open problems and opportunities for PLS in the UAV networks are discussed. In addition, we take a UAV wiretap channel as an example and demonstrate how the design of the secure transmission scheme can effectively improve the secrecy performance of the considered UAV wiretap channel.

To be more specific, we consider a UAV wiretap channel, in which a source communicates with a legitimate UAV in the presence of an eavesdropper. We assume the eavesdropper works in a full-duplex mode. That is, the eavesdropper is capable of performing both eavesdropping and malicious jamming at the same time. The source transmits artificial noise (AN) signals, along with information signals, to interfere with the eavesdropper. We derive the hybrid outage probability, which considers both the transmission outage probability and the secrecy outage probability. We also derive an analytical expression of the asymptotic hybrid outage probability in the regime where the transmit power becomes large. Through analysis and the simulation results, we show how the minimum hybrid outage probability can be achieved by adaptively adjusting the altitude of the UAV, and the power allocation between the AN signals and the information signals.

13.1 Introduction

Unmanned aerial vehicle (UAV) wireless communication has recently been considered to be an effective solution to satisfy various challenging requirements for the

*This chapter has been a part of [1].

[1]School of Information and Communication Engineering, Beijing University of Posts and Telecommunications, China
[2]Department of Electrical and Electronic Engineering, Yonsei University, South Korea
[3]Information Systems Technology and Design Pillar, Singapore University of Technology and Design, Singapore

upcoming sixth-generation (6G) wireless networks, such as reliable and low-latency data transmissions as well as ubiquitous radio coverage [2]. Compared to traditional terrestrial networks or those utilizing high-altitude platforms (HAPs), low-altitude and highly mobile UAVs can achieve capacity enhancement and on-demand coverage in a more cost-effective and flexible manner. Furthermore, UAVs boost the likelihood of establishing short-distance line-of-sight (LoS) communication links, hence improving the link qualities [3]. In terms of whether the UAVs operate as the user equipment (UE) or the aerial base stations (BSs), the UAV wireless networks can be classified into two categories, namely, the cellular-connected UAV networks and the aerial wireless networks. In particular, in cellular-connected UAV networks, the terrestrial cellular networks provide data services to the UAV-UEs, while in the aerial wireless networks, each UAV is mounted with a BS and therefore provides data services to the (terrestrial and/or aerial) UEs. In these two categories of UAV wireless networks, the deployment and the corresponding resource allocation are two key design issues. For example, in [4–6], the deployment of the UAV was optimized to enhance the throughput, successful transmission, and coverage performances, respectively. In [7–9], the transmission power and bandwidth allocation schemes were proposed to improve the energy efficiency [7,8] and reduce the latency [9] of UAV wireless networks.

On the other hand, wireless communications are inherently vulnerable to malicious attacks. Therefore, security is an important issue in the design and implementation of current and future wireless networks. Accordingly, physical layer security in wireless networks has been attracting significant research attention [10–12]. Compared to conventional key-based cryptographic techniques implemented in upper layers, physical layer security can improve the secrecy of wireless data transmissions with no secret keys and complex algorithms, thus making physical layer security more appropriate to the decentralized wireless communication systems of massive amount of users. As a pioneer work, the authors of [13] illustrated the principle of physical layer security which analysis the secrecy of a single-input single-output wiretap channel. The analytical framework in [13] was extended to multiple-input multiple-output (MIMO) wiretap channels, inspired by the advantages of MIMO techniques such as high-speed data rate and high resource efficiency [14–23].

Due to the potential of full-duplex-based techniques in spectral efficiency enhancement [24], this technology has attracted great research attention in recent years [25–28]. In [25], the beamforming and jamming scheme for a single-cell cellular system with a full-duplex BS was jointly designed to improve the secrecy rate. Considering the full-duplex-based legitimate receiver, the authors of [26] explored the optimal jamming covariance matrix for a legitimate receiver which improves the secrecy rate. Taking the use of a full-duplex legitimate receiver into consideration, [26] examined the design issue of jamming covariance matrix at the legitimate receiver which enhances the secrecy rate. The authors of [27,28] explored the advantages of full-duplex relays on secrecy performance enhancement in single-hop relay systems as well as multi-hop relay systems, individually. The works in [14–23,25–28] commonly examined the secrecy performance in the presence of a passive eavesdropper. The authors of [29–32] considered the impact of an active eavesdropper and analyzed the secrecy performance of wiretap channels. In particular, by utilizing game theoretic approaches, the authors of [29,30] investigated the physical layer security for communication systems in the presence of a full-duplex eavesdropper, where the legitimate system perfectly knows the channel state information from the eavesdropper. The authors of [31] examined the ergodic secrecy rate in communication systems with a mode-switching half-duplex eavesdropper, where the eavesdropper performed as an active jammer or a passive eavesdropper. In [32], the secrecy performance of wireless networks with a two-antenna full-duplex eavesdropper was analyzed, where it is assumed that the eavesdropper can eliminate the self-interference.

Despite there has been extensive research on the physical layer security in wireless communication systems, the efforts devoted to the security issues for UAV wireless networks were inadequate. Owing to the potential benefits of UAV communication systems in 6G wireless networks, it is critical to investigate physical layer security in UAV wireless communications. The most related works taking the security issue of UAV wireless communication into consideration are those of [33–39], which commonly focused on the scenarios of aerial wireless networks, and adopted the free-path loss model to characterize the ground-to-air channel. In particular, in [33,34], the deployment strategy of UAV-BS was examined for physical layer security enhancement. In [35], the transmission power at the UAV and the user association were jointly optimized to maximize the secrecy rate of UAV wireless networks. Considering the case where the UAVs act as the relay, the authors of [36,37] investigated the secure transmission schemes that improve the secrecy rate and energy efficiency, respectively. Taking the impacts of the UAV's mobility and jittering into consideration, the secrecy outage probability and secrecy capacity were evaluated in [38,39]. In addition, it was shown that the UAV's mobility and jittering may degrade the secrecy performance of the UAV wireless networks, due to the imperfect CSI and the degraded link reliability. Nevertheless, these works commonly assumed that the source perfectly knows the instantaneous channel state information (CSI) of the links from the eavesdropper, which is hard to realize in practice. Moreover, most of these related works only took the single-antenna nodes into consideration.

Inspired by the above observations, in this chapter, a secure transmission scheme for UAV wiretap channels is proposed considering a source communicates with a legitimate UAV in the presence of a full-duplex active eavesdropper. We assume that the *active* eavesdropper can receive signals and transmit malicious jamming signals to degrade the capability of secrecy transmission between the source and legitimate UAV via passive eavesdropping and/or active jamming [31]. To cope with such an eavesdropper, we assume that the source transmits information signals along with artificial noise (AN) signals to disturb the eavesdropper. Moreover, we adopt the ground-to-air channel model and assume that the non-line-of-sight (NLoS) and LoS probability are node-height-dependent, which is more appropriate for UAV wireless communication systems. To illustrate the performance of the communication secrecy of the legitimate link, we define and derive the hybrid outage probability. We jointly optimize the power allocation factor between information signals and AN signals at the source and UAV operating height to minimize the hybrid outage probability of the UAV wireless communication system. The contributions in this chapter are summarized as follows:

- Instead of using the free-path loss model as [33–37], in this chapter, we characterize the ground-to-air channel with node-height-dependent LoS probability. Besides, we examine a worst-case situation where the jamming signals transmitted from the eavesdropper are directed toward the UAV, which degrades the quality of received signals at the UAV most. We take hybrid outage probability into consideration and derive a compact expression for it. Utilizing the derived expression, we further illustrate the impacts of the UAV operating height, the power allocation factor, and the attacking strategy of the eavesdropper on the secrecy performance.
- To minimize the hybrid outage probability, we optimize the power allocation factor between information signals and AN signals at the source and the operating height of the UAV, respectively. We subsequently examine the impact of critical parameters such as the antenna configurations (the number of antennas at the source and eavesdropper) and the transmission power on the hybrid outage probability.
- From the perspective of the eavesdropper, we examine the optimal antenna configuration which maximizes the hybrid outage probability. In particular, for a fixed number of eavesdropper antennas, we optimize the number of antennas for transmitting jamming signals in different UAV system environments, which suggests that utilizing all antennas to transmit jamming signals is not always the best transmission scheme for the eavesdropper. This can provide a significant reference to design UAV communication systems with guaranteeing a certain level of secrecy, especially when the information of eavesdropper configuration is limited.

The remainder of the chapter is organized as follows. We introduce the system model in Section 13.2. In Section 13.3, we analyze and optimize the secrecy performance achieved by the proposed scheme. We show the simulation and numerical results in Section 13.4, which confirms the accuracy of our analysis. We conclude this chapter in Section 13.5.

13.2 System model

In this section, we introduce our system model, which includes ground-to-UAV channel model, transmission of AN signals, and active eavesdropping.

13.2.1 Ground-to-UAV channel model

We take a wiretap channel into consideration, as shown in Figure 13.1, in which a source (S) communicates with a legitimate UAV (U) in the presence of an eavesdropper (E). In this scenario, the source and the eavesdropper are ground equipment. We consider that the source communicates with the UAV in a half-duplex mode, and the active eavesdropper works in a full-duplex mode, that is, the eavesdropper can complete the information transmission and reception simultaneously. The antenna number of the source and eavesdropper is N_s and N_t, respectively, while the UAV is

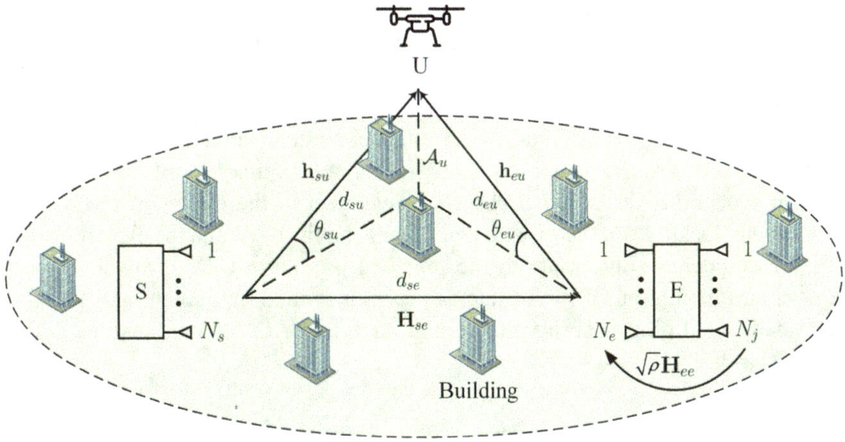

Figure 13.1 Illustration of a wiretap channel consisting of a source, a legitimate UAV, and an eavesdropper. ©IEEE 2019, reprinted with permission from [1].

equipped with a single antenna due to the limited payload. However, we note that our analytical framework can be readily applied to the scenario of multi-antenna UAV. The numbers of receiving antennas and transmitting antennas at the eavesdropper are N_e and N_j, respectively, where $N_e + N_j = N_t$. We assume that the UAV can only perform signal reception when it is static.

Utilizing a three-dimensional Cartesian coordinate, we describe the location of the source, the UAV, and the eavesdropper as $\zeta_s = (0, 0, 0)$, $\zeta_u = (\mu_u, \nu_u, \mathscr{A}_u)$, and $\zeta_e = (\mu_e, \nu_e, 0)$, respectively. We assume that the range of UAV operating height is $0 \leq \mathscr{A}_u \leq \mathscr{A}_u^{\max}$ and the source knows the locations of all nodes.

We consider that the perfect CSI from the source to the UAV is available to the source, and only the statistical character of CSI such as mean and variance of channels between source and eavesdropper, and channels between eavesdropper and UAV are available to the source. We denote $\mathbf{h}_{su}$, $\mathbf{H}_{se}$, and $\mathbf{h}_{eu}$ as the $1 \times N_s$ channel vector from the source to the UAV, the $N_e \times N_s$ channel matrix from the source to the eavesdropper's receiving antennas, and the $1 \times N_j$ channel vector from the eavesdropper's transmitting antennas to the UAV, respectively. We consider that $\mathbf{H}_{se}$ is governed by quasi-static independent and identically distributed (i.i.d.) Rayleigh fading. We assume that $\mathbf{H}_{se}$ is subject to quasi-static independent and identically distributed (i.i.d.) Rayleigh fading. With the emergence of self-interference cancellation techniques, self-interference can be suppressed in the spatial domain [40], digital circuit domain [41,42], and analog circuit domain [43]. We utilize the digital self-interference cancellation in this chapter, which can realize up to 110 dB self-interference cancellation. Hence, the effective self-interference channel from the eavesdropper's transmitting antennas to the eavesdropper's receiving antennas as

$\sqrt{\rho}\mathbf{H}_{ee}$ (as in [26,30]), where $\mathbf{H}_{ee}$ is subject to quasi-static i.i.d. Rayleigh fading and the factor ρ with $0 \leq \rho \leq 1$, characterizes the effect of the digital self-interference cancellation. Specifically, $\rho = 0$ represents no self-interference, and $0 < \rho \leq 1$ characterize the different degree of self-interference.

We now describe the ground-to-UAV channel model involved in this chapter. We assume that each ground equipment can establish a LoS link towards the UAV with a certain probability. The LoS probability is subject to the effects of the wireless environment, locations of the ground equipment and the UAV, and the elevation angle [44]. This model is appropriate to the practical ground-to-UAV channel owing to the random distribution of the obstacles between ground equipment and the UAV. Specifically, the LoS probability between ground device i, $i \in \{s, e\}$, and the UAV is expressed as [44,45]

$$P_{iu}^{\mathrm{L}} = \frac{1}{1 + \varphi \exp\left(-\beta\left(\theta_{iu} - \varphi\right)\right)}, \tag{13.1}$$

in which φ and β are environment-determined constant values, and θ_{iu} represents the elevation angle, that is

$$\theta_{iu} = \frac{180}{\pi} \arcsin\left(\frac{\mathscr{A}_u}{d_{iu}}\right), \tag{13.2}$$

where d_{iu} denotes the distance between ground device i and the UAV, expressed as

$$d_{iu} = \sqrt{\left(\mu_u - \mu_i\right)^2 + \left(\nu_u - \nu_i\right)^2 + \mathscr{A}_u^2}. \tag{13.3}$$

The NLoS probability of the link between node i and the UAV can be presented as $P_{iu}^{\mathrm{NL}} = 1 - P_{iu}^{\mathrm{L}}$. Noting that the LoS probability in (13.1) increases as θ_{iu} increases, i.e., as the height of the UAV increases.

We then denote the path loss exponent model for LoS and NLoS links between the ground device i and the UAV as

$$\eta_{iu} = \begin{cases} \eta_{\mathrm{L}}, & \text{with probability } P_{iu}^{\mathrm{L}} \\ \eta_{\mathrm{NL}}, & \text{with probability } 1 - P_{iu}^{\mathrm{L}} \end{cases}, \tag{13.4}$$

where $\eta_{\mathrm{NL}} > \eta_{\mathrm{L}} > 2$.

We assume that the channel experiences Rician fading for LoS propagation conditions and Rayleigh fading for NLoS propagation conditions. Therefore, the channel between the ground device i and the UAV for LoS propagation is given by

$$\mathbf{h}_{iu} = \sqrt{\frac{K_{iu}}{1 + K_{iu}}}\mathbf{h}_{iu}^{o} + \sqrt{\frac{1}{1 + K_{iu}}}\mathbf{h}_{iu}^{r}, \tag{13.5}$$

where K_{iu} represents the Rician K-factor of the channel between the ground device i and the UAV, $\mathbf{h}_{iu}^{o}$ represents the LoS component of the channel between the ground device i and the UAV, and $\mathbf{h}_{iu}^{r}$ represents the scattered component of the channel between the ground device i and the UAV—the elements of which are assumed to be i.i.d. complex Gaussian random variables with zero mean and unit variance, i.e., $\mathbf{h}_{iu}^{r} \sim \mathcal{CN}\left(\mathbf{0}_{1\times N_s}, \mathbf{I}_{N_s}\right)$. In (13.5), $\mathbf{h}_{iu}^{o}$ is expressed as [46]

$$\mathbf{h}_{iu}^{o} = \left[1, \cdots, \exp\left(-j2\pi\left(N_s - 1\right)\delta_i \sin\beta_{iu}\cos\theta_{iu}\right)\right], \tag{13.6}$$

where δ_i denotes the constant spacing, in wavelengths, between adjacent antennas of the ground device i, and β_{iu} denotes the azimuth angle between the ground device i and the UAV.

13.2.2 Transmission of AN signals

We characterize the transmission strategy between the source and the UAV, where the source transmits AN signals together with information signals, aiming to degrade the eavesdropper's reception quality [16,17,20,21]. Against this backdrop, we denote the signal transmitted from the source as

$$\mathbf{x}_{\mathrm{s}} = \mathbf{W}\mathbf{t}, \tag{13.7}$$

where $\mathbf{W}$ represents an $N_s \times N_s$ beamforming matrix, and $\mathbf{t}$ represents the information signal together with the AN signal. The beamforming matrix $\mathbf{W}$ can be further designed as

$$\mathbf{W} = \left[\mathbf{w}_{\mathrm{s}}, \mathbf{W}_{\mathrm{san}}\right], \tag{13.8}$$

where $\mathbf{w}_{\mathrm{s}}$ and $\mathbf{W}_{\mathrm{san}}$ are used for the information signal and the AN signal at the source, aiming at confusing the eavesdropper, and preserving the reception quality at the UAV, respectively. Specifically, we consider $\mathbf{w}_{\mathrm{s}} = \frac{\mathbf{h}_{su}^{H}}{\|\mathbf{h}_{su}\|}$ and assume $\mathbf{W}_{\mathrm{san}}$ is the projection matrix onto the null space of $\mathbf{h}_{su}$. Hence, we have $\mathbf{h}_{su}\mathbf{W}_{\mathrm{san}} = \mathbf{0}$ and the columns of $\mathbf{W}$ form an orthonormal basis. We subsequently express $\mathbf{t}$ as

$$\mathbf{t} = \begin{bmatrix} t_{\mathrm{s}} \\ \mathbf{t}_{\mathrm{san}} \end{bmatrix}, \tag{13.9}$$

where t_{s} and $\mathbf{t}_{\mathrm{san}}$ represent the information signal and AN signal transmitted from the source, where the dimension of $\mathbf{t}_{\mathrm{san}}$ is $(N_s - 1) \times 1$. We design a factor donated as α_s which represents the proportion of the power allocated to the information signal at the source. Therefore, we have $\mathbb{E}\left[|t_{\mathrm{s}}|^2\right] = \alpha_s$ and $\mathbb{E}\left[\mathbf{t}_{\mathrm{san}}\mathbf{t}_{\mathrm{san}}^{H}\right] = \frac{1-\alpha_s}{N_s-1}\mathbf{I}_{N_s-1}$.

13.2.3 Active eavesdropping

Then, we illustrate the eavesdropping scheme utilized by the active eavesdropper. We consider that the knowledge of $\mathbf{h}_{eu}$ is available to the full-duplex eavesdropper, which can be used to realize directional jamming transmission to impact the quality of the signal reception at the UAV. Concurrently, the eavesdropper receives signals from the source. Specifically, we denote the directional jamming signal transmitted by the eavesdropper as

$$\mathbf{x}_{\mathrm{eu}} = \mathbf{w}_{\mathrm{eu}} t_{\mathrm{eu}}, \tag{13.10}$$

where $\mathbf{w}_{\mathrm{eu}}$ is designed for realizing the transmission of directional jamming signal, t_{eu}, from the eavesdropper. We assume $\mathbb{E}\left[|t_{\mathrm{eu}}|^2\right] = 1$. Moreover, we consider that

$\mathbf{w}_{\text{eu}} = \frac{\mathbf{h}_{eu}^H}{\|\mathbf{h}_{eu}\|}$, thus the quality of received signal at the UAV is degraded most. Hence, the received signal at the UAV can be expressed as

$$y_u = \sqrt{P_s d_{su}^{-\eta_{su}}} \mathbf{h}_{su} \mathbf{w}_{\text{s}} t_{\text{s}} + \sqrt{P_e d_{eu}^{-\eta_{eu}}} \mathbf{h}_{eu} \mathbf{w}_{\text{eu}} t_{\text{eu}} + n_u, \tag{13.11}$$

where P_s and P_e represent the transmit power at the source and the eavesdropper, respectively, and n_u denotes the thermal noise at the UAV, which is assumed to be a complex Gaussian random variable with zero mean and variance σ_u^2, i.e., $n_u \sim \mathcal{CN}\left(0, \sigma_u^2\right)$. We also express the received signal at the eavesdropper as

$$\mathbf{y}_e = \sqrt{P_s d_{se}^{-\eta_{se}}} \mathbf{H}_{se} \left(\mathbf{w}_{\text{s}} t_{\text{s}} + \mathbf{W}_{\text{san}} \mathbf{t}_{\text{san}}\right) + \sqrt{\rho P_e} \mathbf{H}_{ee} \mathbf{x}_{\text{e}} + \mathbf{n}_e, \tag{13.12}$$

where $\mathbf{n}_e$ denotes the thermal noise vector at the eavesdropper's receiving antennas, the elements of which are assumed to be i.i.d. complex Gaussian random variables with zero mean and variance σ_e^2, i.e., $\mathbf{n}_e \sim \mathcal{CN}\left(\mathbf{0}_{1\times N_e}, \sigma_e^2 \mathbf{I}_{N_e}\right)$.

Based on (13.11), we express the received signal-to-interference-plus-noise ratio (SINR) at the UAV as

$$\gamma_u = \frac{\alpha_s \tilde{\gamma}_{su}}{\tilde{\gamma}_{eu} + 1}, \tag{13.13}$$

where $\tilde{\gamma}_{su} = \overline{\gamma}_{su} \|\mathbf{h}_{su}\|^2$ with $\overline{\gamma}_{su} = P_s d_{su}^{-\eta_{su}} / \sigma_u^2$ and $\tilde{\gamma}_{eu} = \overline{\gamma}_{eu} \|\mathbf{h}_{eu}\|^2$ with $\overline{\gamma}_{eu} = P_e d_{eu}^{-\eta_{eu}} / \sigma_u^2$. We clarify that in (13.13) only the value of $\tilde{\gamma}_{su}$ is available at the source since we consider that the instantaneous knowledge of $\mathbf{h}_{eu}$ is not available at the source.

We assume that the eavesdropper applies the minimum mean square error (MMSE) combining to obtain the maximum instantaneous SINR [47]. The MMSE combiner at the eavesdropper is expressed as

$$\mathbf{v} = \mathbf{R}^{-1} \mathbf{H}_{se} \mathbf{w}_{\text{s}}, \tag{13.14}$$

where

$$\mathbf{R} = \frac{1 - \alpha_s}{N_s - 1} \overline{\gamma}_{se} \mathbf{H}_{se} \mathbf{W}_{\text{san}} \mathbf{W}_{\text{san}}^H \mathbf{H}_{se}^H + \overline{\gamma}_{ee} \mathbf{H}_{ee} \mathbf{x}_{\text{e}} \mathbf{x}_{\text{e}}^H \mathbf{H}_{ee}^H + \mathbf{I}_{N_e} \tag{13.15}$$

with $\overline{\gamma}_{se} = P_s d_{se}^{-\eta_{se}} / \sigma_e^2$ and $\overline{\gamma}_{ee} = \rho P_e / \sigma_e^2$. Using the MMSE combining, the SINR at the eavesdropper is expressed as [47]

$$\gamma_e = \alpha_s \overline{\gamma}_{se} \mathbf{w}_{\text{s}}^H \mathbf{H}_{se}^H \mathbf{R}^{-1} \mathbf{H}_{se} \mathbf{w}_{\text{s}}. \tag{13.16}$$

To achieve the MMSE combining at the eavesdropper, we consider that the eavesdropper has the perfect knowledge of $\mathbf{H}_{se}$, $\mathbf{w}_{\text{s}}$, and $\mathbf{R}$, which refers to the worst-case scenario in terms of the secrecy.† We state that the source cannot obtain the value of γ_e, due to the instantaneous knowledge of $\mathbf{H}_{se}$ and $\mathbf{H}_{ee}$ is not available to the source. We utilize the statistical CSI of source-eavesdropper and eavesdropper-UAV links to characterize $\tilde{\gamma}_{eu}$ and γ_e, which will be adopted to derive the communication secrecy performance in Section 13.3.

†Note that this assumption has been commonly adopted in existing works on the physical layer security (e.g., [18,19]).

13.3 Hybrid outage probability analysis

In this section, the communication secrecy performance of the transmission scheme proposed in Section 13.2 is illustrated. By jointly considering the transmission outage probability and the secrecy outage probability, a compact expression is derived to characterize the hybrid outage probability. We further explore the optimal power allocation factor α_s and optimize the operating height of the UAV to minimize the hybrid outage probability. We subsequently obtain a simple yet valuable expression for the asymptotic hybrid outage probability in the high SNR regime. Furthermore, we explore the hybrid outage probability in the presence of a passive eavesdropper, which can be a special case of the proposed analytical framework.

13.3.1 Statistics of SINRs

The statistics of γ_u and γ_e for the proposed system are illustrated in this subsection, which will be utilized to derive the expressions of the hybrid outage probability. Owing to perfect CSI of the main link is available to the source, such as the value of $\tilde{\gamma}_{su}$, the conditional cumulative distribution function (CDF) of γ_u for a given $\tilde{\gamma}_{su}$ can be derived in the following lemma.

Lemma 1. *For a given $\tilde{\gamma}_{su}$, the conditional CDF of the SINR at UAV, γ_u, is derived as*

$$F_{\gamma_u}(z|\tilde{\gamma}_{su}) = \begin{cases} F_{\gamma_u}^{(1)}(z|\tilde{\gamma}_{su}), \text{ with probability } P_{eu}^{L}, \\ F_{\gamma_u}^{(2)}(z|\tilde{\gamma}_{su}), \text{ with probability } 1 - P_{eu}^{L}, \end{cases} \tag{13.17}$$

where $F_{\gamma_u}^{(1)}(z|\tilde{\gamma}_{su})$ and $F_{\gamma_u}^{(2)}(z|\tilde{\gamma}_{su})$ are given by

$$F_{\gamma_u}^{(1)}(z|\tilde{\gamma}_{su}) = 1 - \frac{\gamma\left(N_j m_{eu}, \frac{\frac{\alpha_s \tilde{\gamma}_{su}}{z} - 1}{m_{eu}^{-1}\overline{\gamma}_{eu}}\right)}{\Gamma\left(N_j m_{eu}\right)}, \tag{13.18}$$

$$F_{\gamma_u}^{(2)}(z|\tilde{\gamma}_{su}) = \exp\left(-\frac{\frac{\alpha_s \tilde{\gamma}_{su}}{z} - 1}{\overline{\gamma}_{eu}}\right) \sum_{p=1}^{N_j} \frac{\left(\frac{\frac{\alpha_s \tilde{\gamma}_{su}}{z} - 1}{\overline{\gamma}_{eu}}\right)^{p-1}}{\Gamma(p)}. \tag{13.19}$$

Here, $m_{eu} = \frac{(K_{eu}+1)^2}{2K_{eu}+1}$, $\gamma(\cdot,\cdot)$ is the lower incomplete gamma function, defined as $\gamma(\mu,\nu) = \int_0^{\nu} \exp(-t)\, t^{\mu-1} dt$ [48, Eq. (8.350)], and $\Gamma(\cdot)$ is the Gamma function, defined as $\Gamma(z) = \int_0^{\infty} \exp(-t)\, t^{z-1} dt$ [48, Equation (8.310)].

With the aid of [49], we derive the CDF of the SINR at the eavesdropper, γ_e, as

$$F_{\gamma_e}(z) = 1 - \frac{\exp\left(-\frac{z}{\alpha_s \overline{\gamma}_{se}}\right)}{(1+\kappa_1 z)^{N_s-1}(1+\kappa_2 z)^{N_j}} \times \sum_{m=1}^{N_e} \frac{1}{\Gamma(m)} \left(\frac{z}{\alpha_s \overline{\gamma}_{se}}\right)^{m-1} \sum_{n=0}^{A} \binom{N_j}{n} (\kappa_2 z)^n \times \sum_{t=0}^{N_e-m-n} \binom{N_s-1}{t} (\kappa_1 z)^t, \tag{13.20}$$

where $A = \min\left(N_j, N_e - m\right)$, $\kappa_1 = \frac{1-\alpha_s}{\alpha_s(N_s-1)}$, and $\kappa_2 = \frac{\overline{\gamma}_{ee}}{\alpha_s\overline{\gamma}_{se}N_j}$. Based on (13.20), we then obtain the probability density function (PDF) of γ_e as

$$f_{\gamma_e}(z) = \frac{\exp\left(-\frac{z}{\alpha_s\overline{\gamma}_{se}}\right)}{(1+\kappa_1 z)^{N_s-1}(1+\kappa_2 z)^{N_j}} \sum_{m=1}^{N_e} \frac{\left(\frac{z}{\alpha_s\overline{\gamma}_{se}}\right)^{m-1}}{\Gamma(m)}$$
$$\times \sum_{n=0}^{A} \binom{N_j}{n} (\kappa_2 z)^n \sum_{t=0}^{N_e-m-n} \binom{N_s-1}{t} (\kappa_1 z)^t$$
$$\times \left(\frac{1}{\alpha_s\overline{\gamma}_{se}} - \frac{m-1+n+t}{z} + \frac{1-\alpha_s}{\alpha_s(1+\kappa_1 z)} + \frac{\kappa_2 N_j}{(1+\kappa_2 z)}\right). \quad (13.21)$$

13.3.2 Hybrid outage probability

In this subsection, we clarify the definition of the transmission outage event and the secrecy outage event. Subsequently, by jointly considering the transmission outage probability and the secrecy outage probability, we derive the expression of hybrid outage probability for the proposed transmission scheme.

Based on (13.13)–(13.15), the achievable secrecy rate of the system is expressed as [14]

$$C_s = \begin{cases} C_u - C_e, & \gamma_u > \gamma_e \\ 0, & \gamma_u \le \gamma_e, \end{cases} \quad (13.22)$$

where $C_u = \log_2(1+\gamma_u)$ is the capacity of the main channel between the source and the UAV, and $C_e = \log_2(1+\gamma_e)$ is the capacity of the eavesdropper's channel. In order to secure the transmissions, we consider the use of a wiretap code [13] with the parameter pair (R_u, R_e), where R_u denotes the transmission rate of the wiretap code, and R_e denotes the redundancy rate of the wiretap code, representing the cost of preventing the transmitted signals from being eavesdropped. In addition, we note that $R_s = R_b - R_e$ denotes the target secrecy rate. During the transmission, there are three mutually exclusive events that can occur, namely, the secure transmission event, the transmission outage event, and the secrecy outage event. Specifically, the secure transmission event occurs if $C_u \ge R_u$ and $C_e < R_e$. In this case, the perfect secrecy is guaranteed. If $C_u < R_u$ and $C_e < R_e$, the transmission outage occurs. In this case, the main channel between the source and the UAV cannot support the transmission of the wiretap code. If $C_e \ge R_e$, the secrecy outage event occurs. In this case, the information on the transmitted signal is leaked to Eve, and the secrecy is compromised. We cannot guarantee perfect secrecy in the proposed system, due to the instantaneous CSI of the eavesdropper-source link is not available to the source. Hence, to assess the impact of the eavesdropper on the secrecy performance of our proposed system, we utilize the hybrid outage probability as a performance metric, expressed as

$$P_{\text{ho}}\left(\alpha_s, \mathscr{A}_u, R_u, R_e\right) = \underbrace{\Pr\left(C_u \ < R_u, C_e < R_e|\tilde{\gamma}_{su}\right)}_{P_{\text{to}}\left(\alpha_s, \mathscr{A}_u, R_u, R_e\right)} + \underbrace{\Pr\left(C_e \geq R_e|\tilde{\gamma}_{su}\right)}_{P_{\text{so}}\left(\alpha_s, \mathscr{A}_u, R_u, R_e\right)}, \quad (13.23)$$

where $P_{\text{to}}\left(\alpha_s, \mathscr{A}_u, R_u, R_e\right)$ denotes the transmission outage probability and $P_{\text{so}}\left(\alpha_s, \mathscr{A}_u, R_u, R_e\right)$ denotes the secrecy outage probability. We highlight that the hybrid outage probability is an appropriate performance metric for our system, due to the fact we consider the full-duplex eavesdropper who can perform eavesdropping (that affects secrecy outage) and malicious jamming (that affects transmission outage) simultaneously. Note that in our system the hybrid outage probability is a function of the power allocation factor at the source α_s, the operating height of UAV $\mathscr{A}_u$, and the wiretap code rates (R_u, R_e).

Based on (13.17), (13.20), and (13.21), we now examine the hybrid outage probability of the system in the following theorem.‡

Theorem 1. *The hybrid outage probability of the system is derived as*

$$\begin{aligned} P_{ho}\left(\alpha_s, \mathscr{A}_u, R_u, R_e\right) &= P_{eu}^{L} P_{ho}^{(1)}\left(\alpha_s, \mathscr{A}_u, R_u, R_e\right) \\ &\quad + \left(1 - P_{eu}^{L}\right) P_{ho}^{(2)}\left(\alpha_s, \mathscr{A}_u, R_u, R_e\right), \end{aligned} \quad (13.24)$$

where $P_{ho}^{(1)}\left(\alpha_s, \mathscr{A}_u, R_u, R_e\right)$ and $P_{ho}^{(2)}\left(\alpha_s, \mathscr{A}_u, R_u, R_e\right)$ are the hybrid outage probabilities when the eavesdropper-UAV channel is LoS and NLoS, given by (13.25) *and* (13.26), *shown at the top of the next page, where $\tau_u = 2^{R_u} - 1$ and $\tau_e = 2^{R_e} - 1$.*

$$\begin{aligned} &P_{ho}^{(1)}\left(\alpha_s, \mathscr{A}_u, R_u, R_e\right) \\ &= 1 - \frac{\gamma\left(N_j m_{eu}, \frac{\frac{\alpha_s \tilde{\gamma}_{su}}{\tau_u} - 1}{m_{eu}^{-1} \overline{\gamma}_{eu}}\right)}{\Gamma\left(N_j m_{eu}\right)} \left(1 - \frac{\exp\left(-\frac{\tau_e}{\alpha_s \overline{\gamma}_{se}}\right)}{\left(1 + \kappa_1 \tau_e\right)^{N_s - 1} \left(1 + \kappa_2 \tau_e\right)^{N_j}}\right. \\ &\left.\times \sum_{m=1}^{N_e} \frac{1}{\Gamma(m)} \left(\frac{\tau_e}{\alpha_s \overline{\gamma}_{se}}\right)^{m-1} \sum_{n=0}^{A} \binom{N_j}{n} \left(\kappa_2 \tau_e\right)^n \sum_{t=0}^{N_e - m - n} \binom{N_s - 1}{t} \left(\kappa_1 \tau_e\right)^t\right). \end{aligned} \quad (13.25)$$

‡Here, the analysis of the hybrid outage probability is non-trivial as we consider the full-duplex eavesdropper. As such, in addition to the AN signals from the source, the eavesdropper also suffers from the self-interference. Consequently, the distribution of the SINR at the eavesdropper changes and results in existing works that use AN only (e.g., [16–18]) cannot be applied directly to our analysis.

$$
\begin{aligned}
&P_{ho}^{(2)}\left(\alpha_s, \mathscr{A}_u, R_u, R_e\right) \\
&= \exp\left(-\frac{\frac{\alpha_s \tilde{\gamma}_{su}}{\tau_u} - 1}{\overline{\gamma}_{eu}}\right) \sum_{p=1}^{N_j} \frac{\left(\frac{\frac{\alpha_s \tilde{\gamma}_{su}}{\tau_u} - 1}{\overline{\gamma}_{eu}}\right)^{p-1}}{\Gamma(p)} + \left(1 - \exp\left(-\frac{\frac{\alpha_s \tilde{\gamma}_{su}}{\tau_u} - 1}{\overline{\gamma}_{eu}}\right)\right. \\
&\left.\times \sum_{p=1}^{N_j} \frac{\left(\frac{\frac{\alpha_s \tilde{\gamma}_{su}}{\tau_u} - 1}{\overline{\gamma}_{eu}}\right)^{p-1}}{\Gamma(p)}\right) \frac{\exp\left(-\frac{\tau_e}{\alpha_s \overline{\gamma}_{se}}\right)}{\left(1 + \kappa_1 \tau_e\right)^{N_s - 1}\left(1 + \kappa_2 \tau_e\right)^{N_j}} \sum_{m=1}^{N_e} \frac{1}{\Gamma(m)} \left(\frac{\tau_e}{\alpha_s \overline{\gamma}_{se}}\right)^{m-1} \\
&\times \sum_{n=0}^{A} \binom{N_j}{n} \left(\kappa_2 \tau_e\right)^n \sum_{t=0}^{N_e - m - n} \binom{N_s - 1}{t} \left(\kappa_1 \tau_e\right)^t . \qquad (13.26)
\end{aligned}
$$

Note that Theorem 1 provides an efficient tool for evaluating the hybrid outage probability, and the derived expression is valid for arbitrary values of N_s, $\overline{\gamma}_{su}$, $\overline{\gamma}_{eu}$, K_{su}, and K_{eu}. In this chapter, we adopt the adaptive transmission scheme proposed in [15], of which the secrecy performance can be optimized for each specific realization of the main channel by judiciously selecting transmission parameters. Hence, the hybrid outage probability is analyzed for a realization of the main channel as in [15–18]. We also note that our analysis of the hybrid outage probability for a given $\tilde{\gamma}_{su}$ can be readily extended to obtain the overall hybrid outage probability, i.e., $\mathbb{E}_{\tilde{\gamma}_{su}}\left[P_{\text{ho}}\left(\alpha_s, \mathscr{A}_u, R_u, R_e\right)\right]$, by using the PDF of $\tilde{\gamma}_{su}$, given by

$$
f_{\tilde{\gamma}_{su}}(\gamma) = \begin{cases} f_{\tilde{\gamma}_{su}}^{(1)}(\gamma), \text{ with probability } P_{su}^{\mathrm{L}}, \\ f_{\tilde{\gamma}_{su}}^{(2)}(\gamma), \text{ with probability } 1 - P_{su}^{\mathrm{L}}, \end{cases} \qquad (13.27)
$$

In (13.27), $f_{\tilde{\gamma}_{su}}^{(1)}(\gamma)$ and $f_{\tilde{\gamma}_{su}}^{(2)}(\gamma)$ are given by

$$
f_{\tilde{\gamma}_{su}}^{(1)}(\gamma) = \left(\frac{m_{su}}{\overline{\gamma}_{su}}\right)^{N_s m_{su}} \frac{\gamma^{N_s m_{su} - 1}}{\Gamma\left(N_s m_{su}\right)} \exp\left(\frac{-m_{su} \gamma}{\overline{\gamma}_{su}}\right), \qquad (13.28)
$$

$$
f_{\tilde{\gamma}_{su}}^{(2)}(\gamma) = \frac{\exp\left(-\frac{\gamma}{\overline{\gamma}_{su}}\right) \gamma^{N_s - 1}}{\overline{\gamma}_{su}^{N_s} \Gamma\left(N_s\right)}. \qquad (13.29)
$$

Here, $m_{su} = \frac{(K_{su}+1)^2}{2K_{su}+1}$.

Our key goal is to determine the optimal parameter pair $\left(\alpha_s^{*\circ}, \mathscr{A}_u^{*\circ}\right)$ that minimizes the hybrid outage probability. Accordingly, the optimization problem can be formulated as

$$
\min_{\alpha_s, \mathscr{A}_u} \quad P_{\text{ho}}\left(\alpha_s, \mathscr{A}_u, R_u, R_e\right), \qquad (13.30a)
$$

$$
\text{s.t.} \quad 0 < \alpha_s \leq 1, \qquad (13.30b)
$$

$$
0 \leq \mathscr{A}_u \leq \mathscr{A}_u^{\max}. \qquad (13.30c)
$$

In (13.30), it is difficult to analyze the convexity of the objective function, i.e., $P_{\rm ho}(\alpha_s, \mathscr{A}_u, R_u, R_e)$, since $\mathscr{A}_u$ is implicitly embedded into the elevation angles and the distance between terrestrial devices and the UAV. Consequently, it is difficult to obtain the optimal α_s and $\mathscr{A}_u$ mathematically. However, we can handle the problem (13.30) in two steps:

1. First, for a fixed operating height of UAV $\mathscr{A}_u$, we find the power allocation factor, $\alpha_s^*(\mathscr{A}_u)$, that minimizes the hybrid outage probability. The hybrid outage probability achieved by $\alpha_s^*(\mathscr{A}_u)$ is denoted by $P_{\rm ho}\left(\alpha_s^*(\mathscr{A}_u), \mathscr{A}_u, R_u, R_e\right)$.
2. Then, we find the optimal height of UAV, $\mathscr{A}_u^{*\circ}$, that minimizes $P_{\rm ho}(\alpha_s^*(\mathscr{A}_u), \mathscr{A}_u, R_u, R_e)$ for $0 \le \mathscr{A}_u \le \mathscr{A}_u^{\max}$. Accordingly, the value of $\alpha_s^*(\mathscr{A}_u)$ associated with $\mathscr{A}_u^{*\circ}$ is defined as $\alpha^{*\circ}$.

Finally, the minimum hybrid outage probability can be achieved by $\left(\alpha_s^{*\circ}, \mathscr{A}_u^{*\circ}\right)$, and the hybrid outage probability becomes $P_{\rm ho}^{*\circ}(\alpha_s, \mathscr{A}_u, R_u, R_e)$. We note that although obtaining a closed-form of $\left(\alpha_s^{*\circ}, \mathscr{A}_u^{*\circ}\right)$ is mathematically intractable, we are able to determine $\left(\alpha_s^{*\circ}, \mathscr{A}_u^{*\circ}\right)$ numerically, as will be shown in Section 13.4.

13.3.3 Asymptotic analysis of hybrid outage probability

In this subsection, we examine the asymptotic behavior of the hybrid outage probability. To this end, we consider that $P_s \to \infty$, $P_e \to \infty$, and $P_s/P_e = \xi$. We now quantize the asymptotic behavior of $P_{\rm ho}(\alpha_s, \mathscr{A}_u, R_u, R_e)$ in the following corollary.

Corollary 1. *The asymptotic hybrid outage probability when $P_s \to \infty$, $P_e \to \infty$, and $P_s/P_e = \xi$ is given by*

$$P_{ho}^{\infty}(\alpha_s, \mathscr{A}_u, R_u, R_e) = P_{eu}^{L} P_{ho}^{(1)\infty}(\alpha_s, \mathscr{A}_u, R_u, R_e) + \left(1 - P_{eu}^{L}\right) P_{ho}^{(2)\infty}(\alpha_s, \mathscr{A}_u, R_u, R_e), \tag{13.31}$$

where $P_{ho}^{(1)\infty}(\alpha_s, \mathscr{A}_u, R_u, R_e)$ and $P_{ho}^{(2)\infty}(\alpha_s, \mathscr{A}_u, R_u, R_e)$ are given by (13.32) *and* (13.33)*, shown at the top of the next page, where $\kappa_3 = \xi \phi_u N_j$ and $\phi_u = \frac{d_{eu}^{\eta_{eu}}}{d_{su}^{\eta_{su}}} \|\mathbf{h}_{su}\|^2$.*

$$\begin{aligned} P_{ho}^{(1)\infty}(\alpha_s, \mathscr{A}_u, R_u, R_e) = 1 - &\frac{\gamma\left(N_j m_{eu}, \frac{\alpha_s m_{eu} \kappa_3}{\tau_u}\right)}{\Gamma\left(N_j m_{eu}\right)} \left(1 - (1 + \kappa_1 \tau_e)^{-(N_s - 1)}\right. \\ &\times \left(1 + \frac{d_{se}^{\eta_{se}} \tau_e}{\alpha_s \xi N_j}\right)^{-N_j} \sum_{n=0}^{B} \binom{N_j}{n} \left(\frac{d_{se}^{\eta_{se}} \tau_e}{\alpha_s \xi N_j}\right)^n \\ &\times \left.\sum_{t=0}^{N_e - 1 - n} \binom{N_s - 1}{t} (\kappa_1 \tau_e)^t \right). \end{aligned} \tag{13.32}$$

$$P_{ho}^{(2)\infty}\left(\alpha_s,\mathscr{A}_u,R_u,R_e\right)=\exp\left(-\frac{\alpha_s\kappa_3}{\tau_u}\right)\sum_{p=1}^{N_j}\frac{\left(\frac{\alpha_s\kappa_3}{\tau_u}\right)^{p-1}}{\Gamma(p)}$$
$$+\left(1-\exp\left(-\frac{\alpha_s\kappa_3}{\tau_u}\right)\sum_{p=1}^{N_j}\frac{\left(\frac{\alpha_s\kappa_3}{\tau_u}\right)^{p-1}}{\Gamma(p)}\right)\left(1+\kappa_1\tau_e\right)^{-(N_s-1)}$$
$$\times\left(1+\frac{d_{se}^{\eta_{se}}\tau_e}{\alpha_s\xi N_j}\right)^{-N_j}\sum_{n=0}^{B}\binom{N_j}{n}\left(\frac{d_{se}^{\eta_{se}}\tau_e}{\alpha_s\xi N_j}\right)^n$$
$$\times\sum_{t=0}^{N_e-1-n}\binom{N_s-1}{t}\left(\kappa_1\tau_e\right)^t. \tag{13.33}$$

We find that $P_{\text{ho}}\left(\alpha_s,\mathscr{A}_u,R_u,R_e\right)$ approaches a certain value that is independent of P_s and P_e when $P_s\to\infty$, $P_e\to\infty$ and $P_s/P_e=\xi$. This value is determined by ξ, and decreases as ξ increases.

13.3.4 Passive eavesdropping

So far, both the exact and the asymptotic hybrid outage probability of UAV wiretap channels in the presence of a full-duplex active eavesdropper have been analyzed. However, the question: "What is the most harmful eavesdropper setting: active eavesdropping or passive eavesdropping?" is still confusing. To explore the answer to the question, in this subsection, we take a passive eavesdropper into consideration and analyze the hybrid outage probability for UAV wiretap channels. For the sake of fairness, we assume that the number of passive eavesdropper's antenna is N_t, where $N_t=N_e+N_j$.

According to [17], the hybrid outage probability for the considered system is given by

$$P_{\text{ho}}^{(p)}\left(\alpha_s,\mathscr{A}_u,R_s\right)$$
$$=\frac{\exp\left(-\frac{2^{-R_s}\kappa_4}{\alpha_s\overline{\gamma}_{se}}\right)}{\left(1+2^{-R_s}\kappa_1\kappa_4\right)^{N_s-1}}\sum_{p=1}^{N_t}\frac{1}{\Gamma(p)}\left(\frac{2^{-R_s}\kappa_4}{\alpha_s\overline{\gamma}_{se}}\right)^{p-1}\times\sum_{q=0}^{N_t-p}\binom{N_s-1}{q}\left(2^{-R_s}\kappa_1\kappa_4\right)^q, \tag{13.34}$$

where $\kappa_4=1+\alpha_s\tilde{\gamma}_{su}-2^{R_s}$.

We note that the expression in (13.34) enables us to compare the secrecy performance of the system in the presence of a full-duplex active eavesdropper with that in the presence of a passive eavesdropper.

13.4 Simulations and numerical analysis

In this section, the simulation results are presented to validate our analytical framework and evaluate the benefits of the proposed transmission scheme. We also explore the effect of eavesdropping manner (e.g., active eavesdropping and passive eavesdropping) on the hybrid outage probability. Furthermore, we illustrate the significant impact of UAV operating height on the hybrid outage probability. Without loss of generality, we take normalized distances between nodes into consideration in this section. In particular, we assume that the distance between the source and the eavesdropper is the reference distance and normalize other distances between nodes accordingly. We also consider that $\sigma_d^2 = \sigma_e^2 = 1$. In addition, we consider an urban environment with $\varphi = 11.95$ and $\beta = 0.14$ [44].

13.4.1 Optimal configuration of legitimate system

By utilizing Monte Carlo simulations, we validate the accuracy of the hybrid outage probability. In Figure 13.2, we plot $P_{\text{ho}}\left(\alpha_s, \mathscr{A}_u, R_u, R_e\right)$ versus P_s for different values of N_s with $N_e = 1$, $N_j = 2$, $\alpha_s = 0.5$, $P_e = 10$ dB, $\theta_{su} = \pi/6$, $\theta_{eu} = \pi/4$, $\zeta_e = (1,0,0)$, $\zeta_u = \left(3/4, \sqrt{3}/4, 1/2\right)$, $R_u = 2$ bits/Hz/s, and $R_e = 1$ bits/Hz/s. Figure 13.2 illustrates that the analytical curves which are generated from Theorem 1 precisely match the Monte Carlo simulation results marked by red circles, thus validating the correctness of the analytical expression for $P_{\text{ho}}\left(\alpha_s, \mathscr{A}_u, R_u, R_e\right)$ in Theorem 1. Subsequently, we observe that $P_{\text{ho}}\left(\alpha_s, \mathscr{A}_u, R_u, R_e\right)$ decreases as P_s increases. Moreover, we see that $P_{\text{ho}}\left(\alpha_s, \mathscr{A}_u, R_u, R_e\right)$ decreases significantly as N_s increases. These results imply that the secrecy performance of the proposed system can be enhanced by increasing the transmission power of the source and/or equipping more antennas at the source.

Figure 13.3 illustrates the $P_{\text{ho}}\left(\alpha_s, \mathscr{A}_u, R_u, R_e\right)$ versus α_s for different configurations of antenna number at the source. The solid and dotted lines represent the hybrid outage probability of the considered system in the presence of a full-duplex active eavesdropper with different values of ρ. As the value of α_s changes from 0 to 1, the hybrid outage probability first decreases and then increases. Therefore, we can find a unique $\alpha_s^*(\mathscr{A}_u)$ that minimizes the hybrid outage probability. The explanation for this observation is as follows. For a small α_s, the power assigned to transmit information signal is little, which degrades the quality of information reception at the UAV and results in a high hybrid outage probability. As α_s become larger, the hybrid outage probability is reduced. Nevertheless, as the value of α_s larger than $\alpha_s^*(\mathscr{A}_u)$, the benefit of increasing the transmit power to information signals is limited, while the impact of reducing the transmit power of AN on the secrecy performance becomes dominant which results in an improved link quality of the eavesdropper's channel. Consequently, the hybrid outage probability becomes larger. Moreover, we also observe that the hybrid outage probability increases as ρ reduces, owing to less self-interference at the eavesdropper. Furthermore, we observe that $\alpha_s^*(\mathscr{A}_u)$ reduces as the value of ρ reduces or N_s increases. The reason is that as the value of ρ reduces or N_s increases, the performance of eavesdropping at the eavesdropper is enhanced.

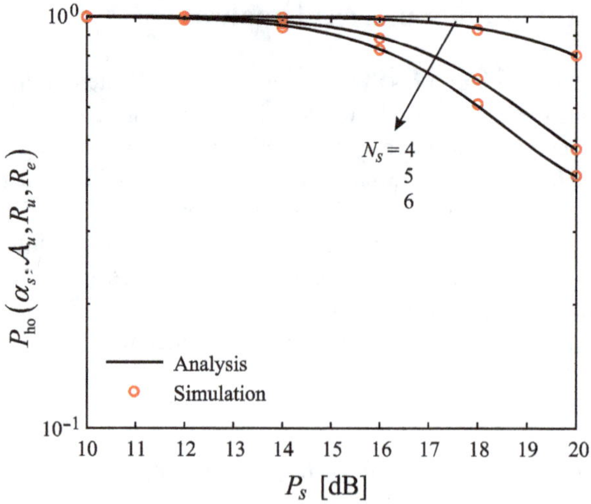

Figure 13.2 $P_{ho}\,(\alpha_s, \mathscr{A}_u, R_u, R_e)$ *versus* P_s *for different values of* N_s *with* $N_e = 1$, $N_j = 2$, $\alpha_s = 0.5$, $\rho = 1$, $P_e = 10$ *dB*, $\beta_{su} = \pi/3$, $\theta_{su} = \pi/6$, $\beta_{eu} = \pi/6$, $\theta_{eu} = \pi/4$, $\zeta_e = (1, 0, 0)$, $\zeta_u = \left(3/4, \sqrt{3}/4, 1/2\right)$, $R_u = 2$ *bits/Hz/s, and* $R_e = 1$ *bits/Hz/s. ©IEEE 2019, reprinted with permission from [1].*

Therefore, to constrain the reception quality at the eavesdropper, more transmit power needs to be assigned to AN signals.

To explore the effect of the operating height of UAV on the hybrid outage probability, Figure 13.4 illustrates $P_{\mathrm{ho}}\left(\alpha_s^*(\mathscr{A}_u), \mathscr{A}_u, R_u, R_e\right)$ versus $\mathscr{A}_u$ for different values of P_s and ρ for a given $\alpha_s^*(\mathscr{A}_u)$. We clarify that when the operating height of the UAV $\mathscr{A}_u = 0$, the channels for source-UAV and UAV-eavesdropper links are both subject to Rayleigh fading. We see that the value of $P_{\mathrm{ho}}\left(\alpha_s^*(\mathscr{A}_u), \mathscr{A}_u, R_u, R_e\right)$ increases as ρ reduces. As $\mathscr{A}_u$ becomes larger, the value of $P_{\mathrm{ho}}\left(\alpha_s^*(\mathscr{A}_u), \mathscr{A}_u, R_u, R_e\right)$ first increases and then decreases. The reason is that, as the $\mathscr{A}_u$ is in a small value regime, the hybrid outage probability decreases with $\mathscr{A}_u$ even though the legitimate communication distance increase. The reason is that the jamming signal received at the UAV decreases as $\mathscr{A}_u$ becomes larger, and the legitimate channel is more likely to establish LoS link as the $\mathscr{A}_u$ increase. Nevertheless, as $\mathscr{A}_u$ works in a large regime and keeps increasing, the transmission outage probability becomes larger as the legitimate link distance increases while the impact of jamming signal is subtle. Therefore, the hybrid outage probability becomes smaller as $\mathscr{A}_u$ increases when $\mathscr{A}_u$ is in a large value regime. This observation reveals that by judiciously selecting the operating height of the UAV, the communication secrecy performance of the system is able to be enhanced, which is consistent with our analysis.

Figure 13.5 illustrates the minimum hybrid outage probability $P_{\mathrm{ho}}^{*\circ}\,(\alpha_s, \mathscr{A}_u, R_u, R_e)$ versus P_s for different eavesdropper configurations and different values of N_s. We

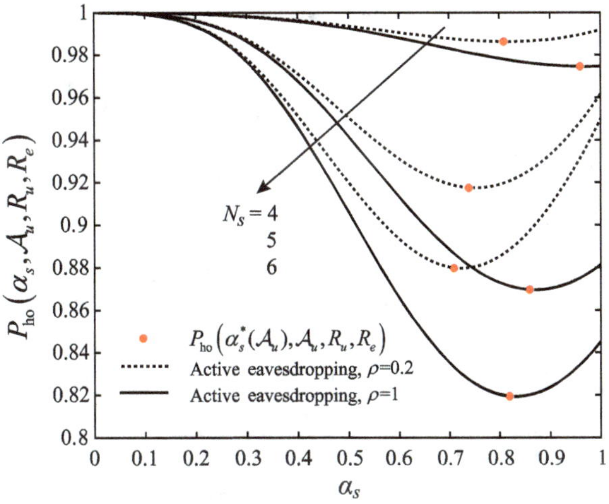

Figure 13.3 $P_{ho}\left(\alpha_s, \mathscr{A}_u, R_u, R_e\right)$ *versus* α_s *for different values of* N_s *with* $N_e = 1$, $N_j = 2$, $P_s = 15$ *dB,* $P_e = 10$ *dB,* $\beta_{su} = \pi/3$, $\theta_{su} = \pi/6$, $\beta_{eu} = \pi/6$, $\theta_{eu} = \pi/4$, $\zeta_e = (1, 0, 0)$, $\zeta_u = \left(3/4, \sqrt{3}/4, 1/2\right)$, $R_u = 2$ *bits/Hz/s, and* $R_e = 1$ *bits/Hz/s. ©IEEE 2019, reprinted with permission from [1].*

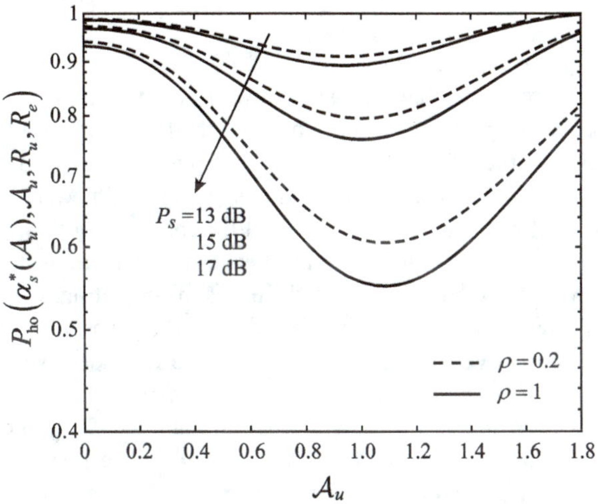

Figure 13.4 $P_{ho}\left(\alpha_s^*(\mathscr{A}_u), \mathscr{A}_u, R_u, R_e\right)$ *versus* $\mathscr{A}_u$ *for different values of* P_s *with* $N_s = 6$, $N_e = 1$, $N_j = 2$, $P_e = 10$ *dB,* $\zeta_e = (1, 0, 0)$, $\mu_u = 3/4$, $\nu_u = \sqrt{3}/4$, $R_u = 2$ *bits/Hz/s, and* $R_e = 1$ *bits/Hz/s. ©IEEE 2019, reprinted with permission from [1].*

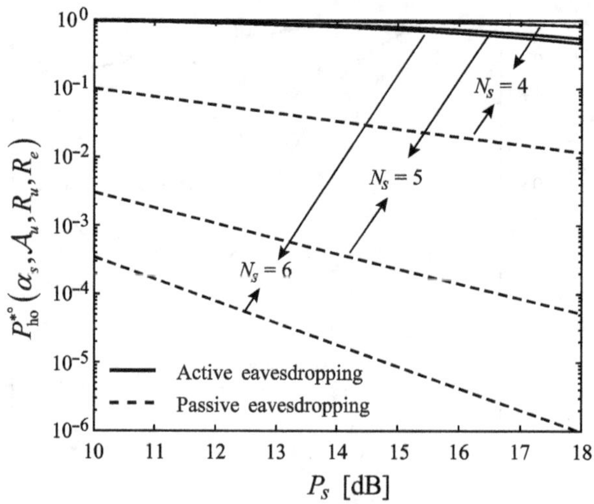

Figure 13.5 $P_{ho}^{*o}(\alpha_s, \mathscr{A}_u, R_u, R_e)$ *versus* P_s *for different values of* N_s *with* $N_e = 1$, $N_j = 2$, $\rho = 1$, $P_e = 10$ *dB,* $\zeta_e = (1, 0, 0)$, $\zeta_u = \left(3/4, \sqrt{3}/4, \mathscr{A}_u\right)$, $R_u = 2$ *bits/Hz/s, and* $R_e = 1$ *bits/Hz/s. ©IEEE 2019, reprinted with permission from [1].*

see that, for a given N_s, the value of $P_{\text{ho}}^{*o}(\alpha_s, \mathscr{A}_u, R_u, R_e)$ for a system with active eavesdropping is larger than that for a system with passive eavesdropping, revealing that the active eavesdropper can do more harm to the legitimate system.

Then, we explore the effect of N_e on the optimal power allocation factor $\alpha_s^*(\mathscr{A}_u)$ for a given N_t in Figure 13.6. Figure 13.6 illustrates the optimal power allocation factor, $\alpha_s^*(\mathscr{A}_u)$, versus N_e for fixed N_t and $\mathscr{A}_u$ and different values of P_s, where we take $N_s = 6$ and $N_t = 3$. Therefore, the number of eavesdropper's transmitting antenna is $N_j = N_t - N_e$. We clarify that, as $N_e = 0$, the eavesdropper utilizes all its antennas to transmit jamming signals, while when $N_e = N_t = 3$, the eavesdropper configuration transforms from an active eavesdropper to a passive eavesdropper. When $N_e > 1$, for both the cases of $\mathscr{A}_u = 1/2$ and $\mathscr{A}_u = 1$, we observe that the value of $\alpha_s^*(\mathscr{A}_u)$ becomes smaller as P_s increases, showing that, to maintain $P_{\text{ho}}\left(\alpha_s^*(\mathscr{A}_u), \mathscr{A}_u, R_u, R_e\right)$ when P_s becomes larger, the source needs to assign more power to transmit AN signals. In addition, we also observe that, when $N_e > 1$, for a given N_t, the value of $\alpha_s^*(\mathscr{A}_u)$ becomes smaller as N_e increases, which reveals that, when the capability of the eavesdropper increases, the source needs to assign more power to transmit AN signal. We observe that $\alpha_s^*(\mathscr{A}_u) = 1$, when $N_e = 1$, for both the cases of $P_s = 10$ dB and $P_s = 13$ dB. The reason is that the directional jamming from the eavesdropper degrades the received SINR at the UAV severely, which requires the source to allocate more power to transmit information signals to maintain $P_{\text{ho}}\left(\alpha_s^*(\mathscr{A}_u), \mathscr{A}_u, R_u, R_e\right)$ when the transmit power at the source is relatively small. Moreover, we observe that $\alpha_s^*(\mathscr{A}_u) = 1$

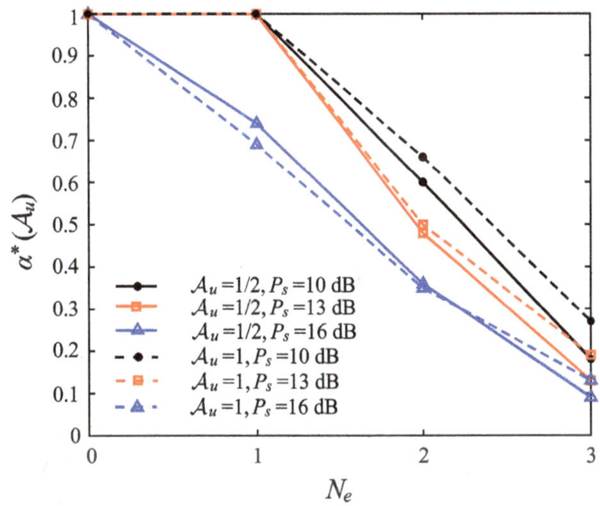

Figure 13.6 $\alpha_s^*(\mathscr{A}_u)$ *versus* N_e *for different values of* P_s *with* $N_s = 6$, $N_t = 3$, $P_e = 10$ *dB*, $\zeta_e = (1, 0, 0)$, $\zeta_u = \left(3/4, \sqrt{3}/4, \mathscr{A}_u\right)$, $R_u = 2$ *bits/Hz/s, and* $R_e = 1$ *bits/Hz/s. ©IEEE 2019, reprinted with permission from [1].*

when $N_e = 0$ regardless of the values of P_s. This is because the eavesdropper operates purely malicious jamming. Thus, the optimal scheme for the source is to allocate all its power to transmit information signals.

13.4.2 Optimal configuration of eavesdropper

Eventually, from the perspective of the eavesdropper, we explore the optimal antenna configuration of the eavesdropper which maximizes the hybrid outage probability in Figures 13.7 and 13.8, where the configuration of the eavesdropper is most harmful to the legitimate system. To this end, we consider that each eavesdropper's antenna is equipped with both transmit and receive radio frequency chains, thus the eavesdropper can adjust the antenna's configuration to degrade the communication secrecy. In particular, for a given N_s and N_t, we explore the optimal number of transmit antennas, N_e^*, which maximizes the hybrid outage probability at the UAV. We clarify that the curves in Figures 13.7 and 13.8 are plotted with the optimal $\alpha_s^*(\mathscr{A}_u)$.

Figure 13.7 illustrates N_e^* versus $\mathscr{A}_u$ for different values of P_s with a given $P_e = 10$ dB. We observe that $N_e^* = 0$ when the value of P_s is in a small regime. This reveals that the most harmful case is that the eavesdropper assigns all its antennas to conduct malicious jamming when the source's transmit power is small. In addition, we also observe that when P_s becomes larger, the value of N_e^* is nonzero, which indicates that, when the source increases its transmit power, purely jamming is less effective

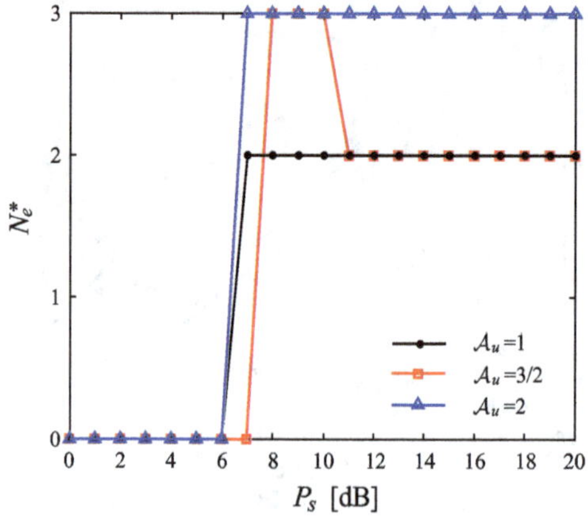

Figure 13.7 N_e^* *versus* P_s *for different values of* $\mathscr{A}_u$ *with* $N_s = 6$, $N_t = 3$, $\rho = 1$, $P_e = 10$ *dB*, $\zeta_e = (1, 0, 0)$, $\zeta_u = \left(3/4, \sqrt{3}/4, \mathscr{A}_u\right)$, $R_u = 2$ *bits/Hz/s, and* $R_e = 1$ *bits/Hz/s. ©IEEE 2019, reprinted with permission from [1].*

in decreasing the secrecy performance. Moreover, the value of N_e^* is different for the different operating heights of the UAV. Specifically, when the UAV operates at a large height, e.g., $\mathscr{A}_u = 2$, the impact of the jamming becomes less, owing to the distance between the eavesdropper and the UAV is large, thus N_e^* becomes three (i.e., no jamming) as P_s increases. In the case of moderate UAV operating height, i.e., $\mathscr{A}_u = 3/2$ case, $N_e^* = 3$ for small P_s while $N_e^* = 2$ for large P_s. When the value of P_s is small, the eavesdropping capability is small and the transmission outage probability is already high. Therefore, the impact of the jamming signal on the transmission outage probability is subtle, so it is more sensible to allocate all antennas for eavesdropping, i.e., $N_e = 3$. Nevertheless, jamming becomes more effective when the value of P_s is large, thus N_e^* becomes two. In the case of low UAV operating height, i.e., $\mathscr{A}_u = 1$ case, the impact of the jamming on the transmission outage probability is significant, owing to the distance between the eavesdropper and the UAV is short, thus the optimal number of jamming antennas is nonzero (i.e., $N_e^* < 3$) in the most range of P_s. Moreover, the effect of ρ on N_e^* is also revealed in the case of $\mathscr{A}_u = 2$ with different values of ρ. We can observe that, owing to smaller ρ, N_e^* changes from 3 to 1 for P_s ranging from 12 dB to 14 dB, suggesting that when the residual self-interference is relatively small, a more effective antenna configuration for the eavesdropper is to allocate more antennas for jamming.

Figure 13.8 illustrates N_e^* versus P_s for different values of $\mathscr{A}_u$. We can see that it has similar trends shown in Figure 13.7. Particularly, when $P_s = 0$ dB, we observe that $N_e^* = 0$ for $\mathscr{A}_u$ ranging from 0 to 2, which validates that the optimal transmit scheme for the eavesdropper is to utilize all its antennas to transmit jamming signals. We also observe that when the UAV operating height $\mathscr{A}_u$ becomes larger, utilizing all antennas to transmit jamming signals is no longer the optimal scheme for the eavesdropper.

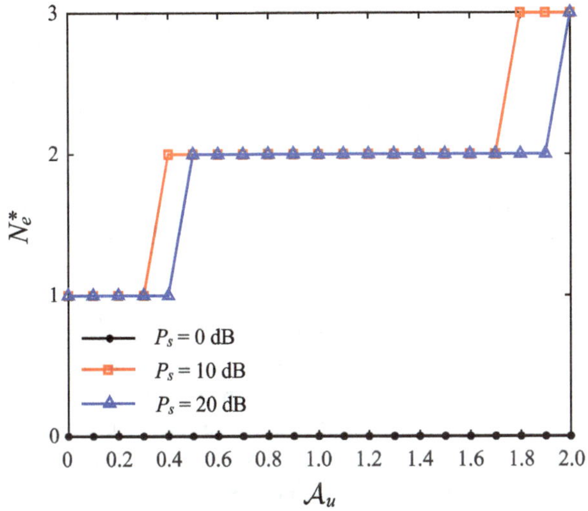

Figure 13.8 N_e^* *versus* $\mathscr{A}_u$ *for different values of* P_s *with* $N_s = 6$, $N_t = 3$, $\rho = 1$, $P_e = 10$ *dB*, $\zeta_e = (1, 0, 0)$, $\zeta_u = \left(3/4, \sqrt{3}/4, \mathscr{A}_u\right)$, $R_u = 2$ *bits/Hz/s, and* $R_e = 1$ *bits/Hz/s. ©IEEE 2019, reprinted with permission from [1].*

13.5 Conclusions and future works

In this chapter, we design a secure transmission strategy for a UAV wiretap channel, where a multi-antenna source communicates with a single-antenna UAV in the presence of a full-duplex active eavesdropper. We assume that the source transmits AN signals along with the information signals to degrade the reception quality of the eavesdropper. We obtain an expression in a compact form to characterize the hybrid outage probability. In addition, we derive a new expression for the asymptotic hybrid outage probability. Utilizing our analytical framework, the optimal power allocation factor for the source is explored, which can minimize the hybrid outage probability. Moreover, we illustrate how to judiciously select the operating height of the UAV, which can further enhance the secrecy performance of the system. We also explore the impact of the antenna configuration at the eavesdropper on the secrecy performance, and obtain the optimal number of the transmitting antennas for the eavesdropper which can maximize the hybrid outage probability, revealing that the eavesdropper transmits jamming signals with all its antennas can be most harmful to the legitimate system when the source's transmit power is relatively small.

In addition to the PLS transmission scheme presented for UAV wireless networks in this chapter, we identify future challenges and opportunities for PLS in UAV wireless networks, as follows.

PLS for air–space–ground integrated networks: With the vision of realizing global seamless coverage in future 6G networks, especially for the remote areas such as marine and desert regions, air–space–ground integrated networks have been proposed as a promising framework to support ubiquitous connections. With the aid of satellites and UAVs, the air–space–ground integrated networks can efficiently facilitate the heavy burden of conventional terrestrial networks, thus has attracted significant attention in both industry and academic communities. Nevertheless, the open transmission environment degrades the communication secrecy, resulting in information leakage and malicious attacks. Moreover, due to the integration of terrestrial and non-terrestrial facilities, the heterogeneity of the networks complicates the PLS design. Therefore, the PLS enhancement techniques for air–space–ground integrated networks need to be investigated. Besides, the cooperative scheme among different types of nodes for secrecy transmissions in the air–space–ground integrated networks, such as signal processing and data exchange, play a vital role in PLS for air–space–ground integrated networks, and worthwhile investigation.

PLS for massive MIMO-enabled UAV networks: Massive MIMO has been investigated extensively over the past decades, which can provide a high transmission rate, while guaranteeing the spectral efficiency and energy efficiency. By deploying massive number of antennas at both transmitters and receivers, as well as utilizing the spatial multiplexing techniques, massive MIMO techniques can significantly improve the channel capacity. Meanwhile, with the extension of bandwidth, millimeter-wave and terahertz techniques can mitigate the scarcity of spectrum, while achieving the miniaturization of antennas. Through integrating massive MIMO techniques with UAVs, a better adaptation to the complex transmission environment and the secrecy enhancement can be achieved. However, due to the massive number of antennas, the complexity of channel estimation and beamforming designing increases accordingly. Therefore, it is essential to design low-complexity channel estimation algorithms and precoding schemes for PLS enhancement in massive MIMO-enabled UAV networks.

PLS for IRS-aided UAV networks: Intelligent reflecting surface (IRS) is a promising technology for boosting the secrecy performance, since the IRS can tune the signal reflection and dynamically alter the wireless channel. In PLS for IRS-aided UAV networks, one important open problem is to the joint optimization of the UAV's trajectory and IRS's passive beamforming such that the secrecy performance can be maximized. In addition, the effectiveness of the IRS relies on the assumption of perfect CSI, while the UAV's mobility and jitter may lead to significant channel estimation error. As such, designing novel secure transmission schemes that are robust to the mobility and jitter of the UAV is also a challenging and interesting problem.

References

[1] Liu C, Lee J, and Quek TQS. Safeguarding UAV communications against full-duplex active eavesdropper. *IEEE Transactions on Wireless Communications*. 2019;18(6):2919–2931.

[2] Zeng Y, Zhang R, and Lim TJ. Wireless communications with unmanned aerial vehicles: opportunities and challenges. *IEEE Communications Magazine*. 2016;54(5):36–42.

[3] Mozaffari M, Saad W, Bennis M, *et al.* Mobile unmanned aerial vehicles (UAVs) for energy-efficient Internet of Things communications. *IEEE Transactions on Wireless Communications*. 2017;16(11):7574–7589.

[4] Li P, Xie L, Yao J, *et al.* Cellular-connected UAV with adaptive air-to-ground interference cancellation and trajectory optimization. *IEEE Communications Letters*. 2022;26(6):1368–1372.

[5] Gao Y, Xiao L, Wu F, *et al.* Cellular-connected UAV trajectory design with connectivity constraint: a deep reinforcement learning approach. *IEEE Transactions on Green Communications and Networking*. 2021;5(3):1369–1380.

[6] Mou Z, Zhang Y, Gao F, *et al.* Deep reinforcement learning based three-dimensional area coverage with UAV swarm. *IEEE Journal on Selected Areas in Communications*. 2021;39(10):3160–3176.

[7] Zeng F, Hu Z, Xiao Z, *et al.* Resource allocation and trajectory optimization for QoE provisioning in energy-efficient UAV-enabled wireless networks. *IEEE Transactions on Vehicular Technology*. 2020;69(7):7634–7647.

[8] Chen K, Wang Y, Zhao J, *et al.* URLLC-oriented joint power control and resource allocation in UAV-assisted networks. *IEEE Internet of Things Journal*. 2021;8(12):10103–10116.

[9] Horani M and Hasna MO. Latency analysis of UAV based communication networks. In: *2018 International Conference on Information and Communication Technology Convergence (ICTC)*; 2018. p. 385–390.

[10] Hong YWP, Lan PC, and Kuo CCJ. Enhancing physical-layer secrecy in multi-antenna wireless systems: an overview of signal processing approaches. *IEEE Signal Processing Magazine*. 2013;30(5):29–40.

[11] Mukherjee A, Fakoorian SAA, Huang J, *et al.* Principles of physical layer security in multiuser wireless networks: a survey. *IEEE Communications Surveys & Tutorials*. 2014;16(3):1550–1573.

[12] Yang N, Wang L, Geraci G, *et al.* Safeguarding 5G wireless communication networks using physical layer security. *IEEE Communications Magazine*. 2015;53(4):20–27.

[13] Wyner AD. The wire-tap channel. *The Bell System Technical Journal*. 1975;54(8):1355–1387.

[14] Oggier F and Hassibi B. The secrecy capacity of the MIMO wiretap channel. *IEEE Transactions on Information Theory*. 2011;57(8):4961–4972.

[15] Zhou X, McKay MR, Maham B, *et al.* Rethinking the secrecy outage formulation: a secure transmission design perspective. *IEEE Communications Letters*. 2011;15(3):302–304.

[16] Zhang X, Zhou X, and McKay MR. On the design of artificial-noise-aided secure multi-antenna transmission in slow fading channels. *IEEE Transactions on Vehicular Technology*. 2013;62(5):2170–2181.

[17] Yang N, Yan S, Yuan J, *et al.* Artificial noise: transmission optimization in multi-input single-output wiretap channels. *IEEE Transactions on Communications*. 2015;63(5):1771–1783.

[18] Yang N, Elkashlan M, Duong TQ, *et al.* Optimal transmission with artificial noise in MISOME wiretap channels. *IEEE Transactions on Vehicular Technology*. 2016;65(4):2170–2181.

[19] Zhou X and McKay MR. Secure transmission with artificial noise over fading channels: achievable rate and optimal power allocation. *IEEE Transactions on Vehicular Technology*. 2010;59(8):3831–3842.

[20] Liu C, Yang N, Yuan J, *et al.* Location-based secure transmission for wiretap channels. *IEEE Journal on Selected Areas in Communications*. 2015;33(7):1458–1470.

[21] Wang J, Lee J, Wang F, *et al.* Jamming-aided secure communication in massive MIMO Rician channels. *IEEE Transactions on Wireless Communications*. 2015;14(12):6854–6868.

[22] Yan S and Malaney R. Location-based beamforming for enhancing secrecy in Rician wiretap channels. *IEEE Transactions on Wireless Communications*. 2016;15(4):2780–2791.

[23] Liu C and Malaney R. Location-based beamforming and physical layer security in Rician wiretap channels. *IEEE Transactions on Wireless Communications*. 2016;15(11):7847–7857.

[24] Liu G, Yu FR, Ji H, *et al.* In-band full-duplex relaying: a survey, research issues and challenges. *IEEE Communications Surveys & Tutorials*. 2015;17(2): 500–524.

[25] Zhu F, Gao F, Yao M, *et al.* Joint information- and jamming-beamforming for physical layer security with full duplex base station. *IEEE Transactions on Signal Processing*. 2014;62(24):6391–6401.

[26] Zheng G, Krikidis I, Li J, *et al.* Improving physical layer secrecy using full-duplex jamming receivers. *IEEE Transactions on Signal Processing*. 2013;61(20):4962–4974.

[27] Chen G, Gong Y, Xiao P, *et al.* Physical layer network security in the full-duplex relay system. *IEEE Transactions on Information Forensics and Security*. 2015;10(3):574–583.

[28] Lee JH. Full-duplex relay for enhancing physical layer security in multi-hop relaying systems. *IEEE Communications Letters*. 2015;19(4):525–528.

[29] Huang W, Chen W, Bai B, *et al.* Wiretap Channel with full-duplex proactive eavesdropper: a game theoretic approach. *IEEE Transactions on Vehicular Technology*. 2018;67(8):7658–7663.

[30] Mukherjee A and Swindlehurst AL. A full-duplex active eavesdropper in mimo wiretap channels: construction and countermeasures. In: *2011 Conference Record of the Forty-Fifth Asilomar Conference on Signals, Systems and Computers (ASILOMAR)*; 2011. p. 265–269.

[31] Mukherjee A and Swindlehurst AL. Jamming games in the MIMO wiretap channel with an active eavesdropper. *IEEE Transactions on Signal Processing*. 2013;61(1):82–91.

[32] Kong L, He J, Kaddoum G, *et al.* Secrecy analysis of a MIMO full-duplex active eavesdropper with channel estimation errors. In: *2016 IEEE 84th Vehicular Technology Conference (VTC-Fall)*; 2016. p. 1–5.

[33] Zhang G, Wu Q, Cui M, *et al.* Securing UAV communications via joint trajectory and power control. *IEEE Transactions on Wireless Communications*. 2019;18(2):1376–1389.

[34] Cui M, Zhang G, Wu Q, *et al.* Robust trajectory and transmit power design for secure UAV communications. *IEEE Transactions on Vehicular Technology*. 2018;67(9):9042–9046.
[35] Li R, Wei Z, Yang L, *et al.* Resource allocation for secure multi-UAV communication systems with multi-eavesdropper. *IEEE Transactions on Communications*. 2020;68(7):4490–4506.
[36] Xiao L, Xu Y, Yang D, *et al.* Secrecy energy efficiency maximization for UAV-enabled mobile relaying. *IEEE Transactions on Green Communications and Networking*. 2020;4(1):180–193.
[37] Shen L, Zhu Z, Wang N, *et al.* Trajectory optimization for physical layer secure buffer-aided UAV mobile relaying. In: *2019 IEEE 90th Vehicular Technology Conference (VTC2019-Fall)*; 2019. p. 1–6.
[38] Wu H, Li H, Wei Z, *et al.* Secrecy performance analysis of air-to-ground communication with UAV jitter and multiple random walking eavesdroppers. *IEEE Transactions on Vehicular Technology*. 2021;70(1):572–584.
[39] Yuan X, Feng Z, Ni W, *et al.* Secrecy performance of terrestrial radio links under collaborative aerial eavesdropping. *IEEE Transactions on Information Forensics and Security*. 2020;15:604–619.
[40] Riihonen T, Werner S, and Wichman R. Mitigation of loopback self-interference in full-duplex MIMO relays. *IEEE Transactions on Signal Processing*. 2011;59(12):5983–5993.
[41] Korpi D, Tamminen J, Turunen M, *et al.* Full-duplex mobile device: pushing the limits. *IEEE Communications Magazine*. 2016;54(9):80–87.
[42] Sim MS, Chung M, Kim D, *et al.* Nonlinear self-interference cancellation for full-duplex radios: from link-level and system-level performance perspectives. *IEEE Communications Magazine*. 2017;55(9):158–167.
[43] Duarte M, Dick C, and Sabharwal A. Experiment-driven characterization of full-duplex wireless systems. *IEEE Transactions on Wireless Communications*. 2012;11(12):4296–4307.
[44] Al-Hourani A, Kandeepan S, and Lardner S. Optimal LAP altitude for maximum coverage. *IEEE Wireless Communications Letters*. 2014;3(6):569–572.
[45] Bor-Yaliniz RI, El-Keyi A, and Yanikomeroglu H. Efficient 3-D placement of an aerial base station in next generation cellular networks. In: *2016 IEEE International Conference on Communications (ICC)*; 2016. p. 1–5.
[46] Tsai JA, Buehrer RM, and Woerner BD. BER performance of a uniform circular array versus a uniform linear array in a mobile radio environment. *IEEE Transactions on Wireless Communications*. 2004;3(3):695–700.
[47] Tse D and Viswanath P. *Fundamentals of Wireless Communication*. Cambridge University Press; 2005.
[48] Zwillinger D, Moll V, Gradshteyn IS, *et al.* *Table of Integrals*. Academic Press; 2014.
[49] Liu C, Yang N, Malaney R, *et al.* Artificial-noise-aided transmission in multi-antenna relay wiretap channels with spatially random eavesdroppers. *IEEE Transactions on Wireless Communications*. 2016;15(11):7444–7456.

Chapter 14

Secure communication in mobile edge computing networks with RF energy harvesting

Dac-Binh Ha[1], Van-Long Nguyen[2], Van-Truong Truong[1] and Anand Nayyar[2]

In this chapter, we study the physical layer security (PLS) of a multiuser non-orthogonal multiple access (NOMA) radio frequency energy harvesting (RF EH) relaying system. The user selection scheme, decode-and-forward (DF) relay selection solution, users' quality of service (QoS) requirements, and time switching ratio of RF EH are considered for secure NOMA transmission. Specifically, we investigate the scenario that two legitimate user (LU) clusters with different priorities depend on their QoS requirements to offload their tasks to a server located at a base station by applying uplink/downlink NOMA with the help of multiple relays in the presence of a passive eavesdropper. To enhance the secrecy performance, we examine a relay selection scheme and a user selection method. In this context, to evaluate the secrecy performance, closed-form expressions of the existence probability of secrecy capacity and secrecy outage probability for selected LUs and relay are derived. Furthermore, we propose an optimization algorithm based on particle swarm optimization (PSO) to obtain an optimal parameter set, e.g., time switching ratio and power allocation ratio, to achieve the best secrecy performance for this considered system. To investigate the system's PLS performance, we simulate the system with essential parameters such as transmit power, power allocation coefficient, time switching ratio, number of users in clusters, and number of relays. Our theoretical results confirmed by simulation show that PLS performance can be improved by using an optimization algorithm or/and enlarging transmit power or/and increasing the number of users and relays. Moreover, the system performance under the proposed optimal algorithm and protocol is superior to that of the traditional OMA solution. Finally, these results also indicate that our proposed optimal algorithm achieves the best secrecy performance.

[1]Faculty of Electrical-Electronic Engineering, Duy Tan University, Vietnam
[2]Faculty of Information Technology, Duy Tan University, Vietnam

14.1 Introduction

The rapid growth of the number of wireless devices and their functional diversity in the next-generation networks (5G–6G) leads to many challenging problems, e.g., mobility management, network coverage, data rate, energy, and security issues [1–3]. Recently, a non-orthogonal multiple access (NOMA) technique with outstanding spectral efficiency, multi-user support, and user fairness compared with traditional multiple access has been considered a good candidate for a 5G network solution [2]. However, strong co-channel interference and high signal processing complexity in the multi-user NOMA model are significant challenges to this technique. Thus, a novel solution is proposed that users will be classified into several subgroups, where they can be served by utilizing different time slots or subcarriers, and NOMA is carried out between the users in each group [4–6]. Specifically, a practical scenario was considered in [7], in which two users are paired to perform NOMA.

The authors of [8] examine users' QoS requirements in a two-user NOMA network as an alternative to the traditional approach to their channel conditions. Specifically, the authors investigated secrecy performance in terms of the exact and approximate expressions of secrecy outage probability (SOP). However, a proposed simple NOMA system model consisting of one source and two users in the presence of one eavesdropper presented has not highlighted the network properties. Very recently, some studies have been made regarding radio frequency energy harvesting (RF EH) and relaying technique to NOMA systems [9–11]. RF EH technology was introduced to alleviate the energy consumption problem for limited-energy devices in wireless systems. By allocating the proper energy collection time in the NOMA network, RF EH ensures devices have enough power to operate and ensure a significant increase in system performance [12]. Meanwhile, the relaying technique is proposed in NOMA networks, which helps obtain high data rate coverage, saving spectrum resources, and more reliable transmission [13,14].

Furthermore, mobile edge computing (MEC) is a technique often closely combined with NOMA to create a system that can meet large computational tasks with minimal delay [15,16]. It is a network model considered the next generation of Cloud computing when computing or storage services are brought closer to the data source [15]. The computation tasks of NOMA users are transmitted to the small-scale access point (AP) or base station (BS) in the uplink so that the MEC server there supports decoding and computation [5]. Meanwhile, under the downlink NOMA model, user data can be supported by multiple APs [11].

However, the fact that information is broadcast in the NOMA-MEC network presents a considerable security challenge. Therefore, besides the NOMA MEC system performance survey in recent years, the physical layer security (PLS) has also received significant attention from the research community [17–20]. Security methods such as cryptographic techniques have been designed at the network layer based on the hypothesis that an error-free link in the physical layer is observed [18]. However, this hypothesis does not hold up in experimental conditions, so PLS is a physical layer technique exploiting the dynamic characteristics of wireless channels that have been studied as an additional solution to improve information security for 5G networks.

Considering the combination of NOMA, MEC, and PLS, many studies have evaluated systems to enhance the information security capability of these networks [21–24] as follows. In more detail, the work [21] investigated the multi-user NOMA MEC network offloading part of the task to the BS with the appearance of a malicious eavesdropper. The authors used SOP to determine the security performance of the offloading process and as a constraint in the problem of minimizing the sum energy consumption of all the NOMA users. The sequential convex approximation and penalty dual decomposition technique proposed to solve the optimization problem has been proven superior energy efficiency with secrecy requirements. In [22], secure communication in a single-input, single-output (SISO) NOMA network was considered by deriving the closed-form expression of an optimal power allocation policy to provide the maximum secrecy sum rate of the NOMA system subject to the users' QoS requirements. However, this work has not considered multiple antennas scheme and a general channel model, i.e., Nakagami-*m*.

Unlike the previous works, in this paper, we aim at proposing a secure communication protocol in two clusters uplink–downlink RF EH NOMA MEC network in the presence of an eavesdropper. In this system, users' QoS requirements are considered instead of their channel conditions to perform NOMA. In our scenario, a decode-and-forward (DF) relay is selected based on the channel information transmitted from the relay cluster to the base station (BS) to support communications with two legitimate user (LU) clusters, in which the QoS requirements of users in the first cluster are higher than that in the second one, in the presence of an passive eavesdropper. In particular, to raise information security, a relay-user selection method is evaluated, where one user in each cluster with the best channel condition is chosen for NOMA-based transmission. In this context, the main contributions of this chapter are as follows:

- We investigate the secrecy performance of an uplink–downlink RF EH NOMA system consisting of one base station, multiple DF relays, one high-priority user cluster, and one low-priority user cluster in the presence of an eavesdropper. Users' QoS requirements, Rayleigh fading, and relay-user selection solutions are considered in the studied network.
- We derive the closed-form expressions of the existence probability of secrecy capacity and secrecy outage probability of LUs in the considered network.
- We evaluate the impact of key system parameters such as the average transmit signal-to-noise ratio (SNR), the number of relays in the relay cluster, and the number of users in two clusters on the secrecy performance of the considered network.
- We propose an algorithm based on PSO, namely MSOP-PSO, to determine the optimal set of parameters, i.e., time switching ratio and power allocation ratio, to optimize the secrecy performance of each user.

The rest of this chapter is organized as follows. Section 14.2 presents the literature review. The system model is presented in Section 14.3. Secrecy performance analysis of the considered system is provided in Section 14.4. The numerical results are shown in Section 14.5. Finally, Section 14.6 is the conclusion of our chapter.

Notation: $\mathscr{CN}(0, \sigma^2)$ indicates a scalar complex Gaussian distribution with zero mean and variance σ^2, $\mathbb{E}[\cdot]$ denotes the expectation operator, $\binom{K}{k} = \frac{K!}{k!(K-k)!}$.

14.2 Literature review

This section summarizes the related works on the PLS to NOMA systems [25,26], relaying transmission [24,27], the application of RF EH [28,29], and MEC paradigm [21,30–33], which is represented in the following.

In [26], randomly deployed users were considered to investigate the PLS of NOMA in large-scale networks. The exact and asymptotic expressions of secrecy outage probability were derived to characterize the secrecy performance. Moreover, to improve the information security of large-scale NOMA networks, the application of multiple antennas, artificial noise, and protected zone were introduced and analyzed clearly. The obtained results indicated that invoking the protected zone and generating artificial noise at the base station (BS) can help to raise the secrecy performance of NOMA systems.

In [24], Hoang *et al.* presented a cooperative relay network in which a single source communicates to multiple destinations via a set of multi-trusted relays, with the presence of an eavesdropper. The authors derived an upper bound on the ergodic secrecy rate in two cases: SIMOSE and MISOSE. The author evaluated the security performance of the proposed system in terms of the adapted cooperative diversity gain. In [27], Nguyen *et al.* proposed the system model of the PLS NOMA relaying network where two energy-constrained clusters harvest RF energy from the specified relay to communicate to BS in the presence of a passive eavesdropper. The authors propose a meta-heuristic algorithm to determine the best system parameter set that can bring the system's best security performance.

In [28], Nasir *et al.* proposed simple RF EH NOMA, in which one BS serves one near-user and one far-user. The authors found that energy transmission requires more energy than conventional information transmission, making it challenging to accommodate the QoS in terms of throughput for all users. In [29], Rauniyar *et al.* proposed an RF EH relaying NOMA system, which contains one source node that communicates with destination nodes with the help of a power-constrained IoT relay node. The authors derived analytical expressions for two case relaying protocols for the system. An optimal algorithm was introduced to find the optimal time switching and power splitting ratios that maximize the sum throughput for RF EH NOMA-based system.

In [30], Li *et al.* investigated the security performance of the cooperative NOMA MEC system. Specifically, the authors considered a scenario in which an edge-computing device offloads its tasks to the edge-computing server against the security attack of a malicious eavesdropper. The cooperative jamming solution and the stochastic learning-based three-layered optimization algorithm were proposed to solve the problem of non-convex, minimizing the total energy consumption. In [21], Wu *et al.* studied a secure offloading in uplink NOMA MEC network without knowing the

Table 14.1 Mathematical notations used in this chapter

Notation	Meaning
g_0, g_1, g_2, g_3	Channel power gain of the wireless links of R^*-S, A^*-R^*, B^*-R^*, R^*-E
T	Transmission block time
α	Time switching ratio
ρ	Power allocation coefficient
η	Energy conversion efficiency
M	Number of users in cluster **A**
N	Number of users in cluster **B**
K	Number of relays
P_0	Transmit power of S
P_U	Total transmit power of user pair (A^*, B^*)
R^*	The selected relay
(A^*, B^*)	The NOMA MEC users pair
max	The maximum function
arg max	Arguments of the maximum function
$\mathcal{CN}(0, \sigma^2)$	The complex Gaussian distribution with zero mean and variance σ^2
$\mathbb{E}[\cdot]$	The expectation operator

eavesdropper's channel state information (CSI). The authors stated optimal problem and proposed an iterative algorithm to maximize the minimum anti-eavesdropping ability of users with constraints on secrecy rate and transmission power budgets.

Through the literature review process, we realize that many issues still need further research in RF EH relaying NOMA-based MEC systems. Several studies analyze the PLS behavior of single-user systems, while the optimization algorithms have not been mentioned or use loops with high algorithmic complexity. Our research focuses on a multi-user system suitable for reality, and, at the same time, proposes an algorithm based on the PSO that can bring high performance while having relatively low complexity. That is the motivation for us to do this study.

14.3 System channel and model

We define the notations used in the next part of this work in 14.1.

This chapter considers PLS in a multiuser uplink/downlink relaying NOMA MEC system. Specifically, a cluster of M MEC-users (MUs) with high priority, denoted by $\mathbf{A} = \{A_1, \ldots, A_M\}$, and a cluster of N MUs with low priority, denoted by $\mathbf{B} = \{B_1, \ldots, B_N\}$ offload their tasks to the MEC server located at the base station (BS), denoted by $\mathbf{S} = \{S\}$, with the help of a cluster of K relays, denoted by $\mathbf{R} = \{R_1, \ldots, R_K\}$ in the presence of a passive eavesdropper, denoted by $\mathbf{E} = \{E\}$, as depicted in Figure 14.1. In this system, the MUs in clusters **A** and **B** are considered to be high-priority and low-priority users, respectively, i.e., the QoS requirements of the users in cluster **A** are higher than that of the users in cluster **B** [22,34]. Contrary to the

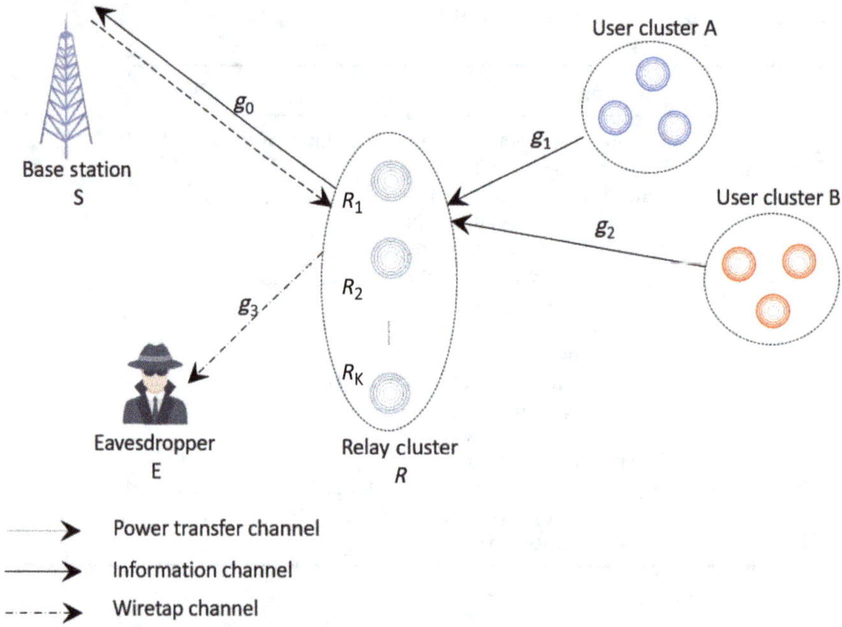

Figure 14.1 Model of PLS in a multiuser uplink/downlink NOMA relaying MEC system with RF HE

hypothesis in [27], **R** are energy-constrained devices, thus they need power support from **S** to relay offload information from clusters **A** and **B**. Assume that there is no direct link between the users and the **S** and the **E** and all channels of this network experience independent and identically Rayleigh fading. Furthermore, each device is equipped with a single antenna and operated in half-duplex mode.

Let h_{XY} is the channel coefficient of the link from **A** or **B** to **R**, $X \in \{A_m, B_n\}$, $Y \in \{R_k\}$, $1 \leq m \leq M, 1 \leq n \leq N, 1 \leq k \leq K$. Let h_{YZ} is the channel coefficient of the link from **R** to **S** (and vice versa) or from **R** to **E**, $Z \in \{S, E\}$. In addition, the distance and path loss exponent of $X \to Y$ or $Y \to Z$ channel are denoted by d_{XY} or d_{YZ}, and θ_{XY} or θ_{YZ}, respectively.

In our study, a selected relay R^* harvests the RF energy from **S** to help NOMA user pair transmit information using DF scheme. Specifically, prior to information transmission, **S** transfers the power to **R** in duration αT, where T denotes as transmission time block, α stands for time switching ratio $(0 \leq \alpha \leq 1)$. Next, the best MUs in the clusters **A** and **B**, denoted by A^* and B^*, are paired to perform uplink NOMA based on the users' QoS requirements to transmit information to **R**. Finally, **R** retransmits the signal to **S** by using downlink NOMA. Meanwhile, the **E** tries to extract information transmitted from **R** only without any attack. Note that in our proposed model, the best relay is selected in phase 1, which received the most energy for relaying data in phase 2. This selection ensures that the cooperative relay criterion must collect enough power to operate [35,36].

We propose a protocol for this considered system as Algorithm 1.

Algorithm 1: Proposed schemes for RF EH uplink/downlink NOMA relaying MEC system

---

Initialization Selection: relay R^*, user pair (A^*, B^*), α, ρ, setup P_0, and P_U.
while True **do**
    R^* harvests energy from **S** in duration αT.
    User pair (A^*, B^*) offloads own task to R^* during the duration $\frac{(1-\alpha)T}{2}$.
    R^* decodes and retransmits to **S** in the remained duration $\frac{(1-\alpha)T}{2}$.
end while

14.3.1 Relay and user selection solution

Given the above discussion, in this subsection, we present the solution of relay and user selection for secure communication in the considered RF energy harvesting NOMA-based MEC system.

14.3.1.1 Relay selection

Due to passive wiretap [37,38], it is difficult to obtain the CSI of illegitimate link. Thus, in this context, the best relay is selected based on the best channel of link **S**–**R**. Mathematically, the selected relay R^* in cluster **R** is written as:

$$R^* = arg \max_{1 \le k \le K} \left\{ \left| h_{SR_k} \right|^2 \right\}, \tag{14.1}$$

i.e., the channel power gain of this selected channel is given by

$$g_0 = \max_{1 \le k \le K} \left\{ \left| h_{SR_k} \right|^2 \right\}. \tag{14.2}$$

14.3.1.2 User selection

The best MUs in clusters **A** and **B** are selected for transmission based on the channels from the users to the selected relay, i.e., R^*. Given this context, a pair of users in cluster **A** and **B** are selected to maximize the channel power gains of the links from the user to the chosen relay. Thus, the selected MUs in cluster **A** and **B** are, respectively, given by:

$$A^* = arg \max_{1 \le m \le M} \left\{ \left| h_{A_m R^*} \right|^2 \right\}, \tag{14.3}$$

$$B^* = arg \max_{1 \le n \le N} \left\{ \left| h_{B_n R^*} \right|^2 \right\}. \tag{14.4}$$

The channel power gains of this selected channel are, respectively, expressed as

$$g_1 = \max_{1 \le m \le M} \left\{ \left| h_{A_m R*} \right|^2 \right\}, \tag{14.5}$$

$$g_2 = \max_{1 \le n \le N} \left\{ |h_{B_n R*}|^2 \right\}. \tag{14.6}$$

Based on (14.2), (14.5), and (14.6), under Rayleigh fading, the cumulative distribution function (CDF) and the probability density function (PDF) of $|h_{XY}|^2$ ($X \in \{A_m, B_n\}, Y \in \{R_k\}, 1 \le m \le M, 1 \le n \le N, 1 \le k \le K$) are, respectively, given by

$$F_{XY}(x) = 1 - e^{-\frac{x}{\lambda_{XY}}}, \tag{14.7}$$

$$f_{XY}(x) = \frac{e^{-\frac{lx}{\lambda_{XY}}}}{\lambda_{XY}}, \tag{14.8}$$

where $\lambda_{XY} = \mathbb{E}\left[|h_{XY}|^2\right]$.

Accordingly, the CDF and PDF of g_i, ($i \in \{0, 1, 2\}$) are, respectively, obtained by [18]:

$$F_{g_i}(x) = \left(1 - e^{-\frac{x}{\lambda_i}}\right)^L = \sum_{l=0}^{L} \binom{L}{l} (-1)^l e^{-\frac{lx}{\lambda_i}}, \tag{14.9}$$

$$f_{g_i}(x) = \sum_{l=1}^{L} \binom{L}{l} \frac{(-1)^{l+1} l}{\lambda_i} e^{-\frac{lx}{\lambda_i}}, \tag{14.10}$$

where $L \in \{K, M, N\}, \lambda_i = \mathbb{E}[g_i]$.

For eavesdropper E, the PDF and CDF of $g_3 = |h_{R*E}|^2$ are, respectively, expressed as

$$f_{g_3}(x) = \frac{1}{\lambda_3} e^{-\frac{x}{\lambda_3}}, \tag{14.11}$$

$$F_{g_3}(x) = 1 - e^{-\frac{x}{\lambda_3}}, \tag{14.12}$$

where $\lambda_3 = \mathbb{E}[g_3]$.

14.3.2 NOMA-based DF relay offloading with RF EH description

Given the above discussion, the overall energy harvesting and information communication process can be mathematically described as follows. First, the energy-constrained relay harvests the RF energy from **S** at the time of αT. The harvested energy at R^* is given by

$$E_H = \eta \alpha T P_0 g_0, \tag{14.13}$$

where η stands for the energy conversion efficiency of the energy receiver ($0 < \eta \le 1$), P_0 is the transmit power of **S**. Then, following the principle of uplink NOMA, MU pair (A^*, B^*) transmit information to R^* during the duration $\frac{(1-\alpha)T}{2}$. The received signal at R^* is obtained by

$$y = \sqrt{P_U \rho} x_A h_{A^* R^*} + \sqrt{(1-\rho) P_U} x_B h_{B^* R^*} + n_1, \tag{14.14}$$

where P_U is the total transmit power of user pair (A^*, B^*), ρ denotes the power allocation coefficient, x_A and x_B represent the messages of users A^* and B^*, respectively, as well as $n_1 \sim \mathscr{CN}(0, \sigma^2)$ denotes the additive white Gaussian noise (AWGN) at relay R^*. Notice that since user A^* has a higher priority than user B^*, this implies $0.5 < \rho < 1$. The successive interference cancellation (SIC) is applied at relay R^* to detect x_A and x_B [39]. Specifically, they first decode the message x_A, then remove this component from the received signal to detect the message x_B. Accordingly, the instantaneous signal-to-interference-and-noise ratio (SINR) at relay R^* to detect x_A and x_B is, respectively, written as

$$\gamma_{R*}^{x_A} = \frac{\rho \gamma_U g_1}{(1-\rho)\gamma_U g_2 + 1}, \tag{14.15}$$

$$\gamma_{R*}^{x_B} = (1-\rho)\gamma_U g_2, \tag{14.16}$$

where $\gamma_U = \frac{P_U}{\sigma^2}$ is the average transmit SNR. Finally, relay retransmits the decoded messages to **S** by applying downlink NOMA with all harvested RF energy in the remaining time of $\frac{(1-\alpha)T}{2}$.

The received signal at **S** is obtained by

$$z = \sqrt{\rho P_H} x_A h_{R^*S} + \sqrt{(1-\rho)P_H} x_B h_{R^*S} + n_2, \tag{14.17}$$

where $P_H = \frac{E_H}{(1-\alpha)T/2}$ is the transmit power of selected relay R^*, and $n_2 \sim \mathscr{CN}(0, \sigma^2)$.

Accordingly, by using SIC to detect received multi-user data stream, the SINR at **S** to detect x_A and x_B is, respectively, written as

$$\gamma_S^{x_A} = \frac{\rho b \gamma_0 g_0^2}{(1-\rho) b \gamma_0 g_0^2 + 1}, \tag{14.18}$$

$$\gamma_S^{x_B} = (1-\rho) b \gamma_0 g_0^2, \tag{14.19}$$

where $b = \frac{2\eta\alpha}{(1-\alpha)}$, and $\gamma_0 = \frac{P_0}{\sigma^2}$ is the average transmit SNR.

At eavesdropper E, the signal received from R^* has the following form:

$$z_E = \sqrt{\rho P_H} x_A h_{R^*E} + \sqrt{(1-\rho)P_H} x_B h_{R^*E} + n_E, \tag{14.20}$$

where $n_E \sim \mathscr{CN}(0, \sigma_E^2)$. It is assumed that eavesdropper also uses SIC to detect received multi-user data streams. Thus, the instantaneous SINR and SNR at E to detect x_A and x_B are, respectively, given by

$$\gamma_E^{x_A} = \frac{\rho b \gamma_E g_0 g_3}{(1-\rho) b \gamma_E g_0 g_3 + 1}, \tag{14.21}$$

$$\gamma_E^{x_B} = (1-\rho) b \gamma_E g_0 g_3, \tag{14.22}$$

where $\gamma_E = \frac{P_0}{\sigma_E^2}$ stands for the average transmit SNR according to eavesdropper.

14.3.3 Derivation for the joint CDF of SINRs and SNRs

From (14.18), (14.19), (14.21), and (14.22), the joint CDFs of $\left(\gamma_S^{x_A}, \gamma_E^{x_A}\right)$, and $\left(\gamma_S^{x_B}, \gamma_E^{x_B}\right)$ can be derived through Lemmas 1 and 2 as follows.

Lemma 1. *Under Rayleigh fading, the joint CDF of* $\left(\gamma_S^{x_A}, \gamma_E^{x_A}\right)$ *is given by*

$$F_{\gamma_S^{x_A},\gamma_E^{x_A}}(x,y)=\begin{cases} 1, & if\, x \ge \frac{\rho}{1-\rho},\, y \ge \frac{\rho}{1-\rho} \\ \left(1-e^{-\beta_{1x}/\lambda_0}\right)^K, & if\, x < \frac{\rho}{1-\rho},\, y \ge \frac{\rho}{1-\rho} \\ \sum\limits_{k=1}^{K}\binom{K}{k}(-1)^{k+1}\left[1-\mu_{1y}\mathbf{K}_1(\mu_{1y})\right], & if\, x \ge \frac{\rho}{1-\rho},\, y < \frac{\rho}{1-\rho} \\ \sum\limits_{k=1}^{K}\binom{K}{k}(-1)^{k+1}[1-e^{-\frac{k\beta_{1x}}{\lambda_0}}-\Xi], & if\, x < \frac{\rho}{1-\rho},\, y < \frac{\rho}{1-\rho} \end{cases} \quad (14.23)$$

where $\mu_{1y}=2\sqrt{\frac{ky}{b\lambda_0\lambda_3\gamma_E[\rho-(1-\rho)y]}}$, $\Xi=\frac{k}{\lambda_0}\frac{\pi\beta_{1x}}{2P}\sum\limits_{p=1}^{P}e^{\mu_{2y}}\sqrt{1-\varphi_p^2}$, $\beta_{1x}=\sqrt{\frac{x}{b\gamma_0[\rho-(1-\rho)x]}}$, $\mu_{2y}=-\frac{y}{b\lambda_3\gamma_E[\rho-(1-\rho)y]\theta_p}-\frac{k\theta_p}{\lambda_0}$, $\mathbf{K}_v(.)$ *is the modified Bessel function of the second kind and* vth *order, P are the complexity-vs-accuracy trade-off coefficients,* $\theta_p=\frac{(\varphi_p+1)\beta_{1x}}{2}$, $\varphi_p=\cos\left(\frac{2p-1}{2P}\pi\right)$.

Proof. See in Appendix A.

Lemma 2. *Under Rayleigh fading, the joint CDF of* $\left(\gamma_S^{x_B}, \gamma_E^{x_B}\right)$ *is obtained by*

$$F_{\gamma_S^{x_B},\gamma_E^{x_B}}(x,y)=\sum_{k=1}^{K}\binom{K}{k}(-1)^{k+1}$$

$$\times\left[1-e^{-\frac{k\beta_{2x}}{\lambda_0}}-\frac{\pi k\beta_{2x}}{2P\lambda_0}\sum_{p=1}^{P}e^{-\frac{y}{(1-\rho)\lambda_3 b\gamma_E\theta_p}-\frac{k\theta_p}{\lambda_0}}\sqrt{1-\varphi_p^2}\right] \quad (14.24)$$

where $\beta_{2x}=\sqrt{\frac{x}{(1-\rho)b\gamma_0}}$, $\theta_p=\frac{(\varphi_p+1)\beta_{2x}}{2}$, $\varphi_p=\cos\left(\frac{2p-1}{2P}\pi\right)$, *P are the complexity-vs-accuracy trade-off coefficients.*

Proof. See in Appendix B.

14.4 Secrecy performance analysis

In this section, the secrecy performance of the considered system is analyzed in terms of the existence probability of secrecy capacity (Ψ_{ep}) and secrecy outage probability (Ψ_{sop}).

14.4.1 Preliminaries

Existence probability of secrecy capacity (Ψ_{ep}) is an important measure that is useful for system designers to evaluate the secrecy performance of a wireless communication system. The existence probability of secrecy capacity is defined as

$$\Psi_{ep} = \Pr\left(Cs > 0\right). \tag{14.25}$$

The secrecy outage probability of a legitimate node, denoted as (Ψ_{sop}), is defined as the probability that the instantaneous secrecy capacity, e.g., Cs, falls below a predetermined secrecy rate threshold $Cs_{th} > 0$ and expressed as

$$\Psi_{sop} = \Pr\left(Cs < Cs_{th}\right). \tag{14.26}$$

In the considered network, the capacity of **S** and that of eavesdropper E to detect $x_V, V \in \{A, B\}$ are, respectively, written as

$$C_{R^*S} = \frac{1-\alpha}{2}\log_2\left(1+\gamma_S^{x_V}\right), \tag{14.27}$$

and

$$C_{R^*E} = \frac{1-\alpha}{2}\log_2\left(1+\gamma_E^{x_V}\right). \tag{14.28}$$

Thus, the instantaneous secrecy capacity for the selected relay R^* is obtained by

$$\begin{aligned} Cs^{x_V} &= \max\left\{0, C_{R*S} - C_{R*E}\right\} \\ &= \max\left\{0, \frac{1-\alpha}{2}\log_2\left(\frac{1+\gamma_S^{x_V}}{1+\gamma_E^{x_V}}\right)\right\}. \end{aligned} \tag{14.29}$$

Here, we can express the secrecy outage probability of R^* to detect x_V as follows:

$$\Psi_{sop}^{x_V} = \Pr\left(Cs_{x_V} < Cs_{th}^{x_V}\right), \tag{14.30}$$

where $Cs_{th}^{x_V}$ stands for a predetermined secrecy rate threshold to hear x_V.

Considering the transmission process with the DF scheme, the SOP of this system to detect x_V is expressed as

$$\Omega_{sop}^{x_V} = \Phi_{out}^{x_V} + \left(1-\Phi_{out}^{x_V}\right).\Psi_{sop}^{x_V}, \tag{14.31}$$

where $\Phi_{out}^{x_V}$ is the outage probability to detect x_V defined as follows:

$$\Phi_{out}^{x_V} = \Pr\left(C_{x_V} < R_{x_V}\right). \tag{14.32}$$

where $C_{x_V} = \frac{1-\alpha}{2}\log_2\left(1+\gamma_{R*}^{x_V}\right)$, and R_{x_V} stands for target rate for V^*, $V \in \{A, B\}$.

In order to analyze secrecy performance of this system, we obtain the following theorems.

14.4.2 Existence probability of secrecy capacity analysis

Theorem 1. *Under Rayleigh fading channel, the closed-form expression of the existence probability of secrecy capacity $\Omega_{ep}^{x_A}$ of the considered system to detect x_A is as follows:*

$$\Omega_{ep}^{x_A} = \left(1-\Phi_1^{x_A}\right).\Psi_{ep}^{x_A}, \tag{14.33}$$

where

$$\Phi_1^{x_A} = 1 + \sum_{m=1}^{M}\sum_{n=1}^{N}\binom{M}{m}\binom{N}{n}\frac{(-1)^{m+n+1} n\rho\lambda_1 e^{-\frac{m\gamma_{th}^A}{\rho\gamma_U\lambda_1}}}{m\gamma_{th}^A(1-\rho)\lambda_2 + n\rho\lambda_1},$$

$$\Psi_{ep}^{x_A} = \sum_{k=1}^{K}\binom{K}{k}(-1)^{k+1}\left[e^{-\frac{k\beta_\gamma^A}{\lambda_0}} - \frac{\pi^2\rho\left(a-\gamma_{th}^A\right)k}{4PQ\lambda_0}\right.$$
$$\times\sum_{q=1}^{Q}\sum_{p=1}^{P}\frac{e^{-\frac{\chi_a}{b\lambda_3\gamma_E[\rho-(1-\rho)\chi_a]\theta_p}-\frac{k\theta_p}{\lambda_0}}}{\sqrt{b\lambda_0\chi_a}[\rho-(1-\rho)\chi_a]^{\frac{3}{2}}}$$
$$\left.\times\left(1+\frac{\chi_a}{b\lambda_3\gamma_E\left[\rho-(1-\rho)\chi_a\right]\theta_p}-\frac{k\theta_p}{\lambda_0}\right)\times\sqrt{\left(1-\varphi_p^2\right)\left(1-\phi_q^2\right)}\right],$$

$a = \frac{\rho}{1-\rho}$, $\beta_\gamma^A = \sqrt{\frac{\gamma_{th}^A}{b\gamma_0[\rho-(1-\rho)\gamma_{th}^A]}}$, $\theta_p = \frac{(\varphi_p+1)\beta_q}{2}$, $\varphi_p = \cos\left(\frac{2p-1}{2P}\pi\right)$, $\beta_q = \sqrt{\frac{\chi_q}{b\gamma_0[\rho-(1-\rho)\chi_q]}}$, $\chi_a = \frac{(\phi_q+1)a}{2}$, $\phi_q = \cos\left(\frac{2q-1}{2Q}\pi\right)$, *P and Q are the complexity-vs-accuracy trade-off coefficients.*

Proof. See in Appendix C.

Theorem 2. *Under Rayleigh fading channel, the closed-form expression of the existence probability of secrecy capacity* $\Omega_{ep}^{x_B}$ *of the considered system to detect* x_B *is as follows:*

$$\Omega_{ep}^{x_B} = \left(1-\Phi_1^{x_B}\right).\Psi_{ep}^{x_B}, \tag{14.34}$$

where

$$\Phi_1^{x_B} = \left(1 - e^{-\frac{\gamma_{th}^B}{(1-\rho)\gamma_U\lambda_2}}\right)^N,$$

$$\Psi_{ep}^{x_B} = \sum_{k=1}^{K}\binom{K}{k}(-1)^{k+1}\left[e^{-\frac{k\beta_\gamma^B}{\lambda_0}} - \frac{\pi^2 k e^{-\gamma_{th}^B}}{4PQ\lambda_0}\times\sum_{q=1}^{Q}\sum_{p=1}^{P}\frac{e^{-\frac{\chi_b}{(1-\rho)\lambda_3 b\gamma_E\theta_p}-\frac{k\theta_p}{\lambda_0}}}{\sqrt{(1-\rho)b\gamma_0\chi_b}u_q}\right.$$
$$\left.\times\left(1+\frac{\chi_b}{(1-\rho)\lambda_3 b\gamma_E\theta_p}-\frac{k\theta_p}{\lambda_0}\right)\times\sqrt{\left(1-\varphi_p^2\right)\left(1-\phi_q^2\right)}\right],$$

$\beta_\gamma^B = \sqrt{\frac{\gamma_{th}^B}{(1-\rho)b\gamma_0}}$, $\theta_p = \frac{(\varphi_p+1)\beta_q}{2}$, $\varphi_p = \cos\left(\frac{2p-1}{2P}\pi\right)$, $\beta_q = \sqrt{\frac{\chi_b}{(1-\rho)b\gamma_0}}$, $\chi_b = -\ln(u_q)$, $u_q = \frac{(\phi_q+1)e^{-\gamma_{th}^B}}{2}$, $\phi_q = \cos\left(\frac{2q-1}{2Q}\pi\right)$, *P and Q are the complexity-vs-accuracy trade-off coefficients.*

Proof. See in Appendix D.

14.4.3 Secrecy outage probability analysis

Theorem 3. *Under Rayleigh fading channel, the secrecy outage probability of this RF EH QoS-based multiuser uplink/downlink NOMA communication network without CSI of eavesdropper to detect x_A can be obtained by*

$$\Omega_{sop}^{x_A} = \Phi_1^{x_A} + \left(1 - \Phi_1^{x_A}\right).\Phi_2^{x_A} + \left(1 - \Phi_1^{x_A}\right).\Psi_{sop}^{x_A}, \tag{14.35}$$

where

$$\Phi_1^{x_A} = 1 + \sum_{m=1}^{M}\sum_{n=1}^{N}\binom{M}{m}\binom{N}{n}\frac{(-1)^{m+n+1} n\rho\lambda_1 e^{-\frac{m\gamma_{th}^A}{\rho\gamma_U\lambda_1}}}{m\gamma_{th}^A(1-\rho)\lambda_2 + n\rho\lambda_1},$$

$$\Phi_2^{x_A} = \begin{cases} 1, & \gamma_{th}^A > \frac{\rho}{1-\rho} \\ \left(1 - e^{-\frac{1}{\lambda_0}\sqrt{\frac{\gamma_{th}^A}{b\gamma_0\left[\rho-(1-\rho)\gamma_{th}^A\right]}}}\right)^K, & \gamma_{th}^A < \frac{\rho}{1-\rho} \end{cases},$$

$$\begin{aligned}\Psi_{sop}^{x_A} = \frac{\pi^2\rho}{2PQ\lambda_0 b\lambda_3\gamma_E}\Bigg\{ & a_1\sum_{k=1}^{K}\sum_{p=1}^{P}\sum_{q=1}^{Q}\binom{K}{k}\frac{(-1)^{k+1} k e^{-\frac{\xi_a}{b\lambda_3\gamma_E[\rho-(1-\rho)\xi_a]\theta_q}-\frac{k\theta_q}{\lambda_0}}}{[\rho-(1-\rho)\xi_a]^2(\varphi_p+1)}\Phi \\ & +(a_2-a_1)\sum_{k=1}^{K}\sum_{p=1}^{P}\sum_{q=1}^{Q}\binom{K}{k}\frac{(-1)^{k+1} k e^{-\frac{\psi_q}{b\lambda_3\gamma_E\left[\rho-(1-\rho)\psi_q\right]\theta_a}-\frac{k\theta_a}{\lambda_0}}}{\left[\rho-(1-\rho)\psi_q\right]^2(\varphi_p+1)}\Phi \\ & -(a_2-a_0)\sum_{k=1}^{K}\sum_{p=1}^{P}\sum_{q=1}^{Q}\binom{K}{k}\frac{(-1)^{k+1} k e^{-\frac{\psi_q}{b\lambda_3\gamma_E\left[\rho-(1-\rho)\psi_q\right]\theta_\gamma}-\frac{k\theta_\gamma}{\lambda_0}}}{\left[\rho-(1-\rho)\psi_q\right]^2(\varphi_p+1)}\Phi\Bigg\},\end{aligned}$$

$\Phi = \sqrt{\left(1-\varphi_p^2\right)\left(1-\phi_q^2\right)}$, $a_0 = \frac{\gamma_{th}^A+1}{2^{Rs_{x_A}}} - 1$, $a_1 = \frac{1}{(1-\rho)2^{Rs_{x_A}}} - 1$, $a_2 = \frac{\rho}{1-\rho} - \delta$, δ *is the accuracy coefficient,* $\beta_q = \sqrt{\frac{2^{Rs_{x_A}}(\xi_a+1)-1}{b\gamma_0\left\{\rho-(1-\rho)\left[2^{Rs_{x_A}}(\xi_a+1)-1\right]\right\}}}$, $\beta_a = \sqrt{\frac{a_2}{b\gamma_0[\rho-(1-\rho)a_2]}}$, $\beta_\gamma = \sqrt{\frac{\gamma_{th}^A}{b\gamma_0\left[\rho-(1-\rho)\gamma_{th}^A\right]}}$, $\theta_a = \frac{(\varphi_p+1)\beta_a}{2}$, $\varphi_p = \cos\left(\frac{2p-1}{2P}\pi\right)$, $\xi_a = \frac{(\phi_q+1)a_1}{2}$, $\theta_q = \frac{(\varphi_p+1)\beta_q}{2}$, $\psi_a = \frac{(\phi_q+1)(a_2-a_1)}{2} + a_1$, $\phi_q = \cos\left(\frac{2q-1}{2Q}\pi\right)$, $Rs_{x_A} = \frac{2Cs_{th}^{x_A}}{1-\alpha}$, *P and Q are the complexity-vs-accuracy trade-off coefficients.*

Proof. See in Appendix E.

Theorem 4. *Under Rayleigh fading channel, the secrecy outage probability of this RF EH QoS-based multiuser uplink/downlink NOMA communication network without CSI of eavesdropper to detect x_B can be written as*

$$\Omega_{sop}^{x_B} = \Phi_1^{x_B} + \left(1 - \Phi_1^{x_B}\right).\Phi_2^{x_B} + \left(1 - \Phi_1^{x_B}\right).\Psi_{sop}^{x_B}, \tag{14.36}$$

where $\Phi_1^{\chi B} = \left(1 - e^{-\frac{\gamma_{th}^B}{(1-\rho)\gamma_U\lambda_2}}\right)^N$, $\Phi_2^{\chi B} = \left(1 - e^{-\frac{\gamma_{th}^B}{(1-\rho)b\gamma_0\lambda_0}}\right)^K$,

$$\Psi_{sop}^{\chi B} = \frac{\pi^2 e^{-c}}{2PQ}\sum_{k=1}^{K}\sum_{q=1}^{Q}\sum_{p=1}^{P}\binom{K}{k}\frac{(-1)^{k+1}k\sqrt{(1-\varphi_p^2)(1-\phi_q^2)}}{(1-\rho)\lambda_0\lambda_3 b\gamma_E u_q(\varphi_p+1)}$$
$$\times\left[e^{-\frac{\chi_q}{(1-\rho)\lambda_3 b\gamma_E\theta_q}-\frac{k\theta_q}{\lambda_0}} - e^{-\frac{\chi_q}{(1-\rho)\lambda_3 b\gamma_E\theta_\gamma}-\frac{k\theta_\gamma}{\lambda_0}}\right],$$

$c = \frac{\gamma_{th}^B+1}{2^{R_{S\chi B}}} - 1$, $\beta_y = \sqrt{\frac{2^{R_{S\chi B}}(y+1)-1}{(1-\rho)b\gamma_0}}$, $\theta_y = \frac{(\varphi_p+1)\beta_y}{2}$, $\beta_\gamma = \sqrt{\frac{\gamma_{th}^B}{(1-\rho)b\gamma_0}}$, $\theta_y = \frac{(\varphi_p+1)\beta_\gamma}{2}$,
$\beta_q = \sqrt{\frac{2^{R_{S\chi B}}(\chi_q+1)-1}{(1-\rho)b\gamma_0}}$, $\theta_q = \frac{(\varphi_p+1)\beta_q}{2}$, $\chi_q = -\ln(u_q)$, $a = e^{-c}$, $u_q = \frac{(\phi_q+1)a}{2}$, $\phi_q = \cos\left(\frac{2q-1}{2Q}\pi\right)$, *P and Q are the complexity-vs-accuracy trade-off coefficients.*

Proof. See in Appendix F.

14.4.4 Secrecy performance optimization

In this section, we present the problem of minimizing the secrecy outage probability (Ω_{sop}) for the proposed system, namely MSOP. Specifically, the mathematical description of the optimization problem is expressed by the following formulas:

$$\text{(MSOP)} \quad : \min_{\alpha,\rho} \left(\Omega_{sop}\right) \tag{14.37a}$$

$$\text{subject to} \quad 0 < \alpha \leq 1 \tag{14.37b}$$

$$0.5 < \rho \leq 1 \tag{14.37c}$$

where conditions (14.37b) describe the constraint of the time switching ratio, and conditions (14.37c) describe the constraint of the power allocation ratio.

We propose to use a particle swarm optimization (PSO)-based algorithm [40–42] to solve the MSOP, namely MSOP-PSO, as Algorithm 2. PSO results from modeling the birds searching for food and is classified as a swarm intelligence class. At the beginning of the search, the swarm flies in some very random direction and speed. The position and the velocity of the ith particle at time t are given by:

$$x_i(t) = (\alpha_i(t), \rho_i(t)) \tag{14.38}$$

$$v_i(t) = (v_i^\alpha(t), v_i^\rho(t)) \tag{14.39}$$

After searching in the solution space, some particles began to find a place to store food. Then, they send signals, and they are the knowledge that they accumulate in the process of searching, to other particles in swarm, i.e., $pBest_i(t)$ and $gBest(t)$. Based on the information received, each particle will adjust its flight direction and speed in the direction of the most food [41]. This mechanism helps particles find out where the most food is in the vast search space.

The process of updating the particle's velocity and coordinates is described by the following formula [43]:

$$v_i\,(t+1) = \chi(v_i\,(t) + e_1.r_1.\,(pBest_i\,(t) - x_i\,(t)) + e_2.r_2.\,(gBest\,(t) - x_i\,(t))\,) \tag{14.40}$$

$$x_i\,(t+1) = x_i\,(t) + v_i\,(t+1) \tag{14.41}$$

where v_i and $v_i\,(t+1)$ are the velocity of the particle ith at time t and time $t+1$ respectively, x_i and $x_i\,(t+1)$ are the coordinates of the particle ith at time t and time $t+1$, respectively. In [43], Clerc *et al.* proposed constriction factor $\chi = \frac{2}{\left|2-\phi-\sqrt{\phi^2-4\phi}\right|}$, with $\phi = \phi_1 + \phi_2 > 4$, r_1 and r_2 are random number follow uniform distribution with a range of 0 to 1, $e_1 = \chi\phi_1, e_2 = \chi\phi_2$ are two acceleration coefficients and are, respectively, cognitive and social characteristics.

We use the maximum number of evolutionary iterations as the stopping condition for the algorithm, denoted as *MaxIt*.

Algorithm 2: Minimal the secrecy outage probability based on particle swarm optimization (MSOP-PSO)

Input K, M, N, P, Q, SOP, α, ρ
Output α^*, ρ^*
Initialize population size *nPop* with random position $x_i(\alpha, \rho)$, $v_i(0) \leftarrow 0$, $\phi_1 = 2.05$, $\phi_2 = 2.05$, and *MaxIt*.
for $t = 1 : MaxIt$ **do**
    for $i = 1 : nPop$ **do**
        Update x_i and v_i using (14.40) and (14.41)
        Compute particles fitness value using the SOP function
        Update best particles $pBest_i(t)$ and $gBest(t)$
    end for
end for
Return (α^*, ρ^*)

The MSOP-PSO algorithm complexity is analyzed as follows: the use of formulas (14.38) and (14.40) to update the position and speed of particles is simple addition and multiplication operations of complexity, thus, it has complexity $\mathcal{O}(1)$. Similarly, comparisons for selection $pBest_i(t)$ and $gBest(t)$ have complexity $\mathcal{O}(1)$. Calculating fitness function Ω_{sop} requires two nested loops to calculate $\Phi_1^{x_A}$ and three nested loops to calculate $\Psi_{sop}^{x_A}$. This whole process is executed through a *MaxIt* loop over *nPop* particles. Thus the algorithmic complexity of MSOP-PSO is:

$$\mathcal{T} = MaxIt.nPOp(\mathcal{O}(1) + M.N + K.P.Q + \mathcal{O}(1)) \tag{14.42}$$

$$= MaxIt.nPOp(M.N + K.P.Q) \tag{14.43}$$

14.5 Numerical results and discussion

This section describes the numerical results to verify the analytical expression of the existence probability of secrecy capacity and the secrecy outage probability for the multiuser RF EH NOMA relaying MEC system. We present the Monte Carlo simulation parameters of this work in Table 14.2. In the simulations, we assume that the **A** and **B** are allocated $1m$ away from **R**, and **S** is fixed at the edge $1m$ away from the relay cluster in the opposite direction. According to the assumption presented in Section 14.3, we set up **E** to be distributed on the **S** side and $1m$ away from **R**. Without loss of generality, we assume that the path loss exponents for all of the devices equals 2. Applying the exponential distribution model, the channels are modeled as standard Rayleigh fading components with zero mean and unit variance.

In this experiment, we investigate the effect of the time switching ratio (α) on the existence probability of secrecy capacity and SOP of the proposed system. We investigate the average SNR levels of 10 dB and 20 dB, respectively. The results in Figures 14.2 and 14.3 show that as α increases and Ω_{ep} decreases, while Ω_{sop} increases. Another observation is that as the average SNR increases, Ω_{ep} increases, and Ω_{sop} decreases. In other words, increasing the transmit power of **S** can help improve the security performance of the system. Furthermore, Figures 14.2 and 14.3, respectively, show the comparison of Ω_{ep} and Ω_{sop} between NOMA and OMA schemes. As expected, we observe that the secrecy performance of the NOMA scheme significantly outperforms that of the OMA scheme. This trend is inevitable and will reappear in our following investigations; thus, we will not present them in the remaining figures.

In the next experiment, we examined Ω_{ep} and Ω_{sop} under the impact of the number of relays in relay cluster. We fixed the number of users in cluster **A** as $M = 3$ and the number of users in cluster **B** as $N = 4$, then gradually increased the number of relays K. The results are in Figures 14.4 and 14.5 show that Ω_{ep} of two users increases and Ω_{sop} decreases with increase gradually the number of relays. In other words, increasing the number of relays can improve physical layer security performance for users in the proposed system. This result coincides with the system's operating

Table 14.2 Simulation parameters

Parameters	Notation	Typical values
Environment		Rayleigh
Number of users in cluster **A**	M	3
Number of users in cluster **B**	N	4
Number of users in relay cluster **R**	K	2
Average transmit SNR	γ_0	0–20 dB
Time switching ratio	α	0.2
Energy conversion efficiency	η	0.8
Power allocation ratio	ρ	0.8
The population size	*nPop*	50
The maximum evolutionary generation	*MaxIt*	50

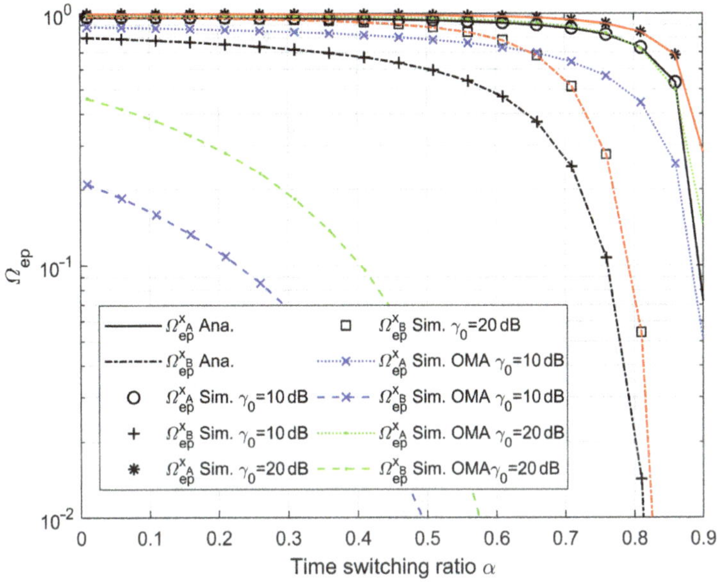

Figure 14.2 Impact of time switching ratio on existence probability of secrecy capacity with different average transmit SNR values

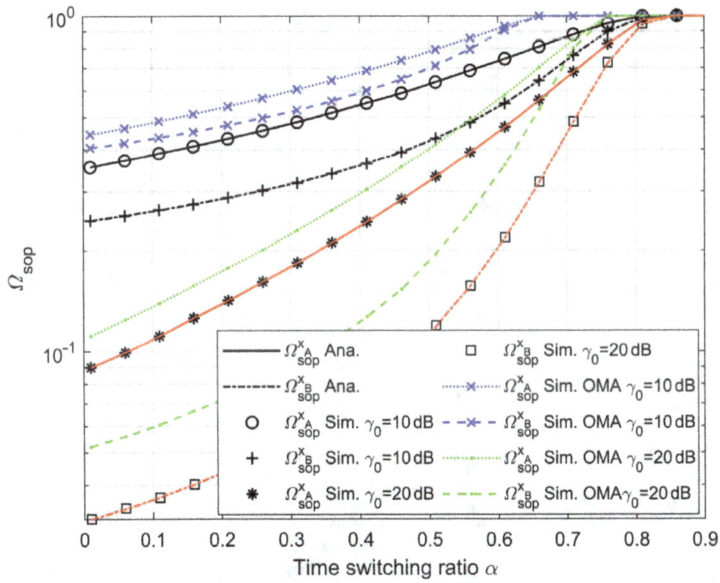

Figure 14.3 Impact of time switching ratio on secrecy outage probability with different average transmit SNR values

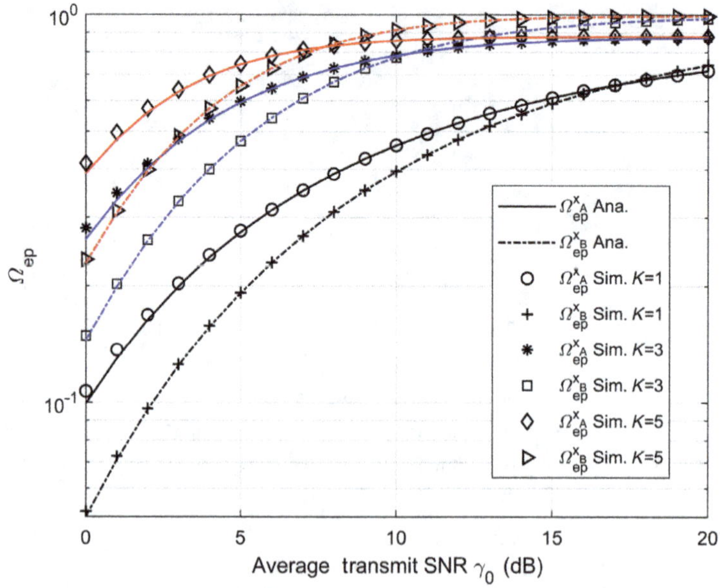

Figure 14.4 Impact of number of relays on existence probability of secrecy capacity

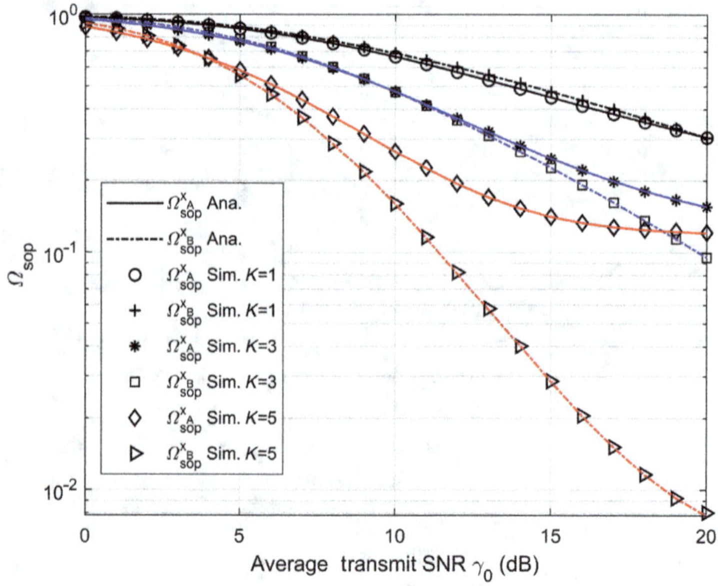

Figure 14.5 Impact of number of relays on secrecy outage probability

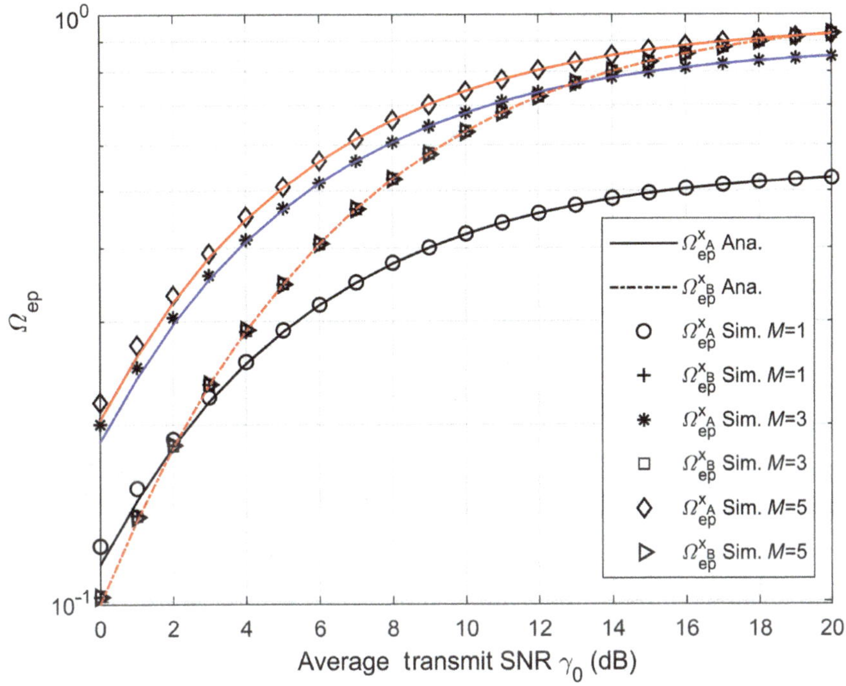

Figure 14.6 Impact of number of users in cluster ***A*** *on existence probability of secrecy capacity*

protocol; the more relays support, the more likely **S** is to choose the best relay R^* to support the communication of A^* and B^*. Another observation is that when the transmit power is higher than 25 dB, Ω_{ep} and Ω_{sop} of A^* and B^* tend to be saturated. It shows that for the proposed system, increasing the transmit power too high is not necessary.

We investigated the physical layer security performance of the proposed system under the impact of the number of users in the cluster. We fixed the number of users in cluster **B** to $N = 4$, the number of relays $K = 2$, and changed the number of users in cluster **A** to $M = \{1, 3, 5\}$, respectively. Figures 14.6 and 14.7 describe the user's Ω_{ep} and Ω_{sop} when changing the number of users in cluster **A** M. The results show that as M increases and Ω_{ep} increases, while Ω_{sop} decreases. It shows that the higher the number of users in cluster **A**, the better the physical layer security performance of the proposed system.

We continue to similarly examine the impact of the number of users in cluster **B** on the system's physical layer security performance. The simulation parameters are set up as follows: $K = 2$, $M = 3$ and $N = \{1, 3, 5\}$. The results obtained in Figures 14.8 and 14.9 show that as N increases the user's Ω_{ep} increases while Ω_{sop} decreases.

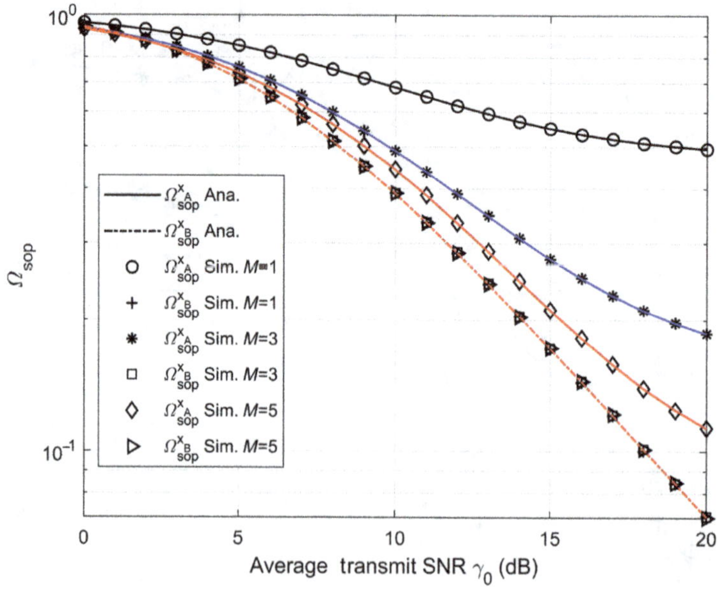

Figure 14.7 Impact of number of users in cluster ***A*** *on secrecy outage probability*

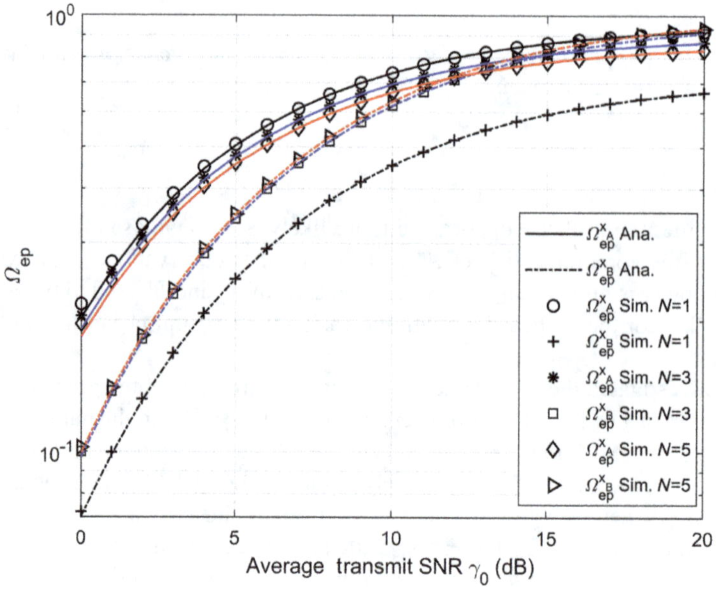

Figure 14.8 Impact of number of users in cluster ***B*** *on existence probability of secrecy capacity*

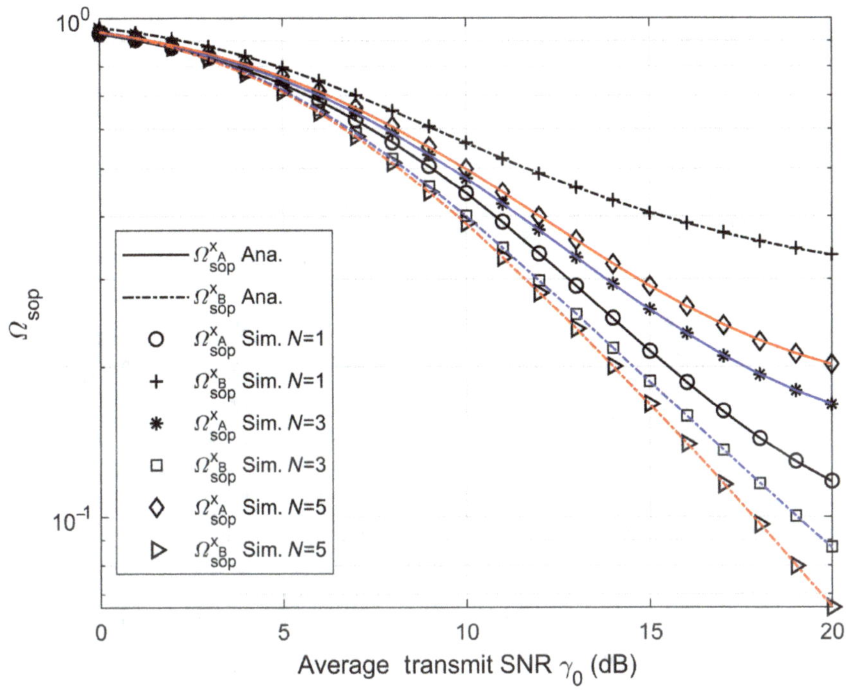

Figure 14.9 Impact of number of users in cluster ***B*** *on secrecy outage probability*

This result once again confirms that increasing the number of users can improve the physical layer performance of the system. These simulation results are completely consistent with the communication protocol in the proposed system.

In this experiment, we investigate the effect of the power allocation ratio ρ on the existence probability of secrecy capacity of the proposed system. The results in Figure 14.10 show that as ρ increases, the Ω_{ep} of A^* increases, while the Ω_{ep} of B^* decreased. This result is completely consistent with hypothesis $0.5 < \rho < 1$ because cluster **A** has a higher priority than cluster **B**. As ρ increases, that means more energy is allocated to A^*, helping A^* improve Ω_{ep}. However, at the same time, this negatively affects B^*'s Ω_{ep} due to its reduced energy allocation.

The results in Figure 14.11 describe the effect of the power allocation ratio ρ on the secrecy outage probability of the proposed system. As ρ increases, user A*'s Ω_{sop} decreases, and user B*'s Ω_{sop} increases. This result is entirely logical with the result obtained in Figure 14.10. In other words, changing the power allocation ratio can improve the physical layer security performance of users in a cluster but degrade physical layer security performance for the user in another cluster.

In the next section, we evaluate the convergence of the MSOP-PSO algorithm using different transmit SNR. The results obtained in Figure 14.12 show that the

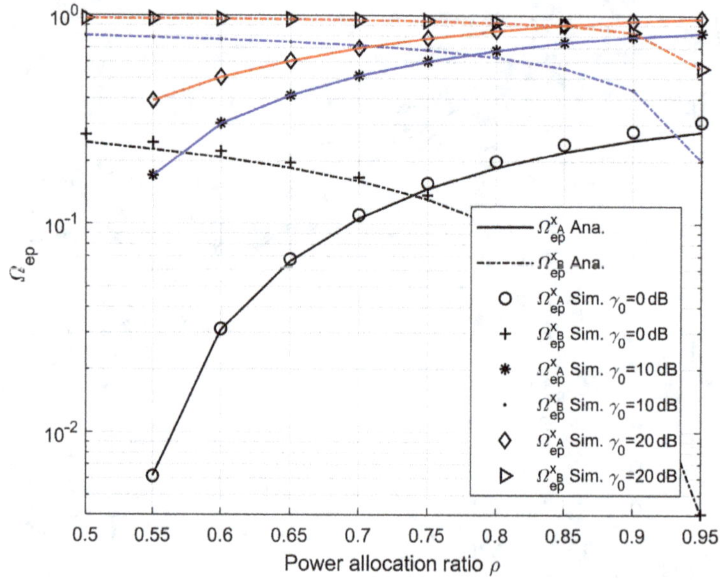

Figure 14.10 Impact of the power allocation ratio on the existence probability of secrecy capacity

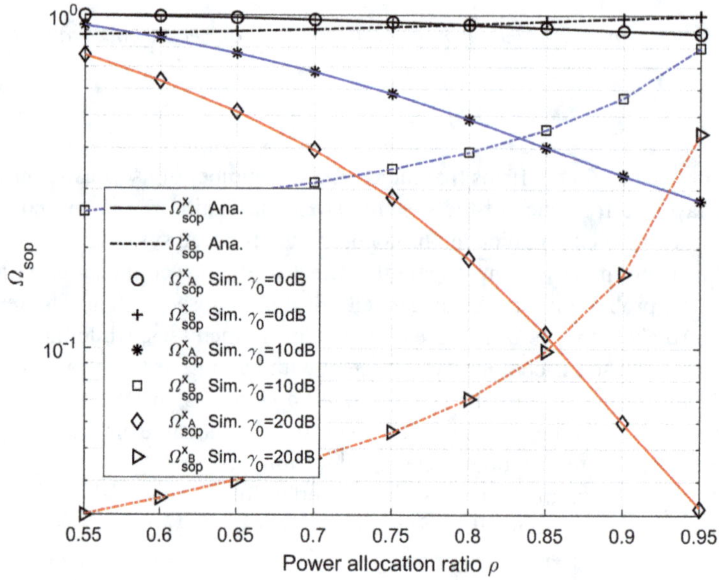

Figure 14.11 Impact of the power allocation ratio on the secrecy outage probability

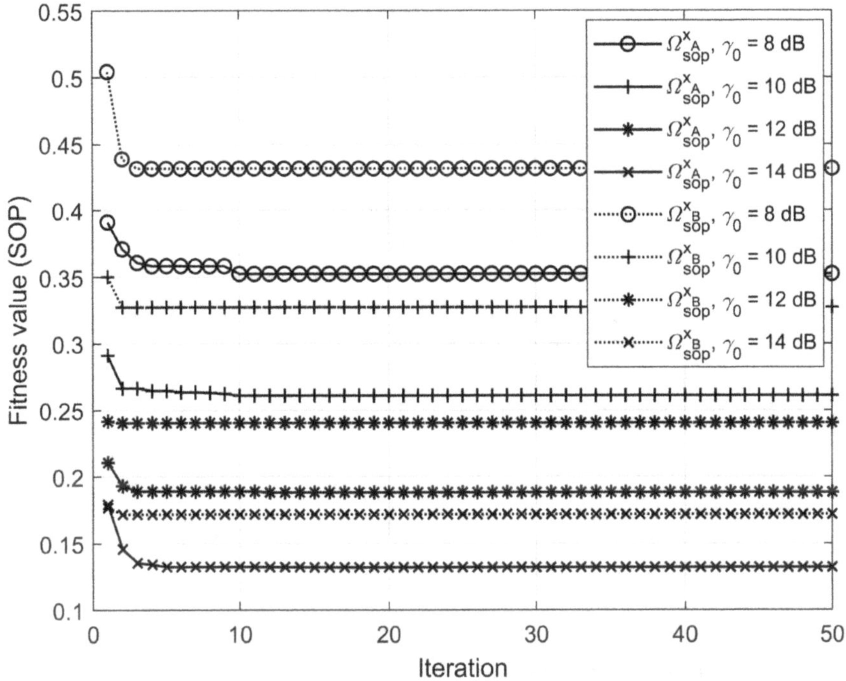

Figure 14.12 The convergence of MSOP-PSO with different average transmit SNR values

algorithm quickly converges in the established terminating condition. It proves that the proposed PSO-based algorithm can solve the MSOP optimization problem well.

With the optimal parameter sets (α^*, ρ^*) obtained from the MSOP-PSO algorithm, we continue to investigate the SOP of each user according to different power levels with three cases: (i) apply MSOP-PSO to user A^* (MSOPA), (ii) apply MSOP-PSO to user B^* (MSOPB), and (iii) do not use optimization algorithm, as shown in Figure 14.13. Using MSOP-PSO can significantly decrease the user's SOP than not using this algorithm. It shows the efficiency of our proposed algorithm. However, when the SOP of one user decreases, the SOP of the other user will be very high, even higher than the case without using the optimal algorithm. It is entirely consistent with the results obtained in the above experiments.

Finally, we examine the best relay selection (BRS) scheme with the random relay selection (RRS) scheme as a benchmark [44], as shown in Figure 14.14. Note that the proposed protocol selects the best relay to collect the most energy to guarantee power for the transition phase. Meanwhile, the RRS scheme can be used to ensure the most straightforward operating protocol for the system. Simulation results show that BRS offers outperform RRS in terms of SOP.

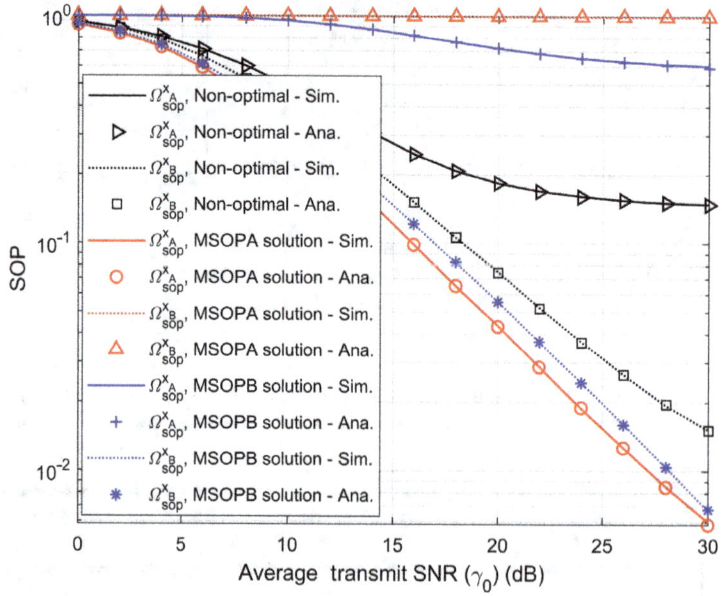

Figure 14.13 Comparison SOP in the case of using the optimal algorithm and not using the optimal algorithm

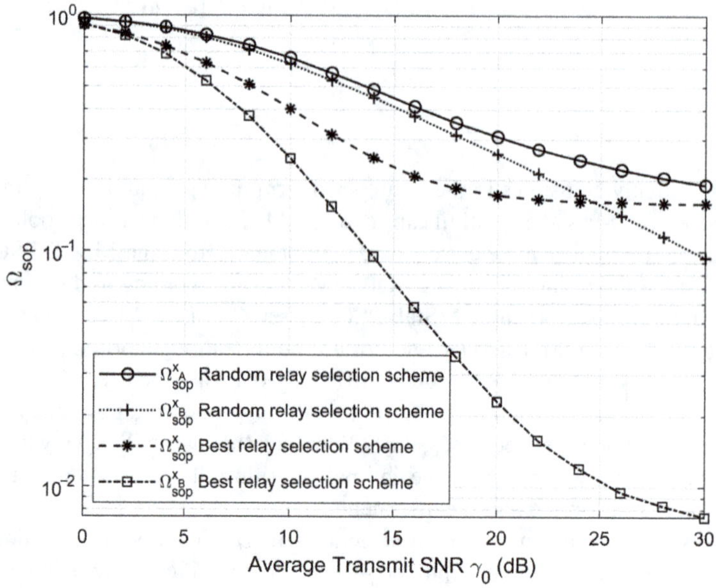

Figure 14.14 Comparison SOP in the case of different relay selection scheme

The simulation results (Sim.) coincide entirely with the calculated results (Ana.) in the relevant parameters, showing that the approach and calculation model are correct.

14.6 Conclusion and future scope

This work evaluated the physical layer security of a multiuser RF EH NOMA MEC system in the presence of a passive eavesdropper over the Rayleigh fading channel. Specifically, a relay with the best channel coefficient is selected to receive RF energy from the BS and use that energy to support communication for the cluster. Accordingly, two users in two clusters with different priorities are paired to apply NOMA and communicate with the selected relay. Furthermore, we have investigated the physical layer security performance of the system in terms of the existence probability of secrecy capacity and the secrecy outage probability with essential parameters. The results show that the proposed model and protocol using the NOMA scheme significantly outperform that of the OMA scheme in terms of secrecy performance. Finally, we state the secrecy performance optimization problem for each user and propose the MSOP-PSO algorithm to solve those problems. The results show that the proposed algorithm can help the user's SOP achieve a better value than the optimal algorithm. The obtained simulation results are entirely consistent with our theoretical analysis, proving the correctness of this study.

We will extend the system model with multi-antenna BS over a general channel in our future studies. Furthermore, we will examine the system protocol in the amplify-and-forward relays scheme, and E can eavesdrop in both phases. Finally, the improved algorithms based on PSO will also be considered in the next studies.

Appendix A: Proof of Lemma 1

The joint CDF of $\gamma_S^{x_A}$ and $\gamma_E^{x_A}$ is derived as follows:

$$\begin{aligned}
F_{\gamma_S^{x_A},\gamma_E^{x_A}}(x,y) \\
&= \Pr\left(\gamma_S^{x_A} < x,\ \gamma_E^{x_A} < y\right) \\
&= \Pr\left(\frac{\rho b\gamma_0 g_0^2}{(1-\rho)b\gamma_0 g_0^2+1} < x, \frac{\rho b\gamma_E g_0 g_3}{(1-\rho)b\gamma_E g_0 g_3+1} < y\right) \\
&= \begin{cases}
1, \text{if } x \geq \frac{\rho}{1-\rho}, y \geq \frac{\rho}{1-\rho} \\
\Pr\left(g_0 < \sqrt{\frac{x}{b\gamma_0[\rho-(1-\rho)x]}}\right), \text{if } x < \frac{\rho}{1-\rho}, y \geq \frac{\rho}{1-\rho} \\
\Pr\left(g_3 < \frac{y}{b\gamma_E[\rho-(1-\rho)y]g_0}\right), \text{if } x \geq \frac{\rho}{1-\rho}, y < \frac{\rho}{1-\rho} \\
\Pr\left(g_0 < \sqrt{\frac{x}{b\gamma_0[\rho-(1-\rho)x]}}, g_3 < \frac{y}{b\gamma_E[\rho-(1-\rho)y]g_0}\right), \\
\quad \text{if } x < \frac{\rho}{1-\rho}, y < \frac{\rho}{1-\rho}
\end{cases}
\end{aligned}$$

$$
= \begin{cases} 1, \text{if}\, x \geq \frac{\rho}{1-\rho},\, y \geq \frac{\rho}{1-\rho} \\ \left(1 - e^{-\beta x/\lambda_0}\right)^K, \text{if}\, x < \frac{\rho}{1-\rho},\, y \geq \frac{\rho}{1-\rho} \\ \int_0^\infty \left[1 - e^{-\frac{y}{b\lambda_3\gamma_E[\rho-(1-\rho)y]t}}\right] \sum \frac{(-1)^{k+1} k e^{-\frac{kt}{\lambda_0}}}{\lambda_0} dt, \\ \text{if}\, x \geq \frac{\rho}{1-\rho},\, y < \frac{\rho}{1-\rho} \\ \int_0^{\beta_1} \left[1 \quad e^{-\frac{y}{b\lambda_3\gamma_E[\rho-(1-\rho)y]t}}\right] \sum \frac{(-1)^{k+1} k e^{-\frac{kt}{\lambda_0}}}{\lambda_0} dl, \\ \text{if}\, x < \frac{\rho}{1-\rho},\, y < \frac{\rho}{1-\rho} \end{cases}
$$

$$
\stackrel{(a)}{=} \begin{cases} 1, & \text{if}\, x \geq \frac{\rho}{1-\rho},\, y \geq \frac{\rho}{1-\rho} \\ \left(1 - e^{-\beta_{1x}/\lambda_0}\right)^K, & \text{if}\, x < \frac{\rho}{1-\rho},\, y \geq \frac{\rho}{1-\rho} \\ \sum (-1)^{k+1} \left[1 - \mu_{1y} \mathbf{K}_1(\mu_{1y})\right], & \text{if}\, x \geq \frac{\rho}{1-\rho},\, y < \frac{\rho}{1-\rho} \\ \sum (-1)^{k+1} [1 - e^{-\frac{k\beta_{1x}}{\lambda_0}} - \Xi], & \text{if}\, x < \frac{\rho}{1-\rho},\, y < \frac{\rho}{1-\rho} \end{cases} \tag{14.44}
$$

where

$\sum \triangleq \sum_{k=1}^{K} \binom{K}{k}$, $\Xi = \frac{k}{\lambda_0} \frac{\pi \beta_{1x}}{2P} \sum_{p=1}^{P} e^{\mu_{2y}} \sqrt{1 - \varphi_p^2}$, $\beta_{1x} = \sqrt{\frac{x}{b\gamma_0[\rho-(1-\rho)x]}}$,

$\mu_{1y} = 2\sqrt{\frac{ky}{b\lambda_0\lambda_3\gamma_E[\rho-(1-\rho)y]}}$, $\mu_{2y} = -\frac{y}{b\lambda_3\gamma_E[\rho-(1-\rho)y]\theta_p} - \frac{k\theta_p}{\lambda_0}$,

$\mathbf{K}_\nu(\cdot)$ is the modified Bessel function of the second kind and νth order.

Note that step (a) is obtained by applying again the Gaussian–Chebyshev quadrature method with P is the complexity-vs-accuracy trade-off coefficient, $\theta_p = \frac{(\varphi_p+1)\beta_{1x}}{2}$, $\varphi_p = \cos\left(\frac{2p-1}{2P}\pi\right)$ [45]. The proof is completed.

Appendix B: Proof of Lemma 2

The joint CDF of γ_S^{xB} and γ_E^{xB} is derived as follows:

$$
\begin{aligned}
F_{\gamma_S^{xB},\gamma_E^{xB}}(x,y) &= \Pr\left(g_0 < \sqrt{\frac{x}{(1-\rho)b\gamma_0}}, g_3 < \frac{y}{(1-\rho)b\gamma_E g_0}\right) \\
&= \sum_{k=1}^{K} \binom{K}{k} \frac{(-1)^{k+1} k}{\lambda_0} \int_0^{\beta_{2x}} \left[1 - e^{-\frac{y}{\lambda_3(1-\rho)b\gamma_E t}}\right] e^{-\frac{kt}{\lambda_0}} dt \\
&\stackrel{(b)}{=} \sum_{k=1}^{K} \binom{K}{k} (-1)^{k+1} \left[1 - e^{-\frac{k\beta_{2x}}{\lambda_0}} - \frac{\pi k \beta_{2x}}{2P\lambda_0} \right. \\
&\quad \left. \times \sum_{p=1}^{P} e^{-\frac{y}{(1-\rho)\lambda_3 b\gamma_E\theta_p} - \frac{k\theta_p}{\lambda_0}} \sqrt{1 - \varphi_p^2}\right]
\end{aligned} \tag{14.45}
$$

where $\beta_{2x} = \sqrt{\frac{x}{(1-\rho)b\gamma_0}}$, $\theta_p = \frac{(\varphi_p+1)\beta_{2x}}{2}$, $\varphi_p = \cos\left(\frac{2p-1}{2P}\pi\right)$. Note that step (b) is obtained by applying again the Gaussian–Chebyshev quadrature method with P is the complexity-vs-accuracy trade-off coefficient. The proof is completed.

Appendix C: Proof of Theorem 3

Here, we derived the closed-form expressions of $\Phi_1^{x_A}$ and $\Psi_{ep}^{x_A}$ as follows:

$$\begin{aligned}\Phi_1^{x_A} &= \Pr\left(\gamma_{R*}^{x_A} < \gamma_{th}^A\right) = \Pr\left(\frac{\rho\gamma_U g_1}{(1-\rho)\gamma_U g_2 + 1} < \gamma_{th}^A\right)\\ &= \Pr\left(g_1 < \frac{\gamma_{th}^A\left[(1-\rho)\gamma_U g_2 + 1\right]}{\rho\gamma_U}\right)\\ &= 1 + \sum_{m=1}^{M}\sum_{n=1}^{N}\binom{M}{m}\binom{N}{n}\frac{(-1)^{m+n+1}n\rho\lambda_1 e^{-\frac{m\gamma_{th}^A}{\rho\gamma_U\lambda_1}}}{m\gamma_{th}^A(1-\rho)\lambda_2 + n\rho\lambda_1}.\end{aligned} \tag{14.46}$$

$$\begin{aligned}\Psi_{ep}^{x_A} &= \int_{\gamma_{th}^A}^{\infty}\int_{0}^{x} f_{\gamma_S^{x_A},\gamma_E^{x_A}}(x,y)\,dy\,dx = \int_{\gamma_{th}^A}^{a}\left[\frac{\partial F_{\gamma_S^{x_A},\gamma_E^{x_A}}(x,y)}{\partial x}\right]_{y=x} dx\\ &= \sum_{k=1}^{K}\binom{K}{k}(-1)^{k+1}\left\{e^{-\frac{k\beta_\gamma^A}{\lambda_0}} - \frac{\pi\rho k}{2P\lambda_0}\times\int_{\gamma_{th}^A}^{a}\sum_{p=1}^{P}\frac{e^{-\frac{x}{b\lambda_3\gamma_E[\rho-(1-\rho)x]\theta_x}-\frac{k\theta_x}{\lambda_0}}}{\sqrt{b\lambda_0 x}[\rho-(1-\rho)x]^{\frac{3}{2}}}\right.\\ &\quad\left.\times\left\{1+\frac{x}{b\lambda_3\gamma_E\left[\rho-(1-\rho)x\right]\theta_x} - \frac{k\theta_x}{\lambda_0}\right\}\sqrt{1-\varphi_p^2dx}\right\}\\ &\overset{(c)}{=} \sum_{k=1}^{K}\binom{K}{k}(-1)^{k+1}\left\{e^{-\frac{k\beta_\gamma^A}{\lambda_0}} - \frac{\pi^2\rho\left(a-\gamma_{th}^A\right)k}{4PQ\lambda_0}\right.\\ &\quad\times\sum_{q=1}^{Q}\sum_{p=1}^{P}\frac{e^{-\frac{\chi_a}{b\lambda_3\gamma_E[\rho-(1-\rho)\chi_a]\theta_p}-\frac{k\theta_p}{\lambda_0}}}{\sqrt{b\lambda_0\chi_a}[\rho-(1-\rho)\chi_a]^{\frac{3}{2}}}\\ &\quad\times\left(1+\frac{\chi_a}{b\lambda_3\gamma_E\left[\rho-(1-\rho)\chi_a\right]\theta_p} - \frac{k\theta_p}{\lambda_0}\right)\times\sqrt{\left(1-\varphi_p^2\right)\left(1-\phi_q^2\right)}\end{aligned} \tag{14.47}$$

where $a = \frac{\rho}{1-\rho}$, $\theta_x = \frac{(\varphi_p+1)\beta_1}{2}$, $\beta_\gamma^A = \sqrt{\frac{\gamma_{th}^A}{b\gamma_0[\rho-(1-\rho)\gamma_{th}^A]}}$, $\theta_p = \frac{(\varphi_p+1)\beta_q}{2}$, $\varphi_p = \cos\left(\frac{2p-1}{2P}\pi\right)$, $\beta_q = \sqrt{\frac{\chi_q}{b\gamma_0[\rho-(1-\rho)\chi_q]}}$, $\chi_a = \frac{(\phi_q+1)a}{2}$, $\phi_q = \cos\left(\frac{2q-1}{2Q}\pi\right)$. Note that step (c) is obtained by applying again the Gaussian–Chebyshev quadrature method with P and Q are the complexity-vs-accuracy trade-off coefficients. The proof is completed.

Appendix D: Proof of Theorem 4

Here, we derived the closed-form expressions of Φ_1^{xB} and Ψ_{ep}^{xB} as follows:

$$\begin{aligned}\Phi_1^{xB} &= \Pr\left(\gamma_{R*}^{xB} < \gamma_{th}^{B}\right) = \Pr\left(g_2 < \frac{\gamma_{th}^{B}}{(1-\rho)\gamma_U}\right)\\ &= \left(1 - e^{-\frac{\gamma_{th}^{B}}{(1-\rho)\gamma_U\lambda_2}}\right)^N. \end{aligned} \tag{14.48}$$

$$\begin{aligned}\Psi_{ep}^{xB} &= \int_{\gamma_{th}^{B}}^{\infty}\left[\frac{\partial F_{\gamma_S^{xA},\gamma_E^{xA}}(x,y)}{\partial x}\right]_{y=x} dx\\ &= \sum_{k=1}^{K}\binom{K}{k}(-1)^{k+1}\left[e^{-\frac{k\beta_\gamma^B}{\lambda_0}} - \frac{\pi k}{2P\lambda_0}\sum_{p=1}^{P}\int_{\gamma_{th}^{B}}^{\infty}\frac{e^{-\frac{x}{(1-\rho)\lambda_3 b\gamma_E\theta_p}-\frac{k\theta_p}{\lambda_0}}}{\sqrt{(1-\rho)b\gamma_0 x}}\right.\\ &\quad\times\left[1+\frac{x}{(1-\rho)\lambda_3 b\gamma_E\theta_p}-\frac{k\theta_p}{\lambda_0}\right]dx.\sqrt{\left(1-\varphi_p^2\right)}\\ &\overset{(d)}{=} \sum_{k=1}^{K}\binom{K}{k}(-1)^{k+1}\left[e^{-\frac{k\beta_\gamma^B}{\lambda_0}} - \frac{\pi^2 k e^{-\gamma_{th}^{B}}}{4PQ\lambda_0}\sum_{q=1}^{Q}\sum_{p=1}^{P}\frac{e^{-\frac{\chi_b}{(1-\rho)\lambda_3 b\gamma_E\theta_p}-\frac{k\theta_p}{\lambda_0}}}{\sqrt{(1-\rho)b\gamma_0\chi_b u_q}}\right.\\ &\quad\left.\times\left[1+\frac{\chi_b}{(1-\rho)\lambda_3 b\gamma_E\theta_p}-\frac{k\theta_p}{\lambda_0}\right]\sqrt{\left(1-\varphi_p^2\right)\left(1-\phi_q^2\right)}\right],\end{aligned} \tag{14.49}$$

where $\beta_\gamma^B = \sqrt{\frac{\gamma_{th}^{B}}{(1-\rho)b\gamma_0}}$, $\theta_p = \frac{(\varphi_p+1)\beta_q}{2}$, $\varphi_p = \cos\left(\frac{2p-1}{2P}\pi\right)$, $\beta_q = \sqrt{\frac{\chi_b}{(1-\rho)b\gamma_0}}$, $\chi_b = -\ln(u_q)$, $u_q = \frac{(\phi_q+1)e^{-\gamma_{th}^{B}}}{2}$, $\phi_q = \cos\left(\frac{2q-1}{2Q}\pi\right)$. Note that step (d) is obtained by applying again the Gaussian–Chebyshev quadrature method with P and Q are the complexity-vs-accuracy trade-off coefficients. The proof is completed.

Appendix E: Proof of Theorem 5

Here, we derived the closed-form expression of Ψ_{sop}^{xA} as follows:

$$\begin{aligned}\Phi_2^{xA} &= \Pr\left(\gamma_S^{xA} < \gamma_{th}^{A}\right) = \Pr\left(\frac{\rho b\gamma_0 g_0^2}{(1-\rho)b\gamma_0 g_0^2+1} < \gamma_{th}^{A}\right)\\ &= \begin{cases} 1, \gamma_{th}^{A} > \frac{\rho}{1-\rho} \\ \left(1 - e^{-\frac{1}{\lambda_0}\sqrt{\frac{\gamma_{th}^{A}}{b\gamma_0\left[\rho-(1-\rho)\gamma_{th}^{A}\right]}}}\right)^K, \gamma_{th}^{A} < \frac{\rho}{1-\rho}\end{cases}\end{aligned}$$

$$\begin{aligned}
\Psi_{sop}^{x_A} &= \int_{0}^{\infty} \int_{\gamma_{th}^{A}}^{2^{Rs_{x_A}}(y+1)-1} f_{\gamma_S^{x_A},\,\gamma_E^{x_A}}(x,y)\,dx\,dy \\
&= \int_{0}^{a_1} \int_{\gamma_{th}^{A}}^{2^{Rs_{x_A}}(y+1)-1} f_{\gamma_S^{x_A},\,\gamma_E^{x_A}}(x,y)\,dx\,dy + \int_{a_1}^{a_2} \int_{\gamma_{th}^{A}}^{a_2} f_{\gamma_S^{x_A},\,\gamma_E^{x_A}}(x,y)\,dx\,dy \\
&= \int_{0}^{a_1} \left[\frac{\partial F_{\gamma_S^{x_A},\gamma_E^{x_A}}(x,y)}{\partial y} \right]_{x=\gamma_{th}^{A}}^{x=2^{Rs_{x_A}}(y+1)-1} dy + \int_{a_1}^{a_2} \left[\frac{\partial F_{\gamma_S^{x_A},\gamma_E^{x_A}}(x,y)}{\partial y} \right]_{x=\gamma_{th}^{A}}^{x=a_2} dy \\
&\overset{(e)}{=} \frac{\pi^2 \rho}{2PQ\lambda_0 b\lambda_3 \gamma_E} \Bigg\{ a_1 \sum_{k=1}^{K} \sum_{p=1}^{P} \sum_{q=1}^{Q} \binom{K}{k} \\
&\quad \times \frac{(-1)^{k+1} k e^{-\frac{\xi_a}{b\lambda_3\gamma_E[\rho-(1-\rho)\xi_a]\theta_q} - \frac{k\theta_q}{\lambda_0}}}{[\rho-(1-\rho)\xi_a]^2 (\varphi_p+1)} \sqrt{\left(1-\varphi_p^2\right)\left(1-\phi_q^2\right)} \\
&\quad + (a_2 - a_1) \sum_{k=1}^{K} \sum_{p=1}^{P} \sum_{q=1}^{Q} \binom{K}{k} \\
&\quad \times \frac{(-1)^{k+1} k e^{-\frac{\psi_q}{b\lambda_3\gamma_E\left[\rho-(1-\rho)\psi_q\right]\theta_a} - \frac{k\theta_a}{\lambda_0}}}{\left[\rho-(1-\rho)\psi_q\right]^2 (\varphi_p+1)} \sqrt{\left(1-\varphi_p^2\right)\left(1-\phi_q^2\right)} \\
&\quad - (a_2 - a_0) \sum_{k=1}^{K} \sum_{p=1}^{P} \sum_{q=1}^{Q} \binom{K}{k} \\
&\quad \times \frac{(-1)^{k+1} k e^{-\frac{\psi_q}{b\lambda_3\gamma_E\left[\rho-(1-\rho)\psi_q\right]\theta_\gamma} - \frac{k\theta_\gamma}{\lambda_0}}}{\left[\rho-(1-\rho)\psi_q\right]^2 (\varphi_p+1)} \sqrt{\left(1-\varphi_p^2\right)\left(1-\phi_q^2\right)} \Bigg\}, \qquad (14.50)
\end{aligned}$$

where $a_0 = \frac{\gamma_{th}^{A}+1}{2^{Rs_{x_A}}} - 1$, $a_1 = \frac{1}{(1-\rho)2^{Rs_{x_A}}} - 1$, $a_2 = \frac{\rho}{1-\rho} - \delta$ (δ denotes the known accuracy coefficient), $\beta_y = \sqrt{\frac{2^{Rs_{x_A}}(y+1)-1}{b\gamma_0\left\{\rho-(1-\rho)\left[2^{Rs_{x_A}}(y+1)-1\right]\right\}}}$, $\beta_q = \sqrt{\frac{2^{Rs_{x_A}}(\xi_a+1)-1}{b\gamma_0\left\{\rho-(1-\rho)\left[2^{Rs_{x_A}}(\xi_a+1)-1\right]\right\}}}$, $\beta_a = \sqrt{\frac{a_2}{b\gamma_0[\rho-(1-\rho)a_2]}}$, $\beta_\gamma = \sqrt{\frac{\gamma_{th}^{A}}{b\gamma_0\left[\rho-(1-\rho)\gamma_{th}^{A}\right]}}$, $\theta_y = \frac{(\varphi_p+1)\beta_y}{2}$, $\theta_\gamma = \frac{(\varphi_p+1)\beta_\gamma}{2}$, $\theta_a = \frac{(\varphi_p+1)\beta_a}{2}$, $\varphi_p = \cos\left(\frac{2p-1}{2P}\pi\right)$, $\xi_a = \frac{(\phi_q+1)a_1}{2}$, $\theta_q = \frac{(\varphi_p+1)\beta_q}{2}$, $\psi_a = \frac{(\phi_q+1)(a_2-a_1)}{2} + a_1$, $\phi_q = \cos\left(\frac{2q-1}{2Q}\pi\right)$, $Rs_{x_A} = \frac{2Cs_{th}^{x_A}}{1-\alpha}$. Note that step (e) is obtained by applying again the Gaussian–Chebyshev quadrature method with P and Q are the complexity-vs-accuracy trade-off coefficients. The proof is completed.

Appendix F: Proof of Theorem 6

Here, we derived the closed-form expression of Ψ_{sop}^{xB} as follows:

$$\Phi_2^{xB} = \Pr\left(\gamma_S^{xB} < \gamma_{th}^B\right) = \Pr\left(g_2 < \frac{\gamma_{th}^B}{(1-\rho)b\gamma_0}\right)$$
$$= \left(1 - e^{-\frac{\gamma_{th}^B}{(1-\rho)b\gamma_0\lambda_0}}\right)^K. \tag{14.51}$$

$$\Psi_{sop}^{xB} = \int_0^{\infty} \int_{\gamma_{th}^B}^{2^{R_{S}xB}(y+1)-1} f_{\gamma_S^{xB},\gamma_E^{xB}}(x,y)dxdy$$
$$= \int_0^{\infty} \left[\frac{\partial F_{\gamma_S^{xB},\gamma_E^{xB}}(x,y)}{\partial y}\right]_{x=\gamma_{th}^B}^{x=2^{R_{S}xB}(y+1)-1} dy$$
$$= \sum_{k=1}^{K} \binom{K}{k} (-1)^{k+1} \frac{\pi k}{P\lambda_0} \sum_{p=1}^{P} \frac{\sqrt{1-\varphi_p^2}}{(1-\rho)\lambda_3 b\gamma_E(\varphi_p+1)}$$
$$\times \int_c^{\infty} \left[e^{-\frac{y}{(1-\rho)\lambda_3 b\gamma_E\theta_y} - \frac{k\theta_y}{\lambda_0}} - e^{-\frac{y}{(1-\rho)\lambda_3 b\gamma_E\theta_Y} - \frac{k\theta_Y}{\lambda_0}}\right]dy$$
$$\overset{(f)}{=} \frac{\pi^2 e^{-c}}{2PQ} \sum_{k=1}^{K}\sum_{q=1}^{Q}\sum_{p=1}^{P} \binom{K}{k} \frac{(-1)^{k+1}k\sqrt{\left(1-\varphi_p^2\right)\left(1-\phi_q^2\right)}}{(1-\rho)\lambda_0\lambda_3 b\gamma_E u_q(\varphi_p+1)}$$
$$\times \left[e^{-\frac{\chi_q}{(1-\rho)\lambda_3 b\gamma_E\theta_q} - \frac{k\theta_q}{\lambda_0}} - e^{-\frac{\chi_q}{(1-\rho)\lambda_3 b\gamma_E\theta_Y} - \frac{k\theta_Y}{\lambda_0}}\right], \tag{14.52}$$

where $c = \frac{\gamma_{th}^B+1}{2^{R_{S}xB}} - 1$, $\beta_y = \sqrt{\frac{2^{R_{S}xB}(y+1)-1}{(1-\rho)b\gamma_0}}$, $\theta_y = \frac{(\varphi_p+1)\beta_y}{2}$, $\beta_Y = \sqrt{\frac{\gamma_{th}^B}{(1-\rho)b\gamma_0}}$, $\theta_y = \frac{(\varphi_p+1)\beta_Y}{2}$, $\beta_q = \sqrt{\frac{2^{R_{S}xB}(\chi_q+1)-1}{(1-\rho)b\gamma_0}}$, $\theta_q = \frac{(\varphi_p+1)\beta_q}{2}$, $\chi_q = -\ln(u_q), a = e^{-c}$, $u_q = \frac{(\phi_q+1)a}{2}$, $\phi_q = \cos\left(\frac{2q-1}{2Q}\pi\right)$. Note that step (f) is obtained by applying again the Gaussian–Chebyshev quadrature method with P and Q are the complexity-vs-accuracy trade-off coefficients. The proof is completed.

References

[1] Y. Saito, Y. Kishiyama, A. Benjebbour, T. Nakamura, A. Li, and K. Higuchi, "Non-orthogonal multiple access (NOMA) for cellular future radio access," in *Proc. 2013 IEEE 77th Veh. Technol. Conf. (VTC Spring)*. IEEE, 2013, pp. 1–5.

[2] L. Dai, B. Wang, Y. Yuan, S. Han, I. Chih-Lin, and Z. Wang, "Non-orthogonal multiple access for 5G: solutions, challenges, opportunities, and future research trends," *IEEE Commun. Mag.*, vol. 53, no. 9, pp. 74–81, 2015.

[3] Y. Liu, Z. Ding, M. Elkashlan, and H. V. Poor, "Cooperative non-orthogonal multiple access with simultaneous wireless information and power transfer," *IEEE J. Selected Areas Commun.*, vol. 34, no. 4, pp. 938–953, 2016.

[4] A. Masaracchia, D. B. Da Costa, T. Q. Duong, M.-N. Nguyen, and M. T. Nguyen, "A PSO-based approach for user-pairing schemes in NOMA systems: Theory and applications," *IEEE Access*, vol. 7, pp. 90550–90564, 2019.

[5] X. Zhang, X. Zhu, and H. Zhu, "Joint user clustering and multi-dimensional resource allocation in downlink MIMO-NOMA networks," *IEEE Access*, vol. 7, pp. 81783–81793, 2019.

[6] Z. Ding, M. Peng, and H. V. Poor, "Cooperative non-orthogonal multiple access in 5G systems," *IEEE Commun. Lett.*, vol. 19, no. 8, pp. 1462–1465, 2015.

[7] Z. Ding, P. Fan, and H. V. Poor, "Impact of user pairing on 5G nonorthogonal multiple-access downlink transmissions," *IEEE Trans. Veh. Technol.*, vol. 65, no. 8, pp. 6010–6023, 2015.

[8] H. Lei, J. Zhang, K.-H. Park, P. Xu, I. S. Ansari, G. Pan, B. Alomair, and M.-S. Alouini, "On secure NOMA systems with transmit antenna selection schemes," *IEEE Access*, vol. 5, pp. 17450–17464, 2017.

[9] A. Rauniyar, P. E. Engelstad, and O. N. Osterbo, "On the performance of bidirectional NOMA-SWIPT enabled IoT relay networks," *IEEE Sensors J.*, vol. 21, no. 2, pp. 2299–2315, 2020.

[10] M. Waqas, W. Ejaz, G. A. Sidhu, and S. Aslam, "Resource optimization of D2D-Assisted CR network with NOMA for 5G and beyond systems," *IEEE Internet Things J.*, vol. 9, no. 21, pp. 21 232–21 245, 2022.

[11] D.-B. Ha, V.-T. Truong, and Y. Lee, "Performance analysis for RF energy harvesting Mobile Edge Computing networks with SIMO/MISO-NOMA schemes," *EAI Endorsed Trans. Ind. Netw. Intell. Syst.*, vol. 8, no. 27, p. e2, 2021.

[12] Y. Luo, C. Wu, Y. Leng, N. Huang, L. Mao, and J. Tang, "Throughput optimization for NOMA cognitive relay network with RF energy harvesting based on improved bat algorithm," *Mathematics*, vol. 10, no. 22, p. 4357, 2022.

[13] N. T. Luan and D.-T. Do, "A new look at AF two-way relaying networks: energy harvesting architecture and impact of co-channel interference," *Annals Telecommun.*, vol. 72, pp. 669–678, 2017.

[14] T. N. Do, D. B. da Costa, T. Q. Duong, and B. An, "Improving the performance of cell-edge users in NOMA systems using cooperative relaying," *IEEE Trans. Commun.*, vol. 66, no. 5, pp. 1883–1901, 2018.

[15] M. Hua, H. Tian, X. Lyu, W. Ni, and G. Nie, "Online offloading scheduling for NOMA-Aided MEC under partial device knowledge," *IEEE Internet Things J.*, vol. 9, no. 3, pp. 2227–2241, 2021.

[16] T. P. Truong, T.-V. Nguyen, W. Noh, S. Cho *et al.*, "Partial computation offloading in NOMA-assisted Mobile-Edge Computing systems using deep reinforcement learning," *IEEE Internet Things J.*, vol. 8, no. 17, pp. 13196–13208, 2021.

[17] M. Bloch, J. Barros, M. R. Rodrigues, and S. W. McLaughlin, "Wireless information-theoretic security," *IEEE Trans. Inform. Theory*, vol. 54, no. 6, pp. 2515–2534, 2008.

[18] N. Yang, P. L. Yeoh, M. Elkashlan, R. Schober, and I. B. Collings, "Transmit antenna selection for security enhancement in MIMO wiretap channels," *IEEE Trans. Commun.*, vol. 61, no. 1, pp. 144–154, 2012.

[19] J. D. V. Sanchez, L. Urquiza-Aguiar, M. C. P. Paredes, and D. P. M. Osorio, "Survey on physical layer security for 5G wireless networks," *Annals Telecommun.*, vol. 76, pp. 155–174, 2021.

[20] Z. Ding, H. Dai, and H. V. Poor, "Relay selection for cooperative NOMA," *IEEE Wireless Commun. Lett.*, vol. 5, no. 4, pp. 416–419, 2016.

[21] Y. Wen, T.-X. Zheng, Y. Tong, H.-W. Liu, X. Chen, P. Mu, and H.-M. Wang, "Physical layer security for NOMA-Enabled Multi-Access Edge Computing wireless networks," in *Proc. 2021 IEEE/CIC Int. Conf. Commun. China (ICCC)*. IEEE, 2021, pp. 41–46.

[22] Z. Qin, Y. Liu, Z. Ding, Y. Gao, and M. Elkashlan, "Physical layer security for 5G non-orthogonal multiple access in large-scale networks," in *Proc. 2016 IEEE Int. Conf. Commun. (ICC)*. IEEE, 2016, pp. 1–6.

[23] X. Li, Y. Pei, H. Jiang, X. Yue, Y. Wang, and Y. Suo, "Performance analysis of MEC based on NOMA under imperfect CSI with eavesdropper," *Wireless Commun. Mobile Comput.*, vol. 2021, 2021.

[24] T. M. Hoang, T. Q. Duong, H. A. Suraweera, C. Tellambura, and H. V. Poor, "Cooperative beamforming and user selection for physical layer security in relay systems," in *Proc. 2015 IEEE Global Commun. Conf. (GLOBECOM)*. IEEE, 2015, pp. 1–6.

[25] H. Chamkhia, A. Erbad, A. Al-Ali, A. Mohamed, A. Refaey, and M. Guizani, "Pls performance analysis of a hybrid NOMA-OMA based IoT system with mobile sensors," in *proc. 2022 IEEE Wireless Commun. Netw. Conf. (WCNC)*. IEEE, 2022, pp. 1419–1424.

[26] C. Gong, X. Yue, Z. Zhang, X. Wang, and X. Dai, "Enhancing physical layer security with artificial noise in large-scale NOMA networks," *IEEE Trans. Veh. Technol.*, vol. 70, no. 3, pp. 2349–2361, 2021.

[27] T. M. Hoang, B. C. Nguyen, X. N. Tran, T. Kim *et al.*, "Secrecy outage performance of FD-NOMA relay system with multiple non-colluding eavesdroppers," *IEEE Trans. Veh. Technol.*, vol. 70, no. 12, pp. 12 985–12 997, 2021.

[28] A. Nasir, H. Tuan, T. Duong, and M. Debbah, "NOMA for throughput and EE maximization in energy harvesting enabled networks," *arXiv preprint arXiv:1803.09225*, 2018.

[29] A. Rauniyar, P. Engelstad, and O. N. Osterbo, "RF energy harvesting and information transmission based on NOMA for wireless powered IoT relay systems," *Sensors*, vol. 18, no. 10, p. 3254, 2018.

[30] B. Li, W. Wu, W. Zhao, and H. Zhang, "Security enhancement with a hybrid cooperative NOMA scheme for MEC system," *IEEE Trans. Veh. Technol.*, vol. 70, no. 3, pp. 2635–2648, 2021.

[31] H. Lin, Y. Cao, Y. Zhong, and P. Liu, "Secure computation efficiency maximization in NOMA-Enabled Mobile Edge Computing networks," *IEEE Access*, vol. 7, pp. 87504–87512, 2019.

[32] W. Wu, X. Wang, F. Zhou, K.-K. Wong, C. Li, and B. Wang, "Resource allocation for enhancing offloading security in NOMA-Enabled MEC networks," *IEEE Systems J.*, vol. 15, no. 3, pp. 3789–3792, 2020.

[33] D.-D. Tran and D.-B. Ha, "Secrecy performance analysis of QoS-based Non-orthogonal Multiple Access networks over Nakagami-*m* fading," in *Proc. Int. Conf. Recent Adv. Signal Process., Telecommun. Comput. (SigTelCom)*, HCMC, Vietnam, 2018.

[34] H. Xing, L. Liu, and R. Zhang, "Secrecy wireless information and power transfer in fading wiretap channel," *IEEE Trans. Veh. Technol.*, vol. 65, no. 1, pp. 180–190, 2015.

[35] I. Ahmed, A. Ikhlef, R. Schober, and R. K. Mallik, "Joint power allocation and relay selection in energy harvesting AF relay systems," *IEEE Wireless Commun. Lett.*, vol. 2, no. 2, pp. 239–242, 2013.

[36] K.-H. Liu, "Performance analysis of relay selection for cooperative relays based on wireless power transfer with finite energy storage," *IEEE Trans. Veh. Technol.*, vol. 65, no. 7, pp. 5110–5121, 2015.

[37] C. Guo, L. Zhao, C. Feng, Z. Ding, and H.-M. Wang, "Secrecy performance of NOMA systems with energy harvesting and full-duplex relaying," *IEEE Trans. Veh. Technol.*, vol. 69, no. 10, pp. 12 301–12 305, 2020.

[38] Z. Cao, X. Ji, J. Wang, W. Wang, K. Cumanan, Z. Ding, and O. A. Dobre, "Artificial noise aided secure communications for cooperative NOMA networks," *IEEE Trans. Cognitive Commun. Netw.*, vol. 8, no. 2, pp. 946–963, 2021.

[39] L. Fan, N. Yang, T. Q. Duong, M. Elkashlan, and G. K. Karagiannidis, "Exploiting direct links for physical layer security in multiuser multirelay networks," *IEEE Trans. Wireless Commun.*, vol. 15, no. 6, pp. 3856–3867, 2016.

[40] H.-G. T. Pham, Q.-V. Pham, A. T. Pham, and C. T. Nguyen, "Joint task offloading and resource management in NOMA-based MEC systems: A swarm intelligence approach," *IEEE Access*, vol. 8, pp. 190463–190474, 2020.

[41] J. Kennedy and R. Eberhart, "Particle swarm optimization," in *Proc. ICNN'95-Int. Conf. Neural Netw.*, vol. 4. IEEE, 1995, pp. 1942–1948.

[42] R. Poli, J. Kennedy, and T. Blackwell, "Particle swarm optimization: An overview," *Swarm Intelligence*, vol. 1, pp. 33–57, 2007.

[43] M. Clerc and J. Kennedy, "The particle swarm-explosion, stability, and convergence in a multidimensional complex space," *IEEE Trans. Evolutionary Comput.*, vol. 6, no. 1, pp. 58–73, 2002.

[44] C. Yin, N.-P. Nguyen, E. Garcia-Palacios, X. N. Tran, and T. Le-Tien, "Secure energy harvesting communications with relay selection over Nakagami-*m* fading channels," *Mobile Netw. App.*, vol. 23, pp. 1555–1562, 2018.

[45] I. S. Gradshteyn and I. M. Ryzhik, *Table of integrals, series, and products*, 8th ed., I. G. I. R. D. Zwillinger, V. Moll, Ed. Academic Press, 2014.

Chapter 15
Secure and private localization in 6G networks*

Stefano Tomasin[1]*, Shihao Yan*[2] *and Robert Malaney*[3]

A wireless vehicular network (WVN) is a particular type of intelligent transportation system that utilizes vehicle-to-everything (e.g., inter-vehicle and infrastructure-vehicle) communications to carry out different network operations. In the context of communicating cars, with the aid of a WVN, we can optimize traffic routing, minimize traffic congestion, improve road tolling infrastructure, assist smart city traffic planning, aid in road infrastructure expansions, achieve seamless in-vehicle entertainment services, increase electric vehicle charging performance [2], and most importantly enhance road safety. As per a survey conducted by the World Health Organization, there are annually over one million causalities worldwide due to car accidents. Eliminating such a high number of fatalities, even partially, is a challenge that motivates an increasing amount of research on WVNs. To enable the full functions of a WVN, we require the following three desirable features: ultra-high data rates, ultra-reliability, and ultra-low latency. These desirable features cannot be fully met by existing general wireless communication systems such as fifth-generation (5G) cellular systems or the dedicated short-range communication (DSRC) system. However, the forthcoming fifth-generation (5G) (B5G) or the sixth-generation (6G) technology is specifically designed with these desirable features in mind.

Location information plays an important role in WVNs and even serves a foundation role in several key functional areas [3–5]. For example, many network operations of WVNs, such as location-based routing, depend critically on the location information of each vehicle in the network. Also, location information serves as the enabler of many services provided by WVNs, e.g., collision avoidance and location-based advertising.

Considering the critical uses of localization information, its security becomes of pivotal importance. However, the reliability and correctness of a vehicle's reported location information cannot be fully guaranteed by existing techniques. Meanwhile, some privacy concerns on location information are emerging. This chapter addresses two key features for 6G location security, i.e., the security associated with localization

*This chapter has been a part of [1].

[1]Department of Information Engineering, University of Padova, Italy
[2]School of Science (Cyber Security), Edith Cowan University, Australia
[3]Institute for Cybersecurity (IFCYBER), The University of New South Wales, Australia

services: (a) location verification systems (LVSs) [6], aiming at ensuring that the reportedly derived location is correct, and (b) location privacy techniques, aiming at preventing the disclosure of the position information to other entities. Lastly, we shed some light on new solutions based on quantum security to improve location privacy and verification.

15.1 Location verification

Different from positioning systems, an LVS aims at confirming whether a user (e.g., a vehicle) is physically at its reported/claimed location. Mathematically, location verification is a binary detection problem, while localization is an estimation problem. The importance of an LVS can be evidenced by the negative effects of fake location information on WVNs, e.g., dramatically reduced packet delivery in position-based routing protocols or even life-threatening road accidents. In principle, an LVS can remove these negative impacts on WVN operations. We also note that verified location information can improve the performance of communications based on millimeter wave (mmWave), which is a key enabler of the B5G wireless networks, since this location information can aid in the determination of accurate channel state information (CSI). To fully support the services of emerging WVNs, location verification should be embedded into emerging B5G technology.

In this section, we first review location verification frameworks and techniques, and then present the state-of-the-art LVS algorithms. We then provide insight into the optimal use of the B5G technology to address location verification. In addition, noting the recently emerging use of machine learning (ML) in the context of WVNs, we summarize how ML can enhance the performance and be deployed in real-world LVS systems. Some critical aspects of ML-based LVS that can only be deployed with the aid of the B5G technology are also identified. We also identify an opportunity of using mmWave to incorporate location verification into a communication system with a low implementation cost. We note that this section has reproduced text and figures from [1].

15.1.1 Physical-layer-based location verification

Physical-layer-based LVSs utilize the inherent properties of the wireless medium to verify a user's reported location information and thus infer whether the user is legitimate or malicious. Their performance is based on the fundamental theories of detection, which serves as the performance limit of location verification in wireless networks.

In the LVS literature, the work [7], for the first time, developed an information-theoretic framework for location verification, proving that the likelihood ratio test is optimal in terms of maximizing the mutual information between the input and the output of an LVS. This information-theoretic framework was deployed and examined in an LVS with received signal strength (RSS) measurements. The work [8] extended this framework by considering directional antennas under realistic dual-slope large-scale fading channel models in vehicular networks, showing that the directional antenna can significantly increase the capability of an LVS in correctly detecting malicious vehicles. Also in the context of vehicular networks, the authors of [9] considered location verification with multiple antennas under Rician fading channels. Interestingly, it was proved that the performance limit of LVSs is independent of the parameters of the channel from a malicious vehicle to a base station. In addition, as proved, the malicious vehicle's number of antennas does not affect this limit once it is greater than the legitimate vehicle's number of antennas. This is mainly due to the consideration of the worst-case scenario, where the malicious vehicle optimizes all the system parameters under its control. Because of this optimization, a further increase in the number of antennas held by a malicious vehicle has little additional impact.

LVSs also draw an increasing amount of research attention in Internet of Things (IoT). We note that WVN is one type of IoT, where the "things" are vehicles. For example, the work [10] developed an enhanced location verification using audibility and two-way time-of-arrival information to guarantee the reliability of location information for geo-spatial tagging and location-based services in IoT. Based on both synthetic and real-world datasets, it demonstrated that the LVS performance is improved when audibility is used. Considering the low-complexity hardware used to collect RSS measurements, they have been widely used in the context of location estimation and verification. To eliminate the requirement of knowing transmit power in RSS-based systems, the RSS (DRSS) have been used as measurements as well. For the first time, the authors of [11] proved the identity of using RSS and DRSS in location verification for known and unknown transmit power, respectively.

15.1.2 Machine-learning-based location verification

Although LVSs based on information or statistical detection theory can offer performance limit of location verification, they require ideal operating conditions (e.g., known channel parameters), which may not be fully satisfied in practice. One motivation for using ML in practical LVSs is to eliminate this requirement. In this subsection, we briefly review some recent works on ML-based location verifications.

The authors of [12] developed an LVS based on neural network and demonstrated its efficient performance in the absence of prior knowledge on the proportion of legitimate or malicious vehicles among all the vehicles. It was shown that the LVS based on neural network outperforms an information-theoretic LVS when the signals from the vehicles are under the impact of non-line-of-sight biases. The work [13] resorted to ML solutions for in-region location verification to solve the issues of lacking channel feature statistics. It was shown that the developed solution based on neural network and support vector machines, using typical loss functions, becomes the most powerful test at learning convergence for sufficiently complex learning machines and large training datasets, even in the absence of communication channel statistics. We note that the aforementioned typical loss functions refer to those used during the training phase of various ML algorithms, e.g., the cross-entropy loss function and the mean squared error loss function. Different from most aforementioned physical-layer-based LVSs assuming known channel models, the authors of [14] used logistic regression algorithm of the ML family for conducting location verification. This algorithm eliminates the assumption of known channel models or parameters and thus is applicable to more general scenarios of wireless networks. The performance of the proposed algorithm was examined based on RSS measurements at multiple landmarks (each is equipped with multiple antennas), where a distributed Frank–Wolfe-based verification is also developed to reduce the communication overhead. The detection accuracy of the proposed algorithm was confirmed by simulation and experimental results.

Many physical-layer-based LVSs will function optimally if the system model they assumed is a reality. However, this is nearly impossible in practice due to abrupt changes in channel conditions, changing noise parameters, diverse transceiver characteristics, and environment-based positioning errors. These issues may lead to a circumstance where an LVS designed for one application scenario may not function properly in other scenarios (e.g., with different channel conditions). For instance, an LVS designed for an urban environment may not function properly in a rural area (and vice versa). To accommodate all the above issues, we desire an LVS that can retune itself to the changing environments. The reviewed works mentioned in Section 15.2 confirm that ML plays a pivotal role toward the development of such an LVS within the context of WVNs.

It is part of our own ongoing work to seamlessly integrate ML with optimal decision theory to develop an LVS solution for WVNs that is practically deployable. A schematic of our approach to development of such a solution is shown in Figure 15.1. We aim to integrate the information-theoretic LVS framework into neural networks. As shown in phase 1 of Figure 15.1, we use the data considered for the information-theoretic LVS as training data for the LVS based on neural networks, where we aim

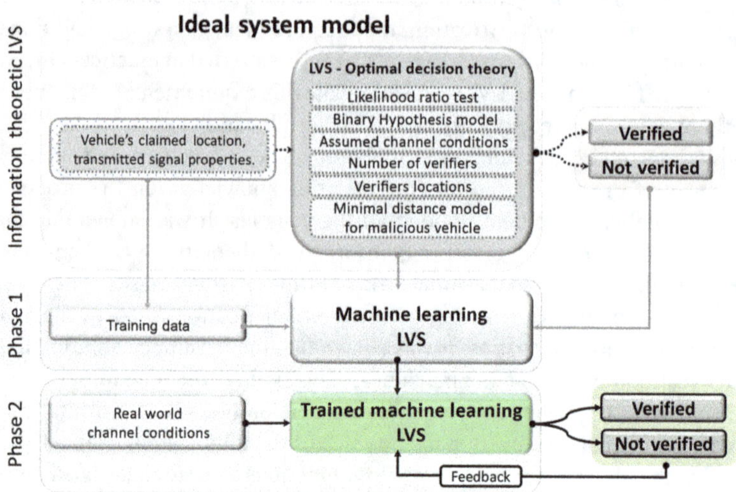

Figure 15.1 A schematic of integrating ML with optimal decision theory to produce an LVS that is practically functioning in real-world channel conditions

to achieve similar location verification performance by neural networks to that of the information-theoretic LVS. In addition, the information-theoretic framework is able to provide a benchmark performance that can aid the training of the neural networks, e.g., to avoid the over-training issues. In phase 2, we introduce real-world channel conditions to the ML-based LVS. Specifically, we plan to supply RSS, time of arrival (ToA), and angle of arrival (AoA) measurements of the vehicles' transmitted signals, and their claimed locations, as input to the LVS. Also, with the aid of existing and newly identified features in the input data and by adjusting the neural network architectures, we plan to further enhance the performance for the ML-based LVS. An active feedback from the verification output assists in continuously tuning the hyperparameters of the neural network, which leads to a scalable, adoptable, and up-to-date LVS in this context.

Some preliminary results achieved by our framework detailed above are presented in Figure 15.2, where the total error is adopted as the performance metric. This total error is defined as $P_0\alpha + P_1\beta$, where P_0 and P_1 are the proportions of the legitimate and malicious vehicles, respectively, and α and β are the false positive and missed detection rates, respectively. To spoof the LVS, we consider that a malicious vehicle adopts an attacking strategy, where it randomly claims its attack location at a predetermined distance (e.g., 100 m) away from its true location. We recall that, as discussed in Section 15.3.1, the malicious vehicle not only reports a false location but also intends to modify the measurements used by an LVS for verifying the reported location (to deceive the LVS). To plot Figure 15.2, we assumed that the

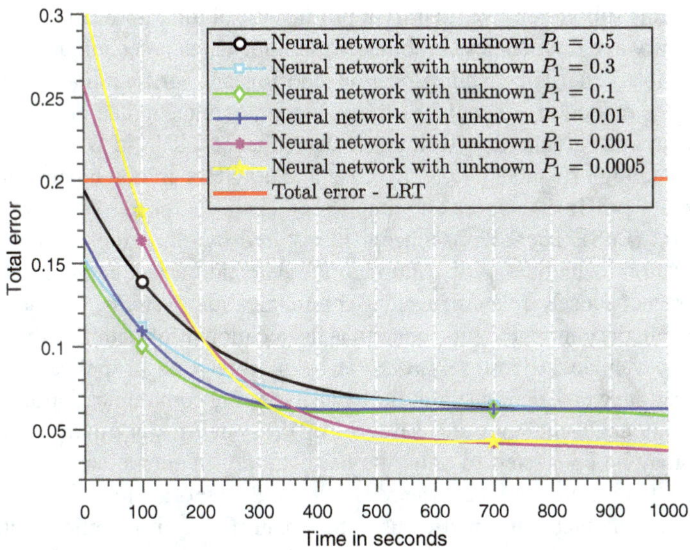

Figure 15.2 Preliminary comparison results between an LVS-based neural networks with unknown P_1 and an information-theoretic LVS based on LRT with known P_1

malicious vehicle has optimized its transmit power using the methodology presented in [11]. This is the worst-case scenario where the malicious vehicle optimizes all the parameters under its control.

An information-theoretic LVS [15], which is developed based on a likelihood ratio test (LRT), needs some *a prior* knowledge (e.g., P_1) for operation, while an LVS based on neural networks may not need such knowledge [12]. In the simulations, we train the neural network with incremental data as the time increases before testing the performance of the neural network by using the test data. Specifically, the number of training examples at 1 s is 1 and this number increases by 1 in each second. Therefore, we have severe fluctuations in the total error at the start, and polynomial fitting is used to smooth the total error curves. In this figure, we show that the ML-based LVS can still outperform the former, even when the ML-based LVS does not know P_1 (the values of P_1 shown in the legend of Figure 15.2 are unknown). However, for the calculations in Figure 15.2, it is assumed the LRT-based LVS knows the value of P_1. To examine the robustness of the ML-based LVS, we consider different values of P_1, ranging from very high (e.g., 50%) to very low (i.e., 0.05%). This result explicitly demonstrates the superiority of the ML-based LVS relative to the LRT-based LVS. We also see that the total error for the ML-based LVS converges once the framework is trained for approximately 800 s. The slight variation in the curves after 800 s is due to the applied polynomial fitting.

15.1.3 Simultaneous location reporting and verification

B5G wireless networks bring new opportunities to fully address the requirements of WVNs. In this subsection, we identify a promising solution to location verification by using mmWave frequencies combined with multi-antenna techniques.

An obvious benefit of enabling location reporting and verification simultaneously (LRVS) is the reduction of needed hardware resources and entailed processing delay. Communications via mmWave techniques can enhance LRVS in WVNs mainly due to the utilized high frequency (e.g., 28 GHz and 60 GHz) and high bandwidth (e.g., 400 MHz and 1 GHz). Also, the communication channels for mmWave are typically line of sight (LoS). The LRVS concept is inspired by emerging combined radar-communication techniques [16]. Although the detection target and communication content are not closely related in radar-communication systems, in our proposed LRVS system, the communication content is the location information of the detection target (e.g., a vehicle). Therefore, LRVS can be interpreted as an application of radar-communication, which demonstrates the benefits of joint communication and sensing. The multi-antenna techniques are useful for LRVS systems, since in many scenarios, a vehicle may be connected to only one base station. In such a scenario, a single-antenna system cannot guarantee the reliability of the reported location information. This is due to the fact that a malicious vehicle can modify its location metric, e.g., transmit power or transmission time, together with a false reported location to deceive an LVS [11].

We now present two frameworks of using mmWave combined with multi-antenna techniques to enable LRVS in WVNs as shown in Figure 15.3. The first framework

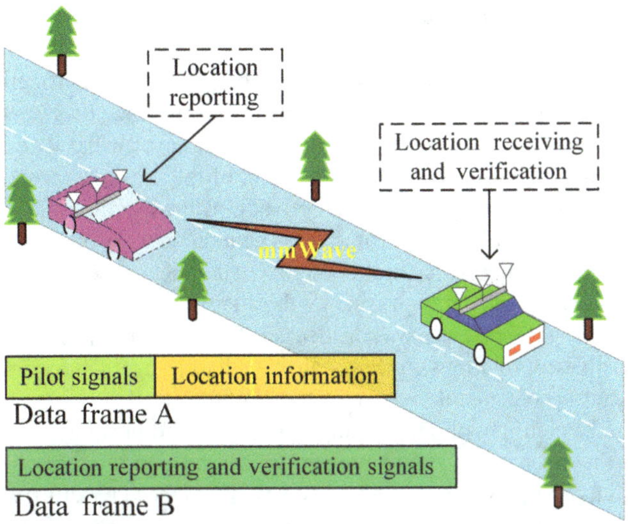

Figure 15.3 The frameworks of location reporting and verification simultaneously with the aid of mmWave and multi-antenna techniques in B5G wireless vehicular networks

involves the use of the data frame A, shown in Figure 15.3. The frame form is widely used in wireless communications using the full CSI (coherent communications). In the context of LRVS, the pilot signals of this frame are used for channel estimation and the location information part (indicated by the yellow shading in Figure 15.3) is used simply to transfer the encoded location coordinates. In addition to channel estimation, the pilot signals can also be used for location verification [9], since one benefit of using data frame A for LRVS is a "double-dipping" on the use of pilot signals—a technique that can significantly reduce latency.

A second framework for LRVS involves the use of the data frame B, shown in Figure 15.3. In this frame, only statistical information of the channel is used for encoding (non-coherent communications). In this case, a combined encoding of the claimed location coordinates and the verification information (e.g., RSS and ToA) can be utilized. We note that the aforementioned two data frame structure does not require specific size, and, in general, a commonly used transmission control protocol (TCP) data size (e.g., 1500 bytes) is acceptable.

15.2 Location privacy

While confidentiality is a security target to prevent the disclosure of a message *content* to an eavesdropper, privacy aims at preventing the disclosure of *side information* associated with the transmission of a message. With location privacy, we refer in particular to the protection of the information about the location and movements of a transmitting device.

Any privacy problem is intrinsically more complex than a confidentiality problem, since the information to be protected is not in the message itself, and the attacks may exploit several features of the transmission. Moreover, what is to be protected by privacy is often more blurred than the content of the message (protected by confidentiality). This occurs also for location privacy, as the location of a device can be established with a desired precision, or more in general with a probability distribution function: the appropriate privacy metric on these statistical observations depends on the use of the location information done by the attacker. For example, if the attacker aims at determining if the device was inside a specific area or to which shop it entered (within a possible set of shops), privacy aims at disrupting such hypothesis testing (or classification) problems. If instead the target of the attacker is simply to track the movements of the victim (without focusing on precise relevant points), the localization precision (as the error variance on the position) can be a good privacy metric. Thus, several location privacy metrics have been investigated (see [17–20] for surveys).

Moreover, privacy protection on message transmission should still ensure the correct reception of the message, which may not immediately related to the location privacy metric. Thus, another metric measuring the possible degradation of the communication must be taken into account when designing and evaluating location privacy mechanisms.

15.2.1 Location privacy in mobile networks

Mobile networks are designed to provide communication link to mobile devices and are therefore natural contexts where the location and movements of users can be estimated and tracked. The localization accuracy has improved over the various network generations, thanks to specific technologies that have been deployed to provide this service. Indeed, location-based services (LBSs) have drawn increasing attention by network operators and manufacturers.

At the same time, concerns about the location privacy have risen, also indicating that the perception of weak privacy on our location may reduce our trust in mobile communications. In 2001, a study [21] already warned that LBSs "often may be implemented in a way that exposes sensitive personal information, there are several privacy issues to consider," and defined the main concepts of privacy policies on the data access inside and outside the network. In 2008, an overview of location privacy issues in cellular networks has been provided in [22], also indicating several privacy-preserving approaches. A collaborative solution among devices for location privacy was proposed in 20211 [23], where devices keep their context information in a buffer, until it expires, and they pass it to other users seeking such information, thus reducing the need to contact the LBS server. In 2012, the evolution of LBSs was explored in [24], including the use of Apple's and Google's crowd-sourced data and the development of anonymous LBSs was advocated. The deployment of measures to improve an individual's location privacy through a user-controllable global system for mobile communications (GSM) software stack was considered in 2013 [25], and the effects of the interaction between the subscribers and the providers were examined. In the same year, the authors of [26] indicated a possible attack against anonymization solutions for location privacy. In 2015, a location privacy attack against the long-term evolution (LTE) of the third-generation partnership project (3GPP) cellular standard was proposed [27], based on the timing advance parameter. In 2021, the use position information was considered for contact-tracing in the spread of diseases, and a solution based on the separation between the knowledge of the user identity (by the network operator) and the user location (by an external company handing anonymized data) has been advocated [28]. In 2022, a power control scheme was proposed in [29] for mitigating user location tracking in LTE networks.

The relevance of location privacy has been confirmed also by the legislation of several countries. In the United States of America, for example, the 2018 case *United States v. Carpenter* [30] led the Supreme Court to warn against the fundamental danger to privacy due to the knowledge and storage by the mobile network operators (MNOs) of the cell location information. Indeed, the observation that user equipments (UEs) are almost always within 2 m from its owner, make the UE localization and use of positional data a very sensitive topic for our privacy. The European Union's General Data Protection Regulation (GDPR) [31] protects the collection and storage or personal data to support privacy: it specifically identifies (in Art. 4) as personal data to be protected "any information relating to an identified or identifiable natural person ('data subject'); an identifiable natural person is one who can be identified, directly or indirectly, in particular by reference to an identifier such as a name, an identification number, location data, …." For an in-depth analysis of legal issues related to location privacy, see also [32,33].

Despite both the repeated alarms on the scarce location privacy of cellular networks and the various technical solutions that have been identified to mitigate the threats, limited actions have been taken, mostly concentrated on limiting the use of location information by applications in smartphones.

A comprehensive approach to location privacy in 5G and beyond cellular networks has been advocated in [34], where the idea of the location-private mobile network (LPMN) has been first introduced.† The LPMN comprises trusted UEs that communicate directly in a device-to-device (D2D) mode and collectively transmit to the new-generation node base (gNB) only through selected gateway nodes. This prevents the gNB from receiving signals directly from the UEs, thus hiding their position, as the network will only know the position of the gateways. In addition, the LPMN provides that all data pertaining to the identities of the UEs are stored by a trusted *authentication and billing authority* and are not accessible to the MNO. The MNO and the UE use temporary ID numbers for authentication, with the support of the billing authority. The idea has been further elaborated in [35], where a routing protocol for the LPMN has been designed, to prevent the cellular network from inferring the UE position from the statistics of the gateway usage.

†In [34], the LPMN is denoted virtual private mobile network.

On the same line of preventing an accurate localization by the network, in [36], it is proposed to modify the transmitted I/Q samples by leveraging an irreversible randomized sequence, which distorts the location-specific information in the resulting channel, thus preventing localization without affecting the communication performance.

15.2.2 5G and beyond networks

Starting from release 16 of the third-generation partnership project (3GPP) cellular standard, the network has been designed as a reconfigurable set of interconnected software functions with much higher capabilities in handling information on the UE. Table 15.1 summarizes the main functions involved in the localization process and that are relevant for location privacy.

A function obtains the location of the user by doing a request to the access and mobility management function (AMF), which sends the location–service request to the location management function (LMF). This function is specifically in charge for localizing the UE, and, in fact, it schedules the resources needed for such service. Operations of the LMF may also include a request to the UE to provide its own location by its GNSS receiver. Basically, this procedure is similar to what happened in previous-generation networks, now organized as software functions. The main innovation is instead in the use of the location information.

Two new functions can handle the location information of the UEs. One is the network data analytic function (NWDAF) that aggregates data from multiple UEs over time and is able to infer and predict the UE location and movements. The NWDAF can in fact exploit not only the information on the movement of the target UE but also of the other (neighboring) UEs and will use powerful AI techniques. Since the NWDAF has a global picture of all the UEs moving in the network, it can also infer the relations among the users, i.e., checking who, when, and for how long was in

Table 15.1 5G network functions involved in location privacy

Function	**Relation to location privacy**
Location management function (LMF)	Interacts with the access and mobility management function (AMF) to track UE movements
Access and mobility management function (AMF)	Starts the location process upon request by other functions, interacting with the gNB and UE
New-generation node base (gNB)	Implements the localization algorithms based on the signals at the physical layer
Network data analytic function (NWDAF)	Provides inferences on location data of all the UEs
Net work exposure function (NEF)	Provides information on the UEs (possibly including their location) to external applications of over-the-top (OTT) operators
User equipment (UE)	Provides additional location information, e.g., obtained by global navigation satellite systems (GNSSs)

proximity. With such function, the MNO may obtain lateral information on the user (beyond its location), which is clearly relevant for privacy.

The second function handling location information is the network exposure function (NEF), which gives network information to OTT operators: the location of UEs is among the data that can be passed outside of the network. By this function, the leakage of information is extended from the MNO to the OTT operators and the effects of the lack of privacy can be much more disruptive.

The GMLC function operates as the gateway between other functions and the AMF, which actually locates the UE. The purpose of the GMLC is to check if the requesting function has the necessary authorization to access the location information. This function could be strengthened to implement to more sophisticated privacy-preserving techniques.

We should also remark that other entities beyond MNOs and OTTs may obtain information on the UE position by intercepting the wireless signal transmitted by the UE, since the wireless medium intrinsically let several devices to receive the transmitted signals. As remarked in [34], suitable techniques should be deployed (a) to make wireless signals as anonymous as possible to external entities, thus making them unusable to track devices and persons; (b) to avoid leakage of location information through the control signals exchanged between the UE and the gNB. Indeed, target (a) may be difficult to achieve, due to the space–temporal correlation of signals and the possibility to merge a significant amount of data (e.g., by AI techniques). Target (b) instead requires the encryption of control signals, thus increasing the communication overhead, but with limited computational effort. A localization attack based on the exchanged control signals in 5G networks has been investigated in [37], by noting that, according to the 3GPP standard, UEs feedback to the gNB the precoder (as an index to a shared codebook) to be used for downlink transmissions. Since the precoder is strongly related to the user channel, which in turn contains information on the user position within the cell, it is possible to localize the UE by this feedback signal.

The trends for beyond-5G and 6G networks follow the path of the previous generations on localization and tracking. The need for a more accurate localization is confirmed and extended to both outdoor and indoor scenarios. To this end, several technologies [38] now under study for adoption in the 6G networks will support localization services. First, transmission in the THz band will allow for a boost of the location accuracy at the millimeter level. The integration of sensing and communications will also provide a network knowledge of the physical environment that can be further exploited for localization, tracking, and inference of other location-based information on the UEs. Even the integration of visible-light communication (VLC) will enhance localization, in particular indoor. Moreover, drones deployed to increase coverage can be also exploited to improve localization and tracking. Lastly, the boost of use of AI at all levels of the network will further enhance the ability of MNO to extract additional data and information from their observation, posing a significant threat to our privacy in several domains.

Looking at the use cases targeted by 6G networks, we also realize the relevance of techniques to protect privacy, since e-Health applications [38] are of specific interest, opening several concerns on the privacy of our data, including our location.

In the following, we will consider (a) location-privacy-preserving techniques operating at the single UE to prevent localization by the network, (b) cooperative location-privacy-preserving techniques based on multiple UEs that exchange data in a D2D fashion to hide their location to the gNB, and (c) a procedure to obtain the location of the UE exploiting control signals for channel estimation and techniques to mitigate such attack.

15.2.3 Localization techniques

Before describing various privacy-preserving techniques, we outline here the most important localization strategies by the gNB, some based on the received signal and one based on the exchanged control signals.

About the basic strategies based on the received signals, first, we observe that sensing techniques are suitable to detect the presence of (large) scatterers and are at the moment not suitable to obtain information on the position of passive devices, such as a UE. Hence, localization techniques should be used by the gNB, which are based on the reception of signals actively transmitted by the UE.

Localization may be performed at the gNB by exploiting the signals directly received from the UE in LoS or indirectly through the scatterers, in NLoS (see Figure 15.4).

15.2.3.1 AoA-based localization in LoS conditions

In LoS conditions, the position of the UE can be estimated by a set of K gNBs using the triangulation technique. By estimating the AoA θ_k, $k = 1, \ldots, K$, and knowing

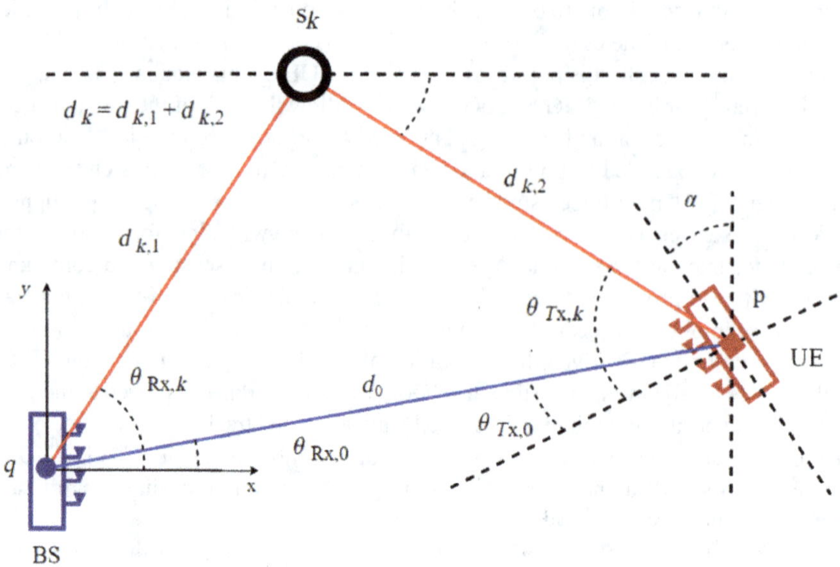

Figure 15.4 LoS and NLoS conditions for localization

the position $\boldsymbol{p}(k) = [p_x(k), p_y(k)]$, $k = 1, \ldots, K$, of each gNB, the position of the UE $\boldsymbol{q} = [q_x, q_y]$ can be estimated through the system of equations

$$(q_y - p_y(k)) = (q_x - p_x(k)) \tan \theta_k, \quad k = 1, \ldots, K, \tag{15.1}$$

which can be solved in general when at least three non-aligned gNBs are available. A similar problem can be written when considering a three-dimensional space (and three coordinates per node). Note that angle measurements are in general affected by errors. Assuming approximately Gaussian errors on the angles, (15.1) should be rewritten to minimize the mean square error (MSE) with respect to the measured angles, i.e.,

$$\min_{\theta_k} \sum_{k=1}^{K} |[\theta_k - \hat{\theta}_k]_{2\pi}|^2, \tag{15.2}$$

s.t. (15.1), where $[\cdot]_{2\pi}$ is the 2π-modulus operation and θ_k, $k = 1, \ldots, K$, are the (noisy) estimated angles. For a gNB equipped with a linear array of M antennas, the AoA can be easily estimated, since the baseband uplink narrowband channel vector at gNB k for a single-antenna UE is $\boldsymbol{h}(k) = G_k \boldsymbol{a}_M(\theta_k)$, where

$$\boldsymbol{a}_M(\theta_k) = [1, e^{j\frac{2\pi}{\lambda} d \sin \theta_k}, \ldots, e^{j(M-1)\frac{2\pi}{\lambda} d \sin \theta_k}]^T, \tag{15.3}$$

G_k is the channel gain (due to the gNB-UE distance), λ is the carrier wavelength, and d is the distance between two adjacent elements in the antenna array.

15.2.3.2 AoA-based localization in NLoS conditions

In NLoS conditions instead, localization can be obtained even with a single gNB, by exploiting the scatterers that operate as a distributed antenna system, providing AoA at different locations, as in a scenario with several gNBs. In this case, however, the gNB (at position $\boldsymbol{p} = [p_x, p_y]$) must also know the angle of departure (AoD) of the signals from the UE. Moreover, the UE should be equipped with multiple antennas, so that the gNB can estimate the AoD from the channel matrix. Consider N scatterers, where scatter n is located at position $\boldsymbol{s}(n) = [s_x(n), s_y(n)]$, $n = 1, \ldots, N$. Let us indicate the AoD from the UE to scatter n as ϕ_n, $n = 1, \ldots, N$ and the AoAs at the gNB from scatterer as θ_n, for $n = 1, \ldots, N$. Note that in this case also the positions of the scatterers are unknown. Thus, we have $2N$ equations into $2N + 2$ unknowns (the positions of both the scatterers and the UE). Hence, exploit the information on the ToA at the gNBs. Let τ_n be the ToA of the signal received through the scatterer $n = 1, \ldots, N$, and let $d_{n,1}$ be the distance between the gNB and the scatterer n. Then we have

$$\boldsymbol{p} = \boldsymbol{q} + d_{n,1} \begin{bmatrix} \cos \theta_n \\ \sin \theta_n \end{bmatrix} + (c\tau_n - d_{n,1}) \begin{bmatrix} \cos (\phi_n + \alpha) \\ \sin (\phi_n + \alpha) \end{bmatrix}, \tag{15.4}$$

where c is the speed of light and α is the unknown orientation angle of the UE array. Also, in this case, the angle estimates are noisy, and, in practical, an MSE approach should be adopted. The AoDs and ToAs can be obtained from the channel matrix. Assume that also the UE has a linear array of P antennas, spaced by d, then the MIMO narrowband channel matrix is

$$\boldsymbol{H} = \boldsymbol{A}_R \boldsymbol{\Gamma} \boldsymbol{A}_T^H, \tag{15.5}$$

where ^{H} denotes the Hermitian operator,

$$A_T = [a_P(\phi_1), \ldots, a_P(\phi_N)], \quad A_R = [a_M(\theta_1), \ldots, a_M(\theta_N)], \tag{15.6}$$

$$\Gamma = G_n \text{diag}\left\{\frac{h_1}{\sqrt{\rho_1}} e^{-j2\pi \frac{\tau_1}{T_s}}, \ldots, \frac{h_N}{\sqrt{\rho_N}} e^{-j2\pi \frac{\tau_N}{T_s}}\right\}, \tag{15.7}$$

T_s is the symbol period, and ρ_n and h_n, $n = 1, \ldots, N$, denote the path loss and the complex channel gain for path n, respectively.

15.2.3.3 RSS-based localization in LoS

When operating in LoS conditions, an alternative to estimate the position of the UE is based on the use of RSS. In this case, the distance from the UE to the gNB k is estimated by taking into account that the RSS decays with the square of the gNB-UE distance, thus, from the estimated RSS $\hat{\gamma}_k$, the estimated distances of UE from gNB are

$$\hat{d}_k = \sqrt{\frac{P_0}{\hat{\gamma}_k}}, \tag{15.8}$$

where P_0 is the UE transmit power. From these distances, the positions can be obtained by solving the following minimization problem:

$$\hat{q} = \text{argmin}_q \sum_{k=1}^{K} \left[||q - p(k)||^2 - \hat{d}_k^2\right], \quad k = 1, \ldots, K. \tag{15.9}$$

15.2.3.4 TDoA-based localization in LoS

Still when operating in LoS conditions, localization can also be performed using the ToA of signals transmitted by the UE to the K gNBs. In particular, the TDoA technique exploits the difference ToA at the different gNBs. Let $\hat{\tau}_k$, $k = 1, \ldots, K$ the estimated time ToA of the same signal transmitted by the UE to the K gNB, and let K be the reference gNB, with respect to which the TDoA

$$\hat{\Delta}_k = \hat{\tau}_k - \hat{\tau}_K \tag{15.10}$$

are computed. Then, the differential distances are defined as $\hat{d}_{k,K} = c\hat{\Delta}_k$, and the position estimation problem can be written as

$$\hat{q} = \text{argmin}_q \sum_{k=1}^{K-1} \left((\hat{d}_{k,K} + ||q - p(k)||^2)^2 - \hat{d}_k^2\right)^2. \tag{15.11}$$

15.2.4 *Singe-UE privacy using AoA-based localization*

A first solution to preserve location privacy in future communication networks is based on a suitable beamforming of the signal transmitted by the UE.

The basic idea is that a suitable beamforming at the UE before signal transmission modifies the estimate of channel parameters (AoA, AoD, and ToA) by the gNB and prevents the gNB from obtaining a good localization. Several approaches can be adopted to this end. In the following, we will refer to the case of a UE equipped with a linear array and a single gNB attempting to estimate the position of the user, as discussed in the previous section. Let $\boldsymbol{x} = [x_1, \ldots, x_P]^T$ be the baseband transmitted vector signal, obtained as

$$\boldsymbol{x} = \boldsymbol{B}\boldsymbol{d}, \tag{15.12}$$

where $\boldsymbol{d} = [d_1, \ldots, d_L]^T$ is the vector of transmitted L data symbols and $\boldsymbol{B}$ is the $P \times P$ beamforming matrix.

A first option is to use a ZF transmit beamformer at the UE, i.e.,

$$\boldsymbol{B} = A\boldsymbol{H}^{\dagger}, \tag{15.13}$$

the pseudo-inverse of matrix $\boldsymbol{H}$, where A is a suitable scaling factor to satisfy the transmit power constraint. When the UE has more antennas than the gNB ($P > M$), all the information on the UE position is removed. Still, such approach suffers from a performance degradation in terms of data rate, especially when $\boldsymbol{H}$ is badly invertible.

A second option is obtained by observing that localization can be altered by modifying only some channel features rather than preventing the estimate of all of them at the gNB. Thus, we can focus on eliminating the information on the link between the UE and the scatterers, which, from the decomposition (15.7), is readily available at transmission (right side of the matrix product). In particular, we can hide only the AoDs: indeed, the AoA and AoD of NLoS paths are weakly correlated and disrupting the AoD estimate already prevents localization. Hence, a privacy-preserving beamformer aims at inverting matrix $\boldsymbol{A}_T^H$ as a normalized pseudoinverse of $\boldsymbol{A}_T^H[n]$, where column n of $\boldsymbol{B}$ is (see [39])

$$(\boldsymbol{B})_k = A(\boldsymbol{A}_T^H)_k^{-1}, \quad k = 1, \ldots, K, \tag{15.14}$$

A is a power-preserving scaling factor, and $(\boldsymbol{A}_T^H)_k^{-1}$ denotes column k of $(\boldsymbol{A}_T^H)^{-1}$. In case an LoS component is present, the UE should not beamform signals in that direction, which is obtained by removing the column corresponding to the LoS path from $\boldsymbol{B}$ (see [39] for more details). Such choice of beamforming significantly increases the achievable data rate, while still ensuring location privacy to the UE.

15.2.5 Singe-UE privacy using RSS-based localization

Considering the trilateration technique based on the RSS, privacy can be preserved by obfuscating the estimated position. A simple approach provides that the transmitter intentionally changes the transmit power P_0 to induce a wrong distance estimation in (15.8), as discussed in [40]. This still requires the use of beamforming, so that the different gNBs obtain a set of received powers consistent with their positions. Indeed, we note that in problem (15.9), we can consider P_0 as an unknown and include its estimate in the minimization process: this will provide an accurate position estimate as long as all the gNBs receive a power only scaled by the square of the distance.

15.2.6 Single-UE privacy using TDoA-based localization

In [41], location-privacy solution has been proposed, when localization is performed using the TDoA. This solution assumes that the gNBs do not cooperate to localize the

UE, a scenario that refers to the case of a curious network operator that still complies with the standard. In turn, the standard should be modified to include the protocol proposed in [41].

In particular, following the algebraic derivations of [41], problem (15.11) can be first solved as

$$\begin{bmatrix} \hat{\boldsymbol{q}} \\ \hat{d}_K \end{bmatrix} = \frac{1}{2} \begin{bmatrix} \boldsymbol{A}_{11} & \boldsymbol{A}_{12} \\ \boldsymbol{A}_{21} & \boldsymbol{A}_{22} \end{bmatrix}^{-1} \begin{bmatrix} \boldsymbol{B}_{11} + \boldsymbol{B}_{12} \\ \boldsymbol{B}_{21} + \boldsymbol{B}_{22} \end{bmatrix}, \tag{15.15}$$

where

$$\begin{aligned} \boldsymbol{A}_{11} &= (K-1)\boldsymbol{p}(K)\boldsymbol{p}^T(K) + \sum_{k=1}^{K-1} \boldsymbol{p}(k)\boldsymbol{p}^T(k) \\ &\quad - \left(\sum_{k=1}^{K-1} \boldsymbol{p}(k)\right)\boldsymbol{p}^T(K) - \boldsymbol{p}(K)\left(\sum_{k=1}^{K-1} \boldsymbol{p}^T(k)\right), \end{aligned} \tag{15.16}$$

$$\begin{aligned} \boldsymbol{A}_{12} &= \sum_{k=1}^{K-1} \boldsymbol{p}(k)\hat{\Delta}_{k,K} - \boldsymbol{p}(K)\left(\sum_{k=1}^{K-1} \hat{\Delta}_{k,K}\right), \\ \boldsymbol{A}_{21} &= \sum_{k=1}^{K-1} \boldsymbol{p}^T(k)\hat{\Delta}_{k,K} - \boldsymbol{p}^T(K)\left(\sum_{k=1}^{K-1} \hat{\Delta}_{k,K}\right), \end{aligned} \tag{15.17}$$

$$\boldsymbol{A}_{22} = \sum_{k=1}^{K-1} \hat{\Delta}_{k,K}^2, \tag{15.18}$$

and

$$\begin{aligned} \boldsymbol{B}_{11} &= (K-1)\boldsymbol{p}(K)\boldsymbol{p}^T(K)\boldsymbol{p}(K) + \sum_{k=1}^{K-1} \boldsymbol{p}(k)\boldsymbol{p}^T(k)\boldsymbol{p}(k) \\ &\quad - \left(\sum_{k=1}^{K-1} \boldsymbol{p}(k)\right)\boldsymbol{p}^T(K)\boldsymbol{p}(K) - \end{aligned} \tag{15.19}$$

$$\boldsymbol{p}(K)\left(\sum_{k=1}^{K-1} \boldsymbol{p}^T(k)\boldsymbol{p}(k)\right), \tag{15.20}$$

$$\boldsymbol{B}_{12} = -\sum_{k=1}^{K-1} \boldsymbol{p}(k)\hat{\Delta}_{k,K}^2 + \boldsymbol{p}(K)\left(\sum_{k=1}^{K-1} \hat{\Delta}_{k,K}^2\right), \tag{15.21}$$

$$\boldsymbol{B}_{21} = \sum_{k=1}^{K-1} \boldsymbol{p}^T(k)\boldsymbol{p}(k)\hat{\Delta}_{k,K} - \boldsymbol{p}^T(K)\boldsymbol{p}(K)\left(\sum_{k=1}^{K-1} \hat{\Delta}_{k,K}\right), \tag{15.22}$$

$$\boldsymbol{B}_{22} = -\sum_{k=1}^{K-1} \hat{\Delta}_{k,K}^3. \tag{15.23}$$

Now, to protect location privacy, the privacy-preserving summation technique is used. To explain this approach, consider that the UE aims at computing $\sum_{k=1}^{K} \boldsymbol{Y}_k$, where matrix $\boldsymbol{Y}_k$ is known to gNB k. We aim at keeping $\boldsymbol{Y}_k$ private to each gNB, i.e., we do not want to share it with the other gNBs. The privacy-preserving summation technique provides that each gNB k generates K matrices, denoted as $\boldsymbol{W}_{k,k'}$, $k' = 1, \ldots, K$, such that $\sum_{k'=1}^{K} \boldsymbol{W}_{k,k'} = \boldsymbol{0}$. Then gNB k keeps $\boldsymbol{W}_{k,k}$ and sends the rest to the other $K-1$ gNBs. By summing all the received random matrices from the other $K-1$ gNBs with $\boldsymbol{W}_{k,k}$, gNB k can get a random matrix $\boldsymbol{W}_k$. Now, if each gNB k transmits $\tilde{\boldsymbol{Y}}_k = \boldsymbol{Y}_k + \boldsymbol{W}_k$, no other gNB can compute the original value of $\boldsymbol{Y}_k$, but anyone can compute the summation of all $\boldsymbol{Y}_k$, since

$$\sum_{k=1}^{K} \boldsymbol{Y}_k = \sum_{k=1}^{K} \tilde{\boldsymbol{Y}}_k. \tag{15.24}$$

The location-privacy-preserving algorithm then works as follows. gNB K shares with the other gNBs its estimated ToA $\hat{\tau}_K$. Then, using their own TDoA estimate $\hat{\tau}_k$, $k = 1, \ldots, K-1$, all the other gNBs compute the TDoA $\hat{\Delta}_{k,K}$ from (15.10). Using the privacy-preserving summation technique, gNB K computes the summations in parenthesis in (15.15)–(15.23) obtaining information from the gNBs $1, \ldots, K-1$. Then, the UE computes $\boldsymbol{A}_{11}$, $\boldsymbol{A}_{21}$, $\boldsymbol{A}_{12}$, $\boldsymbol{A}_{22}$, $\boldsymbol{B}_{11}$, $\boldsymbol{B}_{21}$, $\boldsymbol{B}_{12}$, and $\boldsymbol{B}_{22}$, using (15.15)–(15.23) and receiving the required terms by privacy-preserving summation technique from all gNB (including gNB K).

With this technique, only the UE will know its position. We should remark that this technique requires honest gNBs that do not collude among them to locate the UE (since, in any case, they have all the TDoA needed to locate it). Moreover, other techniques based on cryptography may be used to exchange the information on the ToA with the UE without disclosing its value to other entities, including other gNBs.

15.2.7 Cooperative location-preserving privacy technique

15.2.7.1 The mix zone approach

A first approach to exploit cooperation among UEs to enhance location privacy is the *mix zone* technique [42]. This assumes that a trusted middleware system is available at the interface between the cellular network location system and untrusted parts (e.g., the network operators themselves or the OTT operators). Specific geographical areas are then identified, where the untrusted parts can access the precise location of the UEs (the so-called *application zones*), while for other areas, the location information should be masked (the so-called *mix zones*). UEs are associated to time-varying pseudonyms and untrusted parts do not access a traceable user identity to prevent association with a location, but only the current pseudonym. Whenever a UE enters of exits a mix zone, it changes its pseudonym, preventing tracking UEs moving from the outside to the inside of the mix zone. To avoid the association of different pseudonyms to the same UE observed in close times, the boundary of mix regions is blurred, so that a random time gap (and a random position) is made available to the untrusted application, thus reducing the accuracy of the pseudonym (and location) association.

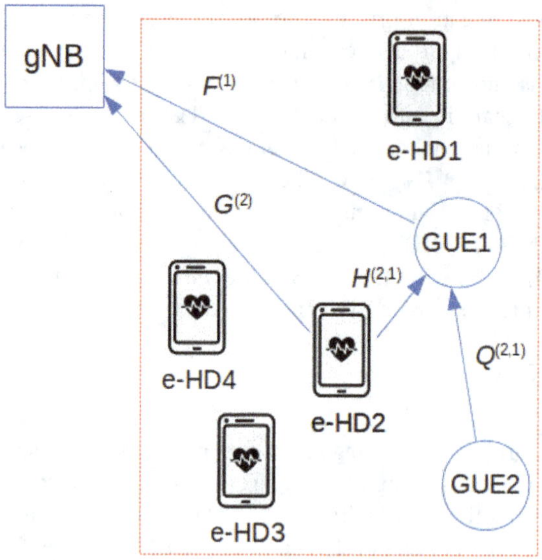

Figure 15.5 Example of LPMN

15.2.7.2 The LPMN approach

Another more comprehensive solution is the LPMN [34]. Consider a scenario with an LPMN with N_P UEs and N_G gateway UEs GUEs (see Figure 15.5): the UEs aim at protecting their location privacy, while GUEs operate as gateways between the UEs and the gNB and their location is known to the cellular network: they can be either devices dedicated to this operation (thus a network of GUEs should be specifically deployed) or UEs that at the moment are not interested to hide their location to the network, e.g., because they need a better localization for specific LBSs and do not have special privacy constraints.

The gNB is equipped with M antennas, while both GUEs and UEs have P antennas each, and we denote with $\boldsymbol{H}^{(i,j)}$ the $P \times P$ channel matrix from UE i, $i = 1, \dots, N_P$, to GUE $j = 1, \dots, N_G$, while $\boldsymbol{G}^{(i)}$ is the $M \times P$ channel matrix from UE i to the gNB, $\boldsymbol{F}^{(j)}$ is the $M \times P$ channel matrix from GUE j to the gNB, and $\boldsymbol{Q}^{(i,j)}$ is the $P \times P$ channel matrix from GUE i to GUE j. All channels are known to the GUEs and UEs, while the gNB knows only the channel matrix to the GUE, thus the UE cannot be localized by the gNB.

We now focus on the protection of the communication link between the UE and the GUE, with the objective of preventing the gNB from localizing the UE through overhearing the signal transmitted by the UE to the GUE [43]. In the previous section, we have seen two design criteria for the transmit beamformer to improve location privacy, namely the ZF beamformer and the AoD-information removing beamformer. Now we consider a third option, wherein the transmit beamformer is such that the signal is weak enough at the gNB to prevent an accurate localization, while still being strong enough at the GUE for data transmission. Moreover, we let the other GUE transmit an AN signal to the gNB to further reduce the signal-to-noise ratio (SNR) there. Indeed, by providing an SNR of about −20 dB, we can significantly degrade the localization accuracy to about 100 m [44].

We will assume that UE i transmits a single data stream to GUE j, using the degrees of freedom of its multiple antennas to beamform this signal. We denote with $\boldsymbol{w}^{(i)}$ the transmit beamformer of UE i. Furthermore, we will design the AN transmitted by the other GUEs such that it does not affect GUE j, which then can use the maximum-ratio combining (MRC) receiver on the resulting channel $\boldsymbol{H}^{(i,j)}\boldsymbol{w}^{(i)}$, cascade of the LPB of UE i and its channel $\boldsymbol{H}^{(i,j)}$.

15.2.7.3 AN transmission

As the other GUEs transmit AN, GUE k, $k = 1, \ldots, N_G, k \neq j$ generates the AN vector $\boldsymbol{x}^{(k)}$ of L independent zero-mean Gaussian variables with power σ_k^2. The transmitted signal is then $\boldsymbol{V}^{(k)}\boldsymbol{x}^{(k)}$, where $\boldsymbol{V}^{(k)}$ is the AN beamformer with unitary power constraint. The signal received at the gNB is

$$\boldsymbol{r}_B = \boldsymbol{G}^{(i)}\boldsymbol{w}^{(i)}d^{(i)} + \underbrace{\sum_{k=1}^{N_U} \boldsymbol{F}^{(k)}\boldsymbol{V}^{(k)}\boldsymbol{x}^{(k)}}_{\boldsymbol{n}_{\text{AN}}} + \boldsymbol{n}_B, \tag{15.25}$$

where $\boldsymbol{n}_{\text{AN}}$ is the received AN vector, $d^{(i)}$ is the transmitted data symbol, and $\boldsymbol{n}_B$ is the noise vector. To avoid interference of the AN with the signal received by GUE j, we impose

$$\boldsymbol{w}^{(i)H}\boldsymbol{H}^{(i,j)H}\boldsymbol{Q}^{(k,j)}\boldsymbol{V}^{(k)} = 0, \quad k = 1, \ldots, N_G, k \neq j, \tag{15.26}$$

thus we must have

$$\boldsymbol{V}^{(k)} = \bar{\boldsymbol{Q}}^{(k,j)}\boldsymbol{V}^{(k)'}, \quad k = 1, \ldots, N_G, k \neq j, \tag{15.27}$$

where $\bar{\boldsymbol{Q}}^{(k,j)}$ is a matrix, whose columns are an orthonormal basis of the space orthogonal to $\boldsymbol{w}^{(i)H}\boldsymbol{H}^{(i,j)H}\boldsymbol{Q}^{(k,j)}$.

15.2.7.4 SNR at the gNB

The gNB knows the correlation matrix $\boldsymbol{R}^{(j)}$of the artificial and natural noise, since it depends only on the channels between the gNB and the GUEs, and we have

$$\boldsymbol{R}^{(j)} = \sigma_G^2\boldsymbol{I} + \sum_{k=1,k\neq j}^{N_G} \boldsymbol{F}^{(k)}\boldsymbol{V}^{(k)}\boldsymbol{V}^{(k)^H}\boldsymbol{F}^{(k)H}\sigma_k^2. \tag{15.28}$$

In an attempt to reduce the effects of the AN, the gNB first whitens the noise (including the AN) by multiplying $\boldsymbol{r}_B$ by $\boldsymbol{R}^{(j)-1/2}$, where $\boldsymbol{R}^{(j)-1/2} = \boldsymbol{\Lambda}^{-1/2}\boldsymbol{U}^H$ is computed from the SVD $\boldsymbol{R}^{(j)} = \boldsymbol{U}\boldsymbol{\Lambda}\boldsymbol{U}^H$. Then, the gNB applies the MRC combiner $\boldsymbol{w}^{(i)H}\boldsymbol{G}^{(i)H}\boldsymbol{R}^{(j)-1/2}$. The resulting channel between the gNB and the UE is therefore $\boldsymbol{R}^{(j)-1/2}\boldsymbol{G}^{(i)}\boldsymbol{w}^{(i)}$. Hence, we can define the SNR at the gNB as

$$\Upsilon^{(i,j)} = \boldsymbol{w}^{(i)H}\boldsymbol{G}^{(i)H}\boldsymbol{R}^{(j)-1}\boldsymbol{G}^{(i)}\boldsymbol{w}^{(i)}. \tag{15.29}$$

The location privacy is obtained by imposing that the SNR at the gNB remains below a threshold value ε, i.e.,

$$\Upsilon^{(i,j)} \leq \varepsilon. \tag{15.30}$$

The general optimization of the transmit beamformer $\boldsymbol{w}^{(i)}$ and the AN beamformers $\boldsymbol{V}^{(k)}$ cannot be performed in close form, but can be addressed by numerical methods (see [43] for more details).

15.3 Quantum-enhanced location privacy and authentication

At the physical layer, the ultimate technique for securing communication is based on quantum communication—an exciting technology whose mainstream adoption is just commencing. Commercially available quantum key distribution (QKD) systems for point-to-point fiber optic links and line-of-sight wireless links are now available, and, in August 2016, the first quantum-enabled satellite, Micius, was launched by the Chinese Space Agency [45]. The importance of this latter event cannot be overstated as it represents the first step toward the so-called Quantum Internet—a paradigm dependent on the viability of reliable ground-satellite quantum communication links. Following on from its launch into Low Earth Orbit, Micius has now confirmed all aspects of quantum communication via low Earth orbit satellites is possible, including quantum entanglement distribution [45], QKD [46], and quantum teleportation [47]. Quantum technology is, therefore, "here and now" and the Quantum Internet is imminent. A detailed tutorial on quantum technology, as required for the emerging Quantum Internet, is given elsewhere [48].

Excitement over the "arrival" of quantum technology has led many in the engineering community to consider the integration of such technology with next-generation 6G systems, e.g., [49–52]. In the review of [52], a road-map for the integration of quantum-enabled systems within 6G was outlined as follows: near-term (2–3 years) secure communications for back-haul (via fiber QKD); middle-term (3–5 years) secure satellite communications (via satellite QKD); long-term (5–10 years) data computing with privacy preservation (via quantum computing) and real-time wireless AI (via quantum machine learning). While some of these developments may need to await 7G deployments, the scope for integration of quantum communications, at least at some level, within maturing 6G standards is entirely plausible.

In this review, we are particularly interested in the use of quantum communications for localization purposes. We are especially interested in the privacy and authentication of location information.

Privacy issues are largely delivered using QKD techniques. Current authentication mechanisms within communication systems are of course purely classical—i.e., they do not include the use of any quantum technology. Going forward, these current systems will become redundant, especially when quantum computers are prevalent. Secure communication requires, ideally, a system whereby the entity (the sending device) and the messages sent are both verified at an "information-theoretic" security level. It is known that classical physical layer security techniques cannot provide this information-theoretic security at the same level as quantum techniques can—classical techniques are information-theoretic secure only under additional restrictions on the adversary model relative to quantum techniques [53]. It is well established that some QKD protocols can, in principle, deliver information-theoretic security without any assumptions on the adversary model—the one-time-pad (the key) distributed being completely random with vanishing leakage to the adversary (the probability of key leakage is bounded to be less than some value). As such, ultimately privacy of the location information will be enabled by quantum techniques that encrypt the transfer of the quantum data between entities in the network. We will not delve further into the details of QKD here—as there are many excellent reviews and tutorials available on this well-traveled subject. In terms of 6G networks, the review of [54] should be of particular value as it focuses on optical networking elements of QKD. In 6G networks, most quantum communication links will be free space optical links—although short-range quantum communication (10–20 m range) should be possible at microwave frequencies [55,56]. It is also perhaps worth mentioning another close link between QKD and classical-only wireless communications—both need similarly advanced and efficient classical error-correcting codes operating close to the Shannon limit. In the classical reconciliation step of QKD—finite-size key effects can drive the code block length to be set at very high values (sometimes 10^{10}, and above)—a scenario that necessitates the use of sophisticated optimization and parallel processing [57].

Perhaps more challenging in the context of localization within the context of 6G is the authentication of location information within a network. In classical communications, information-theoretic authentication of entities and messages is provided by a Wegman–Carter Message Authentication Code (MAC) [58]. However, these codes have some deployment-related shortcomings. MACs require long pre-shared secrets and the use of pre-assigned hash functions—rendering their implementation in a quantum-enabled system unwieldy, e.g., [59]. Also, as is the case in most protocols, MACs are susceptible to side-channel attacks [60]. However, even worse is the fact that adversaries can easily spoof their location, rendering the transmitted location information at best useless, and, at worst, a hazard to the entire network. This issue requires the intervention of an LVS.

The goal of an LVS is to verify a user's claimed position (e.g., a claimed GNSS position) using wireless network measurements, so as to perform a binary decision on whether the user is legitimate—i.e., claims to be at his true position, or malicious—i.e., deliberately spoofs his claimed position, e.g., [15,61–63]. Researchers have formulated a number of information-theoretic LVSs over the past few years which utilize available vehicle signal metrics, such as RSS, time of arrival (ToA), and the angle of arrival (AoA).

The main purpose of an LVS is to detect location spoofing by obtaining a low false positive rate for legitimate users and a high detection rate for malicious users, leading to a trade-off best illustrated by a receiver operating characteristic (ROC) curve. However, it is well established that ROCs are not always perfect when it comes to comparing separate systems, e.g., [64]. We also note that the use of an ROC does not inform us where the optimal operating point of an LVS is (in terms of system parameters). Optimization of an LVS, however, may be possible through the use of

Bayesian Decision Theory or via Generalized Likelihood Ratio Tests. These tests minimize the input/output classification error in the scenario where the costs of all types of misclassifications are equal. If the misclassification costs are unequal, then the Bayes criterion may be used as a variation of a more traditional decision rule. However, a key weakness of most of these approaches is that their decision criteria are prone to sensitivity arising from subjective *a priori* specifications on beliefs of key quantities in the decision framework. Inclusion of such subjective specifications leads in turn to a subjective measure of system performance, a generally undesirable trait for any LVS. Although much progress has been made in recent years to optimize the performance of classical LVSs—for the reasons outlined above (and many more), it is impossible for such systems to operate at a level where the authentication of any claimed location is *fully guaranteed* to be correct.

However, fortunately, quantum techniques come to the rescue again through quantum location verification (QLV)—a technique that can fully guarantee the location authentication. This quantum-technique first appeared in the literature in 2010 in the works of [65,66], with some aspects of these works appearing in a 2006 patent (Kent *et al.*, US7075438)—see [63] for details on the historical development of QLV. In general terms, the anti-location-spoofing properties of QLV can be attributed to the quantum no-cloning theorem, coupled with the no-signaling principle of special relativity. Since the works of [65,66], QLV has been further studied in numerous other works with its deployment and security under a wide range of conditions and assumptions investigated (see [63] and refs therein). In [63], arguments for QLV's security against any known attack are presented. Notwithstanding these defenses against attacks, it is well established that QLV is provably unconditionally secure when an *a priori* secret key is used within the protocol—since then the data required for encryption within the QLV is information-theoretic secured. An *a priori* established secret key is already required for all QKD protocols, and, as such, can be easily shared for QLV [63]. A schematic of QLV is given in Figure 15.6.

Building on the notion of QLV, it was later pointed out that by monitoring the location of a device via classical radio communications, and utilizing quantum information embedded in the device, a system could be developed whereby real-time data could be delivered *only* to a specific geographic location [67]. In this system of quantum-enabled geo-encryption, data is sent to the receiver only while the QLV is validated. By obfuscating the quantum information needed by both the quantum communication and the QLV protocols until the time of the decryption and location verification, it becomes possible to make the decryption of data itself conditional upon the location of the receiver [67]. This new form of geo-encryption is free from the vulnerabilities that plague classical-only geo-encryption protocols. We do note, however, the quantum information to be used in geo-encryption is best pre-stored in the device that will invoke the decryption—an issue for emerging 6G systems, as we now discuss.

The benefits of quantum technology within the context of 6G have one major *Achilles heel*—they largely depend on the availability of quantum memory within the network. Ideally, this memory should be operational at room temperature, have a memory-hold time of at least a few seconds, and be available in a reasonable-sized form factor. Such a quantum memory is currently being worked on by many groups

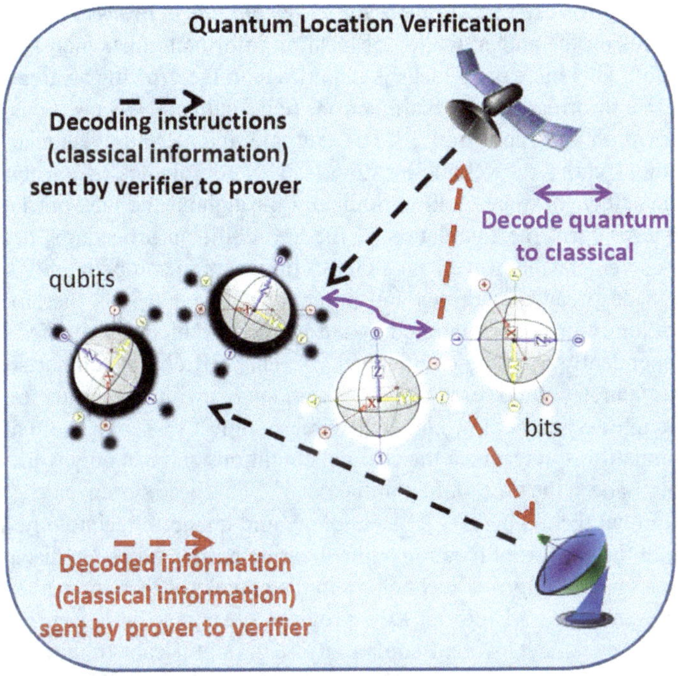

Figure 15.6 *In one configuration of QLV, multiple verifiers can send classical information to a prover (a user whose claimed location is to be proven), who must use this received classical information in order to decode stored quantum information into some new classical information. The new classical information is then sent to the verifiers.*

throughout the world but is not currently available. A detailed review of this important research area can be found elsewhere, e.g., [68–71], where it can be gathered that the development of the required quantum memory is certainly achievable prior to the ultimate demise of any new 6G standard.

A possible candidate for the desired form of quantum memory is that based on solid-state devices as described in [68]. This technique already has allowed storage of quantum light for 0.5 s and holds much promise for longer storage times. In principle, memory timescales equal to the round-trip communication time (milliseconds) are required for quantum geo-encryption. However, more realistically, quantum memory timescales of the order of minutes will likely be required, with even longer timescales greatly expanding the real-world deployment scenarios. The required memory timescale of minutes is predicted to be widely deployable in about 5 years [70]. The current quantum memory records in controlled laboratory settings are 1 min at room temperature and 6 h at low temperature [68–71]. These issues, as well as other related challenges facing the development of the Quantum Internet, are discussed in more detail elsewhere [48].

Case study: 6G location authentication—vehicular networks:
Research focus on the authentication of location information has increased over the past few years. This increasing focus is in part due to the growing awareness of how vulnerable that information is to malfunctions (e.g., faulty GNSS positions), or malicious falsification and spoofing [72]. For example, in emerging vehicular networks for autonomous cars—the vehicle location (GNSS coordinates) is an input relayed in all of the safety messages sent periodically throughout the network [73]. These safety messages form the foundation of the anti-collision procedures deployed by the network's vehicles. However, such GNSS information can be readily spoofed—leading to widespread disruptions and potential collisions. It is therefore critical that all location information supplied by autonomous vehicles is verified (authenticated) independently by some connected LVS. This difficult but important task for autonomous vehicles could form a key application area within emerging 6G applications. Although easy to deploy, classical-wireless-only LVSs may well have certain operation limitations in terms of the channel conditions and can only work under the assumptions made at the time of their initial design. This base system can be enhanced by two additional technologies—AI technology and quantum technology. The latter technology in the context of location authentication has already been discussed—we briefly discuss next the former technology in the context of location authentication.

The integration of AI into an LVS provides clear realizable improvements at the cost of increased deployment complexity (e.g., significant training epochs). A combined architecture of a classical-wireless LVS, an AI-enabled LVS, and a quantum-enabled LVS, that could be embedded within a 6G network, is schematically shown in Figure 15.7. Here, the artificial intelligence component (the AI component) will most likely be machine learning in the form of neural-network (NN) algorithms. However, once trained to a certain limit, an NN-based LVS (NN-LVS) can perform well in the channel conditions they are designed for. Due to their adaptability, the NN-LVS can

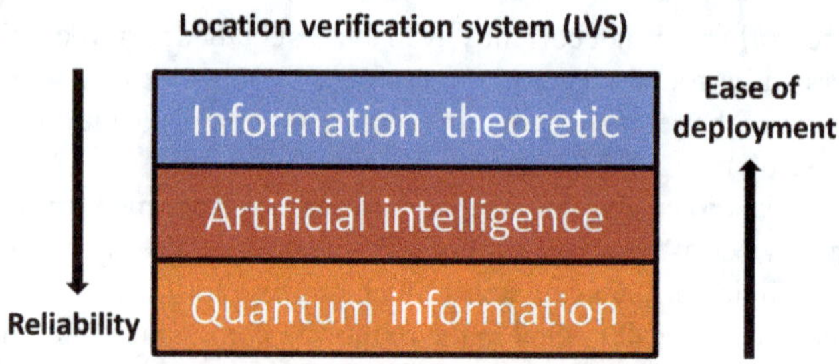

Figure 15.7 In 6G, a suite of LVS solutions can be deployed, depending on the level of performance required. The first solution (information-theoretic) is easiest to deploy and is purely classical, but has the least reliability. The third solution (quantum information) is the most reliable but the most complex to deploy. The other solution, AI, is classical only but can be considered, in terms of complexity, to lie between the top and bottom extremes.

also accommodate changing channel and environment conditions, something which the information-theoretic LVSs lack. Further, the NN-LVS is believed to address multiple threat situations, thus eliminating the need for designing threat-specific LVSs [74]. The NN-LVS also has the potential to update itself through continuous learning thus avoiding the risk of becoming obsolete. The use of AI has recently been shown [75] to lead to dramatic improvements in performance in certain circumstances. However, the same work also highlights the danger of over training and the difficulty in optimally designing NN architectures fit for purpose.

The combined classical-AI-quantum architecture we depict most likely entails the main characteristics of any quantum-enabled system in a 6G deployment. That is, any quantum deployment in 6G will always have at least one additional system that it can revert to should the quantum operations fail or the required quantum signaling (or stored data) not be available. AI will be ubiquitous in 6G systems, so its inclusion within the context of location authentication is natural.

References

[1] S. Yan, U. Ihsan, R. Malaney, L. Sun, and S. Tomasin, "Location verification for future wireless vehicular networks: research directions and challenges," *IEEE Network*, vol. 36, no. 6, pp. 60–66, 2022.

[2] Y. Cao, T. Jiang, O. Kaiwartya, H. Sun, H. Zhou, and R. Wang, "Toward pre-empted EV charging recommendation through V2V-based reservation system," *IEEE Trans. Syst. Man Cybern.*, vol. 51, no. 5, pp. 3026–3039, 2021.

[3] G. Yan, S. Olariu, and M. Weigle, "Providing location security in vehicular ad hoc networks," *IEEE Wirel. Commun.*, vol. 16, no. 6, pp. 48–55, 2009.

[4] R. Lu, X. Lin, X. Liang, and X. Shen, "A dynamic privacy-preserving key management scheme for location-based services in VANETs," *IEEE Trans. Intell. Transp. Syst.*, vol. 13, no. 1, pp. 127–139, 2012.

[5] Y. Cao, X. Zhang, B. Zhou, X. Duan, D. Tian, and X. Dai, "MEC intelligence driven electro-mobility management for battery switch service," *IEEE Trans. Intell. Transp. Syst.*, vol. 22, no. 7, pp. 4016–4029, 2021.

[6] J. Yang, Y. Chen, W. Trappe, and J. Cheng, *Pervasive Wireless Environments: Detecting and Localizing User Spoofing*. New York, NY: Springer, 2014.

[7] S. Yan, R. Malaney, I. Nevat, and G. W. Peters, "Optimal information-theoretic wireless location verification," *IEEE Trans. Veh. Technol.*, vol. 63, no. 7, pp. 3410–3422, 2014.

[8] M. Monteiro, J. Rebelatto, and R. Souza, "Information-theoretic location verification system with directional antennas for vehicular networks," *IEEE Trans. Intell. Transp. Syst.*, vol. 17, no. 1, pp. 93–103, 2016.

[9] S. Yan, R. Malaney, I. Nevat, and G. Peters, "Location verification systems for VANETs in Rician fading channels," *IEEE Trans. Veh. Technol.*, vol. 65, no. 7, pp. 4132–4144, 2016.

[10] J. Y. Koh, I. Nevat, D. Leong, and W.-C. Wong, "Geo-spatial location spoofing detection for Internet of Things," *IEEE Internet Things J.*, vol. 3, no. 6, pp. 971–978, 2016.

[11] S. Yan, I. Nevat, G. Peters, and R. Malaney, "Location verification systems under spatially correlated shadowing," *IEEE Trans. Wireless Commun.*, vol. 15, no. 6, pp. 4132–4144, 2016.

[12] U. Ihsan, S. Yan, and R. Malaney, "Location verification for emerging wireless vehicular networks," *IEEE Internet Things J.*, vol. 6, no. 6, pp. 10261–10272, 2019.

[13] A. Brighente, F. Formaggio, G. Nunzio, and S. Tomasin, "Machine learning for in-region location verification in wireless networks," *IEEE J. Sel. Areas Commun.*, vol. 37, no. 11, pp. 2490–2502, 2019.

[14] L. Xiao, X. Wan, and Z. Han, "PHY-layer authentication with multiple landmarks with reduced overhead," *IEEE Trans. Wireless Commun.*, vol. 17, no. 3, pp. 1676–1687, 2018.

[15] S. Yan, R. Malaney, I. Nevat, and G. Peters, "Optimal information-theoretic wireless location verification," *IEEE Trans. Veh. Technol.*, vol. 63, no. 7, pp. 3410–3422, 2014.

[16] F. Liu, C. Masouros, A. Li, H. Sun, and L. Hanzo, "MU-MIMO communications with MIMO radar: from co-existence to joint transmission," *IEEE Trans. Wirel. Commun.*, vol. 17, no. 4, pp. 2755–2770, 2018.

[17] I. Wagner and D. Eckhoff, "Technical privacy metrics: a systematic survey," *ACM Comput. Surv.*, vol. 51, no. 3, pp. 1–38, 2018.

[18] R. Shokri, G. Theodorakopoulos, J.-Y. Le Boudec, and J.-P. Hubaux, "Quantifying location privacy," in *Proceedings of the 2011 IEEE Symposium on Security and Privacy*, 2011, pp. 247–262.

[19] B. Liu, W. Zhou, T. Zhu, L. Gao, and Y. Xiang, "Location privacy and its applications: a systematic study," *IEEE Access*, vol. 6, pp. 17606–17624, 2018.

[20] H. Jiang, J. Li, P. Zhao, F. Zeng, Z. Xiao, and A. Iyengar, "Location privacy-preserving mechanisms in location-based services: a comprehensive survey," *ACM Comput. Surv.*, vol. 54, no. 1, pp. 1–36, 2021.

[21] E. Snekkenes, "Concepts for personal location privacy policies," in *Proceedings of the 3rd ACM Conference on Electronic Commerce*, ser. EC '01. New York, NY: Association for Computing Machinery, 2001, p. 48–57.

[22] M. Decker, "Location privacy—an overview," in *Proceedings of the 2008 7th International Conference on Mobile Business*, 2008, pp. 221–230.

[23] R. Shokri, P. Papadimitratos, G. Theodorakopoulos, and J.-P. Hubaux, "Collaborative location privacy," in *Proceedings of the 2011 IEEE Eighth International Conference on Mobile Ad-Hoc and Sensor Systems*, 2011, pp. 500–509.

[24] S. B. Wicker, "The loss of location privacy in the cellular age," *Commun. ACM*, vol. 55, no. 8, pp. 60–68, 2012.

[25] K. Rechert, K. Meier, R. Zahoransky, *et al.*, "Reclaiming location privacy in mobile telephony networks—effects and consequences for providers and subscribers," *IEEE Syst. J.*, vol. 7, no. 2, pp. 211–222, 2013.

[26] J. Freudiger, M. H. Manshaei, J.-P. Hubaux, and D. C. Parkes, "Non-cooperative location privacy," *IEEE Trans. Depend. Secure Comput.*, vol. 10, no. 2, pp. 84–98, 2013.

[27] J. D. Roth, M. Tummala, J. C. McEachen, and J. W. Scrofani, "On location privacy in LTE networks," *IEEE Trans. Inf. Forens. Secur.*, vol. 12, no. 6, pp. 1358–1368, 2017.

[28] D. Giustiniano, G. Bianchi, A. Conti, S. Bartoletti, and N. B. Melazzi, "5G and beyond for contact tracing," *IEEE Commun. Mag.*, vol. 59, no. 9, pp. 36–41, 2021.

[29] I. Bang, T. Kim, H. S. Jang, and D. K. Sung, "An opportunistic power control scheme for mitigating user location tracking attacks in cellular networks," *IEEE Trans. Inf. Forens. Secur.*, vol. 17, pp. 1131–1144, 2022.

[30] Court Sentence, United States v. Carpenter. 585 U.S. 2206 (US), 2018.

[31] EU Parliament, *Directive 95/46/EC of the European Parliament and of the Council of 24 October 1995 on the protection of individuals with regard to the processing of personal data and on the free movement of such data*, 1995.

[32] Y. Georgiadou, R. A. de By, and O. Kounadi, "Location privacy in the wake of the GDPR," *ISPRS Int. J. Geo-Inf.*, vol. 8, no. 3, pp. 1–17, 2019.

[33] M. M. Losavio, *Legal Issues and the Need to Engineer Positioning Systems for Protection of Privacy and Personal Security*. Cham: Springer International Publishing, 2021, pp. 151–169.

[34] S. Tomasin, M. Centenaro, G. Seco-Granados, S. Roth, and A. Sezgin, "Location-privacy leakage and integrated solutions for 5G cellular networks and beyond," *Sensors*, vol. 21, no. 15, pp. 1–14, 2021.

[35] S. Tomasin and J. G. L. Hidalgo, "Virtual private mobile network with multiple gateways for B5G location privacy," in *Proceedings of the 2021 IEEE 94th Vehicular Technology Conference (VTC2021-Fall)*, 2021, pp. 1–6.

[36] M. Cominelli, F. Kosterhon, F. Gringoli, R. Lo Cigno, and A. Asadi, "IEEE 802.11 CSI randomization to preserve location privacy: an empirical evaluation in different scenarios," *Comput. Netw.*, vol. 191, p. 107970, 2021.

[37] S. Roth, S. Tomasin, M. Maso, and A. Sezgin, "Localization attack by pre-coder feedback overhearing in 5G networks and countermeasures," *IEEE Trans. Wirel. Commun.*, vol. 20, no. 7, pp. 4100–4112, 2021.

[38] M. Giordani, M. Polese, M. Mezzavilla, S. Rangan, and M. Zorzi, "Toward 6G networks: use cases and technologies," *IEEE Commun. Mag.*, vol. 58, no. 3, pp. 55–61, 2020.

[39] J. J. Checa and S. Tomasin, "Location-privacy-preserving technique for 5G mmWave devices," *IEEE Commun. Lett.*, vol. 24, no. 12, pp. 2692–2695, 2020.

[40] V. Shubina, A. Ometov, S. Andreev, D. Niculescu, and E. S. Lohan, "Privacy versus location accuracy in opportunistic wearable networks," in *Proceedings of the 2020 International Conference on Localization and GNSS (ICL-GNSS)*, 2020, pp. 1–6.

[41] X. Shi and J. Wu, "To hide private position information in localization using time difference of arrival," *IEEE Trans. Signal Process.*, vol. 66, no. 18, pp. 4946–4956, 2018.

[42] A. Beresford and F. Stajano, "Mix zones: user privacy in location-aware services," in *Proceedings of the Second IEEE Annual Conference on Pervasive Computing and Communications Workshops, 2004*, 2004, pp. 127–131.

[43] S. Tomasin, "Beamforming and artificial noise for cross-layer location privacy of e-health cellular devices," in *Proceedings of the 2022 IEEE International Conference on Communications Workshops (ICC Workshops)*, 2022, pp. 568–573.

[44] A. Shahmansoori, G. E. Garcia, G. Destino, G. Seco-Granados, and H. Wymeersch, "Position and orientation estimation through millimeter-wave MIMO in 5G systems," *IEEE Trans. Wirel. Commun.*, vol. 17, no. 3, pp. 1822–1835, 2018.

[45] J. Yin, Y. Cao, Y.-H. Li, *et al.*, "Satellite-based entanglement distribution over 1200 kilometers," *Science*, vol. 356, no. 6343, pp. 1140–1144, 2017.

[46] S.-K. Liao, W.-Q. Cai, W.-Y. Liu, *et al.*, "Satellite-to-ground quantum key distribution," *Nature*, vol. 549, no. 7670, pp. 43–47, 2017.

[47] J.-G. Ren, P. Xu, H.-L. Yong, *et al.*, "Ground-to-satellite quantum teleportation," *Nature*, vol. 549, no. 7670, pp. 70–73, 2017.

[48] A. Singh, K. Dev, H. Siljak, H. D. Joshi, and M. Magarini, "Quantum Internet—applications, functionalities, enabling technologies, challenges, and research directions," *IEEE Commun. Surv. Tuts.*, vol. 23, no. 4, pp. 2218–2247, 2021.

[49] S. J. Nawaz, S. K. Sharma, S. Wyne, M. N. Patwary, and M. Asaduzzaman, "Quantum machine learning for 6G communication networks: state-of-the-art and vision for the future," *IEEE Access*, vol. 7, pp. 46317–46350, 2019.

[50] S. Mumtaz and M. Guizani, "An overview of quantum computing and quantum communication systems," *IET Quantum Commun.*, vol. 2, no. 3, pp. 136–138, 2021.

[51] K. G. Shenoy, *Role of Quantum Technology in 6G*. Cham: Springer International Publishing, 2021, pp. 419–429.

[52] C. Wang and A. Rahman, "Quantum-enabled 6G wireless networks: opportunities and challenges," *IEEE Wirel. Commun.*, vol. 29, no. 1, pp. 58–69, 2022.

[53] R. Malaney, "A note on the information-theoretic-(in)security of fading generated secret keys," in *Proceedings of the 2018 IEEE Global Communications Conference (GLOBECOM)*, 2018, pp. 1–6.

[54] P. Sharma, A. Agrawal, V. Bhatia, S. Prakash, and A. K. Mishra, "Quantum key distribution secured optical networks: a survey," *IEEE Open J. Commun. Soc.*, vol. 2, pp. 2049–2083, 2021.

[55] C. Weedbrook, S. Pirandola, and T. C. Ralph, "Continuous-variable quantum key distribution using thermal states," *Phys. Rev. A*, vol. 86, p. 022318, 2012.

[56] N. Hosseinidehaj and R. Malaney, "Quantum entanglement distribution in next-generation wireless communication systems," in *Proceedings of the 2017 IEEE 85th Vehicular Technology Conference (VTC Spring)*, 2017, pp. 1–7.

[57] X. Ai and R. Malaney, "Optimised multithreaded CV-QKD reconciliation for global quantum networks," *IEEE Trans. Commun.*, vol. 70, no. 9, pp. 6122–6132, 2022.

[58] J. Carter and M. N. Wegman, "Universal classes of hash functions," *J. Comput. Syst. Sci.*, vol. 18, no. 2, pp. 143–154, 1979.

[59] M. Curty and D. J. Santos, "Quantum authentication of classical messages," *Phys. Rev. A*, vol. 64, p. 062309, 2001.

[60] K. Okeya and T. Iwata, "Side channel attacks on message authentication codes," in R. Molva, G. Tsudik, and D. Westhoff, Eds, *Security and Privacy in Ad-hoc and Sensor Networks*. Berlin, Heidelberg: Springer Berlin Heidelberg, 2005, pp. 205–217.

[61] S. Capkun, K. Rasmussen, M. Cagalj, and M. Srivastava, "Secure location verification with hidden and mobile base stations," *IEEE Trans. Mobile Comput.*, vol. 7, no. 4, pp. 470–483, 2008.

[62] S. Yan, R. Malaney, I. Nevat, and G. W. Peters, "Location verification systems for VANETs in Rician fading channels," *IEEE Trans. Veh. Technol.*, vol. 65, no. 7, pp. 5652–5664, 2016.

[63] R. Malaney, "The quantum car," *IEEE Wirel. Commun. Lett.*, vol. 5, no. 6, pp. 624–627, 2016.

[64] G. Gu, P. Fogla, D. Dagon, W. Lee, and B. Skorić, "Measuring intrusion detection capability: an information-theoretic approach," in *Proceedings of the 2006 ACM Symposium on Information, Computer and Communications Security*, ser. ASIACCS'06. New York, NY: Association for Computing Machinery, 2006, p. 90–101.

[65] R. A. Malaney, "Location-dependent communications using quantum entanglement," *Phys. Rev. A*, vol. 81, p. 042319, 2010.

[66] R. A. Malaney, "Quantum location verification in noisy channels," in *Proceedings of the 2010 IEEE Global Telecommunications Conference GLOBECOM 2010*, 2010, pp. 1–6.

[67] R. Malaney, "Quantum geo-encryption," in *Proceedings of the 2016 IEEE Global Communications Conference (GLOBECOM)*, 2016, pp. 1–6.

[68] A. Holzäpfel, J. Etesse, K. T. Kaczmarek, A. Tiranov, N. Gisin, and M. Afzelius, "Optical storage for 0.53 s in a solid-state atomic frequency comb memory using dynamical decoupling," *New J. Phys.*, vol. 22, no. 6, p. 063009, 2020.

[69] X.-L. Pang, A.-L. Yang, J.-P. Dou, *et al.*, "A hybrid quantum memory-enabled network at room temperature," *Sci. Adv.*, vol. 6, no. 6, p. eaax1425, 2020.

[70] L. Ma, X. Tang, and O. Slattery, "Optical quantum memory applications in quantum communication," in K. S. Deacon, Ed., *Quantum Communications and Quantum Imaging XVII*, vol. 11134, International Society for Optics and Photonics. SPIE, 2019, pp. 1 – 9.

[71] M. Zhong, R. L. Ahlefeldt, and M. J. Sellars, "Quantum information processing using frozen core y^{3+} spins in eu^{3+}:y_2sio_5," *N.w J. Phys.*, vol. 21, no. 3, p. 033019, 2019.

[72] M. Amoozadeh, A. Raghuramu, C.-N. Chuah, D. Ghosal, H. M. Zhang, J. Rowe, and K. Levitt, "Security vulnerabilities of connected vehicle streams and their impact on cooperative driving," *IEEE Commun. Mag.*, vol. 53, no. 6, pp. 126–132, 2015.

[73] T. Leinmuller, E. Schoch, and F. Kargl, "Position verification approaches for vehicular ad hoc networks," *IEEE Wirel. Commun.*, vol. 13, no. 5, pp. 16–21, 2006.

[74] S. Hamdan, A. Hudaib, and A. Awajan, "Detecting sybil attacks in vehicular ad hoc networks," *Int. J. Parallel Emergent Distrib. Syst.*, vol. 36, no. 2, pp. 69–79, 2021.

[75] U. Ihsan, S. Yan, and R. Malaney, "Location verification for emerging wireless vehicular networks," *IEEE Internet Things J.*, vol. 6, no. 6, pp. 10261–10272, 2019.

Chapter 16

Radio frequency informed physical layer security—an augmented padlock to wireless transmission secrecy

Jayakrishnan M. Purushothama[1]*, Jiayu Hou*[1]*, Yuan Ding*[1] *and Yue Xiao*[2]

16.1 What is physical layer wireless security, and why it is important

Electromagnetic induction, the earliest form of wireless energy transfer, was demonstrated by Michael Faraday in 1831, happening to be the infancy of radio communication. Since then, this technique in various scales was used for communications, eventually converging to a democratic name "Radio" in the early 1900s as we know it today. Radio frequency or RF communications are collectively and broadly referred to the communication of wireless nature using electromagnetic waves with a frequency of roughly a few kHz to several THz. RF communications are used for several applications ranging from leisure, meteorology, public broadcast, commercial, domestic and defense communications, and so forth. However, when information is communicated, often there is a need to distinguish among the intended receivers and eavesdroppers, for reasons of secrecy, confidentiality, and integrity, like for requirements from personal privacy, till national security. At present, wireless communications are mostly secured in the top level through mathematic encryption—the Advanced Encryption Standard (AES) and the Rivest–Shamir–Adleman (RSA) methods are some examples of this, and in smaller levels by techniques like range limiting, frequency diversity and similar. Anyhow the prominent mode of security is mathematic-based software encryption.

Now it is apparent that researchers (ethical or unethical) are working hard like the research community in software encryption to find techniques to crack into the information transmitted. Very often a scenario is encountered such that a transmitted spectrum is recorded and is then cracked over time using several reverse engineering

[1]Institute of Sensors, Signals and Systems, School of Engineering and Physical Sciences, Heriot-Watt University, UK
[2]National Key Laboratory of Science and Technology on Communications, University of Electronic Science and Technology of China (UESTC), China

and computational approaches. After all, software encryption is nothing but a complicated mathematical function that, in principle, is always solvable, provided enough time and computational effort is applied to the challenge. This is like eavesdropping an unknown language speaker, recording the conversation, and then transliterating it later with the help of a proficient translator.

The current advances in computing and artificial intelligence, like the advent of quantum computing makes this plot worse as computation power and competence is evolving day by day [1–4].

An added fortress to wireless communication in the shadow of such an impending vulnerability is hardware security, which is interchangeably called physical layer network security. This is synonymous to adding an extra padlock alongside a combination lock that extends the security over time and effort. Thus, physical layer wireless security is a set of hardware control approaches which makes sure that an effective reception of transmitted data constellation (often software encrypted) could be received in the right format only along pre-defined secure directions or locations. The transmission along all other directions or locations is either scrambled or prevented. This is a relevant concern and is backed by the fact that in a conventional antenna-based transmission, even though the directional gain of the antenna can be controlled, the constellation pattern of transmitted data is distributed in free space almost omnidirectionally around it. Thus, a receiver of adequate sensitivity would be able to record the transmitted data even though it is encrypted mathematically.

In this chapter, we broadly classify this concept, discuss several types of reported technologies for physical layer wireless security, and outline their comparisons. Finally, a future perspective of the same is drawn for readers who are interested in this promising research topic.

16.2 Broad classification of physical layer wireless security

Physical layer wireless security could be very broadly classified into two cursory categories, namely, secret key agreement and keyless secure communication. In secret key agreement-based secure communication, a common hardware key is shared among friendly communication nodes. Such a hardware layer secret key is formed by techniques such as exploiting the common randomness of hardware systems [5,6], and similar approaches. Moreover, in this approach, often the secret key is never transmitted among the nodes but is generated or applied locally based on hardware-based mathematical approaches, thus effectively mitigating the abduction of the same.

An alternative and more practical method of physical layer network security is the keyless approach. In keyless hardware-secured communication, a spatial diversity system is used—either to direct the communication beam only to the required receiver and place potential nulls to the eavesdroppers, or to direct the right constellation only to the intended receiver and scramble the constellation along all other directions. This is achieved through a vice management of several techniques including antenna array and beamforming, with or without combined signal processing techniques, involving the addition of direction sensitive artificial noise (AN) into the transmission [7–11].

Among these techniques, the keyless approach is more attractive owing to its receiver independence. This is more a transmitter side technology and is easier to implement and modify on a higher level at lower cost owing to no alteration in the receivers.

At present, the most discussed and researched technique among the keyless physical layer network security is directional modulation (DM) and is reviewed in more detail in this chapter. Like the name indicates DM is a technique in which a data transmission beam is encrypted using hardware techniques, sensitive to spatial directions to make sure that the data constellation is recovered only along the pre-defined legitimate or secure dimensional inclinations. Different DM techniques, its classifications and further discussions are summarized in the following sections.

16.3 Directional modulation (DM)

The term DM was initially coined in [12] as a physical layer wireless security technique. In the early stages of research, this terminology was used for several applications that are, however, different from the scope of physical layer wireless security. Anyhow at present the term DM is often linked to the right application domain and is identified synonymous to hardware-secured network security. DM is a transmitter side technique, which projects digitally modulated constellation signals to a pre-defined spatial direction (the legitimate secure communication direction) while simultaneously scrambling the constellation elsewhere around in free space, as shown in Figure 16.1.

The core idea of achieving DM in a conventional transmitter could be summarized as follows. In most RF transmitter systems, an antenna array-based front end is used

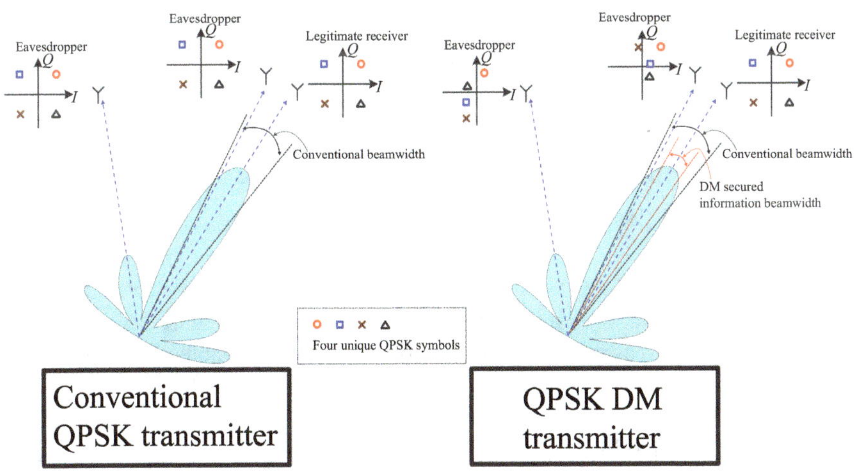

Figure 16.1 Comparison of a physical layer secured DM transmitter to a classic transmitter of similar system architecture except DM. Here QPSK is adopted as an example of digital modulation technique.

for several reasons including achievable gain, spatial directivity, and controllable directional agility. Now if we look at the requirements of DM, the array should be able to project the digitally modulated information into a pre-specified spatially secured direction in free space. The antenna array characteristics that could be used to control a radiation pattern as said herewith could be listed as follows.

- Spatial configuration of antenna elements.
- Locations of excited antenna elements in an array.
- Relative amplitudes of excitation of antenna elements.
- Phases of excitation of antenna elements.
- Radiation pattern of antenna active element patterns.

There are several studies reported around the utilization of these degrees of freedom for DM till date. However, from the authors' point of view, the techniques employing control of relative amplitudes and phases of excitation of antenna elements are more comprehensive and straightforward to implement/upgrade in an existing system (including non-DM conventional transmitters), owing to the flexibility of implementation through a computational logic appendage.

Early form of DM includes several active and passive transmitter architectures. Some notable works include the Walsh waveform-enabled pulse position modulation-enabled DM transmitters [13] that exploit the orthogonal properties of Walsh waveforms for spatially sensitive transmission beams. This was one of the first reported DM-enabled hardware secured wireless transmitter architecture. Other efficient architectures include near field antenna coupling-based systems [14–16], reconfigurable phase shifter integrated active antenna array based [12,17,18], active RF switch array controlled antenna subset modulation based [19], I (in-phase signal) and Q (quadrature signal) segregated transmitter systems [20,21], synthesis-free systems utilizing hardware properties like Fourier lenses [22,23], antenna spatial location based [24], and several others. DM could be achieved in a transmitter through hardware or software-centered approaches. The most relevant of these architectures and techniques are discussed in detail in the following sections.

16.3.1 A general mathematical description of DM with reference to conventional transmission

In a conventional uniform linear array-based transmitter, for a one-dimensional (1D) antenna array of N elements as shown in Figure 16.2, the radiated electric field at a given arbitrary observation point (say along the direction θ in Figure 16.2) in the far field could be defined as in (16.1). Here the operator "$[\bullet]^T$" denotes matrix or vector transpose.

$$\boldsymbol{E}(\boldsymbol{\theta}) = \frac{e^{-j\vec{k}\cdot\vec{r}}}{|\vec{r}|} \begin{bmatrix} \boldsymbol{AP}_1(\boldsymbol{\theta}) \\ \boldsymbol{AP}_2(\boldsymbol{\theta}) \\ \vdots \\ \boldsymbol{AP}_N(\boldsymbol{\theta}) \end{bmatrix}^T \begin{bmatrix} A_1 \cdot e^{-j\vec{k}\cdot\vec{x}_1} \\ A_2 \cdot e^{-j\vec{k}\cdot\vec{x}_2} \\ \vdots \\ A_N \cdot e^{-j\vec{k}\cdot\vec{x}_N} \end{bmatrix} \tag{16.1}$$

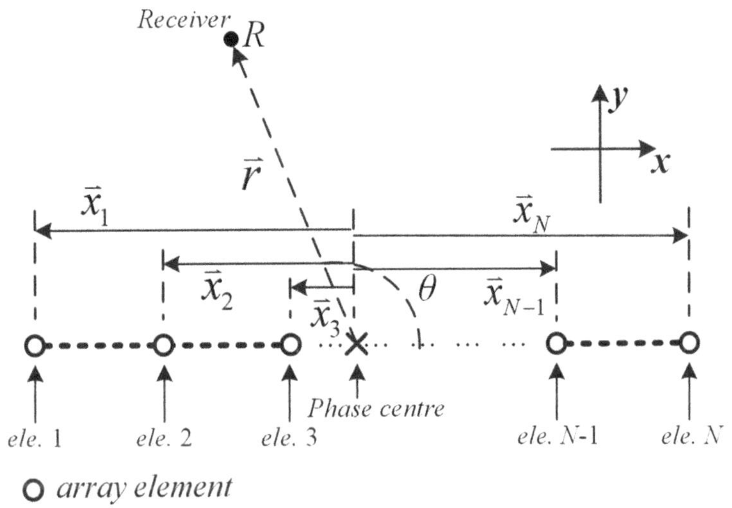

Figure 16.2 An N-element uniform linear array

Here $\vec{x}_n$ and $\vec{r}$ represent the location vectors of the nth array element and the receiver, respectively, relative to the array phase center. And $\vec{k}$ is the wave number vector along the spatial transmission direction, and $\boldsymbol{AP_n}(\theta)$ is the radiation pattern of the nth array element.

Now if we assume that all the radiators have an identical and isotropic radiation pattern and that the element spacing among adjacent elements in the array (i.e., $|\vec{x}_{n+1} - \vec{x}_n|$) is one-half wavelength, the far field electric field component could be approximated as

$$\boldsymbol{E(\theta)} = \frac{e^{-j\vec{k}\cdot\vec{r}}}{|\vec{r}|}\sum_{n=1}^{N}\left(A_n\cdot e^{j\pi\left(n-\frac{N+1}{2}\right)cos(\theta)}\right) \tag{16.2}$$

In (16.2), the term $\frac{e^{-j\vec{k}\cdot\vec{r}}}{|\vec{r}|}$ is a constant that scales, in magnitude and phase, the electric field in any direction and hence could be ignored. Thus, the instantaneous electric filed in any given direction θ could be described as

$$\boldsymbol{E(\theta)} = \sum_{n=1}^{N}\left(A_n\cdot e^{j\pi\left(n-\frac{N+1}{2}\right)cos(\theta)}\right) \tag{16.3}$$

With reference to (16.3) and Figure 16.3, commonly in a conventional non-DM transmitter, all antenna elements are actively excited, and the information to be transmitted is typically modulated to a digital baseband format, and then upconverted using a mixer stage to the desired transmission carrier frequency. This upconverted RF signal stream is then fed to a beamformer network and transferred to the connected antenna array with a chosen beamforming parameter. Here $\boldsymbol{D_m}$ is the modulated

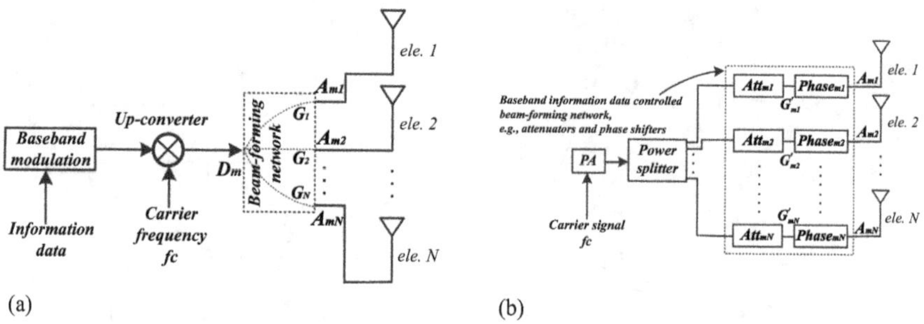

Figure 16.3 (a) A conventional non-DM transmitter array. (b) An active DM transmitter array system consisting of baseband data-controlled RF attenuators and phase shifters.

complex data symbol and $\boldsymbol{G_n}$ for each branch n is the complex gain of the beamformer and is usually a constant with respect to the modulation rate. Thus, the radiated far field electric field $\boldsymbol{E_m}$ is a scaled version with a complex weight $\boldsymbol{M}$ relative of the symbol $\boldsymbol{D_m}$, in the given direction θ (here the subscript m corresponds to the mth symbol in the data stream), as shown in (16.4). Note here that the operator "(•)*" denotes complex conjugation.

$$\begin{aligned}\boldsymbol{E_m}(\theta) &= \sum_{n=1}^{N}\left(\underbrace{\boldsymbol{D_m}\cdot\boldsymbol{G_n^*}}_{A_{mn}}\cdot e^{j\pi\left(n-\frac{N+1}{2}\right)cos(\theta)}\right)\\ &= \boldsymbol{D_m}\cdot\underbrace{\sum_{n=1}^{N}\left(\boldsymbol{G_n^*}\cdot e^{j\pi\left(n-\frac{N+1}{2}\right)cos(\theta)}\right)}_{M}\end{aligned} \tag{16.4}$$

This scaled nature of the electric field in a conventional transmission allows a receiver of adequate sensitivity to capture and recover the symbol constellations in the right format even though the radiation pattern gain may be suppressed along this direction. In essence, the only difference in the radiated electric field along a different direction is a magnitude scaling, and a phase rotation due to distance of wave travel and channel conditions, which could be easily corrected through standard reception practices.

Now if we think in the DM perspective, what is required is a direction-dependent signal format transmission, thus limiting the transmission of right constellation pattern only along the chosen secure communication direction and scrambling elsewhere. An astute solution to this problem is varying the $\boldsymbol{G_n}$, i.e., the $\boldsymbol{M}$ in (16.4) for each symbol to secure the $\boldsymbol{E_m}$ from being just a scaled version of $\boldsymbol{D_m}$ in every direction. Thus, the new $\boldsymbol{G_n}$ and $\boldsymbol{M}$ should be denoted as $\boldsymbol{G_{mn}}$ and $\boldsymbol{M_m}$, respectively, updated at the modulation rate (i.e., for each symbol m). This could be represented mathematically as

$$\boldsymbol{E_m}(\theta) = \sum_{n=1}^{N} \left(\boldsymbol{D_m G^*_{mn}} \cdot e^{j\pi\left(n-\frac{N+1}{2}\right)cos(\theta)} \right) \tag{16.5}$$

$$\boldsymbol{E_m}(\theta) = \begin{bmatrix} \boldsymbol{D_m G^*_{m1}} \\ \boldsymbol{D_m G^*_{m2}} \\ \vdots \\ \boldsymbol{D_m G^*_{mN}} \end{bmatrix}^T \begin{bmatrix} 1 \cdot e^{j\pi\left(1-\frac{N+1}{2}\right)cos(\theta)} \\ 1 \cdot e^{j\pi\left(2-\frac{N+1}{2}\right)cos(\theta)} \\ \vdots \\ 1 \cdot e^{j\pi\left(N-\frac{N+1}{2}\right)cos(\theta)} \end{bmatrix} \tag{16.6}$$

Thus, in a DM transmitter similar to as illustrated in Figure 16.3(b), the carrier signal is split using a power splitter to N branches corresponding to the number of antenna elements in the array. These branches each are then controlled using amplitude and phase weights A_{mn} and $Phase_{mn}$, respectively, which collectively form the complex gain $\boldsymbol{G_{mn}}$. Here m corresponds to the mth transmitted signal symbol ($m = 1, 2, 3 \ldots, M$) and n refers to the nth array element.

16.3.2 Software-centered approaches for DM

DM approaches based on mathematical synthesis of relative amplitudes and phases of excitation of antenna elements, relative to the baseband information to be transmitted, and that could be achieved by software processing the inputs (carrier wave or baseband information) are referred to in this chapter as software-centered approaches. The advantage of this approach is that it could be easily integrated to an existing system with little extra investment, most often with the help of some additional computing facilities like an MCU or FPGA. Future upgradations or modifications using this approach is also facile. Some state-of-the-art software-centered approaches to achieve DM in conventional RF transmitters are discussed here.

16.3.2.1 Orthogonal vector (OV) synthesis-based directional modulation (DM) techniques

In this section a unique and versatile approach to DM through orthogonal vector (OV) synthesis is discussed in detail. The basic idea of OV-based DM is to radiate some extra amount of energy through a conventional array-based transmitter such that the radiated OV waveforms (a complex number—with an arbitrary magnitude and phase) are orthogonal to the chosen secure direction and cancels out, and, at the same time, they combine with the symbol information in other directions in a random fashion [25,26].

The free space channel vector along any arbitrary spatial angle θ for an N-element uniform linear array as illustrated in Figure 16.2, represented as $\vec{\boldsymbol{H}}_{\boldsymbol{\theta}}$ is defined as in (16.7):

$$\vec{\boldsymbol{H}}_{\boldsymbol{\theta}} = \left[e^{jkx_1 cos(\theta)} \; e^{jkx_2 cos(\theta)} \; e^{jkx_3 cos(\theta)} \; \ldots \; e^{jkx_N cos(\theta)} \right]^T \tag{16.7}$$

$$\vec{\boldsymbol{\Lambda}} = \frac{1}{\sqrt{N}}\vec{\boldsymbol{H}}_{\theta_0}\boldsymbol{D}_m \tag{16.8}$$

In the conventional non-DM transmission, an information symbol $\boldsymbol{D}_m$ to be transmitted is scaled across the channel vector for a chosen communication direction θ_0 and the resulting excitation vector $\vec{\boldsymbol{\Lambda}}$, given in (16.8), is fed to the array. This excitation vector is radiated through the array and combines in free space. This is equivalent to multiplication by the conjugated channel, i.e., $\vec{\boldsymbol{H}}_{\theta_0^*}$, achieving a beamforming gain of $10log_{10}(N)$. At the reception side, this information could be received as

$$\vec{\boldsymbol{R}}_{\theta_0} = \vec{\boldsymbol{H}}^*_{\theta_0}\vec{\boldsymbol{\Lambda}} = \frac{1}{\sqrt{N}}\vec{\boldsymbol{H}}^{\dagger}_{\theta_0}\vec{\boldsymbol{H}}_{\theta_0}\boldsymbol{D}_m = \sqrt{N}\boldsymbol{D}_m + \sigma \tag{16.9}$$

Here σ is the added complex channel noise and the operator "$[\bullet]^{\dagger}$" is the Hermitian operator. Thus here, indisputably along all spatial directions, the signals can be calculated as the dot product of the excitation vector and the conjugated channel vector. The only factor that affects these signals is the radiation pattern of the array which is an attribute of the active element patterns. For simplicity, we consider these to be isotropic in this analysis.

Now as indicated above, such a transmission is not spatially secure and could be received around the radiation aperture of the array, with any receiver (legitimate or eavesdropping) having adequate sensitivity adaptive to the radiation pattern intensity. This happens due to the fact that often the product $\vec{\boldsymbol{H}}^{\dagger}_{\theta}\vec{\boldsymbol{H}}_{\theta_0}$ is some value other than zero.

One DM technique to mitigate this radiation leakage is the OV-based synthesis [25,26]. Here an additional vector $\vec{W}$ is added to the excitation vector $\vec{\boldsymbol{\Lambda}}$ prior to feeding this new secure excitation vector $\vec{S}$ to the array elements. This vector $\vec{W}$ is uniquely synthesized such that it is orthogonal to the channel vector along the chosen secure communication direction as shown in (16.10) and (16.11):

$$\vec{\boldsymbol{S}} = \vec{\boldsymbol{\Lambda}} + \vec{\boldsymbol{W}} \tag{16.10}$$

$$\vec{\boldsymbol{W}} = \frac{1}{N-1}\sum_{p=1}^{N-1}\left(\vec{\boldsymbol{Q}}_p v_p\right) \tag{16.11}$$

Here, $\vec{\boldsymbol{W}}$ is chosen from the null space of the channel vector $\vec{\boldsymbol{H}}^*_{\theta_0}$ as shown in (16.11). $\vec{\boldsymbol{Q}}_p$ is the orthonormal basis to the null space of $\vec{\boldsymbol{H}}^*_{\theta_0}$. This orthonormal basis is scaled using a variable for each symbol, to introduce some degree of uniqueness using a random number v_p with a given statistical distribution [27].

Now the physical mechanism for OV-based DM is discussed for a better understanding of the readers. From the receiver point of view in a signal processing perspective, the reception of the transmission facilitated by the conventional excitation vector $\vec{\boldsymbol{\Lambda}}$ in (16.9), along the chosen communication direction θ_0 could be represented vectorially as

$$\boldsymbol{X}_n = \boldsymbol{H}^*_{\theta_0}(n)\boldsymbol{\Lambda}_n = \frac{\boldsymbol{D}_m}{\sqrt{N}} \tag{16.12}$$

In (16.12), $\boldsymbol{H}_{\theta_0}^*(n)$ and $\boldsymbol{\Lambda}_n$ are correspondingly the nth elements of the vectors $\vec{\boldsymbol{H}}_{\theta_0}^*$ and $\vec{\boldsymbol{\Lambda}}$. Similarly, in the DM perspective given in (16.10), the reception along any arbitrary direction could be expressed as (16.13):

$$\boldsymbol{R}_\theta = \vec{\boldsymbol{H}}_\theta^\dagger \vec{\boldsymbol{S}} = \vec{\boldsymbol{H}}_\theta^\dagger \vec{\boldsymbol{\Lambda}} + \vec{\boldsymbol{H}}_\theta^\dagger \vec{\boldsymbol{W}} = \frac{1}{\sqrt{N}} \vec{\boldsymbol{H}}_\theta^\dagger \vec{\boldsymbol{H}}_{\theta_0} \boldsymbol{D}_m + \vec{\boldsymbol{H}}_\theta^\dagger \vec{\boldsymbol{W}} \tag{16.13}$$

In (16.13), along any directions other than the secure communication direction θ_0, the term $\vec{\boldsymbol{H}}_\theta^\dagger \vec{\boldsymbol{W}}$ is a non-zero complex number and thus interferes with the right reception of the intended constellation pattern. This technique also makes sure that the received constellation pattern along two arbitrary directions (other than the secure communication direction) are also different owing to the fact that in the second term $\vec{\boldsymbol{H}}_\theta^\dagger \vec{\boldsymbol{W}}$, the channel vector $\vec{\boldsymbol{H}}_\theta^\dagger$ is different for different θ. However, along the secure communication direction θ_0, this added energy term tends to be zero, as $\vec{\boldsymbol{W}}$ is in the null space of $\vec{\boldsymbol{H}}_{\theta_0}^\dagger$ ensuring proper reception of the transmitted constellation patterns. Now from the receiver's point of view, the reception could be represented vectorially along the secure communication direction as

$$\boldsymbol{Y}_n = \boldsymbol{H}_{\theta_0}^*(n) \boldsymbol{S}_n = \frac{\boldsymbol{D}_m}{\sqrt{N}} \tag{16.14}$$

On comparing (16.13) and (16.14), $\boldsymbol{X}_n$ and $\boldsymbol{Y}_n$ look alike. However, this only means that they converge to the same point. Here in fact, $\boldsymbol{Y}_n$ carries more energy and travels a longer vector path than $\boldsymbol{X}_n$ to converge to the point $\frac{\boldsymbol{D}_m}{\sqrt{N}}$, as here the excitation vector $\vec{\boldsymbol{S}}$ is doped with more energy in the form of the orthogonal "Artificial Noise" as shown in (16.10) and (16.11). This concept is pictorially depicted in Figure 16.4.

Figure 16.4 gives a clear perception of the physical mechanism of OV-based DM explained herewith. Now from this explanation, we could also define two classes of OV-based DM—known as static OV-DM and dynamic OV-DM.

The implemented DM is termed "static" when the updating process of the $\vec{\boldsymbol{W}}$ is static with respect to time, and "dynamic" when the updating rate is concordant to the symbol rate [26]. In the static DM case, usually a $\vec{\boldsymbol{W}}$, different to each unique symbol, is fixed in the entire signal transmission. Whereas in dynamic DM, a new $\vec{\boldsymbol{W}}$ is synthesized for each transmitted symbol in the information trail. Figures 16.5, 16.6, and 16.7 show, respectively, the far field magnitude and phase patterns correspondingly for a conventional, static OV DM, and dynamic OV DM transmissions, for 100 random QPSK symbols, using a uniform linear array with five elements. It is assumed the main beam (either radiation or information) is steered to 45°.

In Figure 16.5, one could see that for all the random symbols, the magnitude is conformal to the same beam shape along the spatial angular domain, and only the phase varies for each of QPSK symbols as expected. Whereas in Figure 16.6 for the static DM case, the right phase difference and magnitude to distinguish the four unique QPSK symbols are retained only along the secure communication direction 45°. However, one could notice here that this scrambling is static with time and could be decoded by spatially distributed multiple receivers eavesdropping strategy. Now in contrast to both of these techniques, in Figure 16.7, the transmission security is unique, such that the proper constellation of QPSK symbols is maintained only along

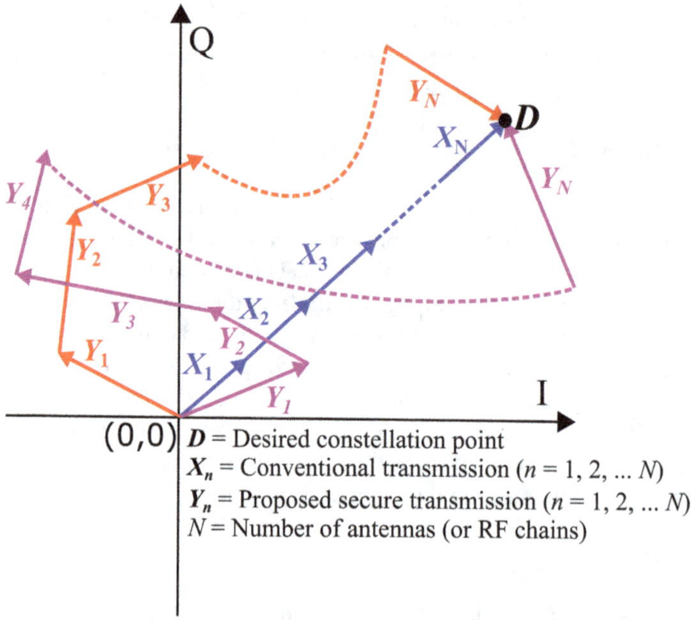

Figure 16.4 Vector representation of (16.12) and (16.14) in the IQ plane, in a receiver point of view in a signal processing perspective. This depiction shows the vector path for a conventional reception in accordance with (16.12) in blue trace, and example of two different vector path for OV DM-enabled transmission on similar hardware in red and pink traces as described in (16.14).

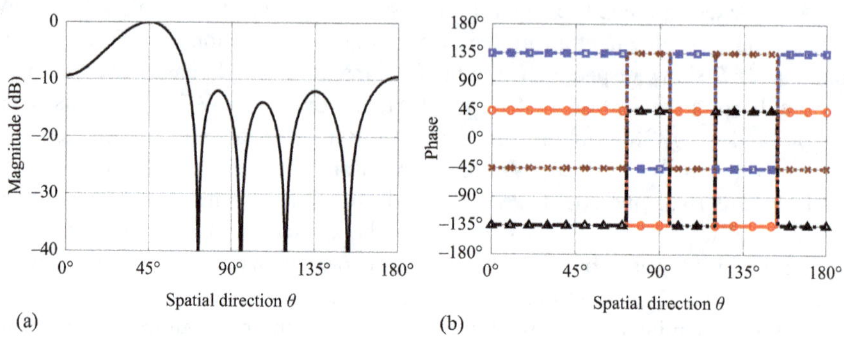

Figure 16.5 Far-field (a) magnitude and (b) phase patterns for 100 random QPSK symbol transmissions using a conventional uniform linear array with 5 elements and the main beam steered to 45°. © 2014 IEEE. Reprinted, with permission, from [26]. Doi: 10.1109/TAP.2013.2287001.

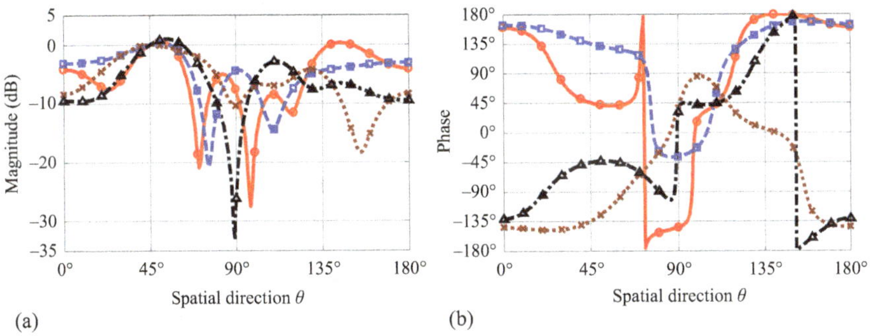

Figure 16.6 Far-field (a) magnitude and (b) phase patterns for 100 random QPSK symbol transmissions using a static DM-enabled uniform linear array with five elements and the main beam steered to 45°. © 2014 IEEE. Reprinted, with permission, from [26]. Doi: 10.1109/TAP.2013.2287001.

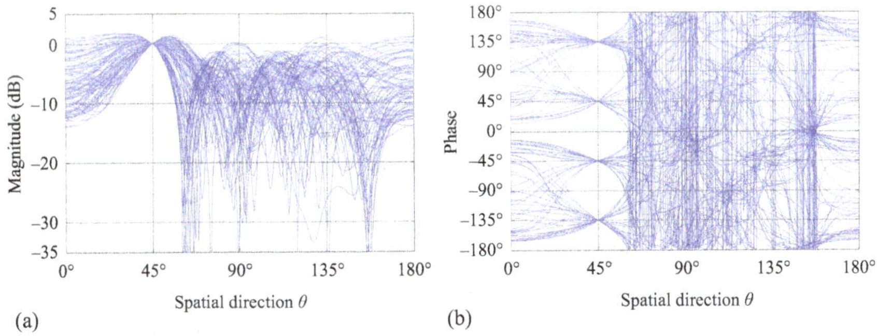

Figure 16.7 Far-field (a) magnitude and (b) phase patterns for 100 random QPSK symbol transmissions using a dynamic DM-enabled uniform linear array with five elements and the main beam steered to 45°. © 2014 IEEE. Reprinted, with permission, from [26]. Doi: 10.1109/TAP.2013.2287001.

45° and both magnitude and phase are scrambled elsewhere at the modulation symbol rate, offering utter chaos for eavesdropping receivers.

The OV-based DM approach could be applied to any type of digital transmissions including constant-envelope techniques like BPSK, QPSK etc., and non-uniform envelope techniques like M-QAM and APSK. The authors of [25] show the application of this technique for multi-beam independent transmissions, for multiuser scenarios. Here spatially secured simultaneous independent data streams modulated to QPSK and BPSK are shown to be transmitted using a same array-based transmitter hardware, and received securely only in chosen directions, while scrambling elsewhere.

16.3.2.2 Time-modulated array-based directional modulation techniques

First proposed in [28], time-modulated array (TMA), also known as four-dimensional (4D) antenna array, explores time as an additional dimension to provide an extra degree of freedom to conventional phased array antennas. Time modulation can be realized, for example, by periodically connecting and disconnecting selected array elements to the signal source. This is often achieved with the help of high-speed RF switches. This periodical signal used to control the switches, usually modeled as time domain square wave, expands its harmonic components to the whole frequency domain following coefficients in the Fourier series. By manipulating this additional time dimension, the radiated power can be concentrated to certain direction (beam steering) and significantly low sidelobe levels (SLLs) can be achieved [28–30].

Notably in physical layer network security aspects, TMA is an elegant candidate to extend DM technology from single-carrier transmission to orthogonal frequency division multiplexing (OFDM). An artificial interference could be generated by optimizing the switching waveform (i.e., ON/OFF waveforms used to control if each antenna is activated or not) to match the frequency spacing among two consecutive OFDM subcarriers [31]. To illustrate, the harmonic frequency components of the periodic signal will shift the original OFDM signal, and the shifted signals will overlap and add constructively at certain directions and destructively at others. Thus, DM functionality is achieved. The proposed transmitter structure in [31] is redrawn in Figure 16.8. In this architecture, the N-element linear antenna array-based transmitter is constructed by connecting the array to a fixed equal-splitting power divider (FPD), and a bank of RF switches. Each RF switch is connected to one antenna element to provide excitation control for time modulation, which can be described as a periodic square wave function $U_n(t)$, where n refers to the nth antenna element ($n = 1, 2, \ldots, N$). The phase shifter of value φ_n is applied to the split OFDM signal at the nth antenna branch for beamforming purpose, similar to those in conventional phased arrays.

Similarly, another DM transmitter configuration was proposed in [32], which, instead of using single pole single throw (SPST) switches, employs single pole triple throw (SP3T) switches. This SP3T switch has three states, namely "ON," "OFF," and "flipping." Besides, only a single RF carrier is injected to the beamforming networks. The OFDM signal is synthesized with the SP3T switching waveform. However, one disadvantage of the proposed OFDM DM transmitter in [31,32] is their poor energy efficiency, which originates from the FPD. To elaborate this, when the switch is at "OFF" state, the power split to the corresponding antenna branch will be absorbed by this RF switch, which inevitably causes power loss. The shorter the proportion of time for "ON" state in one period, the poorer the energy efficiency of the transmitter would be. An approach to improve the energy efficiency of such TMA-based DM was introduced in [33,34], where the absorptive RF switches and FPD were replaced by reconfigurable power dividers (RPDs) proposed in [35,36]. The RPD design uses p–i–n diodes as switches to control power flow and it has an additional impedance converter network to ensure impedance matching. This type of RPD can be configured into different modes and states, which correspond to different number of activated

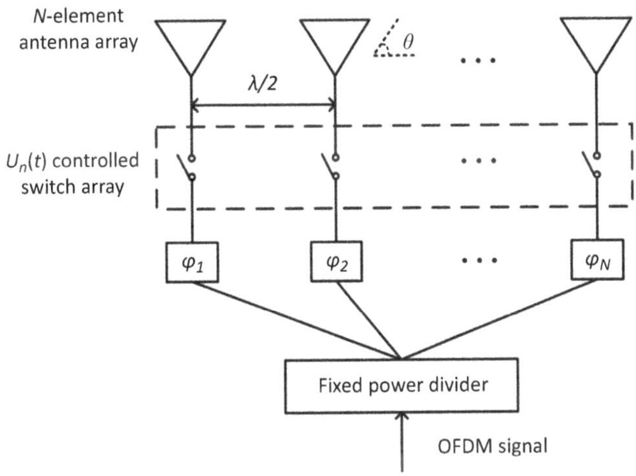

Figure 16.8 Block diagram of time-modulated OFDM transmitter; adapted and redrawn from [31]

antenna elements and their combinations. The RPDs are hence able to recycle the power at deactivated antenna branches, instead of absorbing the same, and then reroute this power to other active antenna branches. As a consequence, the TMA efficiency can be enhanced. For example, around 34% power efficiency increase, or equivalently 2.5 dB power gain, was obtained in [33]. A step further, our research group in Heriot-Watt University has successfully combined works in [31,33,34], and achieving an RPD-enabled efficiency-improved TMA OFDM DM transmitter. In our work, employing a 4-way RPD, a power gain of 7.8 dB, 3.7 dB, and 1.5 dB is achieved when the number of activated antennas is set to 3, 2, and 1, respectively [37].

16.3.3 Hardware-centered approaches for DM

In the previous Section 16.3.2, algorithm-enabled orthogonality-based DM was explained in detail. In this section, a different perspective of hardware enabled DM is elaborated. Here, emphasis is given on the studies that report physical layer wireless security through DM enabled utilizing the hardware properties of different structures like Fourier lenses, Butler matrices, retro-directive arrays, and several others. In this approach as well, the essential idea of DM is achieved as explained in (16.4) and (16.5) by selecting or updating the complex scaling factor $\boldsymbol{M}$ of the transmitted symbol in a specific fashion, through hardware means.

16.3.3.1 DM utilizing the beam orthogonality of Fourier transform beamforming networks

Fourier transforms are extremely useful in signal processing, and equally in RF and microwave engineering. Fourier transform is a multi-versatile tool for defining, analyzing, and achieving several physical and electromagnetic properties. To mention some, the far-field radiation pattern of an antenna is basically the Fourier transform of the current distribution on the antenna, integrated over its length (or surface whichever applicable). The same idea could be applied to phased arrays. Say for a linear-phased array (for simplicity), the far field radiation pattern (array factor) is the Fourier transform of the current distribution (integrated over the normalized array length). Similarly, it is pretty straightforward to synthesize the array input variables for a given radiation pattern or for a given null steering condition by simply finding the inverse Fourier transform of the same [38].

In addition to this, the use of Fourier transform beamforming networks (FT-BFNs) such as Butler matrices and Fourier Rotman lenses as array feed networks is well known for decades [39–44]. A microwave Fourier lens could be used to provide a unique set of progressive phase shifts and approximately equal amplitudes at the output ports, for a given input excitation. The most attractive feature is that switching the input beam ports results in beam-steering of the connected phased array along a discrete set of spatial directions, and these steered beams are orthogonal to all the other beams generated using the same Fourier lens, owing to the orthogonality of Fourier basis functions definitions [45]. For an $N \times N$ FT-BFN, the transfer function among the nth input port and pth output port is given as

$$\boldsymbol{T}_{mn} = \frac{1}{\sqrt{N}} e^{-j\frac{2\pi}{N}\left(p-\frac{N+1}{2}\right)\left(n-\frac{N+1}{2}\right)} \quad (p, n = 1, 2, \cdots, N) \tag{16.15}$$

Thus, the signal vector $\vec{\boldsymbol{A}}$ at the output ports p of the FT-BFN for an excitation vector $\vec{\boldsymbol{B}}$ at the input ports n could be represented in matrix form as

$$\vec{\boldsymbol{A}} = [\boldsymbol{T}_{pn}]\vec{\boldsymbol{B}} \tag{16.16}$$

The radiation pattern (electric field) $\boldsymbol{E}(\theta)$ of a half-wavelength-spaced uniform linear array with isotropically radiating elements, for θ ranging from 0° to 180° with boresight at 90°, is

$$\boldsymbol{E(\theta)} = \sum_{p=1}^{N} \left[\boldsymbol{A}_m \cdot e^{j\pi\left(p-\frac{N+1}{2}\right)cos(\theta)}\right] \tag{16.17}$$

Now if the above FT-BFN is used to feed this array, the radiation pattern $\boldsymbol{E(\theta)}$ for the given excitation $\boldsymbol{B}_n$ at the nth input port of the FT-BFN can be expressed as

$$\begin{aligned}\boldsymbol{E}_n\boldsymbol{(\theta)} &= \sum_{p=1}^{N} \left[(\boldsymbol{T}_{pn} \cdot \boldsymbol{B}_n) \cdot e^{j\pi\left(p-\frac{N+1}{2}\right)cos(\theta)}\right] \\ &= \sum_{p=1}^{N} \left[\left(\frac{\boldsymbol{B}_n}{\sqrt{N}} \cdot e^{-j\frac{2\pi}{N}\left(p-\frac{N+1}{2}\right)\left(n-\frac{N+1}{2}\right)}\right) \cdot e^{j\pi\left(p-\frac{N+1}{2}\right)cos(\theta)}\right]\end{aligned} \tag{16.18}$$

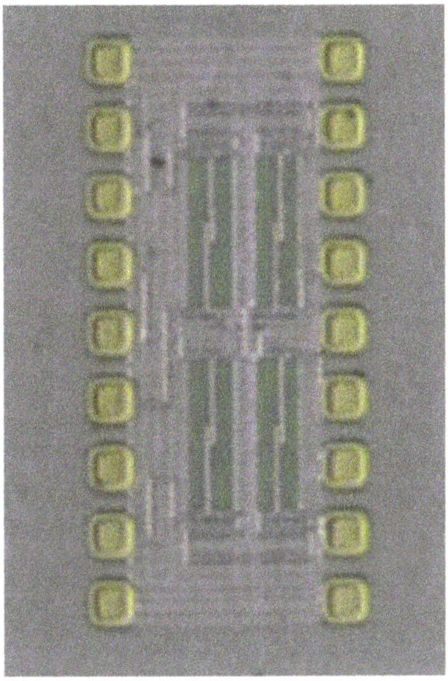

4 x 4 Butler matrix designed for V-band operation via 0.35 μm SiGe bipolar process.

Figure 16.9 Microphotograph of Butler matrix designed for 60 GHz operation [46]

where the main beam pointing direction α_n of $\boldsymbol{E_n}(\theta)$ can be calculated following:

$$-\frac{2\pi}{N}\left(p-\frac{N+1}{2}\right)\left(n-\frac{N+1}{2}\right)+\pi\left(p-\frac{N+1}{2}\right)\cdot cos(\alpha_n)=0$$

$$\Rightarrow cos(\alpha_n)=\frac{2}{N}\left(n-\frac{N+1}{2}\right)=\frac{2n-N-1}{N} \tag{16.19}$$

The electric filed $\boldsymbol{E_q}(\theta)$ pointing toward α_q generated by exciting the qth input port of the FT-BFN could be represented by replacing n in (16.18) and (16.19) with q. By the inherent orthogonality property of FT-BFN, it is always ensured that $\boldsymbol{E_n}(\alpha_q)=0$ when $n \neq q$ [46]. This means that the different radiation patterns generated by the FT-BFN fed antenna array are orthogonal to each other, i.e., each radiation pattern places a null at the beam maxima of every other radiation pattern generated from the same FT-BFN. This orthogonality property of the FT-BFN network hardware is utilized in several studies for achieving synthesis-free DM functionalities. Notable works include [22,23,47]. In [22], a Butler matrix designed for 60 GHz operation, shown in Figure 16.9, is used as the FT-BFN, to feed a 1D array to achieve DM as shown in Figure 16.10.

Two different RF chains are used in this technique, one for injecting the genuine information, connected to the FT-BFN port pointing to the desired secure communication direction θ_0, and the second for generating the scrambling sidelobes. Due

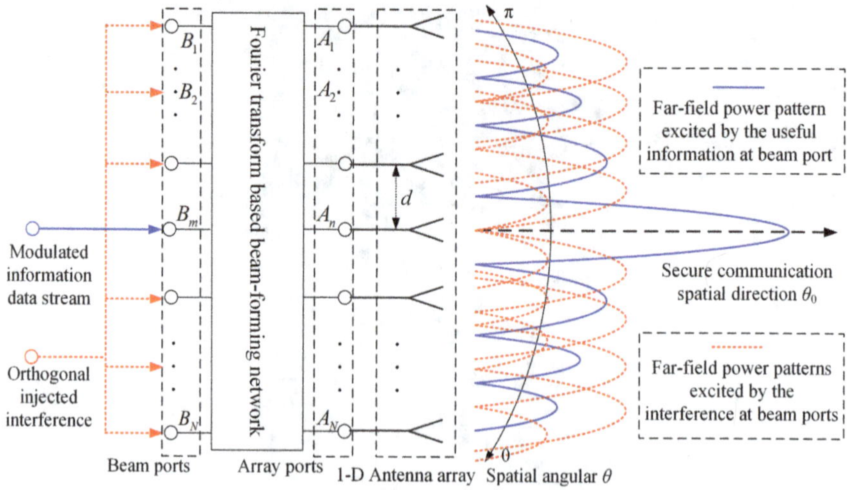

Figure 16.10 FT-BFN-based DM transmitter architecture; adapted and redrawn from [47]

to the orthogonality property of the FT-BFN, the information radiated around θ_0 is unaffected by the interference injected. However, these interference combines in every other direction and thus distorts the signal amplitude and phase, enabling DM. Using this technique, beam steering to a new secure communication direction could be achieved by selecting a new input port on the FT-BFN corresponding to the new direction and using another spare port for the interference injection.

The maximum steering angular resolution that could be achieved using this technique is determined by the number of ports in the FT-BFN. In this regard, one of the limitations faced by the Butler matrices is the complicated design and fabrication process. Thus, DM transmitters with super-fine spatial resolution using Butler matrix hardware are usually very hard to implement practically. Another easy to implement counterpart to Butler matrix is a Fourier Rotman lens-based FT-BFN. In a Fourier Rotman lens-enabled DM, the theory of FT-BFN remains the same with a difference that it is a 2D structure with flexible choice of input and output ports numbers [42,44,48]. In [46], the realization of dynamic DM enabled wireless transmitter based on a 13×13 Fourier Rotman lens FT-BFN was reported. The hardware block diagram and photograph of the implemented hardware are reproduced in Figure 16.11. The video recording of demonstration of DM in this experiment can be accessed in [49].

16.3.3.2 Retro directive array dynamics-based DM

Retro-directive array (RDA) is a well-known concept that enables auto-beam steering transmitter and receiver systems [9,11,50–52]. RDAs are capable of re-transmitting a signal back to the location of a received pilot, without an a priori knowledge of the direction of arrival (DoA) of the signal and without utilizing any signal processing.

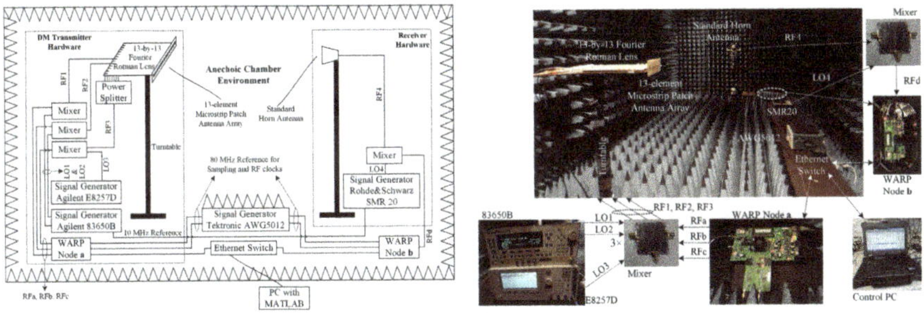

(a) (b)

Figure 16.11 *(a) Hardware block diagram of 13 × 13 Fourier Rotman lens-based DM transmission experiment, adapted and redrawn from [46]. (b) Photograph of implemented hardware [46].*

RDAs could achieve this despite the presence of any spatial or temporal variations in the propagation path. An RDA samples a received pilot tone and conjugates its phase with the help of phase-conjugating networks and re-transmits the signal back. Applied phase conjugation automatically steers the array beam to the location where the pilot signal is originated. Among this process, the conjugated signals associated with each antenna could be encoded with identical data streams. This information is then spatially combined during the re-transmission.

The hardware security property of RDAs has been discussed and demonstrated in several studies [11,53,54]. However, such a spatial security feature of an RDA fails if the multipath channel is static with respect to time or loosens if the dynamism of the multipath channel is reduced. In such a scenario, studies have shown that adding DM concepts to an RDA ensures an extra layer of protection to the spatial security offered by the RDA [46,55–58]. Often at times, it also demonstrates that such a DM concept association with an RDA offers a better security than both of these concepts could offer alone. The authors of [46] show the security enhancement of and RDA performance using OV-DM technique. Here a seven-element half-wavelength-spaced linear RDA in a static multipath environment is DM secured, by synthesizing secure excitations based on OVs generated from the channel vector corresponding to received pilot tone. The RDA and the static multipath environment, composed of a two-ray model, is depicted in Figure 16.12.

Here the pilot tone is generated from an intended receiver placed 100 wavelengths (λ) along the boresight (θ_1) of the RDA, the second reflected path of the pilot tone is generated from an ideal electromagnetic reflector placed perpendicular to the array axis at a distance of 100λ from the array phase center (also the geometric center) and the pilot location (receiver). The reflector thus reflects the pilot tone originating from the intended receiver to generate a new quasi-pilot tone that incidents at an angle θ_2 = 150° to the RDA. OV DM can be applied to the channel vector of the RDA owing to the reception of the pilot tone and its reflection, similar to as explained in

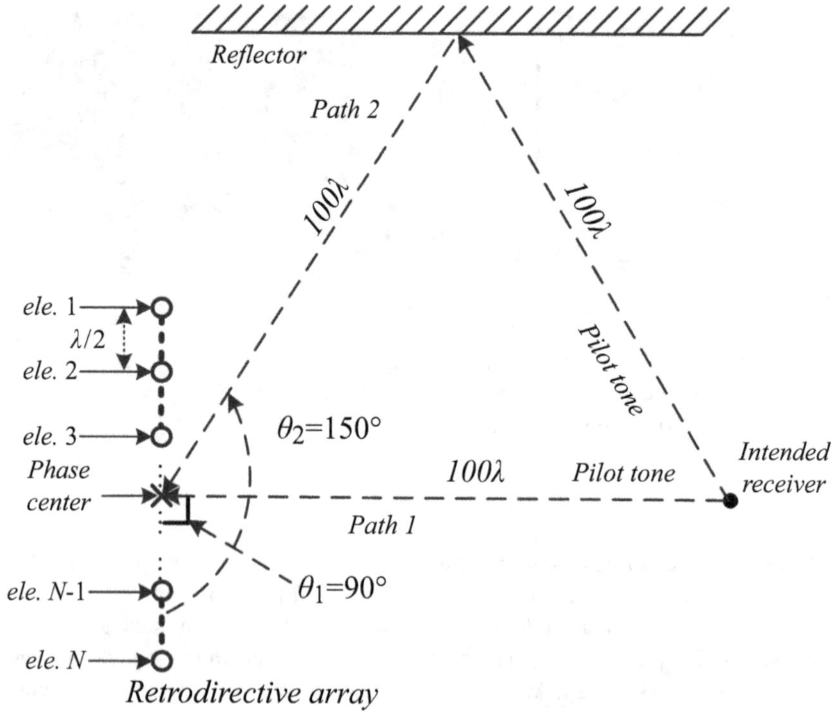

Figure 16.12 Depiction of a seven-element half wavelength-spaced linear RDA in a two-ray static multipath environment model

Section 16.3.2.1. Now Figure 16.13 shows the bit error rate (BER) simulation of the received data along the 0°–180° angular domain for a radial distance of 90λ–200λ from the RDA.

From Figure 16.13, one can see that the error-free zone is limited to a small area around the intended receiver and is a function of the channels signal-to-noise-ratio (SNR) and the amount of OV energy allocated to the applied DM (termed as PE_{DM} and explained in detail in a later Section 16.4.2). This result proves the effectivity of using DM along with RDA techniques for physical layer spatial security. In addition, the authors of [55] present a more mature discussion of the same concept in a better multipath-rich environment with outstanding security features seen in the BER computations. Further, the authors of [56] demonstrate DM through TMA-DM techniques applied to an RDA. This study also shows that this concept works even when there are two legitimate receivers. In summary, RDA-based DM is an effective technique for physical layer network security, as this method extends the traditional free-space DM to multipath environments, and it can automatically adapt to dynamic channels.

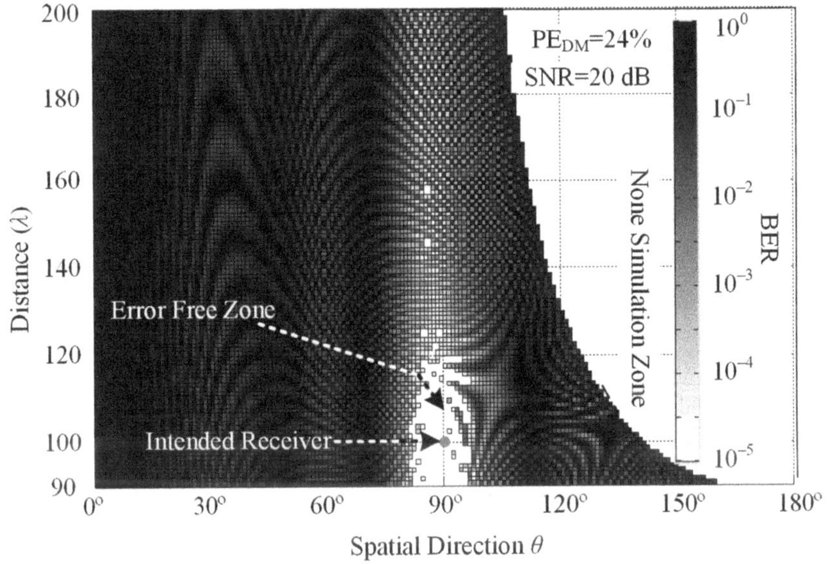

Figure 16.13 BER spatial distributions along the beam-space of the RDA model in Figure 16.12, with OV-based DM enabled

16.3.3.3 Antenna hardware dynamics-based DM techniques

There are a multitude of versatile and genius techniques reported in the past and present decades for achieving physical layer network security through DM based on intelligent rearrangement of antenna or antenna array hardware dynamics. These include controlling the number and position of active radiators in an antenna array with redundant number of elements, controlling the phase shift of active radiators, modifying the near field radiating boundary conditions of antenna elements, and similar. Indeed, different parameters are controlled in different studies. However, what brings them under one-umbrella in this section is the fact that in all the systems grouped herewith, the modulation is taking place at the RF front end stage and not in the baseband, and, in most cases, the classic baseband section in the RF chain start is replaced by a simple continuous wave (CW) local oscillator [12,14,18,19,59–61]. Notable works discussed here include the antenna subset modulation [19] and near-field direct antenna modulation [14].

The authors of [19] report an antenna array hardware-based DM technique titled as antenna subset modulation (ASM), at millimeter wave frequencies. The block diagram of the hardware used in this study is depicted in Figure 16.14. In this transmitter, there are M RF chains and N antennas such that $N > M$. A CW local oscillator producing a sinusoid connects to the antenna array through an array of high-speed RF switches. The CW oscillator is split to M chains each passing through a phase shifter and an RF power amplifier (PA). Now, transmission is done here by connecting the M

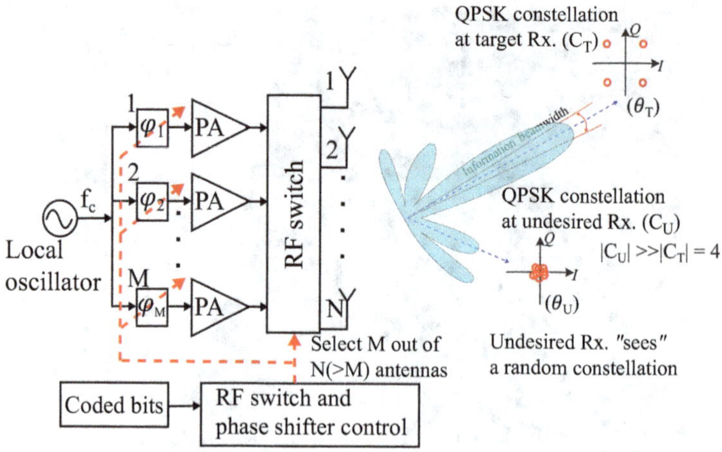

Figure 16.14 Hardware block diagram of antenna subset modulation (ASM)-based DM transmitter; adapted and redrawn from [19]

RF chains to M antennas out of the N array elements, in a synthesized fashion, at the modulation rate to achieve the DM functionality. Here the baseband-coded bits data is used to control the phase shifters and the RF switch simultaneously, according to the synthesized order. Thus, the modulation takes place at the RF end and not in the baseband part. Through this technique, the selected M antennas beam form in the far field to produce the right magnitude and phase in the pre-defined secure communication direction θ_0, and is scrambled, or combined to exhibit a chaotic magnitude and phase elsewhere. This scenario is in some limited sense similar to the vector mapping shown in Figure 16.4, where the different subsets of antennas each time choose a different vector path to reach the constellation point along the secure communication, and this path combination happens to be out of tune in other unsecure directions. ASM technique applies only to uniform envelope modulation techniques like QPSK and BPSK.

A similar DM technique in which the baseband modulation is done directly at the antenna array level was reported in [14]. The technique in this study is titled as near-field direct antenna modulation (NFDAM). Here, antenna arrays are constructed using unique antenna base elements, and each antenna element is loaded with a large number of passive reflectors in its nearfield boundary. These reflectors are exploited to reconfigure its electrical length using several high-speed switches as shown in Figure 16.15. When a base antenna element is considered for analysis, different far field magnitude and phase patterns could be created for a CW feed by controlling the switches. These far-field patterns could be translated to different constellation points in the in-phase and quadrature (IQ) plane, detected along each angular direction along the beam space of the antenna. The key here is that the IQ constellation point for a particular switch configuration is different along different angular directions. Thus,

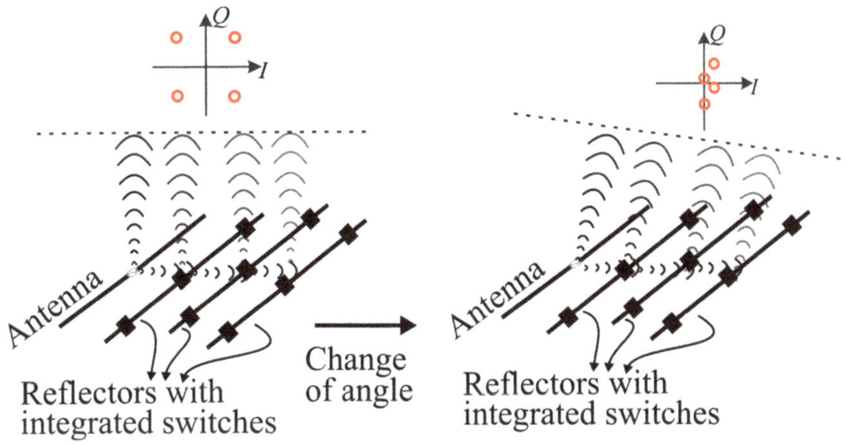

Figure 16.15 Concept of DM through near-field direct antenna modulation; adapted and redrawn from [14]

for arrays composed with such loaded elements, a large number of IQ points could be generated for each angular direction, by selecting all possible combinations of a large number of switches. From this switch combination, a subset could be selected for each direction for a given constellation pattern. The achievable degrees of freedom or the size of this subset is a function of the antenna near-field arrangement and the number of switches embedded. This technique is straight forward, but complicated to implement due to the exhaustive requirement of computation of all possible switch combinations beforehand. This complexity increases with the number of switches or loaded antenna elements in the array. The technique was originally demonstrated for millimeter wave frequencies, but conceptually is applicable to microwave frequencies as well, provided suitable hardware downscaling is achieved.

16.3.4 Emerging DM techniques for low-cost and high-efficiency applications

So far, we have discussed several techniques to achieve DM through both software and hardware methods. However, essentially, this is only a rough grouping to identify how in principle the DM functionality is achieved in a given communication system through several available techniques. In practical application scenarios, very often the best performance is achieved in a system is through an ingenious combination of both hardware and software approaches [62–67]. Some such applications, clearly outlining the interdependence of the two broad areas discussed above are discussed herewith. This is around the adaptation of DM functionality through hybrid techniques for low cost and compact application like for emerging Internet of Things (IoT) networks and/or subsystems.

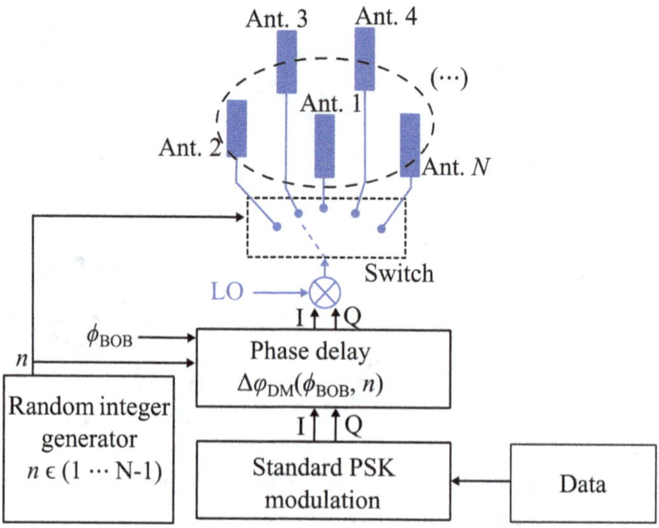

Figure 16.16 Block diagram of energy-efficient DM system with electrically small antennas; reported in [63] (adapted and redrawn)

Narbudowicz *et al.* in [63] present a DM transmitter technique, with similarities to the TMA method. This system uses a single radio frequency chain connected at a time to a single antenna of a uniform circular array of monopole radiators, as shown in Figure 16.16. The radiating antenna among the array is selected at random and the excitation is applied to this antenna, scaled with a phase shift $\Delta\varphi_{DM}$ corresponding to the chosen receiver location and the random antenna number as in (16.20):

$$\Delta\varphi_{DM} = -phase(P_n(\phi_{BOB})) \tag{16.20}$$

In (16.20) $P_n(\phi)$ is the radiation pattern of the nth antenna and ϕ_{BOB} is the direction of the legitimate receiver. This way the correct constellation phase is preserved only along the chosen communication direction and the system does not produce any artificial noise, preventing any side-lobes or interference to other systems. This technique was practically demonstrated on an array of diameter of 0.6 wavelength and has additional advantage that it is suitable for space and power-constrained compact devices, and here the control switching sequence could be generated without the requirement of a well-calibrated synchronization with the transmitter, and at random.

Next, the authors of [62] proposed DM, using a compact switchless reconfigurable two-port antenna which was also reported in [68]. This is a considerable achievement using a compact antenna in place of an extensive antenna array, and this is demonstrated for the first time according to the authors and is suitable for compact gadgets including for IoT applications. This antenna has two ports that produce two unique radiation patterns, where the phase is constant over the spatial angle for one of the patterns and constantly rotating along the spatial domain for the second. Here these antenna properties are combined in a simple relation to achieve directional modulation, without the need for any complex optimizations. In essence here, the I and Q components of the symbol to be transmitted are radiated separately using the two ports, applying a scaling factor which is a function of the radiation pattern of the individual ports and the intended secure transmission angle. 8-PSK modulation was demonstrated herewith, and this technique achieves DM with decent secrecy. Then, DM using a compact multiport antenna, which is basically a modification of the technique explained in [62], is given in [64]. Here a four-port-stacked circular patch antenna is used for this technique, with orthogonal radiation patterns for excitation among the ports, owing to four radiating modes. This technique reports a front to back ratio of at least 10 dB and 360° spatial steering. The basic OV DM theory in [26] is used to explain this technique. The authors demonstrate DM through a transmission of QPSK-modulated data symbols, using this compact multiport antenna with an overall diameter of 0.7 wavelength. The authors claimed in this study that the presented compact antenna with a significantly reduced size can provide equivalent performance to a conventional four-element array with a spacing of half-wavelength.

Another simple and low-cost technique to achieve physical layer security in an existing wireless system is through the use of intelligent reflecting surfaces (IRSs), placed along with the standard transmitter. IRS is a versatile low-cost technique enabling to boost the energy and spectrum efficiencies of wireless communication networks, and is now adapted and optimized to provide DM. The possibility of achieving DM through the use of an IRS, placed alongside a conventional transmitter hardware as shown in Figure 16.17, is reported in [65]. Here the spatial security functionality is achieved through optimization algorithms based on two different techniques, the Rayleigh–Ritz theorem and derivation operation in the former, and a zero-forcing algorithm in the latter. The authors claim that this technique is a good fit to apply the DM functionality to future communications networks including unmanned aerial vehicles (UAVs), satellite, and vehicular communications.

Another important aspect of DM techniques is the possibility of multicarrier transmitters. This way one could increase the efficiency of the transmitter, in terms of power, information, and spectrum. Several techniques have reported till date to achieve these multiuser scenarios. A hint of achieving multiuser functionality by transmitting BPSK and QPSK simultaneously toward two directions is explained in Section 16.3.2.1. Now we look at some practical examples.

Demonstration of multiuser dynamic DM using a simple software-defined radio hardware was reported in [69]. Two uncorrelated signals along two different observation angles are transmitted, showing effectiveness of the concept. One channel transmits QPSK and other 16-QAM. The theory of DM and transmission is based on the channel combination method using pseudo-inverse matrix projection techniques, given in [25]. Similarly, the authors of [70] present multi-beam DM synthesis using zero-forcing synthesis, which is claimed by the authors to be simpler than previous techniques. The zero-forcing method is also closely related to the above-described pseudo-inverse technique.

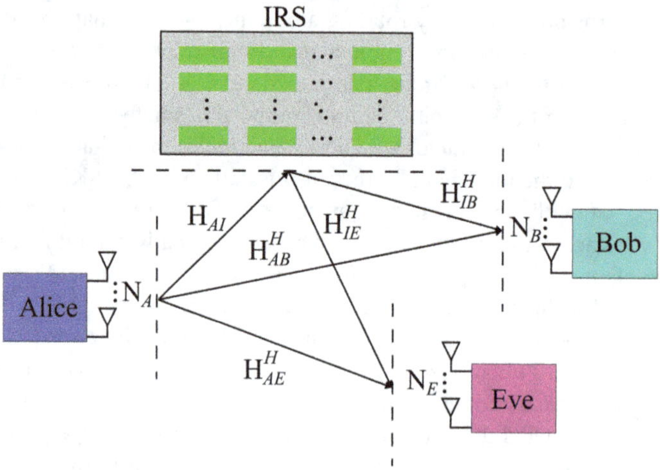

Figure 16.17 Block diagram of IRS-based DM system, reported in [65] (adapted and redrawn)

16.4 Inter-system benchmarking and other considerations

A comprehensive survey of reported DM techniques is summarized in the previous sections. Now it is important to state some bench marking information for inter-comparison of different techniques and for qualitative and quantitative analysis. In the following sub-sections, two relevant aspects are discussed which help toward this deliberation.

16.4.1 Metrics for DM evaluation

Adequate metrics are necessary in any domain to evaluate the effectiveness of a strategy. This is not different for DM systems. Several metrics are reported in open literature till date for assessing the performance of DM systems. These include several gauges like error vector magnitude (EVM) type metrices, BER, secrecy rate, and similar. To pick a universal metric that works well under all constrains is difficult due to the width of the domain, with extensive mathematical and signal processing prospects, as well as emerging counter measures, in physical layer information security perspective. Anyhow, a very agreeable metric often seen popularly in most studies is the BER spatial distribution. The authors of [27,46] outline a comprehensive discussion on the most used metrices and their applicability in static and dynamic DM systems. An in-depth mathematical analysis of the multitude of these techniques established as metrices is very exhaustive and is beyond the scope of this chapter and hence a very brief summary of comparison of these metrices, given in [46], is re-created herewith in a scale of applicability format in Table 16.1.

Table 16.1 *Inter comparison of different metrices to evaluate the effectivity of various DM techniques*

Type of metric		Static DM	Dynamic DM
EVM-based metric		☹	😐
BER	Closed form equation based	😐	😐
	Calculated from data stream	☺	☺
Secrecy rate	Numerical calculus based	☺	😐
	Bit-wise calculations	☹	😐

☹ Metric cannot be calculated or does not work.
😐 Metric works with limited applicability.
☺ Metric works well.

16.4.2 DM power efficiency

In DM, it is obvious that some extra energy is added to the transmission than in a conventional case. It is this energy's cost that enables the information security. This added energy could be in different forms, like the addition of OVs into the excitation as in the OV DM cases [25,26], energy for control of relays, phase shifters or switches, as in the hardware-based approaches based on antenna subset modulation, near-field direct antenna modulation [14,16,19], etc., or other extra hardware processing requirements like in RDA or TMA DM approaches [32,55,57]. If we look closely into these implementations, one could see that the achieved information security quality is always a function of the added extra energy, thus PE_{DM} (DM power efficiency) is defined in (16.21) to facilitate a fair DM system comparison:

$$PE_{DM} = \frac{\sum_{i=1}^{I} \textit{Power transmitted in a conventional transmission}}{\sum_{i=1}^{I} \textit{Power transmitted in DM transmission}} \times 100\% \tag{16.21}$$

Here, I is the number of modulation states (i.e., 4 in QPSK) in static DM or the total number of transmitted symbols in dynamic DM. In a practical application example, PE_{DM} can be computed using (16.22) and (16.23), defined in [26,46] for OV-based DM systems explained in Section 16.3.2.1:

$$PE_{DM\_i} = \frac{1}{1 + \sum_{p=1}^{N-1} \left(\frac{1}{N-1} \cdot v_p\right)^2} \times 100\% \tag{16.22}$$

and

$$PE_{DM} = \frac{1}{I} \cdot \sum_{i=1}^{I} PE_{DM\_i} \tag{16.23}$$

16.5 Conclusion, future perspectives, and other discussions

In this chapter, we have summarized the basic idea of physical layer wireless security, in particular DM techniques. Discussions on the evolution and present state of the art in DM are contemplated herewith. The relevance and demand of physical layer wireless security is increasing at present in the light of rapid evolution in computing in the current decade. This trend can, in the not-too-distant future, compromise the current software encryption-based security. The main objective of physical layer security is to define a well-defined physical transmission boundary for wireless networks, which otherwise are not specifically spatially bounded [71]. DM or physical layer wireless security could literally be an augmented padlock to wireless transmission secrecy.

Some of the potential future directions and the works in progress among our group at present are discussed herewith. As apparent from the prior discussion in this chapter, DM-enabled physical layer security is achieved at the cost of extra energy pumped into the transmission in comparison to a conventional system. In the OV DM case, this energy is in the form of additional OVs added to the symbol excitation. Often there is a scenario in OV DM that this OV data overloads the RF PAs in the chain due to high peak to average power ratio (PAPR). Hence, the output backoff point (OPBO point) of the RF PA must be raised to prevent pushing the PA to nonlinear region than what is used in a conventional case. This leads to the degradation of the entire transmission efficiency. Techniques reported till date to mitigate this effect include the ones similar to adding synthesized digital pre-distortion (DPD) signals into the transmission prior to the RF power amplification for system linearization, peak clipping schemes, and similar [72–75]. However, these methods are computationally exhaustive, or schemes like clipping, compromises signal fidelity.

One technique to lower the PAPR is to normalize the orthogonal vectors, prior to pumping into the system. In this technique, the OVs are synthesized such that the final excitation vector is always mapped to a fixed amplitude. Hence, the magnitude of the OVs is analytically ensured to be the difference of the symbol excitation to be transmitted and the ceiling magnitude set for a given transmission. This idea could be applied without distinction to uniform as well as non-uniform envelope modulation types. Such an approach enables the PAs to work at a constant PAPR and even at 0 dB PAPR, enabling the choice of a lower OPBO point and higher system efficiency, as reported in [76].

Another domain for future work is the realization of physical layer security based on frequency diverse arrays (FDAs). FDA is a technique in which an incremental or decremental propagating frequency shift is added in addition to the propagating phase shift in a conventional antenna array. This additional propagating frequency shift enables the array to focus radiation patterns as a function of range, time, and angle [77]. Some groups have published the utilization of FDA techniques for achieving DM functionalities. However, there are some criticisms and shortcomings on the effectiveness of this approach [78]. A potential future perspective in DM technology would be the investigation of possible mitigation techniques toward these limitations and enabling physical layer security through this concept.

Other emerging techniques that claim to provide efficient DM accomplishments include sparse array-based and metasurface-based solutions [79–84]. Sparse arrays need fewer antenna elements than classic antenna arrays to realize a given aperture or radiation pattern, and this is achieved through advanced numerical techniques like genetic algorithm and convex optimizations. Some research groups have reported techniques that combine sparse array and metasurface to obtain classic spatial angle secure DM, with additional advantage of range or position secrecy, like in [81]. Which the authors call positional modulation (PM) using a combination of a reflecting metasurface and a linear antenna array. Here in order to reduce the number of active elements, a sparsity is introduced in both the antenna array and metasurface. The $l1$ norm minimization technique is used herewith to pursue the design goals of the sparse antenna array and sparse metasurface. These are of course promising directions and assure a great future potential for physical layer wireless security.

Intersectoral research extensions of DM to several other heads like for radar and sensing applications is also a hot topic of discussion at present [85,86]. The emerging machine learning and artificial intelligence (AI)-based optimization concepts have a lot of potential in physical layer security optimizations. For example, in converging to the right type of OV for a balanced OV DM transmission adhering to the channel hostility and pilot location, choosing the right type of switch combination in NFDAM DM techniques, waveform optimizations in TMA DM techniques, and similar.

Acknowledgments

This work was supported by the Engineering and Physical Sciences Research Council (EPSRC), UK, under Grant no. EP/V002635/1.

References

[1] P. Pacheco, *An Introduction to Parallel Programming*. Amsterdam: Elsevier, 2011.

[2] I. A. Contributor, "Intel AI brandvoice: The rise in computing power: Why ubiquitous artificial intelligence is now a reality," *Forbes*. https://www.forbes.com/sites/intelai/201807/17/the-rise-in-computing-power-why-ubiquitous-artificial-intelligence-is-now-a-reality/(accessed Jun. 27, 2022).

[3] C. Ferrie, "Quantum computing for all?," *Nat. Electron.*, vol. 2, no. 3, Art. no. 3, 2019, doi: 10.1038/s41928-019-0223-4.

[4] A. Mehonic and A. J. Kenyon, "Brain-inspired computing needs a master plan," *Nature*, vol. 604, no. 7905, Art. no. 7905, 2022, doi: 10.1038/s41586-021-04362-w.

[5] H. Ahmadi and R. Safavi-Naini, "Secret key establishment over a pair of independent broadcast channels," in *2010 International Symposium on Information Theory & Applications*, Oct. 2010, pp. 185–190, doi: 10.1109/ISITA.2010.5650100.

[6] C. Chen and M. A. Jensen, "Secret key establishment using temporally and spatially correlated wireless channel coefficients," *IEEE Trans. Mob. Comput.*, vol. 10, no. 2, pp. 205–215, 2011, doi: 10.1109/TMC.2010.114.

[7] O. Malyuskin and V. Fusco, "Amplitude modulated signal transmission through a phase conjugating lens," *IET Microw. Antennas Propag.*, vol. 6, no. 9, pp. 1026–1033, 2012, doi: 10.1049/iet-map.2011.0543.

[8] O. Malyuskin and V. Fusco, "Spatial data encryption using phase conjugating lenses," *IEEE Trans. Antennas Propag.*, vol. 60, no. 6, pp. 2913–2920, 2012, doi: 10.1109/TAP.2012.2194661.

[9] N. B. Buchanan and V. Fusco, "Bit error rate performance enhancement of a retrodirective array over a conventional fixed beam array in a dynamic multipath environment," *IEEE Trans. Microw. Theory Tech.*, vol. 58, no. 4, pp. 757–763, 2010, doi: 10.1109/TMTT.2010.2041518.

[10] O. N. Alrabadi and G. F. Pedersen, "Directional space-time modulation: a novel approach for secured wireless communication," in *2012 IEEE International Conference on Communications (ICC)*, Jun. 2012, pp. 3554–3558, doi: 10.1109/ICC.2012.6364574.

[11] V. F. Fusco and N. B. Buchanan, "Retrodirective antenna spatial data protection," *IEEE Antennas Wirel. Propag. Lett.*, vol. 8, pp. 490–493, 2009, doi: 10.1109/LAWP.2009.2021078.

[12] M. P. Daly and J. T. Bernhard, "Directional modulation technique for phased arrays," *IEEE Trans. Antennas Propag.*, vol. 57, no. 9, pp. 2633–2640, 2009, doi: 10.1109/TAP.2009.2027047.

[13] C. M. Elam and D. A. Leavy, "Secure communication using array transmitter," US6275679B1, Aug. 14, 2001 Accessed: Jun. 15, 2022. [Online]. Available: https://patents.google.com/patent/US6275679B1/en

[14] A. Babakhani, D. B. Rutledge, and A. Hajimiri, "Transmitter architectures based on near-field direct antenna modulation," *IEEE J. Solid-State Circuits*, vol. 43, no. 12, pp. 2674–2692, 2008, doi: 10.1109/JSSC.2008.2004864.

[15] A. Babakhani, D. B. Rutledge, and A. Hajimiri, "Near-field direct antenna modulation," *IEEE Microw. Mag.*, vol. 10, no. 1, pp. 36–46, 2009, doi: 10.1109/MMM.2008.930674.

[16] A. H. Chang, A. Babakhani, and A. Hajimiri, "Near-field direct antenna modulation (NFDAM) transmitter at 2.4 GHz," in *2009 IEEE Antennas Propagation Society International Symposium*, Jun. 2009, pp. 1–4, doi: 10.1109/APS.2009.5171550.

[17] D. M.P., "Physical layer encryption using fixed and reconfigurable antennas," PhD Thesis, University of Illinois at Urbana-Champaign, USA, 2013.

[18] M. P. Daly, E. L. Daly, and J. T. Bernhard, "Demonstration of directional modulation using a phased array," *IEEE Trans. Antennas Propag.*, vol. 58, no. 5, pp. 1545–1550, 2010, doi: 10.1109/TAP.2010.2044357.

[19] N. Valliappan, A. Lozano, and R. W. Heath, "Antenna subset modulation for secure millimeter-wave wireless communication," *IEEE Trans. Commun.*, vol. 61, no. 8, pp. 3231–3245, 2013, doi: 10.1109/TCOMM.2013.061013.120459.

[20] T. Hong, M.-Z. Song, and Y. Liu, "Dual-beam directional modulation technique for physical-layer secure communication," *IEEE Antennas Wirel. Propag. Lett.*, vol. 10, pp. 1417–1420, 2011, doi: 10.1109/LAWP.2011.2178384.

[21] Y. Ding and V. Fusco, "Polarization distortion as a means for securing wireless communication," in *8th European Conference on Antennas Propagation (EuCAP 2014)*, Apr. 2014, pp. 1203–1207, doi: 10.1109/EuCAP.2014.6901990.

[22] Y. Ding and V. Fusco, "Sidelobe manipulation using Butler matrix for 60 GHz physical layer secure wireless communication," in *2013 Loughborough Antennas Propagation Conference (LAPC)*, Nov. 2013, pp. 61–65, doi: 10.1109/LAPC.2013.6711852.

[23] Y. Ding, Y. Zhang, and V. Fusco, "Fourier Rotman lens enabled directional modulation transmitter," *Int. J. Antennas Propag.*, vol. 2015, p. e285986, 2015, doi: 10.1155/2015/285986.

[24] S. M. Ellison, J. M. Merlo, and J. A. Nanzer, "Distributed antenna array dynamics for secure wireless communication," *IEEE Trans. Antennas Propag.*, vol. 70, no. 4, pp. 2740–2749, 2022, doi: 10.1109/TAP.2021.3137449.

[25] Y. Ding and V. Fusco, "Orthogonal vector approach for synthesis of multi-beam directional modulation transmitters," *IEEE Antennas Wirel. Propag. Lett.*, vol. 14, pp. 1330–1333, 2015, doi: 10.1109/LAWP.2015.2404818.

[26] Y. Ding and V. F. Fusco, "A vector approach for the analysis and synthesis of directional modulation transmitters," *IEEE Trans. Antennas Propag.*, vol. 62, no. 1, pp. 361–370, 2014, doi: 10.1109/TAP.2013.2287001.

[27] Y. Ding and V. F. Fusco, "Establishing metrics for assessing the performance of directional modulation systems," *IEEE Trans. Antennas Propag.*, vol. 62, no. 5, pp. 2745–2755, 2014, doi: 10.1109/TAP.2014.2307318.

[28] H. E. Shanks and R. W. Bickmore, "Four-dimensional electromagnetic radiators," *Can. J. Phys.*, vol. 37, no. 3, pp. 263–275, 1959, doi: 10.1139/p59-031.

[29] W. Kummer, A. Villeneuve, T. Fong, and F. Terrio, "Ultra-low sidelobes from time-modulated arrays," *IEEE Trans. Antennas Propag.*, vol. 11, no. 6, pp. 633–639, 1963, doi: 10.1109/TAP.1963.1138102.

[30] H. Shanks, "A new technique for electronic scanning," *IRE Trans. Antennas Propag.*, vol. 9, no. 2, pp. 162–166, 1961, doi: 10.1109/TAP.1961.1144965.

[31] Y. Ding, V. Fusco, J. Zhang, and W.-Q. Wang, "Time-modulated OFDM directional modulation transmitters," *IEEE Trans. Veh. Technol.*, vol. 68, no. 8, pp. 8249–8253, 2019, doi: 10.1109/TVT.2019.2924543.

[32] G. Huang, Y. Ding, and S. Ouyang, "Multicarrier directional modulation symbol synthesis using time-modulated phased arrays," *IEEE Antennas Wirel. Propag. Lett.*, vol. 20, no. 4, pp. 567–571, 2021, doi: 10.1109/LAWP.2021.3056962.

[33] J. Chen, X. Liang, C. He, *et al.*, "Efficiency improvement of time modulated array with reconfigurable power divider/combiner," *IEEE Trans. Antennas Propag.*, vol. 65, no. 8, pp. 4027–4037, 2017, doi: 10.1109/TAP.2017.2712811.

[34] J. Chen, X. Liang, C. He, *et al.*, "Instantaneous gain optimization in time modulated array using reconfigurable power divide/combiner," *IEEE Antennas Wirel. Propag. Lett.*, vol. 17, no. 4, pp. 530–533, 2018, doi: 10.1109/LAWP.2018.2795008.

[35] H. Fan, J. Geng, X. Liang, R. Jin, and X. Zhou, "A three-way reconfigurable power divider/combiner," *IEEE Trans. Microw. Theory Tech.*, vol. 63, no. 3, pp. 986–998, 2015, doi: 10.1109/TMTT.2015.2395421.

[36] H. Fan, X. Liang, J. Geng, L. Liu, and R. Jin, "An N-way reconfigurable power divider," *IEEE Trans. Microw. Theory Tech.*, vol. 65, no. 11, pp. 4122–4137, 2017, doi: 10.1109/TMTT.2017.2702115.

[37] J. Hou, J. Methapettyparambu Purushothama, H. Fan, C. Song, Y. Ding, and M. Sellathurai, "Energy efficient time-modulated OFDM directional modulation transmitters," *Microw. Opt. Technol. Lett.*, vol. 65, pp. 5–13, 2022.

[38] C. A. Balanis, *Antenna Theory: Analysis and Design*, 3rd ed. Hoboken, NJ: John Wiley, 2005.

[39] W. Jaeckle, "Systematic design of a matrix network used for antenna beam steering," *IEEE Trans. Antennas Propag.*, vol. 15, no. 2, pp. 314–316, 1967, doi: 10.1109/TAP.1967.1138908.

[40] M. Ueno, "A systematic design formulation for Butler matrix applied FFT algorithm," *IEEE Trans. Antennas Propag.*, vol. 29, no. 3, pp. 496–501, 1981, doi: 10.1109/TAP.1981.1142601.

[41] Y. Gao, M. Khaliel, F. Zheng, and T. Kaiser, "Rotman lens based hybrid analog–digital beamforming in massive MIMO systems: array architectures, beam selection algorithms and experiments," *IEEE Trans. Veh. Technol.*, vol. 66, no. 10, pp. 9134–9148, 2017, doi: 10.1109/TVT.2017.2714693.

[42] I. Munina, P. Turalchuk, A. Shitvov, and D. Kozlov, "Simplified spatial division analog beamforming based on Fourier Rotman lens," in *12th European Conference on Antennas Propagation (EuCAP 2018)*, Apr. 2018, pp. 1–4, doi: 10.1049/cp.2018.1163.

[43] P. Turalchuk, I. Munina, and A. Shitvov, "Analog beamforming based on Fourier Rotman lens for multibeam applications," in *2018 48th European Microwave Conference (EuMC)*, Sep. 2018, pp. 1573–1576, doi: 10.23919/EuMC.2018.8541739.

[44] Y. Zhang, S. Christie, V. Fusco, R. Cahill, G. Goussetis, and D. Linton, "Reconfigurable beam forming using phase-aligned Rotman lens," *IET Microw. Antennas Propag.*, vol. 6, no. 3, pp. 326–330, 2012, doi: 10.1049/iet-map.2011.0379.

[45] A. V. Oppenheim, A. S. Willsky, and S. H. Nawab, *Signals & Systems*. Prentice-Hall of India Private Limited, 2008.

[46] Y. Ding, "Directional modulation secured wireless communication," PhD Thesis, Queen's University Belfast, Belfast, UK, 2014.

[47] Y. Zhang, Y. Ding, and V. Fusco, "Sidelobe modulation scrambling transmitter using Fourier Rotman lens," *IEEE Trans. Antennas Propag.*, vol. 61, no. 7, pp. 3900–3904, 2013, doi: 10.1109/TAP.2013.2254453.

[48] Y. Zhang and V. Fusco, "Fourier transform using a Rotman lens," in *2012 9th European Radar Conference*, Oct. 2012, pp. 389–392.

[49] Fourier Rotman lens directional modulation demonstrator experiment, (Jun. 30, 2014). Accessed: Jun. 21, 2022. [Online Video]. Available: https://www.youtube.com/ watch?v=FsmCcxo-TPE

[50] N. B. Buchanan, V. F. Fusco, and M. van der Vorst, "SATCOM retrodirective array," *IEEE Trans. Microw. Theory Tech.*, vol. 64, no. 5, pp. 1614–1621, 2016, doi: 10.1109/TMTT.2016.2541121.

[51] D. S. Goshi, K. M. K. H. Leong, and T. Itoh, "A secure high-speed retrodirective communication link," *IEEE Trans. Microw. Theory Tech.*, vol. 53, no. 11, pp. 3548–3556, 2005, doi: 10.1109/TMTT.2005.857107.

[52] L. D. DiDomenico and G. M. Rebeiz, "Digital communications using self-phased arrays," *IEEE Trans. Microw. Theory Tech.*, vol. 49, no. 4, pp. 677–684, 2001, doi: 10.1109/22.915442.

[53] V. Fusco and N. Buchanan, "Developments in retrodirective array technology," *IET Microw. Antennas Propag.*, vol. 7, no. 2, pp. 131–140, 2013, doi: 10.1049/iet-map.2012.0565.

[54] Y. Ding, J. Zhang, and V. F. Fusco, "Retrodirective-assisted secure wireless key establishment," *IEEE Trans. Commun.*, vol. 65, no. 1, pp. 320–334, 2017, doi: 10.1109/TCOMM.2016.2616406.

[55] Y. Ding and V. Fusco, "Directional modulation-enhanced retrodirective array," *Electron. Lett.*, vol. 51, no. 1, pp. 118–120, 2015, doi: https://doi.org/10.1049/el.2014.3188.

[56] C. Sun, S. Yang, Y. Chen, J. Guo, S. Qu, and J. Hu, "4-D retro-directive antenna arrays for secure communication based on improved directional modulation," *IEEE Trans. Antennas Propag.*, vol. 66, no. 11, pp. 5926–5933, 2018, doi: 10.1109/TAP.2018.2866536.

[57] S. Y. Nusenu, S. Huaizong, Y. Pan, and A. Basit, "Directional modulation with precise legitimate location using time-modulation retrodirective frequency diversity array for secure IoT communications," *IEEE Syst. J.*, vol. 15, no. 1, pp. 1109–1119, 2021, doi: 10.1109/JSYST.2020.3010787.

[58] Y. Ding and V. Fusco, "A synthesis-free directional modulation transmitter using retrodirective array," *IEEE J. Sel. Top. Signal Process.*, vol. 11, no. 2, pp. 428–441, 2017, doi: 10.1109/JSTSP.2016.2605066.

[59] H. Shi and A. Tennant, "Secure physical-layer communication based on directly modulated antenna arrays," in *2012 Loughborough Antennas Propagation Conference (LAPC)*, Nov. 2012, pp. 1–4, doi: 10.1109/LAPC.2012.6403072.

[60] H. Shi and A. Tennant, "Enhancing the security of communication via directly modulated antenna arrays," *IET Microw. Antennas Amp. Propag.*, vol. 7, no. 8, pp. 606–611, 2013, doi: 10.1049/iet-map.2012.0644.

[61] H. Shi and A. Tennant, "Simultaneous, multichannel, spatially directive data transmission using direct antenna modulation," *IEEE Trans. Antennas Propag.*, vol. 62, no. 1, pp. 403–410, 2014, doi: 10.1109/TAP.2013.2287284.

[62] A. Narbudowicz, M. J. Ammann, and D. Heberling, "Directional modulation for compact devices," *IEEE Antennas Wirel. Propag. Lett.*, vol. 16, pp. 2094–2097, 2017, doi: 10.1109/LAWP.2017.2697720.

[63] A. Narbudowicz, A. Zandamela, N. Marchetti, and M. J. Ammann, "Energy-efficient dynamic directional modulation with electrically small antennas," *IEEE Antennas Wirel. Propag. Lett.*, vol. 21, no. 4, pp. 681–684, 2022, doi: 10.1109/LAWP.2022.3141212.

[64] J. Parron, E. A. Cabrera-Hernandez, A. Tennant, and P. de Paco, "Multiport compact stacked patch antenna with 360° beam steering for generating dynamic directional modulation," *IEEE Trans. Antennas Propag.*, vol. 69, no. 2, pp. 1162–1167, 2021, doi: 10.1109/TAP.2020.3008065.

[65] R. Dong, S. Jiang, X. Hua, Y. Teng, F. Shu, and J. Wang, "Low-complexity joint phase adjustment and receive beamforming for directional modulation networks via IRS," *IEEE Open J. Commun. Soc.*, vol. 3, pp. 1234–1243, 2022, doi: 10.1109/OJCOMS.2022.3195050.

[66] X. Ai and L. Gan, "Directional modulation for secure IoT networks via accurate phase response control," *IEEE Internet Things J.*, vol. 9, no. 21, pp. 21537–21547, 2022, doi: 10.1109/JIOT.2022.3181474.

[67] N. Ye, X. Zhuo, J. Li, B. Di, and J. An, "Secure directional modulation in RIS-aided networks: a low-sidelobe hybrid beamforming approach," *IEEE Wirel. Commun. Lett.*, vol. 11, no. 8, pp. 1753–1757, 2022, doi: 10.1109/LWC.2022.3180931.

[68] A. Narbudowicz, M. J. Ammann, and D. Heberling, "Switchless reconfigurable antenna with 360° steering," *IEEE Antennas Wirel. Propag. Lett.*, vol. 15, pp. 1689–1692, 2016, doi: 10.1109/LAWP.2016.2524199.

[69] E. A. Cabrera-Hernández, J. P. Granados, and A. Tennant, "Multichannel dynamic directional modulation with software defined radio," in *2020 14th European Conference on Antennas Propagation (EuCAP)*, Mar. 2020, pp. 1–4, doi: 10.23919/EuCAP48036.2020.9135816.

[70] T. Xie, J. Zhu, and Y. Li, "Artificial-noise-aided zero-forcing synthesis approach for secure multi-beam directional modulation," *IEEE Commun. Lett.*, vol. 22, no. 2, pp. 276–279, 2018, doi: 10.1109/LCOMM.2017.2772228.

[71] X. Li, J. Hwu, and E. P. Ratazzi, "Using antenna array redundancy and channel diversity for secure wireless transmissions," *J. Commun.*, vol. 2, pp. 24–32, 2007.

[72] L. Chen, W. Chen, Y.-J. Liu, *et al.*, "Linearization of a directional modulation transmitter using low-complexity cascaded digital predistortion," *IEEE Trans. Microw. Theory Tech.*, vol. 67, no. 11, pp. 4467–4478, 2019, doi: 10.1109/TMTT.2019.2921346.

[73] E. Tollefson, B. R. Jordan, and J. D. Gaeddert, "Out-phased array linearized signaling (OPALS): a practical approach to physical layer encryption," in *MILCOM 2015–2015 IEEE Military Communications Conference*, Oct. 2015, pp. 294–299, doi: 10.1109/MILCOM.2015.7357458.

[74] L. Chen, W. Chen, Y. Liu, C. Yang, and Z. Feng, "An efficient directional modulation transmitter with novel crest factor reduction technique," *IEEE Microw.*

Wirel. Compon. Lett., vol. 29, no. 8, pp. 554–556, 2019, doi: 10.1109/LMWC.2019.2923361.

[75] Y. He, Y. Liu, N. Du, and N. Xie, "Directional modulation for QAM signals with PAPR reduction," in *2016 46th European Microwave Conference (EuMC)*, Oct. 2016, pp. 1307–1310, doi: 10.1109/EuMC.2016.7824591.

[76] J. M. Purushothama, Y. Ding, G. Goussetis, G. Huang, and Y. Xiao, "Synthesis of energy efficiency enhanced directional modulation transmitters," *IEEE Trans. Green Commun. Netw.*, vol. 7, pp. 1–1, 2022, doi: 10.1109/TGCN.2022.3208023.

[77] P. Antonik, "An investigation of a frequency diverse array," Doctoral, UCL (University College London), 2009. Accessed: May 9, 2021. [Online]. Available: https://discovery.ucl.ac.uk/id/eprint/18746/

[78] Y. Ding, A. Narbudowicz, and G. Goussetis, "Physical limitation of range-domain secrecy using frequency diverse arrays," *IEEE Access*, vol. 8, pp. 63302–63309, 2020, doi: 10.1109/ACCESS.2020.2984643.

[79] T. Hong, X.-P. Shi, and X.-S. Liang, "Synthesis of sparse linear array for directional modulation via convex optimization," *IEEE Trans. Antennas Propag.*, vol. 66, no. 8, pp. 3959–3972, 2018, doi: 10.1109/TAP.2018.2835641.

[80] B. Zhang, W. Liu, Q. Li, *et al.*, "Metasurface based positional modulation design," *IEEE Access*, vol. 8, pp. 113807–113813, 2020, doi: 10.1109/ACCESS.2020.3003495.

[81] B. Zhang, W. Liu, J. Ma, *et al.*, "Sparse antenna array based positional modulation design with a low-complexity metasurface," *IEEE Access*, vol. 8, pp. 177640–177646, 2020, doi: 10.1109/ACCESS.2020.3027285.

[82] K. Chen and Y. Feng, "A review of recent progress on directional metasurfaces: concept, design, and application," *J. Phys. Appl. Phys.*, vol. 55, no. 38, p. 383001, 2022, doi: 10.1088/1361-6463/ac7e04.

[83] N.-Y. Liu, X. Yang, W.-B. Sun, Y.-Y. Gong, and R. Wang, "Multicast directional modulation based on sparse array optimization," in *2021 International Wireless Communications and Mobile Computing (IWCMC)*, Jun. 2021, pp. 985–990. doi: 10.1109/IWCMC51323.2021.9498909.

[84] F. Liu, L. Wang, and J. Xie, "Directional modulation technique for linear sparse arrays," *IEEE Access*, vol. 7, pp. 13230–13240, 2019, doi: 10.1109/ACCESS.2019.2894219.

[85] V. F. Fusco, A. Chepala, and M. A. B. Abbasi, "Target location using dual-beam directional modulated circular array," *IEEE Trans. Antennas Propag.*, vol. 66, no. 12, pp. 7525–7529, 2018, doi: 10.1109/TAP.2018.2869257.

[86] G. Huang, Y. Ding, S. Ouyang, and J. M. Purushothama, "Target localization using time-modulated directional modulated transmitters," *IEEE Sens. J.*, vol. 22, pp. 1–1, 2022, doi: 10.1109/JSEN.2022.3178699.

Chapter 17
Physical layer security for non-orthogonal multiple access toward 6G*

Na Li[1] *and Yuanwei Liu*[2]

The sixth-generation (6G) networks are expected to support an extremely high data rate, a very large number of users, and a variety of applications and services. In order to achieve these stringent targets, one of the most fundamental issues is the design of sophisticated multiple access techniques, intending to enable a tremendous number of users/devices to be efficiently and flexibly connected with the network over the given wireless radio resources. Non-orthogonal multiple access (NOMA) provides significant benefits in supporting massive connectivity, achieving high spectrum efficiency (SE) and energy efficiency (EE), guaranteeing user fairness, being compatible with many other techniques, and thus is one of the promising candidates for 6G [1,2]. On the one hand, NOMA empowers other promising techniques for 6G, like terahertz (THz), grant-free (GF) access, orthogonal time–frequency space (OTFS) modulation, visible light communication (VLC), backscatter communication (BC), etc., which can significantly maximize the achievable rates, SE and number of connectivity to support diverse applications in 6G. On the other hand, NOMA can be empowered by the emerging 6G techniques, like reconfigurable intelligent surface (RIS), OTFS, etc., which can motivate NOMA to support flexible quality of service (QoS) requirements.

Moreover, physical layer security (PLS) has drawn extensive attention in recent years. Abundant studies have been carried out for various scenarios including traditional NOMA. However, new challenges and requirements arise when NOMA is being applied in 6G. In this chapter, we first review the PLS for NOMA toward 6G. Then we take the RIS-aided NOMA as case studies to show its PLS against internal or external eavesdropping attacks.

17.1 Brief review of PLS for NOMA toward 6G

In the following, we first clarify how NOMA interplays with other enabling techniques of 6G to achieve better connectivity or efficiency. Then we review the current research

*This chapter has been a part of [3,4].

[1]National Engineering Lab for Mobile Network Technologies, Beijing University of Posts and Telecommunications, China
[2]School of Electronic Engineering and Computer Science, Queen Mary University of London, UK

about the PLS for these NOMA-aided new scenarios toward 6G. Finally, we discuss the corresponding open problems and possible future research directions.

17.1.1 NOMA with new modulation: OTFS-NOMA

Supporting users with different mobility profiles is an important feature of 6G. OTFS modulation has been proposed as an effective physical layer alternative to traditional orthogonal frequency division multiplexing (OFDM) modulation in high-mobility environments [5]. The key idea of OTFS is to modulate the information bits in both the delay and Doppler (D–D) domains, which can effectively convert a rapidly time-varying channel in the time–frequency (T–F) domain into a quasi-stationary channel model in the D–D domain. To further improve the SE and support massive connectivity, the OTFS-NOMA has been proposed in two basic schemes [6]. In the first scheme, multiple users with different mobile profiles share the same D–D resources and are distinguished by either different power levels or sparse codewords. In the second scheme, the high-mobility users using OTFS in the D–D domain are paired with low-mobility users using OFDM in the T–F domain for non-orthogonal channel sharing.

PLS of OTFS-based communications attracted only limited attention. In [7], the security performance of OTFS-based uplink satellite communication was investigated. A cooperative unmanned aerial vehicle (UAV) is employed to send jamming signals against the reconnaissance satellite. The closed-form expression for the secrecy outage probability (SOP) was derived, and the security advantage of OTFS over OFDM was analyzed. In [8], the OTFS waveform was proposed to be rotated by the singular value decomposition (SVD)-based precoding to achieve both secure and reliable transmissions.

There is still a lack of research contributions on the PLS for OTFS-NOMA systems. The main challenge comes from the fact that the resource degrees of freedoms (DoFs) are extended from the traditional T–F domain to the D–D domain or hybrid domain, which brings alternative advantages and new chances for external or internal eavesdropping attacks. For example, the eavesdroppers with various mobility profiles may be paired with the legitimate ones, and thus cause much more serious information leakage.

17.1.2 NOMA with new spectrums: mmWave/THz/VLC-NOMA

17.1.2.1 mmWave/THz-NOMA

Utilizing the huge bandwidths offered in the mmWave/THz bands is a promising approach for boosting the data rate in 6G. To obtain better throughput and enhanced SE, NOMA can be incorporated to increase the number of supportable users by exploiting the power-domain or code-domain multiplexing gain.

PLS of mmWave-NOMA communications has drawn increasing attention. Secrecy performance in terms of SOP or secrecy throughput was evaluated by deriving the corresponding closed-form expressions [9–17]. Resource allocations including user pairing, beamforming, cooperative jamming (CJ), and/or artificial noise (AN) were optimized to improve the secrecy performance [18–21]. Pilot contamination attack detection for mmWave-NOMA was also preliminary studied in [22,23]. These studies focused on the external eavesdropping problem. The internal eavesdropping issue for mmWave-NOMA is also critical to be studied.

THz-band signals are often assumed to be more secure than lower frequency signals due to the more directional transmission and the narrower beams. However, eavesdroppers located in the sharp sector of the beam coverage can still endanger information security. Information confidentiality or communication covertness under the threat of the in-band eavesdropping has been studied in various communication scenarios with a single receiver or multiple users based on orthogonal multiple access (OMA) scheme [24], while PLS for THz-NOMA is still to be explored.

17.1.2.2 VLC-NOMA

VLC is a potential technology for complementing and/or offloading radio frequency (RF) communication systems for various indoor user-dense scenarios in 6G networks, because of its usage of unlicensed spectrum, efficient implementation, high confidentiality, and low energy consumption [25]. One main drawback of the current VLC systems is that the modulation bandwidth is small, which significantly restricts the achievable data rate. Therefore, NOMA is an attractive technique indispensable for VLC to enhance the SE [5,25,26].

PLS has gained significant consideration in VLC communications [27–29]. In [30], the exact SOP expressions were derived for the single light-emitting diode (LED) VLC-NOMA network with a single or multiple eavesdroppers. In [31], the outer and inner bounds of the secrecy capacity were first derived for the multi-LED VLC-NOMA system, based on which the optimal beamforming was proposed to minimize the total LED power or to maximize the secrecy capacity. In [32,33], the optimal precoding for the multi-LED NOMA-VLC network was studied, where AN was introduced to suppress the eavesdroppers. In [32], collusive eavesdroppers were considered. The secrecy capacity maximization problems were formulated respectively for cases with known or unknown eavesdroppers' locations. In [33], robust resource allocation was proposed to minimize the total transmit power under the assumption of imperfect channel state information (CSI). In [34], the optimal and robust beamforming was proposed for the simultaneous light-wave information and power transfer (SLIPT) multi-LED VLC-NOMA network. The transmit power was minimized or the secrecy rate was maximized under the assumption of perfect or imperfect CSI. In [35], a novel precoding scheme reinforced with random constellation coding was proposed for the multi-LED VLC-NOMA network. In [36,37], PLS was studied in a cooperative NOMA-aided hybrid VLC/RF system, where the near user (NU) can help the far user (FU) by using RF signals. In [36], the secrecy rate of the FU is maximized under the QoS constraints of the NU by optimizing the beamforming and power allocation. In [37], RIS was further introduced to enhance the secrecy, and the closed-form expression of the SOP was derived. In [38], an underwater multi-LED NOMA-VLC network was studied. Both optimal and suboptimal LED selection schemes were proposed to select an LED that can transmit the information with the highest secrecy rate. And the closed-form SOP was derived for both single-LED and multi-LED transmission strategies with known or unknown CSI of the eavesdropper. In [39,40], the dynamic resource allocation was optimized to maximize the secrecy performance of the VLC-NOMA network with moving users. In [39], the users are assumed to move uniformly within the cells, while in [40], the users are assumed to move based on the random way-point (RWP) mobility model.

The above studies focused mainly on one legitimate user suffering security threats from one or more external eavesdroppers, while only little research has been developed on the PLS for VLC-NOMA systems, and the internal eavesdropping issue is also critical to be studied.

17.1.3 NOMA with new random access: GF-NOMA

Massive connections and strict QoS requirements are key performance indicators for 6G machine-type communication (MTC), which requires a novel design of random-access protocols. GF access has been identified as a key medium access control technique for providing massive connectivity [41]. However, the inability of multi-user detection caused by frequent collisions among massive devices is the most essential challenge for the GF strategy. NOMA is able to handle this problem by serving multiple devices within the same resource blocks. Moreover, as a further step to improve the SE, semi-grant-free (SGF) NOMA was proposed [42,43], where the GF users are admitted to share the same spectrum resources of the grant-based (GB) users.

The outage performance analysis, resource allocation optimization, protocol design for GF-NOMA, and SGF-NOMA systems have been studied in several scenarios despite still being in their infancy. However, research on the PLS for GF-NOMA or SGF-NOMA systems has attracted only limited attention. In [44], the authors first studied the secrecy performance of the SGF-NOMA system with single or multiple GF users and a single GB user, in the presence of an external eavesdropper aiming to wiretap the information of GF user. The exact analytical expression for the SOP was derived.

17.1.4 NOMA with new schemes: ISAC/BC-NOMA

17.1.4.1 ISAC-NOMA

ISAC can provide new sensing capabilities and reduce the weight of the equipment, and thus has been recognized as an important technique for 6G [45]. However, ISAC may suffer from severe inter-functionality interference due to the hardware platform and radio resource sharing. The prominent features of NOMA in efficient interference management and flexible resource allocation match well with the requirements encountered in ISAC [46–50].

Due to the spectrum sharing between sensing targets and information receivers, ISAC raises unique security challenges [51]. On the one hand, the sensing target may act as a potential eavesdropper for the information users. Meanwhile, the reflected echo also includes the environmental information, which might be obtained by the radar eavesdropper. On the other hand, the sensing functionality of ISAC also brings a new opportunity for secure design by proactively obtaining the location and CSI of the eavesdropping targets, thus resolving the eavesdroppers' CSI issue faced in conventional secrecy communication systems.

To settle the above inter-functionality eavesdropping issues, a few studies of PLS have been carried out for the typical ISAC systems or those integrating sensing and transmission functions like ISAC. In [51], the challenges and contradictory objectives involved in securing ISAC transmission were discussed. In [52–54], a unified joint system of passive radar and communication systems was considered, where the communication user (CU) might be eavesdropped by the target of passive radar. In [52], an optimal resource allocation strategy was proposed to maximize the signal-to-noise ratio (SNR) at the radar target while keeping the secrecy rate above a certain threshold. In [53], AN was adopted to enhance the secrecy, and optimization problems were formulated by minimizing the SNR received at radar target, while guaranteeing the signal-to-interference-plus-noise ratio (SINR) requirement at legitimate users. In [54], an extra signal carrying false information was adopted to confuse the targets instead of AN. And optimization problems were studied to improve the secrecy of CU, including secrecy rate maximization, target return SNR maximization, and transmit power minimization. In [55], a secure ISAC system was studied, where AN signal is used to confuse and sense the targets. The matching error between the transmit beampattern and a desired sensing beampattern was minimized by jointly optimizing the transmit information and sensing beamforming, subject to the minimum secrecy rate requirement at the CU. In [56], an integrated sensing, jamming, and communications (ISJC) framework was proposed, where AN was used to jam and sense the eavesdropper. Based on the sensing location and velocity information obtained in the previous time slot, a channel correlation-based online resource allocation algorithm was developed to determine the user scheduling and precoding policy for the current time slot.

The inter-functionality eavesdropping problem in NOMA-ISAC can be significantly magnified due to the internal eavesdropping issue between NOMA receivers. The corresponding research remains to be explored.

17.1.4.2 BC-NOMA

BC is one of the revolutionizing techniques to support low-powered Internet of Things (IoTs) in 6G [5,57,58], since it enables low-power transmission of information by backscatter devices (BDs) through the modulation of information over incident RF carriers, without generating any power by the BDs. The interplay between BC and NOMA could boost the performance of data conveying, EE and SE for massive IoT networks [5,59]. First, NOMA can potentially be a great enabler to BC to support a large number of BDs through power domain multiplexing with different backscattered power levels. Second, BC can be applied to enhance the cooperative NOMA transmission, by backscattering the surplus power of the received downlink signals at one user to enhance the reception of the user who cannot recover its information. Third, NOMA and BC can cooperate in a symbiotic radio (SR) network, in which a RF source transmits signals to NOMA users, while a BD transmits its own messages to the users by modulating them over its received RF signals.

Due to the promising and exciting prospects, BC-NOMA has triggered an overwhelming research interest. The research efforts have been mainly focused on the outage performance evaluation and throughput maximization. Research of PLS for BC-NOMA has also drawn increasing attention, but still in its infancy. Only few works have studied the secrecy performance of BC-NOMA in the SR network. In [60], PLS was studied for the BC-NOMA-aided SR network with a potential eavesdropper trying to intercept the signals transmitted from the RF source as well as those from the BD. The objective is to maximize the η-outage secrecy rate from the BD to the NU under the achievable secrecy rate constraints from the BS to the NU and the FU. In [61,62], the outage probability (OP) and intercept probability (IP) were obtained in closed forms by assuming that all nodes are with a single antenna, and the impact of in-phase and quadrature-phase imbalance (IQI) was considered. In [63], the authors further derived the analytical expressions for the OP and the IP, under the realistic assumptions of residual hardware impairments (RHIs), channel estimation errors and imperfect successive interference cancellation. AN is injected into the RF signal to improve the secrecy. In [64], the secrecy performance of an overlay cognitive BC-NOMA network was evaluated by deriving the SOP among the primary network, overlay secondary network, and the eavesdroppers, respectively. In [65], a multi-cell BC-NOMA network was studied in the presence of multiple concluding eavesdroppers trying to overhear the transmission from the BD to the NU. The secrecy rate was maximized by optimizing the reflection coefficient of the backscatter node in each cell.

17.1.5 NOMA with RIS: RIS-NOMA

RIS has recently emerged as one of the promising and revolutionizing techniques for enhancing the SE and EE of wireless networks. RISs are capable of reconfiguring the wireless propagation environment by carefully tuning the phase shifts of a large number of low-cost reflecting elements. Hence, RISs provide additional DoFs to further improve the system performance by interplaying with other physical layer techniques. RIS and NOMA, the two communication techniques, are perfectly complementary to each other and can be integrated to cope with the challenging efficiency and connection requirements envisioned for 6G networks [59,66]. Specifically, NOMA can improve the SE and connectivity of RIS systems and RISs ensure that the propagation environments can be effectively and intelligently customized for the implementation of NOMA.

Extensive research efforts for RIS-NOMA have been devoted to investigating the performance evaluation and resource allocations. PLS for RIS-NOMA is also drawing increasing attention.

For the external eavesdropping case, RIS can be utilized to enhance the channel gains of legitimate users (LUs) or degrade the channel gains of eavesdroppers, by adding an extra reconfigurable wireless link. For secrecy performance evaluation, in [67], the SOP was derived and analyzed for the two-user downlink RIS-NOMA. In [68], HDI was further considered, and the SOP was re-derived. In [69], the direct link between the source and users/Eve was assumed to exist, and the average secrecy rate (ASR) was derived. In [70,71], the RIS was designed to maximize the channel gain of the FU. In [70], the SOP was derived for the multi-user downlink NOMA, where each FU is assisted by a specific RIS. In [71], the authors considered that RIS assists the transmission to FU, and derived the SOP and ASR. In [72], RIS was designed to minimize the channel gain of Eve, and the SOP was derived. As for optimal resource allocation studies to maximize the secrecy performance, in [73], AN was added to degrade the unknown eavesdropper in the RIS-aided NOMA network. To avoid its impact on LUs, the jamming power should be the highest at all receivers, so that it can be decoded and eliminated in the first step of serial interference cancellation (SIC). An optimization problem was formulated to maximize the sum rate of all LUs, subject to the QoS requirements, the RIS reflecting constraint, and the designed SIC decoding condition. In [74], the max–min fair hybrid beamforming for secure uplink RIS-NOMA system was studied. Both users send messages to the base station (BS) while emitting AN that confuses the eavesdropper. By the joint optimization of the active beamforming at users and the passive beamforming at RIS, the minimal secrecy rate among users was maximized. In [75], a robust and secure RIS-NOMA transmission was studied by assuming imperfect eavesdropping CSI with bounded errors. A worst-case power minimization problem was investigated, by jointly optimizing the active and passive beamforming, subject to the minimum QoS constraints at users, the maximum eavesdropping rates, and a given SIC decoding order constraint. In [66,76], a robust beamforming scheme was proposed to guarantee secure NOMA transmission with multiple distributed RISs by assuming imperfect eavesdropping CSI with statistical errors. The minimum secrecy rate was maximized subject to the total power constraint at the BS, the phase shifts constraints of RISs, and the SOP/SIC constraints.

As for the internal eavesdropping case, RIS is utilized to re-order users' channel gains according to their secrecy requirements. On the one hand, the NU may be untrusted and try to intercept the information of the FU. To protect the information of FU, the SIC order should be switched that the FU's information symbol is not the first one to be decoded. To achieve this, we have to modify the equivalent channel gain of the FU to be the larger one and allocate less power to the FU. In other words, the users' channel gains should be re-ordered in such a way that the roles of FU and NU are switched. In [77], the RIS phase shifts and active beamforming vectors were jointly optimized to maximize the FU's equivalent channel gain such that a positive secrecy rate can be guaranteed with high probability. In [78], AN generated in the null space of FU's channel was further added to jam the NU. On the other hand, the FU may be untrusted and try to intercept the information of the NU, after detecting its own information symbol and by subtracting it from the superposed signal. In this way, a positive secrecy rate can always be guaranteed as long as the NU has a larger channel gain. However, to enlarge the secrecy capacity, the phase shifts and active beamforming vectors should be jointly optimized under the constraint that the order of channel gains keeps unchanged [79].

17.1.6 Conclusion in open problems and research directions

Based on the above analysis, NOMA can interplay with many new techniques to meet the efficiency and QoS requirements toward 6G. To cope with the security challenges arise from the aforementioned NOMA-aided new scenarios, preliminary research efforts have been carried out in academia, which however is still in its infancy.

First, the new attack models drawn from the integration of NOMA and other techniques are not fully recognized. For example, the eavesdropping behaviors may become more intelligent in the ISAC- and RIS-NOMA systems, since the attackers may obtain and manipulate more DoFs of side information and radio resources to create the eavesdropping advantages or hide the behaviors. And the eavesdroppers may become more active in the GF-NOMA system by deliberately destroying the GF accesses of legitimate users.

Second, the distinct features of the new waveform, high-frequency spectrum and hybrid communication modes have not been well utilized for the PLS of NOMA. For example, the AN and/or CJ may be created in the D–D domain in the OTFS-NOMA system, or multiplexed with the sensing beam or BC signals in the ISAC- or BC-NOMA system.

Third, many new techniques are still in their exploring stage. Certain practical challenges are waiting to be settled. For example, the hardware imperfections of the THz and RIS systems, the CSI acquisition, interference problems in the ISAC and BC systems, etc. These practical issues should be considered in advance when studying the corresponding PLS.

17.2 Motivation

Like NOMA, RIS can be easily integrated with other 6G techniques. Moreover, RIS and NOMA are perfectly complementary to each other. The integration of RIS and NOMA is expected to be the most powerful solution to cope with the challenging requirements envisioned for 6G networks. Therefore, in this chapter, we focus on the PLS for RIS-NOMA systems.

From the security perspective, RIS introduces two aspects of advantages for NOMA. First, RIS provides an extra wireless link which can be used to improve the LU's equivalent channel and/or suppress the eavesdropping channel. In this way, RIS-NOMA is believed to achieve better and efficient secrecy. Second, RIS is able to flexibly adjust the SIC orders of the NOMA users according to their individual QoS and secrecy requirements. This unique function is crucial for 6G, which is expected to support a variety of applications and services with diverse QoS requirements. How to effectively and efficiently exploit these advantages of RIS-NOMA is one of the fundamental studying problems.

Based on the aforementioned PLS researches for RIS-NOMA, the joint optimal design of the active beamforming at the transmitter and the passive beamforming at the RIS is the fundamental strategy to guarantee the security of RIS-NOMA against external or internal eavesdroppers. AN can be applied to further improve the secrecy performance. In this chapter, we introduce two general secure RIS-NOMA communication models against the external or internal eavesdropper respectively, based on which we aim to verify the security advantages of RIS-NOMA, and show how to efficiently exploit the advantages brought by RIS to NOMA.

17.2.1 RIS-NOMA under external eavesdropping

For the external eavesdropping case, we extend the aforementioned conventional RIS (C-RIS)-aided NOMA to the general simultaneously transmitting/refracting and reflecting (STAR)-RIS-aided NOMA [3]. Compared with the C-RIS, STAR-RIS is able to provide a full-space coverage and more flexible deployment, by transmitting and reflecting incident signals toward the users on both the same and opposite sides of the source, simultaneously. However, this coverage advantage can also be exploited by adversaries to launch a full-space eavesdropping, which requires a novel secrecy design framework. An initial exploration on PLS for STAR-RIS-NOMA was studied in [80], where the indoor user and outdoor user send their signals to the BS through the reflection/transmission links provided by STAR-RIS in the presence of an external eavesdropper. Joint beamforming or power optimization problems was formulated for minimum secrecy capacity maximization or maximum SOP minimization, subject to the SIC decoding constraints and reflection/transmission coefficient constraints. It showed that compared with C-RIS, STAR-RIS is capable of providing better secrecy performance.

In this chapter, the secure communication problem is studied for the STAR-RIS-assisted downlink NOMA system [3]. AN is introduced to further combat the performance gain brought by STAR-RIS to eavesdroppers. The secrecy rate maximization problem is formulated under the constraints of the SIC decoding order and individual secrecy requirements of users, by optimizing the passive beamforming and the AN model.

17.2.2 RIS-NOMA under internal eavesdropping

According to the NOMA principle, the user with a higher SIC order is allowed to decode the information symbols of the users with lower orders, and thus can achieve a higher-level secrecy, while the user with a lower SIC order achieves lower-level secrecy or no secrecy. In practice, the secrecy levels and requirements of users may not be consistent with the SIC orders which are generally decided by their channel conditions. In other words, a user with a weaker channel may require a higher-level secrecy. Based on the above researches, RIS can help to adjust the users' channels and SIC orders according to their secrecy requirements. Specifically, when the FU requires a higher secrecy level than the NU, the system can re-configure the SIC orders and change the FU to be the NU. When it comes to the more-than-two-user case, where users have diverse secrecy levels and secrecy requirements, a general model to realize the internal secrecy for RIS-NOMA should be explored.

In this chapter, we propose a novel RIS-aided strategy to satisfy the internal secrecy among NOMA users [4]. The transmit power minimization problem is formulated under the individual secrecy rate constraints, by jointly optimizing the beamforming vectors and the RIS coefficients.

Notations: The lower-case letter x, bold-face lower-case letter $\mathbf{a}$, and upper-case letter $\mathbf{A}$ denote the scalar, vector, and matrix, respectively. $\mathbb{C}^{M\times N}$ denotes the complex matrix with the space $N \times M$. $\mathbb{H}^{N}$ represents the N-dimensional Hermitian matrix. $\text{rank}\{\mathbf{A}\}$ and $\text{tr}\{\mathbf{A}\}$ indicate the rank and trace of matrix $\mathbf{A}$, respectively. $\mathbf{A} \succeq 0$

means that **A** is a positive semi-definite matrix. $\mathbf{a}^T$ and $\mathbf{a}^H$ denotes the transpose and conjugate transpose of $\mathbf{a}$. $a_{\{j,k\}}$ means a_j and a_k. $\Re(x)$ denotes the real value of x.

17.3 Secure NOMA communication against the external eavesdropper assisted by STAR-RIS

In this chapter, the STAR-RIS-assisted secure NOMA communication against the external eavesdropper is studied [3].

17.3.1 System model and problem formulation

As shown in Figure 17.1, we consider a STAR-RIS-aided downlink NOMA communication system serving $K \geq 2$ users. An external eavesdropper denoted as Eve tries to eavesdrop on the users' information. The STAR-RIS with M reflection elements is deployed, which divides the communication space into the reflecting area and the transmitting (refracting) area. The energy splitting (ES) protocol is employed at the STAR-RIS, since the ES protocol entails a lower hardware implementation complexity than time switching (TS) and can achieve higher performance than mode switching (MS) in the downlink transmissions [81,82]. The users and Eve locate in the reflecting area or the transmitting area respectively and each is equipped with single antenna. Considering the 360° coverage of the STAR-RIS, we assume a single Eve who may locate in the reflecting or transmitting area. In practice, multiple eavesdroppers may exist on both sides of the STAR-RIS, and the eavesdroppers may or may not collude with each other. In the colluding scenario, the final eavesdropping SINR could be the sum of all SINRs. While in the non-colluding scenario, the final eavesdropping SINR could be the maximum one of all SINRs. The BS has $N \geq 2$ antennas. The channels from BS to RIS, BS to user k or Eve, and RIS to user k or Eve are denoted by $\mathbf{G} \in \mathbb{C}^{M\times N}$, $\mathbf{h}_{\{k,e\}} \in \mathbb{C}^{1\times N}$ and $\mathbf{r}_{\{k,e\}} \in \mathbb{C}^{1\times M}$, respectively. All channels are assumed to follow independent block fading and perfectly known. Note here we

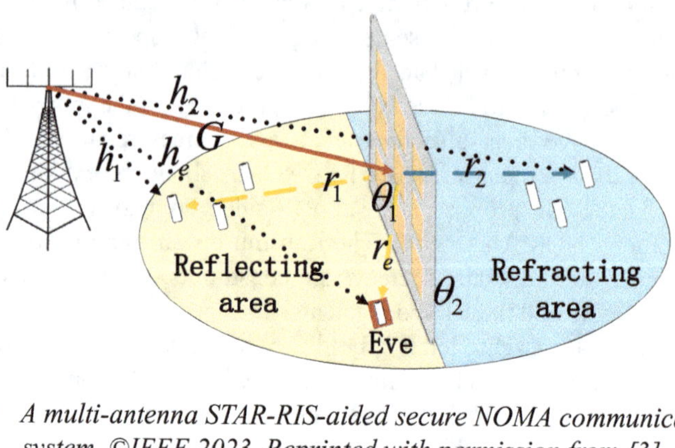

Figure 17.1 A multi-antenna STAR-RIS-aided secure NOMA communication system. ©IEEE 2023. Reprinted with permission from [3].

focus on studying the value of the STAR-RIS in improving the secrecy performance of the NOMA communications. As a result, the CSI is simply assumed to be perfectly known. In practice, the CSI is never perfect and hard to be obtained correctly. In this case, the SOP metric is usually adopted instead of the secrecy rate as given in the references, which will not be discussed in this chapter.

AN is utilized to enhance the secrecy performance. The signal sent by BS can be denoted by $\mathbf{x} = \sum_{k=1}^{K} \mathbf{w}_k s_k + \mathbf{z}$, where $\mathbf{w}_k \in \mathbb{C}^{N\times 1}$ denotes the beamforming vector of user k, $\mathbf{z} \in \mathbb{C}^{N\times 1}$ represents the AN vector, jointly satisfying the power constraint $\mathbf{z}^H\mathbf{z} + \sum_{k=1}^{K} \mathbf{w}_k^H \mathbf{w}_k \leq P$ with P as the maximum total transmit power. $s_k \in \mathbb{C}^{1\times 1}$ is the information symbol for user k satisfying $\mathbb{E}|s_k|^2 = 1$. The received signal at user k or Eve is

$$y_{\{k,e\}} = \mathbf{v}_{l_{\{k,e\}}} \mathbf{H}_{\{k,e\}} \left(\sum_{k=1}^{K} \mathbf{w}_k s_k + \mathbf{z} \right) + n_{\{k,e\}}, \tag{17.1}$$

where $l_{\{k,e\}}$ indicates that user k or Eve locates in the reflecting ($l = 1$) or transmitting area ($l = 2$). Moreover, $\mathbf{v}_l = [\mathbf{u}_l \ 1]$, $\mathbf{u}_l = [u_{l,1}, u_{l,2}, \ldots, u_{l,M}]$, with $u_{l,m} = \sqrt{\alpha_{l,m}} e^{j\theta_{l,m}}$, $\alpha_{l,m} \in [0,1]$, $\theta_{l,m} \in [0, 2\pi)$, $m = 1, 2, \ldots, M$, $\mathbf{H}_{\{k,e\}} = \left[\mathbf{G}^H \operatorname{diag}\left\{\mathbf{r}_{\{k,e\}}\right\}^H, \ \mathbf{h}_{\{k,e\}}^H \right]^H$, and $n_{\{k,e\}}$ is the complex additive Gaussian white noise (AWGN) with variance σ^2.

NOMA is adopted for simultaneous information transmission to multiple users. Let $\pi(k)$ denote the decoding order of user k. Then the SINR for the jth user to decode symbol s_k can be obtained by

$$\gamma_{k\to j} = \frac{\varsigma_{k\to j}}{\sum_{\pi(k)<\pi(t)} \varsigma_{t\to j} + \varsigma_{\mathbf{z}\to j} + \sigma_j^2}, \forall \pi(k) \leq \pi(j), \tag{17.2}$$

where $\varsigma_{t\to j} \triangleq \left|\mathbf{v}_{l_j} \mathbf{H}_j \mathbf{w}_t\right|^2$ for $\forall \pi(k) \leq \pi(t)$, $\varsigma_{\mathbf{z}\to j} \triangleq \left|\mathbf{v}_{l_j} \mathbf{H}_j \mathbf{z}\right|^2$. And the SINR for Eve to decode symbol s_k can be written by

$$\gamma_{k\to e} = \frac{\varsigma_{k\to e}}{\sum_{t\neq k} \varsigma_{t\to e} + \varsigma_{\mathbf{z}\to e} + \sigma_e^2}, \forall k, \tag{17.3}$$

where $\varsigma_{t\to e} \triangleq \left|\mathbf{v}_{l_e} \mathbf{H}_e \mathbf{w}_t\right|^2$, $\varsigma_{\mathbf{z}\to e} \triangleq \left|\mathbf{v}_{l_e} \mathbf{H}_e \mathbf{z}\right|^2$. Note here we assume Eve cannot decode any user's information symbol since the following secrecy constraint should be satisfied:

$$\log_2(1 + \gamma_{k\to k}) - \log_2(1 + \gamma_{k\to e}) \geq 0, \forall k. \tag{17.4}$$

17.3.2 Problem formulation and proposed solution

We aim to maximize the sum secrecy rate via a joint design of the beamforming vectors $\mathbf{w}_k$ and AN vector $\mathbf{z}$ at BS, reflecting and transmitting beamforming vectors $\mathbf{u}_1, \mathbf{u}_2$ at STAR-RIS, and the decoding orders π of NOMA. There are $K!$ different decoding orders for K users, which is usually a small value since the number of users supported by NOMA is restricted by the complexity of SIC [4]. Then the optimal decoding order can be easily obtained by searching over all combinations. In order to simplify the description, we formulate the problem with a given decoding order as below.

For a certain decoding order π, the secrecy rate maximization problem is formulated as

$$\textbf{(P1)}: \max_{\substack{\{\mathbf{w}_k, R_{k\to k}\},\\ \mathbf{z},\mathbf{v}_1,\mathbf{v}_2}} \sum_{k=1}^{K} \left(R_{k\to k} - \log_2\left(1+\gamma_{k\to e}\right)\right) \tag{17.5}$$

$$\text{s.t. } \mathbf{z}^H\mathbf{z} + \sum_{k=1}^{K} \mathbf{w}_k^H\mathbf{w}_k \le P, \tag{17.5a}$$

$$R_{k\to k} \ge \log_2\left(1+\gamma_{k\to e}\right), \forall k, \tag{17.5b}$$

$$\left|\mathbf{v}_{l_i}^H\mathbf{H}_i\mathbf{w}_j\right|^2 \le \left|\mathbf{v}_{l_i}^H\mathbf{H}_i\mathbf{w}_k\right|^2, \forall\pi(k) < \pi(j), \forall i \tag{17.5c}$$

$$R_{k\to k} \le \min_{\pi(k)\le\pi(j)} \log_2\left(1+\gamma_{k\to j}\right), \forall k, \tag{17.5d}$$

$$0 \le \left|\mathbf{v}_{1,m}\right|^2, \left|\mathbf{v}_{2,m}\right|^2, \left|\mathbf{v}_{1,m}\right|^2 + \left|\mathbf{v}_{2,m}\right|^2 \le 1 \tag{17.5e}$$

$$\left|\mathbf{v}_{1,M+1}\right|^2 = \left|\mathbf{v}_{2,M+1}\right|^2 = 1, \tag{17.5f}$$

where $R_{k\to k}$ denotes the data rate chosen for user k, (17.5a) restricts the total transmit power, (17.5b) guarantees the secrecy communication, (17.5c) guarantees rate fairness among all users, and (17.5d) guarantees that SIC can be carried out successfully. It is worth to mention that the above problem also applies to the conventional reflecting-only or transmitting-only RIS scenario by setting $\left|\mathbf{v}_{2,m}\right|^2 = 0$ or $\left|\mathbf{v}_{1,m}\right|^2 = 0, \forall m$.

The original problem **(P1)** is not jointly convex due to the coupling variables $\{\mathbf{w}_k\}$, $\mathbf{z}$ and $\mathbf{v}_1, \mathbf{v}_2$, which cannot be solved optimally and directly. To make this problem tractable, we first reformulate the original problem by applying the successive convex approximation (SCA) technique. Specifically, we first introduce slack variables $\{R_{k\to e}\}$ to replace the logarithmic functions in the objective, then add a new constraint

$$R_{k\to e} \ge \log_2\left(1+\gamma_{k\to e}\right), \forall k. \tag{17.6}$$

Then, we further introduce slack variables $\{c_e, d_{k\to e}\}$ and $\{a_{k\to j}, b_{k\to j}\}$ to replace the numerators and denominators of the fractional expressions in the logarithmic functions of (17.5d) and (17.6), respectively. Finally, Problem **(P1)** can be reformulated as

$$\textbf{(P2)}: \max_{\Delta} \sum_{k=1}^{K} \left(R_{k\to k} - R_{k\to e}\right) \tag{17.7}$$

$$\text{s.t. } R_{k\to k} \ge R_{k\to e}, \forall k, \tag{17.7a}$$

$$R_{k\to e} \ge c_e - d_{k\to e}, \forall k, \tag{17.7b}$$

$$R_{k\to k} \le a_{k\to j} - b_{k\to j}, \forall\pi(k) \le \pi(j), \tag{17.7c}$$

$$2^{a_{k\to j}} \le \sum_{\pi(k)\le\pi(t)}^{K} \varsigma_{t\to j} + \varsigma_{\mathbf{z}\to j} + \sigma_j^2, \forall\pi(k) \le \pi(j), \tag{17.7d}$$

$$2^{d_{k\to e}} \leq \sum\nolimits_{t\neq k}^{K} \varsigma_{t\to e} + \varsigma_{\mathbf{z}\to e} + \sigma_e^2, \forall k \tag{17.7e}$$

$$2^{b_{k\to j}} \geq \sum\nolimits_{\pi(k)<\pi(t)}^{K} \varsigma_{t\to j} + \varsigma_{\mathbf{z}\to j} + \sigma_j^2, \forall \pi(k) \leq \pi(j), \tag{17.7f}$$

$$2^{c_e} \geq \sum\nolimits_{t=1}^{K} \varsigma_{t\to e} + \varsigma_{\mathbf{z}\to e} + \sigma_e^2, \tag{17.7g}$$

$$(17.5a), (17.5c), (17.5e), (17.5f), \tag{17.7h}$$

where $\Delta \triangleq \left\{\left\{\mathbf{w}_k, R_{k\to k}, R_{k\to e}, c_e, d_{k\to e}, a_{k\to j}, b_{k\to j}\right\}, \mathbf{v}_1, \mathbf{v}_2, \mathbf{z}\right\}$ denotes the feasible parameters of the above problem.

To deal with the non-convexity of (17.7f) and (17.7g), we replace the left-hand side (LHS) with their lower bounds derived from the first-order Taylor expansions, denoted by $\alpha_{k,j} \triangleq 2^{b_{k\to j}^{(i-1)}} ln2 + 2^{b_{k\to j}^{(i-1)}}(b_{k\to j} - b_{k\to j}^{(i-1)})ln2, \forall \pi(k) \leq \pi(j)$ and $\kappa_e \triangleq 2^{c_e^{(i-1)}} ln2 + 2^{c_e^{(i-1)}}(c^e - c_e^{(i-1)})ln2$, respectively, where $b_{k\to j}^{(i-1)}$ and $c_e^{(i-1)}$ indicate the values of $b_{k\to j}$ and c_e in the $(i-1)$th iteration. Then constraints (17.7f) and (17.7g) become

$$\alpha_{k,j} \geq \sum\nolimits_{\pi(k)<\pi(t)}^{K} \varsigma_{t\to j} + \varsigma_{\mathbf{z}\to j} + \sigma_j^2, \forall \pi(k) \leq \pi(j), \tag{17.8}$$

$$\kappa_e \geq \sum\nolimits_{t=1}^{K} \varsigma_{t\to e} + \varsigma_{\mathbf{z}\to e} + \sigma_e^2. \tag{17.9}$$

Problem (**P2**) is still hard to be solved directly because variables $\mathbf{w}_k$, $\mathbf{z}$ and $\mathbf{v}_1, \mathbf{v}_2$ are still highly coupled. In the following, we decompose Problem (**P2**) into two sub-problems by optimizing the active beamforming vectors $\{\mathbf{w}_k\}, \mathbf{z}$ and the passive beamforming vectors $\mathbf{v}_1, \mathbf{v}_2$, separately, and propose an alternating optimization (AO) based iterative algorithm, where the two subproblems are alternately solved by utilizing the semi-definite relaxation (SDR) and SCA techniques, respectively.

17.3.2.1 Active beamforming and AN optimization

Given $\mathbf{v}_1$ and $\mathbf{v}_2$, we apply the rank lifting method by setting $\mathbf{W}_k = \mathbf{w}_k\mathbf{w}_k^H$, $\mathbf{Z} = \mathbf{z}\mathbf{z}^H$, $\mathbf{V}_1 = \mathbf{v}_1\mathbf{v}_1^H$, $\mathbf{V}_2 = \mathbf{v}_2\mathbf{v}_2^H$, then $\varsigma_{k\to j}, \varsigma_{\mathbf{z}\to j}, \varsigma_{k\to e}, \varsigma_{\mathbf{z}\to e}$ can be rewritten as $\varsigma_{k\to j} = \text{tr}\left\{\mathbf{W}_k\mathbf{H}_j^H\mathbf{V}_{l_j}\mathbf{H}_j\right\}$, $\varsigma_{\mathbf{z}\to j} = \text{tr}\left\{\mathbf{Z}\mathbf{H}_j^H\mathbf{V}_{l_j}\mathbf{H}_j\right\}$, $\varsigma_{k\to e} = \text{tr}\left\{\mathbf{W}_k\mathbf{H}_e^H\mathbf{V}_{l_e}\mathbf{H}_e\right\}$, $\varsigma_{\mathbf{z}\to e} = \text{tr}\left\{\mathbf{Z}\mathbf{H}_e^H\mathbf{V}_{l_e}\mathbf{H}_e\right\}$. We rewrite the corresponding constraints as, respectively

$$\text{tr}\left\{\mathbf{Z}\right\} + \sum\nolimits_{k=1}^{K} \text{tr}\left\{\mathbf{W}_k\right\} \leq P, \tag{17.10}$$

$$\text{tr}\left\{\mathbf{W}_j\mathbf{H}_i^H\mathbf{V}_{l_i}\mathbf{H}_i\right\} \leq \text{tr}\left\{\mathbf{W}_k\mathbf{H}_i^H\mathbf{V}_{l_i}\mathbf{H}_i\right\}, \forall \pi(k) < \pi(j), \forall i. \tag{17.11}$$

Then, the subproblem (**P2.1**) is formulated as

$$\textbf{(P2.1)}: \max_{\Delta_1} \sum_{k=1}^{K} (R_{k\to k} - R_{k\to e}) \tag{17.12}$$

$$\text{s.t. } (17.7) \text{ - } (17.7e), (17.8) \text{ - } (17.11), \tag{17.12a}$$

$$\mathbf{W}_k, \mathbf{Z} \succeq 0, \mathbf{W}_k, \mathbf{Z} \in \mathbf{H}^N, \forall k, \tag{17.12b}$$

$$\text{rank}\{\mathbf{W}_k\} = 1, \forall k, \tag{17.12c}$$

where $\Delta_1 = \left\{ \left\{ \mathbf{W}_k, R_{k\to k}, R_{k\to e}, c_e, d_{k\to e}, a_{k\to j}, b_{k\to j} \right\}, \mathbf{Z} \right\}$.

Remark 1. *Due to the rank-one constraints in Problem (**P2.1**), the above problem cannot be solved directly. By using the SDR approach and dropping the rank-one constraint [4], it can be reformulated as a convex semidefinite program (SDP) problem and efficiently solved. The optimal beamforming vectors can be obtained with* $\mathbf{W}_k^* = \mathbf{w}_k^* \mathbf{w}_k^{*H}$, $\mathbf{Z}^* = \mathbf{z}^* \mathbf{z}^{*H}$ *by using the Cholesky decomposition.*

17.3.2.2 Passive beamforming optimization

Given $\{\mathbf{w}_k\}$ and $\mathbf{z}$, the passive beamforming optimization subproblem is formulated as

$$\textbf{(P2.2)}: \max_{\Delta_2} \sum_{k=1}^{K} (R_{k\to k} - R_{k\to e}) \tag{17.13}$$

$$\text{s.t. } (17.7) \text{ - } (17.7e), (17.8) \text{ - } (17.11), \tag{17.13a}$$

$$0 \le [\mathbf{V}_1]_{m,m}, [\mathbf{V}_2]_{m,m}, [\mathbf{V}_1]_{m,m} + [\mathbf{V}_2]_{m,m} \le 1, \tag{17.13b}$$

$$[\mathbf{V}_1]_{M+1,M+1} = [\mathbf{V}_2]_{M+1,M+1} = 1, \tag{17.13c}$$

$$\mathbf{V}_1, \mathbf{V}_2 \succeq 0, \mathbf{V}_1, \mathbf{V}_2 \in \mathbf{H}^{M+1}, \forall k, \tag{17.13d}$$

$$\text{rank}\{\mathbf{V}_1\} = 1, \text{rank}\{\mathbf{V}_2\} = 1, \forall k, \tag{17.13e}$$

where $\Delta_2 = \left\{ \left\{ R_{k\to k}, R_{k\to e}, c_e, d_{k\to e}, a_{k\to j}, b_{k\to j} \right\}, \mathbf{V}_1, \mathbf{V}_2 \right\}$.

We cannot simply remove the rank-one constraint in Problem (P2.2) since the ranks of $\mathbf{V}_1, \mathbf{V}_2$ may not be 1. To deal with this problem, we adopt the sequential rank-one constraint relaxation (SROCR) algorithm [83] by replacing the rank-one constraint (17.13e) with its relaxed convex constraints (17.14a) and (17.14b), which control the largest eigenvalue to the trace ratios of $\mathbf{V}_1, \mathbf{V}_2$ with parameters $\omega_1, \omega_2 \in [0, 1]$, and provide a locally optimal rank-one solution. The relaxed problem is given by

$$\textbf{(P2.3)}: \max_{\Delta_2} \sum_{k=1}^{K} (R_{k\to k} - R_{k\to e}) \tag{17.14}$$

$$\text{s.t. } \mathbf{u}_{\max}\left(\mathbf{V}_1^{(i)}\right)^H \mathbf{V}_1 \, \mathbf{u}_{\max}\left(\mathbf{V}_1^{(i)}\right) \ge \omega_1^{(i)} \text{tr}\{\mathbf{V}_1\}, \tag{17.14a}$$

Algorithm 1: Iterative algorithm for solving Problem (**P2.2**)

Initialize convergence thresholds ξ_1, ξ_2 and feasible $\mathbf{V}^0$;
Solve Problem (**P2.3**) with $\omega_1^{(0)} = \omega_2^{(0)} = 0$ for $\mathbf{V}_1^{(0)}, \mathbf{V}_2^{(0)}$;
Set $i = 1$, initialize the step size $\delta^{(i)}$.

repeat
  Solve Problem (**P2.3**) with given $\omega_1^{(i)}, \mathbf{V}_1^{(i)}, \omega_2^{(i)}, \mathbf{V}_2^{(i)}$;
  if Problem (**P2.3**) is solvable **then**
    The optimal solution is denoted by $\mathbf{V}_1^{(i+1)}, \mathbf{V}_2^{(i+1)}$;
    $\delta^{(i+1)} = \delta^{(i)}$;
  else
    $\delta^{(i+1)} = \delta^{(i)}/2, \mathbf{V}_1^{(i+1)} = \mathbf{V}_1^{(i)}, \mathbf{V}_2^{(i+1)} = \mathbf{V}_2^{(i)}$;
  end if
  $\omega_{\{1,2\}}^{(i+1)} = \min\left(1, \frac{\lambda_{\max}\left(\mathbf{V}_{\{1,2\}}^{(i+1)}\right)}{\mathrm{tr}\left(\mathbf{V}_{\{1,2\}}^{(i+1)}\right)} + \delta^{(i+1)}\right)$;
  $i = i + 1$;
until $\left|1 - \omega^{(i-1)}\right| \leq \xi_1$ and the fractional reduce of the objective value is below the threshold ξ_2

$$\mathbf{u}_{\max}\left(\mathbf{V}_2^{(i)}\right)^H \mathbf{V}_2\, \mathbf{u}_{\max}\left(\mathbf{V}_2^{(i)}\right) \geq \omega_2^{(i)} \mathrm{tr}\left\{\mathbf{V}_2\right\}, \tag{17.14b}$$

$$\text{(17.7)-(17.7e), (17.8)-(17.11), (17.13b)-(17.13d),} \tag{17.14c}$$

where $\mathbf{V}_1^{(i)}, \mathbf{V}_2^{(i)}$ are the optimal solutions in the ith iteration with $\omega_1^{(i)}$, $\omega_2^{(i)}$, and $\mathbf{u}_{\max}\left(\mathbf{V}_1^{(i)}\right)$ and $\mathbf{u}_{\max}\left(\mathbf{V}_2^{(i)}\right)$ denote the eigenvectors corresponding to the maximal eigenvalues $\lambda_{\max}\left(\mathbf{V}_1^{(i)}\right)$ and $\lambda_{\max}\left(\mathbf{V}_2^{(i)}\right)$.

17.3.2.3 Overall iterative algorithm and complexity analysis

The proposed iterative optimization algorithm is summarized in Algorithm 2. According to [84], the total complexity is $O\left(I_{ite}^N\left(A + I_{ite}^S B\right)\right)$, where $A = \max\{N, m\}^4\sqrt{N}\log\frac{1}{\xi}$, $B = \max\{M+1, m\}^4\sqrt{M+1}\log\frac{1}{\xi}$, $m = \frac{K^2(K+1)}{2} + K + 1$ and ξ denotes the accuracy, I_{ite}^S and I_{ite}^N denote the number of iterations for Algorithms 1 and 2, respectively. This indicates that the complexity of the proposed algorithm is pretty low.

17.3.3 Simulation results

In this section, numerical results are presented to evaluate the performance of the proposed algorithm. The locations of BS, STAR-RIS, and users are illustrated in Figure 17.2, where $K = 2$ users are considered and two different setups for Eve is studied according to Eve's location, respectively, defined as Cases 1 and 2.

Algorithm 2: Overall iterative algorithm

Initialize convergence threshold $\xi > 0$, maximum iterations ε
Find a feasible set $\left\{\mathbf{W}_k^{(0)}\right\}, \mathbf{Z}^{(0)}, \mathbf{V}_1^{(0)}, \mathbf{V}_2^{(0)}$ to Problem (**P1**);
Set iteration round $l = 1$, calculate $b_{k\to j}$ and c_e;
repeat
    Solve subproblem (**P2.1**) for $\left\{\mathbf{W}_k^{(l+1)}\right\}, \mathbf{Z}^{(l+1)}$;
    Solve subproblem (**P2.2**) for $\mathbf{V}_1^{(l+1)}, \mathbf{V}_2^{(l+1)}$ applying Algorithm 1;
    Update the parameters $c_e^{(l)} = c_e, b_{k\to j}^{(l)} = b_{k\to j}$;
    $l = l + 1$
until the fractional reduce of the objective value is below a threshold ξ or $l \geq \varepsilon$

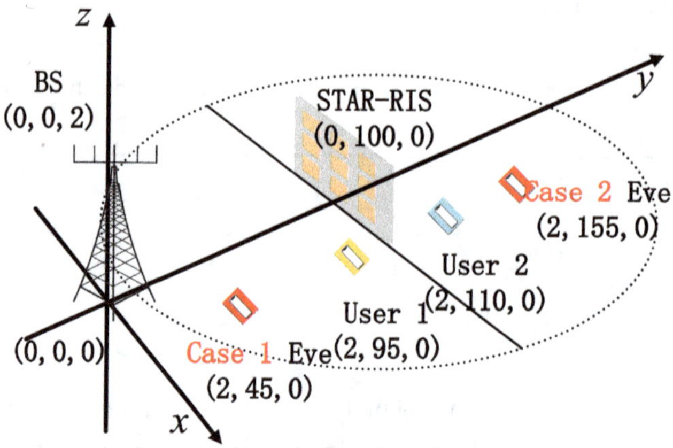

Figure 17.2 Simulation setups. ©2023. Reprinted with permission from [3].

The wireless channels are modeled as $\mathbf{h} = \sqrt{L_0 d^{\alpha}}\tilde{\mathbf{h}}$, where L_0 is the large-scale path loss at $d = 1$ m, α is the path loss exponent, d denotes the distance between the sender and the receiver. $\tilde{\mathbf{h}}$ denotes the small-scale channels, which follows Rayleigh fading for the direct channels from BS to the users and Eve, while follows Rician fading for the STAR-RIS related channels. The Rician fading channel is modeled as $\tilde{\mathbf{h}} = \sqrt{\beta/(1+\beta)}\tilde{\mathbf{h}}^{\mathrm{LoS}} + \sqrt{1/(1+\beta)}\tilde{\mathbf{h}}^{\mathrm{NLoS}}$, where $\beta \geq 0$ is the Rician factor, $\tilde{\mathbf{h}}^{\mathrm{LoS}}$ denotes the deterministic line-of-sight (LoS) components, and $\tilde{\mathbf{h}}^{\mathrm{NLoS}}$ represents the non-LoS (NLoS) components, which is modeled as Rayleigh fading.

If not specified, the simulation parameters are set as follows. The path loss exponents of the BS-RIS, RIS-user/Eve, and BS-user/Eve channels are $L_0 = -30$ dB, $\alpha = 2.2$, the Rician factors of BS-RIS and RIS-user channels are set as $\beta = 3$ dB, and the noise power is $\sigma^2 = BN_0$ with bandwidth $B = 100$ MHz, the effective noise power density of $N_0 = -150$ dBm/Hz. The total transmit power is $P = 30$ dBm. The

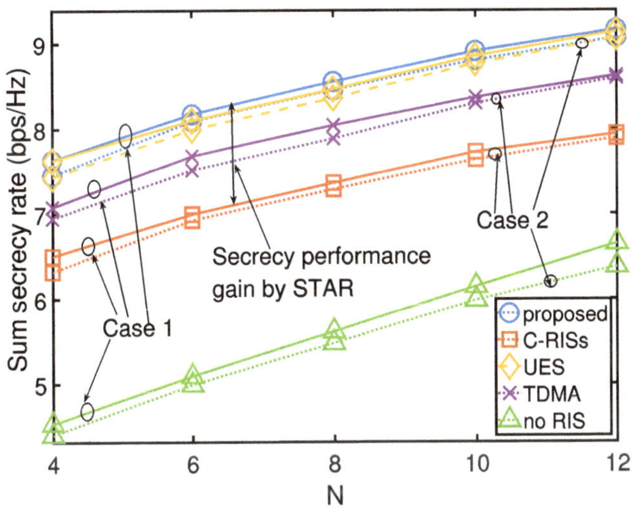

Figure 17.3 Secrecy performance versus the number of antennas at BS, $M = 30$*. ©IEEE 2023. Reprinted with permission from [3].*

location of Eve is as Case 1. All results are obtained over 200 independent channel realizations. The convergence threshold is set to be $\xi = 10^{-2}$.

First, we compare the performance of our proposed AN aided STAR-RIS NOMA scheme with three benchmark schemes: (1) C-RISs: The STAR-RIS is replaced by two conventional RISs with $M/2$ elements, where M is assumed to be even for simplicity. One of the two RISs is reflecting-only, the other one is transmitting-only. (2) uniform energy splitting scheme (also referred to as UES): In this case, we assume that all elements of the ES STAR-RIS employ the same amplitude coefficients for reflecting and transmitting, respectively. (3) TDMA scheme (also referred to as the TS mode of the STAR-RIS): BS and RIS serve each user with a minimum secrecy rate requirement in different orthogonal time slots.

From Figure 17.3, one can observe that the sum secrecy rate increases with the number of antennas at BS. This is expected since larger N enables a higher active beamforming gain, which in turn increases the secrecy performance. In addition, the secrecy performance of the proposed scheme is obviously better than other schemes. This is because the STAR-RIS can deploy flexible numbers of reflecting and transmitting elements, which provides more DoFs to enhance the desired signal strength than conventional RISs.

In Figure 17.4, we compare the secrecy performance of our proposed algorithm with baseline naive null-space AN scheme (also referred to as naive null), in which the AN is generated uniformly over the null space of the equivalent channels of users. We observe that our proposed algorithm can provide a much better AN signal model. However, the simple null space-based AN can approximately approach the optimal

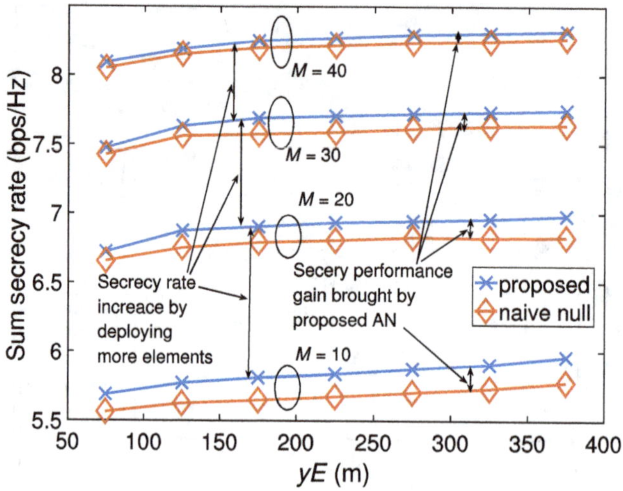

Figure 17.4 Secrecy performance versus yE, N = 4, the distance between BS and Eve increases with yE. ©IEEE 2023. Reprinted with permission from [3].

solution when the number of RIS elements is sufficiently large. This is because the naive null AN can efficiently degrade the reception at Eve with larger M, thus more power can be allocated to send the users' signals. Additionally, we can also find that when M is sufficiently large, the location of Eve has negligible impact on the sum secrecy rate, since the eavesdropping rate is always small for large values of M. These results imply that the introduction of STAR-RIS can effectively enhance the robustness of the sum secrecy rate with respect to (w.r.t.) Eve's locations and the strategy of AN, especially when M is large.

In Figure 17.5, we add extra constraints in Problem (**P1**) to ensure that the minimum data rates for users should be larger than 3 bps/Hz or 0 bps/Hz, respectively. It is interesting to observe that after exchanging the minimum data rate requirements between users, the achievable secrecy sum rate is almost unchanged in the STAR-RIS case, while relatively large gap happens in other cases. This implies that STAR-RIS can more effectively reconfigure the users' channels according to their QoS requirements.

It is further observed in Figure 17.6 that the proposed scheme requires less AN power but can achieve better secrecy performance than C-RISs scheme. One can also observe that the power required for AN in Case 2 is smaller than that in Case 1 since Eve in Case 2 is much further from BS and STAR-RIS. Additionally, the power ratio of AN decreases with the number of antennas at BS in Case 1 but increases in Case 2. This is because more DoFs become available for the beamforming of BS with larger N to degrade the reception at Eve, which thus renders the power required of AN decreases. However, when Eve is far away from BS in Case 2, the beamforming of BS is fully exploited to enhance the desired signals at the users, as a result, the

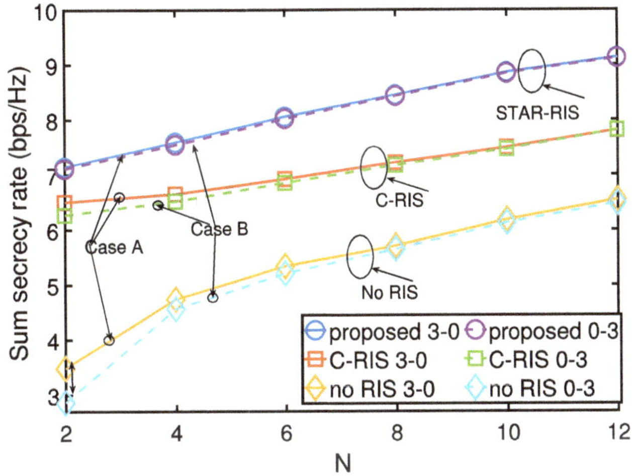

Figure 17.5 *Secrecy performance under different QoS requirements, M = 30. ©IEEE 2023. Reprinted with permission from [3].*

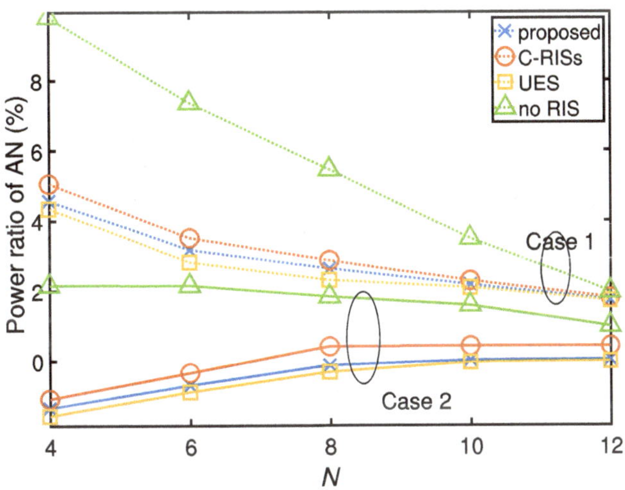

Figure 17.6 *The power required of AN versus the number of antennas, M = 30. ©IEEE 2023. Reprinted with permission from [3].*

power required by AN increases with N [85]. This insight provides useful guidelines for practical STAR-RIS implementation.

In Figure 17.7, we can find that the power ratio of AN can rapidly reach a fixed value with the increase of the number of elements at STAR-RIS, but decrease slowly

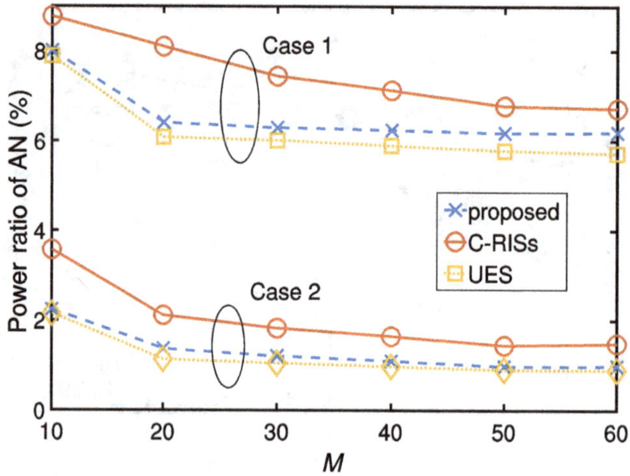

Figure 17.7 The power required of AN versus the number of elements, N = 4. ©IEEE 2023. Reprinted with permission from [3].

in the C-RIS scheme. It is reasonable since the STAR-RIS provides more DoFs than the C-RIS to decline the SINR at Eve with the help of AN. Moreover, we observe from Figures 17.3 and 17.7 that the RIS-assisted schemes achieve much higher secrecy sum rate with much lower AN power compared with the traditional system without RIS. This further proves that RIS can provide much greener secure transmission strategy for wireless networks. This advantage is more prominent when STAR-RIS is adopted.

17.3.4 Conclusion

The sum secrecy rate maximization problem in a new scenario of STAR-RIS-assisted NOMA transmission system was studied. Simulation results validated that the secrecy performance of the proposed scheme was better than the benchmark schemes. The optimal AN strategy generated by the proposed scheme could significantly improve the secrecy rate compared with the null-space-based AN. However, as the number of RIS elements increases, the proposed scheme becomes robust to the AN strategy. When Eve is close to the BS, the AN power increases with the number of antennas at BS, while it decreases when Eve is close to the STAR-RIS. These insights can provide practical guidance for the design of AN assisted STAR-RIS network.

In this chapter, we assume that the reflecting and transmitting phase shifts can be independently and continuously adjusted, which requires complex RIS hardware. A more practical and low-cost model is that the phase-shift coefficients are drawn from a discrete set and the transmission and reflection phase shifts are coupled with each other[86], which introduces new design challenges for future work.

17.4 Heterogeneous internal secrecy for NOMA assisted by RIS

In this chapter, the RIS-assisted secure NOMA communication with heterogeneous internal secrecy requirements is studied [4].

17.4.1 System model

As shown in Figure 17.8, we consider an RIS-assisted downlink multiple-input–multiple-output (MISO) system serving $K \geq 2$ users with heterogeneous internal secrecy requirements. In particular, we assume legitimate but curious users who may act as internal eavesdroppers. Each user k is at a certain secrecy level, denoted by $\mathscr{L}(k) \geq 0, \forall k$. The user at a higher level does not trust those at lower levels and treats them as internal eavesdroppers. Moreover, each user requires a minimum secrecy rate $r_{s,k} \geq 0, \forall k$. Note for users with no secrecy demands, we set $\mathscr{L}(k) = 0$ and $r_{s,k} = 0$.

There are $N \geq 1$ antennas at the BS, $M \geq 1$ reflecting elements at the RIS, and all users are equipped with a single antenna. The channels from BS to user k, RIS to user k and BS to RIS are denoted by $\mathbf{h}_k \in \mathbb{C}^{1\times N}$, $\mathbf{r}_k \in \mathbb{C}^{1\times M}$, and $\mathbf{G} \in \mathbb{C}^{M\times N}$, respectively. All channels follow independent block fading and are perfectly known.

NOMA is adopted for simultaneous information transmission to multiple users. Let s_k denote the information symbol for user k with $\mathbb{E}|s_k|^2 = 1$ and $\mathbf{w}_k \in \mathbb{C}^{N\times 1}$ denote the corresponding active beamforming vector. Then the combined transmit signal for NOMA is denoted as $\mathbf{x} = \sum_{k=1}^{K} \mathbf{w}_k s_k$, and the received signal at user k can be expressed as [83]

$$y_k = \mathbf{v}\mathbf{H}_k \sum_{i=1}^{K} \mathbf{w}_i s_i + n_k, \tag{17.15}$$

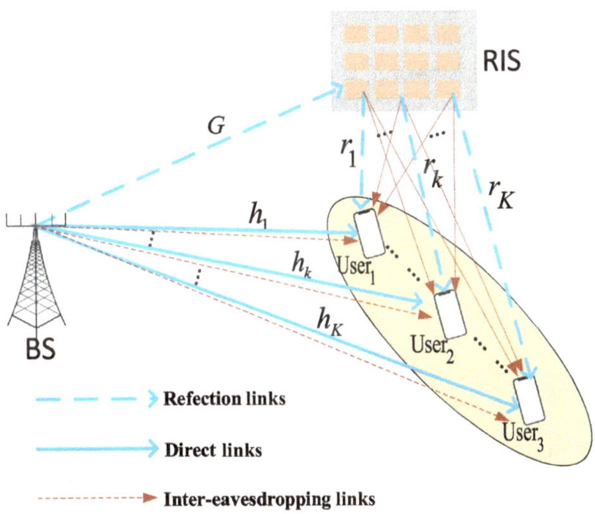

Figure 17.8 RIS-assisted NOMA with internal secrecy requirements. ©IEEE 2023. Reprinted with permission from [4].

where $\mathbf{v} = [\mathbf{u}\ 1]$, $\mathbf{u} = [\mu_1, \ldots, \mu_M]$, $\mu_m = \alpha_m e^{j\theta_m}$, $\alpha_m \in [0, 1]$, $\theta_m \in [0, 2\pi)$, $\mathbf{H}_k = \left[\mathbf{G}^H \operatorname{diag}\{\mathbf{r}_k\}^H,\ \mathbf{h}_k^H\right]^H$, and n_k is the AWGN with variance σ^2.

Let $\pi(k)$ denote the decoding order of user k. The SINR for user j to decode symbol s_k is expressed as

$$\gamma_{k\to j} = \frac{|\mathbf{v}\mathbf{H}_j\mathbf{w}_k|^2}{\sum_{\pi(i)>\pi(k)} |\mathbf{v}\mathbf{H}_j\mathbf{w}_i|^2 + \sigma^2}, \forall \pi(j) \geq \pi(k). \tag{17.16}$$

Here perfect SIC is assumed, so that we can focus our study on the internal secrecy of the RIS-assisted NOMA. In fact, the derivations listed in the following also apply to the imperfect SIC scenario, by adding $\sum_{\pi(i)<\pi(k)} \rho_j |\mathbf{v}\mathbf{H}_j\mathbf{w}_i|^2$ to the denominator [87], where $\rho_j \in [0, 1]$ denotes the residual interference factor due to imperfect SIC at user j.

To guarantee the internal secrecy, the user at higher secrecy level should be configured with a higher decoding order. That is, $\pi(k) < \pi(i)$ if $\mathscr{L}(k) < \mathscr{L}(i)$. The eavesdropping SINR of user i to the information of user k is

$$\hat{\gamma}_{k\to i} = \frac{|\mathbf{v}\mathbf{H}_i\mathbf{w}_k|^2}{\sum_{m\in\Psi_{k,i}\backslash k} |\mathbf{v}\mathbf{H}_i\mathbf{w}_m|^2 + \sigma^2}, \forall \pi(i) < \pi(k), \tag{17.17}$$

where $\Psi_{k,i} = \{m : \pi(m) > \pi(i), \mathscr{L}(m) > 0\}$ denotes the set of users whose decoding order is higher than that of user i. Note here we assume a worst case that the symbols of the users at $\mathscr{L}(m) = 0$ can always be successfully decoded and canceled, since they have no secrecy constraints.

According to the well-known Wyner's encoding scheme [88], the transmitter sets two rates for each user, i.e., the transmission rate $R_{t,k}$ and the secrecy rate $R_{s,k}$. The data rate $R_{t,k}$ chosen by user k should satisfy $r_{t,k} \leq R_{t,k} \leq \min_{\pi(j)\geq\pi(k)} \log_2(1 + \gamma_{k\to j})$, where $r_{t,k}$ indicates the minimum QoS requirement. Perfect secrecy is achieved when $C_{e,k} \leq R_{t,k} - R_{s,k}$, where $C_{e,k} = \max_{\pi(i)<\pi(k)} \log_2\left(1 + \hat{\gamma}_{k\to i}\right), \forall \mathscr{L}(k)$ denotes the internal eavesdropping rate of user k. That is, $R_{s,k}$ should satisfy $R_{s,k} \leq R_{t,k} - \max_{\pi(i)<\pi(k)} \log_2\left(1 + \hat{\gamma}_{k\to i}\right), \forall \mathscr{L}(k) > 0$.

17.4.2 Problem formulation and derivations

Our goal is to minimize the total transmit power under users' individual secrecy requirements, by optimizing the beamforming vectors $\{\mathbf{w}_k\}$, the passive reflecting vector $\mathbf{u}$, and the decoding orders π. Set $\bar{K}$ as the number of users with secrecy requirements. Based on the above analysis, there are $\left(K - \bar{K}\right)!$ different orders, which is usually a small value since the number of users supported by NOMA is restricted by the complexity of SIC. As a result, the optimal order can be easily obtained by searching over all combinations.

For a certain decoding order π, the power minimization problem is defined as

$$\textbf{(P3)}: \min_{\mathbf{v},\{\mathbf{w}_k\},\{R_{t,k}\},\{R_{s,k}\}} \sum\nolimits_{k=1}^{K} \|\mathbf{w}_k\|^2 \tag{17.18a}$$

$$\text{s.t.} \quad R_{t,k} \geq r_{t,k}, \forall k, \tag{17.18b}$$

$$R_{s,k} \geq r_{s,k}, \forall \mathscr{L}(k) > 0, \tag{17.18c}$$

$$R_{t,k} - R_{s,k} \geq \max_{\pi(i)<\pi(k)} \log_2 (1 + \hat{\gamma}_{k,i}), \forall \mathscr{L}(k) > 0, \tag{17.18d}$$

$$R_{t,k} \leq \min_{\pi(i)\geq\pi(k)} \log_2 (1 + \gamma_{k,i}), \forall k \tag{17.18e}$$

$$|\mathbf{v}\mathbf{H}_k\mathbf{w}_i|^2 \leq |\mathbf{v}\mathbf{H}_k\mathbf{w}_j|^2, \forall \pi(i) > \pi(j), \forall k, \tag{17.18f}$$

$$0 \leq |\mathbf{v}(l)|^2 \leq 1, 1 \leq l \leq M, \mathbf{v}(M+1) = 1, \tag{17.18g}$$

where (17.18b) restricts the minimum QoS, (17.18c) and (17.18d) guarantee the internal secrecy, and (17.18e) and (17.18f) are drawn from the NOMA principle.

Problem **(P3)** is non-convex and hard to be solved optimally. In this paper, we propose an efficient sub-optimal algorithm by using the SCA technique. Specifically, we introduce slack variables $\{c_{k,i}, d_{k,i}\}$, $\forall \pi(i) < \pi(k), \mathscr{L}(k) > 0$, and $\{a_{k,i}, b_{k,i}\}$, $\forall \pi(i) \geq \pi(k)$ to replace the numerators and denominators of the fractional expressions in the logarithmic functions of (17.18d) and (17.18e), respectively. Then **(P3)** is transformed into

$$\textbf{(P4)}: \min_{\Delta,\mathbf{v},\{\mathbf{w}_k\}} \sum\nolimits_{k=1}^{K} \|\mathbf{w}_k\|^2 \tag{17.19a}$$

$$\text{s.t.} \quad R_{t,k} - R_{s,k} \geq c_{k,i} - d_{k,i}, \forall \mathscr{L}(k) > 0, \forall \pi(i) < \pi(k), \tag{17.19b}$$

$$e^{c_{k,i}} \geq \sum\nolimits_{j\in\Psi_{k,i}} |\mathbf{v}\mathbf{H}_i\mathbf{w}_j|^2 + \sigma^2, \forall \mathscr{L}(k) > 0, \forall \pi(i) < \pi(k), \tag{17.19c}$$

$$e^{d_{k,i}} \leq \sum\nolimits_{j\in\Psi_{k,i}\setminus k} |\mathbf{v}\mathbf{H}_i\mathbf{w}_j|^2 + \sigma^2, \forall \mathscr{L}(k) > 0, \forall \pi(i) < \pi(k), \tag{17.19d}$$

$$R_{t,k} \leq a_{k,i} - b_{k,i}, \forall k, \forall \pi(i) \geq \pi(k), \tag{17.19e}$$

$$e^{a_{k,i}} \leq \sum\nolimits_{\pi(j)\geq\pi(k)} |\mathbf{v}\mathbf{H}_i\mathbf{w}_j|^2 + \sigma^2, \forall \pi(i) \geq \pi(k), \tag{17.19f}$$

$$e^{b_{k,i}} \geq \sum\nolimits_{\pi(j)>\pi(k)} |\mathbf{v}\mathbf{H}_i\mathbf{w}_j|^2 + \sigma^2, \forall \pi(i) \geq \pi(k), \tag{17.19g}$$

$$(17.18b), (17.18c), (17.18f), (17.18g), \tag{17.19h}$$

where Δ denotes the parameter set consisting of $\{R_{t,k}\}$, $\{R_{s,k}\}$, $\{a_{k,i}, b_{k,i}\}$, and $\{c_{k,i}, d_{k,i}\}$.

To deal with the non-convexity of constraints (17.19c) and (17.19g), we further replace the LHS with the lower bounds derived from the first-order Taylor expansions at given local points $c_{k,i}^{(t-1)}$ and $b_{k,i}^{(t-1)}$, respectively

$$e^{c_{k,i}} \geq \lambda_{k,i} \triangleq e^{c_{k,i}^{(t-1)}} + e^{c_{k,i}^{(t-1)}}(c^{k,i} - c_{k,i}^{(t-1)}), \tag{17.20}$$

$$e^{b_{k,i}} \geq \kappa_{k,i} \triangleq e^{b_{k,i}^{(t-1)}} + e^{b_{k,i}^{(t-1)}}(b_{k,i} - b_{k,i}^{(t-1)}). \tag{17.21}$$

Problem **(P4)** is still hard to be solved directly because variables $\mathbf{v}$ and $\{\mathbf{w}_k\}$ are highly coupled. In the following, we decompose Problem **(P4)** into the beamforming optimization subproblem and the phase shifts optimization subproblem, and propose an iterative algorithm using AO, where the two subproblems are alternately solved by utilizing the SDR and SCA techniques, respectively.

17.4.2.1 Beamforming optimization

For a given $\mathbf{v}$, we reformulate **(P4)** as a SDP problem by using the matrix lifting technique. Specifically, we define $\mathbf{W}_k = \mathbf{w}_k\mathbf{w}_k^H, \forall k$, and rewrite the constraints as, respectively

$$\text{tr}\left\{\mathbf{W}_i\mathbf{H}_k^H\mathbf{V}\mathbf{H}_k\right\} \le \text{tr}\left\{\mathbf{W}_j\mathbf{H}_k^H\mathbf{V}\mathbf{H}_k\right\}, \forall\pi(i) > \pi(j), \forall k, \tag{17.22}$$

$$\lambda_{k,i} \ge \sum\nolimits_{j\in\Psi_{k,i}} \text{tr}\left\{\mathbf{W}_j\mathbf{H}_i^H\mathbf{V}\mathbf{H}_i\right\} + \sigma^2, \forall\mathscr{L}(k) > 0, \forall\pi(i) < \pi(k), \tag{17.23}$$

$$e^{d_{k,i}} \le \sum\nolimits_{j\in\Psi_{k,i}\backslash k} \text{tr}\left\{\mathbf{W}_j\mathbf{H}_i^H\mathbf{V}\mathbf{H}_i\right\} + \sigma^2, \forall\mathscr{L}(k) > 0, \forall\pi(i) < \pi(k), \tag{17.24}$$

$$e^{a_{k,i}} \le \sum\nolimits_{\pi(j)\ge\pi(k)} \text{tr}\left\{\mathbf{W}_j\mathbf{H}_i^H\mathbf{V}\mathbf{H}_i\right\} + \sigma^2, \forall\pi(i) \ge \pi(k), \tag{17.25}$$

$$\kappa_{k,i} \ge \sum\nolimits_{\pi(j)>\pi(k)} \text{tr}\left\{\mathbf{W}_j\mathbf{H}_i^H\mathbf{V}\mathbf{H}_i\right\} + \sigma^2, \forall\pi(i) \ge \pi(k), \tag{17.26}$$

which are convex w.r.t. $\mathbf{W}_k$. Based on the above, the subproblem **(P4.1)** is defined as

$$(\mathbf{P4.1}): \min_{\Delta,\{\mathbf{W}_k\}} \sum\nolimits_{k=1}^{K} \text{tr}\left\{\mathbf{W}_k\right\} \tag{17.27a}$$

$$\text{s.t.}\quad (17.18b), (17.18c), (17.19b), (17.19e), (17.22)-(17.26), \tag{17.27b}$$

$$\mathbf{W}_i \succeq 0, \mathbf{W}_i \in \mathbb{H}^N, \forall i \tag{17.27c}$$

$$\text{rank}\left\{\mathbf{W}_i\right\} = 1, \forall i. \tag{17.27d}$$

Note **(P4.1)** is still non-convex due to the rank-one constraint (17.27d). By using the SDR approach and dropping the rank-one constraint according to Lemma 1 given in [4], which indicates that the rank-one solution of **(P4.1)** can be always obtained, it can be reformulated as a convex SDP problem and efficiently solved with readily available software packages. The optimal beamforming vectors can be obtained with $\mathbf{W}_k^* = \mathbf{w}_k^*\mathbf{w}_k^{*H}$ by using the Cholesky decomposition.

17.4.2.2 Phase shifts optimization

For given $\{\mathbf{w}_k\}$, the non-convex constraints (17.18f), (17.19d), and (17.19f) of **(P4)** are transformed into their convex approximation expressions by using SCA. Specifically, the RHS is replaced with its lower bound derived from the first-order Taylor expansion at the given local point $\mathbf{v}^{(t)}$:

$$|\mathbf{v}\mathbf{H}_k\mathbf{w}_i|^2 \ge \vartheta_{i,k} \triangleq |\mathbf{v}^{(t)}\mathbf{H}_k\mathbf{w}_i|^2 + 2\Re\left\{\mathbf{v}^{(t)}\mathbf{H}_k\mathbf{W}_i\mathbf{H}_k^H\left(\mathbf{v}-\mathbf{v}^{(t)}\right)\right\}, \tag{17.28}$$

and the constraints become

$$|\mathbf{v}\mathbf{H}_k\mathbf{w}_i|^2 \le \vartheta_{i,k}, \forall\pi(i) > \pi(j), \forall k, \tag{17.29}$$

Algorithm 3: Proposed iterative algorithm for solving **(P4)**

Initialization:
Find a feasible set $\{\mathbf{w}_k^{(0)}\}$ and $\mathbf{v}^{(0)}$ to **(P3)**;
Calculate $\{b_{k,i}^{(0)}\}$, $\{c_{m,k}^{(0)}\}$ according to (17.19c) and (17.19g);
Set $t = 1$.

1: **repeat**
Calculate $\{\lambda_{k,i}^{(t)}\}$ and $\{\kappa_{k,i}^{(t)}\}$ according to (17.20) and (17.21).
Solve **(P4.1)** with given $\mathbf{v}^{(t)}$ for $\{\mathbf{w}_k^{(t+1)}\}$.
Calculate $\{\vartheta_{i,k}^{(t)}\}$ according (17.28).
Find $\mathbf{v}^{(t+1)}$ by solving **(P4.2)** for given $\{\mathbf{w}_k^{(t+1)}\}$.
$t = t + 1$.
Update $\{b_{k,i}^{(t)}\}$, $\{c_{m,k}^{(t)}\}$.
2: **until** the fractional reduce of the objective value is below a threshold $\xi > 0$.

$$e^{d_{k,i}} \leq \sum\nolimits_{j \in \Psi_{k,i} \backslash k} \vartheta_{j,i} + \sigma^2, \mathscr{L}(k) > 0, \forall \pi(i) < \pi(k), \tag{17.30}$$

$$e^{a_{k,i}} \leq \sum\nolimits_{\pi(j) \geq \pi(k)} \vartheta_{j,i} + \sigma^2, \forall \pi(i) \geq \pi(k). \tag{17.31}$$

The second subproblem **(P4.2)** is defined as

$$(\mathbf{P4.2}) : \textit{find} \;\; \mathbf{v} \tag{17.32a}$$

$$\text{s.t.} \;\; (17.18a), (17.18c), (17.18g), (17.19b), (17.19c), (17.19e), (17.19g), (17.29) - (17.31), \tag{17.32b}$$

which is convex and can be solved efficiently.

17.4.2.3 Iterative algorithm

The proposed iterative optimization algorithm is summarized in Algorithm 3. The initial feasible set is obtained from **(P1)** by setting $R_{t,k} = r_{t,k}, \forall k$ and $R_{s,k} = r_{s,k}, \forall \mathscr{L}(k) > 0$. The complexity of Algorithm 3 is $o\left(I_{ite}\left(\max\{m, N\}^4 \sqrt{N} \log\left(\frac{1}{\xi}\right) + n^{3.5}\right)\right)$, where $m = K\left(\frac{K(K+1)}{2} + 1\right) + \bar{K}(2K - \bar{K} - 1)$, $n = M + K(K+2) + \bar{K}(2K - \bar{K})$, ξ denotes the accuracy, and I_{ite} denotes the number of iterations.

It is worth to mention that Algorithm 3 can be easily extended to the non-ideal RIS scenarios with constant reflection amplitudes or discrete phase shifts by two steps. First, get the optimal $\theta_m = \text{angle}(\mathbf{v}(m)), \forall \leq M$, from Algorithm 3 by setting $|\mathbf{v}(m)|^2 = \alpha_m \in (0, 1], \forall m \leq M$ in (17.18g). Second, quantize θ_m to the nearest point [83] $\hat{\theta}_m = \arg\min_{\hat{\theta}_m \in \mathscr{D}} |\hat{\theta}_m - \theta_m|$, where $\mathscr{D} \triangleq \left\{\frac{2\pi n}{2^B}, n = 0, 1, \ldots, 2^B - 1\right\}$ and B denotes the number of resolution bits.

17.4.3 Numerical results

We consider a three-dimensional coordinate network, where the transmitter and the RIS are at (0, 0, 0) and (0, 50*m*, 5*m*), respectively. The users are randomly distributed in a region centered at (8*m*, 50*m*, 0). We set $K = 4$ and $\mathscr{L}(1) = \mathscr{L}(2) = 0$, $\mathscr{L}(4) > \mathscr{L}(3) > 0$. Since only users 3 and 4 have secrecy requirements, the decoding orders could be set as $\pi_1 = \{\pi(1) < \pi(2)\}$ or $\pi_2 = \{\pi(2) < \pi(1)\}$. The four users' distances to the BS are set as $\mathbf{d}_{BU} = [d_{BU,1}\ d_{BU,2}\ d_{BU,3}\ d_{BU,4}]$, where $d_{BU,k}, k = 1, ..., 4$ means the distance between the BS and user k. The noise power σ^2 is chosen as 10^{-7} dBm. All simulation results are obtained with the same randomly generated channel realization and user locations, and the convergence threshold is set to be $\xi = 10^{-2}$.

For comprehensive performance comparison, two baseline schemes are considered: (1) C-NOMA, i.e., NOMA without RIS, where the active beamforming vectors are optimized by solving (**P4.1**) with $\mathbf{v}(m) = 0, \forall m \leq M$; (2) OMA with or without RIS, where the users are served via time division multiple access. We propose a similar algorithm by alternatively optimizing the beamforming and the RIS shifts. Moreover, to better show the internal secrecy performance for either near users or far users, two user distribution scenarios are considered. One is defined as "C1" with $\mathbf{d}_{BU} = [53m\ 48m\ 55m\ 56m]$, which indicates that the far users 3 and 4 have secrecy requirements. The other is defined as "C2" with $\mathbf{d}_{BU} = [65m\ 65m\ 50m\ 30m]$, which means that near users 3 and 4 have secrecy requirements. If not specified, C1 is adopted in the simulations.

Figure 17.9 shows the minimum transmit power required to satisfy the internal secrecy requirements versus the number of RIS elements. It can be seen that the internal secrecy of either near users in C2 or the far users in C1 can be effectively

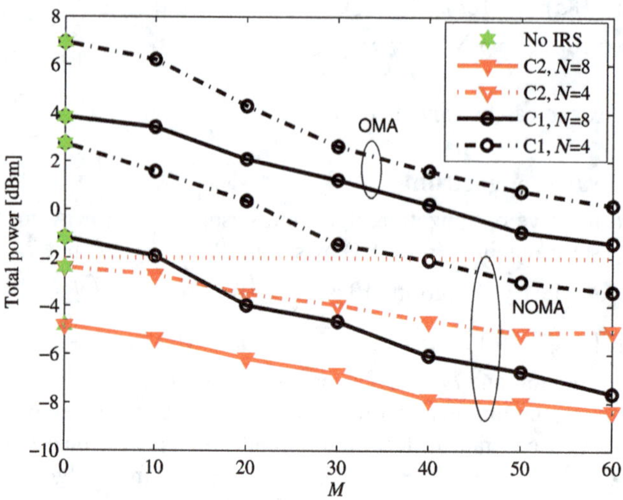

Figure 17.9 Total power versus N. We set $r_{t,k} = 1$ bps/Hz, $\forall k$, $r_{s,3} = r_{s,4} = 0.5$ bps/Hz. ©IEEE 2023. Reprinted with permission from [4].

guaranteed if RIS or multiple antennas at the BS can provide sufficient DoFs. For a given minimum transmit power constraint, e.g., −2 dBm, only the near users' internal secrecy can be guaranteed with more than 10 RIS elements if $N = 4$ and 40 RIS elements if $N = 8$. While when the minimum transmit power is restricted to 4 dBm, the internal secrecy for the far users can also be guaranteed. It is obvious that more power is required to guarantee the secrecy of the far users, since extra power is cost for the reflecting links to counteract the direct link disadvantage. However, the power gap between C1 and C2 decreases with M. This is because that the diversity gain produced by the RIS with a sufficiently large number of elements gradually dominates the channel gain of the direct link. This result indicates that RIS can help to provide much more efficient and flexible secrecy assurance.

Figure 17.9 also compares the secrecy performance of the proposed RIS-assisted NOMA with the baseline schemes. It shows that the RIS-assisted NOMA/OMA shows better performance than the C-NOMA/OMA. The performance gain created by RIS with $M = 10$ can reach about 100% for C1 with $N = 4$. Compared with OMA, NOMA shows higher power efficiency under the same secrecy constraints.

Figure 17.10 shows the transmit power required to satisfy the secrecy requirement of the far users. The required power increases with the secrecy requirement, while decreases with N. When the total power is restricted to be no larger than −2 dBm, C-NOMA (i.e., $M = 0$) cannot support the secrecy requirement of the far users, even though multiple antennas ($N = 4$) is available. Assisted by $M \geq 20$ or $M \geq 60$ RIS elements, the secrecy of the far users becomes feasible, and a minimum $r_{s,4} = 0.2$ bps/Hz or $r_{s,4} = 1$ bps/Hz secrecy rate can be guaranteed, respectively.

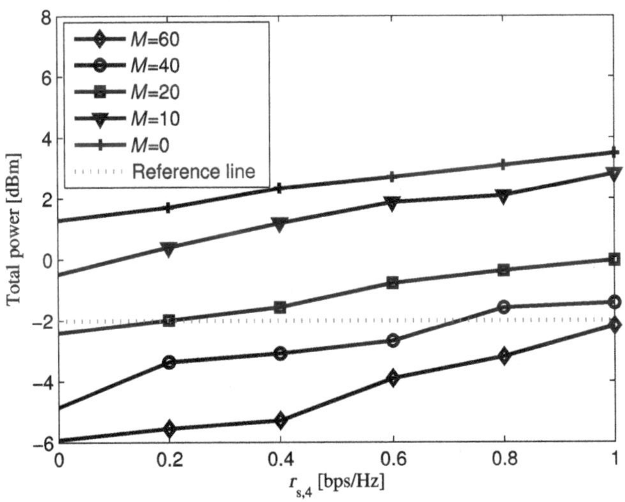

Figure 17.10 Total power versus secrecy rate requirement of FU. We set $K = 4$, $N = 4$, $r_{t,k} = 1$ bps/Hz, $\forall k$. ©IEEE 2023. Reprinted with permission from [4].

17.4.4 Conclusion

In this chapter, RIS has been adopted to enhance the internal secrecy of NOMA users with heterogeneous secrecy requirements. The active beamforming and the passive shifts have been jointly optimized to minimize the power consumption. It has been found that the proposed RIS-assisted NOMA algorithm can significantly reduce the power consumption under the users' individual secrecy constraints. With a small number of RIS elements, significant power reduction can be achieved compared with the C-NOMA. Moreover, the proposed RIS-assisted strategy can effectively enable the internal secrecy of any users while C-NOMA may fail in the far user cases. These results indicate the good efficiency and effectiveness of our proposed RIS-assisted algorithm in satisfying the heterogeneous internal secrecy of NOMA users.

To focus on the secrecy gain brought by RIS to NOMA, we made certain ideal assumptions, including perfectly known CSI, continuous RIS phase shifts, and perfect SIC capability at the receiver. It is necessary to further consider the practical imperfections, which is remained for future work.

References

[1] Liu Y, Zhang S, Mu X, *et al.* Evolution of NOMA toward next generation multiple access (NGMA) for 6G. *IEEE J Select Areas Commun.* 2022;40(4):1037–1071.

[2] Liu Y, Qin Z, Elkashlan M, *et al.* Enhancing the physical layer security of non-orthogonal multiple access in large-scale networks. *IEEE Trans Wirel Commun.* 2017;16(3):1656–1672.

[3] Han Y, Li N, Liu Y, *et al.* Artificial noise aided secure NOMA Communications in STAR-RIS networks. *IEEE Wirel Commun Lett.* 2022;11(6):1191–1195.

[4] Li N, Li M, Liu Y, *et al.* Intelligent reflecting surface assisted NOMA with heterogeneous internal secrecy requirements. *IEEE Wireless Commun Lett.* 2021;10(5):1103–1107.

[5] Hadani R, Rakib S, Tsatsanis M, *et al.* Orthogonal time frequency space modulation. In: *2017 IEEE Wireless Communications and Networking Conference (WCNC)*. San Francisco, CA, USA: IEEE; 2017. p. 1–6.

[6] Ding Z, Schober R, Fan P, *et al.* OTFS-NOMA: an efficient approach for exploiting heterogenous user mobility profiles. *IEEE Trans Commun.* 2019;67(11):7950–7965.

[7] Hu J, Shi J, Ma S, *et al.* Secrecy analysis for orthogonal time frequency space scheme based uplink LEO satellite communication. *IEEE Wirel Commun Lett.* 2021;10(8):1623–1627.

[8] Sun J, Wang Z, and Huang Q. Secure precoded orthogonal time frequency space modulation. In: *2021 13th International Conference on Wireless Communications and Signal Processing (WCSP)*. Changsha, China: IEEE; 2021. p. 1–5.

[9] Zhu Y, Wang L, Wong KK, *et al.* Secure communications in millimeter wave ad hoc networks. *IEEE Trans Wirel Commun.* 2017;16(5):3205–3217.

[10] Song Y, Yang W, Xiang Z, *et al.* Secure transmission in mmWave NOMA networks with cognitive power allocation. *IEEE Access*. 2019;7:76104–76119.

[11] Zhao H, Zhang J, Yang L, *et al.* Secure mmWave communications in cognitive radio networks. *IEEE Wireless Commun Lett*. 2019;8(4):1171–1174.

[12] Song Y, Yang W, Xiang Z, *et al.* An analysis on secure millimeter wave NOMA communications in cognitive radio networks. *IEEE Access*. 2020;8: 78965–78978.

[13] Sun X, Yang W, and Cai Y. Secure communication in NOMA-assisted millimeter-wave SWIPT UAV networks. *IEEE Internet Things J*. 2020;7(3): 1884–1897.

[14] Huang S, Xiao M, and Poor HV. On the physical layer security of millimeter wave NOMA networks. *IEEE Trans Veh Technol*. 2020;69(10):11697–11711.

[15] Song Y, Yang W, Xiang Z, *et al.* Research on cognitive power allocation for secure millimeter-wave NOMA networks. *IEEE Trans Veh Technol*. 2020;69(11):13424–13436.

[16] Sun X, Yang W, and Cai Y. Secure and reliable transmission in mmWave NOMA relay networks With SWIPT. *IEEE Syst J*. 2022;16(3):4861–4872.

[17] Solanki S, Gurjar DS, Sharma PK, *et al.* On the secrecy-reliability performance trade-off for NOMA-enabled 5G mmWave networks. In: *2021 IEEE 32nd Annual International Symposium on Personal, Indoor and Mobile Radio Communications (PIMRC)*. Helsinki, Finland: IEEE; 2021. p. 720–725.

[18] Yapici Y, Rupasinghe N, Guvenc I, *et al.* Physical layer security for NOMA transmission in mmWave drone networks. *IEEE Trans Veh Technol*. 2021;70(4):3568–3582.

[19] Zhao F, Hao W, Shen L, *et al.* Secure energy efficiency transmission for mmWave-NOMA system. *IEEE Syst J*. 2021;15(2):2226–2229.

[20] Zhao F, Hao W, Guo H, *et al.* Secure energy efficiency for mmWave-NOMA cognitive satellite terrestrial network. *IEEE Commun Lett*. 2023;27(1): 283–287.

[21] Cao Y, Wang S, Jin M, *et al.* Joint user grouping and power optimization for secure mmWave-NOMA systems. *IEEE Trans Wireless Commun*. 2022;21(5):3307–3320.

[22] Wang N, Jiao L, and Zeng K. Pilot contamination attack detection for NOMA in Mm-Wave and massive MIMO 5G communication. In: *2018 IEEE Conference on Communications and Network Security (CNS)*. Beijing, China: IEEE; 2018. p. 1–9.

[23] Wang N, Jiao L, Alipour-Fanid A, *et al.* Pilot contamination attack detection for NOMA in 5G Mm-Wave massive MIMO networks. *IEEE Trans Inform Forens Secur*. 2020;15:1363–1378.

[24] Nguyen VL, Lin PC, Cheng BC, *et al.* Security and privacy for 6G: a survey on prospective technologies and challenges. *IEEE Commun Surv Tutorials*. 2021;23(4):2384–2428.

[25] Obeed M, Salhab AM, Alouini MS, *et al.* On optimizing VLC networks for downlink multi-user transmission: a survey. *IEEE Commun Surv Tutorials*. 2019;21(3):2947–2976.

[26] Vaezi M, Aruma Baduge GA, Liu Y, *et al.* Interplay between NOMA and other emerging technologies: a survey. *IEEE Trans Cogn Commun Netw*. 2019;5(4):900–919.

[27] Obeed M, Salhab AM, Alouini MS, *et al.* Survey on physical layer security in optical wireless communication systems. In: *2018 Seventh International Conference on Communications and Networking (ComNet)*. Hammamet, Tunisia: IEEE; 2018. p. 1–5.

[28] Arfaoui MA, Soltani MD, Tavakkolnia I, *et al.* Physical layer security for visible light communication systems: a survey. *IEEE Commun Surv Tuts*. 2020;22(3):1887–1908.

[29] Yesilkaya A, Cogalan T, Erkucuk S, *et al.* Physical-layer security in visible light communications. In: *2020 2nd 6G Wireless Summit (6G SUMMIT)*. Levi, Finland: IEEE; 2020. p. 1–5.

[30] Zhao X, Chen H, and Sun J. On physical-layer security in multiuser visible light communication systems with non-orthogonal multiple access. *IEEE Access*. 2018;6:34004–34017.

[31] Du C, Zhang F, Ma S, *et al.* Secure transmission for downlink NOMA visible light communication networks. *IEEE Access*. 2019;7:65332–65341.

[32] Peng H, Wang Z, Han S, *et al.* Physical layer security for MISO NOMA VLC system under eavesdropper collusion. *IEEE Trans Veh Technol*. 2021;70(6):6249–6254.

[33] Liu X, Chen Z, Wang Y, *et al.* Robust artificial noise-aided beamforming for a secure MISO-NOMA visible light communication system. *China Commun*. 2020;17(11):42–53.

[34] Liu X, Wang Y, Zhou F, *et al.* Beamforming design for secure MISO visible light communication networks with SLIPT. *IEEE Trans Commun*. 2020;68(12):7795–7809.

[35] Su N, Panayirci E, Koca M, *et al.* Transmit precoding for physical layer security of MIMO-NOMA-based visible light communications. In: *2021 17th International Symposium on Wireless Communication Systems (ISWCS)*. Berlin, Germany: IEEE; 2021. p. 1–6.

[36] Obeed M, Chaaban A, Salhab AM, *et al.* Physical Layer Security in Cooperative NOMA Hybrid VLC/RF Systems. arXiv; 2020.

[37] Zhao X and Sun J. Secure reconfigurable intelligent surface aided heterogeneous VLC–RF cooperative NOMA networks. *Opt Commun*. 2022;511:127983.

[38] Kumar A and Jayakody DNK. Secure NOMA assisted multi-LED underwater visible light communication. *IEEE Trans Veh Technol*. 2022;71(7):7769–7779.

[39] Zhao X and Sun J. Physical-layer security for mobile users in NOMA-enabled visible light communication networks. *IEEE Access*. 2020;8: 205411–205423.

[40] Bahadori F, Sajad Sadough SM, and Ghassemlooy Z. The sum secrecy rate of NOMA-enabled VLC network with the random-way point mobility model. In: *2021 Third South American Colloquium on Visible Light Communications (SACVLC)*. Toledo, Brazil: IEEE; 2021. p. 01–06.

[41] Shahab MB, Abbas R, Shirvanimoghaddam M, *et al.* Grant-free non-orthogonal multiple access for IoT: a survey. *IEEE Commun Surv Tuts.* 2020;22(3):1805–1838.
[42] Zhang C, Liu Y, and Ding Z. Semi-grant-free NOMA: a stochastic geometry model. *IEEE Trans Wireless Commun.* 2022;21(2):1197–1213.
[43] Ding Z, Schober R, Fan P, *et al.* Simple semi-grant-free transmission strategies assisted by non-orthogonal multiple access. *IEEE Trans Commun.* 2019;67(6):4464–4478.
[44] Lei H, Yang F, Liu H, *et al.* On Secure NOMA-Aided Semi-Grant-Free Systems. arXiv; 2022.
[45] Liu F, Cui Y, Masouros C, *et al.* Integrated sensing and communications: toward dual-functional wireless networks for 6G and beyond. *IEEE J Select Areas Commun.* 2022;40(6):1728–1767.
[46] Mu X, Wang Z, Liu Y. NOMA for Integrating Sensing and Communications towards 6G: A Multiple Access Perspective. arXiv; 2022.
[47] Mu X, Liu Y, Guo L, *et al.* NOMA-aided joint radar and multicast-unicast communication systems. *IEEE J Select Areas Commun.* 2022;40(6):1978–1992.
[48] Wang Z, Liu Y, Mu X, *et al.* NOMA Inspired Interference Cancellation for Integrated Sensing and Communication. arXiv; 2021.
[49] Wang Z, Liu Y, Mu X, *et al.* NOMA empowered integrated sensing and communication. *IEEE Commun Lett.* 2022;26(3):677–681.
[50] Zhang C, Yi W, Liu Y, *et al.* Semi-Integrated-Sensing-and-Communication (Semi-ISaC): From OMA to NOMA. arXiv; 2022.
[51] Wei Z, Liu F, Masouros C, *et al.* Toward multi-functional 6G wireless networks: integrating sensing, communication, and security. *IEEE Commun Mag.* 2022;60(4):65–71.
[52] Chalise BK and Amin MG. Performance tradeoff in a unified system of communications and passive radar: a secrecy capacity approach. *Digit Signal Process.* 2018;82:282–293.
[53] Su N, Liu F, and Masouros C. Secure radar-communication systems with malicious targets: integrating radar, communications and jamming functionalities. *IEEE Trans Wireless Commun.* 2021;20(1):83–95.
[54] Deligiannis A, Daniyan A, Lambotharan S, *et al.* Secrecy rate optimizations for MIMO communication radar. *IEEE Trans Aerosp Electron Syst.* 2018;54(5):2481–2492.
[55] Ren Z, Qiu L, and Xu J. Optimal Transmit Beamforming for Secrecy Integrated Sensing and Communication. arXiv; 2021.
[56] Wei Z, Liu F, Kwan Ng DW, *et al.* Safeguarding UAV networks through integrated sensing, jamming, and communications. In: *ICASSP 2022 – 2022 IEEE International Conference on Acoustics, Speech and Signal Processing (ICASSP).* Singapore, Singapore: IEEE; 2022. p. 8737–8741.
[57] Khan WU, Ihsan A, Nguyen TN, *et al.* NOMA-enabled backscatter communications for green transportation in automotive-industry 5.0. *IEEE Trans Ind Inf.* 2022;18(11):7862–7874.

[58] Jameel F, Zeb S, Khan WU, *et al.* NOMA-enabled backscatter communications: toward battery-free IoT networks. *IEEE Internet of Things M*. 2020;3(4):95–101.

[59] Ding Z, Lv L, Fang F, *et al.* A state-of-the-art survey on reconfigurable intelligent surface-assisted non-orthogonal multiple access networks. *Proc IEEE*. 2022;110(9):1358–1379.

[60] Li Y, Jiang M, Zhang Q, *et al.* Secure Beamforming in MISO NOMA Backscatter Device Aided Symbiotic Radio Networks. arXiv; 2019.

[61] Li X, Zheng Y, Alshehri MD, *et al.* Cognitive AmBC-NOMA IoV-MTS networks With IQI: reliability and security analysis. *IEEE Trans Intell Transport Syst*. 2023;24(2):2596–2607.

[62] Li X, Zhao M, Liu Y, *et al.* Secrecy analysis of ambient backscatter NOMA systems under I/Q imbalance. *IEEE Trans Veh Technol*. 2020;69(10): 12286–12290.

[63] Li X, Zhao M, Zeng M, *et al.* Hardware impaired ambient backscatter NOMA systems: reliability and security. *IEEE Trans Commun*. 2021;69(4):2723–2736.

[64] Zheng Y, Li X, Zhang H, *et al.* Overlay cognitive ABCom-NOMA-based ITS: an in-depth secrecy analysis. *IEEE Trans Intell Transport Syst*. 2023;24(2):2271–2228.

[65] Khan WU, Liu J, Jameel F, *et al.* Secure backscatter communications in multi-cell NOMA networks: enabling link security for massive IoT networks. In: *IEEE INFOCOM 2020 - IEEE Conference on Computer Communications Workshops (INFOCOM WKSHPS)*. Toronto, ON, Canada: IEEE; 2020. p. 213–218.

[66] Zhang Z, Chen J, Wu Q, *et al.* Securing NOMA networks by exploiting intelligent reflecting surface. *IEEE Trans Commun*. 2022;70(2):1096–1111.

[67] Nguyen ND, Van Nguyen MS, and Munochiveyi M. Empowering reconfigurable intelligent surfaces for security of downlink NOMA. *Wirel Commun Mobile Comput*. 2022;2022:1–15.

[68] Chen Q, Li M, Yang X, *et al.* Impact of residual hardware impairment on the IoT secrecy performance of RIS-assisted NOMA networks. *IEEE Access*. 2021;9:42583–42592.

[69] Tuan VP and Hong IP. Enhancing secrecy performance for NOMA systems with intelligent reflecting surface: analysis and optimization. *IEEE Access*. 2021;9:99060–99072.

[70] Yang L and Yuan Y. Secrecy outage probability analysis for RIS-assisted NOMA systems. *Electron Lett*. 2020;56(23):1254–1256.

[71] Tang Z, Hou T, Liu Y, *et al.* Physical layer security of intelligent reflective surface aided NOMA n. *IEEE Trans Veh Technol*. 2022;71(7):7821–7834.

[72] Tang Z, Hou T, Liu Y, *et al.* A novel design of RIS for enhancing the physical layer security for RIS-aided NOMA networks. *IEEE Wireless Commun Lett*. 2021;10(11):2398–2401.

[73] Wang W, Liu X, Tang J, *et al.* Beamforming and jamming optimization for IRS-aided secure NOMA networks. *IEEE Trans Wireless Commun*. 2022;21(3):1557–1569.

[74] Feng Y, Chen J, Xue X, *et al.* Max–min fair beamforming for IRS-aided secure NOMA systems. *IEEE Commun Lett.* 2022;26(2):234–238.

[75] Zhang Z, Lv L, Wu Q, *et al.* Robust and secure communications in intelligent reflecting surface assisted NOMA networks. *IEEE Commun Lett.* 2021;25(3):739–743.

[76] Zhang Z, Chen J, Wu Q, *et al.* Enhancing security of NOMA networks via distributed intelligent reflecting surfaces. In: *2021 IEEE Global Communications Conference (GLOBECOM)*. Madrid, Spain: IEEE; 2021. p. 1–6.

[77] Zhang Z, Zhang C, Jiang C, *et al.* Improving physical layer security for reconfigurable intelligent surface aided NOMA 6G networks. *IEEE Trans Veh Technol.* 2021;70(5):4451–4463.

[78] Jiang C, Zhang C, Hao P, *et al.* Robust secure design for RIS-aided NOMA network against internal near-end eavesdropping. *IEEE Access.* 2021;9:142105–142113.

[79] Gong C, Yue X, Wang X, *et al.* Intelligent reflecting surface aided secure communications for NOMA networks. *IEEE Trans Veh Technol.* 2022;71(3): 2761–2773.

[80] Zhang Z, Chen J, Liu Y, *et al.* On the secrecy design of STAR-RIS assisted uplink NOMA networks. *IEEE Trans Wireless Commun.* 2022;21(12):11207–11221.

[81] Liu Y, Mu X, Xu J, *et al.* STAR: simultaneous transmission and reflection for 360° coverage by intelligent surfaces. *IEEE Wirel Commun.* 2021;28(6): 102–109.

[82] Mu X, Liu Y, Guo L, *et al.* Simultaneously transmitting and reflecting (STAR) RIS aided wireless communications. *IEEE Trans Wirel Commun.* 2022;21(5):3083–3098.

[83] Mu X, Liu Y, Guo L, *et al.* Exploiting intelligent reflecting surfaces in NOMA networks: joint beamforming optimization. *IEEE Trans Wirel Commun.* 2020;19(10):6884–6898.

[84] Luo ZQ, Ma WK, So A, *et al.* Semidefinite relaxation of quadratic optimization problems. *IEEE Signal Process Mag.* 2010;27(3):20–34.

[85] Guan X, Wu Q, and Zhang R. Intelligent reflecting surface assisted secrecy communication: is artificial noise helpful or not? *IEEE Wireless Commun Lett.* 2020;9(6):778–782.

[86] Xu J, Liu Y, Mu X, *et al.* STAR-RISs: a correlated T&R phase-shift model and practical phase-shift configuration strategies. *IEEE J Select Top Signal Process.* 2022;16(5):1097–1111.

[87] Han S, Xu X, Liu Z, *et al.* Energy-efficient short packet communications for uplink NOMA-based massive MTC networks. *IEEE Trans Veh Technol.* 2019;68(12):12066–12078.

[88] Wyner AD. The wire-tap channel. *Bell Syst Tech J.* 1975;54(8):1355–1387.

Chapter 18
Detecting unpredictable adversaries in the industrial network with blockchain

Vishal Sharma[1], Aman Kataria[2], John McAllister[1], Amie Weedon[3] and Robert Bennett[4]

A cyber-physical system (CPS) is integral to Industry 4.0 and beyond, depending heavily on the Internet of Things (IoT). CPS can be related to Industrial Control Systems, Supervisory Control and Data Acquisition, and automation—collectively termed industrial cyber-physical systems (ICPSs). Because of their critical role and requirement of all-time availability, the safety of devices, processes, and workflows in ICPS require a resilient communication setup for safer data exchange. ICPS combines physical and computational systems through sophisticated networking controlled through internal decision modules. Issues related to external and insider threats are most prominent in ICPS because of their large-scale critical deployment and devolved network infrastructure. These threats are scaled up if the nature and the type of adversary are unknown, and such unpredictable adversaries are challenging to detect. With the motivation of controlling unexpected adversaries, this chapter discusses the security of ICPS from digital warfare and vandalism points of view, particularly looking towards network-initiated activities. Furthermore, challenges associated with detecting unpredictable adversaries are listed, followed by scenario-specific simulated discussions on impact and event validation, highlighting the adoption of blockchain and research opportunities for securing future Industrial IoT networks.

18.1 Introduction

Industrial cyber-physical systems (ICPSs) combine network, physical, and logical components to control processes and associated equipment [1]. ICPS allows automating tasks in several industries, such as autonomous automobile systems, medical monitoring, nuclear plants, metros and railways, smart grids, and intelligent transportation and is pivotal in enhancing the applications of the Internet of Things (IoT) [2]. A basic ICPS is divisible into three layers, namely, external, internal CPS, and physical [3]. An external environment is the physical resources, including

[1]School of Electronics, Electrical Engineering and Computer Science, Queen's University Belfast, UK
[2]Amity Institute of Defence Technology, Amity University, India
[3]Research and Innovation Department, MASS, Cohort PLC, UK
[4]Solutions Directorate, Lone Star Analysis, UK

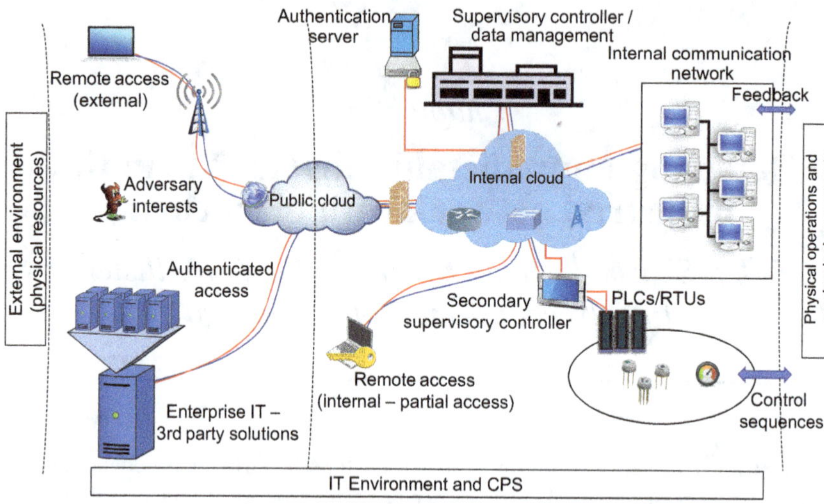

Figure 18.1 An exemplary illustration of the accessibility in the ICPS

external remote accessibility or third-party solutions, usually enterprise solutions offered over the cloud. IT environment and CPS are the intermediate layers and core of the entire system, including the internal authentication server, supervisory systems, data managers, the Programmable Logic Controller (PLC), Remote Terminal Unit (RTU), intelligent devices, and an internal communication network manageable through an internal cloud [4]. Physical operations and mechanical components are the third layer and have mechanical components that are responsible for industrial-end operations. Every element in the ICPS has a specific role that depends on the nature of the application. The initial schedulers filter the external requests, and the external firewall manages the flow. Figure 18.1 illustrates an exemplary scenario of components in an ICPS. The flow of information is in two ways—internal and external requests depending on the implementation and network architecture.

In contrast, the internal requests require local authentication, and there are internal routers and switches to handle the traffic. The information and control sequences are shared using communication and security protocols. These protocols can either be applied device-to-device or multi-device to multi-device communications. In the last decade, the focus of the ICPS from a closed setup has changed to make it more distributed by including many IoT devices that have raised the demand for competitive resource allocation [5] and security and rely heavily on the performance of the network infrastructure.

The security components of an ICPS are a direct result of unpatched flaws, which remain unnoticed when these systems and the associated network components are deployed. Non-standard security evaluation and vulnerability analysis practices often account for significant security threats. These attacks can happen at lower levels, such as PLCs and RTUs, or higher levels at the application peripheral interface (API). Most recently, the lack of access control policies and the unavailability of well-evaluated authenticated protocols are the primary reasons for cyber threats in ICPS. Weak password security and lack of multifactor authentication are other reasons for attacks against such systems. Adversaries use low-level privileges to initiate attacks based on unpatched vulnerabilities for longer to gain access to crucial systems and launch attacks such as Man-In-the-Middle, Denial-of-Service, or malware like ransomware [4]. Information security practices and using IDS can help control these attacks to an extent, but not if the adversaries evolve rapidly and have unpredictable properties. It is tedious to identify such adversaries instantly because of the lack of a pattern, and adapting mitigating strategies and having a solid event validation system can be of much value in those circumstances. Alongside these, authenticated software updates [6] can help overcome eavesdropping on security patches when attacks are recognized.

18.2 Digital warfare and cyber-vandalism

Digital or cyber warfare is an Internet-enabled attack on digital infrastructure, such as computer viruses, hacking, or denial-of-service attacks [7]. It is a crucial aspect of modern cybersecurity and covers defensive and offensive mechanisms. Covert and overt attacks have found their way into digital warfare, making it a top priority for security agencies to build strategies to protect national infrastructure. Industries like smart grids, nuclear plants, and intelligent transportation systems have been critical targets of digital warfare. The attacks on information disrupt information processing and decision-making and threaten nation-states with destruction, disruption, or loss of life. Using technology against technology has been the prime motto of adversaries aiming at digital warfare. Robust authentication mechanisms must secure components in the industries, and validation must ensure the accurate flow of information and resultant action in the network space [8]. Checkpoint mechanisms with distributed technology are the critical enablers in such a scenario [9,10]. It is advised against using a centralized model of detection and mitigation as intelligence is a crucial feature of devices in industries, and adversaries can utilize this feature for composing an attack by impacting several possible single points of failure. Violating rules, gaining privileges, feeding incorrect reports, and bypassing risk mitigation and due diligence are the essential areas to look for when dealing with adversaries operating in an intelligent environment [11].

With the evolution of adversaries, the cost of attack is no longer a concern. Adversaries backed by some of the most notorious groups invest excessive resources for harming the systems, leading to cyber-vandalism [11], e.g., abusive adversaries [12]. To date, the solutions aiming at preventing and mitigating adversaries consider threat models with limited resources for adversaries or operate at a threshold, thus undermining the abilities of adversaries looking for cyber-vandalism. The desired requirement is that the solutions for handling adversaries must look beyond the scenario of proving that their solution is ideal in a limited set of configurations. A more open setting evaluation for validating detection systems is the need of the hour. It also raises the need for new verification, validation, and assessment methods. It must start with non-relaxed threat modelling and better strategies for studying systems under adversarial attacks.

18.3 Unpredictable adversaries

Adversaries intend to exploit the features of AI to hide their intentions and approach to attack [13,14]. Such adversaries are unpredictable and do not exhibit any fixed pattern of attack. Because of uncertain behaviour, maintaining the attack corpus is tedious. Computational logic is a go-to approach for scenarios where adversaries exhibit multiple or diverse patterns. However, the issue is that logical evaluation may require evaluating the system in a step-by-step mechanism. It will slow down the process flow rate and increase the system overhead. Additionally, if state machines operate alongside the computational logic, the delay caused by step-by-step evaluation appears as an anomaly that shows high false positives in the system. Limitation to classify the unpredictable adversaries into active or passive categories prevents detection using an available attack corpus.

18.3.1 Zero-click adversaries

Zero-click adversaries aim to exploit zero-click vulnerabilities that lead to zero-click attacks [15] for compromising the system without subsequent action from the user. For ICPS, the system is unlikely to fall prey to a fresh zero-click adversary. The sequential nature of attacks or unpatched vulnerabilities can be the reasoning behind the zero-click attack that results from the system's response without any action by the user. For example, initiate a transmission channel between the devices by sending a random SMS irrespective of whether the user responds. Most likely, adversaries can take advantage and launch zero-click attacks with modern technologies, where the original equipment manufacturers (OEMs) have a non-user interaction-based updating system [16]. Zero-click adversaries are unpredictable and non-traceable in the existing signature or attack corpus. Thus, it is complicated for detection strategies to trace these adversaries proactively, and mitigation depends on the threat level from such adversaries. Consistent and regular testing for bugs by the OEMs appears to prevent these adversaries in networked components of the ICPS.

18.3.2 Challenges in detecting unpredictable adversaries

Unpredictable adversaries are like unknown territories where the detection systems have little idea of the features and expectations in terms of an adversary's behaviour towards the system. Figure 18.2 lists some impact-to-consequences entanglement that needs further exploration to understand the issues with unpredictable adversaries. Several challenges associated with the detection of unpredictable adversaries are as follows:

- **Random events**: The devices in ICPS follow a dedicated workflow with processes initiated based on events or inputs. Event failures are manageable as a part of risk management. However, failure of random processes (such as those occurring with human experts [17]) is difficult to detect while keeping the system alive and usually requires halting all operations. It makes an adversarial attack a success.

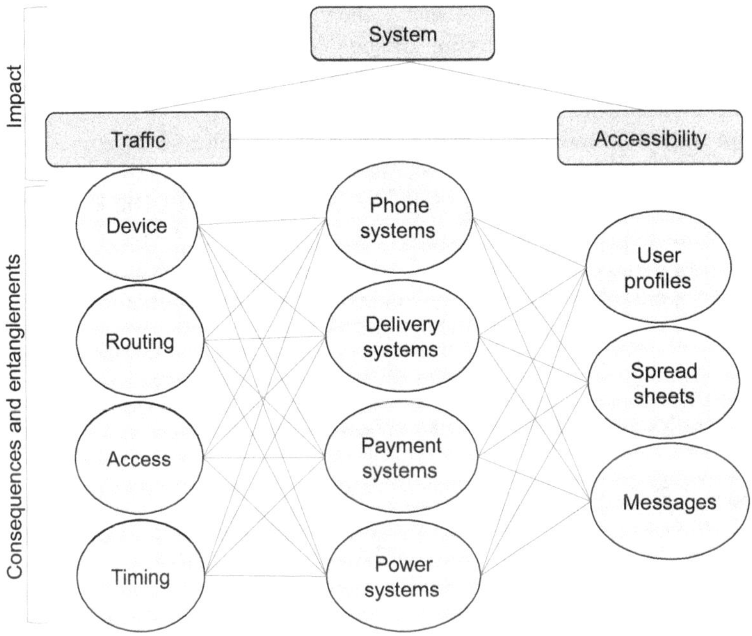

Figure 18.2 Impact-to-consequences entanglement for unpredictable adversaries

Detecting random events requires a high density of redundant systems, which increases the cost of operation and is computationally expensive.

- **Limited attack corpus**: Unpredictable adversaries may lead to zero-day attacks and have a lower record time. The discovery and security patching phases are too close for immediate mitigation, making it difficult to act against these adversaries. The response time plays a critical role when the attack corpus is unavailable. Moreover, automated systems can help track uncertain events. However, these systems require consistent upgrading and rely on an in-depth corpus of attack indicators.
- **Complex and interdependent processes**: ICPS combines large-scale networking with industrial and computing infrastructure, which may have hundreds of dedicated equipment operating in proximity to each other [18]. The complexity of the architecture often leads to high false positives and negatives, as immediate detection of adversarial processes is not practically feasible. Several components may rely on multiple sources for inputs, and fault at any of these components requires backtracking all associated past events and devices. It makes detection highly complex and requires more time, giving an advantage to the adversary for covering its track and bypassing detection.
- **Devices diversity**: ICPS does not have a similar kind of setup and configuration for all types of processes. Devices in ICPS are large-scale and diverse in make,

configuration, and operations. It means the detection process cannot rely on one sole strategy. Instead, a collection of suitable approaches needs to be in place to check each device distinctively. However, maintaining synergy between the distinct mechanisms of adversary detection is an additional challenge to resolve.

- **Large-scale data imputation**: Data imputation is a solution to generate/replace missing data when some of the components fail to report promptly [19]. It helps to ensure that techniques requiring data must have the complete input to check on adversaries. With various devices and large-scale deployments, the pattern of missing data is difficult to trace, which works favouring adversaries. Thus, detecting approaches based on data imputations do not account for unpredictable adversaries.

18.4 Attacks by unpredictable adversaries

Untraceable events that do not exhibit patterns go unnoticed and require sophisticated prevention strategies rather than detection to build a foolproof system against unpredictable adversaries. The properties relevant to adversary detection include confidentiality, availability, integrity, and authenticity. The design principles of the ICPS enable security systems to incorporate strategies for evaluating malicious intentions to predict an attack. Each activity of the devices in ICPS is markable as a factor affecting one of the properties. Based on these properties, vulnerability and attack analyses can help understand if the system illustrates unpredictable behaviour. Over the years, the issues associated with vulnerabilities and attacks remain the same [11,20]. However, the adversaries have become more intelligent, and the attack's impact has increased. Figure 18.3 shows some of the most common vulnerabilities and attacks in ICPS, with the majority falling into the predictable category. Distributing the adversarial detection strategies and validating them across the system can be a helpful solution against unpredictable adversaries.

To further highlight the impact of unpredictable adversaries, we have considered a specific scenario where the adversary sits in the ICPS's networking components and acts as a watchdog to perform a selective state replication attack. It is a type of state estimation attack using the Markov Chain where the correctness of the systems is observed based on Weibull survivability. Standard deviation method (STDM) [21] was used for shape and scale estimation. The assumption is that the adversary controls PLC/RTU to initiate processes from states it can control. Specifically, the number of processes assists in labelling the state and associated probability in the settings. It replicates an event-based system where the adversary aims to control the events without getting detected and exhibiting any pattern. The strategy used by the adversary, attack selection cafeteria and observations are illustrated in Figure 18.3. The adversary relies on the process observation probability by tracking the events triggered in the ICPS and makes a probabilistic graph between the states. An adversary may fetch this information from the logs. Once a graph is available, the adversary classifies the graph using the Markov Chain into different survivable classes to form actual transition probabilities. Here, the adversary can take different approaches to control

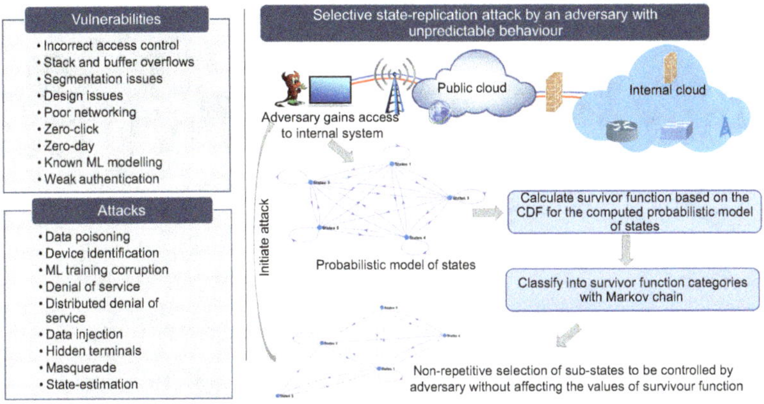

Figure 18.3 Common vulnerabilities and attacks in ICPS [11] and an exemplary strategy for an adversary to launch a selective state-replication attack with unpredictable behaviour

the states. This work relies on selective-state replication as an approach for adversarial settings. The adversary calculates each state's survival function and the state location where max is observed for the survival function. Next, the adversary takes a subgraph of states based on this maximum value and manipulates the probability to ensure that these states' net change of survival function remains zero. With this, the adversary can make multiple strategies to launch non-pattern-based attacks by initiating different states in different orders, making it challenging to predict. Figure 18.4 illustrates the earlier process with an exemplary probability of state transition and the selection of favourable states for the adversary to control.

18.5 Validating ICPS workflow using blockchain

The problem with detecting unpredictable adversaries is that it is quite tedious, given the randomness associated with timely identification. Due to the nature of the adversary, the detection systems can evaluate different settings using decentralized techniques, such as blockchain [22], which can help validate the processes in the ICPS, making it more secure even against the unpredictable behaviour of adversaries. In this part, the exemplary attack situation of Figure 18.5 is reconsidered, and blockchain is adopted to validate the workflow in a distributed manner.

18.5.1 Entities

The key entities involved in the workflow validation are listed below:

- Process requester: It is the requesting process that has completed its execution on a given state of the system and requests validation and transfer of control to the upcoming procedure.

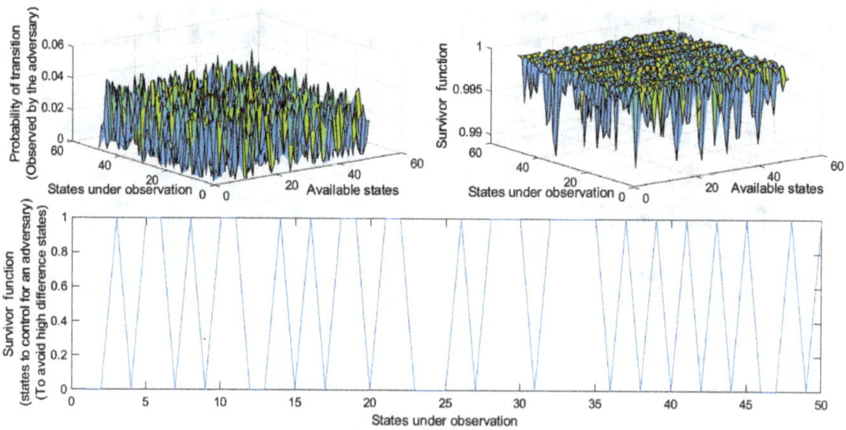

Figure 18.4 *An exemplary presentation of the selective state-replication attack by exploiting system's survivor functionality by an unpredictable adversary to pick states based on its attacking capacity*

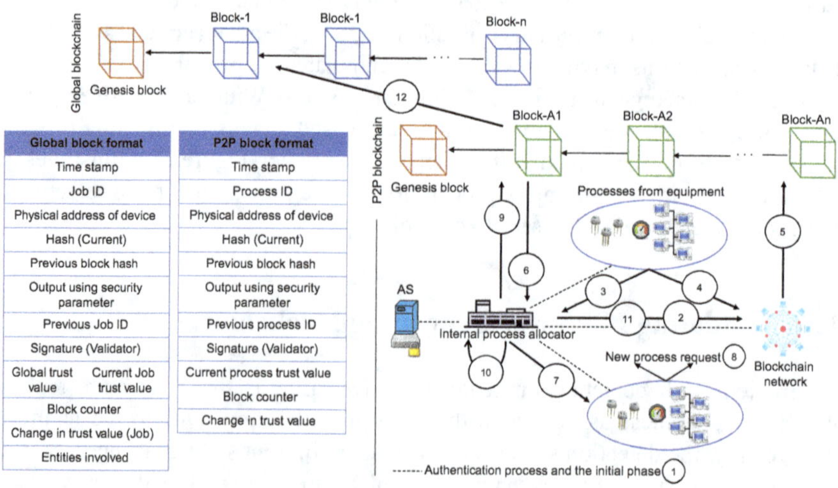

Figure 18.5 *An exemplary illustration of validating ICPS workflow with blockchain (block formats and procedures)*

- Internal process allocator: The supervisory controller is responsible for managing the process allocation and regulating the completed processes.
- Blockchain network: This is responsible for updating the transaction via a public ledger and validating the transaction via various consensus mechanisms. Specifically, event validation as a part of workflow control relies on this blockchain

network. The blockchain network helps to process and record all information and process details on the shared ledger using a consensus scheme. In this scenario, Proof-of-Trust is used, which is driven by Proof-of-Stake like consensus.

- Validators: These are responsible for writing transactions into the block and going through the blockchain network to allow validated blocks to be appended to the blockchain. Validators go through the process of leader selection to pick the signing node in each round. These components work similarly to miners in Proof-of-work, except that these rely on stakes for selecting leaders rather than the hash-calculating competition. Validators can be picked by installing secondary devices alongside each process to help speed up the process and lower the number of authentications. These can be several cost-effective IoT devices that can serve the validator's purpose.
- Ledger: This is the global record accessible to every node in the internal network of the ICPS. External nodes can query it based on permissioned or permissionless scenario. With many components and blocks to be appended, the ledger size does not favour the ICPS. Thus, the network considers having a global blockchain and process-to-process (P2P) blockchain initiated by network components, not bounded by performance issues. Whenever the ICPS completes a job (task), the P2P blockchain submits its appended blocks to the global blockchain. The validators, the memory, and energy-restricted devices reset their settings before the initialization of the next job.

18.5.2 Model for event validation

The resultant event validation via blockchain will evaluate each device and process in an ICPS and provide detailed reporting while preventing potential threats to the components, as expressed in Figure 18.5. Trust validation will ensure that all devices and processes comply with accurate working and consensus is not broken, assuring that the system does not fall prey to any unpredictable event.

The validation of blocks accounts for evaluating the submitted blocks and the value of the transaction submitted alongside them. Each validator calculates a trust score based on the input values received by the associated processes in the ICPS. There are several ways to calculate the trust by the validator [23]. The survivor function keeps it consistent with the example in Figures 18.3 and 18.4. Validators must commit several stakes in every round to initiate the validation process and attempt to be the leader. On completion, the internal process allocator issues incentives based on the committed ratio of stakes. Here, an interesting concept is that the internal process allocator can reset the stakes for every job. Suppose the internal process allocator is in a trusted zone; in that case, it can predetermine the number of stakes each validator may require depending on the number of processes executed and validated for completing the assigned job. Such a strategy can help counteract the adversary's attacks to compromise the blockchain with several known attacks, such as feather forking and Eclipse Attacks. The stakes here can be assigned initially based on the role of the components in the initial phase. The procedures are as follows:

① **Initial phase**: In the initial phase, the supervisory controllers are authenticated using long-term shared keys that help to derive several subkeys as well as form the part of the public and private key pairs used by each of the entities. Next, the system relies on the same AS to calculate the long-term keys shared with each entity to operate the processes required to complete the job. It helps to secure the network before the processes are executed. The secret keys derived from the shared keys are used for creating anonymous wallets for the involved validators. A global blockchain is formed and rolled with the genesis block. Next, a P2P blockchain is initiated with the address of its genesis block.
②–⑧ **Intermediary phase**: The intermediary phase is the event validation phase, where the P2P blockchain helps to ensure that the events taking place as a part of the process execution are correct. The following steps are executed to attain this phase:

- ② Process call: The internal process allocator first submits the first process call to the blockchain network. It is the initialization request for the processes and contains the hash of the genesis block of the P2P blockchain. This call means the new job is submitted, and any previous stakes are not allowed. With this, the wallets are updated with this request. As stated earlier, if a trusted zone is available, these wallet updates are performed by computing the expected number of rounds to complete the submitted job.
- ③ and ④ Process request: In this phase, the requesting device authenticates itself to the internal process allocator and ensures that it is the legitimate source of the request. The process requester can be an equipment that wants to initiate requests for another subprocess. The process requester obtains available processes for execution as a part of the submitted job. The process requester relies on the blockchain network to validate the request by submitting a block containing the transaction and showcasing the upcoming process period and the execution site.
- ⑤–⑦ Request validation and allocation: Once the block is submitted to the blockchain network, the validators get initiated. Each listed validator selects a stake proportional to the connectivity between the request process and the traces it can make to validate the process. Once the validators invest the stakes, the trust is calculated to ensure that the system has not violated any property and the processes are executed in order. A leader is deterministically selected based on the invested stakes and the degree of connectivity between it and the following executable process. Next, the leader signs the block and submits it for validation, which is appended to the P2P blockchain. If the trust score submitted by the leader is above a specific value (e.g. threshold value), the internal process allocator awards a fee to each of the processes. Next, with the block contents of the P2P blockchain, the internal process allocator associates the request to handle the process. The fee in the model is the trust score for the validators. The trust score of the validator increases the chances of selecting that device as a leader in the next round. An adversary will target the most recent leader chosen to compromise the consensus. Here, a solution like non-deterministic leader selection is helpful to prevent an adversary from trailing the network.

- ⑧ Next process call: Once the block is appended to the P2P blockchain, the system moves to the next step and waits for the next process requester. The intermediary phase continues until a submitted job is completed.

⑨ **De-registration phase**: Once the P2P blockchain is appended with the final desired result or observation, it marks the completion of the job, and the internal process allocator initiates the de-registration phase. Now, irrespective of the number of times a secondary device has validated the process in the previous job, the stakes are re-initiated.
⑩–⑫ **Total incentives and trust score**: The internal process allocator is a supervisory controller that checks for the total incentives earned by the validators during a job as well as the trust score associated with the process and the validating system. The internal process allocator has precalculated values of the required number of processes and the associated incentives, which it uses to check any violations in the job. Moreover, this system can validate the processes in real-time as an internal supervisory controller can evaluate the results of every block each time these are appended. Once the incentives are decided, a block is created and validated by the blockchain network to append to the global blockchain.

The devices can validate the events, and the ICPS workflow can be managed using the above blockchain-dependant solution, preventing unpredictable adversaries from compromising the events during job submission. However, this strategy is not a complete solution, and it needs further investigation on several aspects, as expressed in Figure 18.6, which covers parts of remote code execution to flexible and adaptive authentication and availability of nodes for a better turnaround time for such a decentralized solution to work seamlessly.

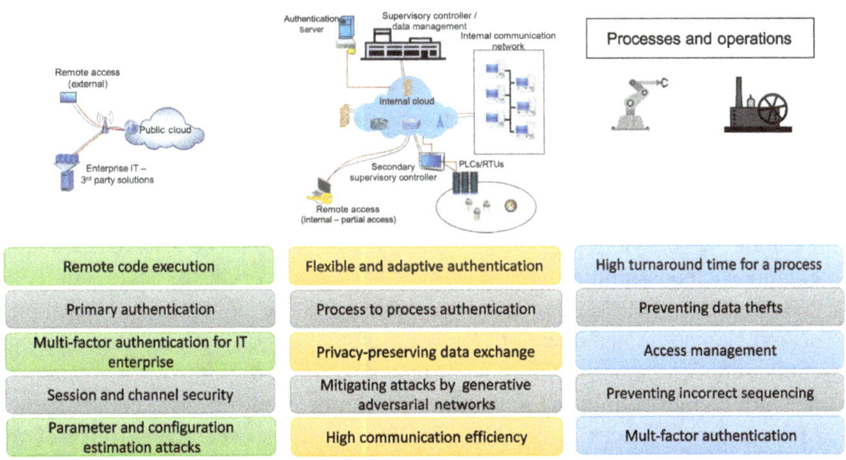

Figure 18.6 An illustration of research challenges and opportunities for preventing unpredictable adversaries in ICPS

18.6 Future research trends

Below are some prominent future research trends associated with security operations in ICPS and the associated Industrial IoT devices.

- **Big graph evaluations**: Activity charts and graphs are vital in detecting adversaries in large-scale complex networks [24]. If the detection system deploys a rule or activity-based mechanism, large and complex graphs are generated as the system passes through different states and IoT devices. It further complicates the structure when multiple events can trigger the same condition. The system must process many vertices and edges for identity conflicts and possible issues, which requires practical tools to deduce the output in less time, such as query mechanisms [25].
- **Secure AI**: AI has been a prominent role player in the intelligent detection of adversaries in ICPS as it helps protect the dynamic enterprise attack surface and provides much-needed analysis and threat identification. However, the ethical concerns associated with the safe use of AI are significant [26,27]. The approaches need to consider if the evaluation rules used as a part of the AI are correct and verified as the system progresses or the configurations are updated. Auto-updates of the system may expose zero-click vulnerabilities, which can lead to attacks such as ransomware.
- **Model-based control and attack indicators**: Identifying the attack properties and matching these to the systems' metrics are critical to detecting unpredictable adversaries. Attack indicators help to confirm situations where the properties of the system are violated. These indicators are responsible for ensuring the system's confidentiality, integrity, and availability. Distributed control models [18] and threat models [3] are used along with attack indicators to give an exact idea of the systems' issues in light of known and unknown attacks.
- **Digital twinning of the systems:** Every system working using the principles of computing exhibits some patterns, which can be studied as a part of behaviour modelling [28] and digital twins [29]. This modelling allows an understanding of how the system responds to specific changes in the process flow and what automated remedies are adapted to counter unknown (foreign) changes in the workflow. Behaviour modelling can be studied using state-based and non-state-based solutions. In the literature, rule-based mechanisms have been dominantly used for the behaviour modelling of computing systems. The researchers should explore such possibilities for managing the accurate working of devices in ICPS.
- **Network-based authentication**: Devices in ICPS use authentication based on a shared channel established with key-exchange protocols. The instruments with similar configurations can follow group authentication. With new technologies, network slicing with group authentication can be used for faster authentication of a group of devices in distributed ICPS [30]. However, the security of the key exchange protocols and ensuring that the ICPS workflows are safe from network attacks are other areas to look into, and solutions are required that can rely on cryptographic keys to provide more robust encryption mechanisms for

the devices. However, the efficiency of the solution, alongside security, is a considerable challenge given the time requirements of the devices to complete the process without affecting the workflow.

18.7 Conclusion

This chapter discusses the decentralized security of ICPSs, focusing on unpredictable adversaries that can be detected using blockchain. The chapter discusses how digital warfare and vandalism are critical motivators for adversaries. The adversaries can easily fool the detection using intelligent techniques and prevent themselves from being logged in the attack corpus. Decentralized technologies such as blockchain can help tame these adversaries to an extent. However, these technologies have their associated research aspects and challenges. Several open issues discussed as part of understanding unpredictable adversaries show that this area of research is essential, specifically with the large-scale deployment of the IoT across Industry 4.0 and beyond.

References

[1] Colombo AW, Karnouskos S, Kaynak O, *et al.* Industrial cyberphysical systems: a backbone of the fourth industrial revolution. *IEEE Industrial Electronics Magazine*. 2017;11(1):6–16.

[2] Zhang K, Shi Y, Karnouskos S, *et al.* Advancements in industrial cyber-physical systems: an overview and perspectives. *IEEE Transactions on Industrial Informatics*. 2022;19(1):716–729.

[3] Lee CC, Tan TG, Sharma V, *et al.* Quantum computing threat modelling on a generic CPS setup. In: *International Conference on Applied Cryptography and Network Security*. New York, NY: Springer; 2021. p. 171–190.

[4] Franco J, Aris A, Canberk B, *et al.* A survey of honeypots and honeynets for Internet of Things, industrial Internet of Things, and cyber-physical systems. *IEEE Communications Surveys Tutorials*. 2021;23(4):2351–2383.

[5] Seid AM, Abishu HN, Erbad A, *et al.* HDFRL-empowered energy efficient resource allocation for aerial MEC-enabled smart city cyber physical system in 6G. In: *2023 International Wireless Communications and Mobile Computing (IWCMC)*. IEEE; 2023. p. 836–841.

[6] Magnanini F, Ferretti L, and Colajanni M. Scalable, confidential and survivable software updates. *IEEE Transactions on Parallel and Distributed Systems*. 2022;33(1):176–191.

[7] Robinson M, Jones K, and Janicke H. Cyber warfare: issues and challenges. *Computers & Security*. 2015;49:70–94.

[8] Babu PR, Amin R, Reddy AG, *et al.* Robust authentication protocol for dynamic charging system of electric vehicles. *IEEE Transactions on Vehicular Technology*. 2021;70(11):11338–11351.

[9] Sari A and Akkaya M. Fault tolerance mechanisms in distributed systems. *International Journal of Communications, Network and System Sciences*. 2015;8(12):471.

[10] Javaid U and Sikdar B. A checkpoint enabled scalable blockchain architecture for industrial Internet of Things. *IEEE Transactions on Industrial Informatics*. 2020;17(11):7679–7687.

[11] Maleh Y, Shojafar M, Darwish A, *et al. Cybersecurity and Privacy in Cyber Physical Systems*. London: CRC Press; 1st ed., 2019, 10.1201/978042 9263897.

[12] Sharma V, Varghese B, McAllister J, *et al.* Abusive adversaries in 5G and beyond IoT. *IEEE Consumer Electronics Magazine*. 2021;11(4):11–20, 10.1109/MCE.2021. 3079998.

[13] Qammar A, Ding J, and Ning H. Federated learning attack surface: taxonomy, cyber defences, challenges, and future directions. *Artificial Intelligence Review*. 2022;55:3569–3606.

[14] Hoffman W. AI and the future of cyber competition. In: *CSET Issue Brief*. 2021;1–35.

[15] Ali S, Anantharaman P, Lucas Z, *et al.* What we have here is failure to validate: summer of LangSec. *IEEE Security & Privacy*. 2021;19(3):17–23.

[16] Teague V, Schink M, Ising F, *et al.* When is a proof actually not? In: *14th {USENIX} Workshop on Offensive Technologies ({WOOT} 20)*; 2020.

[17] Drašar M, Moskal S, Yang S, *et al.* Session-level adversary intent-driven cyberattack simulator. In: *2020 IEEE/ACM 24th International Symposium on Distributed Simulation and Real Time Applications (DS-RT)*. IEEE; 2020. p. 1–9.

[18] Ding D, Han QL, Wang Z, *et al.* A survey on model-based distributed control and filtering for industrial cyber-physical systems. *IEEE Transactions on Industrial Informatics*. 2019;15(5):2483–2499.

[19] Dunin-Kęplicz B, Jankowski A, Skowron A, *et al.* Data security and null value imputation in distributed information systems. In: *Monitoring, Security, and Rescue Techniques in Multiagent Systems*. New York, NY: Springer; 2005. p. 133–146.

[20] Sharma V, You I, Andersson K, *et al.* Security, privacy and trust for smart mobile-Internet of Things (M-IoT): a survey. *IEEE Access*. 2020;8: 167123–167163.

[21] Shaban AH, Resen AK, and Bassil N. Weibull parameters evaluation by different methods for windmills farms. *Energy Reports*. 2020;6:188–199.

[22] Sharma V, You I, Palmieri F, *et al.* Secure and energy-efficient handover in fog networks using blockchain-based DMM. *IEEE Communications Magazine*. 2018;56(5):22–31.

[23] Najib W and Sulistyo S. Survey on trust calculation methods in Internet of Things. *Procedia Computer Science*. 2019;161:1300–1307.

[24] Gebreegziabher F and Patgiri R. The major challenges of big graph and their solutions: a review. *Advances in Computers*. 2023;128:399–421.

[25] Fan W and Hu C. Big graph analyses: from queries to dependencies and association rules. *Data Science and Engineering*. 2017;2:36–55.

[26] Fulton N and Platzer A. Safe AI for CPS. In: *2018 IEEE International Test Conference (ITC)*. IEEE; 2018. p. 1–7.

[27] Armstrong S and O'Rorke X. Good and Safe Uses of AI Oracles; 2017. arXiv preprint arXiv:171105541.

[28] Cherskikh E and Saveliev A. Survey on behavioral strategies of cyber-physical systems in case of loss of integrity. In: *Electromechanics and Robotics*. New York, NY: Springer; 2022. p. 463–474.

[29] Flammini F. Digital twins as run-time predictive models for the resilience of cyber-physical systems: a conceptual framework. *Philosophical Transactions of the Royal Society A*. 2021;379(2207):20200369.

[30] Ni J, Lin X, and Shen XS. Efficient and secure service-oriented authentication supporting network slicing for 5G-enabled IoT. *IEEE Journal on Selected Areas in Communications*. 2018;36(3):644–657.

Chapter 19

Using support vector machines for detecting active spoofing attacks

Tiep M. Hoang[1] and Trung Q. Duong[2,3]

19.1 Introduction

Physical layer security (PLS) has become an important topic in wireless communications for decades. Different from security solutions operating at the upper ISO layers, PLS solutions rely on the randomness of wireless channels to cope with eavesdroppers/spoofers [1–3]. In general, there are two common types of PLS concerns: (i) passive eavesdroppers conceal their presence while listening to the transmitted signals and (ii) active spoofers impersonate some legitimate users to break into the system. An active spoofer (i.e., an active E) is more malicious than a passive eavesdropper (i.e., a passive E) because the amount of information leaked to the active E is higher [3,4]. Thus, many works pay special attention to active spoofing, e.g., [3,5–7]. Since spoofers actively send their spoofing signals, there is a chance for us to detect their presence based on received signals.*

If a transmitter is aware of security risks, it can generate artificial noise and inject the AN into the transmitted signals to deteriorate the decoding process at the spoofers' side. However, the generation of AN requires the transmitter to sacrifice a portion of the power budget [2]. On the other hand, if the transmitter can detect the presence of spoofers, then it can be easier for the transmitter to deal with the spoofers by providing efficient beamforming designs rather than the low-efficient AN. In short, there is a need to address intrusion detection at the physical layer. Also concerning intrusion detection, machine learning is emerging as a powerful method of detecting the presence of spoofers. However, most of the previous works emphasize the use of machine learning at the upper ISO layers, where the personal datasets of users are available [9–11]. However, there are a few contributions on physical layer authentication [12–21]. In general, physical layer authentication based on machine learning is still a largely open area of research.

*This chapter has been published partly in [8].

[1]Department of Electrical and Microelectronic Engineering, Rochester Institute of Technology, US
[2]Faculty of Engineering and Applied Science, Memorial University of Newfoundland, Canada
[3]School of Electronics, Electrical Engineering and Computer Science, Queen's University Belfast, UK

In [12], the authors suggested using carrier frequency offset (CFO) and Kalman filtering for authentication. Also concerning, the authors of [13] proposed channel probing and hypothesis testing. However, these contributions did not consider machine learning algorithms. The usefulness of combining Q-learning and game theory was presented in [14]. Also using Q-learning, the authors of [15] considered PLS concerns in ad-hoc networks. On the other hand, the authors of [16] showed the potential of using extreme learning machines for providing new authentication methods. Gaussian mixture models were performed in the work of [17] to identify spoofing attacks, while the authors of [18] harnessed industrial datasets and compared the detection performance of different algorithms, including a twin-class support vector machine (TC-SVM) algorithm, the classic k-nearest neighbors (k-NN) and decision trees. Both TC-SVM and Fisher's linear discriminant analysis (Fisher's LDA) were examined in [19] for detecting the channels compromised by spoofers. Furthermore, to cope with PLS problems of the smart grid, TC-SVM was used [20]. Meanwhile, the so-called in-situ machine learning was studied in [21]. In general, various different machine learning techniques have been studied for various aspects of PLS, but the detection of spoofing attacks by active E has received limited attention.

Despite recent efforts, there remain many PLS concerns and solutions needed to be further addressed. For example, the influence of propagation on modulated signals was ignored from [14–16,19–21]. As an exception, the authors of [22] employed machine learning techniques to look into the nature of realistic signals. Similar to [22], this chapter will also consider wireless signal models and employ machine learning for detecting abnormal signals associated with spoofing attacks. However, different from [22] that uses Gaussian mixture models to classify signals, we will show the potential of the TC-SVM and of the single-class SVM (SC-SVM) models in this chapter. To be more specific, TC-SVM and SC-SVM are not dependent on probabilistic assumptions as seen in Gaussian mixture models, but on the distances among data points. Thus, once the distribution of the data is unavailable, it may be unreasonable to employ Gaussian mixture models.

Focusing on extracting features from wireless signals, the authors of [15,16,19–21] suggested advanced techniques for detecting active E. For example, the authors of [15] used the received signal strength (RSS) indicator and time of arrival (TOA) of a packet as a pair of features of a data point. Meanwhile, the authors of [19] used the TOA, the RSS and the Z-transform as a triple of features to describe a data point. In [16], the authors relied on the Pearson correlation and Euclidean distance to form the features of a dataset. On the other hand, the data features studied in [20] were supposed to obey some distribution that is fitted to the pre-processed data. In [22], the authors constructed datasets based on the normalized magnitudes of estimated channels. Moreover, according to [21], the features of data can be formulated by the in-phase and quadrature (I–Q) components and the CFO. However, there is no winner for constructing the features of a dataset, because of a great deal of variety of solutions. Thus, it is a new challenge to convert wireless signals into datasets whose features are suitable for detection/classification tasks performed by machine learning algorithms. Indeed, feature selection is necessary to facilitate physical-layer authentication systems and make them more reliable [23].

In this chapter, we will propose a framework for creating our own dataset from wireless signals so that all the features of the data have some statistical properties. Based on those datasets, we aim to detect spoofing attacks with the aid of both TC-SVM and SC-SVM. Our contributions can be summarized as follows.

- The received signals will be converted into a *structured dataset* that has two distinct features (namely, MEAN and RATIO). These features are formulated in a way so that when the number of samples is large enough, they are associated with some constants. Moreover, the more data points we have, the closer MEAN and RATIO reach their associated constants.
- As for the training process, we build the so-called *artificial training data* (ATD). To be more specific, two types of ATD will be built based on the knowledge of the channel state information (CSI). If the CSI of an active E's channel is known, then the ATD has two classes (or two labels), which will then be trained by TC-SVM classifiers. By contrast, if the CSI of a passive eavesdropper's channel is unknown, the ATD has a single class (or a single label), which will be trained by SC-SVM classifiers.
- The spoofing detection performance of TC-SVM and that of SC-SVM are evaluated through metrics like accuracy, sensitivity and specificity. Moreover, given that the heart of an SVM classifier is a kernel function, we compare the effectiveness of four different kernel functions.

The remainder of the chapter is organized as follows. The principles of TC-SVM and SC-SVM are introduced in Section 19.2. The process of constructing the features of data from wireless signals is presented in Section 19.3. Section 19.4 presents two different types of ATD that are trained by TC-SVM and SC-SVM, respectively. Finally, Sections 19.5 and 19.6 present numerical results and conclusions, respectively.

Notations: $\text{sign}(z) = 1$ if $z > 0$ and $\text{sign}(z) = -1$ if $z \leq 0$. Vectors and matrices are represented by lowercase boldface and uppercase boldface, respectively; $[\cdot]^T$ and $[\cdot]^\dagger$ denote the transpose operator, and Hermitian operator, respectively; $\mathbf{I}_L$ is the $L \times L$ identity matrix; $\|\cdot\|$ denotes the Euclidean norm; $\mathbf{z} \sim \mathscr{CN}(\mathbf{0}, \mathbf{I}_L)$ denotes a complex Gaussian random vector $\mathbf{z} \in \mathbb{C}^{L\times 1}$ with zero-mean and covariance $\mathbf{I}_L$; $\mathbb{E}_{\mathbf{v}_1,\ldots,\mathbf{v}_M}\{\cdot\}$ denotes the expectation over $\mathbf{v}_1, \ldots, \mathbf{v}_M$. Some important symbols are defined in Table 19.1, while the remaining symbols will be defined whenever used.

19.2 A brief introduction to TC-SVM and SC-SVM

19.2.1 The development of typical models

The concept of perceptron is used for describing the functional relationship $l = \text{sign}(\langle \mathbf{w}, \mathbf{r}\rangle + b)$ between an input vector $\mathbf{r} \in \mathscr{R} \subseteq \mathbb{C}^{n\times 1}$ and an output value l. Herein, l is referred to as the label for the input vector $\mathbf{r}$ and, assigning a value (say 1 or -1) to l represents the classification of the input vector. For example, if $l = 1$ means that $\mathbf{r}$ has some attribute A, then $l = -1$ can be used to show that $\mathbf{r}$ is not of that attribute A. Thus, the input space $\mathscr{R}$ is separated into two regions by the hyperplane

Table 19.1 Notation

Symbols	Meanings
$\mathscr{R}, \mathscr{H}$	Input space and feature space, respectively.
$\phi(\mathbf{x}_s)$	The image of a training sample $\mathbf{x}_s$ in $\mathscr{H}$.
$\mathscr{K}(\mathbf{x}_s, \mathbf{x})$	The kernel function that computes the inner products between $\phi(\mathbf{x}_s)$ and $\phi(\mathbf{x})$ in $\mathscr{H}$.
C	The regularization parameter in SVM.
γ	The parameter that appears in kernel functions, e.g., radius basis function (RBF), polynomial, and sigmoid kernels.
$\left\{\varphi_k^{(i)}\right\}_{i=1}^{2}$	The *ideal* features of training data are used to check if a spoofer is attacking user k. These two features are theoretically formulated.
$\left\{f_k^{(i)}[t]\right\}_{i=1}^{3}$	The practical features of the tth training sample in training data. These two features are formulated based on post-processing signals.
$\left\{\widehat{f}_k^{(i)}\left[\widehat{t}\right]\right\}_{i=1}^{3}$	The features of the $\widehat{t}$th training sample in ATD.
T	When a certain pilot is requested for authentication during the uplink phase, it is repeatedly sent T times.
$\widehat{T}$	The number of *artificial* training samples that are stuck with the label (#1) (or the label (#0)).
T_{tot}	The total number of training samples that are fed into SVM algorithms. In numerical results, $T_{tot} = 2\widehat{T}$.
(#1), $\mathscr{H}_1$	There is a spoofing attack.
(#0), $\mathscr{H}_0$	There is not any spoofing attack.

$\langle \mathbf{w}, \mathbf{r}\rangle - b = 0$. In one of the two regions, l can take the value of 1, while in the other region, the value of l is (-1). Since the binary classification relies on the position of the separating hyperplane, during the training process the perceptron looks for the coefficients $\mathbf{w}$ and b so that assigning a value to l (i.e., classifying $\mathbf{r}$) is as close to optimal as possible.

Upon replacing sign$(\cdot)$ by a continuous function, such as tanh$(\cdot)$, *multi-layer perceptrons* (MLPs) were formed. In many cases [24–26], support vector machines (SVMs) can yield better performance than classical MLPs. Different types of SVM classifiers basically rely on different kernel functions, such as the *radial basis function* (RBF) kernel and the sigmoid kernel. Noticeably, SVM classifiers can perform well even on small datasets.

19.2.2 Kernel methods

Prior to describing the operation of SVMs, we will clarify the role of kernel functions in SVM classifiers. Let $\phi(\cdot) : \mathscr{R} \rightarrow \mathscr{H}$ be a nonlinear mapping that casts the *input space* $\mathscr{R}$ to a higher-dimensional space $\mathscr{H}$ (the *feature space*). If $\mathbf{x}_t \in \mathscr{R}$ and $\mathbf{x}_{t'} \in \mathscr{R}$, then the inner product $\mathscr{K}(\mathbf{x}_t, \mathbf{x}_{t'}) = \langle\phi(\mathbf{x}_t), \phi(\mathbf{x}_{t'})\rangle$ is a kernel function [27,28]. The main rationale for using the kernel function is that once it has been given beforehand, one can directly compute $\mathscr{K}(\mathbf{x}_t, \mathbf{x}_{t'})$ from $\mathbf{x}_t$ and $\mathbf{x}_{t'}$ without necessarily having

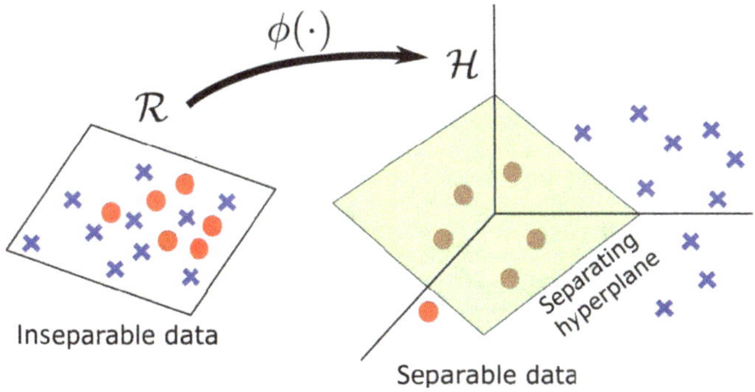

Figure 19.1 Using kernel methods, the data in the input space can become more linearly separable in a higher-dimensional space. ©IEEE 2021. Reprinted with permission from [8].

explicit expressions for $\phi(\mathbf{x}_t)$ and $\phi(\mathbf{x}_{t'})$ during learning. That helps us to accelerate the computational speed. Moreover, the kernel function allows us to avoid working in $\mathscr{R}$; instead, we only evaluate the input samples using the inner product in $\mathscr{H}$. From the viewpoint of data classification, the map $\phi(\cdot)$ can help us to convert *linearly inseparable* data (in $\mathscr{R}$) into linearly separable structures (in $\mathscr{H}$) [28]. As an example, Figure 19.1 illustrates the reason for mapping data points in $\mathscr{R}$ to the associated data points in $\mathscr{H}$ in a way that the crosses and circles become separable by the hyperplane.

19.2.3 Classification based on twin-class SVM

The binary classification problem based on SVM can be formulated as follows [28,29]:

$$\min_{\mathbf{w},b} \frac{1}{2}\|\mathbf{w}\|^2 \tag{19.1a}$$

$$\text{s.t.} \ \ l_\omega \left(\langle \mathbf{w}, \mathbf{x}_\omega \rangle + b\right) \geq 1, \tag{19.1b}$$

where $\mathbf{x}_\omega$ is the position vector of the ωth sample, $\omega \in \Omega \triangleq \{1, 2, \ldots, T_{tot}\}$, and T_{tot} is the total number of training samples. In (19.1b), $l_\omega = +1$ if $\mathbf{x}_\omega$ is labeled as (#1); otherwise, $l_\omega = -1$ if $\mathbf{x}_\omega$ is labeled as (#0).

As illustrated in Figure 19.2, the samples are separated by a hyperplane that takes the form of $h(\mathbf{x}) = \langle \mathbf{w}, \mathbf{x} \rangle + b = 0$. The separating hyperplane is in the middle of two margins, and the Euclidean distance between the two margins is equal to $2/\|\mathbf{w}\|$. The goal of (19.1) is to maximize the margin width $2/\|\mathbf{w}\|$, so that the samples are correctly separated. Note that maximizing $2/\|\mathbf{w}\|$ is equivalent to (19.1a), while the correct separation is equivalent to (19.1b).

Let $\Omega_{(\#0)}$ and $\Omega_{(\#1)}$ denote the set of indices $\{\omega \in \Omega | l_\omega = -1\}$ and the set of indices $\{\omega \in \Omega | l_\omega = 1\}$, respectively. If (19.1) has optimal solutions of $\mathbf{w} = \mathbf{w}_\star$ and $b = b_\star$, then we say that $\mathscr{X} = \{(\mathbf{x}_\omega, l_\omega) \, | \omega \in \Omega\}$ are *linearly separable*. In this case,

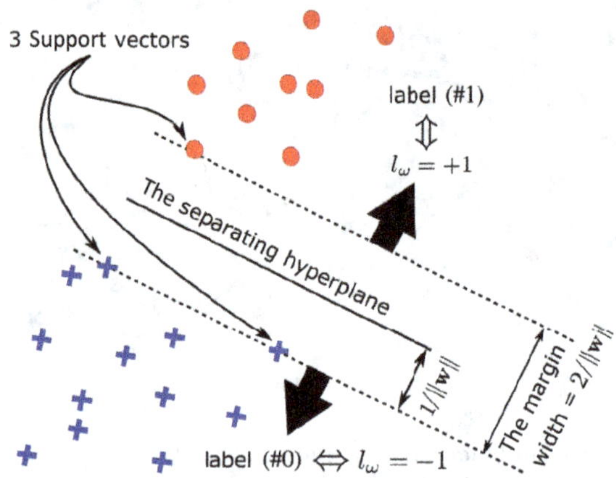

Figure 19.2 A dataset, with two types of samples, is separated by a separating hyperplane. The ωth sample, which is positioned at $\mathbf{x}_\omega$, will be labeled as (#1) (or $l_w = +1$) if it lies above the separating hyperplane. Otherwise, it will be labeled as (#0) (or $l_w = -1$). ©IEEE 2021. Reprinted with permission from [8].

there exists at least one separating hyperplane that separates the samples into two sub-sets, i.e., $\mathscr{X}_{(\#0)} = \left\{(\mathbf{x}_\omega, l_\omega) \,|\omega \in \Omega_{(\#0)}\right\}$ and $\mathscr{X}_{(\#1)} = \mathscr{X} \setminus \mathscr{X}_{(\#0)}$. Furthermore, the optimal separating hyperplane among all possible hyperplanes will have the following equation:

$$h_\star(\mathbf{x}) = \langle \mathbf{w}_\star, \mathbf{x} \rangle + b_\star = 0. \tag{19.2}$$

Note that $\mathbf{x}$ in (19.2) is not necessarily the same as $\mathbf{x}_\omega$ ($\omega \in \Omega$). A certain labeled sample $\mathbf{x}_\omega$ ($\omega \in \Omega$) only lies on the optimal hyperplane if it satisfies $h_\star(\mathbf{x}_t) = 0$. Given the hyperplane $h_\star(\mathbf{x})$, the input space is divided into two sub-spaces, where one of the sub-spaces contains all samples $\mathbf{x}_\omega \in \mathscr{X}_{(\#0)}$, while the other contains all samples $\mathbf{x}_\omega \in \mathscr{X}_{(\#1)}$.

It should be noted that (19.1) may not work in numerous practical scenarios, because the labeled samples are linearly inseparable. In this case, the classification tasks simply fail to work and no hyperplane can be found. For this reason, the use of kernel methods will make classification tasks easier when the inseparable data can be translated into a higher-dimensional feature space in which the structure of data becomes more separable [27–29].[†] Thus, we will replace $\mathbf{x}_\omega$ in (19.1) by the function $\phi(\mathbf{x}_\omega)$ so as to avoid working directly in the input space. However, the problem is that the transformed data may remain inseparable. To overcome this difficulty, slack variables $\xi_\omega \geq 0$ are added to the classification problems (see [30] and [31, Chapter 7]).

[†] Once again, Figure 19.1 illustrates the role of kernel methods in translating inseparable data points in the two-dimensional space into separable data points in the three-dimensional space.

This breakthrough method has become the background of SVM-based classification methods [30]. Following this method, one can change the original problem (19.1) to the following:

$$\min_{\mathbf{w},b,\xi_t} \underbrace{\frac{1}{2}\|\mathbf{w}\|^2}_{\text{regularizer}} + \underbrace{C\sum_{\omega=1}^{T_{tot}} \xi_\omega}_{\text{error}} \tag{19.3a}$$

$$\text{s.t.} \quad l_\omega \left(\langle \mathbf{w}, \phi\left(\mathbf{x}_\omega\right)\rangle + b\right) \geq 1 - \xi_\omega, \tag{19.3b}$$

where C is the regularization/margin parameter. To elaborate, (19.3) is called a $L1$ soft-margin in SVM terminology. If $\xi_\omega = 0$ for $\forall\omega \in \Omega$, then (19.3) reduces to the hard-margin SVM (19.1). Otherwise, with $\xi_\omega \neq 0$, we accept that the ωth sample $\mathbf{x}_\omega$ can be misclassified [27, Chapter 2]. According to [27], (19.3) may be transformed into an equivalent optimization problem through the use of the Lagrangian function and Karush–Kuhn–Tucker conditions, which is formulated as

$$\max_{a_1,\ldots,a_T} \sum_{\omega=1}^{T_{tot}} a_\omega - \frac{1}{2}\sum_{\substack{\omega=1\\ \omega'=1}}^{T_{tot}} a_\omega a_{\omega'} l_\omega l_{\omega'} \mathscr{K}\left(\mathbf{x}_\omega, \mathbf{x}_{\omega'}\right) \tag{19.4a}$$

$$\text{s.t.} \quad \sum_{\omega=1}^{T} a_\omega l_\omega = 0 \tag{19.4b}$$

$$0 \leq \alpha_\omega \leq C, \ \omega \in \Omega \tag{19.4c}$$

where $\mathscr{K}\left(\mathbf{x}_\omega, \mathbf{x}_{\omega'}\right) \triangleq \langle\phi(\mathbf{x}_\omega), \phi(\mathbf{x}_{\omega'})\rangle$ is the kernel function.

Remark 1. *Any data point for which $a_\omega = 0$ will not contribute to the prediction of new data points, thus only the data points for which $a_\omega \neq 0$ are called support vectors (SVs). Furthermore, any data point for which $0 < a_\omega < C$ is called an unbounded SV. Also, any data point for which $a_\omega = C$ is called a bounded SV. While the unbounded SVs lie on the margins, the bounded SVs are between the two margins (see [27, p. 24]). Additionally, we denote $\mathscr{S}$ as the set of indices of all SVs, $\mathscr{U}$ as the set of unbounded SV indices, and $\mathscr{S} \setminus \mathscr{U}$ as the set of bounded SV indices. Note that $\mathscr{U} \subseteq \mathscr{S} \subseteq \Omega$.*

The equation of the optimal hyperplane is given by

$$h_{(\star|\text{SVM})}(\mathbf{x}) = \sum_{s\in S} a_s l_s \mathscr{K}\left(\mathbf{x}_s, \mathbf{x}\right) + b, \tag{19.5}$$

where b is calculated for unbounded SVs (see [27, Chapter 2] or [31, (7.37)])

$$b = \frac{1}{|U|}\sum_{u\in U}\left(l_u - \sum_{s\in S} a_s l_s \mathscr{K}\left(\mathbf{x}_s, \mathbf{x}_u\right)\right). \tag{19.6}$$

Note that in (19.5)–(19.6), $\mathbf{x}_s$ is the sth SV, while $\mathbf{x}_u$ is the uth unbounded SV. Due to $\mathscr{U} \subseteq \mathscr{S}$, we have $\{\mathbf{x}_u | u \in \mathscr{U}\} \subseteq \{\mathbf{x}_s | s \in \mathscr{S}\}$. For the kernel $\mathscr{K}\left(\mathbf{x}_s, \mathbf{x}\right)$, we will consider four specific functions in Section 19.5 for comparison purposes.

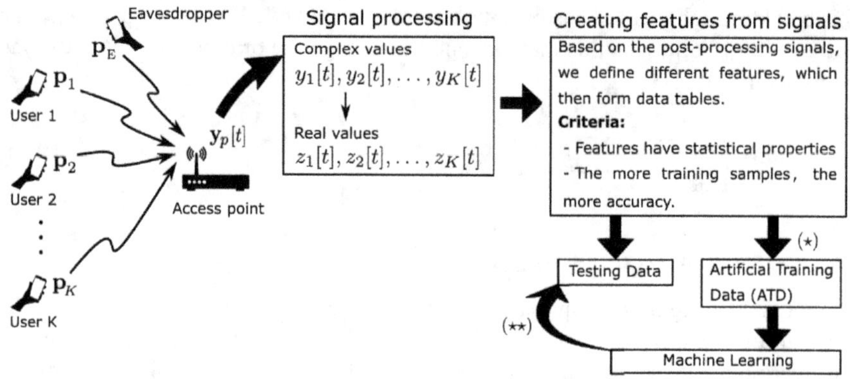

Figure 19.3 System model. At the stage (⋆), the AP forms the ATD by imitating the uplink phase and simulating signal processing. Using machine learning algorithms, the AP has the trained models and uses them for predicting the testing data at the stage (⋆ ⋆). ©IEEE 2021. Reprinted with permission from [8].

19.2.4 Classification based on SC-SVM

Slightly differently from classical TC-SVM models, a SC-SVM model is described as follows [32–34]:

$$\underset{\mathbf{w},\rho,\xi_i}{\text{minimize}} \quad \underbrace{\frac{1}{2}\|\mathbf{w}\|^2}_{\text{regularizer}} + \underbrace{\frac{1}{\nu T_{tot}}\sum_{w=1}^{T_{tot}} \xi_w - \rho}_{\text{error}} \tag{19.7a}$$

$$\text{subject to} \quad \langle \mathbf{w}, \phi(\mathbf{x}_w)\rangle \geq \rho - \xi_w. \tag{19.7b}$$

Herein, ρ is an offset parameter determining the distance from the origin to the hyperplane, while $\nu \in (0, 1]$ is a parameter balancing the maximal distance from the origin and the number of data points in the region created by the hyperplane [33]. Each $\xi_w \geq 0$ is a slack variable, and $\phi(\cdot) : \mathscr{X} \to \mathscr{F}$ is a mapping that casts the input space $\mathscr{X}$ to a higher-dimensional space $\mathscr{F}$ (namely, the feature space) (Figure 19.3). Once this minimization problem has been solved, the decision function will be given by

$$y(\mathbf{x}) = \text{sign}\left(\mathbf{w}^T\phi(\mathbf{x}) - \rho\right). \tag{19.8}$$

According to [34], most of the data points are put into a region (e.g., a ball) and these data points in the set $\{\mathbf{x}_1, \ldots, \mathbf{x}_{T_{tot}}\}$ will be labeled by $+1$. At the same time, any data points outside the region will be referred to as *outliers*. Consequently, some true outliers might be potentially misclassified when being put inside the region. In short, the objective of SC-SVM is to create a region for most of the data points in $\mathscr{X}$, while the remaining ones lie outside the region and are considered as being associated with spoofing attacks.

19.3 Collecting wireless signals and creating/defining features

While the previous section presented useful tools for classifying new data, this section shows the process of creating relevant features and then forming datasets.

19.3.1 Collecting wireless data

We assume that a single access point (AP) supports K legitimate users in the presence of an active E. Each node is equipped with a single antenna and all nodes are randomly positioned. Let the channel between the AP and the kth user be g_k. Similarly, the channel between the AP and E is denoted by g_{E}. We have two phases: uplink phase for authentication and downlink phase for confidential data transmission.

19.3.1.1 Uplink phase

In the uplink, the AP requests users to send pilot signals so that it can estimate channels as well as authenticate legitimate users. The kth user (or user k) is assumed to send some pilot vector $\mathbf{p}_k$ to the AP. Herein, $\mathbf{p}_k \in \mathbb{C}^{\mathscr{L}\times 1}$ is a column vector with $\mathscr{L}$ entries. By assuming $\mathscr{L} \geq K$, we can design K orthogonal pilot vectors satisfying that $\mathbf{p}_k^\dagger \mathbf{p}_{k'} = 0$ for $k \neq k'$ and $\|\mathbf{p}_k\|^2 = 1$. If E wants to overhear the signal s_k that is intended for the kth user, E will design her pilot sequence $\mathbf{p}_{\mathrm{E}}$ to be the same as $\mathbf{p}_k$ and will send $\mathbf{p}_{\mathrm{E}}$ to the AP (see [7,35]). Upon receiving pilots from the legitimate users and E, the AP interprets these pilots as requests for information. As a result, the AP will send confidential downlink signals s_k, which is intended for user k, to both user k and E. Obviously, due to the misunderstanding of the AP, the confidential information between the AP and the kth user will be leaked to E.

At some instance t, the AP receives the following signal:

$$\mathbf{y}_p[t] = \begin{cases} \sqrt{\mathscr{L}\rho_u}\sum_{k=1}^{K} \mathbf{p}_k g_k[t] + \mathbf{n}[t], & \text{non-attack} \\ \sqrt{\mathscr{L}\rho_u}\sum_{k=1}^{K} \mathbf{p}_k g_k[t] + \sqrt{\mathscr{L}\rho_{\mathrm{E}}}\mathbf{p}_{\mathrm{E}} g_{\mathrm{E}}[t] + \mathbf{n}[t], & \text{attack,} \end{cases} \tag{19.9}$$

where $\rho_u \triangleq P_u/N_0$ and $\rho_{\mathrm{E}} \triangleq P_{\mathrm{E}}/N_0$. Herein, P_u and P_{E} are the average transmit power of each user and that of E, respectively; while N_0 is the average noise power per receive antenna; $\mathbf{n}$ is an additive white Gaussian noise (AWGN) vector with $\mathbf{n} \sim \mathscr{CN}(\mathbf{0}, \mathbf{I}_{\mathscr{L}})$. Note that $\mathbf{y}_p[t]$, $g_k[t]$, $g_{\mathrm{E}}[t]$, and $\mathbf{n}[t]$ are the realizations of $\mathbf{y}_p$, g_k, g_{E}, and $\mathbf{n}$ at time t, respectively.

19.3.1.2 Downlink phase

We assume that the AP transmits confidential messages to the legitimate users in the downlink phase. Due to the broadcast nature of wireless propagation, E can receive the confidential messages that are intended for a certain user k. Moreover, given that in the previous uplink phase, the AP has received the identical pilot sequences from user k and E, then in this downlink phase the AP can misunderstand that a pair of legitimate users are requesting the same messages. In such a situation, the AP is unaware of the presence of E and her spoofing attack.

Indeed, the data rate of user k can decrease when E impersonates user k in the uplink phase [7]. Moreover, the signal-to-noise ratio (SNR) of the kth user, denoted as snr_k, reduces with the increase of E's power. Hence, the difference between the data rate of user k and the data rate of E, i.e. achievable security rate, also becomes lower. It is important that we need to detect the spoofing attacks from E in the uplink phase. If the AP succeeds in detecting a spoofing attack, it can find a suitable transmission scheme to deal with E, e.g. the AP can combine beamforming and artificial noise in transmitted signals. However, we do not investigate those transmission schemes in the downlink phase. Instead, we will focus on the authentication process during the uplink phase. To be more specific, SVM classifiers will be used for classifying normal and abnormal signals.

19.3.2 Creating features/attributes

By projecting $\mathbf{y}_p[t]$ along the pilot vector $\mathbf{p}_k^\dagger$, we have the post-processed signal $y_k[t] = \mathbf{p}_k^\dagger \mathbf{y}_p[t]$, i.e.

$$y_k[t] = \begin{cases} \sqrt{\mathscr{L}\rho_u} g_k[t] + \mathbf{p}_k^\dagger \mathbf{n}[t], & \text{non-attack} \\ \sqrt{\mathscr{L}\rho_u} g_k[t] + \sqrt{\mathscr{L}\rho_{\mathrm{E}}} g_{\mathrm{E}}[t] + \mathbf{p}_k^\dagger \mathbf{n}[t], & \text{attack} \end{cases}. \tag{19.10}$$

Proceeding by defining $z_k[t] \triangleq |y_k[t]|^2$, the AP can calculate

$$\varphi_k^{(1)} \triangleq \mathbb{E}_t\left\{z_k[t]\right\}, \tag{19.11}$$

$$\varphi_k^{(2)} \triangleq \frac{\mathbb{E}_t\left\{z_k[t]\right\} - \mathbb{E}_t\left\{\left|\mathbf{p}_k^\dagger \mathbf{n}[t]\right|^2\right\}}{\mathbb{E}_t\left\{\left|\mathbf{p}_k^\dagger \mathbf{n}[t]\right|^2\right\}} \tag{19.12}$$

based on sufficient statistical knowledge of $\{g_k\}_{k=1}^K$ and g_{E}. Note that $\mathbb{E}_t\{\cdot\} \equiv \mathbb{E}_{\{g_k\}_{k=1}^K, g_{\mathrm{E}}, \mathbf{n}}\{\cdot\}$ due to the dependence of $\{g_k\}_{k=1}^K$, g_{E} and $\mathbf{n}$ on t.

In practice, if user k sends his/her pilot vector $\mathbf{p}_k$ to the AP T times,[‡] then the AP will have T different values $z_k[1], \ldots, z_k[T]$. Using these values, the AP will create a structured dataset that consists of the following two features:

- Feature 1 (MEAN):

$$f_k^{(1)}[T] \triangleq \frac{1}{T}\sum_{t=1}^{T} z_k[t] = \begin{cases} f_{k|\mathscr{H}_0}^{(1)}[T], & \text{non-attack} \\ f_{k|\mathscr{H}_1}^{(1)}[T], & \text{attack} \end{cases}. \tag{19.13}$$

Herein, $\mathscr{H}_0$ implies that there is no spoofing attack, while $\mathscr{H}_0$ implies spoofing.

‡Herein, $\mathbf{p}_k$ is repeatedly sent for authentication.

- Feature 2 (RATIO):

$$f_k^{(2)}[T] \triangleq \frac{\sum_{t=1}^{T} z_k[t] - \sum_{t=1}^{T} |\mathbf{p}_k^{\dagger}\mathbf{n}[t]|^2}{\sum_{t=1}^{T} |\mathbf{p}_k^{\dagger}\mathbf{n}[t]|^2} = \begin{cases} f_{k|\mathcal{H}_0}^{(2)}[T], \text{ non-attack} \\ f_{k|\mathcal{H}_1}^{(2)}[T], \text{ attack} \end{cases}. \tag{19.14}$$

Statistically, if the AP receives a sufficient number of samples (i.e., if T is large enough), we have

$$f_k^{(1)}[T] \overset{\text{large } T}{\approx} \varphi_k^{(1)}, \tag{19.15}$$

$$f_k^{(2)}[T] \overset{\text{large } T}{\approx} \varphi_k^{(2)}. \tag{19.16}$$

Table 19.2 illustrates the two features of the testing data used in this chapter. The testing dataset begins with the T_0th data point, which is related to the T_0th time slot, $T_0 \in \{1, 2, \ldots, T\}$. If $T_0 = 1$, we exploit all the T data samples collected during T time slots in the uplink. By contrast, if $T_0 > 1$, we only consider $T - T_0$ data samples from the T_0th time slot to the Tth time slot. The position of the tth data point in the 2-dimensional space is $(f_k^{(1)}[t], f_k^{(2)}[t])$. Note that the subscript k indicates that we are checking whether the kth user is under attack or not. Moreover, our testing data has not been labeled yet. In Figure 19.4, the impact of T on the distribution of data points is depicted. It is readily seen that the first data samples are close together, making them hard to be separated. On the other hand, the larger T, the more separable the data becomes.

Table 19.2 Testing data: T points are associated with the label (#0) and T points are associated with the label (#1)

Feature 1 (MEAN)	**Feature 2 (RATIO)**
$f_{k\|\mathcal{H}_0}^{(1)}[T_0]$	$f_{k\|\mathcal{H}_0}^{(2)}[T_0]$
$\vdots$	$\vdots$
$f_{k\|\mathcal{H}_0}^{(1)}[T]$	$f_{k\|\mathcal{H}_0}^{(2)}[T]$
$f_{k\|\mathcal{H}_1}^{(1)}[T_0]$	$f_{k\|\mathcal{H}_1}^{(2)}[T_0]$
$\vdots$	$\vdots$
$f_{k\|\mathcal{H}_0}^{(1)}[T]$	$f_{k\|\mathcal{H}_0}^{(2)}[T]$

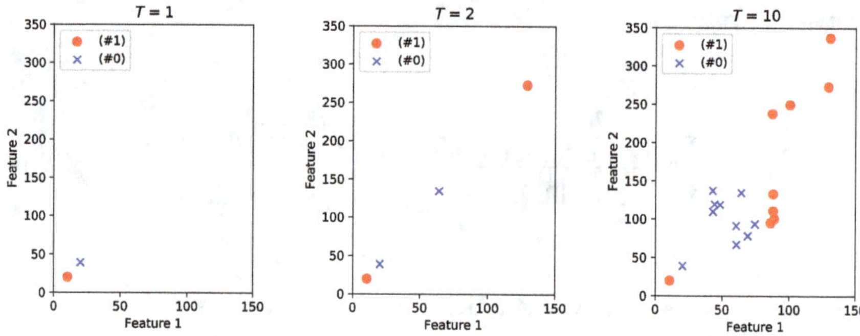

Figure 19.4 The distribution of data points in $\mathbb{R}^2$. The 1st red dot and the 1st blue cross are, respectively, positioned at $\left(f^{(1)}_{k|\mathcal{H}_1}[1], f^{(2)}_{k|\mathcal{H}_1}[1]\right)$ and at $\left(f^{(1)}_{k|\mathcal{H}_0}[1], f^{(2)}_{k|\mathcal{H}_0}[1]\right)$. Similarly, the 2nd red dot and the 2nd blue cross are, respectively, positioned at $\left(f^{(1)}_{k|\mathcal{H}_1}[2], f^{(2)}_{k|\mathcal{H}_1}[2]\right)$ and at $\left(f^{(1)}_{k|\mathcal{H}_0}[2], f^{(2)}_{k|\mathcal{H}_0}[2]\right)$. When the value of T increases, red dots tend to cluster together and become separable from blue crosses. ©IEEE 2021. Reprinted with permission from [8].

19.4 Artificial training data

In this section, we create the so-called artificial training data (ATD) and use it for the training purposes. The ATD, which will be used for TC-SVM, corresponds to the case of having both the legitimate users' CSI and the active E's CSI. By contrast, the ATD used for SC-SVM corresponds to the scenario of only having the legitimate users' CSI.

19.4.1 ATD for twin-class SVM

The AP can create the ATD by imitating the uplink phase by performing the following steps:

- Step 1: Start with $k = 1$.
- Step 2: Start with $\widehat{t} = 1$.
- Step 3: Generate $\widehat{\mathbf{g}}_k \overset{dist}{=} \mathbf{g}_k$, $\widehat{\mathbf{g}}_{\mathrm{E}} \overset{dist}{=} \mathbf{g}_{\mathrm{E}}$ and $\widehat{\mathbf{n}} \sim \mathcal{CN}(\mathbf{0}, \mathbf{I}_{\mathcal{L}})$. Note that the notation $X \overset{dist}{=} Y$ implies that X and Y have the same distribution.
- Step 4: Calculate

$$\widehat{z}_k\left[\widehat{t}\right] = \begin{cases} \left|\sqrt{\mathcal{L}\rho_u}\widehat{\mathbf{g}}_k\left[\widehat{t}\right] + \mathbf{p}_k^{\dagger}\widehat{\mathbf{n}}\left[\widehat{t}\right]\right|^2, & \text{non-attack} \\ \left|\sqrt{\mathcal{L}\rho_u}\widehat{\mathbf{g}}_k\left[\widehat{t}\right] + \sqrt{\mathcal{L}\rho_{\mathrm{E}}}\widehat{\mathbf{g}}_{\mathrm{E}}\left[\widehat{t}\right] + \mathbf{p}_k^{\dagger}\widehat{\mathbf{n}}\left[\widehat{t}\right]\right|^2, & \text{attack} \end{cases} \quad (19.17)$$

$$\widehat{f}_k^{(1)}\left[\widehat{t}\,\right] = \frac{1}{\widehat{t}}\sum_{\varsigma=1}^{\widehat{t}} \widehat{z}_k[\varsigma] = \begin{cases} \widehat{f}_{k|\mathcal{H}_0}^{(1)}\left[\widehat{t}\,\right], & \text{non-attack} \\ \widehat{f}_{k|\mathcal{H}_1}^{(1)}\left[\widehat{t}\,\right], & \text{attack} \end{cases}, \tag{19.18}$$

$$\widehat{f}_k^{(2)}\left[\widehat{t}\,\right] = \frac{\sum_{\varsigma=1}^{\widehat{t}} \widehat{z}_k[\varsigma] - \sum_{\varsigma=1}^{\widehat{t}} |\mathbf{p}_k^{\dagger}\widehat{\mathbf{n}}[\varsigma]|^2}{\sum_{\varsigma=1}^{\widehat{t}} |\mathbf{p}_k^{\dagger}\widehat{\mathbf{n}}[\varsigma]|^2} = \begin{cases} \widehat{f}_{k|\mathcal{H}_0}^{(2)}\left[\widehat{t}\,\right], & \text{non-attack} \\ \widehat{f}_{k|\mathcal{H}_1}^{(2)}\left[\widehat{t}\,\right], & \text{attack} \end{cases} \tag{19.19}$$

- Step 5: When $\widehat{t} \geq T_0$, we stick (#0) to the $\widehat{t}$th data point $\left(\widehat{f}_{k|\mathcal{H}_0}^{(1)}\left[\widehat{t}\,\right], \widehat{f}_{k|\mathcal{H}_0}^{(2)}\left[\widehat{t}\,\right]\right)$ in order to imply that user k is not under attack. By contrast, we stick the label (#1) to the $\widehat{t}$th data point $\left(\widehat{f}_{k|\mathcal{H}_1}^{(1)}\left[\widehat{t}\,\right], \widehat{f}_{k|\mathcal{H}_1}^{(2)}\left[\widehat{t}\,\right]\right)$ in order to imply that user k is under attack.
- Step 6: Set $\widehat{t} = \widehat{t} + 1$ and repeat Steps 3–5. Go to Step 7 if $\widehat{t} > \widehat{T}$. Herein, $\widehat{T}$ is a large number that can be freely determined by the designers. To make the ATD statistically reliable, we choose $\widehat{T} \gg T$.
- Step 7: Set $k = k + 1$ and repeat Steps 2–6. Stop the process if $k > K$.

Remark 2. *Table 19.3 describes the structure of the ATD used for classical TC-SVM. Since the ATD in Table 19.3 will be fed into classical TC-SVM models for training. Then, the trained models are applied to the testing dataset in Table 19.2, whereby we can classify data samples and point out the spoofing attacks from E.*

Remark 3. *Unlike the parameter T, we can denote $\widehat{T}$ as a very large number because $\widehat{T}$ belongs to an artificial process. For example, in the uplink, the AP can request the kth user and E to send the same pilot signal $\mathbf{p}_k$ ten times (i.e., $T = 10$), but cannot*

Table 19.3 Artificial training data (ATD) used for classical TC-SVM

MEAN	RATIO	Labels
$\widehat{f}_{k\|\mathcal{H}_1}^{(1)}[T_0]$	$\widehat{f}_{k\|\mathcal{H}_1}^{(2)}[T_0]$	(#1)
$\vdots$	$\vdots$	$\vdots$
$\widehat{f}_{k\|\mathcal{H}_1}^{(1)}[\widehat{T}]$	$\widehat{f}_{k\|\mathcal{H}_1}^{(2)}[\widehat{T}]$	(#1)
$\widehat{f}_{k\|\mathcal{H}_0}^{(1)}[T_0]$	$\widehat{f}_{k\|\mathcal{H}_0}^{(2)}[T_0]$	(#0)
$\vdots$	$\vdots$	$\vdots$
$\widehat{f}_{k\|\mathcal{H}_0}^{(1)}[\widehat{T}]$	$\widehat{f}_{k\|\mathcal{H}_0}^{(2)}[\widehat{T}]$	(#0)

Table 19.4 *Artificial training data (ATD) used for SC-SVM*

MEAN	RATIO	Labels
$\widehat{f}^{(1)}_{k\|\mathcal{H}_0}[T_0]$	$\widehat{f}^{(2)}_{k\|\mathcal{H}_0}[T_0]$	(#0)
$\vdots$	$\vdots$	$\vdots$
$\widehat{f}^{(1)}_{k\|\mathcal{H}_0}[\widehat{T}]$	$\widehat{f}^{(2)}_{k\|\mathcal{H}_0}[\widehat{T}]$	(#0)

request them to send $\mathbf{p}_k$ too many times (e.g., $T = 1000$). Fortunately, as for an artificial process, we can simulate the reception of a large number of copies of $\mathbf{p}_k$ as desired. Given sufficiently large $\widehat{T}$, we can reach $\widehat{f}^{(1)}_k[\widehat{T}]/\varphi^{(1)}_k \approx 1$ and $\widehat{f}^{(2)}_k[\widehat{T}]/\varphi^{(2)}_k \approx 1$.

19.4.2 *ATD for single-class SVM*

While the ATD used for TC-SVM has two labels corresponding to normal and abnormal data samples, the ATD used for SC-SVM has only a single label corresponding to normal data samples. Indeed, the ATD used for SC-SVM can be extracted from the ATD used for SVM by removing one of the labels and keeping the other one.

Noticeably, SC-SVM is related to the scenario of not having the CSI of E. Hence, the ATD used for SC-SVM only contains (#0)-labeled data samples (i.e., non-attack). Obviously, if the AP only knows the perfect CSI of legitimate users, we can only simulate the virtual process of transmitting wireless signals from the legitimate users to the AP and create (#0)-labeled data points.

Table 19.4, which is constructed of the last $\widehat{T}$ rows of Table 19.3, shows the ATD used for SC-SVM. Herein, Table 19.4 does not contain any (#1)-labeled data samples associated with spoofing attacks.

19.4.3 *ATD normalization/whitening*

Normally, SVM classifiers work with data within the range of [0, 1]. It is necessary for the AP to normalize all features in the training datasets. The normalization of the data makes all the values fall within the range of [0, 1]. Also concerning the data normalization, if a certain feature column consists of values $u_1, u_2, \ldots$, and $u_{T_{tot}}$, then the AP will have to perform the following steps:

- Step 1: Find $u_{min} = \min\{u_1, u_2, \ldots, u_{T_{tot}}\}$ and $u_{max} = \max\{u_1, u_2, \ldots, u_{T_{tot}}\}$.
- Step 2: Compute $u_k^{temp} = \dfrac{u_k - u_{min}}{u_{max} - u_{min}}$ where $k \in \{1, 2, \ldots, T_{tot}\}$. Assign the temporary value u_k^{temp} to u_k by letting $u_k = u_k^{temp}$.

The execution of the above steps allows us to normalize the ATD before entering the ATD into the SVM classifiers. Moreover, if the training data (i.e., the ATD in this chapter) is normalized, then the testing data also must be normalized accordingly (Table 19.5).

Table 19.5 The kernel function, which is mentioned in (19.5)*, is the RBF kernel, i.e.,* $\mathcal{K}(\boldsymbol{x}_s, \boldsymbol{x}) = exp\{-\gamma \|\boldsymbol{x}_s - \boldsymbol{x}\|\}$

		The accuracy of algorithms (%)	
$\widehat{\mathrm{T}}$	T	**TC-SVM**	**SC-SVM**
200	10	76.86	99.37
	20	81.81	98.6
	50	89.23	97.78
2000	10	76.76	98.76
	20	81.87	98.21
	50	91.28	97.26
4000	10	77.38	99.39
	20	82.76	98.93
	50	91.44	98.31

The SVM parameters are $\gamma = 0.5$ and $C = 1$. The SC-SVM parameters are $\nu = 0.02, \gamma = 0.5$. System parameters: $K = 4$, $\mathcal{L} = 10$, $\rho_u = \rho_E = 5$, $\beta_k = 1$, $\beta_E = 1$, and $T_0 = 1$.
Note: This table is in [8].

19.5 Numerical results

In this section, we present several numerical examples for characterizing the proposed framework under specific settings. As for the channel's fading, we set $g_k = \sqrt{\beta_k} h_k$ and $g_E = \sqrt{\beta_E} h_E$. Herein, β_k and β_E represent the path loss, while $h_k \sim \mathcal{CN}(0, 1)$ and $h_E \sim \mathcal{CN}(0, 1)$ represent the Rayleigh fading.[§]

According to [36], the complexity relies on the selection of the loss function and the optimization method. Using the big-O notation, it can be experimentally confirmed that if the implementation is based on LIBSVM, the run-time of TC-SVM as well as SC-SVM is on the order of about $\mathcal{O}(\widehat{T}^3_{tot})$, where $\widehat{T}_{tot}$ is the total number of training samples in the ATD [37–39]. Our experiments are carried out in Python and the execution of each TC-SVM/SC-SVM classifier takes around 0.09 s for training an ATD-based model and for labeling the data points in the testing data.

19.5.1 Examining $\widehat{T}$*,* T*, and* T_0

Table 19.5 shows the accuracy of RBF-based TC-SVM (and SC-SVM) classifiers for different values of $\widehat{T}$ and T. Observe that the accuracy of 99% is achievable, provided that the parameters are appropriately chosen. Moreover, the results show that in the case of TC-SVM, the accuracy increases both with T as well as with $\widehat{T}$. By contrast, in the case of SC-SVM, there is little change in the accuracy. Explicitly, the change of $\widehat{T}$, as well as T does not affect much how SC-SVM learns the decision

[§]Although Rayleigh fading is used to model h_k and h_E, other types of fading can also be used because the framework, presented in Sections 19.3 and 19.4, is not limited to any specific type of fading.

boundary. Additionally, although TC-SVM offers higher accuracy than SC-SVM, we have to note that a comparison between these two may be unfair due to the differences inherent in these two approaches (e.g., (19.3) uses the regularization parameter C, while (19.7) uses the parameter ν), and that SC-SVM requires only the CSI of the legitimate users.

19.5.2 Examining four different kernel functions

Figure 19.5 shows the accuracy of four different classifiers versus T for TC-SVM. These four classifiers are based on

- linear kernel: $\mathscr{K}(\mathbf{x}_s, \mathbf{x}) = \langle \mathbf{x}_s, \mathbf{x} \rangle$,
- RBF kernel: $\mathscr{K}(\mathbf{x}_s, \mathbf{x}) = \exp\{-\gamma \|\mathbf{x}_s - \mathbf{x}\|\}$,
- polynomial kernel: $\mathscr{K}(\mathbf{x}_s, \mathbf{x}) = (\gamma \langle \mathbf{x}_s, \mathbf{x} \rangle + r)^d$,
- sigmoid kernel: $\mathscr{K}(\mathbf{x}_s, \mathbf{x}) = \tanh(\gamma \langle \mathbf{x}_s, \mathbf{x} \rangle + r)$.

Herein, it should also be noted that we set r to 0, so that the RBF, polynomial, and sigmoid kernels are fairly treated. The other parameters are as follows: $K = 4$, $\mathscr{L} = 10$, $\rho_u = 5$, $\rho_{\mathrm{E}} = 4$, $\beta_k = 1$, $\beta_{\mathrm{E}} = 1$, $\widehat{T} = 2000$, $T_0 = 5$, and $\gamma = 0.001$. Also in Figure 19.5, the accuracy increases with T. For example, when $T \geq 80$, the accuracy exceeds 95% for the linear kernel. Among the four kernels considered, the linear kernel has the highest accuracy (e.g., over 95%), while the polynomial kernel has the lowest accuracy (e.g., less than 75%). The accuracy of four different SC-SVM classifiers versus T indicates no accuracy improvements with T, hence it has been omitted. However, it should be noted that among the SC-SVM models, the sigmoid

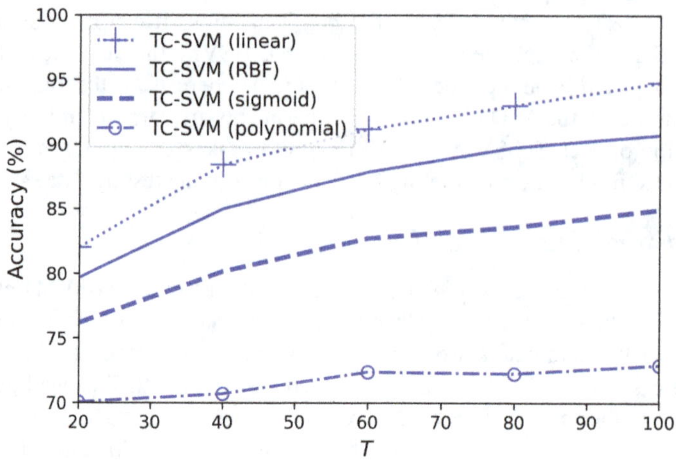

Figure 19.5 An illustration of four trained TC-SVM models that correspond to four different kernel functions (i.e., linear, RBF, polynomial (degree 2), and sigmoid kernels). ©IEEE 2021. Reprinted with permission from [8].

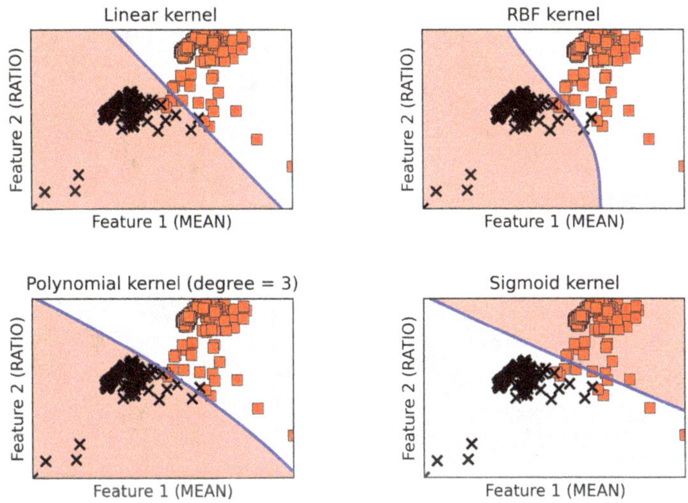

Figure 19.6 An illustration of 4 trained TC-SVM models that correspond to four different kernel functions (i.e., linear, RBF, 3-degree polynomial, and sigmoid kernels). The black crosses relate to (#0), while the red squares relate to (#1). ©IEEE 2021. Reprinted with permission from [8].

kernel has the highest accuracy (over 95%), while the RBF kernel has the lowest accuracy (i.e., around 90%). Additionally, there is no obvious difference between the polynomial kernel and the linear kernel (i.e., around 93%).

For characterizing the above-mentioned kernel functions in the case of TC-SVM, we provide an intuitive visual comparison in Figure 19.6. The total number of training samples is $T_{tot} = 2\widehat{T} = 400$, i.e., 200 samples with the label (#1) and 200 samples with the label (#0). Similarly, Figure 19.7 illustrates four different SC-SVM models with $\widehat{T} = 200$ samples. Observe from Figure 19.7 that there are only ATD points corresponding to the label (#0), because the SC-SVM models are trained on the data having a single class.

19.5.3 Receiver operating characteristic curves and sensitivity-specificity trade-off

According to [40–42], the true positive rate (or probability of detection) is defined as

$$TPR = \frac{\text{True positive (TP)}}{\text{Positive (Pos)}},$$

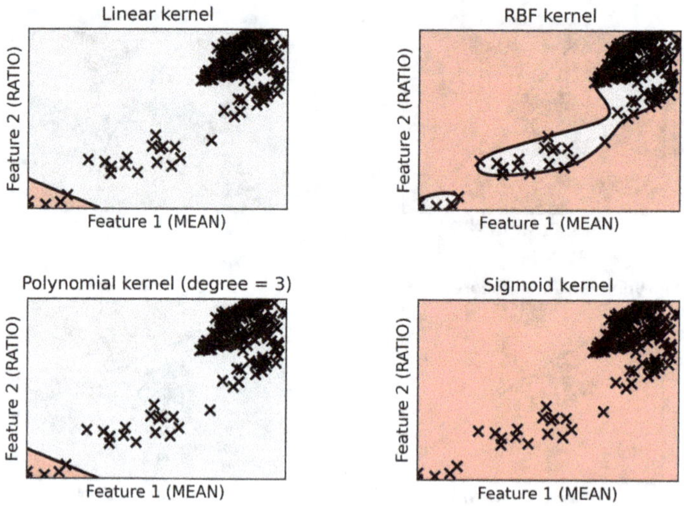

Figure 19.7 An illustration of 4 trained SC-SVM models that correspond to linear, RBF, polynomial (degree 3), and sigmoid kernels. The black crosses are the ATD points associated with the label (#0). The white area contains the ATD points that are correctly detected. ©IEEE 2021. Reprinted with permission from [8].

where TP represents the number of (#1)-labeled samples *correctly classified as (#1)*, and Pos is the total number of *actually* (#1)-labeled samples. By contrast, the false positive rate (or probability of false alarm) is defined as [42]

$$FPR = \frac{\text{False positive (FP)}}{\text{Negative (Neg)}},$$

where FP is the number of (#0)-labeled samples *incorrectly classified as (#1)*, and Neg is the total number of *actually* (#0)-labeled samples.

To plot the *TPR* against the *FPR*, we offer Figure 19.8 that contains two receiver operating characteristic (ROC) curves. The kernel used for Figure 19.8 is the RBF kernel. The other parameters are as follows: $K = 4$, $\mathscr{L} = 10$, $\rho_u = 5$, $\rho_{\mathrm{E}} = 2$, $\beta_k = 1$, $\beta_{\mathrm{E}} = 1$, $C = 1$, $\nu = 0.02$, $\gamma = 0.001$, $\widehat{T} = 1000$, $T = 20$, and $T_0 = 1$. The area under the ROC curve (AUC) is 0.95 in the case of TC-SVM and 0.8 in the case of SC-SVM. These two numbers reveal that TC-SVM may be better than SC-SVM, when taking into account both *TPR* and *FPR*. Since the *TPR* increases with the *FPR*, the careful choice of parameters is required to maintain a low *FPR* and an acceptable *TPR* (e.g., $FPR \leq 0.2$ and $TPR \geq 0.8$).

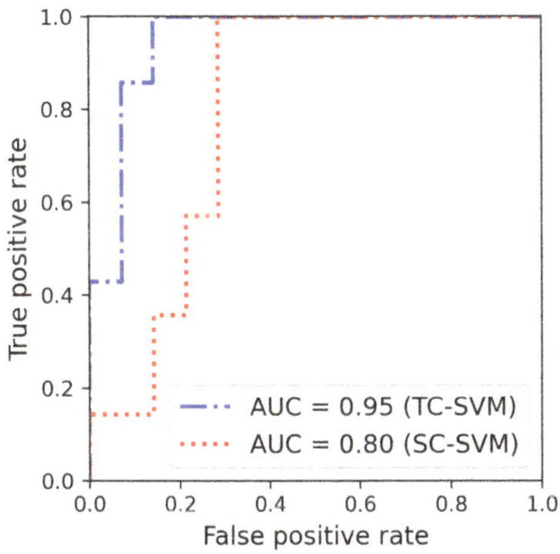

Figure 19.8 ROC curves and the areas under the corresponding ROC curves in two different regimes. ©IEEE 2021. Reprinted with permission from [8].

Apart from the *TPR* and *FPR*, there are two other similar rates referred to as false negative rate (*FNR*) and true negative rate (*TNR*), which are formulated as:

$$FNR = \frac{\text{False negative (FN)}}{\text{Positive (Pos)}} = 1 - TPR,$$

and

$$TNR = \frac{\text{True negative (TN)}}{\text{Negative (Neg)}} = 1 - FPR,$$

where FN is the number of (#1)-labeled samples *incorrectly identified as (#0)*, and TN is the number of (#0)-labeled samples *correctly identified as (#0)*. Note that *TPR* is also known as (a.k.a.) the *sensitivity* that quantifies how well a test can identify true positives, while *TNR* is a.k.a. the *specificity* that measures how well a test can identify true negatives. The trade-off between the sensitivity and the specificity is illustrated in Figure 19.9. We can see from Figure 19.9 that the sensitivity and specificity of the TC-SVM model are higher than those of the SC-SVM model. This confirms again that the TC-SVM model outperforms the SC-SVM model. Additionally, it is readily seen that the higher the sensitivity, the lower the specificity becomes. Explicitly, this means that neither the TC-SVM model nor the SC-SVM can improve the sensitivity and the specificity at the same time. In other words, if a model can improve its probability of detecting malicious samples associated with actual spoofing attacks, then its proportion of admitting benevolent samples associated with the situation of non-attack will go down.

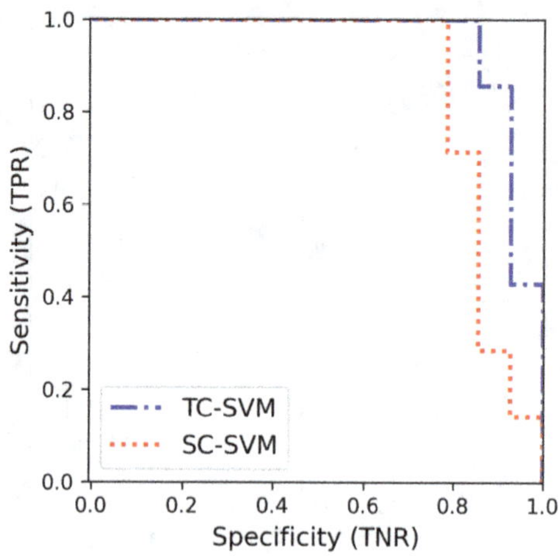

Figure 19.9 The trade-off between the sensitivity and the specificity in two different regimes. ©IEEE 2021. Reprinted with permission from [8].

19.5.4 Examining the impact of ρ_E and γ

Figure 19.10 shows the accuracy versus the ratio ρ_E/ρ_u in two sub-cases corresponding to $\gamma = \{0.001, 1\}$. Herein, we fix $\rho_u = 5$ and change ρ_E. The kernel used in Figure 19.10 is the RBF kernel. The parameter C of TC-SVM is equal to $C = 1$, while the parameter ν of SC-SVM is equal to $\nu = 0.02$. The common system parameters are as follows: $K = 4$, $\mathcal{L} = 10$, $\beta_k = 1$, $\beta_E = 1$, $(\widehat{T}, T, T_0) = (2000, 50, 15)$.

In general, the accuracy of the TC-SVM model increases from about 58% at $\rho_E = 0.1\rho_u$ to about 92% at $\rho_E = \rho_u$. This is because the TC-SVM model learns from the data belonging to two classes in which class (#1) is related to the spoofer's power. When ρ_E increases, the data points in class (#1) become more readily separable from the data points in class (#0), thereby yielding higher accuracy. Additionally, the change of γ has little impact on the accuracy of the TC-SVM model.

By contrast, the accuracy of the SC-SVM does not change significantly with ρ_E. This is because the SC-SVM model learns from the data having the label (#0) only. As a result, the SC-SVM classifier is less sensitive to ρ_E that inherently affects only the class (#1). Noticeably, γ has a significant influence on the accuracy of the SC-SVM classifier. For example, when $\gamma = 0.001$, the accuracy can exceed 97%, even when ρ_E is small. This means that γ is an influential parameter in allowing the SC-SVM classifier to learn the characteristics of the normal data samples, thereby detecting the abnormal outliers that are associated with E.

19.5.5 Over-fitting problems: the impact of γ

In Figure 19.11, we make a comparison among four different sub-cases corresponding to $\gamma = \{0.05, 1, 20, 400\}$. The total number of training samples is equal to $T_{tot} = 2\widehat{T} = 400$. Figure 19.11 illustrates the transformation of models, when γ increases. With

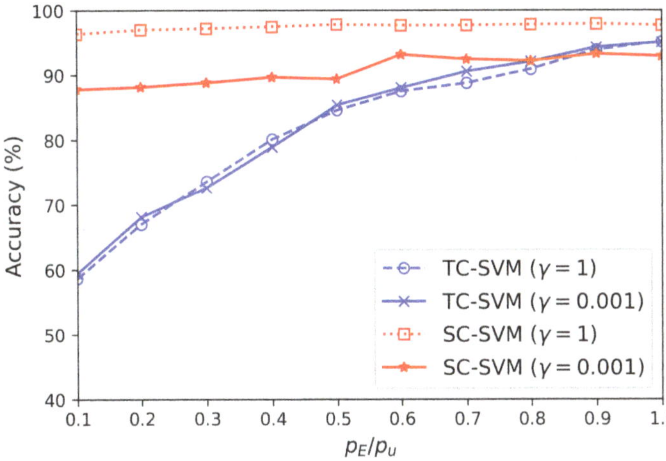

Figure 19.10 An illustration of the trained TC-SVM and SC-SVM models that correspond to the RBF kernel.

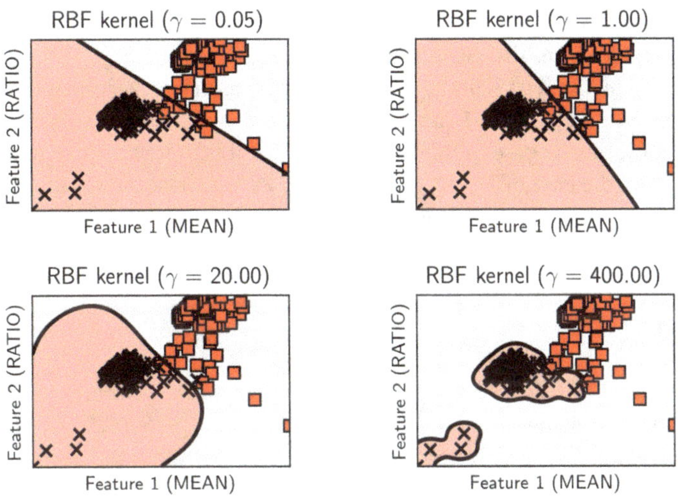

Figure 19.11 The four trained TC-SVM models correspond to four different values of γ in the case of the RBF kernel (with $C = 1$). The black crosses relate to (#0), while the red squares relate to (#1).

$\gamma = 400$, we face an over-fitting problem due to the fact that the fourth model learns so well and becomes too detailed.

19.6 Conclusions

We have considered the uplink phase of a wireless system, where mobile users send pilot signals to the base station for authentication. Under the assumption that E impersonates a legitimate user, we have applied two types of SVM classifiers to identify the presence of E. While TC-SVM is used for the case the CSI of E's channel is known, SC-SVM is used for the case that there is no CSI of E's channel. Through the numerical results, the following insights have been observed:

(i) The conversion of wireless signals into structured datasets with specific features will facilitate SVM classifiers (as well as any other machine-learning classifiers) to learn the characteristics of normal and abnormal data.
(ii) Since the kernel function is the heart of an SVM classifier, there is a significant difference in accuracy when choosing different kernel functions. For example, it is suggested not to use the polynomial kernel function for our datasets because of the low level of accuracy. By contrast, the linear and RBF functions are better candidates for the kernel function of an SVM classifier.
(iii) It is shown that the length of a dataset has a considerable impact on the detection performance, i.e. the more data points, the more accuracy. This implies that working on big data will improve our SVM classifiers' accuracy.
(iv) TC-SVM and SC-SVM models are affected by the value of γ in terms of accuracy, but the case of SC-SVM shows that it is more sensitive to the change of γ than the case of TC-SVM. Moreover, we have also shown the impact of γ on the associated over-fitting problems.

References

[1] T. M. Hoang, T. Q. Duong, and S. Lambotharan, "Secure wireless communication using support vector machines," in *2019 IEEE Conference on Communication and Network Security (CNS)*, 2019, pp. 1–5.
[2] Y. Zou, J. Zhu, X. Wang, and L. Hanzo, "A survey on wireless security: Technical challenges, recent advances, and future trends," *Proceedings of the IEEE*, vol. 104, no. 9, pp. 1727–1765, 2016.
[3] D. Kapetanovic, G. Zheng, and F. Rusek, "Physical layer security for massive MIMO: an overview on passive eavesdropping and active attacks," *IEEE Commun. Mag.*, vol. 53, no. 6, pp. 21–27, 2015.
[4] W. Wang, K. C. Teh, K. H. Li, and S. Luo, "On the impact of adaptive eavesdroppers in multi-antenna cellular networks," *IEEE Trans. Info. Forens. Secur.*, vol. 13, no. 2, pp. 269–279, 2018.

[5] A. Al-Nahari, "Physical layer security using massive multiple-input and multiple-output: passive and active eavesdroppers," *IET Commun.*, vol. 10, no. 1, pp. 50–56, 2016.

[6] Y. Wu, R. Schober, D. W. K. Ng, C. Xiao, and G. Caire, "Secure massive MIMO transmission with an active eavesdropper," *IEEE Trans. Info. Theory*, vol. 62, no. 7, pp. 3880–3900, 2016.

[7] T. M. Hoang, H. Q. Ngo, T. Q. Duong, H. D. Tuan, and A. Marshall, "Cell-free massive MIMO networks: optimal power control against active eavesdropping," *IEEE Trans. Commun.*, vol. 66, no. 10, pp. 4724–4737, 2018.

[8] T. M. Hoang, T. Q. Duong, H. D. Tuan, S. Lambotharan, and L. Hanzo, "Physical layer security: detection of active eavesdropping attacks by support vector machines," *IEEE Access*, vol. 9, pp. 31595–31607, 2021.

[9] H. Li, P. He, S. Wang, A. Rocha, X. Jiang, and A. C. Kot, "Learning generalized deep feature representation for face anti-spoofing," *IEEE Trans. Inf. Forens. Secur.*, vol. 13, no. 10, pp. 2639–2652, 2018.

[10] L. Wu, J. Yang, M. Zhou, Y. Chen, and Q. Wang, "LVID: A multimodal biometrics authentication system on smartphones," *IEEE Trans. Inf. Forens. Secur.*, vol. 15, pp. 1572–1585, 2020.

[11] H. Zhang, J. Liu, K. Li, H. Tan, and G. Wang, "Gait learning based authentication for intelligent things," *IEEE Trans. Veh. Tech.*, vol. 69, no. 4, pp. 4450–4459, 2020.

[12] W. Hou, X. Wang, J. Chouinard, and A. Refaey, "Physical layer authentication for mobile systems with time-varying carrier frequency offsets," *IEEE Trans. Commun.*, vol. 62, no. 5, pp. 1658–1667, 2014.

[13] L. Xiao, L. J. Greenstein, N. B. Mandayam, and W. Trappe, "Using the physical layer for wireless authentication in time-variant channels," *IEEE Trans. Wirel. Commun.*, vol. 7, no. 7, pp. 2571–2579, 2008.

[14] L. Xiao, Y. Li, G. Han, G. Liu, and W. Zhuang, "PHY-layer spoofing detection with reinforcement learning in wireless networks," *IEEE Trans. Veh. Tech.*, vol. 65, no. 12, pp. 10037–10047, 2016.

[15] X. Lu, L. Xiao, T. Xu, Y. Zhao, Y. Tang, and W. Zhuang, "Reinforcement learning based PHY authentication for VANETs," *IEEE Trans. Veh. Tech.*, vol. 69, no. 3, pp. 3068–3079, 2020.

[16] N. Wang, T. Jiang, S. Lv, and L. Xiao, "Physical-layer authentication based on extreme learning machine," *IEEE Commun. Lett.*, vol. 21, no. 7, pp. 1557–1560, 2017.

[17] X. Qiu, T. Jiang, S. Wu, and M. Hayes, "Physical layer authentication enhancement using a Gaussian mixture model," *IEEE Access*, vol. 6, pp. 53583–53592, 2018.

[18] F. Pan, Z. Pang, H. Wen, M. Luvisotto, M. Xiao, R. Liao, and J. Chen, "Threshold-free physical layer authentication based on machine learning for industrial wireless CPS," *IEEE Trans. Ind. Informat.*, vol. 15, no. 12, pp. 6481–6491, 2019.

[19] C. Pei, N. Zhang, X. S. Shen, and J. W. Mark, "Channel-based physical layer authentication," in *IEEE Global Communication Conference*, Dec. 2014, pp. 4114–4119.

[20] M. Esmalifalak, L. Liu, N. Nguyen, R. Zheng, and Z. Han, "Detecting stealthy false data injection using machine learning in smart grid," *IEEE Syst. J.*, vol. 11, no. 3, pp. 1644–1652, 2017.

[21] B. Chatterjee, D. Das, S. Maity, and S. Sen, "RF-PUF: Enhancing IoT security through authentication of wireless nodes using in-situ machine learning," *IEEE Internet Things J.*, vol. 6, no. 1, pp. 388–398, 2019.

[22] A. Weinand, M. Karrenbauer, R. Sattiraju, and H. Schotten, "Application of machine learning for channel based message authentication in mission critical machine type communication," in *The 23rd European Wireless 2017*, Dresden, Germany, May 2017, pp. 342–346.

[23] H. Fang, X. Wang, and S. Tomasin, "Machine learning for intelligent authentication in 5G and beyond wireless networks," *IEEE Wirel. Commun.*, vol. 26, no. 5, pp. 55–61, 2019.

[24] V. N. Vapnik, *The Nature of Statistical Learning Theory*. New York, NY: Springer-Verlag, 1995.

[25] N. I. Sapankevych and R. Sankar, "Time series prediction using support vector machines: a survey," *IEEE Compu. Intell. Mag.*, vol. 4, no. 2, pp. 24–38, 2009.

[26] C. J. C. Burges, "A tutorial on support vector machines for pattern recognition," *Knowl. Discov. Data Min.*, vol. 2, no. 2, pp. 121–167, 1998.

[27] S. Abe, *Support Vector Machines for Pattern Classification*. New York, NY: Springer-Verlag, 2005.

[28] F. Perez-Cruz and O. Bousquet, "Kernel methods and their potential use in signal processing," *IEEE Sig. Process. Mag.*, vol. 21, no. 3, pp. 57–65, 2004.

[29] B. Scholkopf and A. J. Smola, *Learning with Kernels: Support Vector Machines, Regularization, Optimization, and Beyond*. New York, NY: The MIT Press, 2002.

[30] C. Cortes and V. Vapnik, "Support-vector networks," *Mach. Learn., Springer*, vol. 3, no. 20, pp. 273–29, 1995.

[31] C. M. Bishop, *Pattern Recognition and Machine Learning*. New York, NY: Springer-Verlag, 2006.

[32] L. M. Manevitz and M. Yousef, "One-class SVMs for document classification," *J. Mach. Learn. Res.*, vol. 2, pp. 139–154, 2001.

[33] Y. Wang, J. Wong, and A. Miner, "Anomaly intrusion detection using one class SVM," in *The fifth annual IEEE SMC Info. Assurance Workshop*, West Point, NY, June 2004, pp. 358–364.

[34] B. Scholkopf, J. C. Platt, J. Shawe-Taylor, A. J. Smola, and R. C. Williamson, "Estimating the support of a high-dimensional distribution," *Neural Comput.*, vol. 13, no. 7, pp. 1443–1471, 2001.

[35] Y. Wu, R. Schober, D. W. K. Ng, C. Xiao, and G. Caire, "Secure massive MIMO transmission in presence of an active eavesdropper," in *IEEE International Conference on Communication (ICC)*, London, UK, Jun. 2015, pp. 1434–1440.

[36] O. Chapelle, "Training a support vector machine in the primal," *Neural Comput.*, vol. 19, no. 5, pp. 1155–1178, 2007.

[37] C.-C. Chang and C.-J. Lin, "LIBSVM: a library for support vector machines," *ACM Trans. Intell. Syst. Technol.*, vol. 2, no. 3, 2011. Available: https://www.csie.ntu.edu.tw/ cjlin/papers/libsvm

[38] A. Abdiansah and R. Wardoyo, "Time complexity analysis of support vector machines (SVM) in LibSVM," *Int. J. Comput. Appl.*, vol. 128, no. 3, pp. 28–34, 2015.

[39] G. L. Grinblat, L. C. Uzal, and P. M. Granitto, "Abrupt change detection with one-class time-adaptive support vector machines," *Expert Syst. Appl.*, vol. 40, no. 18, pp. 7242–7249, 2013.

[40] J. Davis and M. Goadrich, "The relationship between precision-recall and ROC curves," in *Proceedings of the 23rd International Conference on Machine Learning*, ser. ICML '06. New York, NY: ACM, 2006, pp. 233–240.

[41] T. Fawcett, "An introduction to ROC analysis," *Pattern Recogn. Lett.*, vol. 27, no. 8, pp. 861–874, 2006, ROC Analysis in Pattern Recognition. Available: http://www.sciencedirect.com/science/article/pii/S016786550500303X

[42] Y. Tang, Y. Zhang, N. V. Chawla, and S. Krasser, "SVMs modeling for highly imbalanced classification," *IEEE Trans. Syst. Man Cyber. Part B (Cyber.)*, vol. 39, no. 1, pp. 281–288, 2009.

Index

www.ingramcontent.com/pod-product-compliance
Lightning Source LLC
LaVergne TN
LVHW011538200926
848484LV00001B/282